Contents

STUDENT'S
SOLUTIONS MANUAL

JUDITH A. PENNA
Indiana University Purdue University Indianapolis

INTRODUCTORY ALGEBRA
TENTH EDITION

Marvin L. Bittinger
Indiana University Purdue University Indianapolis

Boston San Francisco New York
London Toronto Sydney Tokyo Singapore Madrid
Mexico City Munich Paris Cape Town Hong Kong Montreal

Reproduced by Pearson Addison-Wesley from electronic files supplied by the author.

Copyright © 2007 Pearson Education, Inc.
Publishing as Pearson Addison-Wesley, 75 Arlington Street, Boston, MA 02116.

ISBN   0-321-30599-X

3 4 5 6 BB 09 08 07

Chapter R

Prealgebra Review

1. We first find some factorizations:

   $20 = 1 \cdot 20, \ 20 = 2 \cdot 10, \ 20 = 4 \cdot 5$

   The factors of 20 are 1, 2, 4, 5, 10, and 20.

3. We first find some factorizations:

   $72 = 1 \cdot 72, \ 72 = 2 \cdot 36, \ 72 = 3 \cdot 24,$

   $72 = 4 \cdot 18, \ 72 = 6 \cdot 12, \ 72 = 8 \cdot 9$

   The factors of 72 are 1, 2, 3, 4, 6, 8, 9, 12, 18, 24, 36 and 72.

5. $15 = 3 \cdot 5$

   Both factors are prime, so we have the prime factorization of 15.

7. $22 = 2 \cdot 11$

   Both factors are prime, so we have the prime factorization of 22.

9. $9 = 3 \cdot 3$

   Both factors are prime, so we have the prime factorization of 9.

11. $49 = 7 \cdot 7$

    Both factors are prime, so we have the prime factorization of 49.

13. We begin by factoring 18 in any way that we can and continue factoring until each factor is prime.

    $18 = 2 \cdot 9 = 2 \cdot 3 \cdot 3$

15. We begin by factoring 40 in any way that we can and continue factoring until each factor is prime.

    $40 = 4 \cdot 10 = 2 \cdot 2 \cdot 2 \cdot 5$

17. We begin by factoring 90 in any way that we can and continue factoring until each factor is prime.

    $90 = 2 \cdot 45 = 2 \cdot 9 \cdot 5 = 2 \cdot 3 \cdot 3 \cdot 5$

19. We go through the table of primes until we find a prime that divides 210. We continue dividing by that prime until it is not possible to do so any longer. We continue this process until each factor is prime.

    $210 = 2 \cdot 105 = 2 \cdot 3 \cdot 35 = 2 \cdot 3 \cdot 5 \cdot 7$

21. We go through the table of primes until we find a prime that divides 91. The first such prime is 7.

    $91 = 7 \cdot 13$

    Both factors are prime, so this is the prime factorization.

23. We go through the table of primes until we find a prime that divides 119. The first such prime is 7.

    $119 = 7 \cdot 17$

    Both factors are prime, so this is the prime factorization.

25. a) We find the prime factorizations:

    $4 = 2 \cdot 2$

    $5 = 5$ (5 is prime)

    b) We write 2 as a factor two times (the greatest number of times it occurs in any one factorization). We write 5 as a factor one time (the greatest number of times it occurs in any one factorization). The LCM is $2 \cdot 2 \cdot 5$, or 20.

27. a) We find the prime factorizations:

    $24 = 2 \cdot 2 \cdot 2 \cdot 3$

    $36 = 2 \cdot 2 \cdot 3 \cdot 3$

    b) We write 2 as a factor three times (the greatest number of times it occurs in any one factorization). We write 3 as a factor two times (the greatest number of times it occurs in any one factorization). The LCM is $2 \cdot 2 \cdot 2 \cdot 3 \cdot 3$, or 72.

29. $3 = 3$ (3 is prime)
    $15 = 3 \cdot 5$

    The LCM is $3 \cdot 5$, or 15.

31. $30 = 2 \cdot 3 \cdot 5$
    $40 = 2 \cdot 2 \cdot 2 \cdot 5$

    The LCM is $2 \cdot 2 \cdot 2 \cdot 3 \cdot 5$, or 120.

33. 13 and 23 are both prime. The LCM is $13 \cdot 23$, or 299.

35. $18 = 2 \cdot 3 \cdot 3$
    $30 = 2 \cdot 3 \cdot 5$

    The LCM is $2 \cdot 3 \cdot 3 \cdot 5$, or 90.

37. $30 = 2 \cdot 3 \cdot 5$
    $36 = 2 \cdot 2 \cdot 3 \cdot 3$

    The LCM is $2 \cdot 2 \cdot 3 \cdot 3 \cdot 5$, or 180.

39. $24 = 2 \cdot 2 \cdot 2 \cdot 3$
    $30 = 2 \cdot 3 \cdot 5$

    The LCM is $2 \cdot 2 \cdot 2 \cdot 3 \cdot 5$, or 120.

41. 17 and 29 are both prime. The LCM is $17 \cdot 29$, or 493.

43. $12 = 2 \cdot 2 \cdot 3$
    $28 = 2 \cdot 2 \cdot 7$

    The LCM is $2 \cdot 2 \cdot 3 \cdot 7$, or 84.

45. 2, 3, and 5 are all prime. The LCM is $2 \cdot 3 \cdot 5$, or 30.

47. $24 = 2 \cdot 2 \cdot 2 \cdot 3$

$36 = 2 \cdot 2 \cdot 3 \cdot 3$

$12 = 2 \cdot 2 \cdot 3$

The LCM is $2 \cdot 2 \cdot 2 \cdot 3 \cdot 3$, or 72.

49. $5 = 5$ (5 is prime)

$12 = 2 \cdot 2 \cdot 3$

$15 = 3 \cdot 5$

The LCM is $2 \cdot 2 \cdot 3 \cdot 5$, or 60.

51. $6 = 2 \cdot 3$

$12 = 2 \cdot 2 \cdot 3$

$18 = 2 \cdot 3 \cdot 3$

The LCM is $2 \cdot 2 \cdot 3 \cdot 3$, or 36.

53. Jupiter: $12 = 2 \cdot 2 \cdot 3$

Saturn: $30 = 2 \cdot 3 \cdot 5$

The LCM is $2 \cdot 2 \cdot 3 \cdot 5$, or 60.

Jupiter and Saturn will appear in the same direction in the night sky as seen from the earth every 60 years.

55. Saturn: $30 = 2 \cdot 3 \cdot 5$

Uranus: $84 = 2 \cdot 2 \cdot 3 \cdot 7$

The LCM is $2 \cdot 2 \cdot 3 \cdot 5 \cdot 7$, or 420.

Saturn and Uranus will appear in the same direction in the night sky as seen from the earth every 420 years.

57. The smallest number of strands that can be used is the LCM of 10 and 3.

$10 = 2 \cdot 5$

$3 = 3$

$\text{LCM} = 2 \cdot 5 \cdot 3 = 30$

The smallest number of strands that can be used is 30.

59. $8 = 2 \cdot 2 \cdot 2$

$12 = 2 \cdot 2 \cdot 3$

The greatest number of times 2 occurs in the factorizations of 8 and 12 is three times. The greatest number of times 3 occurs is once. Thus, the LCM must contain exactly three factors of 2 and one factor of 3. The LCM is $2 \cdot 2 \cdot 2 \cdot 3$.

a) No; $2 \cdot 2 \cdot 3 \cdot 3$ is not a multiple of 8.

b) No; $2 \cdot 2 \cdot 2 \cdot 3 \cdot 5$ is a multiple of both 8 and 12, but it is not the least common multiple.

c) No; $2 \cdot 3 \cdot 3$ is not a multiple of 8 or 12.

d) Yes; $2 \cdot 2 \cdot 2 \cdot 3$ is a multiple of both 8 and 12, and it is the smallest such multiple.

61. Use a calculator to find multiples of 7800. Divide each one by 2700 to determine if it is also a multiple of 2700. The first such number, 70,200, is the LCM. ($70,200 = 7800 \cdot 9 = 2700 \cdot 26$)

Exercise Set R.2

1. $\dfrac{3}{4} = \dfrac{3}{4} \cdot 1$ Identity property of 1

$= \dfrac{3}{4} \cdot \dfrac{3}{3}$ Using $\dfrac{3}{3}$ for 1

$= \dfrac{9}{12}$ Multiplying numerators and denominators

3. $\dfrac{3}{5} = \dfrac{3}{5} \cdot 1$ Identity property of 1

$= \dfrac{3}{5} \cdot \dfrac{20}{20}$ Using $\dfrac{20}{20}$ for 1

$= \dfrac{60}{100}$ Multiplying numerators and denominators

5. $\dfrac{13}{20} = \dfrac{13}{20} \cdot 1$ Identity property of 1

$= \dfrac{13}{20} \cdot \dfrac{8}{8}$ Using $\dfrac{8}{8}$ for 1

$= \dfrac{104}{160}$

7. We will use $\dfrac{3}{3}$ for 1 since $24 = 8 \cdot 3$.

$\dfrac{7}{8} = \dfrac{7}{8} \cdot 1 = \dfrac{7}{8} \cdot \dfrac{3}{3} = \dfrac{21}{24}$

9. We will use $\dfrac{4}{4}$ for 1 since $16 = 4 \cdot 4$.

$\dfrac{5}{4} = \dfrac{5}{4} \cdot 1 = \dfrac{5}{4} \cdot \dfrac{4}{4} = \dfrac{20}{16}$

11. We will use $\dfrac{23}{23}$ for 1 since $437 = 19 \cdot 23$.

$\dfrac{17}{19} = \dfrac{17}{19} \cdot 1 = \dfrac{17}{19} \cdot \dfrac{23}{23} = \dfrac{391}{437}$

13. $\dfrac{18}{27} = \dfrac{2 \cdot 9}{3 \cdot 9}$ Factoring numerator and denominator

$= \dfrac{2}{3} \cdot \dfrac{9}{9}$ Factoring the fractional expression

$= \dfrac{2}{3} \cdot 1$

$= \dfrac{2}{3}$ Identity property of 1

15. $\dfrac{56}{14} = \dfrac{4 \cdot 14}{1 \cdot 14}$ Factoring and inserting a factor of 1 in the numerator

$= \dfrac{4}{1} \cdot \dfrac{14}{14}$ Factoring the fractional expression

$= \dfrac{4}{1} \cdot 1$

$= 4$ Identity property of 1

17. $\dfrac{6}{42} = \dfrac{1 \cdot 6}{7 \cdot 6}$ Factoring and inserting a factor of 1 in the numerator

$\quad\quad = \dfrac{1}{7} \cdot \dfrac{6}{6}$

$\quad\quad = \dfrac{1}{7} \cdot 1$

$\quad\quad = \dfrac{1}{7}$

19. $\dfrac{56}{7} = \dfrac{8 \cdot 7}{1 \cdot 7} = \dfrac{8}{1} \cdot \dfrac{7}{7} = \dfrac{8}{1} \cdot 1 = 8$

21. $\dfrac{19}{76} = \dfrac{1 \cdot 19}{4 \cdot 19}$ Factoring and inserting a factor of 1 in the numerator

$\quad\quad = \dfrac{1 \cdot \cancel{19}}{4 \cdot \cancel{19}}$ Removing a factor of 1: $\dfrac{19}{19} = 1$

$\quad\quad = \dfrac{1}{4}$

23. $\dfrac{100}{20} = \dfrac{5 \cdot 20}{1 \cdot 20}$ Factoring and inserting a factor of 1 in the denominator

$\quad\quad = \dfrac{5 \cdot \cancel{20}}{1 \cdot \cancel{20}}$ Removing a factor of 1: $\dfrac{20}{20} = 1$

$\quad\quad = \dfrac{5}{1}$

$\quad\quad = 5$ Simplifying

25. $\dfrac{425}{525} = \dfrac{17 \cdot 25}{21 \cdot 25}$ Factoring the numerator and the denominator

$\quad\quad = \dfrac{17 \cdot \cancel{25}}{21 \cdot \cancel{25}}$ Removing a factor of 1: $\dfrac{25}{25} = 1$

$\quad\quad = \dfrac{17}{21}$

27. $\dfrac{2600}{1400} = \dfrac{2 \cdot 13 \cdot 100}{2 \cdot 7 \cdot 100}$ Factoring

$\quad\quad = \dfrac{13 \cdot \cancel{2} \cdot \cancel{100}}{7 \cdot \cancel{2} \cdot \cancel{100}}$ Removing a factor of 1: $\dfrac{2 \cdot 100}{2 \cdot 100} = 1$

$\quad\quad = \dfrac{13}{7}$

29. $\dfrac{8 \cdot x}{6 \cdot x} = \dfrac{2 \cdot 4 \cdot x}{2 \cdot 3 \cdot x}$ Factoring

$\quad\quad = \dfrac{4 \cdot \cancel{2} \cdot \cancel{x}}{3 \cdot \cancel{2} \cdot \cancel{x}}$ Removing a factor of 1: $\dfrac{2 \cdot x}{2 \cdot x} = 1$

$\quad\quad = \dfrac{4}{3}$

31. $\dfrac{1}{3} \cdot \dfrac{1}{4} = \dfrac{1 \cdot 1}{3 \cdot 4}$ Multiplying numerators and denominators

$\quad\quad = \dfrac{1}{12}$

33. $\dfrac{15}{4} \cdot \dfrac{3}{4} = \dfrac{15 \cdot 3}{4 \cdot 4} = \dfrac{45}{16}$

35. $\dfrac{1}{3} + \dfrac{1}{3} = \dfrac{1 + 1}{3}$ Adding numerators; keeping the same denominator

$\quad\quad = \dfrac{2}{3}$

37. $\dfrac{4}{9} + \dfrac{13}{18} = \dfrac{4}{9} \cdot \dfrac{2}{2} + \dfrac{13}{18}$ LCD is 18.

$\quad\quad = \dfrac{8}{18} + \dfrac{13}{18}$

$\quad\quad = \dfrac{21}{18}$

$\quad\quad = \dfrac{7 \cdot \cancel{3}}{6 \cdot \cancel{3}} = \dfrac{7}{6}$ Simplifying

39. $\dfrac{3}{10} + \dfrac{8}{15} = \dfrac{3}{10} \cdot \dfrac{3}{3} + \dfrac{8}{15} \cdot \dfrac{2}{2}$ LCD is 30.

$\quad\quad = \dfrac{9}{30} + \dfrac{16}{30}$

$\quad\quad = \dfrac{25}{30}$

$\quad\quad = \dfrac{5 \cdot \cancel{5}}{6 \cdot \cancel{5}} = \dfrac{5}{6}$ Simplifying

41. $\dfrac{7}{30} + \dfrac{5}{12} = \dfrac{7}{30} \cdot \dfrac{2}{2} + \dfrac{5}{12} \cdot \dfrac{5}{5}$ LCD is 60.

$\quad\quad = \dfrac{14}{60} + \dfrac{25}{60}$

$\quad\quad = \dfrac{39}{60}$

$\quad\quad = \dfrac{\cancel{3} \cdot 13}{\cancel{3} \cdot 20} = \dfrac{13}{20}$ Simplifying

43. $\dfrac{5}{4} - \dfrac{3}{4} = \dfrac{2}{4}$

$\quad\quad = \dfrac{1 \cdot \cancel{2}}{2 \cdot \cancel{2}} = \dfrac{1}{2}$

45. $\dfrac{11}{12} - \dfrac{3}{8} = \dfrac{11}{12} \cdot \dfrac{2}{2} - \dfrac{3}{8} \cdot \dfrac{3}{3}$ LCD is 24.

$\quad\quad = \dfrac{22}{24} - \dfrac{9}{24}$

$\quad\quad = \dfrac{13}{24}$

47. $\dfrac{11}{12} - \dfrac{2}{5} = \dfrac{11}{12} \cdot \dfrac{5}{5} - \dfrac{2}{5} \cdot \dfrac{12}{12}$ LCD is 60.

$\quad\quad = \dfrac{55}{60} - \dfrac{24}{60}$

$\quad\quad = \dfrac{31}{60}$

49. $\dfrac{7}{6} \div \dfrac{3}{5} = \dfrac{7}{6} \cdot \dfrac{5}{3}$ $\left(\dfrac{5}{3} \text{ is the reciprocal of } \dfrac{3}{5}.\right)$

$\quad\quad = \dfrac{35}{18}$ Multiplying

51. $\dfrac{8}{9} \div \dfrac{4}{15} = \dfrac{8}{9} \cdot \dfrac{15}{4} = \dfrac{8 \cdot 15}{9 \cdot 4} = \dfrac{2 \cdot \cancel{4} \cdot \cancel{3} \cdot 5}{\cancel{3} \cdot 3 \cdot \cancel{4}} = \dfrac{10}{3}$

53. $\frac{1}{8} \div \frac{1}{4} = \frac{1}{8} \cdot \frac{4}{1} = \frac{1 \cdot 4}{8 \cdot 1} = \frac{1 \cdot 4}{2 \cdot 4 \cdot 1} = \frac{1}{2}$

55. $\frac{\frac{13}{12}}{\frac{39}{5}} = \frac{13}{12} \div \frac{39}{5} = \frac{13}{12} \cdot \frac{5}{39} = \frac{13 \cdot 5}{12 \cdot 39} = \frac{13 \cdot 5}{12 \cdot 3 \cdot 13} = \frac{5}{36}$

57. $100 \div \frac{1}{5} = \frac{100}{1} \div \frac{1}{5} = \frac{100}{1} \cdot \frac{5}{1} = \frac{100 \cdot 5}{1 \cdot 1} = \frac{500}{1} = 500$

59. $\frac{3}{4} \div 10 = \frac{3}{4} \cdot \frac{1}{10} = \frac{3 \cdot 1}{4 \cdot 10} = \frac{3}{40}$

61. Discussion and Writing Exercise

63. We begin by factoring 28 in any way that we can and continue factoring until each factor is prime.

$28 = 4 \cdot 7 = 2 \cdot 2 \cdot 7$

65. We begin by factoring 1000 in any way that we can and continue factoring until each factor is prime.

$1000 = 4 \cdot 250 = 2 \cdot 2 \cdot 2 \cdot 125 = 2 \cdot 2 \cdot 2 \cdot 5 \cdot 25 =$

$2 \cdot 2 \cdot 2 \cdot 5 \cdot 5 \cdot 5$

67. We begin by factoring 2001 in any way that we can and continue factoring until each factor is prime.

$2001 = 3 \cdot 667 = 3 \cdot 23 \cdot 29$

Since 3, 23 and 29 are prime numbers, the prime factorization is $3 \cdot 23 \cdot 29$.

69. $16 = 2 \cdot 2 \cdot 2 \cdot 2$
$24 = 2 \cdot 2 \cdot 2 \cdot 3$
The LCM = $2 \cdot 2 \cdot 2 \cdot 2 \cdot 3$, or 48.

71. $48 = 2 \cdot 2 \cdot 2 \cdot 2 \cdot 3$
$64 = 2 \cdot 2 \cdot 2 \cdot 2 \cdot 2 \cdot 2$
$96 = 2 \cdot 2 \cdot 2 \cdot 2 \cdot 2 \cdot 3$
The LCM = $2 \cdot 2 \cdot 2 \cdot 2 \cdot 2 \cdot 2 \cdot 3$, or 192.

73. $\frac{192}{256} = \frac{3 \cdot 64}{4 \cdot 64} = \frac{3}{4}$

75. $\frac{64 \cdot a \cdot b}{16 \cdot a \cdot b} = \frac{4 \cdot 16 \cdot a \cdot b}{1 \cdot 16 \cdot a \cdot b} = \frac{4}{1} = 4$

77. $\frac{36 \cdot (2 \cdot h)}{8 \cdot (9 \cdot h)} = \frac{9 \cdot 4 \cdot 2 \cdot h}{2 \cdot 4 \cdot 9 \cdot h} = 1$

Exercise Set R.3

1. 5.3. $5.3 = \frac{53}{10}$
1 place   1 zero

3. 0.67. $0.67 = \frac{67}{100}$
2 places   2 zeros

5. 2.0007. $2.0007 = \frac{20,007}{10,000}$
4 places   4 zeros

7. 7889.8. $7889.8 = \frac{78,898}{10}$
1 place   1 zero

9. $\frac{1}{10}$ 0.1. $\frac{1}{10} = 0.1$
1 zero 1 place

11. $\frac{1}{10,000}$ 0.0001. $\frac{1}{10,000} = 0.0001$
4 zeros   4 places

13. $\frac{9999}{1000}$ 9.999. $\frac{9999}{1000} = 9.999$
3 zeros   3 places

15. $\frac{4578}{10,000}$ 0.4578. $\frac{4578}{10,000} = 0.4578$
4 zeros   4 places

17. 415.78
$+ 29.16$
$\overline{444.94}$

19. 234.000
$+156.617$
$\overline{390.617}$

21. $85 + 67.95 + 2.774$
We have:
$85.$
67.95
$+ 2.774$
$\overline{155.724}$

23. $17.95 + 16.99 + 28.85$
We have:
17.95
16.99
$+28.85$
$\overline{63.79}$

25. 78.110
-45.876
$\overline{32.234}$

27.
```
              16
         7  6  9 10
     3  8 . 7  0  0
   − 1  1 . 8  6  5
   ─────────────────
     2  6 . 8  3  5
```

29.
```
           16
        4  6  18
     5  7 . 8  6
   −    9 . 9  5
   ───────────────
     4  7 . 9  1
```

31.
```
        2  9  9  9 10
     3 . 0  0  0  0
   − 1 . 0  8  0  7
   ──────────────────
     1 . 9  1  9  3
```

33.
```
        2  3
     7 . 3  4  ←── 2 decimal places
   ×    1 . 8  ←── 1 decimal place
   ─────────────
     5  8  7  2
     7  3  4  0
   ─────────────────
   1  3 . 2  1  2  ←── 3 decimal places
```

35.
```
        5
        1
     0 . 8  6  ←── 2 decimal places
   × 0 . 9  3  ←── 2 decimal places
   ─────────────
        2  5  8
     7  7  4  0
   ──────────────────
   0 . 7  9  9  8  ←── 4 decimal places
```

37.
```
     1  7 . 9  5  ←── 2 decimal places
   ×       1  0  ←── 0 decimal places
   ─────────────────
   1  7  9 . 5  0  ←── 2 decimal places
```

39.
```
           1  2
           4  5
     0 . 4  5  7  ←── 3 decimal places
   ×    3 . 0  8  ←── 2 decimal places
   ─────────────────
        3  6  5  6
     1  3  7  1  0  0
   ─────────────────────
   1 . 4  0  7  5  6  ←── 5 decimal places
```

41.
```
           5  3  1
           5  3  1
     3 . 6  4  2  ←── 3 decimal places
   × 0 . 9  9  ←── 2 decimal places
   ─────────────────
     3  2  7  7  8
     3  2  7  7  8  0
   ─────────────────────
   3 . 6  0  5  5  8  ←── 5 decimal places
```

43. Place the decimal point in the quotient directly above the decimal point in the dividend. Then divide as whole numbers.

```
               2 . 3
        7 2 ⟌ 1 6 5 . 6
              1 4 4
              ───────
                2 1 6
                2 1 6
                ──────
                    0
```

45.
```
                 5 . 2
        8.5∧⟌ 4 4.2∧0
               4 2 5
               ──────
                 1 7 0
                 1 7 0
                 ──────
                     0
```

47.
```
                0 . 0 2 3
        9.9∧⟌ 0 . 2∧2 7 7
                 1 9 8
                 ──────
                   2 9 7
                   2 9 7
                   ──────
                       0
```

49.
```
                   1 8 . 7 5
        0.6 4∧⟌ 1 2.0 0∧0 0
                 6 4
                 ──────
                 5 6 0
                 5 1 2
                 ──────
                   4 8 0
                   4 4 8
                   ──────
                     3 2 0
                     3 2 0
                     ──────
                         0
```

51.
```
                 6 6 0 .
        1.0 5∧⟌ 6 9 3.0 0∧
               6 3 0
               ──────
                 6 3 0
                 6 3 0
                 ──────
                     0
                     0
                     ──
                     0
```

53.
```
                 0 . 6 8
        8.6∧⟌ 5.8∧4 8
               5 1 6
               ──────
                 6 8 8
                 6 8 8
                 ──────
                     0
```

55.
```
                 0 . 3 4 3 7 5
        3 2 ⟌ 1 1 . 0 0 0 0 0
               9 6
               ──────
               1 4 0
               1 2 8
               ──────
                 1 2 0
                   9 6
                 ──────
                 2 4 0
                 2 2 4
                 ──────
                   1 6 0
                   1 6 0
                   ──────
                       0
```

Decimal notation for $\frac{11}{32}$ is 0.34375.

57.

```
        1.1 8 1 8
  1 1 [ 1 3.0 0 0 0
        1 1
        ‾‾‾
          2 0
          1 1
          ‾‾‾
            9 0
            8 8
            ‾‾‾
              2 0
              1 1
              ‾‾‾
                9 0
                8 8
                ‾‾‾
                  2
```

Since 2 and 9 alternate as remainders, the sequence of digits following the decimal point in the quotient repeats. Thus, decimal notation for $\frac{13}{11}$ is 1.1818..., or $1.\overline{18}$.

59.

```
       0.5 5
  9 [ 5.0 0
      4 5
      ‾‾‾
        5 0
        4 5
        ‾‾‾
          5
```

The number 5 repeats as a remainder, so the digit 5 will repeat in the quotient. Thus, decimal notation for $\frac{5}{9}$ is 0.55..., or $0.\overline{5}$.

61.

```
       2.1 1
  9 [ 1 9.0 0
      1 8
      ‾‾‾
        1 0
          9
        ‾‾‾
          1
```

The number 1 repeats as a remainder, so the digit 1 will repeat in the quotient. Thus, decimal notation for $\frac{19}{9}$ is 2.11..., or $2.\overline{1}$.

63. 745.06534

Round to the nearest hundredth: The digit in the hundredths place is 6. The next digit to the right, 5, is 5 or higher, so we round up: 745.07

Round to the nearest tenth: The digit in the tenths place is 0. The next digit to the right, 6, is 5 or higher, so we round up: 745.1

Round to the nearest one: The digit in the ones place is 5. The next digit to the right, 0, is less than 5, so we round down: 745

Round to the nearest ten: The digit in the tens place is 4. The next digit to the right, 5, is 5 or higher, so we round up: 750

Round to the nearest hundred: The digit in the hundreds place is 7. The next digit to the right, 4, is less than 5, so we round down: 700

65. 6780.50568

Round to the nearest hundredth: The digit in the hundredths place is 0. The next digit to the right, 5, is 5 or higher, so we round up: 6780.51

Round to the nearest tenth: The digit in the tenths place is 5. The next digit to the right, 0, is less than 5, so we round down: 6780.5

Round to the nearest one: The digit in the ones place is 0. The next digit to the right, 5, is 5 or higher, so we round up: 6781

Round to the nearest ten: The digit in the tens place is 8. The next digit to the right, 0, is less than 5, so we round down: 6780

Round to the nearest hundred: The digit in the hundreds place is 7. The next digit to the right, 8, is 5 or higher, so we round up: 6800

67. $17.988

Round to the nearest cent (nearest hundredth): The digit in the hundredths place is 8. The next digit to the right, 8, is 5 or higher, so we round up: $17.99

Round to the nearest dollar (nearest one): The digit in the ones place is 7. The next digit to the right, 9, is 5 or higher, so we round up: $18

69. $346.075

Round to the nearest cent (nearest hundredth): The digit in the hundredths place is 7. The next digit to the right, 5, is 5 or higher, so we round up: $346.08

Round to the nearest dollar (nearest one): The digit in the ones place is 6. The next digit to the right, 0, is less than 5, so we round down: $346

71. $16.95

The digit in the ones place is 6. The next digit to the right, 9, is 5 or higher, so we round up: $17

73. $189.50

The digit in the ones place is 9. The next digit to the right, 5, is 5 or higher, so we round up: $190

75.

```
          1 2.3 4 5 6 7
  8 1 [ 1 0 0 0.0 0 0 0 0
         8 1
         ‾‾‾
         1 9 0
         1 6 2
         ‾‾‾‾‾
           2 8 0
           2 4 3
           ‾‾‾‾‾
             3 7 0
             3 2 4
             ‾‾‾‾‾
               4 6 0
               4 0 5
               ‾‾‾‾‾
                 5 5 0
                 4 8 6
                 ‾‾‾‾‾
                   6 4 0
                   5 6 7
                   ‾‾‾‾‾
                     7 3
```

$\frac{1000}{81} \approx 12.34567$

We round to the nearest:

ten-thousandth: 12.3457

thousandth: 12.346

hundredth: 12.35

tenth: 12.3

one: 12

77.

$$
\begin{array}{r}
0.5\,8\,9\,7\,4 \\
39\overline{)2\,3.0\,0\,0\,0\,0} \\
1\,9\,5 \\
\hline
3\,5\,0 \\
3\,1\,2 \\
\hline
3\,8\,0 \\
3\,5\,1 \\
\hline
2\,9\,0 \\
2\,7\,3 \\
\hline
1\,7\,0 \\
1\,5\,6 \\
\hline
1\,4
\end{array}
$$

$\dfrac{23}{39} \approx 0.58974$

We round to the nearest:

    ten-thousandth: 0.5897

    thousandth: 0.590

    hundredth: 0.59

    tenth: 0.6

    one: 1

79. Discussion and Writing Exercise

81. $\dfrac{7}{8} + \dfrac{5}{32} = \dfrac{7}{8} \cdot \dfrac{4}{4} + \dfrac{5}{32}$,   LCD is 32.

$\qquad = \dfrac{28}{32} + \dfrac{5}{32}$

$\qquad = \dfrac{33}{32}$

83. $\dfrac{15}{16} \cdot \dfrac{11}{12} = \dfrac{15 \cdot 11}{16 \cdot 12}$

$\qquad = \dfrac{\cancel{3} \cdot 5 \cdot 11}{16 \cdot \cancel{3} \cdot 4}$

$\qquad = \dfrac{55}{64}$

85. The LCD is 210.

$\dfrac{9}{70} + \dfrac{8}{15} = \dfrac{9}{70} \cdot \dfrac{3}{3} + \dfrac{8}{15} \cdot \dfrac{14}{14}$

$\qquad = \dfrac{27}{210} + \dfrac{112}{210}$

$\qquad = \dfrac{139}{210}$

87. The LCD is 1000.

$\dfrac{9}{10} + \dfrac{1}{100} + \dfrac{113}{1000} = \dfrac{9}{10} \cdot \dfrac{100}{100} + \dfrac{1}{100} \cdot \dfrac{10}{10} + \dfrac{113}{1000}$

$\qquad = \dfrac{900}{1000} + \dfrac{10}{1000} + \dfrac{113}{1000}$

$\qquad = \dfrac{1023}{1000}$

89. We begin by factoring 208 in any way we can and continue factoring until each factor is prime.

$$208 = 2 \cdot 104 = 2 \cdot 4 \cdot 26 = 2 \cdot 2 \cdot 2 \cdot 2 \cdot 13$$

91. We begin by factoring 1250 in any way we can and continue factoring until each factor is prime.

$$1250 = 10 \cdot 125 = 2 \cdot 5 \cdot 5 \cdot 25 = 2 \cdot 5 \cdot 5 \cdot 5 \cdot 5$$

Exercise Set R.4

1. 21%   0.21.

Move the decimal point 2 places to the left.

21% = 0.21

3. 3.45%   0.03.45

Move the decimal point 2 places to the left.

3.45% = 0.0345

5. 63%   0.63.

Move the decimal point 2 places to the left.

63% = 0.63

7. 94.1%   0.94.1

Move the decimal point 2 places to the left.

94.1% = 0.941

9. 1%   0.01.

Move the decimal point 2 places to the left.

1% = 0.01

11. 0.61%   0.00.61

Move the decimal point 2 places to the left.

0.61% = 0.0061

13. 240%   2.40.

Move the decimal point 2 places to the left.

240% = 2.4

15. 3.25%   0.03.25

Move the decimal point 2 places to the left.

3.25% = 0.0325

17. $2\% = 2 \times \dfrac{1}{100}$ Replacing % with $\times \dfrac{1}{100}$

$\qquad = \dfrac{2}{100}$

19. $10.9\% = 10.9 \times \dfrac{1}{100}$ Replacing % with $\times \dfrac{1}{100}$

$\qquad = \dfrac{10.9}{100}$

$\qquad = \dfrac{10.9}{100} \cdot \dfrac{10}{10}$ Multiplying by 1 to get a whole number in the numerator

$\qquad = \dfrac{109}{1000}$

21. $77\% = 77 \times \dfrac{1}{100}$ Replacing % with $\times \dfrac{1}{100}$

$\qquad = \dfrac{77}{100}$

23. $60\% = 60 \times \dfrac{1}{100}$ Replacing % with $\times \dfrac{1}{100}$

$\qquad = \dfrac{60}{100}$

25. $28.9\% = 28.9 \times \dfrac{1}{100}$ Replacing % with $\times \dfrac{1}{100}$

$\qquad = \dfrac{28.9}{100}$

$\qquad = \dfrac{28.9}{100} \cdot \dfrac{10}{10}$ Multiplying by 1 to get a whole number in the numerator

$\qquad = \dfrac{289}{1000}$

27. $110\% = 110 \times \dfrac{1}{100} = \dfrac{110}{100}$

29. $0.042\% = 0.042 \times \dfrac{1}{100} = \dfrac{0.042}{100} = \dfrac{0.042}{100} \times \dfrac{1000}{1000} =$

$\qquad \dfrac{42}{100,000}$

31. $250\% = 250 \times \dfrac{1}{100} = \dfrac{250}{100}$

33. $3.47\% = 3.47 \times \dfrac{1}{100} = \dfrac{3.47}{100} = \dfrac{3.47}{100} \times \dfrac{100}{100} = \dfrac{347}{10,000}$

35. 0.73   0.73.

Move the decimal point 2 places to the right.

$0.73 = 73\%$

37. 0.41   0.41.

Move the decimal point 2 places to the right.

$0.41 = 41\%$

39. 1   1.00.

Move the decimal point 2 places to the right.

$1 = 100\%$

41. 0.996   0.99.6

Move the decimal point 2 places to the right.

$0.996 = 99.6\%$

43. 0.0047   0.00.47

Move the decimal point 2 places to the right.

$0.0047 = 0.47\%$

45. 0.072   0.07.2

Move the decimal point 2 places to the right.

$0.072 = 7.2\%$

47. 9.2   9.20.

Move the decimal point 2 places to the right.

$9.2 = 920\%$

49. 0.0068   0.00.68

Move the decimal point 2 places to the right.

$0.0068 = 0.68\%$

51. $\dfrac{1}{6} = 0.16\overline{6} = 16.\overline{6}\%$, or $16\dfrac{2}{3}\%$

53. $\dfrac{13}{20} = 0.65 = 65\%$

55. $\dfrac{29}{100} = 0.29 = 29\%$

57. $\dfrac{8}{10} = 0.8 = 80\%$

59. $\dfrac{3}{5} = 0.6 = 60\%$

61. $\dfrac{2}{3} = 0.66\overline{6} = 66.\overline{6}\%$, or $66\dfrac{2}{3}\%$

63. $\dfrac{7}{4} = 1.75 = 175\%$

65. $\dfrac{3}{4} = 0.75 = 75\%$

67. $\dfrac{118.6}{100,000} = \dfrac{118.6}{100,000} \cdot \dfrac{10}{10} = \dfrac{1186}{1,000,000} =$

$\qquad 0.001186 = 0.1186\%$

69. Decimal notation: Move the decimal point 2 places to the left.

$\qquad 49\% = 0.49$

Fraction notation:

$\qquad 49\% = 49 \times \dfrac{1}{100} = \dfrac{49}{100}$

71. Fraction notation:

$$0.29 = \frac{29}{100}$$

Percent notation: Move the decimal point 2 places to the right.

$$0.29 = 29\%$$

73. Decimal notation:

$$\frac{3}{10} = 0.3$$

Percent notation:

$$\frac{3}{10} = 0.3 = 30\%$$

75. Decimal notation: Move the decimal point 2 places to the left.

$$36\% = 0.36$$

Fraction notation:

$$36\% = 36 \times \frac{1}{100} = \frac{36}{100} = \frac{4 \cdot 9}{4 \cdot 25} = \frac{9}{25}$$

77. Discussion and Writing Exercise

79.
```
   2.2 5
4 ⟌9.0 0
   8
   ─
   1 0
     8
     ─
     2 0
     2 0
     ──
      0
```

Decimal notation for $\frac{9}{4}$ is 2.25.

81.
```
      1.4 1 6 6
12 ⟌1 7.0 0 0 0
     1 2
     ──
      5 0
      4 8
      ──
        2 0
        1 2
        ──
         8 0
         7 2
         ──
          8 0
          7 2
          ──
           8
```

The number 8 repeats as a remainder, so the digit 6 will repeat in the quotient. Thus, $\frac{17}{12} = 1.41\overline{6}$.

83.
```
       0.9 0 9 0 9
11 ⟌1 0.0 0 0 0 0 0
      9 9
      ──
      1 0 0
        9 9
        ──
        1 0 0
          9 9
          ──
          1 0 0
            9 9
            ──
              1
```

The pattern of the remainders repeats, so the sequence of digits in the quotient repeats. Thus, $\frac{10}{11} = 0.\overline{90}$.

85.
```
    2 3.4 5 8  ← 3 decimal places
  ×     7.0 3  ← 2 decimal places
    ─────────
     7 0 3 7 4
  1 6 4 2 0 6 0 0
  ───────────────
  1 6 4.9 0 9 7 4  ← 5 decimal places
```

87.
```
    8 0 9.5 6 9
  +   8 6.9 9
  ─────────────
    8 9 6.5 5 9
```

89. $18\% + 14\% = 0.18 + 0.14 = 0.32 = 32\%$

91. $1 - 30\% = 1 - 0.3 = 0.7 = 70\%$

93. $27 \times 100\% = 27 \times 1 = 27 = 2700\%$

95. $3(1 + 15\%) = 3(1 + 0.15) = 3(1.15) = 3.45 = 345\%$

97. $\frac{100\%}{40} = \frac{1}{40} = 0.025 = 2.5\%$

Exercise Set R.5

1. $\underbrace{5 \times 5 \times 5 \times 5}_{4 \text{ factors}} = 5^4$

3. $\underbrace{10 \cdot 10 \cdot 10}_{3 \text{ factors}} = 10^3$

5. $\underbrace{10 \times 10 \times 10 \times 10 \times 10 \times 10}_{6 \text{ factors}} = 10^6$

7. $7^2 = 7 \cdot 7 = 49$

9. $9^5 = 9 \cdot 9 \cdot 9 \cdot 9 \cdot 9 = 59{,}049$

11. $10^2 = 10 \cdot 10 = 100$

13. $1^4 = 1 \cdot 1 \cdot 1 \cdot 1 = 1$

15. $(2.3)^2 = (2.3)(2.3) = 5.29$

17. $(0.2)^3 = (0.2)(0.2)(0.2) = 0.008$

19. $(20.4)^2 = (20.4)(20.4) = 416.16$

21. $\left(\frac{3}{8}\right)^2 = \left(\frac{3}{8}\right)\left(\frac{3}{8}\right) = \frac{9}{64}$

23. $5^3 = 5 \cdot 5 \cdot 5 = 125$

25. $1000 \times (1.02)^3 = 1000 \times (1.02 \times 1.02 \times 1.02) = 1000 \times 1.061208 = 1061.208$

27. $9 + 2 \times 8 = 9 + 16$ Multiplying
$= 25$ Adding

29. $9(8) + 7(6) = 72 + 42$ Multiplying
$= 114$ Adding

31. $39 - 4 \times 2 + 2 = 39 - 8 + 2$ Multiplying
$= 31 + 2$ Subtracting
$= 33$ Adding

33. $9 \div 3 + 16 \div 8 = 3 + 2$ Dividing
$$= 5 \qquad \text{Adding}$$

35. $7 + 10 - 10 \div 2 = 7 + 10 - 5$ Dividing
$$= 17 - 5 \qquad \text{Adding and subtracting}$$
$$= 12 \qquad \text{from left to right}$$

37. $(6 \cdot 3)^2 = 18^2$ Multiplying within parentheses
$$= 324$$

39. $4 \cdot 5^2 = 4 \cdot 25$ Evaluating the exponential expression
$$= 100$$

41. $(8 + 2)^3 = 10^3$ Adding within parentheses
$$= 1000 \qquad \text{Evaluating the exponential expression}$$

43. $6 + 4^2 = 6 + 16$ Evaluating the exponential expression
$$= 22 \qquad \text{Adding}$$

45. $(3 - 2)^2 = 1^2$ Subtracting within parentheses
$$= 1 \qquad \text{Evaluating the exponential expression}$$

47. $4^3 \div 8 - 4$
$$= 64 \div 8 - 4 \qquad \text{Evaluating the exponential expression}$$
$$= 8 - 4 \qquad \text{Dividing}$$
$$= 4 \qquad \text{Subtracting}$$

49. $120 - 3^3 \cdot 4 \div 6$
$$= 120 - 27 \cdot 4 \div 6 \qquad \text{Evaluating the exponential expression}$$
$$= 120 - 108 \div 6 \qquad \text{Multiplying}$$
$$= 120 - 18 \qquad \text{Dividing}$$
$$= 102 \qquad \text{Subtracting}$$

51. $6[9 + (3 + 4)] = 6[9 + 7]$ Adding inside the parentheses
$$= 6[16] \qquad \text{Adding inside the brackets}$$
$$= 96 \qquad \text{Multiplying}$$

53. $8 + (7 + 9) = 8 + 16$ Adding inside the parentheses
$$= 24 \qquad \text{Adding}$$

55. $15(4 + 2) = 15(6)$ Adding inside the parentheses
$$= 90 \qquad \text{Multiplying}$$

57. $12 - (8 - 4) = 12 - 4$ Subtracting inside the parentheses
$$= 8 \qquad \text{Subtracting}$$

59. $1000 \div 100 \div 10 = 10 \div 10$ Dividing in order from left to right
$$= 1 \qquad \text{Doing the second division}$$

61. $2000 \div \dfrac{3}{50} \cdot \dfrac{3}{2}$

$$= \frac{2000}{1} \cdot \frac{50}{3} \cdot \frac{3}{2} \qquad \text{Dividing}$$

$$= \frac{100,000}{3} \cdot \frac{3}{2} \qquad \text{Completing the division}$$

$$= \frac{100,000 \cdot 3}{3 \cdot 2} \qquad \text{Multiplying}$$

$$= \frac{\cancel{2} \cdot 50,000 \cdot \cancel{3}}{1 \cdot \cancel{3} \cdot \cancel{2}}$$

$$= 50,000 \qquad \text{Simplifying}$$

63. We multiply and divide in order from left to right.
$$75 \div 15 \cdot 4 \cdot 8 \div 32 = 5 \cdot 4 \cdot 8 \div 32$$
$$= 20 \cdot 8 \div 32$$
$$= 160 \div 32$$
$$= 5$$

65. We multiply and divide in order from left to right.
$$16 \cdot 5 \div 80 \div 12 \cdot 36 \cdot 9 = 80 \div 80 \div 12 \cdot 36 \cdot 9$$
$$= 1 \div 12 \cdot 36 \cdot 9$$
$$= \frac{1}{12} \cdot 36 \cdot 9$$
$$= \frac{36}{12} \cdot 9$$
$$= 3 \cdot 9$$
$$= 27$$

67. We will do the calculations in the numerator and in the denominator and then divide the results.
$$\frac{80 - 6^2}{9^2 + 3^2} = \frac{80 - 36}{81 + 9}$$
$$= \frac{44}{90}$$
$$= \frac{22 \cdot \cancel{2}}{45 \cdot \cancel{2}}$$
$$= \frac{22}{45}$$

69. $\dfrac{3(6 + 7) - 5 \cdot 4}{6 \cdot 7 + 8(4 - 1)} = \dfrac{3 \cdot 13 - 5 \cdot 4}{6 \cdot 7 + 8 \cdot 3}$
$$= \frac{39 - 20}{42 + 24}$$
$$= \frac{19}{66}$$

71. $8 \cdot 2 - (12 - 0) \div 3 - (5 - 2)$
$$= 8 \cdot 2 - 12 \div 3 - 3 \qquad \text{Subtracting inside the parentheses}$$
$$= 16 - 4 - 3 \qquad \text{Multiplying and dividing in order from left to right}$$
$$= 12 - 3 \qquad \text{Subtracting in order}$$
$$= 9 \qquad \text{from left to right}$$

73. Discussion and Writing Exercise

75. $\dfrac{5}{16} = 0.3125$ 0.31.25

Move the decimal point two places to the right.

$\dfrac{5}{16} = 0.3125 = 31.25\%$

77. $\dfrac{9}{2001} = \dfrac{\cancel{3} \cdot 3}{\cancel{3} \cdot 667} = \dfrac{3}{667}$

79. We begin by factoring 48 in any way that we can and continue factoring until each factor is prime.

$48 = 2 \cdot 24 = 2 \cdot 4 \cdot 6 = 2 \cdot 2 \cdot 2 \cdot 2 \cdot 3$

81. $\dfrac{10^5}{10^3} = \dfrac{\cancel{10} \cdot \cancel{10} \cdot \cancel{10} \cdot 10 \cdot 10}{\cancel{10} \cdot \cancel{10} \cdot \cancel{10} \cdot 1} = 10 \cdot 10 = 10^2$

83. $5^4 \cdot 5^2 = (5 \cdot 5 \cdot 5 \cdot 5) \cdot (5 \cdot 5) = 5^6$

85. See the answer section in the text.

Exercise Set R.6

1. Perimeter $= 4 \text{ mm} + 6 \text{ mm} + 7 \text{ mm}$
$= (4 + 6 + 7) \text{ mm}$
$= 17 \text{ mm}$

3. Perimeter $= 3.5 \text{ in.} + 3.5 \text{ in.} + 4.25 \text{ in.} +$
$0.5 \text{ in.} + 3.5 \text{ in.}$
$= (3.5 + 3.5 + 4.25 + 0.5 + 3.5) \text{ in.}$
$= 15.25 \text{ in.}$

5. $P = 2 \cdot (l + w)$ Perimeter of a rectangle
$P = 2 \cdot (5 \text{ ft} + 10 \text{ ft})$
$P = 2 \cdot (15 \text{ ft})$
$P = 30 \text{ ft}$

7. $P = 2 \cdot (l + w)$ Perimeter of a rectangle
$P = 2 \cdot (34.67 \text{ cm} + 4.9 \text{ cm})$
$P = 2 \cdot (39.57 \text{ cm})$
$P = 79.14 \text{ cm}$

9. $P = 4 \cdot s$ Perimeter of a square

$P = 4 \cdot 20\dfrac{3}{8} \text{ ft}$

$P = 4 \cdot \dfrac{163}{8} \text{ ft}$

$P = \dfrac{4 \cdot 163}{8} \text{ ft} = \dfrac{\cancel{4} \cdot 163}{2 \cdot \cancel{4}} \text{ ft}$

$P = \dfrac{163}{2} \text{ ft} = 81\dfrac{1}{2} \text{ ft}$

11. $P = 4 \cdot s$ Perimeter of a square
$P = 4 \cdot 45.5 \text{ mm}$
$P = 182 \text{ mm}$

13. *Familiarize.* We label the missing lengths on the drawing and let $P =$ the perimeter.

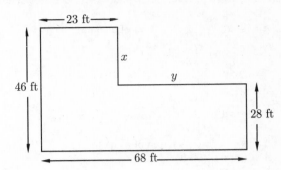

Translate. First we find the missing lengths x and y.

28 ft	plus	how many more ft	is	46 ft
↓	↓	↓	↓	↓
28	+	x	=	46

23 ft	plus	how many more ft	is	68 ft
↓	↓	↓	↓	↓
23	+	y	=	68

Solve. We solve for x and y.

$$28 + x = 46 \qquad\qquad 23 + y = 68$$
$$x = 46 - 28 \qquad\qquad y = 68 - 23$$
$$x = 18 \qquad\qquad\qquad y = 45$$

a) To find the perimeter we add the lengths of the sides of the house.

$P = 23 \text{ ft} + 18 \text{ ft} + 45 \text{ ft} + 28 \text{ ft} + 68 \text{ ft} + 46 \text{ ft}$
$= (23 + 18 + 45 + 28 + 68 + 46) \text{ ft}$
$= 228 \text{ ft}$

b) Next we find t, the total cost of the gutter.

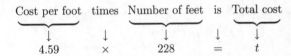

Cost per foot	times	Number of feet	is	Total cost
↓	↓	↓	↓	↓
4.59	×	228	=	t

We carry out the multiplication.

$$
\begin{array}{r}
2\,2\,8 \\
\times\ 4\,.5\,9 \\
\hline
2\,0\,5\,2 \\
1\,1\,4\,0\,0 \\
9\,1\,2\,0\,0 \\
\hline
1\,0\,4\,6\,.5\,2
\end{array}
$$

Thus, $t = 1046.52$.

Check. We can repeat the calculations.

State. (a) The perimeter of the house is 228 ft. (b) The total cost of the gutter is $1046.52.

15. $A = l \cdot w$ Area of a rectangle
$A = (5 \text{ km}) \cdot (3 \text{ km})$
$A = 5 \cdot 3 \cdot \text{ km} \cdot \text{ km}$
$A = 15 \text{ km}^2$

17. $A = l \cdot w$ Area of a rectangle
$A = (2 \text{ in.}) \cdot (0.7 \text{ in.})$
$A = 2 \cdot 0.7 \cdot \text{ in.} \cdot \text{ in.}$
$A = 1.4 \text{ in}^2$

19. $A = s \cdot s$ $\qquad$ Area of a square

$$A = \left(\frac{2}{3}\text{ yd}\right) \cdot \left(\frac{2}{3}\text{ yd}\right)$$

$$A = \frac{2}{3} \cdot \frac{2}{3} \cdot \text{ yd} \cdot \text{ yd}$$

$$A = \frac{4}{9}\text{ yd}^2$$

21. $A = s \cdot s$ $\qquad$ Area of a square

$A = (90\text{ ft}) \cdot (90\text{ ft})$

$A = 90 \cdot 90 \cdot \text{ ft} \cdot \text{ ft}$

$A = 8100\text{ ft}^2$

23. $A = l \cdot w$ $\qquad$ Area of a rectangle

$A = (10\text{ ft}) \cdot (5\text{ ft})$

$A = 10 \cdot 5 \cdot \text{ ft} \cdot \text{ ft}$

$A = 50\text{ ft}^2$

25. $A = l \cdot w$ $\qquad$ Area of a rectangle

$A = (34.67\text{ cm}) \cdot (4.9\text{ cm})$

$A = 34.67 \cdot 4.9 \cdot \text{ cm} \cdot \text{ cm}$

$A = 169.883\text{ cm}^2$

27. $A = l \cdot w$ $\qquad$ Area of a rectangle

$$A = \left(\frac{2}{3}\text{ in.}\right) \cdot \left(\frac{5}{6}\text{ in.}\right)$$

$$A = \frac{2}{3} \cdot \frac{5}{6} \cdot \text{ in.} \cdot \text{ in.}$$

$$A = \frac{2 \cdot 5}{3 \cdot 2 \cdot 3}\text{ in}^2$$

$$A = \frac{2}{2} \cdot \frac{5}{3 \cdot 3}\text{ in}^2$$

$$A = \frac{5}{9}\text{ in}^2$$

29. $A = s \cdot s$ $\qquad$ Area of a square

$$A = \left(22\frac{1}{4}\text{ ft}\right) \cdot \left(22\frac{1}{4}\text{ ft}\right)$$

$$A = 22\frac{1}{4} \cdot 22\frac{1}{4} \cdot \text{ ft} \cdot \text{ ft}$$

$$A = \frac{89}{4} \cdot \frac{89}{4}\text{ ft}^2$$

$$A = \frac{7921}{16}\text{ ft}^2$$

$$A = 495\frac{1}{16}\text{ ft}^2$$

31. $A = s \cdot s$ $\qquad$ Area of a square

$A = (56.9\text{ km}) \cdot (56.9\text{ km})$

$A = 56.9 \cdot 56.9 \cdot \text{ km} \cdot \text{ km}$

$A = 3237.61\text{ km}^2$

33. $A = s \cdot s$ $\qquad$ Area of a square

$$A = \left(\frac{3}{8}\text{ yd}\right) \cdot \left(\frac{3}{8}\text{ yd}\right)$$

$$A = \frac{3}{8} \cdot \frac{3}{8} \cdot \text{ yd} \cdot \text{ yd}$$

$$A = \frac{9}{64}\text{ yd}^2$$

35. *Familiarize.* We use the drawing in the text.

Translate. We let A = the area of the sidewalk.

Area of sidewalk	is	Total area	minus	Area of building
$\downarrow$	$\downarrow$	$\downarrow$	$\downarrow$	$\downarrow$
A	$=$	$(113.4\text{ m}) \times (75.4\text{ m})$	$-$	$(110\text{ m}) \times (72\text{ m})$

Solve. The total area is

$$(113.4\text{ m}) \times (75.4\text{ m}) = 113.4 \times 75.4 \times \text{ m} \times \text{ m} = 8550.36\text{ m}^2.$$

The area of the building is

$$(110\text{ m}) \times (72\text{ m}) = 110 \times 72 \times \text{ m} \times \text{ m} = 7920\text{ m}^2.$$

The area of the sidewalk is

$$A = 8550.36\text{ m}^2 - 7920\text{ m}^2 = 630.36\text{ m}^2.$$

Check. Repeat the calculations.

State. The area of the sidewalk is 630.36 m^2.

37. $A = b \cdot h$ $\qquad$ Area of a parallelogram

$A = 8\text{ cm} \cdot 4\text{ cm}$ $\qquad$ Substituting 8 cm for b and 4 cm for h

$A = 32\text{ cm}^2$

39. $A = \frac{1}{2} \cdot b \cdot h$ $\qquad$ Area of a triangle

$A = \frac{1}{2} \cdot 15\text{ in.} \cdot 8\frac{1}{2}\text{ in.}$ $\qquad$ Substituting 15 in. for b and $8\frac{1}{2}$ in. for h

$$A = \frac{1}{2} \cdot 15 \cdot \frac{17}{2}\text{ in}^2$$

$$A = \frac{255}{4}\text{ in}^2$$

$$A = 63\frac{3}{4}\text{ in}^2$$

41. $A = b \cdot h$ $\qquad$ Area of a parallelogram

$A = 2.3\text{ cm} \cdot 3.5\text{ cm}$ $\qquad$ Substituting 2.3 cm for b and 3.5 cm for h

$A = 8.05\text{ cm}^2$

43. $A = \frac{1}{2} \cdot b \cdot h$ $\qquad$ Area of a triangle

$A = \frac{1}{2} \cdot 4\text{ mi} \cdot 3\frac{1}{2}\text{ mi}$ $\qquad$ Substituting 4 m for b and $3\frac{1}{2}$ mi for h

$$A = \frac{1}{2} \cdot 4 \cdot \frac{7}{2}\text{ mi}^2$$

$$A = \frac{1 \cdot 4 \cdot 7}{2 \cdot 2}\text{ mi}^2$$

$$A = \frac{1 \cdot \cancel{2} \cdot \cancel{2} \cdot 7}{\cancel{2} \cdot \cancel{2} \cdot 1}\text{ mi}^2$$

$$A = 7\text{ mi}^2$$

45. $d = 2 \cdot r$

$d = 2 \cdot 7$ cm $= 14$ cm

$C = 2 \cdot \pi \cdot r$

$C \approx 2 \cdot \dfrac{22}{7} \cdot 7$ cm $= \dfrac{2 \cdot 22 \cdot 7}{7}$ cm $= 44$ cm

$A = \pi \cdot r \cdot r$

$A \approx \dfrac{22}{7} \cdot 7$ cm $\cdot 7$ cm $= \dfrac{22}{7} \cdot 49$ cm$^2 = 154$ cm^2

47. $d = 2 \cdot r$

$d = 2 \cdot \dfrac{3}{4}$ in. $= \dfrac{6}{4}$ in. $= \dfrac{3}{2}$ in., or $1\dfrac{1}{2}$ in.

$C = 2 \cdot \pi \cdot r$

$C \approx 2 \cdot \dfrac{22}{7} \cdot \dfrac{3}{4}$ in. $= \dfrac{2 \cdot 22 \cdot 3}{7 \cdot 4}$ in. $= \dfrac{132}{28}$ in. $= \dfrac{33}{7}$ in.,

or $4\dfrac{5}{7}$ in.

$A = \pi \cdot r \cdot r$

$A \approx \dfrac{22}{7} \cdot \dfrac{3}{4}$ in. $\cdot \dfrac{3}{4}$ in. $= \dfrac{22 \cdot 3 \cdot 3}{7 \cdot 4 \cdot 4}$ in$^2 = \dfrac{99}{56}$ in^2, or $1\dfrac{43}{56}$ in^2

49. $r = \dfrac{d}{2}$

$r = \dfrac{32 \text{ ft}}{2} = 16$ ft

$C = \pi \cdot d$

$C \approx 3.14 \cdot 32$ ft $= 100.48$ ft

$A = \pi \cdot r \cdot r$

$A \approx 3.14 \cdot 16$ ft $\cdot 16$ ft $\quad \left(r = \dfrac{d}{2}; r = \dfrac{32 \text{ ft}}{2} = 16 \text{ ft}\right)$

$A = 3.14 \cdot 256$ ft^2

$A = 803.84$ ft^2

51. $r = \dfrac{d}{2}$

$r = \dfrac{1.4 \text{ cm}}{2} = 0.7$ cm

$C = \pi \cdot d$

$C \approx 3.14 \cdot 1.4$ cm $= 4.396$ cm

$A = \pi \cdot r \cdot r$

$A \approx 3.14 \cdot 0.7$ cm $\cdot 0.7$ cm

$\qquad \left(r = \dfrac{d}{2}; r = \dfrac{1.4 \text{ cm}}{2} = 0.7 \text{ cm}\right)$

$A = 3.14 \cdot 0.49$ cm$^2 = 1.5386$ cm^2

53. $r = \dfrac{d}{2}$

$r = \dfrac{14 \text{ ft}}{2} = 7$ ft

$A = \pi \cdot r \cdot r$

$A \approx \dfrac{22}{7} \cdot 7$ ft $\cdot 7$ ft

$A = \dfrac{22}{7} \cdot 49$ ft$^2 = \dfrac{22 \cdot 49}{7}$ ft^2

$A = 154$ ft^2

55. $V = l \cdot w \cdot h$

$V = 12$ cm $\cdot 8$ cm $\cdot 8$ cm

$V = 12 \cdot 64$ cm^3

$V = 768$ cm^3

57. $V = l \cdot w \cdot h$

$V = 7.5$ in. $\cdot 2$ in. $\cdot 3$ in.

$V = 7.5 \cdot 6$ in^3

$V = 45$ in^3

59. $V = l \cdot w \cdot h$

$V = 10$ m $\cdot 5$ m $\cdot 1.5$ m

$V = 10 \cdot 7.5$ m^3

$V = 75$ m^3

61. $V = l \cdot w \cdot h$

$V = 6.5$ yd $\cdot 5.5$ yd $\cdot 10$ yd

$V = 357.5$ yd^3

63. $V = l \cdot w \cdot h$

$V = 22$ in. $\times 9$ in. $\times 14$ in.

$V = 2772$ in^3

65. Discussion and Writing Exercise

67. 0.875

a) Move the decimal point 2 places to the right. 0.87.5

b) Add a percent symbol. 87.5%

$0.875 = 87.5\%$

69. a) Find decimal notation using long division.

```
      0.3 7 5
  8 ⟌ 3.0 0 0
      2 4
      ───
        6 0
        5 6
        ───
          4 0
          4 0
          ───
            0
```

$\dfrac{3}{8} = 0.375$

b) Convert the decimal notation to percent notation. Move the decimal point two places to the right, and write a % symbol.

0.37.5

$\dfrac{3}{8} = 37.5\%$

71. The LCD is 360.

$\dfrac{17}{72} + \dfrac{13}{60} = \dfrac{17}{72} \cdot \dfrac{5}{5} + \dfrac{13}{60} \cdot \dfrac{6}{6}$

$\qquad = \dfrac{85}{360} + \dfrac{78}{360}$

$\qquad = \dfrac{163}{360}$

73. The LCD is 108.
$$\frac{49}{54} - \frac{19}{36} = \frac{49}{54} \cdot \frac{2}{2} - \frac{19}{36} \cdot \frac{3}{3}$$
$$= \frac{98}{108} - \frac{57}{108}$$
$$= \frac{41}{108}$$

75. $V = Bh = \pi \cdot r^2 \cdot h$
$\approx 3.14 \times 8 \text{ in.} \times 8 \text{ in.} \times 4 \text{ in.}$
$\approx 803.84 \text{ in}^3$

77. $V = Bh = \pi \cdot r^2 \cdot h$
$\approx 3.14 \times 5 \text{ cm} \times 5 \text{ cm} \times 4.5 \text{ cm}$
$\approx 353.25 \text{ cm}^3$

Chapter R Review Exercises

1. We begin by factoring 92 in any way that we can and continue factoring until each factor is prime.
$92 = 2 \cdot 46 = 2 \cdot 2 \cdot 23$

2. We begin by factoring 1400 in any way that we can and continue factoring until each factor is prime.
$1400 = 14 \cdot 100 = 2 \cdot 7 \cdot 4 \cdot 25 = 2 \cdot 7 \cdot 2 \cdot 2 \cdot 5 \cdot 5$, or $2 \cdot 2 \cdot 2 \cdot 5 \cdot 5 \cdot 7$

3. $13 = 13$ (13 is prime.)
$32 = 2 \cdot 2 \cdot 2 \cdot 2 \cdot 2$
The LCM is $2 \cdot 2 \cdot 2 \cdot 2 \cdot 2 \cdot 13$, or 416.

4. $5 = 5$ (5 is prime.)
$18 = 2 \cdot 3 \cdot 3$
$45 = 3 \cdot 3 \cdot 5$
The LCM is $2 \cdot 3 \cdot 3 \cdot 5$, or 90.

5. $\frac{2}{5} = \frac{2}{5} \cdot 1 = \frac{2}{5} \cdot \frac{6}{6} = \frac{12}{30}$

6. $\frac{12}{23} = \frac{12}{23} \cdot 1 = \frac{12}{23} \cdot \frac{8}{8} = \frac{96}{184}$

7. We will use $\frac{8}{8}$ for 1 since $64 = 8 \cdot 8$.
$\frac{5}{8} = \frac{5}{8} \cdot 1 = \frac{5}{8} \cdot \frac{8}{8} = \frac{40}{64}$

8. We will use $\frac{7}{7}$ for 1 since $84 = 12 \cdot 7$.
$\frac{13}{12} = \frac{13}{12} \cdot 1 = \frac{13}{12} \cdot \frac{7}{7} = \frac{91}{84}$

9. $\frac{20}{48} = \frac{4 \cdot 5}{4 \cdot 12} = \frac{4}{4} \cdot \frac{5}{12} = 1 \cdot \frac{5}{12} = \frac{5}{12}$

10. $\frac{1020}{1820} = \frac{2 \cdot 2 \cdot 3 \cdot 5 \cdot 17}{2 \cdot 2 \cdot 5 \cdot 7 \cdot 13}$
$= \frac{\cancel{2} \cdot \cancel{2} \cdot 3 \cdot \cancel{5} \cdot 17}{\cancel{2} \cdot \cancel{2} \cdot \cancel{5} \cdot 7 \cdot 13}$
$= \frac{3 \cdot 17}{7 \cdot 13}$
$= \frac{51}{91}$

11. The LCD is 36.
$$\frac{4}{9} + \frac{5}{12} = \frac{4}{9} \cdot \frac{4}{4} + \frac{5}{12} \cdot \frac{3}{3}$$
$$= \frac{16}{36} + \frac{15}{36}$$
$$= \frac{31}{36}$$

12. $\frac{3}{4} \div 3 = \frac{3}{4} \cdot \frac{1}{3} = \frac{\cancel{3} \cdot 1}{4 \cdot \cancel{3}} = \frac{1}{4}$

13. The LCD is 15.
$$\frac{2}{3} - \frac{1}{15} = \frac{2}{3} \cdot \frac{5}{5} - \frac{1}{15}$$
$$= \frac{10}{15} - \frac{1}{15} = \frac{9}{15}$$
$$= \frac{\cancel{3} \cdot 3}{\cancel{3} \cdot 5} = \frac{3}{5}$$

14. $\frac{9}{10} \cdot \frac{16}{5} = \frac{9 \cdot 16}{10 \cdot 5} = \frac{9 \cdot \cancel{2} \cdot 8}{\cancel{2} \cdot 5 \cdot 5} = \frac{72}{25}$

15. The LCD is 144.
$$\frac{11}{18} + \frac{13}{16} = \frac{11}{18} \cdot \frac{8}{8} + \frac{13}{16} \cdot \frac{9}{9}$$
$$= \frac{88}{144} + \frac{117}{144}$$
$$= \frac{205}{144}$$

16. The LCD is 72.
$$\frac{35}{36} + \frac{23}{24} = \frac{35}{36} \cdot \frac{2}{2} + \frac{23}{24} \cdot \frac{3}{3}$$
$$= \frac{70}{72} + \frac{69}{72}$$
$$= \frac{139}{72}$$

17. The LCD is 54.
$$\frac{25}{27} + \frac{17}{18} = \frac{25}{27} \cdot \frac{2}{2} + \frac{17}{18} \cdot \frac{3}{3}$$
$$= \frac{50}{54} + \frac{51}{54}$$
$$= \frac{101}{54}$$

18. The LCD is 84.
$$\frac{29}{42} + \frac{17}{28} = \frac{29}{42} \cdot \frac{2}{2} + \frac{17}{28} \cdot \frac{3}{3}$$
$$= \frac{58}{84} + \frac{51}{84}$$
$$= \frac{109}{84}$$

19. The LCD is 72.
$$\frac{35}{36} - \frac{19}{24} = \frac{35}{36} \cdot \frac{2}{2} - \frac{19}{24} \cdot \frac{3}{3}$$
$$= \frac{70}{72} - \frac{57}{72}$$
$$= \frac{13}{72}$$

20. The LCD is 144.

$$\frac{13}{16} - \frac{11}{18} = \frac{13}{16} \cdot \frac{9}{9} - \frac{11}{18} \cdot \frac{8}{8}$$

$$= \frac{117}{144} - \frac{88}{144}$$

$$= \frac{29}{144}$$

21. The LCD is 84.

$$\frac{29}{42} - \frac{17}{28} = \frac{29}{42} \cdot \frac{2}{2} - \frac{17}{28} \cdot \frac{3}{3}$$

$$= \frac{58}{84} - \frac{51}{84} = \frac{7}{84}$$

$$= \frac{7 \cdot 1}{7 \cdot 12} = \frac{1}{12}$$

22. The LCD is 180.

$$\frac{11}{36} - \frac{1}{20} = \frac{11}{36} \cdot \frac{5}{5} - \frac{1}{20} \cdot \frac{9}{9}$$

$$= \frac{55}{180} - \frac{9}{180} = \frac{46}{180}$$

$$= \frac{2 \cdot 23}{2 \cdot 90} = \frac{23}{90}$$

23. 17.97. $17.97 = \dfrac{1797}{100}$

2 places   2 zeros

24. $\dfrac{2337}{10,000}$ 0.2337. $\dfrac{2337}{10,000} = 0.2337$

4 zeros   4 places

25.
```
   1 1  1
  2 3 4 4.5 6
+     9 8.3 4 5
 2 4 4 2.9 0 5
```

26.
```
   1 1 1
     6.0 4
    7 8.
+    1.9 8 9 8
   8 6.0 2 9 8
```

27.
```
  1 10 3 9 10
  2 0.4 0 0
- 1 1.0 5 8
   9.3 4 2
```

28.
```
          12
      8 9 2 12
  7 8 9.0 3 2
- 6 5 5.7 6 8
  1 3 3.2 6 4
```

29.
```
    1 7.9 5
  ×     2 4
    7 1 8 0
  3 5 9 0 0
  4 3 0.8 0
```

30.
```
      5 6.9 5
  ×      1.9 4
      2 2 7 8 0
    5 1 2 5 5 0
    5 6 9 5 0 0
  1 1 0.4 8 3 0
```

31.
```
            5 5.6
  2.8 ∧ 1 5 5.6∧8
        1 4 0 0
          1 5 6 8
          1 4 0 0
            1 6 8
            1 6 8
                0
```

32.
```
         0.4 5
  5 2 2 3.4 0
      2 0 8
        2 6 0
        2 6 0
            0
```

33.
```
          1.5 8 3
  1 2 1 9.0 0 0
      1 2
        7 0
        6 0
        1 0 0
          9 6
            4 0
            3 6
              4
```

The number 4 repeats as a remainder, so the digit 3 will repeat in the quotient. Thus, decimal notation for $\dfrac{19}{12}$ is $1.5833\ldots$, or $1.58\overline{3}$.

34. Round 34.067 to the nearest tenth. The digit in the tenths place is 0. The next digit to the right, 6, is 5 or higher, so we round up: 34.1.

35. 15.5%   0.15.5

Move the decimal point 2 places to the left.

$15.5\% = 0.155$

36. 0.127   0.12.7

Move the decimal point 2 places to the right.

$0.127 = 12.7\%$

37. $3.11\% = 3.11 \times \dfrac{1}{100} = \dfrac{3.11}{100} = \dfrac{3.11}{100} \times \dfrac{100}{100} = \dfrac{311}{10,000}$

38. $\dfrac{114.1}{100,000} = \dfrac{114.1}{100,000} \cdot \dfrac{10}{10} = \dfrac{1141}{1,000,000} = 0.001141 = 0.1141\%$

39. $\dfrac{5}{8} = 0.625 = 62.5\%$

40. $\dfrac{29}{25} = 1.16 = 116\%$

41. $\underbrace{6 \cdot 6 \cdot 6}_{\text{3 factors}} = 6^3$

42. $(1.06)^2 = (1.06)(1.06) = 1.1236$

43. $\quad 120 - 6^2 \div 4 + 8$

$= 120 - 36 \div 4 + 8 \quad$ Evaluating the exponential expression

$= 120 - 9 + 8 \quad\quad$ Dividing

$= 111 + 8 \quad\quad\quad$ Subtracting

$= 119 \quad\quad\quad\quad$ Adding

44. We multiply and divide in order from left to right.

$64 \div 16 \cdot 32 \div 48 \div 12 \cdot 18 = 4 \cdot 32 \div 48 \div 12 \cdot 18$

$= 128 \div 48 \div 12 \cdot 18$

$= \dfrac{128}{48} \div 12 \cdot 18$

$= \dfrac{128}{48} \cdot \dfrac{1}{12} \cdot 18$

$= \dfrac{128 \cdot 1}{48 \cdot 12} \cdot 18$

$= \dfrac{128 \cdot 18}{48 \cdot 12}$

$= \dfrac{\cancel{2} \cdot \cancel{2} \cdot \cancel{2} \cdot \cancel{2} \cdot \cancel{2} \cdot \cancel{2} \cdot 2 \cdot 2 \cdot \cancel{3} \cdot \cancel{3}}{\cancel{2} \cdot \cancel{2} \cdot \cancel{2} \cdot \cancel{2} \cdot \cancel{3} \cdot \cancel{2} \cdot \cancel{2} \cdot \cancel{3} \cdot 1}$

$= 4$

45. $(120 - 6^2) \div 4 + 8 = (120 - 36) \div 4 + 8$

$= 84 \div 4 + 8$

$= 21 + 8$

$= 29$

46. We will multiply and divide in order from left to right.

$\quad 64 \cdot 16 \div 32 \div 48 \div 12 \cdot 18$

$= 1024 \div 32 \div 48 \div 12 \cdot 18$

$= 32 \div 48 \div 12 \cdot 18$

$= \dfrac{32}{48} \div 12 \cdot 18$

$= \dfrac{32}{48} \cdot \dfrac{1}{12} \cdot 18$

$= \dfrac{32 \cdot 1}{48 \cdot 12} \cdot 18$

$= \dfrac{32 \cdot 18}{48 \cdot 12}$

$= \dfrac{576}{576}$

$= 1$

47. $(120 - 6^2) \div (4 + 8) = (120 - 36) \div 12 = 84 \div 12 = 7$

48. $\quad 8^2 \cdot 2^4 \div 2^2 \cdot 8 \div 48 \div 12 \cdot 18$

$= 64 \cdot 16 \div 4 \cdot 8 \div 48 \div 12 \cdot 18$

$= 1024 \div 4 \cdot 8 \div 48 \div 12 \cdot 18$

$= 256 \cdot 8 \div 48 \div 12 \cdot 18$

$= 2048 \div 48 \div 12 \cdot 18$

$= \dfrac{2048}{48} \div 12 \cdot 18$

$= \dfrac{2048}{48} \cdot \dfrac{1}{12} \cdot 18$

$= \dfrac{2048 \cdot 1}{48 \cdot 12} \cdot 18$

$= \dfrac{2048 \cdot 18}{48 \cdot 12}$

$= \dfrac{\cancel{4} \cdot \cancel{16} \cdot 32 \cdot 2 \cdot \cancel{3} \cdot \cancel{3}}{\cancel{3} \cdot \cancel{16} \cdot \cancel{3} \cdot \cancel{4} \cdot 1}$

$= 64$

49. $\dfrac{4(18 - 8) + 7 \cdot 9}{9^2 - 8^2} = \dfrac{4(10) + 7 \cdot 9}{81 - 64}$

$= \dfrac{40 + 63}{17}$

$= \dfrac{103}{17}$

50. $\text{Perimeter} = 5 \text{ m} + 7 \text{ m} + 4 \text{ m} + 4 \text{ m} + 3 \text{ m}$

$= (5 + 7 + 4 + 4 + 3) \text{ m}$

$= 23 \text{ m}$

51. $\text{Perimeter} = 0.5 \text{ m} + 1.9 \text{ m} + 1.2 \text{ m} + 0.8 \text{ m}$

$= (0.5 + 1.9 + 1.2 + 0.8) \text{ m}$

$= 4.4 \text{ m}$

52. $P = 2 \cdot (l + w) \quad$ Perimeter of a rectangle

$P = 2 \cdot (78 \text{ ft} + 36 \text{ ft})$

$P = 2 \cdot (114 \text{ ft})$

$P = 228 \text{ ft}$

$A = l \cdot w \quad$ Area of a rectangle

$A = (78 \text{ ft}) \cdot (36 \text{ ft})$

$A = 78 \cdot 36 \cdot \text{ ft} \cdot \text{ ft}$

$A = 2808 \text{ ft}^2$

53. $P = 4 \cdot s \quad$ Perimeter of a square

$P = 4 \cdot 9 \text{ ft}$

$P = 36 \text{ ft}$

$A = s \cdot s \quad$ Area of a square

$A = (9 \text{ ft}) \cdot (9 \text{ ft})$

$A = 9 \cdot 9 \cdot \text{ ft} \cdot \text{ ft}$

$A = 81 \text{ ft}^2$

54. $P = 2 \cdot (l + w) \quad$ Perimeter of a rectangle

$P = 2 \cdot (7 \text{ cm} + 1.8 \text{ cm})$

$P = 2 \cdot (8.8 \text{ cm})$

$P = 17.6 \text{ cm}$

$A = l \cdot w$ Area of a rectangle

$A = (7 \text{ cm}) \cdot (1.8 \text{ cm})$

$A = 7 \cdot 1.8 \cdot \text{ cm} \cdot \text{ cm}$

$A = 12.6 \text{ cm}^2$

55. $A = b \cdot h$ Area of a parallelogram

$A = 12 \text{ cm} \cdot 5 \text{ cm}$

$A = 60 \text{ cm}^2$

56. $A = \dfrac{1}{2} \cdot b \cdot h$ Area of a triangle

$A = \dfrac{1}{2} \cdot 15 \text{ m} \cdot 3 \text{ m}$

$A = \dfrac{15 \cdot 3}{2} \text{ m}^2$

$A = \dfrac{45}{2} \text{ m}^2$, or 22.5 m^2

57. $A = \dfrac{1}{2} \cdot b \cdot h$ Area of a triangle

$A = \dfrac{1}{2} \cdot 11 \text{ cm} \cdot 5 \text{ cm}$

$A = \dfrac{11 \cdot 5}{2} \text{ cm}^2$

$A = \dfrac{55}{2} \text{ cm}^2$, or 27.5 cm^2

58. $A = b \cdot h$ Area of a parallelogram

$A = 21 \text{ in.} \cdot 6 \text{ in.}$

$A = 126 \text{ in}^2$

59. *Familiarize.* We use the drawing in the text. The total area has length 70 ft and width 25 ft + 7 ft, or 32 ft. The building has length 70 ft − 7 ft − 7 ft, or 56 ft, and width 25 ft. Let A = the area of the grassy section.

Translate.

Grassy area	is	Total area	minus	Area of building
↓	↓	↓	↓	↓
A	=	70 ft · 32 ft	−	56 ft · 25 ft

Solve. We carry out the computation.

$A = 70 \text{ ft} \cdot 32 \text{ ft} - 56 \text{ ft} \cdot 25 \text{ ft}$

$A = 2240 \text{ ft}^2 - 1400 \text{ ft}^2$

$A = 840 \text{ ft}^2$

Check. Repeat the calculation.

State. The area of the grassy section is 840 ft^2.

60. $r = \dfrac{d}{2} = \dfrac{16 \text{ m}}{2} = 8 \text{ m}$

61. $r = \dfrac{d}{2} = \dfrac{\frac{28}{11} \text{ in.}}{2} = \dfrac{28}{11} \text{ in.} \cdot \dfrac{1}{2} = \dfrac{28}{11 \cdot 2} \text{ in.} = \dfrac{2 \cdot 14}{11 \cdot 2} \text{ in.} = \dfrac{14}{11} \text{ in.}$

62. $d = 2 \cdot r = 2 \cdot 7 \text{ ft} = 14 \text{ ft}$

63. $d = 2 \cdot r = 2 \cdot 10 \text{ cm} = 20 \text{ cm}$

64. $C = \pi \cdot d$

$C \approx 3.14 \cdot 16 \text{ m} = 50.24 \text{ m}$

65. $C = \pi \cdot d$

$C \approx \dfrac{22}{7} \cdot \dfrac{28}{11} \text{ in.} = \dfrac{22 \cdot 28}{7 \cdot 11} \text{ in.} = \dfrac{2 \cdot 11 \cdot 4 \cdot 7}{7 \cdot 11 \cdot 1} \text{ in.} = 8 \text{ in.}$

66. In Exercise 60 we found that $r = 8 \text{ m}$.

$A = \pi \cdot r \cdot r$

$A \approx 3.14 \cdot 8 \text{ m} \cdot 8 \text{ m} = 200.96 \text{ m}^2$

67. In Exercise 61 we found that $r = \dfrac{14}{11} \text{ in.}$

$A = \pi \cdot r \cdot r$

$A \approx \dfrac{22}{7} \cdot \dfrac{14}{11} \text{ in.} \cdot \dfrac{14}{11} \text{ in.} = \dfrac{22 \cdot 14 \cdot 14}{7 \cdot 11 \cdot 11} \text{ in}^2 = \dfrac{2 \cdot 11 \cdot 2 \cdot 7 \cdot 14}{7 \cdot 11 \cdot 11} \text{ in}^2 = \dfrac{56}{11} \text{ in}^2 = 5\dfrac{1}{11} \text{ in}^2$

68. $V = l \cdot w \cdot h$

$V = 12 \text{ m} \cdot 3 \text{ m} \cdot 2.6 \text{ m} = 93.6 \text{ m}^3$

69. $V = l \cdot w \cdot h$

$V = 4.6 \text{ cm} \cdot 3 \text{ cm} \cdot 14 \text{ cm} = 193.2 \text{ cm}^3$

70. *Discussion and Writing Exercise.* See the formulas for area listed at the beginning of the Summary and Review for Chapter R.

71. *Discussion and Writing Exercise.* All represent the same number. When expressed in simplified fraction notation, the numerator is 11 and the denominator is 16.

72. The shaded area is composed of the area of a square with side 8 ft and the area of 3 semicircles, each with radius $\dfrac{8 \text{ ft}}{2}$, or 4 ft. The total area of the 3 semicircles is equivalent to $1\dfrac{1}{2}$, or 1.5 times the area of a circle with radius 4 ft.

Area $= 8 \text{ ft} \cdot 8 \text{ ft} + 1.5 \cdot \pi \cdot 4 \text{ ft} \cdot 4 \text{ ft}$

$\approx 8 \text{ ft} \cdot 8 \text{ ft} + 1.5 \cdot 3.14 \cdot 4 \text{ ft} \cdot 4 \text{ ft}$

$= 64 \text{ ft}^2 + 75.36 \text{ ft}^2$

$= 139.36 \text{ ft}^2$

73. The shaded area is composed of a square with side 10 yd from which a semicircle with radius $\dfrac{10 \text{ yd}}{2}$, or 5 yd, has been removed.

Area $= 10 \text{ yd} \cdot 10 \text{ yd} - \dfrac{1}{2} \cdot \pi \cdot 5 \text{ yd} \cdot 5 \text{ yd}$

$\approx 10 \text{ yd} \cdot 10 \text{ yd} - \dfrac{1}{2} \cdot 3.14 \cdot 5 \text{ yd} \cdot 5 \text{ yd}$

$= 100 \text{ yd}^2 - 39.25 \text{ yd}^2$

$= 60.75 \text{ yd}^2$

74. The shaded region is composed of 4 semicircles with radius $\frac{8 \text{ in.}}{2}$, or 4 in. The total area of the semicircles is equivalent to the area of 2 circles with radius 4 in.

$$\begin{aligned} A &= 2\cdot\pi\cdot 4 \text{ in.}\cdot 4 \text{ in.}\\ &\approx 2\cdot 3.14\cdot 4 \text{ in.}\cdot 4 \text{ in.}\\ &= 100.48 \text{ in}^2 \end{aligned}$$

75. The shaded region is composed of a circle with radius 10 m from which 2 circles with radius $\frac{10 \text{ m}}{2}$, or 5 m, have been removed.

$$\begin{aligned} A &= \pi\cdot 10 \text{ m}\cdot 10 \text{ m} - 2\cdot\pi\cdot 5 \text{ m}\cdot 5 \text{ m}\\ &\approx 3.14\cdot 10 \text{ m}\cdot 10 \text{ m} - 2\cdot 3.14\cdot 5 \text{ m}\cdot 5 \text{ m}\\ &= 314 \text{ m}^2 - 157 \text{ m}^2\\ &= 157 \text{ m}^2 \end{aligned}$$

76. The shaded region is composed of a semicircle with radius $\frac{4.8 \text{ cm}}{2}$, or 2.4 cm, and a triangle with base and height both 4.8 cm.

$$\begin{aligned} A &= \frac{1}{2}\pi\cdot 2.4 \text{ cm}\cdot 2.4 \text{ cm} + \frac{1}{2}\cdot 4.8 \text{ cm}\cdot 4.8 \text{ cm}\\ &\approx \frac{1}{2}\cdot 3.14\cdot 2.4 \text{ cm}\cdot 2.4 \text{ cm} + \frac{1}{2}\cdot 4.8 \text{ cm}\cdot 4.8 \text{ cm}\\ &= 9.0432 \text{ cm}^2 + 11.52 \text{ cm}^2\\ &= 20.5632 \text{ cm}^2 \end{aligned}$$

77. The shaded region is composed of a square with side 28 cm from which a small square with side 18 mm has been removed at each of the four corners. First we will convert each measure to meters.

$$28 \text{ cm} = 28 \cancel{\text{cm}}\cdot\frac{1 \text{ m}}{100 \cancel{\text{cm}}} = 0.28 \text{ m}$$

$$18 \text{ mm} = 18 \cancel{\text{mm}}\cdot\frac{1 \text{ m}}{1000 \cancel{\text{mm}}} = 0.018 \text{ m}$$

Now we find the area.

$$\begin{aligned} A &= 0.28 \text{ m}\cdot 0.28 \text{ m} - 4\cdot 0.018 \text{ m}\cdot 0.018 \text{ m}\\ &= 0.0784 \text{ m}^2 - 0.001296 \text{ m}^2\\ &= 0.077104 \text{ m}^2 \end{aligned}$$

Chapter R Test

1. We begin by factoring 300 in any way that we can and continue factoring until each factor is prime.

$300 = 3\cdot 100 = 3\cdot 10\cdot 10 = 3\cdot 2\cdot 5\cdot 2\cdot 5$, or $2\cdot 2\cdot 3\cdot 5\cdot 5$

2. $15 = 3\cdot 5$

$24 = 2\cdot 2\cdot 2\cdot 3$

$60 = 2\cdot 2\cdot 3\cdot 5$

The LCM is $2\cdot 2\cdot 2\cdot 3\cdot 5$, or 120.

3. $\frac{3}{7} = \frac{3}{7}\cdot 1 = \frac{3}{7}\cdot\frac{7}{7} = \frac{21}{49}$

4. We will use $\frac{3}{3}$ for 1 since $48 = 16\cdot 3$.

$$\frac{11}{16} = \frac{11}{16}\cdot 1 = \frac{11}{16}\cdot\frac{3}{3} = \frac{33}{48}$$

5. $\frac{16}{24} = \frac{2\cdot 8}{3\cdot 8} = \frac{2}{3}\cdot\frac{8}{8} = \frac{2}{3}\cdot 1 = \frac{2}{3}$

6. $\frac{925}{1525} = \frac{25\cdot 37}{25\cdot 61} = \frac{25}{25}\cdot\frac{37}{61} = 1\cdot\frac{37}{61} = \frac{37}{61}$

7. $\frac{10}{27}\div\frac{8}{3} = \frac{10}{27}\cdot\frac{3}{8} = \frac{10\cdot 3}{27\cdot 8} = \frac{\cancel{2}\cdot 5\cdot\cancel{3}}{\cancel{3}\cdot 9\cdot\cancel{2}\cdot 4} = \frac{5}{9\cdot 4} = \frac{5}{36}$

8. The LCD is 40.

$$\frac{9}{10} - \frac{5}{8} = \frac{9}{10}\cdot\frac{4}{4} - \frac{5}{8}\cdot\frac{5}{5} = \frac{36}{40} - \frac{25}{40} = \frac{11}{40}$$

9. The LCD is 36.

$$\frac{11}{12} + \frac{17}{18} = \frac{11}{12}\cdot\frac{3}{3} + \frac{17}{18}\cdot\frac{2}{2} = \frac{33}{36} + \frac{34}{36} = \frac{67}{36}$$

10. $\frac{10}{27}\cdot\frac{3}{8} = \frac{10\cdot 3}{27\cdot 8} = \frac{\cancel{2}\cdot 5\cdot\cancel{3}}{\cancel{3}\cdot 9\cdot\cancel{2}\cdot 4} = \frac{5}{9\cdot 4} = \frac{5}{36}$

11. 6.78. (2 places) $6.78 = \frac{678}{100}$ (2 zeros)

12. $\frac{1895}{1000}$ (3 zeros) 1.895. (3 places) $\frac{1895}{1000} = 1.895$

13.
$$\begin{array}{r} {}^{1\,1\,1} \\ 7.14 \\ 89. \\ +\ \ 2.8787 \\ \hline 99.0187 \end{array}$$

14.
$$\begin{array}{r} {}^{7\ 9\ 9\ 10} \\ 1\cancel{8}\cancel{0}\cancel{0}.\cancel{0}\,\cancel{0} \\ -\quad 3.42 \\ \hline 1796.58 \end{array}$$

15.
$$\begin{array}{r} 123.6 \\ \times\quad 3.52 \\ \hline 2472 \\ 61800 \\ 370800 \\ \hline 435.072 \end{array}$$

16.
$$\begin{array}{r} 1.6 \\ 7.2_{\wedge}\overline{)\,11.5_{\wedge}2} \\ 720 \\ \hline 432 \\ 432 \\ \hline 0 \end{array}$$

17.
$$\begin{array}{r} 2.09 \\ 11\,\overline{)\,23.00} \\ 22 \\ \hline 100 \\ 99 \\ \hline 1 \end{array}$$

The number 1 repeats as a remainder, so the sequence of digits following the decimal point in the quotient repeats. Thus, decimal notation for $\frac{23}{11}$ is $2.0909\ldots$, or $2.\overline{09}$.

18. Round 234.7284 to the nearest tenth. The digit in the tenths place is 7. The next digit to the right, 2, is less than 5, so we round down: 234.7.

19. Round 234.7284 to the nearest thousandth. The digit in the thousandths place is 8. The next digit to the right, 4, is less than 5, so we round down: 234.728.

20.   0.7%   0.00.7

Move the decimal point 2 places to the left.

$0.7\% = 0.007$

21. $91\% = 91 \times \dfrac{1}{100} = \dfrac{91}{100}$

22. $\dfrac{11}{25} = 0.44 = 44\%$

23. $5^4 = 5 \cdot 5 \cdot 5 \cdot 5 = 625$

24. $(1.2)^2 = 1.2 \cdot 1.2 = 1.44$

25.
$$
\begin{aligned}
200 - 2^3 + 5 \times 10 &= 200 - 8 + 5 \times 10 \\
&= 200 - 8 + 50 \\
&= 192 + 50 \\
&= 242
\end{aligned}
$$

26. $8000 \div 0.16 \div 2.5 = 50{,}000 \div 2.5 = 20{,}000$

27. $\dfrac{101.4}{100{,}000} = \dfrac{101.4}{100{,}000} \cdot \dfrac{10}{10} = \dfrac{1014}{1{,}000{,}000} = 0.001014 = 0.1014\%$

28.   2.14%   0.02.14

Move the decimal point 2 places to the left.

$2.14\% = 0.0214$

29. $P = 2(l + w)$ Perimeter of a rectangle

$P = 2 \cdot (9.4 \text{ cm} + 7.01 \text{ cm})$

$P = 2 \cdot (16.41 \text{ cm})$

$P = 32.82 \text{ cm}$

$A = l \cdot w$ Area of a rectangle

$A = (9.4 \text{ cm}) \cdot (7.01 \text{ cm})$

$A = 9.4 \cdot 7.01 \cdot \text{ cm} \cdot \text{ cm}$

$A = 65.894 \text{ cm}^2$

30. $P = 4 \cdot s$ Perimeter of a square

$P = 4 \cdot 25 \text{ m}$

$P = 100 \text{ m}$

$A = s \cdot s$ Area of a square

$A = (25 \text{ m}) \cdot (25 \text{ m})$

$A = 25 \cdot 25 \cdot \text{ m} \cdot \text{ m}$

$A = 625 \text{ m}^2$

31. $A = b \cdot h$ Area of a parallelogram

$A = (10 \text{ cm}) \cdot (2.5 \text{ cm})$

$A = 25 \text{ cm}^2$

32. $A = \dfrac{1}{2} \cdot b \cdot h$ Area of a triangle

$A = \dfrac{1}{2} \cdot 8 \text{ m} \cdot 3 \text{ m}$

$A = 12 \text{ m}^2$

33. $V = l \cdot w \cdot h$

$V = 4 \text{ cm} \cdot 2 \text{ cm} \cdot 10.5 \text{ cm}$

$V = 84 \text{ cm}^3$

34. $d = 2 \cdot r = 2 \cdot \dfrac{1}{8} \text{ in.} = \dfrac{2}{8} \text{ in.} = \dfrac{\cancel{2} \cdot 1}{\cancel{2} \cdot 4} \text{ in.} = \dfrac{1}{4} \text{ in.}$

$C = 2 \cdot \pi \cdot r \approx 2 \cdot \dfrac{22}{7} \cdot \dfrac{1}{8} \text{ in.} = \dfrac{2 \cdot 22}{7 \cdot 8} \text{ in.} =$

$\dfrac{\cancel{2} \cdot \cancel{2} \cdot 11}{7 \cdot \cancel{2} \cdot \cancel{2} \cdot 2} \text{ in.} = \dfrac{11}{14} \text{ in.}$

$A = \pi \cdot r \cdot r \approx \dfrac{22}{7} \cdot \dfrac{1}{8} \text{ in.} \cdot \dfrac{1}{8} \text{ in.} = \dfrac{22}{7 \cdot 8 \cdot 8} \text{ in}^2 =$

$\dfrac{\cancel{2} \cdot 11}{7 \cdot \cancel{2} \cdot 4 \cdot 8} \text{ in}^2 = \dfrac{11}{224} \text{ in}^2$

35. $r = \dfrac{d}{2} = \dfrac{18 \text{ cm}}{2} = 9 \text{ cm}$

$C = \pi \cdot d \approx 3.14 \cdot 18 \text{ cm} = 56.52 \text{ cm}$

$A = \pi \cdot r \cdot r \approx 3.14 \cdot 9 \text{ cm} \cdot 9 \text{ cm} = 254.34 \text{ cm}^2$

36. The window is composed of a semicircle with radius 2 ft and a rectangle with dimensions 5 ft by $2 \cdot 2$ ft, or 5 ft by 4 ft.

$A = \dfrac{1}{2} \cdot \pi \cdot 2 \text{ ft} \cdot 2 \text{ ft} + 5 \text{ ft} \cdot 4 \text{ ft}$

$\approx \dfrac{1}{2} \cdot 3.14 \cdot 2 \text{ ft} \cdot 2 \text{ ft} + 5 \text{ ft} \cdot 4 \text{ ft}$

$= 6.28 \text{ ft}^2 + 20 \text{ ft}^2$

$= 26.28 \text{ ft}^2$

Chapter 1

Introduction to Real Numbers and Algebraic Expressions

Exercise Set 1.1

1. Substitute 56 for x: $56 - 24 = 32$, so it takes Erin 32 min to get to work if it takes George 56 min.

   Substitute 93 for x: $93 - 24 = 69$, so it takes Erin 69 min to get to work if it takes George 93 min.

   Substitute 105 for x: $105 - 24 = 81$, so it takes Erin 81 min to get to work if it takes George 105 min.

3. Substitute 45 m for b and 86 m for h, and carry out the multiplication:

$$A = \frac{1}{2}bh = \frac{1}{2}(45 \text{ m})(86 \text{ m})$$
$$= \frac{1}{2}(45)(86)(\text{m})(\text{m})$$
$$= 1935 \text{ m}^2$$

5. Substitute 65 for r and 4 for t, and carry out the multiplication:
$$d = rt = 65 \cdot 4 = 260 \text{ mi}$$

7. We substitute 6 ft for l and 4 ft for w in the formula for the area of a rectangle.
$$A = lw = (6 \text{ ft})(4 \text{ ft})$$
$$= (6)(4)(\text{ft})(\text{ft})$$
$$= 24 \text{ ft}^2$$

9. $8x = 8 \cdot 7 = 56$

11. $\dfrac{a}{b} = \dfrac{24}{3} = 8$

13. $\dfrac{3p}{q} = \dfrac{3 \cdot 2}{6} = \dfrac{6}{6} = 1$

15. $\dfrac{x+y}{5} = \dfrac{10+20}{5} = \dfrac{30}{5} = 6$

17. $\dfrac{x-y}{8} = \dfrac{20-4}{8} = \dfrac{16}{8} = 2$

19. $b + 7$, or $7 + b$

21. $c - 12$

23. $4 + q$, or $q + 4$

25. $a + b$, or $b + a$

27. $x \div y$, or $\dfrac{x}{y}$, or x/y, or $x \cdot \dfrac{1}{y}$

29. $x + w$, or $w + x$

31. $n - m$

33. $x + y$, or $y + x$

35. $2z$

37. $3m$

39. $4a + 6$, or $6 + 4a$

41. $xy - 8$

43. $2t - 5$

45. $3n + 11$, or $11 + 3n$

47. $4x + 3y$, or $3y + 4x$

49. Let s represent your salary. Then we have $89\%s$, or $0.89s$.

51. A 5% increase in s is represented by $0.05s$, so we have $s + 0.05s$.

53. The distance traveled is the product of the speed and the time. Thus, Danielle traveled $65t$ miles.

55. $\$50 - x$

57. Discussion and Writing Exercise

59. We use a factor tree.

The prime factorization is $2 \cdot 3 \cdot 3 \cdot 3$.

61. We use the list of primes. The first prime that is a factor of 108 is 2.
$$108 = 2 \cdot 54$$

We keep dividing by 2 until it is no longer possible to do so.
$$108 = 2 \cdot 2 \cdot 27$$

Now we do the same thing for the next prime, 3.
$$108 = 2 \cdot 2 \cdot 3 \cdot 3 \cdot 3$$

This is the prime factorization of 108.

63. We use the list of primes. The first prime number that is a factor of 1023 is 3.
$$1023 = 3 \cdot 341$$

We continue through the list of prime numbers until we have

$1023 = 3 \cdot 11 \cdot 31.$

Since 3, 11, and 31 are prime numbers, the prime factorization of 1023 is $3 \cdot 11 \cdot 31$.

65. $6 = 2 \cdot 3$

$24 = 2 \cdot 2 \cdot 2 \cdot 3$

$32 = 2 \cdot 2 \cdot 2 \cdot 2 \cdot 2$

The LCM is $2 \cdot 2 \cdot 2 \cdot 2 \cdot 2 \cdot 3$, or 96.

67. $16 = 2 \cdot 2 \cdot 2 \cdot 2$

$24 = 2 \cdot 2 \cdot 2 \cdot 3$

$32 = 2 \cdot 2 \cdot 2 \cdot 2 \cdot 2$

The LCM is $2 \cdot 2 \cdot 2 \cdot 2 \cdot 2 \cdot 3$, or 96.

69. $\dfrac{a - 2b + c}{4b - a} = \dfrac{20 - 2 \cdot 10 + 5}{4 \cdot 10 - 20}$

$= \dfrac{20 - 20 + 5}{40 - 20}$

$= \dfrac{0 + 5}{20}$

$= \dfrac{5}{20} = \dfrac{\cancel{5} \cdot 1}{\cancel{5} \cdot 4}$

$= \dfrac{1}{4}$

71. $\dfrac{12 - c}{c + 12b} = \dfrac{12 - 12}{12 + 12 \cdot 1} = \dfrac{0}{12 + 12} = \dfrac{0}{24} = 0$

Exercise Set 1.2

1. The integer $-34,000,000$ corresponds to paying a fine of \$34 million.

3. The integer 24 corresponds to 24° above zero; the integer -2 corresponds to 2° below zero.

5. The integer 950,000,000 corresponds to a temperature of 950,000,000°F; the integer -460 corresponds to a temperature of 460°F below zero.

7. The integer -34 describes the situation from the Alley Cats' point of view. The integer 34 describes the situation from the Strikers' point of view.

9. The number $\dfrac{10}{3}$ can be named $3\dfrac{1}{3}$, or $3.3\overline{3}$. The graph is $\dfrac{1}{3}$ of the way from 3 to 4.

11. The graph of -5.2 is $\dfrac{2}{10}$ of the way from -5 to -6.

13. The graph of $-4\dfrac{2}{5}$ is $\dfrac{2}{5}$ of the way from -4 to -5.

15. We first find decimal notation for $\dfrac{7}{8}$. Since $\dfrac{7}{8}$ means $7 \div 8$, we divide.

$$\begin{array}{r} 0.8\,7\,5 \\ 8\,\overline{)7.0\,0\,0} \\ \underline{6\,4} \\ 6\,0 \\ \underline{5\,6} \\ 4\,0 \\ \underline{4\,0} \\ 0 \end{array}$$

Thus $\dfrac{7}{8} = 0.875$, so $-\dfrac{7}{8} = -0.875$.

17. $\dfrac{5}{6}$ means $5 \div 6$, so we divide.

$$\begin{array}{r} 0.8\,3\,3\ldots \\ 6\,\overline{)5.0\,0\,0} \\ \underline{4\,8} \\ 2\,0 \\ \underline{1\,8} \\ 2\,0 \\ \underline{1\,8} \\ 0 \end{array}$$

We have $\dfrac{5}{6} = 0.8\overline{3}$.

19. First we find decimal notation for $\dfrac{7}{6}$. Since $\dfrac{7}{6}$ means $7 \div 6$, we divide.

$$\begin{array}{r} 1.1\,6\,6\ldots \\ 6\,\overline{)7.0\,0\,0} \\ \underline{6} \\ 1\,0 \\ \underline{6} \\ 4\,0 \\ \underline{3\,6} \\ 4\,0 \\ \underline{3\,6} \\ 4 \end{array}$$

Thus $\dfrac{7}{6} = 1.1\overline{6}$, so $-\dfrac{7}{6} = -1.1\overline{6}$.

21. $\dfrac{2}{3}$ means $2 \div 3$, so we divide.

$$\begin{array}{r} 0.6\,6\,6\ldots \\ 3\,\overline{)2.0\,0\,0} \\ \underline{1\,8} \\ 2\,0 \\ \underline{1\,8} \\ 2\,0 \\ \underline{1\,8} \\ 2 \end{array}$$

We have $\dfrac{2}{3} = 0.\overline{6}$.

23. $\dfrac{1}{10}$ means $1 \div 10$, so we divide.

$$10\overline{\smash{\big)}\,\begin{array}{r} 0.1 \\ 1.0 \\ \underline{1\,0} \\ 0 \end{array}}$$

We have $\dfrac{1}{10} = 0.1$

25. We first find decimal notation for $\dfrac{1}{2}$. Since $\dfrac{1}{2}$ means $1 \div 2$, we divide.

$$2\overline{\smash{\big)}\,\begin{array}{r} 0.5 \\ 1.0 \\ \underline{1\,0} \\ 0 \end{array}}$$

Thus $\dfrac{1}{2} = 0.5$, so $-\dfrac{1}{2} = -0.5$

27. $\dfrac{4}{25}$ means $4 \div 25$, so we divide.

$$25\overline{\smash{\big)}\,\begin{array}{r} 0.1\,6 \\ 4.0\,0 \\ \underline{2\,5} \\ 1\,5\,0 \\ \underline{1\,5\,0} \\ 0 \end{array}}$$

We have $\dfrac{4}{25} = 0.16$.

29. Since 8 is to the right of 0, we have $8 > 0$.

31. Since -8 is to the left of 3, we have $-8 < 3$.

33. Since -8 is to the left of 8, we have $-8 < 8$.

35. Since -8 is to the left of -5, we have $-8 < -5$.

37. Since -5 is to the right of -11, we have $-5 > -11$.

39. Since -6 is to the left of -5, we have $-6 < -5$.

41. Since 2.14 is to the right of 1.24, we have $2.14 > 1.24$.

43. Since -14.5 is to the left of 0.011, we have $-14.5 < 0.011$.

45. Since -12.88 is to the left of -6.45, we have $-12.88 \ < \ -6.45$.

47. $-\dfrac{1}{2} = -\dfrac{1}{2} \cdot \dfrac{3}{3} = -\dfrac{3}{6}$

$-\dfrac{2}{3} = -\dfrac{2}{3} \cdot \dfrac{2}{2} = -\dfrac{4}{6}$

Since $-\dfrac{3}{6}$ is to the right of $-\dfrac{4}{6}$, then $-\dfrac{1}{2}$ is to the right of $-\dfrac{2}{3}$, and we have $-\dfrac{1}{2} > -\dfrac{2}{3}$.

49. Since $-\dfrac{2}{3}$ is to the left of $\dfrac{1}{3}$, we have $-\dfrac{2}{3} < \dfrac{1}{3}$.

51. Convert to decimal notation $\dfrac{5}{12} = 0.4166\ldots$ and $\dfrac{11}{25} = 0.44$. Since $0.4166\ldots$ is to the left of 0.44, $\dfrac{5}{12} < \dfrac{11}{25}$.

53. $-3 \geq -11$ is true since $-3 > -11$ is true.

55. $0 \geq 8$ is false since neither $0 > 8$ nor $0 = 8$ is true.

57. $x < -6$ has the same meaning as $-6 > x$.

59. $y \geq -10$ has the same meaning as $-10 \leq y$.

61. The distance of -3 from 0 is 3, so $|-3| = 3$.

63. The distance of 10 from 0 is 10, so $|10| = 10$.

65. The distance of 0 from 0 is 0, so $|0| = 0$.

67. The distance of -30.4 from 0 is 30.4, so $|-30.4| = 30.4$.

69. The distance of $-\dfrac{2}{3}$ from 0 is $\dfrac{2}{3}$, so $\left|-\dfrac{2}{3}\right| = \dfrac{2}{3}$.

71. The distance of $\dfrac{0}{4}$ from 0 is $\dfrac{0}{4}$, or 0, so $\left|\dfrac{0}{4}\right| = 0$.

73. The distance of $-3\dfrac{5}{8}$ from 0 is $3\dfrac{5}{8}$, so $\left|-3\dfrac{5}{8}\right| = 3\dfrac{5}{8}$.

75. Discussion and Writing Exercise

77. 63%   0.63.

Move the decimal point 2 places to the left.

$63\% = 0.63$

79. 110%   1.10.

Move the decimal point 2 places to the left.

$110\% = 1.1$

81. $\dfrac{13}{25} = 0.52 = 52\%$

83. From Exercise 17 we know that $\dfrac{5}{6} = 0.8\overline{3}$, or $0.83\overline{3}$, so $\dfrac{5}{6} = 83.\overline{3}\%$, or $83\dfrac{1}{3}\%$.

85. $-\dfrac{2}{3}, \dfrac{1}{2}, -\dfrac{3}{4}, -\dfrac{5}{6}, \dfrac{3}{8}, \dfrac{1}{6}$ can be written in decimal notation as $-0.\overline{6}$, 0.5, -0.75, $-0.8\overline{3}$, 0.375, $0.1\overline{6}$, respectively. Listing from least to greatest, we have

$-\dfrac{5}{6}, -\dfrac{3}{4}, -\dfrac{2}{3}, \dfrac{1}{6}, \dfrac{3}{8}, \dfrac{1}{2}$.

87. $-8.76, -5.16, -4.24, -2.13, 1.85, 5.23$

89. $0.\overline{1} = \dfrac{0.\overline{3}}{3} = \dfrac{\frac{1}{3}}{3} = \dfrac{1}{3} \cdot \dfrac{1}{3} = \dfrac{1}{9}$

91. First consider $0.\overline{5}$.

$0.\overline{5} = 0.\overline{3} \cdot \dfrac{5}{3} = \dfrac{1}{3} \cdot \dfrac{5}{3} = \dfrac{5}{9}$

Then, $5.\overline{5} = 5 + 0.\overline{5} = 5 + \dfrac{5}{9} = 5\dfrac{5}{9}$, or $\dfrac{50}{9}$.

Exercise Set 1.3

1. $2 + (-9)$ The absolute values are 2 and 9. The difference is $9 - 2$, or 7. The negative number has the larger absolute value, so the answer is negative. $2 + (-9) = -7$

3. $-11 + 5$ The absolute values are 11 and 5. The difference is $11 - 5$, or 6. The negative number has the larger absolute value, so the answer is negative. $-11 + 5 = -6$

5. $-8 + 8$ A negative and a positive number. The numbers have the same absolute value. The sum is 0. $-8 + 8 = 0$

7. $-3 + (-5)$ Two negatives. Add the absolute values, getting 8. Make the answer negative. $-3 + (-5) = -8$

9. $-7 + 0$ One number is 0. The answer is the other number. $-7 + 0 = -7$

11. $0 + (-27)$ One number is 0. The answer is the other number. $0 + (-27) = -27$

13. $17 + (-17)$ A negative and a positive number. The numbers have the same absolute value. The sum is 0. $17 + (-17) = 0$

15. $-17 + (-25)$ Two negatives. Add the absolute values, getting 42. Make the answer negative. $-17 + (-25) = -42$

17. $18 + (-18)$ A positive and a negative number. The numbers have the same absolute value. The sum is 0. $18 + (-18) = 0$

19. $-28 + 28$ A negative and a positive number. The numbers have the same absolute value. The sum is 0. $-28 + 28 = 0$

21. $8 + (-5)$ The absolute values are 8 and 5. The difference is $8 - 5$, or 3. The positive number has the larger absolute value, so the answer is positive. $8 + (-5) = 3$

23. $-4 + (-5)$ Two negatives. Add the absolute values, getting 9. Make the answer negative. $-4 + (-5) = -9$

25. $13 + (-6)$ The absolute values are 13 and 6. The difference is $13 - 6$, or 7. The positive number has the larger absolute value, so the answer is positive. $13 + (-6) = 7$

27. $-25 + 25$ A negative and a positive number. The numbers have the same absolute value. The sum is 0. $-25 + 25 = 0$

29. $53 + (-18)$ The absolute values are 53 and 18. The difference is $53 - 18$, or 35. The positive number has the larger absolute value, so the answer is positive. $53 + (-18) = 35$

31. $-8.5 + 4.7$ The absolute values are 8.5 and 4.7. The difference is $8.5 - 4.7$, or 3.8. The negative number has the larger absolute value, so the answer is negative. $-8.5 + 4.7 = -3.8$

33. $-2.8 + (-5.3)$ Two negatives. Add the absolute values, getting 8.1. Make the answer negative. $-2.8 + (-5.3) = -8.1$

35. $-\dfrac{3}{5} + \dfrac{2}{5}$ The absolute values are $\dfrac{3}{5}$ and $\dfrac{2}{5}$. The difference is $\dfrac{3}{5} - \dfrac{2}{5}$, or $\dfrac{1}{5}$. The negative number has the larger absolute value, so the answer is negative. $-\dfrac{3}{5} + \dfrac{2}{5} = -\dfrac{1}{5}$

37. $-\dfrac{2}{9} + \left(-\dfrac{5}{9}\right)$ Two negatives. Add the absolute values, getting $\dfrac{7}{9}$. Make the answer negative. $-\dfrac{2}{9} + \left(-\dfrac{5}{9}\right) = -\dfrac{7}{9}$

39. $-\dfrac{5}{8} + \dfrac{1}{4}$ The absolute values are $\dfrac{5}{8}$ and $\dfrac{1}{4}$. The difference is $\dfrac{5}{8} - \dfrac{2}{8}$, or $\dfrac{3}{8}$. The negative number has the larger absolute value, so the answer is negative. $-\dfrac{5}{8} + \dfrac{1}{4} = -\dfrac{3}{8}$

41. $-\dfrac{5}{8} + \left(-\dfrac{1}{6}\right)$ Two negatives. Add the absolute values, getting $\dfrac{15}{24} + \dfrac{4}{24}$, or $\dfrac{19}{24}$. Make the answer negative. $-\dfrac{5}{8} + \left(-\dfrac{1}{6}\right) = -\dfrac{19}{24}$

43. $-\dfrac{3}{8} + \dfrac{5}{12}$ The absolute values are $\dfrac{3}{8}$ and $\dfrac{5}{12}$. The difference is $\dfrac{10}{24} - \dfrac{9}{24}$, or $\dfrac{1}{24}$. The positive number has the larger absolute value, so the answer is positive. $-\dfrac{3}{8} + \dfrac{5}{12} = \dfrac{1}{24}$

45. $-\dfrac{1}{6} + \dfrac{7}{10}$ The absolute values are $\dfrac{1}{6}$ and $\dfrac{7}{10}$. The difference is $\dfrac{21}{30} - \dfrac{5}{30} = \dfrac{16}{30} = \dfrac{2 \cdot 8}{2 \cdot 15} = \dfrac{8}{15}$. The positive number has the larger absolute value, so the answer is positive. $-\dfrac{1}{6} + \dfrac{7}{10} = \dfrac{8}{15}$

47. $\dfrac{7}{15} + \left(-\dfrac{1}{9}\right)$ The absolute values are $\dfrac{7}{15}$ and $\dfrac{1}{9}$. The difference is $\dfrac{21}{45} - \dfrac{5}{45} = \dfrac{16}{45}$. The positive number has the larger absolute value, so the answer is positive. $\dfrac{7}{15} + \left(-\dfrac{1}{9}\right) = \dfrac{16}{45}$

49. $76 + (-15) + (-18) + (-6)$

  a) Add the negative numbers: $-15 + (-18) + (-6) = -39$

  b) Add the results: $76 + (-39) = 37$

51. $-44 + \left(-\dfrac{3}{8}\right) + 95 + \left(-\dfrac{5}{8}\right)$

  a) Add the negative numbers: $-44 + \left(-\dfrac{3}{8}\right) + \left(-\dfrac{5}{8}\right) = -45$

  b) Add the results: $-45 + 95 = 50$

53. We add from left to right.

$$
\begin{aligned}
& 98 + (-54) + 113 + (-998) + 44 + (-612) \\
=\ & 44 + 113 + (-998) + 44 + (-612) \\
=\ & 157 + (-998) + 44 + (-612) \\
=\ & -841 + 44 + (-612) \\
=\ & -797 + (-612) \\
=\ & -1409
\end{aligned}
$$

55. The additive inverse of 24 is -24 because $24 + (-24) = 0$.

57. The additive inverse of -26.9 is 26.9 because $-26.9 + 26.9 = 0$.

59. If $x = 8$, then $-x = -8$. (The opposite of 8 is -8.)

61. If $x = -\dfrac{13}{8}$ then $-x = -\left(-\dfrac{13}{8}\right) = \dfrac{13}{8}$. (The opposite of $-\dfrac{13}{8}$ is $\dfrac{13}{8}$.)

63. If $x = -43$ then $-(-x) = -(-(-43)) = -43$. (The opposite of the opposite of -43 is -43.)

65. If $x = \dfrac{4}{3}$ then $-(-x) = -\left(-\dfrac{4}{3}\right) = \dfrac{4}{3}$. (The opposite of the opposite of $\dfrac{4}{3}$ is $\dfrac{4}{3}$.)

67. $-(-24) = 24$ (The opposite of -24 is 24.)

69. $-\left(-\dfrac{3}{8}\right) = \dfrac{3}{8}$ (The opposite of $-\dfrac{3}{8}$ is $\dfrac{3}{8}$.)

71. Let $E =$ the elevation of Mauna Kea above sea level.

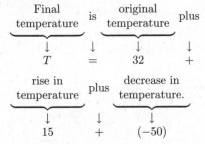

We carry out the addition.

$$E = 33{,}480 + (-19{,}684) = 13{,}796$$

The elevation of Mauna Kea is 13,796 ft above sea level.

73. Let $T =$ the final temperature. We will express the rise in temperature as a positive number and a decrease in the temperature as a negative number.

$$
\begin{array}{ccccc}
\underbrace{\text{Final}}_{} & \text{is} & \underbrace{\text{original}}_{} & \text{plus} \\
\text{temperature} & & \text{temperature} \\
\downarrow & \downarrow & \downarrow & \downarrow \\
T & = & 32 & +
\end{array}
$$

$$
\begin{array}{ccc}
\underbrace{\text{rise in}}_{} & \text{plus} & \underbrace{\text{decrease in}}_{} \\
\text{temperature} & & \text{temperature.} \\
\downarrow & \downarrow & \downarrow \\
15 & + & (-50)
\end{array}
$$

We add from left to right.

$$
\begin{aligned}
T &= 32 + 15 + (-50) \\
&= 47 + (-50) \\
&= -3
\end{aligned}
$$

The final temperature was $-3°$F.

75. Let $S =$ the sum of the profits and losses. We add the five numbers in the bar graph to find S.

$S = \$10{,}500 + (-\$16{,}600) + (-\$12{,}800) + (-\$9600) + \$8200 = -\$20{,}300$

The sum of the profits and losses is $-\$20{,}300$.

77. Let $B =$ the new balance in the account at the end of August. We will express the payments as positive numbers and the original balance and the amount of the new charge as negative numbers.

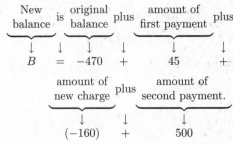

We add from left to right.

$$
\begin{aligned}
B &= -470 + 45 + (-160) + 500 \\
&= -425 + (-160) + 500 \\
&= -585 + 500 \\
&= -85
\end{aligned}
$$

The balance in the account is $-\$85$. Lyle owes \$85.

79. Discussion and Writing Exercise

81.   71.3%   0.71.3

Move the decimal point two places to the left.

$71.3\% = 0.713$

83. $\dfrac{1}{8} = 0.125 = 12.5\%$

85. $\dfrac{2}{3} \div \dfrac{5}{12} = \dfrac{2}{3} \cdot \dfrac{12}{5} = \dfrac{2 \cdot 12}{3 \cdot 5} = \dfrac{2 \cdot 3 \cdot 4}{3 \cdot 5} = \dfrac{8}{5}$

87. When x is positive, the opposite of x, $-x$, is negative, so $-x$ is negative for all positive numbers x.

89. If a is positive, $-a$ is negative. Thus $-a + b$, the sum of two negative numbers, is negative. The correct answer is (b).

Exercise Set 1.4

1. $2 - 9 = 2 + (-9) = -7$

3. $-8 - (-2) = -8 + 2 = -6$

5. $-11 - (-11) = -11 + 11 = 0$

7. $12 - 16 = 12 + (-16) = -4$

9. $20 - 27 = 20 + (-27) = -7$

11. $-9 - (-3) = -9 + 3 = -6$

13. $-40 - (-40) = -40 + 40 = 0$

15. $7 - (-7) = 7 + 7 = 14$

17. $8 - (-3) = 8 + 3 = 11$

19. $-6 - 8 = -6 + (-8) = -14$

21. $-4 - (-9) = -4 + 9 = 5$

23. $-6 - (-5) = -6 + 5 = -1$

25. $8 - (-10) = 8 + 10 = 18$

27. $-5 - (-2) = -5 + 2 = -3$

29. $-7 - 14 = -7 + (-14) = -21$

31. $0 - (-5) = 0 + 5 = 5$

33. $-8 - 0 = -8 + 0 = -8$

35. $7 - (-5) = 7 + 5 = 12$

37. $2 - 25 = 2 + (-25) = -23$

39. $-42 - 26 = -42 + (-26) = -68$

41. $-71 - 2 = -71 + (-2) = -73$

43. $24 - (-92) = 24 + 92 = 116$

45. $-50 - (-50) = -50 + 50 = 0$

47. $-\dfrac{3}{8} - \dfrac{5}{8} = -\dfrac{3}{8} + \left(-\dfrac{5}{8}\right) = -\dfrac{8}{8} = -1$

49. $\dfrac{3}{4} - \dfrac{2}{3} = \dfrac{3}{4} + \left(-\dfrac{2}{3}\right) = \dfrac{9}{12} + \left(-\dfrac{8}{12}\right) = \dfrac{1}{12}$

51. $-\dfrac{3}{4} - \dfrac{2}{3} = -\dfrac{3}{4} + \left(-\dfrac{2}{3}\right) = -\dfrac{9}{12} + \left(-\dfrac{8}{12}\right) = -\dfrac{17}{12}$

53. $-\dfrac{5}{8} - \left(-\dfrac{3}{4}\right) = -\dfrac{5}{8} + \dfrac{3}{4} = -\dfrac{5}{8} + \dfrac{6}{8} = \dfrac{1}{8}$

55. $6.1 - (-13.8) = 6.1 + 13.8 = 19.9$

57. $-2.7 - 5.9 = -2.7 + (-5.9) = -8.6$

59. $0.99 - 1 = 0.99 + (-1) = -0.01$

61. $-79 - 114 = -79 + (-114) = -193$

63. $0 - (-500) = 0 + 500 = 500$

65. $-2.8 - 0 = -2.8 + 0 = -2.8$

67. $7 - 10.53 = 7 + (-10.53) = -3.53$

69. $\dfrac{1}{6} - \dfrac{2}{3} = \dfrac{1}{6} + \left(-\dfrac{2}{3}\right) = \dfrac{1}{6} + \left(-\dfrac{4}{6}\right) = -\dfrac{3}{6}$, or $-\dfrac{1}{2}$

71. $-\dfrac{4}{7} - \left(-\dfrac{10}{7}\right) = -\dfrac{4}{7} + \dfrac{10}{7} = \dfrac{6}{7}$

73. $-\dfrac{7}{10} - \dfrac{10}{15} = -\dfrac{7}{10} + \left(-\dfrac{10}{15}\right) = -\dfrac{21}{30} + \left(-\dfrac{20}{30}\right) = -\dfrac{41}{30}$

75. $\dfrac{1}{5} - \dfrac{1}{3} = \dfrac{1}{5} + \left(-\dfrac{1}{3}\right) = \dfrac{3}{15} + \left(-\dfrac{5}{15}\right) = -\dfrac{2}{15}$

77. $\dfrac{5}{12} - \dfrac{7}{16} = \dfrac{5}{12} + \left(-\dfrac{7}{16}\right) = \dfrac{20}{48} + \left(-\dfrac{21}{48}\right) = -\dfrac{1}{48}$

79. $-\dfrac{2}{15} - \dfrac{7}{12} = -\dfrac{2}{15} + \left(-\dfrac{7}{12}\right) = -\dfrac{8}{60} + \left(-\dfrac{35}{60}\right) = -\dfrac{43}{60}$

81. $18 - (-15) - 3 - (-5) + 2 = 18 + 15 + (-3) + 5 + 2 = 37$

83. $-31 + (-28) - (-14) - 17 = (-31) + (-28) + 14 + (-17) = -62$

85. $-34 - 28 + (-33) - 44 = (-34) + (-28) + (-33) + (-44) = -139$

87. $-93 - (-84) - 41 - (-56) = (-93) + 84 + (-41) + 56 = 6$

89. $-5.4 - (-30.9) + 30.8 + 40.2 - (-12) = -5.4 + 30.9 + 30.8 + 40.2 + 12 = 108.5$

91. $-\dfrac{7}{12} + \dfrac{3}{4} - \left(-\dfrac{5}{8}\right) - \dfrac{13}{24} = -\dfrac{7}{12} + \dfrac{3}{4} + \dfrac{5}{8} + \left(-\dfrac{13}{24}\right) = -\dfrac{28}{48} + \dfrac{36}{48} + \dfrac{30}{48} + \left(-\dfrac{26}{48}\right) = \dfrac{12}{48} = \dfrac{\cancel{12} \cdot 1}{4 \cdot \cancel{12}} = \dfrac{1}{4}$

93. Let D = the difference in elevation.

$$\underbrace{\begin{matrix}\text{Difference}\\\text{in}\\\text{elevation}\end{matrix}}_{\downarrow\,D} \underbrace{\text{is}}_{\downarrow\,=} \underbrace{\begin{matrix}\text{larger}\\\text{depth}\end{matrix}}_{\downarrow\,10,924} \underbrace{\text{minus}}_{\downarrow\,-} \underbrace{\begin{matrix}\text{smaller}\\\text{depth.}\end{matrix}}_{\downarrow\,8605}$$

We carry out the subtraction.

$$D = 10,924 - 8605 = 2319$$

The difference in elevation is 2319 m.

95. Let A = the amount owed.

$$\underbrace{\begin{matrix}\text{Amount}\\\text{owed}\end{matrix}}_{\downarrow\,A} \underbrace{\text{is}}_{\downarrow\,=} \underbrace{\begin{matrix}\text{amount}\\\text{of charge}\end{matrix}}_{\downarrow\,476.89} \underbrace{\text{minus}}_{\downarrow\,-} \underbrace{\begin{matrix}\text{amount}\\\text{of return.}\end{matrix}}_{\downarrow\,128.95}$$

We subtract.

$$A = 476.89 - 128.95 = 347.94$$

Claire owes \$347.94.

97. a) We subtract the number of home runs allowed from the number of home runs hit.

Home run differential $= 197 - 120 = 77$

b) We subtract the number of home runs allowed from the number of home runs hit.

Home run differential $= 153 - 194 = -41$

99. Let D = the difference in elevation.

$$\underbrace{\begin{matrix}\text{Difference}\\\text{in}\\\text{elevation}\end{matrix}}_{\downarrow\,D} \underbrace{\text{is}}_{\downarrow\,=} \underbrace{\begin{matrix}\text{higher}\\\text{elevation}\end{matrix}}_{\downarrow\,-131} \underbrace{\text{minus}}_{\downarrow\,-} \underbrace{\begin{matrix}\text{lower}\\\text{elevation.}\end{matrix}}_{\downarrow\,(-512)}$$

We carry out the subtraction.

$$D = -131 - (-512) = -131 + 512 = 381$$

Lake Assal is 381 ft lower than the Valdes Peninsula.

101. Discussion and Writing Exercise

103. $256 \div 64 \div 2^3 + 100 = 256 \div 64 \div 8 + 100$
$= 4 \div 8 + 100$
$= \dfrac{1}{2} + 100$
$= 100\dfrac{1}{2}$, or 100.5

105. $2^5 \div 4 + 20 \div 2^2 = 32 \div 4 + 20 \div 4 = 8 + 5 = 13$

107. $\dfrac{1}{8} + \dfrac{7}{12} + \dfrac{5}{24} = \dfrac{3}{24} + \dfrac{14}{24} + \dfrac{5}{24} = \dfrac{22}{24} = \dfrac{\cancel{2} \cdot 11}{\cancel{2} \cdot 12} = \dfrac{11}{12}$

109. False. $3 - 0 = 3, 0 - 3 = -3, 3 - 0 \neq 0 - 3$

111. True

113. True by definition of opposites.

Exercise Set 1.5

---

1. -8

3. -48

5. -24

7. -72

9. 16

11. 42

13. -120

15. -238

17. 1200

19. 98

21. -72

23. -12.4

25. 30

27. 21.7

29. $\dfrac{2}{3} \cdot \left(-\dfrac{3}{5}\right) = -\left(\dfrac{2 \cdot 3}{3 \cdot 5}\right) = -\left(\dfrac{2}{5} \cdot \dfrac{3}{3}\right) = -\dfrac{2}{5}$

31. $-\dfrac{3}{8} \cdot \left(-\dfrac{2}{9}\right) = \dfrac{3 \cdot 2}{8 \cdot 9} = \dfrac{3 \cdot 2 \cdot 1}{4 \cdot 2 \cdot 3 \cdot 3} = \dfrac{3 \cdot 2}{3 \cdot 2} \cdot \dfrac{1}{4 \cdot 3} = \dfrac{1}{12}$

33. -17.01

35. $-\dfrac{5}{9} \cdot \dfrac{3}{4} = -\left(\dfrac{5 \cdot 3}{9 \cdot 4}\right) = -\dfrac{5 \cdot 3}{3 \cdot 3 \cdot 4} = -\dfrac{5}{3 \cdot 4} \cdot \dfrac{3}{3} = -\dfrac{5}{12}$

37. $7 \cdot (-4) \cdot (-3) \cdot 5 = 7 \cdot 12 \cdot 5 = 7 \cdot 60 = 420$

39. $-\dfrac{2}{3} \cdot \dfrac{1}{2} \cdot \left(-\dfrac{6}{7}\right) = -\dfrac{2}{6} \cdot \left(-\dfrac{6}{7}\right) = \dfrac{2 \cdot 6}{7 \cdot 6} = \dfrac{2}{7} \cdot \dfrac{6}{6} = \dfrac{2}{7}$

41. $-3 \cdot (-4) \cdot (-5) = 12 \cdot (-5) = -60$

43. $-2 \cdot (-5) \cdot (-3) \cdot (-5) = 10 \cdot 15 = 150$

45. $-\dfrac{2}{45}$

47. $-7 \cdot (-21) \cdot 13 = 147 \cdot 13 = 1911$

49. $-4 \cdot (-1.8) \cdot 7 = (7.2) \cdot 7 = 50.4$

51. $-\dfrac{1}{9} \cdot \left(-\dfrac{2}{3}\right) \cdot \left(\dfrac{5}{7}\right) = \dfrac{2}{27} \cdot \dfrac{5}{7} = \dfrac{10}{189}$

53. $4 \cdot (-4) \cdot (-5) \cdot (-12) = -16 \cdot (60) = -960$

55. $0.07 \cdot (-7) \cdot 6 \cdot (-6) = 0.07 \cdot 6 \cdot (-7) \cdot (-6) = 0.42 \cdot (42) = 17.64$

57. $\left(-\dfrac{5}{6}\right)\left(\dfrac{1}{8}\right)\left(-\dfrac{3}{7}\right)\left(-\dfrac{1}{7}\right) = \left(-\dfrac{5}{48}\right)\left(\dfrac{3}{49}\right) = -\dfrac{5 \cdot 3}{16 \cdot 3 \cdot 49} = -\dfrac{5}{16 \cdot 49} \cdot \dfrac{3}{3} = -\dfrac{5}{784}$

59. 0, The product of 0 and any real number is 0.

61. $(-8)(-9)(-10) = 72(-10) = -720$

63. $(-6)(-7)(-8)(-9)(-10) = 42 \cdot 72 \cdot (-10) = 3024 \cdot (-10) = -30{,}240$

65. $(-1)^{12}$
$= (-1)(-1)(-1)(-1)(-1)(-1)(-1)(-1)(-1)(-1)(-1)(-1)$
$= 1 \cdot 1 \cdot 1 \cdot 1 \cdot 1 \cdot 1 = 1$

67. For $x = 4$:
$$(-x)^2 = (-4)^2 = 16$$
$$-x^2 = -(4)^2 = -(16) = -16$$
For $x = -4$:
$$(-x)^2 = [-(-4)]^2 = [4]^2 = 16$$
$$-x^2 = -(-4)^2 = -(16) = -16$$

69. $(-3x)^2 = (-3 \cdot 7)^2$　　Substituting
$\qquad\quad = (-21)^2$　　Multiplying inside the parentheses
$\qquad\quad = (-21)(-21)$　　Evaluating the power
$\qquad\quad = 441$

$-3x^2 = -3(7)^2$　　Substituting
$\qquad = -3 \cdot 49$　　Evaluating the power
$\qquad = -147$

71. When $x = 2$:　$5x^2 = 5(2)^2$　　Substituting
$\qquad\qquad\qquad = 5 \cdot 4$　　Evaluating the power
$\qquad\qquad\qquad = 20$

When $x = -2$:　$5x^2 = 5(-2)^2$　　Substituting
$\qquad\qquad\qquad = 5 \cdot 4$　　Evaluating the power
$\qquad\qquad\qquad = 20$

73. When $x = 1$, $-2x^3 = -2 \cdot 1^3 = -2 \cdot 1 = -2$.
When $x = -1$, $-2x^3 = -2(-1)^3 = -2(-1) = 2$.

75. Let $w =$ the total weight change. Since Dave's weight decreases 2 lb each week for 10 weeks we have
$$w = 10 \cdot (-2) = -20.$$
Thus, the total weight change is -20 lb.

77. This is a multistep problem. First we find the number of degrees the temperature dropped. Since it dropped $3°\text{C}$ each minute for 18 minutes we have a drop d given by
$$d = 18 \cdot (-3) = -54.$$
Now let $T =$ the temperature at 10:18 AM.
$$T = 0 + (-54) = -54$$
The temperature was $-54°\text{C}$ at 10:18 AM.

79. This is a multistep problem. First we find the total decrease in price. Since it decreased $1.38 each hour for 8 hours we have a decrease in price d given by

$$d = 8(-\$1.38) = -\$11.04.$$

Now let P = the price of the stock after 8 hours.

$$P = \$23.75 + (-\$11.04) = \$12.71$$

After 8 hours the price of the stock was $12.71.

81. This is a multistep problem. First we find the total distance the diver rises. Since the diver rises 7 meters each minute for 9 minutes, the total distance d the diver rises is given by

$$d = 9 \cdot 7 = 63.$$

Now let E = the diver's elevation after 9 minutes.

$$E = -95 + 63 = -32$$

The diver's elevation is -32 m, or 32 m below the surface.

83. Discussion and Writing Exercise

85. $36 = 2 \cdot 2 \cdot 3 \cdot 3$
$60 = 2 \cdot 2 \cdot 3 \cdot 5$
LCM $= 2 \cdot 2 \cdot 3 \cdot 3 \cdot 5$, or 180

87. $\dfrac{26}{39} = \dfrac{2 \cdot \cancel{13}}{3 \cdot \cancel{13}} = \dfrac{2}{3}$

89. $\dfrac{264}{484} = \dfrac{\cancel{2} \cdot \cancel{2} \cdot 2 \cdot 3 \cdot \cancel{11}}{\cancel{2} \cdot \cancel{2} \cdot \cancel{11} \cdot 11} = \dfrac{6}{11}$

91. $\dfrac{275}{800} = \dfrac{\cancel{25} \cdot 11}{\cancel{25} \cdot 32} = \dfrac{11}{32}$

93. $\dfrac{11}{264} = \dfrac{\cancel{11} \cdot 1}{\cancel{11} \cdot 24} = \dfrac{1}{24}$

95. If a is positive and b is negative, then ab is negative and thus $-ab$ is positive. The correct answer is (a).

97. To locate $2x$, start at 0 and measure off two adjacent lengths of x to the right of 0.

To locate $3x$, start at 0 and measure off three adjacent lengths of x to the right of 0.

To locate $2y$, start at 0 and measure off two adjacent lengths of y to the right of 0.

To locate $-x$, start at 0 and measure off the length x to the left of 0.

To locate $-y$, start at 0 and measure off the length y to the left of 0.

To locate $x + y$, start at 0 and measure off the length x to the right of 0 followed by the length y immediately to the right of x. (We could also measure off y followed by x.)

To locate $x - y$, start at 0 and measure off the length x to the right of 0. Then, from that point, measure off the length y going to the left.

To locate $x - 2y$, first locate $x - y$ as described above. Then, from that point, measure off another length y going to the left.

Exercise Set 1.6

1. $48 \div (-6) = -8$ Check: $-8(-6) = 48$

3. $\dfrac{28}{-2} = -14$ Check: $-14(-2) = 28$

5. $\dfrac{-24}{8} = -3$ Check: $-3 \cdot 8 = -24$

7. $\dfrac{-36}{-12} = 3$ Check: $3(-12) = -36$

9. $\dfrac{-72}{9} = -8$ Check: $-8 \cdot 9 = -72$

11. $-100 \div (-50) = 2$ Check: $2(-50) = -100$

13. $-108 \div 9 = -12$ Check: $9(-12) = -108$

15. $\dfrac{200}{-25} = -8$ Check: $-8(-25) = 200$

17. Not defined

19. $\dfrac{0}{-2.6} = 0$ Check: $0(-2.6) = 0$

21. The reciprocal of $\dfrac{15}{7}$ is $\dfrac{7}{15}$ because $\dfrac{15}{7} \cdot \dfrac{7}{15} = 1$.

23. The reciprocal of $-\dfrac{47}{13}$ is $-\dfrac{13}{47}$ because $\left(-\dfrac{47}{13}\right) \cdot \left(-\dfrac{13}{47}\right) = 1$.

25. The reciprocal of 13 is $\dfrac{1}{13}$ because $13 \cdot \dfrac{1}{13} = 1$.

27. The reciprocal of 4.3 is $\dfrac{1}{4.3}$ because $4.3 \cdot \dfrac{1}{4.3} = 1$.

29. The reciprocal of $-\dfrac{1}{7.1}$ is -7.1 because $\left(-\dfrac{1}{7.1}\right)(-7.1) = 1$.

31. The reciprocal of $\dfrac{p}{q}$ is $\dfrac{q}{p}$ because $\dfrac{p}{q} \cdot \dfrac{q}{p} = 1$.

33. The reciprocal of $\dfrac{1}{4y}$ is $4y$ because $\dfrac{1}{4y} \cdot 4y = 1$.

35. The reciprocal of $\dfrac{2a}{3b}$ is $\dfrac{3b}{2a}$ because $\dfrac{2a}{3b} \cdot \dfrac{3b}{2a} = 1$.

37. $4 \cdot \dfrac{1}{17}$

39. $8 \cdot \left(-\dfrac{1}{13}\right)$

41. $13.9 \cdot \left(-\dfrac{1}{1.5}\right)$

43. $x \cdot y$

45. $(3x + 4)\left(\dfrac{1}{5}\right)$

47. $(5a - b)\left(\dfrac{1}{5a + b}\right)$

49. $\dfrac{3}{4} \div \left(-\dfrac{2}{3}\right) = \dfrac{3}{4} \cdot \left(-\dfrac{3}{2}\right) = -\dfrac{9}{8}$

51. $-\dfrac{5}{4} \div \left(-\dfrac{3}{4}\right) = -\dfrac{5}{4} \cdot \left(-\dfrac{4}{3}\right) = \dfrac{20}{12} = \dfrac{5 \cdot 4}{3 \cdot 4} = \dfrac{5}{3}$

53. $-\dfrac{2}{7} \div \left(-\dfrac{4}{9}\right) = -\dfrac{2}{7} \cdot \left(-\dfrac{9}{4}\right) = \dfrac{18}{28} = \dfrac{9 \cdot 2}{14 \cdot 2} = \dfrac{9}{14}$

55. $-\dfrac{3}{8} \div \left(-\dfrac{8}{3}\right) = -\dfrac{3}{8} \cdot \left(-\dfrac{3}{8}\right) = \dfrac{9}{64}$

57. $-6.6 \div 3.3 = -2$ Do the long division. Make the answer negative.

59. $\dfrac{-11}{-13} = \dfrac{11}{13}$ The opposite of a number divided by the opposite of another number is the quotient of the two numbers.

61. $\dfrac{48.6}{-3} = -16.2$ Do the long division. Make the answer negative.

63. $\dfrac{-9}{17 - 17} = \dfrac{-9}{0}$ Division by 0 is not defined.

65. $\dfrac{81}{183} \approx 0.443 \approx 44.3\%$

67. $\dfrac{-22}{428} \approx -0.051 \approx -5.1\%$

69. Discussion and Writing Exercise

71. $2^3 - 5 \cdot 3 + 8 \cdot 10 \div 2$
   $\begin{aligned}
   &= 8 - 5 \cdot 3 + 8 \cdot 10 \div 2 \quad &&\text{Evaluating the power} \\
   &= 8 - 15 + 80 \div 2 \quad &&\text{Multiplying and dividing} \\
   &= 8 - 15 + 40 \quad &&\text{in order from left to right} \\
   &= -7 + 40 \quad &&\text{Adding and subtracting} \\
   &= 33 \quad &&\text{in order from left to right}
   \end{aligned}$

73. $\quad 1000 \div 100 \div 10$
   $\begin{aligned}
   &= 10 \div 10 \quad &&\text{Dividing in order from} \\
   &= 1 \quad &&\text{left to right}
   \end{aligned}$

75. $\dfrac{264}{468} = \dfrac{4 \cdot 66}{4 \cdot 117} = \dfrac{\cancel{4} \cdot \cancel{3} \cdot 22}{\cancel{4} \cdot \cancel{3} \cdot 39} = \dfrac{22}{39}$

77. $\dfrac{7}{8} = 0.875 = 87.5\%$

79. $\dfrac{12}{25} \div \dfrac{32}{75} = \dfrac{12}{25} \cdot \dfrac{75}{32}$
   $\begin{aligned}
   &= \dfrac{12 \cdot 75}{25 \cdot 32} \\
   &= \dfrac{3 \cdot \cancel{4} \cdot 3 \cdot \cancel{25}}{\cancel{25} \cdot \cancel{4} \cdot 8} \\
   &= \dfrac{9}{8}
   \end{aligned}$

81. The reciprocal of -10.5 is $\dfrac{1}{-10.5}$.

   The reciprocal of $\dfrac{1}{-10.5} = -10.5$.

   We see that the reciprocal of the reciprocal is the original number.

83. $-a$ is positive and b is negative, so $\dfrac{-a}{b}$ is the quotient of a positive and a negative number and, thus, is negative.

85. a is negative and $-b$ is positive, so $\dfrac{a}{-b}$ is the quotient of a negative number and a positive number and, thus, is negative. Then $-\left(\dfrac{a}{-b}\right)$ is the opposite of a negative number and, thus, is positive.

87. $-a$ and $-b$ are both positive, so $\dfrac{-a}{-b}$ is the quotient of two positive numbers and, thus, is positive. Then $-\left(\dfrac{-a}{-b}\right)$ is the opposite of a positive number and, thus, is negative.

Exercise Set 1.7

1. Note that $5y = 5 \cdot y$. We multiply by 1, using y/y as an equivalent expression for 1:
$$\dfrac{3}{5} = \dfrac{3}{5} \cdot 1 = \dfrac{3}{5} \cdot \dfrac{y}{y} = \dfrac{3y}{5y}$$

3. Note that $15x = 3 \cdot 5x$. We multiply by 1, using $5x/5x$ as an equivalent expression for 1:
$$\dfrac{2}{3} = \dfrac{2}{3} \cdot 1 = \dfrac{2}{3} \cdot \dfrac{5x}{5x} = \dfrac{10x}{15x}$$

5. Note that $x^2 = x \cdot x$. We multiply by 1, using x/x as an equivalent expression for 1.
$$\dfrac{2}{x} = \dfrac{2}{x} \cdot 1 = \dfrac{2}{x} \cdot \dfrac{x}{x} = \dfrac{2x}{x^2}$$

7. $-\dfrac{24a}{16a} = -\dfrac{3 \cdot 8a}{2 \cdot 8a}$
   $\begin{aligned}
   &= -\dfrac{3}{2} \cdot \dfrac{8a}{8a} \\
   &= -\dfrac{3}{2} \cdot 1 \quad &&\left(\dfrac{8a}{8a} = 1\right) \\
   &= -\dfrac{3}{2} \quad &&\text{Identity property of 1}
   \end{aligned}$

9. $-\dfrac{42ab}{36ab} = -\dfrac{7 \cdot 6ab}{6 \cdot 6ab}$
   $\begin{aligned}
   &= -\dfrac{7}{6} \cdot \dfrac{6ab}{6ab} \\
   &= -\dfrac{7}{6} \cdot 1 \quad &&\left(\dfrac{6ab}{6ab} = 1\right) \\
   &= -\dfrac{7}{6} \quad &&\text{Identity property of 1}
   \end{aligned}$

11. $\dfrac{20st}{15t} = \dfrac{4s \cdot 5t}{3 \cdot 5t}$
   $\begin{aligned}
   &= \dfrac{4s}{3} \cdot \dfrac{5t}{5t} \\
   &= \dfrac{4s}{3} \cdot 1 \quad &&\left(\dfrac{5t}{5t} = 1\right) \\
   &= \dfrac{4s}{3} \quad &&\text{Identity property of 1}
   \end{aligned}$

13. $8 + y$, commutative law of addition

15. nm, commutative law of multiplication

17. $xy + 9$, commutative law of addition

$9 + yx$, commutative law of multiplication

19. $c + ab$, commutative law of addition

$ba + c$, commutative law of multiplication

21. $(a + b) + 2$, associative law of addition

23. $8(xy)$, associative law of multiplication

25. $a + (b + 3)$, associative law of addition

27. $(3a)b$, associative law of multiplication

29. a) $(a + b) + 2 = a + (b + 2)$, associative law of addition

b) $(a + b) + 2 = (b + a) + 2$, commutative law of addition

c) $(a + b) + 2 = (b + a) + 2$ Using the commutative law first,

$= b + (a + 2)$ then the associative law

There are other correct answers.

31. a) $5 + (v + w) = (5 + v) + w$, associative law of addition

b) $5 + (v + w) = 5 + (w + v)$, commutative law of addition

c) $5 + (v + w) = 5 + (w + v)$ Using the commutative law first,

$= (5 + w) + v$ then the associative law

There are other correct answers.

33. a) $(xy)3 = x(y3)$, associative law of multiplication

b) $(xy)3 = (yx)3$, commutative law of multiplication

c) $(xy)3 = (yx)3$ Using the commutative law first,

$= y(x3)$ then the associative law

There are other correct answers.

35. a) $7(ab) = (7a)b$

b) $7(ab) = (7a)b = b(7a)$

c) $7(ab) = 7(ba) = (7b)a$

There are other correct answers.

37. $2(b + 5) = 2 \cdot b + 2 \cdot 5 = 2b + 10$

39. $7(1 + t) = 7 \cdot 1 + 7 \cdot t = 7 + 7t$

41. $6(5x + 2) = 6 \cdot 5x + 6 \cdot 2 = 30x + 12$

43. $7(x + 4 + 6y) = 7 \cdot x + 7 \cdot 4 + 7 \cdot 6y = 7x + 28 + 42y$

45. $7(x - 3) = 7 \cdot x - 7 \cdot 3 = 7x - 21$

47. $-3(x - 7) = -3 \cdot x - (-3) \cdot 7 = -3x - (-21) = -3x + 21$

49. $\frac{2}{3}(b - 6) = \frac{2}{3} \cdot b - \frac{2}{3} \cdot 6 = \frac{2}{3}b - 4$

51. $7.3(x - 2) = 7.3 \cdot x - 7.3 \cdot 2 = 7.3x - 14.6$

53. $-\frac{3}{5}(x - y + 10) = -\frac{3}{5} \cdot x - \left(-\frac{3}{5}\right) \cdot y + \left(-\frac{3}{5}\right) \cdot 10 =$

$-\frac{3}{5}x - \left(-\frac{3}{5}y\right) + (-6) = -\frac{3}{5}x + \frac{3}{5}y - 6$

55. $-9(-5x - 6y + 8) = -9(-5x) - (-9)6y + (-9)8$

$= 45x - (-54y) + (-72) = 45x + 54y - 72$

57. $-4(x - 3y - 2z) = -4 \cdot x - (-4)3y - (-4)2z$

$= -4x - (-12y) - (-8z) = -4x + 12y + 8z$

59. $3.1(-1.2x + 3.2y - 1.1) = 3.1(-1.2x) + (3.1)3.2y - 3.1(1.1)$

$= -3.72x + 9.92y - 3.41$

61. $4x + 3z$ Parts are separated by plus signs. The terms are $4x$ and $3z$.

63. $7x + 8y - 9z = 7x + 8y + (-9z)$ Separating parts with plus signs

The terms are $7x$, $8y$, and $-9z$.

65. $2x + 4 = 2 \cdot x + 2 \cdot 2 = 2(x + 2)$

67. $30 + 5y = 5 \cdot 6 + 5 \cdot y = 5(6 + y)$

69. $14x + 21y = 7 \cdot 2x + 7 \cdot 3y = 7(2x + 3y)$

71. $5x + 10 + 15y = 5 \cdot x + 5 \cdot 2 + 5 \cdot 3y = 5(x + 2 + 3y)$

73. $8x - 24 = 8 \cdot x - 8 \cdot 3 = 8(x - 3)$

75. $-4y + 32 = -4 \cdot y - 4(-8) = -4(y - 8)$

We could also factor this expression as follows:

$-4y + 32 = 4(-y) + 4 \cdot 8 = 4(-y + 8)$

77. $8x + 10y - 22 = 2 \cdot 4x + 2 \cdot 5y - 2 \cdot 11 = 2(4x + 5y - 11)$

79. $ax - a = a \cdot x - a \cdot 1 = a(x - 1)$

81. $ax - ay - az = a \cdot x - a \cdot y - a \cdot z = a(x - y - z)$

83. $-18x + 12y + 6 = -6 \cdot 3x - 6(-2y) - 6(-1) = -6(3x - 2y - 1)$

We could also factor this expression as follows:

$-18x + 12y + 6 = 6(-3x) + 6 \cdot 2y + 6 \cdot 1 = 6(-3x + 2y + 1)$

85. $\frac{2}{3}x - \frac{5}{3}y + \frac{1}{3} = \frac{1}{3} \cdot 2x - \frac{1}{3} \cdot 5y + \frac{1}{3} \cdot 1 =$

$\frac{1}{3}(2x - 5y + 1)$

87. $9a + 10a = (9 + 10)a = 19a$

89. $10a - a = 10a - 1 \cdot a = (10 - 1)a = 9a$

91. $2x + 9z + 6x = 2x + 6x + 9z = (2 + 6)x + 9z = 8x + 9z$

93. $7x + 6y^2 + 9y^2 = 7x + (6 + 9)y^2 = 7x + 15y^2$

95. $41a + 90 - 60a - 2 = 41a - 60a + 90 - 2$

$= (41 - 60)a + (90 - 2)$

$= -19a + 88$

97. $23 + 5t + 7y - t - y - 27$

$= 23 - 27 + 5t - 1 \cdot t + 7y - 1 \cdot y$

$= (23 - 27) + (5 - 1)t + (7 - 1)y$

$= -4 + 4t + 6y$, or $4t + 6y - 4$

99. $\frac{1}{2}b + \frac{1}{2}b = \left(\frac{1}{2} + \frac{1}{2}\right)b = 1b = b$

101. $2y + \frac{1}{4}y + y = 2y + \frac{1}{4}y + 1 \cdot y = \left(2 + \frac{1}{4} + 1\right)y = 3\frac{1}{4}y$, or $\frac{13}{4}y$

103. $11x - 3x = (11 - 3)x = 8x$

105. $6n - n = (6 - 1)n = 5n$

107. $y - 17y = (1 - 17)y = -16y$

109. $\quad -8 + 11a - 5b + 6a - 7b + 7$
$= 11a + 6a - 5b - 7b - 8 + 7$
$= (11 + 6)a + (-5 - 7)b + (-8 + 7)$
$= 17a - 12b - 1$

111. $9x + 2y - 5x = (9 - 5)x + 2y = 4x + 2y$

113. $11x + 2y - 4x - y = (11 - 4)x + (2 - 1)y = 7x + y$

115. $2.7x + 2.3y - 1.9x - 1.8y = (2.7 - 1.9)x + (2.3 - 1.8)y = 0.8x + 0.5y$

117. $\quad \frac{13}{2}a + \frac{9}{5}b - \frac{2}{3}a - \frac{3}{10}b - 42$
$= \left(\frac{13}{2} - \frac{2}{3}\right)a + \left(\frac{9}{5} - \frac{3}{10}\right)b - 42$
$= \left(\frac{39}{6} - \frac{4}{6}\right)a + \left(\frac{18}{10} - \frac{3}{10}\right)b - 42$
$= \frac{35}{6}a + \frac{15}{10}b - 42$
$= \frac{35}{6}a + \frac{3}{2}b - 42$

119. Discussion and Writing Exercise

121. $16 = 2 \cdot 2 \cdot 2 \cdot 2$
$18 = 2 \cdot 3 \cdot 3$
The LCM is $2 \cdot 2 \cdot 2 \cdot 2 \cdot 3 \cdot 3$, or 144.

123. $16 = 2 \cdot 2 \cdot 2 \cdot 2$
$18 = 2 \cdot 3 \cdot 3$
$24 = 2 \cdot 2 \cdot 2 \cdot 3$
The LCM is $2 \cdot 2 \cdot 2 \cdot 2 \cdot 3 \cdot 3$, or 144.

125. $16 = 2 \cdot 2 \cdot 2 \cdot 2$
$32 = 2 \cdot 2 \cdot 2 \cdot 2 \cdot 2$
The LCM is $2 \cdot 2 \cdot 2 \cdot 2 \cdot 2$, or 32.

127. $15 = 3 \cdot 5$
$45 = 3 \cdot 3 \cdot 5$
$90 = 2 \cdot 3 \cdot 3 \cdot 5$
The LCM is $2 \cdot 3 \cdot 3 \cdot 5$, or 90.

129. $\frac{11}{12} + \frac{15}{16} = \frac{11}{12} \cdot \frac{4}{4} + \frac{15}{16} \cdot \frac{3}{3}$ LCD is 48
$= \frac{44}{48} + \frac{45}{48}$
$= \frac{89}{48}$

131. $\frac{1}{8} - \frac{1}{3} = \frac{1}{8} + \left(-\frac{1}{3}\right) = \frac{3}{24} + \left(-\frac{8}{24}\right) = -\frac{5}{24}$

133. No; for any replacement other than 5 the two expressions do not have the same value. For example, let $t = 2$. Then $3 \cdot 2 + 5 = 6 + 5 = 11$, but $3 \cdot 5 + 2 = 15 + 2 = 17$.

135. Yes; commutative law of addition

137. $\quad q + qr + qrs + qrst$ There are no like terms.
$= q \cdot 1 + q \cdot r + q \cdot rs + q \cdot rst$
$= q(1 + r + rs + rst)$ Factoring

Exercise Set 1.8

1. $-(2x + 7) = -2x - 7$ Changing the sign of each term

3. $-(8 - x) = -8 + x$ Changing the sign of each term

5. $-4a + 3b - 7c$

7. $-6x + 8y - 5$

9. $-3x + 5y + 6$

11. $8x + 6y + 43$

13. $9x - (4x + 3) = 9x - 4x - 3$ Removing parentheses by changing the sign of every term
$= 5x - 3$ Collecting like terms

15. $2a - (5a - 9) = 2a - 5a + 9 = -3a + 9$

17. $2x + 7x - (4x + 6) = 2x + 7x - 4x - 6 = 5x - 6$

19. $2x - 4y - 3(7x - 2y) = 2x - 4y - 21x + 6y = -19x + 2y$

21. $\quad 15x - y - 5(3x - 2y + 5z)$
$= 15x - y - 15x + 10y - 25z$ Multiplying each term in parentheses by -5
$= 9y - 25z$

23. $(3x + 2y) - 2(5x - 4y) = 3x + 2y - 10x + 8y = -7x + 10y$

25. $\quad (12a - 3b + 5c) - 5(-5a + 4b - 6c)$
$= 12a - 3b + 5c + 25a - 20b + 30c$
$= 37a - 23b + 35c$

27. $[9 - 2(5 - 4)] = [9 - 2 \cdot 1]$ Computing $5 - 4$
$= [9 - 2]$ Computing $2 \cdot 1$
$= 7$

29. $8[7 - 6(4 - 2)] = 8[7 - 6(2)] = 8[7 - 12] = 8[-5] = -40$

31. $\quad [4(9 - 6) + 11] - [14 - (6 + 4)]$
$= [4(3) + 11] - [14 - 10]$
$= [12 + 11] - [14 - 10]$
$= 23 - 4$
$= 19$

33. $\quad [10(x + 3) - 4] + [2(x - 1) + 6]$
$= [10x + 30 - 4] + [2x - 2 + 6]$
$= [10x + 26] + [2x + 4]$
$= 10x + 26 + 2x + 4$
$= 12x + 30$

35. $[7(x+5)-19]-[4(x-6)+10]$
$=[7x+35-19]-[4x-24+10]$
$=[7x+16]-[4x-14]$
$=7x+16-4x+14$
$=3x+30$

37. $3\{[7(x-2)+4]-[2(2x-5)+6]\}$
$=3\{[7x-14+4]-[4x-10+6]\}$
$=3\{[7x-10]-[4x-4]\}$
$=3\{7x-10-4x+4\}$
$=3\{3x-6\}$
$=9x-18$

39. $4\{[5(x-3)+2]-3[2(x+5)-9]\}$
$=4\{[5x-15+2]-3[2x+10-9]\}$
$=4\{[5x-13]-3[2x+1]\}$
$=4\{5x-13-6x-3\}$
$=4\{-x-16\}$
$=-4x-64$

41. $8-2\cdot3-9=8-6-9$ Multiplying
$\qquad\qquad=2-9$ Doing all additions and
$\qquad\qquad\qquad\qquad$ subtractions in order from
$\qquad\qquad=-7$ left to right

43. $(8-2\cdot3)-9=(8-6)-9$ Multiplying inside the
$\qquad\qquad\qquad\qquad\qquad$ parentheses
$\qquad\qquad\quad=2-9$ Subtracting inside the
$\qquad\qquad\qquad\qquad\qquad$ parentheses
$\qquad\qquad\quad=-7$

45. $[(-24)\div(-3)]\div\left(-\dfrac{1}{2}\right)=8\div\left(-\dfrac{1}{2}\right)=8\cdot(-2)=-16$

47. $16\cdot(-24)+50=-384+50=-334$

49. $2^4+2^3-10=16+8-10=24-10=14$

51. $5^3+26\cdot71-(16+25\cdot3)=5^3+26\cdot71-(16+75)=$
$5^3+26\cdot71-91=125+26\cdot71-91=125+1846-91=$
$1971-91=1880$

53. $4\cdot5-2\cdot6+4=20-12+4=8+4=12$

55. $4^3/8=64/8=8$

57. $8(-7)+6(-5)=-56-30=-86$

59. $19-5(-3)+3=19+15+3=34+3=37$

61. $9\div(-3)+16\div8=-3+2=-1$

63. $-4^2+6=-16+6=-10$

65. $-8^2-3=-64-3=-67$

67. $12-20^3=12-8000=-7988$

69. $2\cdot10^3-5000=2\cdot1000-5000=2000-5000=-3000$

71. $6[9-(3-4)]=6[9-(-1)]=6[9+1]=6[10]=60$

73. $-1000\div(-100)\div10=10\div10=1$

75. $8-(7-9)=8-(-2)=8+2=10$

77. $\dfrac{10-6^2}{9^2+3^2}=\dfrac{10-36}{81+9}=\dfrac{-26}{90}=-\dfrac{13}{45}$

79. $\dfrac{3(6-7)-5\cdot4}{6\cdot7-8(4-1)}=\dfrac{3(-1)-5\cdot4}{42-8\cdot3}=\dfrac{-3-20}{42-24}=-\dfrac{23}{18}$

81. $\dfrac{|2^3-3^2|+|12\cdot5|}{-32\div(-16)\div(-4)}=\dfrac{|8-9|+|12\cdot5|}{-32\div(-16)\div(-4)}=$

$\dfrac{|-1|+|60|}{2\div(-4)}=\dfrac{1+60}{-\frac{1}{2}}=\dfrac{61}{-\frac{1}{2}}=61(-2)=-122$

83.   Discussion and Writing Exercise

85.   The set of integers is
$\{\ldots,-5,\underline{-4,-3},-2,-1,0,1,2,3,\ldots\}.$

87.   The <u>commutative law</u> of addition says that $a+b=b+a$ for any real numbers a and b.

89.   The <u>associative law</u> of addition says that $a+(b+c)=(a+b)+c$ for any real numbers a, b, and c.

91.   Two numbers whose product is 1 are called <u>multiplicative inverses</u> of each other.

93. $6y+2x-3a+c=6y-(-2x)-3a-(-c)=6y-(-2x+3a-c)$

95. $6m+3n-5m+4b=6m-(-3n)-5m-(-4b)=$
$6m-(-3n+5m-4b)$

97. $\{x-[f-(f-x)]+[x-f]\}-3x$
$=\{x-[f-f+x]+[x-f]\}-3x$
$=\{x-[x]+[x-f]\}-3x$
$=\{x-x+x-f\}-3x=x-f-3x=-2x-f$

99.   a) $x^2+3=7^2+3=49+3=52;$
$\qquad x^2+3=(-7)^2+3=49+3=52;$
$\qquad x^2+3=(-5.013)^2+3=25.130169+3=28.130169$

b) $1-x^2=1-5^2=1-25=-24;$
$\qquad 1-x^2=1-(-5)^2=1-25=-24;$
$\qquad 1-x^2=1-(-10.455)^2=1-109.307025=$
$\qquad -108.307025$

101. $\dfrac{-15+20+50+(-82)+(-7)+(-2)}{6}=\dfrac{-36}{6}=-6$

Chapter 1 Review Exercises

1. Substitute 17 for x and 5 for y and carry out the computation.
$$\frac{x-y}{3}=\frac{17-5}{3}=\frac{12}{3}=4$$

2. $19\%x,\ 0.19x$

3. The integer -45 corresponds to a debt of \$45; the integer 72 corresponds to having \$72 in a savings account.

4. The distance of -38 from 0 is 38, so $|-38|=38$.

5. The graph of -2.5 is halfway between -3 and -2.

6. The graph of $\frac{8}{9}$ is $\frac{8}{9}$ of the way from 0 to 1.

7. Since -3 is to the left of 10, we have $-3 < 10$.

8. Since -1 is to the right of -6, we have $-1 > -6$.

9. Since 0.126 is to the right of -12.6, we have $0.126 > -12.6$.

10.
$$-\frac{2}{3} = -\frac{2}{3} \cdot \frac{10}{10} = -\frac{20}{30}$$
$$-\frac{1}{10} = -\frac{1}{10} \cdot \frac{3}{3} = -\frac{3}{30}$$

Since $-\frac{20}{30}$ is to the left of $-\frac{3}{30}$, then $-\frac{2}{3}$ is to the left of $-\frac{1}{10}$ and we have $-\frac{2}{3} < -\frac{1}{10}$.

11. The opposite of 3.8 is -3.8 because $3.8 + (-3.8) = 0$.

12. The opposite of $-\frac{3}{4}$ is $\frac{3}{4}$ because $-\frac{3}{4} + \frac{3}{4} = 0$.

13. The reciprocal of $\frac{3}{8}$ is $\frac{8}{3}$ because $\frac{3}{8} \cdot \frac{8}{3} = 1$.

14. The reciprocal of -7 is $-\frac{1}{7}$ because $-7 \cdot \left(-\frac{1}{7}\right) = 1$.

15. If $x = -34$, then $-x = -(-34) = 34$.

16. If $x = 5$, then $-(-x) = -(-5) = 5$.

17. $4 + (-7)$

The absolute values are 4 and 7. The difference is $7 - 4$, or 3. The negative number has the larger absolute value, so the answer is negative. $4 + (-7) = -3$

18. $6 + (-9) + (-8) + 7$

  a) Add the negative numbers: $-9 + (-8) = -17$

  b) Add the positive numbers: $6 + 7 = 13$

  c) Add the results: $-17 + 13 = -4$

19. $-3.8 + 5.1 + (-12) + (-4.3) + 10$

  a) Add the negative numbers: $-3.8 + (-12) + (-4.3) =$
    -20.1

  b) Add the positive numbers: $5.1 + 10 = 15.1$

  c) Add the results: $-20.1 + 15.1 = -5$

20. $-3 - (-7) = -3 + 7 = 4$

21. $-\frac{9}{10} - \frac{1}{2} = -\frac{9}{10} - \frac{5}{10} = -\frac{9}{10} + \left(-\frac{5}{10}\right) = -\frac{14}{10} =$
$-\frac{7 \cdot 2}{5 \cdot 2} = -\frac{7}{5} \cdot \frac{2}{2} = -\frac{7}{5}$

22. $-3.8 - 4.1 = -3.8 + (-4.1) = -7.9$

23. $-9 \cdot (-6) = 54$

24. $-2.7(3.4) = -9.18$

25. $\frac{2}{3} \cdot \left(-\frac{3}{7}\right) = -\left(\frac{2 \cdot 3}{3 \cdot 7}\right) = -\left(\frac{2}{7} \cdot \frac{3}{3}\right) = -\frac{2}{7}$

26. $3 \cdot (-7) \cdot (-2) \cdot (-5) = -21 \cdot 10 = -210$

27. $35 \div (-5) = -7$ Check: $-7 \cdot (-5) = 35$

28. $-5.1 \div 1.7 = -3$ Check: $-3 \cdot (1.7) = -5.1$

29. $-\frac{3}{11} \div -\frac{4}{11} = -\frac{3}{11} \cdot \left(-\frac{11}{4}\right) = \frac{3 \cdot 11}{11 \cdot 4} = \frac{3}{4} \cdot \frac{11}{11} = \frac{3}{4}$

30. $(-3.4 - 12.2) - 8(-7) = -15.6 - 8(-7)$
$$= -15.6 + 56$$
$$= 40.4$$

31.
$$\frac{-12(-3) - 2^3 - (-9)(-10)}{3 \cdot 10 + 1} = \frac{-12(-3) - 8 - (-9)(-10)}{30 + 1}$$
$$= \frac{36 - 8 - 90}{31}$$
$$= \frac{28 - 90}{31}$$
$$= \frac{-62}{31}$$
$$= -2$$

32. $-16 \div 4 - 30 \div (-5) = -4 - (-6)$
$$= -4 + 6$$
$$= 2$$

33. $\frac{9[(7 - 14) - 13]}{|-2(8) - 4|} = \frac{9[-7 - 13]}{|-16 - 4|} = \frac{9[-20]}{|-20|} = \frac{-180}{20} = -9$

34. Let $t =$ the total gain or loss. We represent the gains as positive numbers and the loss as a negative number. We add the gains and the loss to find t.
$$t = 5 + (-12) + 15 = -7 + 15 = 8$$

There is a total gain of 8 yd.

35. Let $a =$ Kaleb's total assets after he borrows \$300.

Total assets	is	Initial assets	minus	Amount of loan
↓	↓	↓	↓	↓
a	$=$	170	$-$	300

We carry out the subtraction.
$$a = 170 - 300 = -130$$

Kaleb's total assets were $-\$130$.

36. First we multiply to find the total drop d in the price:
$$d = 8(-\$1.63) = -\$13.04$$

Now we add this number to the opening price to find the price p after 8 hr:
$$p = \$17.68 + (-\$13.04) = \$4.64$$

After 8 hr the price of the stock was $\$4.64$ per share.

37. Yuri spent the \$68 in his account plus an additional \$64.65, so he spent a total of $\$68 + \$64.65$, or \$132.65, on seven equally-priced DVDs. Then each DVD cost $\frac{\$132.65}{7}$, or \$18.95.

38. $5(3x - 7) = 5 \cdot 3x - 5 \cdot 7 = 15x - 35$

39. $-2(4x - 5) = -2 \cdot 4x - (-2)(5) = -8x - (-10) = -8x + 10$

40. $10(0.4x + 1.5) = 10 \cdot 0.4x + 10 \cdot 1.5 = 4x + 15$

41. $-8(3 - 6x) = -8 \cdot 3 - (-8)(6x) = -24 - (-48x) = -24 + 48x$

42. $2x - 14 = 2 \cdot x - 2 \cdot 7 = 2(x - 7)$

43. $-6x + 6 = -6 \cdot x - 6(-1) = -6(x - 1)$

The expression can also be factored as follows:

$-6x + 6 = 6(-x) + 6 \cdot 1 = 6(-x + 1)$

44. $5x + 10 = 5 \cdot x + 5 \cdot 2 = 5(x + 2)$

45. $-3x + 12y - 12 = -3 \cdot x - 3(-4y) - 3 \cdot 4 = -3(x - 4y + 4)$

We could also factor this expression as follows:

$-3x + 12y - 12 = 3(-x) + 3 \cdot 4y + 3(-4) = 3(-x + 4y - 4)$

46. $11a + 2b - 4a - 5b = 11a - 4a + 2b - 5b$
$$= (11 - 4)a + (2 - 5)b$$
$$= 7a - 3b$$

47. $7x - 3y - 9x + 8y = 7x - 9x - 3y + 8y$
$$= (7 - 9)x + (-3 + 8)y$$
$$= -2x + 5y$$

48. $6x + 3y - x - 4y = 6x - x + 3y - 4y$
$$= (6 - 1)x + (3 - 4)y$$
$$= 5x - y$$

49. $-3a + 9b + 2a - b = -3a + 2a + 9b - b$
$$= (-3 + 2)a + (9 - 1)b$$
$$= -a + 8b$$

50. $2a - (5a - 9) = 2a - 5a + 9 = -3a + 9$

51. $3(b + 7) - 5b = 3b + 21 - 5b = -2b + 21$

52. $3[11 - 3(4 - 1)] = 3[11 - 3 \cdot 3] = 3[11 - 9] = 3 \cdot 2 = 6$

53. $2[6(y - 4) + 7] = 2[6y - 24 + 7] = 2[6y - 17] = 12y - 34$

54. $[8(x + 4) - 10] - [3(x - 2) + 4]$
$$= [8x + 32 - 10] - [3x - 6 + 4]$$
$$= 8x + 22 - [3x - 2]$$
$$= 8x + 22 - 3x + 2$$
$$= 5x + 24$$

55. $5\{[6(x - 1) + 7] - [3(3x - 4) + 8]\}$
$$= 5\{[6x - 6 + 7] - [9x - 12 + 8]\}$$
$$= 5\{6x + 1 - [9x - 4]\}$$
$$= 5\{6x + 1 - 9x + 4\}$$
$$= 5\{-3x + 5\}$$
$$= -15x + 25$$

56. $-9 \leq 11$ is true since $-9 < 11$ is true.

57. $-11 \geq -3$ is false since neither $-11 > -3$ nor $-11 = -3$ is true.

58. $x > -3$ has the same meaning as $-3 < x$.

59. *Discussion and Writing Exercise.* If the sum of two numbers is 0, they are opposites, or additive inverses of each other. For every real number a, the opposite of a can be named $-a$, and $a + (-a) = (-a) + a = 0$.

60. *Discussion and Writing Exercise.* No; $|0| = 0$, and 0 is not positive.

61. $-\left|\dfrac{7}{8} - \left(-\dfrac{1}{2}\right) - \dfrac{3}{4}\right| = -\left|\dfrac{7}{8} + \dfrac{1}{2} - \dfrac{3}{4}\right|$
$$= -\left|\dfrac{7}{8} + \dfrac{4}{8} - \dfrac{6}{8}\right|$$
$$= -\left|\dfrac{11}{8} - \dfrac{6}{8}\right|$$
$$= -\left|\dfrac{5}{8}\right|$$
$$= -\dfrac{5}{8}$$

62. $(|2.7 - 3| + 3^2 - |-3|) \div (-3)$
$$= (|2.7 - 3| + 9 - |-3|) \div (-3)$$
$$= (|-0.3| + 9 - |-3|) \div (-3)$$
$$= (0.3 + 9 - 3) \div (-3)$$
$$= (9.3 - 3) \div (-3)$$
$$= 6.3 \div (-3)$$
$$= -2.1$$

63. $\underbrace{2000 - 1990}_{\downarrow \atop 10} + \underbrace{1980 - 1970}_{\downarrow \atop 10} + \ldots + \underbrace{20 - 10}_{\downarrow \atop 10}$

Counting by 10's from 10 through 2000 gives us 2000/10, or 200, numbers in the expression. There are 200/2, or 100, pairs of numbers in the expression. Each pair is a difference that is equivalent to 10. Thus, the expression is equal to $100 \cdot 10$, or 1000.

64. Note that the sum of the lengths of the three horizontal segments at the top of the figure, two of which are not labeled and one of which is labeled b, is equivalent to the length of the horizontal segment at the bottom of the figure, a. Then the perimeter is $a + b + b + a + a + a$, or $4a + 2b$.

Chapter 1 Test

---

1. Substitute 10 for x and 5 for y and carry out the computations.
$$\frac{3x}{y} = \frac{3 \cdot 10}{5} = \frac{30}{5} = 6$$

2. Using x for "some number," we have $x - 9$.

3. Substitute 16 ft for b and 30 ft for h in the formula for the area of a triangle and then carry out the multiplication.

$$A = \frac{1}{2}bh = \frac{1}{2}(16 \text{ ft})(30 \text{ ft})$$
$$= \frac{1}{2}(16)(30)(\text{ft})(\text{ft})$$
$$= 240 \text{ ft}^2$$

4. Since -4 is to the left of 0 on the number line, we have $-4 < 0$.

5. Since -3 is to the right of -8 on the number line, we have $-3 > -8$.

6. Since -0.78 is to the right of -0.87 on the number line, we have $-0.78 > -0.87$.

7. Since $-\frac{1}{8}$ is to the left of $\frac{1}{2}$ on the number line, we have $-\frac{1}{8} < \frac{1}{2}$.

8. The distance of -7 from 0 is 7, so $|-7| = 7$.

9. The distance of $\frac{9}{4}$ from 0 is $\frac{9}{4}$, so $\left|\frac{9}{4}\right| = \frac{9}{4}$.

10. The distance of -2.7 from 0 is 2.7, so $|-2.7| = 2.7$.

11. The opposite of $\frac{2}{3}$ is $-\frac{2}{3}$ because $\frac{2}{3} + \left(-\frac{2}{3}\right) = 0$.

12. The opposite of -1.4 is 1.4 because $-1.4 + 1.4 = 0$.

13. If $x = -8$, then $-x = -(-8) = 8$.

14. The reciprocal of -2 is $-\frac{1}{2}$ because $-2\left(-\frac{1}{2}\right) = 1$.

15. The reciprocal of $\frac{4}{7}$ is $\frac{7}{4}$ because $\frac{4}{7} \cdot \frac{7}{4} = 1$.

16. $3.1 - (-4.7) = 3.1 + 4.7 = 7.8$

17. $-8 + 4 + (-7) + 3 = -4 + (-7) + 3$
$$= -11 + 3$$
$$= -8$$

18. $-\frac{1}{5} + \frac{3}{8} = -\frac{1}{5} \cdot \frac{8}{8} + \frac{3}{8} \cdot \frac{5}{5}$
$$= -\frac{8}{40} + \frac{15}{40}$$
$$= \frac{7}{40}$$

19. $2 - (-8) = 2 + 8 = 10$

20. $3.2 - 5.7 = 3.2 + (-5.7) = -2.5$

21. $\frac{1}{8} - \left(-\frac{3}{4}\right) = \frac{1}{8} + \frac{3}{4}$
$$= \frac{1}{8} + \frac{3}{4} \cdot \frac{2}{2}$$
$$= \frac{1}{8} + \frac{6}{8}$$
$$= \frac{7}{8}$$

22. $4 \cdot (-12) = -48$

23. $-\frac{1}{2} \cdot \left(-\frac{3}{8}\right) = \frac{3}{16}$

24. $-45 \div 5 = -9$ Check: $-9 \cdot 5 = -45$

25. $-\frac{3}{5} \div \left(-\frac{4}{5}\right) = -\frac{3}{5} \cdot \left(-\frac{5}{4}\right) = \frac{3 \cdot 5}{5 \cdot 4} = \frac{3 \cdot \cancel{5}}{\cancel{5} \cdot 4} = \frac{3}{4}$

26. $4.864 \div (-0.5) = -9.728$

27. $-2(16) - [2(-8) - 5^3] = -2(16) - [2(-8) - 125]$
$$= -2(16) - [-16 - 125]$$
$$= -2(16) - [-141]$$
$$= -2(16) + 141$$
$$= -32 + 141$$
$$= 109$$

28. $-20 \div (-5) + 36 \div (-4) = 4 + (-9) = -5$

29. Let D = the difference in the temperatures.

Difference in temperature	is	Higher temperature	minus	Lower temperature
↓	↓	↓	↓	↓
D	$=$	-67	$-$	(-81)

We carry out the subtraction.
$$D = -67 - (-81) = -67 + 81 = 14$$

The average high temperature is $14°$F higher than the average low temperature.

30. Let P = the number of points by which the market has changed over the five week period.

Total change	=	Week 1 change	+	Week 2 change	+	Week 3 change	+
↓	↓	↓	↓	↓	↓	↓	
P	$=$	-13	$+$	(-16)	$+$	36	$+$

Week 4 change	+	Week 5 change
↓	↓	↓
(-11)	$+$	19

We carry out the computation.
$$P = -13 + (-16) + 36 + (-11) + 19$$
$$= -29 + 36 + (-11) + 19$$
$$= 7 + (-11) + 19$$
$$= -4 + 19$$
$$= 15$$

The market rose 15 points.

31. First we multiply to find the total decrease d in the population.
$$d = 6 \cdot 420 = 2520$$

The population decreased by 2520 over the six year period.

Now we subtract to find the new population p.
$$18,600 - 2520 = 16,080$$

After 6 yr the population was 16,080.

32. First we subtract to find the total drop in temperature t.

$$t = 16°C - (-17°C) = 16°C + 17°C = 33°C$$

Then we divide to find by how many degrees d the temperature dropped each minute in the 35 minutes from 11:08 A.M. to 11:43 A.M.

$$d = 33 \div 35 = \frac{33}{35}$$

The temperature dropped about $\frac{33}{35}°$C each minute.

33. $3(6 - x) = 3 \cdot 6 - 3 \cdot x = 18 - 3x$

34. $-5(y - 1) = -5 \cdot y - (-5)(1) = -5y - (-5) = -5y + 5$

35. $12 - 22x = 2 \cdot 6 - 2 \cdot 11x = 2(6 - 11x)$

36. $7x + 21 + 14y = 7 \cdot x + 7 \cdot 3 + 7 \cdot 2y = 7(x + 3 + 2y)$

37. $\begin{aligned} 6 + 7 - 4 - (-3) &= 6 + 7 + (-4) + 3 \\ &= 13 + (-4) + 3 \\ &= 9 + 3 \\ &= 12 \end{aligned}$

38. $5x - (3x - 7) = 5x - 3x + 7 = 2x + 7$

39. $\begin{aligned} 4(2a - 3b) + a - 7 &= 8a - 12b + a - 7 \\ &= 9a - 12b - 7 \end{aligned}$

40. $\begin{aligned} &4\{3[5(y - 3) + 9] + 2(y + 8)\} \\ &= 4\{3[5y - 15 + 9] + 2y + 16\} \\ &= 4\{3[5y - 6] + 2y + 16\} \\ &= 4\{15y - 18 + 2y + 16\} \\ &= 4\{17y - 2\} \\ &= 68y - 8 \end{aligned}$

41. $256 \div (-16) \div 4 = -16 \div 4 = -4$

42. $\begin{aligned} 2^3 - 10[4 - (-2 + 18)3] &= 2^3 - 10[4 - (16)3] \\ &= 2^3 - 10[4 - 48] \\ &= 2^3 - 10[-44] \\ &= 8 - 10[-44] \\ &= 8 + 440 \\ &= 448 \end{aligned}$

43. $-2 \geq x$ has the same meaning as $x \leq -2$.

44. $\begin{aligned} &|-27 - 3(4)| - |-36| + |-12| \\ &= |-27 - 12| - |-36| + |-12| \\ &= |-39| - |-36| + |-12| \\ &= 39 - 36 + 12 \\ &= 3 + 12 \\ &= 15 \end{aligned}$

45. $\begin{aligned} &a - \{3a - [4a - (2a - 4a)]\} \\ &= a - \{3a - [4a - (-2a)]\} \\ &= a - \{3a - [4a + 2a]\} \\ &= a - \{3a - 6a\} \\ &= a - \{-3a\} \\ &= a + 3a \\ &= 4a \end{aligned}$

46. The perimeter is equivalent to the perimeter of a square with sides x along with four additional segments of length y. We have $x + x + x + x + y + y + y + y = 4x + 4y$.

Chapter 2
Solving Equations and Inequalities

1. $\underline{x + 17 = 32}$ Writing the equation

$15 + 17 \ ? \ 32$ Substituting 15 for x

$32 \ |$ TRUE

Since the left-hand and right-hand sides are the same, 15 is a solution of the equation.

3. $\underline{x - 7 = 12}$ Writing the equation

$21 - 7 \ ? \ 12$ Substituting 21 for x

$14 \ |$ FALSE

Since the left-hand and right-hand sides are not the same, 21 is not a solution of the equation.

5. $\underline{6x = 54}$ Writing the equation

$6(-7) \ ? \ 54$ Substituting

$-42 \ |$ FALSE

-7 is not a solution of the equation.

7. $\underline{\dfrac{x}{6} = 5}$ Writing the equation

$\dfrac{30}{6} \ ? \ 5$ Substituting

$5 \ |$ TRUE

5 is a solution of the equation.

9. $\underline{5x + 7 = 107}$

$5 \cdot 19 + 7 \ ? \ 107$ Substituting

$95 + 7 \ |$

$102 \ |$ FALSE

19 is not a solution of the equation.

11. $\underline{7(y - 1) = 63}$

$7(-11 - 1) \ ? \ 63$ Substituting

$7(-12) \ |$

$-84 \ |$ FALSE

-11 is not a solution of the equation.

13. $x + 2 = 6$

$x + 2 - 2 = 6 - 2$ Subtracting 2 on both sides

$x = 4$ Simplifying

Check: $\underline{x + 2 = 6}$

$4 + 2 \ ? \ 6$

$6 \ |$ TRUE

The solution is 4.

15. $x + 15 = -5$

$x + 15 - 15 = -5 - 15$ Subtracting 15 on both sides

$x = -20$

Check: $\underline{x + 15 = -5}$

$-20 + 15 \ ? \ -5$

$-5 \ |$ TRUE

The solution is -20.

17. $x + 6 = -8$

$x + 6 - 6 = -8 - 6$

$x = -14$

Check: $\underline{x + 6 = -8}$

$-14 + 6 \ ? \ -8$

$-8 \ |$ TRUE

The solution is -14.

19. $x + 16 = -2$

$x + 16 - 16 = -2 - 16$

$x = -18$

Check: $\underline{x + 16 = -2}$

$-18 + 16 \ ? \ -2$

$-2 \ |$ TRUE

The solution is -18.

21. $x - 9 = 6$

$x - 9 + 9 = 6 + 9$

$x = 15$

Check: $\underline{x - 9 = 6}$

$15 - 9 \ ? \ 6$

$6 \ |$ TRUE

The solution is 15.

23. $x - 7 = -21$

$x - 7 + 7 = -21 + 7$

$x = -14$

Check: $\underline{x - 7 = -21}$

$-14 - 7 \ ? \ -21$

$-21 \ |$ TRUE

The solution is -14.

25. $5 + t = 7$

$-5 + 5 + t = -5 + 7$

$t = 2$

Check: $\underline{5 + t = 7}$

$5 + 2 \ ? \ 7$

$7 \ |$ TRUE

The solution is 2.

27.
$$-7 + y = 13$$
$$7 + (-7) + y = 7 + 13$$
$$y = 20$$

Check:
$$\frac{-7 + y = 13}{}$$
$$-7 + 20 \ ? \ 13$$
$$13 \ | \qquad \text{TRUE}$$

The solution is 20.

29.
$$-3 + t = -9$$
$$3 + (-3) + t = 3 + (-9)$$
$$t = -6$$

Check:
$$\frac{-3 + t = -9}{}$$
$$-3 + (-6) \ ? \ -9$$
$$-9 \ | \qquad \text{TRUE}$$

The solution is −6.

31.
$$x + \frac{1}{2} = 7$$
$$x + \frac{1}{2} - \frac{1}{2} = 7 - \frac{1}{2}$$
$$x = 6\frac{1}{2}$$

Check:
$$\frac{x + \frac{1}{2} = 7}{}$$
$$6\frac{1}{2} + \frac{1}{2} \ ? \ 7$$
$$7 \ | \qquad \text{TRUE}$$

The solution is $6\frac{1}{2}$.

33.
$$12 = a - 7.9$$
$$12 + 7.9 = a - 7.9 + 7.9$$
$$19.9 = a$$

Check:
$$\frac{12 = a - 7.9}{}$$
$$12 \ ? \ 19.9 - 7.9$$
$$| \ 12 \qquad \text{TRUE}$$

The solution is 19.9.

35.
$$r + \frac{1}{3} = \frac{8}{3}$$
$$r + \frac{1}{3} - \frac{1}{3} = \frac{8}{3} - \frac{1}{3}$$
$$r = \frac{7}{3}$$

Check:
$$\frac{r + \frac{1}{3} = \frac{8}{3}}{}$$
$$\frac{7}{3} + \frac{1}{3} \ ? \ \frac{8}{3}$$
$$\frac{8}{3} \ | \qquad \text{TRUE}$$

The solution is $\frac{7}{3}$.

37.
$$m + \frac{5}{6} = -\frac{11}{12}$$
$$m + \frac{5}{6} - \frac{5}{6} = -\frac{11}{12} - \frac{5}{6}$$
$$m = -\frac{11}{12} - \frac{5}{6} \cdot \frac{2}{2}$$
$$m = -\frac{11}{12} - \frac{10}{12}$$
$$m = -\frac{21}{12} = -\frac{\cancel{3} \cdot 7}{\cancel{3} \cdot 4}$$
$$m = -\frac{7}{4}$$

Check:
$$\frac{m + \frac{5}{6} = -\frac{11}{12}}{}$$
$$-\frac{7}{4} + \frac{5}{6} \ ? \ -\frac{11}{12}$$
$$-\frac{21}{12} + \frac{10}{12} \ |$$
$$-\frac{11}{12} \ | \qquad \text{TRUE}$$

The solution is $-\frac{7}{4}$.

39.
$$x - \frac{5}{6} = \frac{7}{8}$$
$$x - \frac{5}{6} + \frac{5}{6} = \frac{7}{8} + \frac{5}{6}$$
$$x = \frac{7}{8} \cdot \frac{3}{3} + \frac{5}{6} \cdot \frac{4}{4}$$
$$x = \frac{21}{24} + \frac{20}{24}$$
$$x = \frac{41}{24}$$

Check:
$$\frac{x - \frac{5}{6} = \frac{7}{8}}{}$$
$$\frac{41}{24} - \frac{5}{6} \ ? \ \frac{7}{8}$$
$$\frac{41}{24} - \frac{20}{24} \ | \ \frac{21}{24}$$
$$\frac{21}{24} \ | \qquad \text{TRUE}$$

The solution is $\frac{41}{24}$.

41.
$$-\frac{1}{5} + z = -\frac{1}{4}$$
$$\frac{1}{5} - \frac{1}{5} + z = \frac{1}{5} - \frac{1}{4}$$
$$z = \frac{1}{5} \cdot \frac{4}{4} - \frac{1}{4} \cdot \frac{5}{5}$$
$$z = \frac{4}{20} - \frac{5}{20}$$
$$z = -\frac{1}{20}$$

Check:
$$-\frac{1}{5} + z = -\frac{1}{4}$$

$$-\frac{1}{5} + \left(-\frac{1}{20}\right) \ ? \ -\frac{1}{4}$$

$$-\frac{4}{20} + \left(-\frac{1}{20}\right) \ \Big| \ -\frac{5}{20}$$

$$-\frac{5}{20} \ \Big| \qquad \text{TRUE}$$

The solution is $-\frac{1}{20}$.

43.
$$x + 2.3 = 7.4$$
$$x + 2.3 - 2.3 = 7.4 - 2.3$$
$$x = 5.1$$
Check:
$$x + 2.3 = 7.4$$
$$5.1 + 2.3 \ ? \ 7.4$$
$$7.4 \ \Big| \qquad \text{TRUE}$$

The solution is 5.1.

45.
$$7.6 = x - 4.8$$
$$7.6 + 4.8 = x - 4.8 + 4.8$$
$$12.4 = x$$
Check:
$$7.6 = x - 4.8$$
$$7.6 \ ? \ 12.4 - 4.8$$
$$\Big| \ 7.6 \qquad \text{TRUE}$$

The solution is 12.4.

47.
$$-9.7 = -4.7 + y$$
$$4.7 + (-9.7) = 4.7 + (-4.7) + y$$
$$-5 = y$$
Check:
$$-9.7 = -4.7 + y$$
$$-9.7 \ ? \ -4.7 + (-5)$$
$$\Big| \ -9.7 \qquad \text{TRUE}$$

The solution is -5.

49.
$$5\frac{1}{6} + x = 7$$
$$-5\frac{1}{6} + 5\frac{1}{6} + x = -5\frac{1}{6} + 7$$
$$x = -\frac{31}{6} + \frac{42}{6}$$
$$x = \frac{11}{6}, \text{ or } 1\frac{5}{6}$$

Check:
$$5\frac{1}{6} + x = 7$$
$$5\frac{1}{6} + 1\frac{5}{6} \ ? \ 7$$
$$7 \ \Big| \qquad \text{TRUE}$$

The solution is $\frac{11}{6}$, or $1\frac{5}{6}$.

51.
$$q + \frac{1}{3} = -\frac{1}{7}$$
$$q + \frac{1}{3} - \frac{1}{3} = -\frac{1}{7} - \frac{1}{3}$$
$$q = -\frac{1}{7} \cdot \frac{3}{3} - \frac{1}{3} \cdot \frac{7}{7}$$
$$q = -\frac{3}{21} - \frac{7}{21}$$
$$q = -\frac{10}{21}$$

Check:
$$q + \frac{1}{3} = -\frac{1}{7}$$
$$-\frac{10}{21} + \frac{1}{3} \ ? \ -\frac{1}{7}$$
$$-\frac{10}{21} + \frac{7}{21} \ \Big| \ -\frac{3}{21}$$
$$-\frac{3}{21} \ \Big| \qquad \text{TRUE}$$

The solution is $-\frac{10}{21}$.

53. Discussion and Writing Exercise

55. $-3 + (-8)$ Two negative numbers. We add the absolute values, getting 11, and make the answer negative.
$$-3 + (-8) = -11$$

57. $-\frac{2}{3} \cdot \frac{5}{8} = -\frac{2 \cdot 5}{3 \cdot 8} = -\frac{\cancel{2} \cdot 5}{3 \cdot \cancel{2} \cdot 4} = -\frac{5}{12}$

59. $\frac{2}{3} \div \left(-\frac{4}{9}\right) = \frac{2}{3} \cdot \left(-\frac{9}{4}\right) = -\frac{2 \cdot 9}{3 \cdot 4} = -\frac{\cancel{2} \cdot \cancel{3} \cdot 3}{\cancel{3} \cdot \cancel{2} \cdot 2} = -\frac{3}{2}$

61.
$$-\frac{2}{3} - \left(-\frac{5}{8}\right) = -\frac{2}{3} + \frac{5}{8}$$
$$= -\frac{2}{3} \cdot \frac{8}{8} + \frac{5}{8} \cdot \frac{3}{3}$$
$$= -\frac{16}{24} + \frac{15}{24}$$
$$= -\frac{1}{24}$$

63. The translation is $\$83 - x$.

65.
$$-356.788 = -699.034 + t$$
$$699.034 + (-356.788) = 699.034 + (-699.034) + t$$
$$342.246 = t$$

The solution is 342.246.

67.
$$x + \frac{4}{5} = -\frac{2}{3} - \frac{4}{15}$$

$$x + \frac{4}{5} = -\frac{2}{3} \cdot \frac{5}{5} - \frac{4}{15} \quad \text{Adding on the right side}$$

$$x + \frac{4}{5} = -\frac{10}{15} - \frac{4}{15}$$

$$x + \frac{4}{5} = -\frac{14}{15}$$

$$x + \frac{4}{5} - \frac{4}{5} = -\frac{14}{15} - \frac{4}{5}$$

$$x = -\frac{14}{15} - \frac{4}{5} \cdot \frac{3}{3}$$

$$x = -\frac{14}{15} - \frac{12}{15}$$

$$x = -\frac{26}{15}$$

The solution is $-\frac{26}{15}$.

69.
$$16 + x - 22 = -16$$

$$x - 6 = -16 \quad \text{Adding on the left side}$$

$$x - 6 + 6 = -16 + 6$$

$$x = -10$$

The solution is -10.

71.
$$x + 3 = 3 + x$$

$$x + 3 - 3 = 3 + x - 3$$

$$x = x$$

$x = x$ is true for all real numbers. Thus the solution is all real numbers.

73.
$$-\frac{3}{2} + x = -\frac{5}{17} - \frac{3}{2}$$

$$\frac{3}{2} - \frac{3}{2} + x = \frac{3}{2} - \frac{5}{17} - \frac{3}{2}$$

$$x = \left(\frac{3}{2} - \frac{3}{2}\right) - \frac{5}{17}$$

$$x = -\frac{5}{17}$$

The solution is $-\frac{5}{17}$.

75.
$$|x| + 6 = 19$$

$$|x| + 6 - 6 = 19 - 6$$

$$|x| = 13$$

x represents a number whose distance from 0 is 13. Thus $x = -13$ or $x = 13$.

The solutions are -13 and 13.

Exercise Set 2.2

1.
$$6x = 36$$

$$\frac{6x}{6} = \frac{36}{6} \quad \text{Dividing by 6 on both sides}$$

$$1 \cdot x = 6 \quad \text{Simplifying}$$

$$x = 6 \quad \text{Identity property of 1}$$

Check:
$$\frac{6x = 36}{6 \cdot 6 \ ? \ 36}$$
$$36 \ | \qquad \text{TRUE}$$

The solution is 6.

3.
$$5x = 45$$

$$\frac{5x}{5} = \frac{45}{5} \quad \text{Dividing by 5 on both sides}$$

$$1 \cdot x = 9 \quad \text{Simplifying}$$

$$x = 9 \quad \text{Identity property of 1}$$

Check:
$$\frac{5x = 45}{5 \cdot 9 \ ? \ 45}$$
$$45 \ | \qquad \text{TRUE}$$

The solution is 9.

5.
$$84 = 7x$$

$$\frac{84}{7} = \frac{7x}{7} \quad \text{Dividing by 7 on both sides}$$

$$12 = 1 \cdot x$$

$$12 = x$$

Check:
$$\frac{84 = 7x}{84 \ ? \ 7 \cdot 12}$$
$$| \ 84 \qquad \text{TRUE}$$

The solution is 12.

7.
$$-x = 40$$

$$-1 \cdot x = 40$$

$$\frac{-1 \cdot x}{-1} = \frac{40}{-1}$$

$$1 \cdot x = -40$$

$$x = -40$$

Check:
$$\frac{-x = 40}{-(-40) \ ? \ 40}$$
$$40 \ | \qquad \text{TRUE}$$

The solution is -40.

9.
$$-x = -1$$

$$-1 \cdot x = -1$$

$$\frac{-1 \cdot x}{-1} = \frac{-1}{-1}$$

$$1 \cdot x = 1$$

$$x = 1$$

Check:
$$\frac{-x = -1}{-(1) \ ? \ -1}$$
$$-1 \ | \qquad \text{TRUE}$$

The solution is 1.

11.
$$7x = -49$$

$$\frac{7x}{7} = \frac{-49}{7}$$

$$1 \cdot x = -7$$

$$x = -7$$

Check:
$$\frac{7x = -49}{7(-7) \ ? \ -49}$$
$$-49 \ | \qquad \text{TRUE}$$

The solution is -7.

13. $\quad -12x = 72$

$$\frac{-12x}{-12} = \frac{72}{-12}$$

$$1 \cdot x = -6$$

$$x = -6$$

Check: $\quad \dfrac{-12x = 72}{\begin{array}{c} -12(-6) \ ? \ 72 \\ 72 \ \big| \end{array}}$ TRUE

The solution is -6.

15. $\quad -21x = -126$

$$\frac{-21x}{-21} = \frac{-126}{-21}$$

$$1 \cdot x = 6$$

$$x = 6$$

Check: $\quad \dfrac{-21x = -126}{\begin{array}{c} -21 \cdot 6 \ ? \ -126 \\ -126 \ \big| \end{array}}$ TRUE

The solution is 6.

17. $\qquad \dfrac{t}{7} = -9$

$$7 \cdot \frac{1}{7}t = 7 \cdot (-9)$$

$$1 \cdot t = -63$$

$$t = -63$$

Check: $\qquad \dfrac{\dfrac{t}{7} = -9}{\begin{array}{c} \dfrac{-63}{7} \ ? \ -9 \\ -9 \ \big| \end{array}}$ TRUE

The solution is -63.

19. $\qquad \dfrac{3}{4}x = 27$

$$\frac{4}{3} \cdot \frac{3}{4}x = \frac{4}{3} \cdot 27$$

$$1 \cdot x = \frac{4 \cdot \cancel{3} \cdot 3 \cdot 3}{\cancel{3} \cdot 1}$$

$$x = 36$$

Check: $\qquad \dfrac{\dfrac{3}{4}x = 27}{\begin{array}{c} \dfrac{3}{4} \cdot 36 \ ? \ 27 \\ 27 \ \big| \end{array}}$ TRUE

The solution is 36.

21. $\qquad \dfrac{-t}{3} = 7$

$$3 \cdot \frac{1}{3} \cdot (-t) = 3 \cdot 7$$

$$-t = 21$$

$$-1 \cdot (-1 \cdot t) = -1 \cdot 21$$

$$1 \cdot t = -21$$

$$t = -21$$

Check: $\qquad \dfrac{\dfrac{-t}{3} = 7}{\begin{array}{c} \dfrac{-(-21)}{3} \ ? \ 7 \\ \dfrac{21}{3} \\ 7 \ \big| \end{array}}$ TRUE

The solution is -21.

23. $\qquad -\dfrac{m}{3} = \dfrac{1}{5}$

$$-\frac{1}{3} \cdot m = \frac{1}{5}$$

$$-3 \cdot \left(-\frac{1}{3} \cdot m \right) = -3 \cdot \frac{1}{5}$$

$$m = -\frac{3}{5}$$

Check: $\qquad \dfrac{-\dfrac{m}{3} = \dfrac{1}{5}}{\begin{array}{c} -\dfrac{\frac{3}{5}}{3} \ ? \ \dfrac{1}{5} \\ -\left(-\dfrac{3}{5} \div 3 \right) \\ -\left(-\dfrac{3}{5} \cdot \dfrac{1}{3} \right) \\ -\left(-\dfrac{1}{5} \right) \\ \dfrac{1}{5} \ \big| \end{array}}$ TRUE

The solution is $-\dfrac{3}{5}$.

25. $\qquad -\dfrac{3}{5}r = \dfrac{9}{10}$

$$-\frac{5}{3} \cdot \left(-\frac{3}{5}r \right) = -\frac{5}{3} \cdot \frac{9}{10}$$

$$1 \cdot r = -\frac{\cancel{5} \cdot \cancel{3} \cdot 3}{\cancel{3} \cdot \cancel{5} \cdot 2}$$

$$r = -\frac{3}{2}$$

Check: $\qquad \dfrac{-\dfrac{3}{5}r = \dfrac{9}{10}}{\begin{array}{c} -\dfrac{3}{5} \cdot \left(-\dfrac{3}{2} \right) \ ? \ \dfrac{9}{10} \\ \dfrac{9}{10} \ \big| \end{array}}$ TRUE

The solution is $-\dfrac{3}{2}$.

27.
$$-\frac{3}{2}r = -\frac{27}{4}$$
$$-\frac{2}{3}\cdot\left(-\frac{3}{2}r\right) = -\frac{2}{3}\cdot\left(-\frac{27}{4}\right)$$
$$1\cdot r = \frac{\cancel{2}\cdot\cancel{3}\cdot 3\cdot 3}{\cancel{3}\cdot\cancel{2}\cdot 2}$$
$$r = \frac{9}{2}$$

Check:
$$-\frac{3}{2}r = -\frac{27}{4}$$
$$\begin{array}{c|c} -\dfrac{3}{2}\cdot\dfrac{9}{2} \ ? \ -\dfrac{27}{4} & \\ -\dfrac{27}{4} & \text{TRUE} \end{array}$$

The solution is $\frac{9}{2}$.

29.　$6.3x = 44.1$
$$\frac{6.3x}{6.3} = \frac{44.1}{6.3}$$
$$1\cdot x = 7$$
$$x = 7$$
Check:
$$6.3x = 44.1$$
$$\begin{array}{c|c} 6.3\cdot 7 \ ? \ 44.1 & \\ 44.1 & \text{TRUE} \end{array}$$
The solution is 7.

31.　$-3.1y = 21.7$
$$\frac{-3.1y}{-3.1} = \frac{21.7}{-3.1}$$
$$1\cdot y = -7$$
$$y = -7$$
Check:
$$3.1y = 21.7$$
$$\begin{array}{c|c} -3.1(-7) \ ? \ 21.7 & \\ 21.7 & \text{TRUE} \end{array}$$
The solution is -7.

33.　$38.7m = 309.6$
$$\frac{38.7m}{38.7} = \frac{309.6}{38.7}$$
$$1\cdot m = 8$$
$$m = 8$$
Check:
$$38.7m = 309.6$$
$$\begin{array}{c|c} 38.7\cdot 8 \ ? \ 309.6 & \\ 309.6 & \text{TRUE} \end{array}$$
The solution is 8.

35.
$$-\frac{2}{3}y = -10.6$$
$$-\frac{3}{2}\cdot\left(-\frac{2}{3}y\right) = -\frac{3}{2}\cdot(-10.6)$$
$$1\cdot y = \frac{31.8}{2}$$
$$y = 15.9$$

Check:
$$-\frac{2}{3}y = -10.6$$
$$\begin{array}{c|c} -\dfrac{2}{3}\cdot(15.9) \ ? \ -10.6 & \\ -\dfrac{31.8}{3} & \\ -10.6 & \text{TRUE} \end{array}$$
The solution is 15.9.

37.
$$\frac{-x}{5} = 10$$
$$5\cdot\frac{-x}{5} = 5\cdot 10$$
$$-x = 50$$
$$-1\cdot(-x) = -1\cdot 50$$
$$x = -50$$
Check:
$$\frac{-x}{5} = 10$$
$$\begin{array}{c|c} \dfrac{-(-50)}{5} \ ? \ 10 & \\ \dfrac{50}{5} & \\ 10 & \text{TRUE} \end{array}$$
The solution is -50.

39.
$$-\frac{t}{2} = 7$$
$$2\cdot\left(-\frac{t}{2}\right) = 2\cdot 7$$
$$-t = 14$$
$$-1\cdot(-t) = -1\cdot 14$$
$$t = -14$$

Check:
$$-\frac{t}{2} = 7$$
$$\begin{array}{c|c} -\dfrac{-14}{2} \ ? \ 7 & \\ -(-7) & \\ 7 & \text{TRUE} \end{array}$$
The solution is -14.

41. Discussion and Writing Exercise

43. $3x + 4x = (3+4)x = 7x$

45. $-4x + 11 - 6x + 18x = (-4-6+18)x + 11 = 8x + 11$

47. $3x - (4+2x) = 3x - 4 - 2x = x - 4$

49. $8y - 6(3y+7) = 8y - 18y - 42 = -10y - 42$

51. The translation is $8r$ miles.

53.　$-0.2344m = 2028.732$
$$\frac{-0.2344m}{-0.2344} = \frac{2028.732}{-0.2344}$$
$$1\cdot m = -8655$$
$$m = -8655$$
The solution is -8655.

55. For all x, $0 \cdot x = 0$. There is no solution to $0 \cdot x = 9$.

57. $2|x| = -12$

$$\frac{2|x|}{2} = \frac{-12}{2}$$

$$1 \cdot |x| = -6$$

$$|x| = -6$$

Absolute value cannot be negative. The equation has no solution.

59. $3x = \dfrac{b}{a}$

$$\frac{1}{3} \cdot 3x = \frac{1}{3} \cdot \frac{b}{a}$$

$$x = \frac{b}{3a}$$

The solution is $\dfrac{b}{3a}$.

61. $\dfrac{a}{b}x = 4$

$$\frac{b}{a} \cdot \frac{a}{b}x = \frac{b}{a} \cdot 4$$

$$x = \frac{4b}{a}$$

The solution is $\dfrac{4b}{a}$.

Exercise Set 2.3

1. $5x + 6 = 31$

$5x + 6 - 6 = 31 - 6$ Subtracting 6 on both sides

$5x = 25$ Simplifying

$\dfrac{5x}{5} = \dfrac{25}{5}$ Dividing by 5 on both sides

$x = 5$ Simplifying

Check: $\dfrac{5x + 6 = 31}{}$

$5 \cdot 5 + 6 \; ? \; 31$

$25 + 6 \;\big|$

$31 \;\big|$ TRUE

The solution is 5.

3. $8x + 4 = 68$

$8x + 4 - 4 = 68 - 4$ Subtracting 4 on both sides

$8x = 64$ Simplifying

$\dfrac{8x}{8} = \dfrac{64}{8}$ Dividing by 8 on both sides

$x = 8$ Simplifying

Check: $\dfrac{8x + 4 = 68}{}$

$8 \cdot 8 + 4 \; ? \; 68$

$64 + 4 \;\big|$

$68 \;\big|$ TRUE

The solution is 8.

5. $4x - 6 = 34$

$4x - 6 + 6 = 34 + 6$ Adding 6 on both sides

$4x = 40$

$\dfrac{4x}{4} = \dfrac{40}{4}$ Dividing by 4 on both sides

$x = 10$

Check: $\dfrac{4x - 6 = 34}{}$

$4 \cdot 10 - 6 \; ? \; 34$

$40 - 6 \;\big|$

$34 \;\big|$ TRUE

The solution is 10.

7. $3x - 9 = 33$

$3x - 9 + 9 = 33 + 9$

$3x = 42$

$\dfrac{3x}{3} = \dfrac{42}{3}$

$x = 14$

Check: $\dfrac{3x - 9 = 33}{}$

$3 \cdot 14 - 9 \; ? \; 33$

$42 - 9 \;\big|$

$33 \;\big|$ TRUE

The solution is 14.

9. $7x + 2 = -54$

$7x + 2 - 2 = -54 - 2$

$7x = -56$

$\dfrac{7x}{7} = \dfrac{-56}{7}$

$x = -8$

Check: $\dfrac{7x + 2 = -54}{}$

$7(-8) + 2 \; ? \; -54$

$-56 + 2 \;\big|$

$-54 \;\big|$ TRUE

The solution is -8.

11. $-45 = 6y + 3$

$-45 - 3 = 6y + 3 - 3$

$-48 = 6y$

$\dfrac{-48}{6} = \dfrac{6y}{6}$

$-8 = y$

Check: $\dfrac{-45 = 6y + 3}{}$

$-45 \; ? \; 6(-8) + 3$

$\big|\; -48 + 3$

$\big|\; -45$ TRUE

The solution is -8.

13. $-4x + 7 = 35$

$-4x + 7 - 7 = 35 - 7$

$-4x = 28$

$\dfrac{-4x}{-4} = \dfrac{28}{-4}$

$x = -7$

Check: $\dfrac{-4x + 7 = 35}{}$

$$-4(-7) + 7 \ ? \ 35$$
$$28 + 7 \ \Big|$$
$$35 \ \Big| \qquad \text{TRUE}$$

The solution is -7.

15. $-8x - 24 = -29\dfrac{1}{3}$

$$-8x - 24 + 24 = -29\dfrac{1}{3} + 24$$

$$-8x = -5\dfrac{1}{3}$$

$$-8x = -\dfrac{16}{3} \qquad \left(-5\dfrac{1}{3} = -\dfrac{16}{3}\right)$$

$$-\dfrac{1}{8}(-8x) = -\dfrac{1}{8}\left(-\dfrac{16}{3}\right)$$

$$x = \dfrac{16}{8 \cdot 3} = \dfrac{2 \cdot \cancel{8}}{\cancel{8} \cdot 3}$$

$$x = \dfrac{2}{3}$$

Check: $\dfrac{-8x - 24 = -29\dfrac{1}{3}}{}$

$$-8 \cdot \dfrac{2}{3} - 24 \ ? \ -29\dfrac{1}{3}$$

$$-\dfrac{16}{3} - 24 \ \Big|$$

$$-5\dfrac{1}{3} - 24 \ \Big|$$

$$-29\dfrac{1}{3} \ \Big| \qquad \text{TRUE}$$

The solution is $\dfrac{2}{3}$.

17. $5x + 7x = 72$

$$12x = 72 \qquad \text{Collecting like terms}$$
$$\dfrac{12x}{12} = \dfrac{72}{12} \qquad \text{Dividing by 12 on both sides}$$
$$x = 6$$

Check: $\dfrac{5x + 7x = 72}{}$

$$5 \cdot 6 + 7 \cdot 6 \ ? \ 72$$
$$30 + 42 \ \Big|$$
$$72 \ \Big| \qquad \text{TRUE}$$

The solution is 6.

19. $8x + 7x = 60$

$$15x = 60 \qquad \text{Collecting like terms}$$
$$\dfrac{15x}{15} = \dfrac{60}{15} \qquad \text{Dividing by 15 on both sides}$$
$$x = 4$$

Check: $\dfrac{8x + 7x = 60}{}$

$$8 \cdot 4 + 7 \cdot 4 \ ? \ 60$$
$$32 + 28 \ \Big|$$
$$60 \ \Big| \qquad \text{TRUE}$$

The solution is 4.

21. $4x + 3x = 42$

$$7x = 42$$
$$\dfrac{7x}{7} = \dfrac{42}{7}$$
$$x = 6$$

Check: $\dfrac{4x + 3x = 42}{}$

$$4 \cdot 6 + 3 \cdot 6 \ ? \ 42$$
$$24 + 18 \ \Big|$$
$$42 \ \Big| \qquad \text{TRUE}$$

The solution is 6.

23. $-6y - 3y = 27$

$$-9y = 27$$
$$\dfrac{-9y}{-9} = \dfrac{27}{-9}$$
$$y = -3$$

Check: $\dfrac{-6y - 3y = 27}{}$

$$-6(-3) - 3(-3) \ ? \ 27$$
$$18 + 9 \ \Big|$$
$$27 \ \Big| \qquad \text{TRUE}$$

The solution is -3.

25. $-7y - 8y = -15$

$$-15y = -15$$
$$\dfrac{-15y}{-15} = \dfrac{-15}{-15}$$
$$y = 1$$

Check: $\dfrac{-7y - 8y = -15}{}$

$$-7 \cdot 1 - 8 \cdot 1 \ ? \ -15$$
$$-7 - 8 \ \Big|$$
$$-15 \ \Big| \qquad \text{TRUE}$$

The solution is 1.

27. $x + \dfrac{1}{3}x = 8$

$$\left(1 + \dfrac{1}{3}\right)x = 8$$

$$\dfrac{4}{3}x = 8$$

$$\dfrac{3}{4} \cdot \dfrac{4}{3}x = \dfrac{3}{4} \cdot 8$$

$$x = 6$$

Check: $\dfrac{x + \dfrac{1}{3}x = 8}{}$

$$6 + \dfrac{1}{3} \cdot 6 \ ? \ 8$$

$$6 + 2 \ \Big|$$

$$8 \ \Big| \qquad \text{TRUE}$$

The solution is 6.

29. $10.2y - 7.3y = -58$

$2.9y = -58$

$\dfrac{2.9y}{2.9} = \dfrac{-58}{2.9}$

$y = -20$

Check: $\dfrac{10.2y - 7.3y = -58}{}$

$10.2(-20) - 7.3(-20) \;?\; -58$

$-204 + 146 \;\Big|\;$

$-58 \;\Big|\;$ TRUE

The solution is -20.

31. $8y - 35 = 3y$

$8y = 3y + 35$ Adding 35 and simplifying

$8y - 3y = 35$ Subtracting $3y$ and simplifying

$5y = 35$ Collecting like terms

$\dfrac{5y}{5} = \dfrac{35}{5}$ Dividing by 5

$y = 7$

Check: $\dfrac{8y - 35 = 3y}{}$

$8 \cdot 7 - 35 \;?\; 3 \cdot 7$

$56 - 35 \;\Big|\; 21$

$21 \;\Big|\;$ TRUE

The solution is 7.

33. $8x - 1 = 23 - 4x$

$8x + 4x = 23 + 1$ Adding 1 and $4x$ and simplifying

$12x = 24$ Collecting like terms

$\dfrac{12x}{12} = \dfrac{24}{12}$ Dividing by 12

$x = 2$

Check: $\dfrac{8x - 1 = 23 - 4x}{}$

$8 \cdot 2 - 1 \;?\; 23 - 4 \cdot 2$

$16 - 1 \;\Big|\; 23 - 8$

$15 \;\Big|\; 15$ TRUE

The solution is 2.

35. $2x - 1 = 4 + x$

$2x - x = 4 + 1$ Adding 1 and $-x$

$x = 5$ Collecting like terms

Check: $\dfrac{2x - 1 = 4 + x}{}$

$2 \cdot 5 - 1 \;?\; 4 + 5$

$10 - 1 \;\Big|\; 9$

$9 \;\Big|\;$ TRUE

The solution is 5.

37. $6x + 3 = 2x + 11$

$6x - 2x = 11 - 3$

$4x = 8$

$\dfrac{4x}{4} = \dfrac{8}{4}$

$x = 2$

Check: $\dfrac{6x + 3 = 2x + 11}{}$

$6 \cdot 2 + 3 \;?\; 2 \cdot 2 + 11$

$12 + 3 \;\Big|\; 4 + 11$

$15 \;\Big|\; 15$ TRUE

The solution is 2.

39. $5 - 2x = 3x - 7x + 25$

$5 - 2x = -4x + 25$

$4x - 2x = 25 - 5$

$2x = 20$

$\dfrac{2x}{2} = \dfrac{20}{2}$

$x = 10$

Check: $\dfrac{5 - 2x = 3x - 7x + 25}{}$

$5 - 2 \cdot 10 \;?\; 3 \cdot 10 - 7 \cdot 10 + 25$

$5 - 20 \;\Big|\; 30 - 70 + 25$

$-15 \;\Big|\; -40 + 25$

$-15 \;\Big|\;$ TRUE

The solution is 10.

41. $4 + 3x - 6 = 3x + 2 - x$

$3x - 2 = 2x + 2$ Collecting like terms on each side

$3x - 2x = 2 + 2$

$x = 4$

Check: $\dfrac{4 + 3x - 6 = 3x + 2 - x}{}$

$4 + 3 \cdot 4 - 6 \;?\; 3 \cdot 4 + 2 - 4$

$4 + 12 - 6 \;\Big|\; 12 + 2 - 4$

$16 - 6 \;\Big|\; 14 - 4$

$10 \;\Big|\; 10$ TRUE

The solution is 4.

43. $4y - 4 + y + 24 = 6y + 20 - 4y$

$5y + 20 = 2y + 20$

$5y - 2y = 20 - 20$

$3y = 0$

$y = 0$

Check: $\dfrac{4y - 4 + y + 24 = 6y + 20 - 4y}{}$

$4 \cdot 0 - 4 + 0 + 24 \;?\; 6 \cdot 0 + 20 - 4 \cdot 0$

$0 - 4 + 0 + 24 \;\Big|\; 0 + 20 - 0$

$20 \;\Big|\; 20$ TRUE

The solution is 0.

45. $\dfrac{7}{2}x + \dfrac{1}{2}x = 3x + \dfrac{3}{2} + \dfrac{5}{2}x$

The least common multiple of all the denominators is 2.
We multiply by 2 on both sides.

$$2\left(\dfrac{7}{2}x + \dfrac{1}{2}x\right) = 2\left(3x + \dfrac{3}{2} + \dfrac{5}{2}x\right)$$

$$2\cdot\dfrac{7}{2}x + 2\cdot\dfrac{1}{2}x = 2\cdot 3x + 2\cdot\dfrac{3}{2} + 2\cdot\dfrac{5}{2}x$$

$$7x + x = 6x + 3 + 5x$$

$$8x = 11x + 3$$

$$8x - 11x = 3$$

$$-3x = 3$$

$$\dfrac{-3x}{-3} = \dfrac{3}{-3}$$

$$x = -1$$

Check:

$$\dfrac{7}{2}x + \dfrac{1}{2}x = 3x + \dfrac{3}{2} + \dfrac{5}{2}x$$

$$\begin{array}{c|c} \dfrac{7}{2}(-1) + \dfrac{1}{2}(-1) \; ? \; 3(-1) + \dfrac{3}{2} + \dfrac{5}{2}(-1) \\[2mm] -\dfrac{7}{2} - \dfrac{1}{2} \;\Big|\; -3 + \dfrac{3}{2} - \dfrac{5}{2} \\[2mm] -4 \;\Big|\; -\dfrac{8}{2} \\[2mm] \Big|\; -4 \qquad \text{TRUE} \end{array}$$

The solution is -1.

47. $\dfrac{2}{3} + \dfrac{1}{4}t = \dfrac{1}{3}$

The least common multiple of all the denominators is 12.
We multiply by 12 on both sides.

$$12\left(\dfrac{2}{3} + \dfrac{1}{4}t\right) = 12\cdot\dfrac{1}{3}$$

$$12\cdot\dfrac{2}{3} + 12\cdot\dfrac{1}{4}t = 12\cdot\dfrac{1}{3}$$

$$8 + 3t = 4$$

$$3t = 4 - 8$$

$$3t = -4$$

$$\dfrac{3t}{3} = \dfrac{-4}{3}$$

$$t = -\dfrac{4}{3}$$

Check:

$$\dfrac{2}{3} + \dfrac{1}{4}t = \dfrac{1}{3}$$

$$\begin{array}{c|c} \dfrac{2}{3} + \dfrac{1}{4}\left(-\dfrac{4}{3}\right) \; ? \; \dfrac{1}{3} \\[2mm] \dfrac{2}{3} - \dfrac{1}{3} \\[2mm] \dfrac{1}{3} \;\Big|\; \text{TRUE} \end{array}$$

The solution is $-\dfrac{4}{3}$.

49. $\dfrac{2}{3} + 3y = 5y - \dfrac{2}{15}, \quad$ LCM is 15

$$15\left(\dfrac{2}{3} + 3y\right) = 15\left(5y - \dfrac{2}{15}\right)$$

$$15\cdot\dfrac{2}{3} + 15\cdot 3y = 15\cdot 5y - 15\cdot\dfrac{2}{15}$$

$$10 + 45y = 75y - 2$$

$$10 + 2 = 75y - 45y$$

$$12 = 30y$$

$$\dfrac{12}{30} = \dfrac{30y}{30}$$

$$\dfrac{2}{5} = y$$

Check:

$$\dfrac{2}{3} + 3y = 5y - \dfrac{2}{15}$$

$$\begin{array}{c|c} \dfrac{2}{3} + 3\cdot\dfrac{2}{5} \; ? \; 5\cdot\dfrac{2}{5} - \dfrac{2}{15} \\[2mm] \dfrac{2}{3} + \dfrac{6}{5} \;\Big|\; 2 - \dfrac{2}{15} \\[2mm] \dfrac{10}{15} + \dfrac{18}{15} \;\Big|\; \dfrac{30}{15} - \dfrac{2}{15} \\[2mm] \dfrac{28}{15} \;\Big|\; \dfrac{28}{15} \quad \text{TRUE} \end{array}$$

The solution is $\dfrac{2}{5}$.

51. $\dfrac{5}{3} + \dfrac{2}{3}x = \dfrac{25}{12} + \dfrac{5}{4}x + \dfrac{3}{4}, \quad$ LCM is 12

$$12\left(\dfrac{5}{3} + \dfrac{2}{3}x\right) = 12\left(\dfrac{25}{12} + \dfrac{5}{4}x + \dfrac{3}{4}\right)$$

$$12\cdot\dfrac{5}{3} + 12\cdot\dfrac{2}{3}x = 12\cdot\dfrac{25}{12} + 12\cdot\dfrac{5}{4}x + 12\cdot\dfrac{3}{4}$$

$$20 + 8x = 25 + 15x + 9$$

$$20 + 8x = 15x + 34$$

$$20 - 34 = 15x - 8x$$

$$-14x = 7x$$

$$\dfrac{-14}{7} = \dfrac{7x}{7}$$

$$-2 = x$$

Check:

$$\dfrac{5}{3} + \dfrac{2}{3}x = \dfrac{25}{12} + \dfrac{5}{4}x + \dfrac{3}{4}$$

$$\begin{array}{c|c} \dfrac{5}{3} + \dfrac{2}{3}(-2) \; ? \; \dfrac{25}{12} + \dfrac{5}{4}(-2) + \dfrac{3}{4} \\[2mm] \dfrac{5}{3} - \dfrac{4}{3} \;\Big|\; \dfrac{25}{12} - \dfrac{5}{2} + \dfrac{3}{4} \\[2mm] \dfrac{1}{3} \;\Big|\; \dfrac{25}{12} - \dfrac{30}{12} + \dfrac{9}{12} \\[2mm] \;\Big|\; \dfrac{4}{12} \\[2mm] \;\Big|\; \dfrac{1}{3} \quad \text{TRUE} \end{array}$$

The solution is -2.

53.
$$2.1x + 45.2 = 3.2 - 8.4x$$

Greatest number of decimal places is 1

$$10(2.1x + 45.2) = 10(3.2 - 8.4x)$$

Multiplying by 10 to clear decimals

$$10(2.1x) + 10(45.2) = 10(3.2) - 10(8.4x)$$

$$21x + 452 = 32 - 84x$$

$$21x + 84x = 32 - 452$$

$$105x = -420$$

$$\frac{105x}{105} = \frac{-420}{105}$$

$$x = -4$$

Check:
$$\frac{2.1x + 45.2 = 3.2 - 8.4x}{}$$

$2.1(-4) + 45.2$? $3.2 - 8.4(-4)$	
$-8.4 + 45.2$	$3.2 + 33.6$
36.8	36.8 TRUE

The solution is -4.

55.
$$1.03 - 0.62x = 0.71 - 0.22x$$

Greatest number of decimal places is 2

$$100(1.03 - 0.62x) = 100(0.71 - 0.22x)$$

Multiplying by 100 to clear decimals

$$100(1.03) - 100(0.62x) = 100(0.71) - 100(0.22x)$$

$$103 - 62x = 71 - 22x$$

$$32 = 40x$$

$$\frac{32}{40} = \frac{40x}{40}$$

$$\frac{4}{5} = x, \text{ or}$$

$$0.8 = x$$

Check:
$$\frac{1.03 - 0.62x = 0.71 - 0.22x}{}$$

$1.03 - 0.62(0.8)$? $0.71 - 0.22(0.8)$	
$1.03 - 0.496$	$0.71 - 0.176$
0.534	0.534 TRUE

The solution is $\frac{4}{5}$, or 0.8.

57.
$$\frac{2}{7}x - \frac{1}{2}x = \frac{3}{4}x + 1, \text{ LCM is 28}$$

$$28\left(\frac{2}{7}x - \frac{1}{2}x\right) = 28\left(\frac{3}{4}x + 1\right)$$

$$28 \cdot \frac{2}{7}x - 28 \cdot \frac{1}{2}x = 28 \cdot \frac{3}{4}x + 28 \cdot 1$$

$$8x - 14x = 21x + 28$$

$$-6x = 21x + 28$$

$$-6x - 21x = 28$$

$$-27x = 28$$

$$x = -\frac{28}{27}$$

Check:
$$\frac{2}{7}x - \frac{1}{2}x = \frac{3}{4}x + 1$$

$$\frac{\frac{2}{7}\left(-\frac{28}{27}\right) - \frac{1}{2}\left(-\frac{28}{27}\right) \ ? \ \frac{3}{4}\left(-\frac{28}{27}\right) + 1}{}$$

$-\frac{8}{27} + \frac{14}{27}$	$-\frac{21}{27} + 1$
$\frac{6}{27}$	$\frac{6}{27}$ TRUE

The solution is $-\dfrac{28}{27}$.

59.
$$3(2y - 3) = 27$$

$$6y - 9 = 27 \qquad \text{Using a distributive law}$$

$$6y = 27 + 9 \qquad \text{Adding 9}$$

$$6y = 36$$

$$y = 6 \qquad \text{Dividing by 6}$$

Check:
$$\frac{3(2y - 3) = 27}{}$$

$3(2 \cdot 6 - 3)$? 27	
$3(12 - 3)$	
$3 \cdot 9$	
27	TRUE

The solution is 6.

61.
$$40 = 5(3x + 2)$$

$$40 = 15x + 10 \qquad \text{Using a distributive law}$$

$$40 - 10 = 15x$$

$$30 = 15x$$

$$2 = x$$

Check:
$$\frac{40 = 5(3x + 2)}{}$$

40 ? $5(3 \cdot 2 + 2)$	
	$5(6 + 2)$
	$5 \cdot 8$
	40 TRUE

The solution is 2.

63.
$$-23 + y = y + 25$$

$$-y - 23 + y = -y + y + 25$$

$$-23 = 25 \qquad \text{FALSE}$$

The equation has no solution.

65.
$$-23 + x = x - 23$$

$$-x - 23 + x = -x + x - 23$$

$$-23 = -23 \qquad \text{TRUE}$$

All real numbers are solutions.

67.
$$2(3 + 4m) - 9 = 45$$

$$6 + 8m - 9 = 45 \qquad \text{Collecting like terms}$$

$$8m - 3 = 45$$

$$8m = 45 + 3$$

$$8m = 48$$

$$m = 6$$

Check: $\dfrac{2(3+4m)-9=45}{}$
$2(3+4\cdot 6)-9 \ ? \ 45$
$2(3+24)-9 \ |$
$2\cdot 27-9 \ |$
$54-9 \ |$
$45 \ |$ \quad TRUE

The solution is 6.

69. $5r-(2r+8)=16$

$5r-2r-8=16$

$3r-8=16$ \qquad Collecting like terms

$3r=16+8$

$3r=24$

$r=8$

Check: $\dfrac{5r-(2r+8)=16}{}$
$5\cdot 8-(2\cdot 8+8) \ ? \ 16$
$40-(16+8) \ |$
$40-24 \ |$
$16 \ |$ \quad TRUE

The solution is 8.

71. $6-2(3x-1)=2$

$6-6x+2=2$

$8-6x=2$

$8-2=6x$

$6=6x$

$1=x$

Check: $\dfrac{6-2(3x-1)=2}{}$
$6-2(3\cdot 1-1) \ ? \ 2$
$6-2(3-1) \ |$
$6-2\cdot 2 \ |$
$6-4 \ |$
$2 \ |$ \quad TRUE

The solution is 1.

73. $5x+5-7x=15-12x+10x-10$

$-2x+5=5-2x$ \quad Collecting like terms

$2x-2x+5=2x+5-2x$ \quad Adding $2x$

$5=5$ \quad TRUE

All real numbers are solutions.

75. $22x-5-15x+3=10x-4-3x+11$

$7x-2=7x+7$ \quad Collecting like terms

$-7x+7x-2=-7x+7x+7$

$-2=7$ \quad FALSE

The equation has no solution.

77. $5(d+4)=7(d-2)$

$5d+20=7d-14$

$20+14=7d-5d$

$34=2d$

$17=d$

Check: $\dfrac{5(d+4)=7(d-2)}{}$
$5(17+4) \ ? \ 7(17-2)$
$5\cdot 21 \ | \ 7\cdot 15$
$105 \ | \ 105$ \quad TRUE

The solution is 17.

79. $8(2t+1)=4(7t+7)$

$16t+8=28t+28$

$16t-28t=28-8$

$-12t=20$

$t=-\dfrac{20}{12}$

$t=-\dfrac{5}{3}$

Check: $\dfrac{8(2t+1)=4(7t+7)}{}$
$8\left(2\left(-\dfrac{5}{3}\right)+1\right) \ ? \ 4\left(7\left(-\dfrac{5}{3}\right)+7\right)$
$8\left(-\dfrac{10}{3}+1\right) \ \Big| \ 4\left(-\dfrac{35}{3}+7\right)$
$8\left(-\dfrac{7}{3}\right) \ \Big| \ 4\left(-\dfrac{14}{3}\right)$
$-\dfrac{56}{3} \ \Big| \ -\dfrac{56}{3}$ \quad TRUE

The solution is $-\dfrac{5}{3}$.

81. $3(r-6)+2=4(r+2)-21$

$3r-18+2=4r+8-21$

$3r-16=4r-13$

$13-16=4r-3r$

$-3=r$

Check: $\dfrac{3(r-6)+2=4(r+2)-21}{}$
$3(-3-6)+2 \ ? \ 4(-3+2)-21$
$3(-9)+2 \ | \ 4(-1)-21$
$-27+2 \ | \ -4-21$
$-25 \ | \ -25$ \quad TRUE

The solution is -3.

83. $19-(2x+3)=2(x+3)+x$

$19-2x-3=2x+6+x$

$16-2x=3x+6$

$16-6=3x+2x$

$10=5x$

$2=x$

Check: $\dfrac{19-(2x+3)=2(x+3)+x}{}$
$19-(2\cdot 2+3) \ ? \ 2(2+3)+2$
$19-(4+3) \ | \ 2\cdot 5+2$
$19-7 \ | \ 10+2$
$12 \ | \ 12$ \quad TRUE

The solution is 2.

85. $2[4 - 2(3 - x)] - 1 = 4[2(4x - 3) + 7] - 25$

$2[4 - 6 + 2x] - 1 = 4[8x - 6 + 7] - 25$

$2[-2 + 2x] - 1 = 4[8x + 1] - 25$

$-4 + 4x - 1 = 32x + 4 - 25$

$4x - 5 = 32x - 21$

$-5 + 21 = 32x - 4x$

$16 = 28x$

$\dfrac{16}{28} = x$

$\dfrac{4}{7} = x$

The check is left to the student.

The solution is $\dfrac{4}{7}$.

87. $11 - 4(x + 1) - 3 = 11 + 2(4 - 2x) - 16$

$11 - 4x - 4 - 3 = 11 + 8 - 4x - 16$

$4 - 4x = 3 - 4x$

$4x + 4 - 4x = 4x + 3 - 4x$

$4 = 3 \qquad \text{FALSE}$

The equation has no solution.

89. $22x - 1 - 12x = 5(2x - 1) + 4$

$22x - 1 - 12x = 10x - 5 + 4$

$10x - 1 = 10x - 1$

$-10x + 10x - 1 = -10x + 10x - 1$

$-1 = -1 \qquad \text{TRUE}$

All real numbers are solutions.

91. $0.7(3x + 6) = 1.1 - (x + 2)$

$2.1x + 4.2 = 1.1 - x - 2$

$10(2.1x + 4.2) = 10(1.1 - x - 2) \quad$ Clearing decimals

$21x + 42 = 11 - 10x - 20$

$21x + 42 = -10x - 9$

$21x + 10x = -9 - 42$

$31x = -51$

$x = -\dfrac{51}{31}$

The check is left to the student.

The solution is $-\dfrac{51}{31}$.

93. Discussion and Writing Exercise

95. Do the long division. The answer is negative.

```
        6 . 5
3. 4⌐| 2 2.1 ⌐0
        2 0 4
        1 7 0
        1 7 0
            0
```

$-22.1 \div 3.4 = -6.5$

97. $7x - 21 - 14y = 7 \cdot x - 7 \cdot 3 - 7 \cdot 2y = 7(x - 3 - 2y)$

99. $-3 + 2(-5)^2(-3) - 7 = -3 + 2(25)(-3) - 7$

$= -3 + 50(-3) - 7$

$= -3 - 150 - 7$

$= -153 - 7$

$= -160$

101. $23(2x - 4) - 15(10 - 3x) = 46x - 92 - 150 + 45x = 91x - 242$

103. First we multiply to remove the parentheses.

$\dfrac{2}{3}\left(\dfrac{7}{8} - 4x\right) - \dfrac{5}{8} = \dfrac{3}{8}$

$\dfrac{7}{12} - \dfrac{8}{3}x - \dfrac{5}{8} = \dfrac{3}{8}, \text{ LCM is 24}$

$24\left(\dfrac{7}{12} - \dfrac{8}{3}x - \dfrac{5}{8}\right) = 24 \cdot \dfrac{3}{8}$

$24 \cdot \dfrac{7}{12} - 24 \cdot \dfrac{8}{3}x - 24 \cdot \dfrac{5}{8} = 9$

$14 - 64x - 15 = 9$

$-1 - 64x = 9$

$-64x = 10$

$x = -\dfrac{10}{64}$

$x = -\dfrac{5}{32}$

The solution is $-\dfrac{5}{32}$.

105. $\dfrac{4 - 3x}{7} = \dfrac{2 + 5x}{49} - \dfrac{x}{14}$

$98\left(\dfrac{4 - 3x}{7}\right) = 98\left(\dfrac{2 + 5x}{49} - \dfrac{x}{14}\right), \text{ LCM is 98}$

$\dfrac{98(4 - 3x)}{7} = 98\left(\dfrac{2 + 5x}{49}\right) - 98 \cdot \dfrac{x}{14}$

$14(4 - 3x) = 2(2 + 5x) - 7x$

$56 - 42x = 4 + 10x - 7x$

$56 - 42x = 4 + 3x$

$56 - 42x + 42x = 4 + 3x + 42x$

$56 = 4 + 45x$

$56 - 4 = 4 + 45x - 4$

$52 = 45x$

$\dfrac{52}{45} = x$

The solution is $\dfrac{52}{45}$.

Exercise Set 2.4

1. a) We substitute 1900 for a and calculate B.

$B = 30a = 30 \cdot 1900 = 57,000$

The minimum furnace output is 57,000 Btu's.

b) $B = 30a$

$\dfrac{B}{30} = \dfrac{30a}{30} \quad$ Dividing by 30

$\dfrac{B}{30} = a$

3. a) We substitute 8 for t and calculate M.

$$M = \frac{1}{5} \cdot 8 = \frac{8}{5}, \text{ or } 1\frac{3}{5}$$

The storm is $1\frac{3}{5}$ miles away.

b)
$$M = \frac{1}{5}t$$
$$5 \cdot M = 5 \cdot \frac{1}{5}t$$
$$5M = t$$

5. a) We substitute 21,345 for n and calculate f.

$$f = \frac{21,345}{15} = 1423$$

There are 1423 full-time equivalent students.

b)
$$f = \frac{n}{15}$$
$$15 \cdot f = 15 \cdot \frac{n}{15}$$
$$15f = n$$

7. We substitute 84 for c and 8 for w and calculate D.

$$D = \frac{c}{w} = \frac{84}{8} = 10.5$$

The calorie density is 10.5 calories per oz.

9. We substitute 7 for n and calculate N.

$$N = n^2 - n = 7^2 - 7 = 49 - 7 = 42$$

42 games are played.

11.
$$y = 5x$$
$$\frac{y}{5} = \frac{5x}{5}$$
$$\frac{y}{5} = x$$

13.
$$a = bc$$
$$\frac{a}{b} = \frac{bc}{b}$$
$$\frac{a}{b} = c$$

15.
$$y = 13 + x$$
$$y - 13 = 13 + x - 13$$
$$y - 13 = x$$

17.
$$y = x + b$$
$$y - b = x + b - b$$
$$y - b = x$$

19.
$$y = 5 - x$$
$$y - 5 = 5 - x - 5$$
$$y - 5 = -x$$
$$-1 \cdot (y - 5) = -1 \cdot (-x)$$
$$-y + 5 = x, \text{ or}$$
$$5 - y = x$$

21.
$$y = a - x$$
$$y - a = a - x - a$$
$$y - a = -x$$
$$-1 \cdot (y - a) = -1 \cdot (-x)$$
$$-y + a = x, \text{ or}$$
$$a - y = x$$

23.
$$8y = 5x$$
$$\frac{8y}{8} = \frac{5x}{8}$$
$$y = \frac{5x}{8}, \text{ or } \frac{5}{8}x$$

25.
$$By = Ax$$
$$\frac{By}{A} = \frac{Ax}{A}$$
$$\frac{By}{A} = x$$

27.
$$W = mt + b$$
$$W - b = mt + b - b$$
$$W - b = mt$$
$$\frac{W - b}{m} = \frac{mt}{m}$$
$$\frac{W - b}{m} = t$$

29.
$$y = bx + c$$
$$y - c = bx + c - c$$
$$y - c = bx$$
$$\frac{y - c}{b} = \frac{bx}{b}$$
$$\frac{y - c}{b} = x$$

31.
$$A = \frac{a + b + c}{3}$$
$$3A = a + b + c \quad \text{Multiplying by 3}$$
$$3A - a - c = b \quad \text{Subtracting } a \text{ and } c$$

33.
$$A = at + b$$
$$A - b = at \quad \text{Subtracting } b$$
$$\frac{A - b}{a} = t \quad \text{Dividing by } a$$

35.
$$A = bh$$
$$\frac{A}{b} = \frac{bh}{b} \quad \text{Dividing by } b$$
$$\frac{A}{b} = h$$

37.
$$P = 2l + 2w$$
$$P - 2l = 2l + 2w - 2l \quad \text{Subtracting } 2l$$
$$P - 2l = 2w$$
$$\frac{P - 2l}{2} = \frac{2w}{2} \quad \text{Dividing by 2}$$
$$\frac{P - 2l}{2} = w, \text{ or}$$
$$\frac{1}{2}P - l = w$$

39. $\quad A = \dfrac{a+b}{2}$

$\qquad 2A = a + b \quad$ Multiplying by 2

$\qquad 2A - b = a \qquad$ Subtracting b

41. $\quad F = ma$

$\qquad \dfrac{F}{m} = \dfrac{ma}{m} \qquad$ Dividing by m

$\qquad \dfrac{F}{m} = a$

43. $\quad E = mc^2$

$\qquad \dfrac{E}{m} = \dfrac{mc^2}{m} \quad$ Dividing by m

$\qquad \dfrac{E}{m} = c^2$

45. $\quad Ax + By = c$

$\qquad Ax = c - By \quad$ Subtracting By

$\qquad \dfrac{Ax}{A} = \dfrac{c - By}{A} \quad$ Dividing by A

$\qquad x = \dfrac{c - By}{A}$

47. $\quad v = \dfrac{3k}{t}$

$\qquad tv = t \cdot \dfrac{3k}{t} \qquad$ Multiplying by t

$\qquad tv = 3k$

$\qquad \dfrac{tv}{v} = \dfrac{3k}{v} \qquad$ Dividing by v

$\qquad t = \dfrac{3k}{v}$

49. Discussion and Writing Exercise

51. We divide:

```
        0.9 2
 2 5 ) 2 3.0 0
       2 2 5
       ─────
         5 0
         5 0
         ───
           0
```

Decimal notation for $\dfrac{23}{25}$ is 0.92.

53. $0.082 + (-9.407) = -9.325$

55. $-45.8 - (-32.6) = -45.8 + 32.6 = -13.2$

57. $3.1\% \qquad 0.\,03.1$

Move the decimal point 2 places to the left.

$3.1\% = 0.031$

59. $-\dfrac{2}{3} + \dfrac{5}{6} = -\dfrac{2}{3} \cdot \dfrac{2}{2} + \dfrac{5}{6}$

$\qquad\qquad = -\dfrac{4}{6} + \dfrac{5}{6}$

$\qquad\qquad = \dfrac{1}{6}$

61. a) We substitute 120 for w, 67 for h, and 23 for a and calculate K.

$\qquad K = 917 + 6(w + h - a)$

$\qquad K = 917 + 6(120 + 67 - 23)$

$\qquad K = 917 + 6(164)$

$\qquad K = 917 + 984$

$\qquad K = 1901$ calories

b) Solve for a:

$\qquad K = 917 + 6(w + h - a)$

$\qquad K = 917 + 6w + 6h - 6a$

$\qquad K + 6a = 917 + 6w + 6h$

$\qquad 6a = 917 + 6w + 6h - K$

$\qquad a = \dfrac{917 + 6w + 6h - K}{6}$

Solve for h:

$\qquad K = 917 + 6(w + h - a)$

$\qquad K = 917 + 6w + 6h - 6a$

$\qquad K - 917 - 6w + 6a = 6h$

$\qquad \dfrac{K - 917 - 6w + 6a}{6} = h$

Solve for w:

$\qquad K = 917 + 6(w + h - a)$

$\qquad K = 917 + 6w + 6h - 6a$

$\qquad K - 917 - 6h + 6a = 6w$

$\qquad \dfrac{K - 917 - 6h + 6a}{6} = w$

63. $\qquad H = \dfrac{2}{a - b}$

$\qquad (a - b)H = (a - b)\left(\dfrac{2}{a - b}\right)$

$\qquad Ha - Hb = 2$

$\qquad Ha - Hb - Ha = 2 - Ha$

$\qquad -Hb = 2 - Ha$

$\qquad -1(-Hb) = -1(2 - Ha)$

$\qquad Hb = -2 + Ha$

$\qquad \dfrac{Hb}{H} = \dfrac{-2 + Ha}{H}$

$\qquad b = \dfrac{-2 + Ha}{H}$, or $\dfrac{Ha - 2}{H}$, or $a - \dfrac{2}{H}$

$\qquad H = \dfrac{2}{a - b}$

$\qquad (a - b)H = (a - b) \cdot \dfrac{2}{a - b}$

$\qquad Ha - Hb = 2$

$\qquad Ha - Hb + Hb = 2 + Hb$

$\qquad Ha = 2 + Hb$

$\qquad \dfrac{Ha}{H} = \dfrac{2 + Hb}{H}$, or $\dfrac{2}{H} + b$

65. $A = lw$

When l and w both double, we have
$$2l \cdot 2w = 4lw = 4A,$$
so A quadruples.

67. $A = \frac{1}{2}bh$

When b increases by 4 units we have
$$\frac{1}{2}(b+4)h = \frac{1}{2}bh + 2h = A + 2h,$$
so A increases by $2h$ units.

Exercise Set 2.5

1. *Translate*.

$$\underbrace{\text{What percent}}_{\downarrow} \text{ of } \underbrace{180}_{\downarrow} \text{ is } \underbrace{36?}_{\downarrow}$$
$$\quad p \qquad\qquad \cdot \quad 180 \ = \ 36$$

Solve. We divide by 36 on both sides and convert the answer to percent notation.

$$p \cdot 180 = 36$$
$$\frac{p \cdot 180}{180} = \frac{36}{180}$$
$$p = 0.2$$
$$p = 20\%$$

Thus, 36 is 20% of 180. The answer is 20%.

3. *Translate*.

$$45 \text{ is } 30\% \text{ of what?}$$
$$\downarrow \ \downarrow \ \downarrow \ \downarrow \ \ \downarrow$$
$$45 \ = \ 30\% \ \cdot \ \ b$$

Solve. We solve the equation.

$$45 = 30\% \cdot b$$
$$45 = 0.3b \qquad \text{Converting to decimal notation}$$
$$\frac{45}{0.3} = \frac{b}{0.3}$$
$$150 = b$$

Thus, 45 is 30% of 150. The answer is 150.

5. *Translate*.

$$\text{What is } 65\% \text{ of } 840?$$
$$\downarrow \quad \downarrow \quad \downarrow \quad \downarrow \quad \downarrow$$
$$a \ \ = \ 65\% \ \cdot \ \ 840$$

Solve. We convert 65% to decimal notation and multiply.

$$a = 65\% \cdot 840$$
$$a = 0.65 \times 840$$
$$a = 546$$

Thus, 546 is 65% of 840. The answer is 546.

7. *Translate*.

$$30 \text{ is } \underbrace{\text{what percent}}_{\downarrow} \text{ of } 125?$$
$$\downarrow \ \downarrow \qquad\qquad\qquad \downarrow \ \downarrow$$
$$30 \ = \qquad p \qquad \cdot \ 125$$

Solve. We solve the equation.

$$30 = p \cdot 125$$
$$\frac{30}{125} = \frac{p \cdot 125}{125}$$
$$0.24 = p$$
$$24\% = p$$

Thus, 30 is 24% of 125. The answer is 24%.

9. *Translate*.

$$12\% \text{ of } \underbrace{\text{what number}}_{\downarrow} \text{ is } 0.3?$$
$$\downarrow \ \downarrow \qquad\qquad\qquad \downarrow \ \downarrow$$
$$12\% \ \cdot \qquad b \qquad = 0.3$$

Solve. We solve the equation.

$$12\% \cdot b = 0.3$$
$$0.12b = 0.3 \qquad \text{Converting to decimal notation}$$
$$\frac{b}{0.12} = \frac{0.3}{0.12}$$
$$b = 2.5$$

Thus, 12% of 2.5 is 0.3. The answer is 2.5.

11. *Translate*.

$$\underbrace{2}_{\downarrow} \text{ is } \underbrace{\text{what percent}}_{\downarrow} \text{ of } \underbrace{40?}_{\downarrow}$$
$$2 \ = \qquad p \qquad \cdot \ 40$$

Solve. We divide by 40 on both sides and convert the answer to percent notation.

$$2 = p \cdot 40$$
$$\frac{2}{40} = \frac{p \cdot 40}{40}$$
$$0.05 = p$$
$$5\% = p$$

Thus, 2 is 5% of 40. The answer is 5%.

13. *Translate*.

$$\underbrace{\text{What percent}}_{\downarrow} \text{ of } 68 \text{ is } 17?$$
$$\quad p \qquad\quad \cdot \ 68 \ = \ 17$$

Solve. We divide by 68 on both sides and then convert to percent notation.

$$p \cdot 68 = 17$$
$$p = \frac{17}{68}$$
$$p = 0.25 = 25\%$$

The answer is 25%.

15. *Translate*.

What is 35% of 240?

$a = 35\% \cdot 240$

Solve. We convert 35% to decimal notation and multiply.

$a = 35\% \cdot 240$

$a = 0.35 \cdot 240$

$a = 84$

The answer is 84.

17. *Translate*.

What percent of 125 is 30?

$p \cdot 125 = 30$

Solve. We divide by 125 on both sides and then convert to percent notation.

$p \cdot 125 = 30$

$p = \dfrac{30}{125}$

$p = 0.24 = 24\%$

The answer is 24%.

19. *Translate*.

What percent of 300 is 48?

$p \cdot 300 = 48$

Solve. We divide by 300 on both sides and then convert to percent notation.

$p \cdot 300 = 48$

$p = \dfrac{48}{300}$

$p = 0.16 = 16\%$

The answer is 16%.

21. *Translate*.

14 is 30% of what number?

$14 = 30\% \cdot b$

Solve. We solve the equation.

$14 = 0.3b \qquad (30\% = 0.3)$

$\dfrac{14}{0.3} = b$

$46.\overline{6} = b$

The answer is $46.\overline{6}$, or $46\frac{2}{3}$, or $\dfrac{140}{3}$.

23. *Translate*.

What is 2% of 40?

$a = 2\% \cdot 40$

Solve. We convert 2% to decimal notation and multiply.

$a = 2\% \cdot 40$

$a = 0.02 \cdot 40$

$a = 0.8$

The answer is 0.8.

25. *Translate*.

0.8 is 16% of what number?

$0.8 = 16\% \cdot b$

Solve. We solve the equation.

$0.8 = 0.16b \qquad (16\% = 0.16)$

$\dfrac{0.8}{0.16} = b$

$5 = b$

The answer is 5.

27. *Translate*.

54 is 135% of what number?

$54 = 135\% \cdot b$

Solve. We solve the equation.

$54 = 1.35b \qquad (135\% = 1.35)$

$\dfrac{54}{1.35} = b$

$40 = b$

The answer is 40.

29. First we reword and translate.

What is 3% of $6600?

$a = 3\% \cdot 6600$

Solve. We convert 3% to decimal notation and multiply.

$a = 3\% \cdot 6600 = 0.03 \cdot 6600 = 198$

The price of the dog is $198.

31. First we reword and translate.

What is 24% of $6600?

$a = 24\% \cdot 6600$

Solve. We convert 24% to decimal notation and multiply.

$a = 24\% \cdot 6600 = 0.24 \cdot 6600 = 1584$

Veterinarian expenses are $1584.

33. First we reword and translate.

What is 8% of $6600?

$a = 8\% \cdot 6600$

Solve. We convert 8% to decimal notation and multiply.

$a = 8\% \cdot 6600 = 0.08 \cdot 6600 = 528$

The cost of supplies is $528.

35. To find the percent of the imported cars that were manufactured in Japan, we first reword and translate.

1,003,745 is what percent of 2,268,093?

$1{,}003{,}745 = p \cdot 2{,}268{,}093$

Solve. We divide by 2,268,093 on both sides and convert to percent notation.

$$1,003,745 = p \cdot 2,268,093$$

$$\frac{1,003,745}{2,268,093} = p$$

$$0.443 \approx p$$

$$44.3\% \approx p$$

About 44.3% of the imported cars were manufactured in Japan.

To find the percent of imported cars that were manufactured in Germany, we first reword and translate.

$$\underbrace{564,910}_{\downarrow} \quad \text{is} \quad \underbrace{\text{what percent}}_{\downarrow} \quad \text{of} \quad \underbrace{2,268,093?}_{\downarrow}$$

$$564,910 = \qquad p \qquad \cdot \quad 2,268,093$$

Solve. We divide by 2,268,093 on both sides and convert to percent notation.

$$564,910 = p \cdot 2,268,093$$

$$\frac{564,910}{2,268,093} = p$$

$$0.249 \approx p$$

$$24.9\% \approx p$$

About 24.9% of the imported cars were manufactured in Germany.

37. First we reword and translate.

$$\underset{\downarrow}{193} \quad \underset{\downarrow}{\text{is}} \quad \underset{\downarrow}{32\%} \quad \underset{\downarrow}{\text{of}} \quad \underbrace{\text{what number?}}_{\downarrow}$$

$$193 = 32\% \cdot \qquad b$$

Solve. We solve the equation.

$$193 = 0.32 \cdot b \quad (32\% = 0.32)$$

$$\frac{193}{0.32} = b$$

$$603 \approx b$$

Sammy Sosa had 603 at-bats.

39. First we reword and translate.

$$\text{What is 3\% of \$6500?}$$

$$\underset{\downarrow}{a} \quad \underset{\downarrow}{=} \quad \underset{\downarrow}{3\%} \quad \underset{\downarrow}{\cdot} \quad \underset{\downarrow}{6500}$$

Solve. We convert 3% to decimal notation and multiply.

$$a = 3\% \cdot 6500 = 0.03 \cdot 6500 = 195$$

Sarah will pay \$195 in interest.

41. a) First we reword and translate.

$$\underbrace{\text{What percent}}_{\downarrow} \text{ of \$25 is \$4?}$$

$$p \qquad \cdot \quad 25 = 4$$

Solve. We divide by 25 on both sides and convert to percent notation.

$$p \cdot 25 = 4$$

$$\frac{p \cdot 25}{25} = \frac{4}{25}$$

$$p = 0.16$$

$$p = 16\%$$

The tip was 16% of the cost of the meal.

b) We add to find the total cost of the meal, including tip:

$$\$25 + \$4 = \$29$$

43. a) First we reword and translate.

$$\text{What is 15\% of \$25?}$$

$$\underset{\downarrow}{a} \quad \underset{\downarrow}{=} \quad \underset{\downarrow}{15\%} \quad \underset{\downarrow}{\cdot} \quad \underset{\downarrow}{25}$$

Solve. We convert 15% to decimal notation and multiply.

$$a = 15\% \cdot 25$$

$$a = 0.15 \times 25$$

$$a = 3.75$$

The tip was \$3.75.

b) We add to find the total cost of the meal, including tip:

$$\$25 + \$3.75 = \$28.75$$

45. a) First we reword and translate.

$$\text{15\% of what is \$4.32?}$$

$$\underset{\downarrow}{15\%} \quad \underset{\downarrow}{\cdot} \quad \underset{\downarrow}{b} \quad \underset{\downarrow}{=} \quad \underset{\downarrow}{4.32}$$

Solve. We solve the equation.

$$15\% \cdot b = 4.32$$

$$0.15 \cdot b = 4.32$$

$$\frac{0.15 \cdot b}{0.15} = \frac{4.32}{0.15}$$

$$b = 28.8$$

The cost of the meal before the tip was \$28.80.

b) We add to find the total cost of the meal, including tip:

$$\$28.80 + \$4.32 = \$33.12$$

47. First we reword and translate.

$$\text{8\% of what is 16?}$$

$$\underset{\downarrow}{8\%} \quad \underset{\downarrow}{\cdot} \quad \underset{\downarrow}{b} \quad \underset{\downarrow}{=} \quad \underset{\downarrow}{16}$$

Solve. We solve the equation.

$$8\% \cdot b = 16$$

$$0.08 \cdot b = 16$$

$$\frac{0.08 \cdot b}{0.08} = \frac{16}{0.08}$$

$$b = 200$$

There were 200 women in the original study.

49. First we reword and translate.

$$\text{What is 16.5\% of 191?}$$

$$\underset{\downarrow}{a} \quad \underset{\downarrow}{=} \quad \underset{\downarrow}{16.5\%} \quad \underset{\downarrow}{\cdot} \quad \underset{\downarrow}{191}$$

Solve. We convert 16.5% to decimal notation and multiply.

$a = 16.5\% \cdot 191$

$a = 0.165 \cdot 191$

$a = 31.515 \approx 31.5$

About 31.5 lb of the author's body weight is fat.

51. We subtract to find the increase.

$\$990 - \$335 = \$655$

The increase is $655.

Now we find the percent of increase.

$655 is what percent of $335?

$\downarrow \quad \downarrow \qquad \downarrow \qquad \downarrow \quad \downarrow$

$655 = \qquad p \qquad \cdot \quad 335$

We divide by 335 on both sides and then convert to percent notation.

$655 = p \cdot 335$

$\dfrac{655}{335} = p$

$1.96 \approx p$

$196\% \approx p$

The percent of increase is about 196%.

53. First we find the increase in the rate for smokers.

Rate increase is 198% of $735.

$\downarrow \qquad \downarrow \quad \downarrow \quad \downarrow \quad \downarrow$

$a \qquad = 198\% \cdot \quad 735$

We convert 198% to decimal notation and multiply.

$a = 198\% \cdot 735 = 1.98 \times 735 \approx 1455$

The rate increase is $1455.

Now we add the rate increase to the rate for nonsmokers to find the rate for smokers.

$\$735 + \$1455 = \$2190$

55. We subtract to find the increase.

$\$5445 - \$1510 = \$3935$

The increase is $3935.

Now we find the percent of increase.

$3935 is what percent of $1510?

$\downarrow \quad \downarrow \qquad \downarrow \qquad \downarrow \quad \downarrow$

$3935 = \qquad p \qquad \cdot \quad 1510$

We divide by 1510 on both sides and then convert to percent notation.

$3935 = p \cdot 1510$

$\dfrac{3935}{1510} = p$

$2.61 \approx p$

$261\% \approx p$

The percent of increase is about 261%.

57. Discussion and Writing Exercise

59.

$$\begin{array}{r} 181.52 \\ 0.05_{\wedge}\overline{)9.07_{\wedge}60} \\ 5 \\ \overline{40} \\ 40 \\ \overline{7} \\ 5 \\ \overline{26} \\ 25 \\ \overline{10} \\ 10 \\ \overline{0} \end{array}$$

The answer is 181.52.

61.

$$\begin{array}{r} \overset{1\ 1\ 1}{} \\ 1.0890 \\ 10.8900 \\ +\ \ 0.1089 \\ \hline 12.0879 \end{array}$$

63. $-5a + 3c - 2(c - 3a)$

$= -5a + 3c - 2 \cdot c - 2(-3a)$

$= -5a + 3c - 2c + 6a$

$= (-5 + 6)a + (3 - 2)c$

$= 1 \cdot a + 1 \cdot c$

$= a + c$

65. $-6.5 + 2.6 = -3.9$ The absolute values are 6.5 and 2.6. The difference is 3.9. The negative number has the larger absolute value, so the answer is negative, -3.9.

67. To simplify the calculation $18 - 24 \div 3 - 48 \div (-4)$, do all the division calculations first, and then the subtraction calculations.

69. Since 6 ft $= 6 \times 1$ ft $= 6 \times 12$ in. $= 72$ in., we can express 6 ft 4 in. as 72 in. $+ 4$ in., or 76 in.

Translate. We reword the problem.

96.1% of what is 76 in.?

$\downarrow \quad \downarrow \quad \downarrow \quad \downarrow \quad \downarrow$

$96.1\% \cdot \quad b \quad = \quad 76$

Solve. We solve the equation.

$96.1\% \cdot b = 76$

$0.961 \cdot b = 76$

$\dfrac{0.961 \cdot b}{0.961} = \dfrac{76}{0.961}$

$b \approx 79$

Note that 79 in. $= 72$ in. $+ 7$ in. $= 6$ ft 7 in.

Jaraan's final adult height will be about 6 ft 7 in.

Exercise Set 2.6

1. ***Familiarize***. Using the labels on the drawing in the text, we let $x =$ the length of the shorter piece, in inches, and $3x =$ the length of the longer piece, in inches.

Translate. We reword the problem.

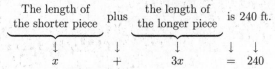

The length of the shorter piece plus the length of the longer piece is 240 ft.

$$x + 3x = 240$$

Solve. We solve the equation.

$$x + 3x = 240$$
$$4x = 240 \quad \text{Collecting like terms}$$
$$\frac{4x}{4} = \frac{240}{4}$$
$$x = 60$$

If x is 60, then $3x = 3 \cdot 60$, or 180.

Check. 180 is three times 60, and $60 + 180 = 240$. The answer checks.

State. The lengths of the pieces are 60 in. and 180 in.

3. *Familiarize.* Let c = the cost of one box of Cinnamon Life cereal.

Translate.

Total cost is Number of boxes times Price of one box

$$17.16 = 4 \cdot c$$

Solve. We solve the equation.

$$17.16 = 4 \cdot c$$
$$\frac{17.16}{4} = c \quad \text{Dividing by 4}$$
$$4.29 = c$$

Check. If one box of Cinnamon Life costs $4.29, then 4 boxes cost $4(\$4.29)$, or $17.16. The answer checks.

State. One box of Cinnamon Life costs $4.29.

5. *Familiarize.* Let d = the amount spent on women's dresses, in billions of dollars.

Translate.

Amount spent on blouses was $0.2 billion more than amount spent on dresses

$$6.5 = 0.2 + d$$

Solve. We solve the equation.

$$6.5 = 0.2 + d$$
$$6.5 - 0.2 = 0.2 + d - 0.2 \quad \text{Subtracting 0.2}$$
$$6.3 = d$$

Check. If we add $0.2 billion to $6.3 billion, we get $6.5 billion. The answer checks.

State. $6.3 billion was spent on dresses.

7. *Familiarize.* Let d = the musher's distance from Nome, in miles. Then $2d$ = the distance from Anchorage, in miles. This is the number of miles the musher has completed. The sum of the two distances is the length of the race, 1049 miles.

Translate.

Distance from Nome plus distance from Anchorage is 1049 mi.

$$d + 2d = 1049$$

Solve. We solve the equation.

$$d + 2d = 1049$$
$$3d = 1049 \quad \text{Collecting like terms}$$
$$\frac{3d}{3} = \frac{1049}{3}$$
$$d = \frac{1049}{3}$$

If $d = \frac{1049}{3}$, then $2d = 2 \cdot \frac{1049}{3} = \frac{2098}{3} = 699\frac{1}{3}$.

Check. $\frac{2098}{3}$ is twice $\frac{1049}{3}$, and $\frac{1049}{3} + \frac{2098}{3} = \frac{3147}{3} = 1049$. The result checks.

State. The musher has traveled $699\frac{1}{3}$ miles.

9. *Familiarize.* Let x = the smaller number and $x + 1$ = the larger number.

Translate. We reword the problem.

First number + second number is 2409

$$x + (x + 1) = 2409$$

Solve. We solve the equation.

$$x + (x + 1) = 2409$$
$$2x + 1 = 2409 \quad \text{Collecting like terms}$$
$$2x + 1 - 1 = 2409 - 1 \quad \text{Subtracting 1}$$
$$2x = 2408$$
$$\frac{2x}{2} = \frac{2408}{2} \quad \text{Dividing by 2}$$
$$x = 1204$$

If x is 1204, then $x + 1$ is 1205.

Check. 1204 and 1205 are consecutive integers, and their sum is 2409. The answer checks.

State. The apartment numbers are 1204 and 1205.

11. *Familiarize.* Let a = the first number. Then $a + 1$ = the second number, and $a + 2$ = the third number.

Translate. We reword the problem.

First number + second number + third number is 126

$$a + (a + 1) + (a + 2) = 114$$

Solve. We solve the equation.

$$a + (a + 1) + (a + 2) = 126$$
$$3a + 3 = 126 \quad \text{Collecting like terms}$$
$$3a + 3 - 3 = 126 - 3$$
$$3a = 123$$
$$\frac{3a}{3} = \frac{123}{3}$$
$$a = 41$$

If a is 41, then $a + 1$ is 42 and $a + 2$ is 43.

Check. 41, 42, and 43 are consecutive integers, and their sum is 126. The answer checks.

State. The numbers are 41, 42, and 43.

13. Familiarize. Let x = the first odd integer. Then $x + 2$ = the next odd integer and $(x + 2) + 2$, or $x + 4$ = the third odd integer.

Translate. We reword the problem.

First odd integer	+	second odd integer	+	third odd integer	is 189
$\downarrow$	$\downarrow$	$\downarrow$	$\downarrow$	$\downarrow$	$\downarrow$ $\downarrow$
x	+	$(x+2)$	+	$(x+4)$	= 189

Solve. We solve the equation.

$$x + (x + 2) + (x + 4) = 189$$
$$3x + 6 = 189 \qquad \text{Collecting like terms}$$
$$3x + 6 - 6 = 189 - 6$$
$$3x = 183$$
$$\frac{3x}{3} = \frac{183}{3}$$
$$x = 61$$

If x is 61, then $x + 2$ is 63 and $x + 4$ is 65.

Check. 61, 63, and 65 are consecutive odd integers, and their sum is 189. The answer checks.

State. The integers are 61, 63, and 65.

15. Familiarize. Using the labels on the drawing in the text, we let w = the width and $3w + 6$ = the length. The perimeter P of a rectangle is given by the formula $2l + 2w = P$, where l = the length and w = the width.

Translate. Substitute $3w + 6$ for l and 124 for P:
$$2l + 2w = P$$
$$2(3w + 6) + 2w = 124$$

Solve. We solve the equation.

$$2(3w + 6) + 2w = 124$$
$$6w + 12 + 2w = 124$$
$$8w + 12 = 124$$
$$8w + 12 - 12 = 124 - 12$$
$$8w = 112$$
$$\frac{8w}{8} = \frac{112}{8}$$
$$w = 14$$

The possible dimensions are $w = 14$ ft and $l = 3w + 6 = 3(14) + 6$, or 48 ft.

Check. The length, 48 ft, is 6 ft more than three times the width, 14 ft. The perimeter is 2(48 ft) + 2(14 ft) = 96 ft + 28 ft = 124 ft. The answer checks.

State. The width is 14 ft, and the length is 48 ft.

17. Familiarize. Let p = the regular price of the shoes. At 15% off, Amy paid 85% of the regular price.

Translate.

$63.75 is 85% of the regular price.

63.75	=	0.85	$\cdot$	p

Solve. We solve the equation.

$$63.75 = 0.85p$$
$$\frac{63.75}{0.08} = p \qquad \text{Dividing both sides by 0.85}$$
$$75 = p$$

Check. 85% of $75, or 0.85($75), is $63.75. The answer checks.

State. The regular price was $75.

19. Familiarize. Let b = the price of the book itself. When the sales tax rate is 5%, the tax paid on the book is 5% of b, or $0.05b$.

Translate.

Price of book plus sales tax is $89.25.

b	+	$0.05b$	=	89.25

Solve. We solve the equation.

$$b + 0.05b = 89.25$$
$$1.05b = 89.25$$
$$b = \frac{89.25}{1.05}$$
$$b = 85$$

Check. 5% of $85, or 0.05($85), is $4.25 and $85 + $4.25 is $89.25, the total cost. The answer checks.

State. The book itself cost $85.

21. Familiarize. Let n = the number of visits required for a total parking cost of $27.00. The parking cost for each $1\frac{1}{2}$ hour visit is $1.50 for the first hour plus $1.00 for part of a second hour, or $2.50. Then the total parking cost for n visits is $2.50n$ dollars.

Translate. We reword the problem.

Total parking cost is $27.00.

$2.50n$	=	27.00

Solve. We solve the equation.

$$2.5n = 27$$
$$10(2.5n) = 10(27) \qquad \text{Clearing the decimal}$$
$$25n = 270$$
$$\frac{25n}{25} = \frac{270}{25}$$
$$n = 10.8$$

If the total parking cost is $27.00 for 10.8 visits, then the cost will be more than $27.00 for 11 or more visits.

Check. The parking cost for 10 visits is $2.50(10), or $25, and the parking cost for 11 visits is $2.50(11), or $27.50. Since 11 is the smallest number for which the parking cost exceeds $27.00, the answer checks.

State. The minimum number of weekly visits for which it is worthwhile to buy a parking pass is 11.

23. Familiarize. Let x = the measure of the first angle. Then $3x$ = the measure of the second angle, and $x + 40$ = the measure of the third angle. Recall that the sum of measures of the angles of a triangle is $180°$.

Translate.

$$\underbrace{\text{Measure of first angle}} + \underbrace{\text{measure of second angle}} + \underbrace{\text{measure of third angle}} \text{ is } 180.$$
$$x \quad + \quad 3x \quad + \quad (x+40) \quad = \quad 180$$

Solve. We solve the equation.

$$x + 3x + (x + 40) = 180$$
$$5x + 40 = 180$$
$$5x + 40 - 40 = 180 - 40$$
$$5x = 140$$
$$\frac{5x}{5} = \frac{140}{5}$$
$$x = 28$$

Possible answers for the angle measures are as follows:

First angle: $x = 28°$

Second angle: $3x = 3(28) = 84°$

Third angle: $x + 40 = 28 + 40 = 68°$

Check. Consider $28°$, $84°$, and $68°$. The second angle is three times the first, and the third is $40°$ more than the first. The sum, $28° + 84° + 68°$, is $180°$. These numbers check.

State. The measures of the angles are $28°$, $84°$, and $68°$.

25. Familiarize. Using the labels on the drawing in the text, we let x = the measure of the first angle, $x + 5$ = the measure of the second angle, and $3x + 10$ = the measure of the third angle. Recall that the sum of measures of the angles of a triangle is $180°$.

Translate.

$$\underbrace{\text{Measure of first angle}} + \underbrace{\text{measure of second angle}} + \underbrace{\text{measure of third angle}} \text{ is } 180.$$
$$x \quad + \quad (x+5) \quad + \quad (3x+10) \quad = \quad 180$$

Solve. We solve the equation.

$$x + (x + 5) + (3x + 10) = 180$$
$$5x + 15 = 180$$
$$5x + 15 - 15 = 180 - 15$$
$$5x = 165$$
$$\frac{5x}{5} = \frac{165}{5}$$
$$x = 33$$

Possible answers for the angle measures are as follows:

First angle: $x = 33°$

Second angle: $x + 5 = 33 + 5 = 38°$

Third angle: $3x + 10 = 3(33) + 10 = 109°$

Check. The second angle is $5°$ more than the first, and the third is $10°$ more than 3 times the first. The sum, $33° + 38° + 109°$, is $180°$. The numbers check.

State. The measures of the angles are $33°$, $38°$, and $109°$.

27. Familiarize. Let a = the amount Sarah invested. The investment grew by 28% of a, or $0.28a$.

Translate.

$$\underbrace{\text{Amount invested}} \text{ plus } \underbrace{\text{amount of growth}} \text{ is \$448.}$$
$$a \quad + \quad 0.28a \quad = \quad 448$$

Solve. We solve the equation.

$$a + 0.28a = 448$$
$$1.28a = 448$$
$$a = 350$$

Check. 28% of \$350 is $0.28(\$350)$, or \$98, and \$350 + \$98 = \$448. The answer checks.

State. Sarah invested \$350.

29. Familiarize. Let b = the balance in the account at the beginning of the month. The balance grew by 2% of b, or $0.02b$.

Translate.

$$\underbrace{\text{Original balance}} \text{ plus } \underbrace{\text{amount of growth}} \text{ is \$870.}$$
$$b \quad + \quad 0.02b \quad = \quad 870$$

Solve. We solve the equation.

$$b + 0.02b = 870$$
$$1.02b = 870$$
$$b \approx \$852.94$$

Check. 2% of \$852.94 is $0.02(\$852.94)$, or \$17.06, and \$852.94 + \$17.06 = \$870. The answer checks.

State. The balance at the beginning of the month was \$852.94.

31. Familiarize. The total cost is the initial charge plus the mileage charge. Let d = the distance, in miles, that Courtney can travel for \$12. The mileage charge is the cost per mile times the number of miles traveled or $0.75d$.

Translate.

$$\underbrace{\text{Initial charge}} \text{ plus } \underbrace{\text{mileage charge}} \text{ is \$12.}$$
$$3 \quad + \quad 0.75d \quad = \quad 12$$

Solve. We solve the equation.

$$3 + 0.75d = 12$$
$$0.75d = 9$$
$$d = 12$$

Check. A 12-mi taxi ride from the airport would cost $\$3 + 12(\$0.75)$, or \$3 + \$9, or \$12. The answer checks.

State. Courtney can travel 12 mi from the airport for \$12.

33. *Familiarize.* Let $c =$ the cost of the meal before the tip. We know that the cost of the meal before the tip plus the tip, 15% of the cost, is the total cost, $41.40.

Translate.

$$\underbrace{\text{Cost of meal}}_{\downarrow} \underset{\downarrow}{\text{plus}} \underset{\downarrow}{\text{tip}} \underset{\downarrow}{\text{is}} \underset{\downarrow}{\$41.40}$$

$$c \qquad + \quad 15\%c = \quad 41.40$$

Solve. We solve the equation.

$$c + 15\%c = 41.40$$
$$c + 0.15c = 41.40$$
$$1c + 0.15c = 41.40$$
$$1.15c = 41.40$$
$$\frac{1.15c}{1.15} = \frac{41.40}{1.15}$$
$$c = 36$$

Check. We find 15% of $36 and add it to $36:

$15\% \times \$36 = 0.15 \times \$36 = \$5.40$ and $\$36 + \$5.40 = \$41.40$.

The answer checks.

State. The cost of the meal before the tip was added was $36.

35. *Familiarize.* Tom paid a total of $3 \cdot \$34$, or $102, for the three ties. Let $t =$ the price of one of the ties. Then $2t =$ the price of another and we are told that the remaining tie cost $27.

Translate.

$$\underbrace{\text{Total cost of the ties}}_{\downarrow} \underset{\downarrow}{\text{is}} \underset{\downarrow}{\$102.}$$

$$t + 2t + 27 \qquad = \quad 102$$

Solve. We solve the equation.

$$t + 2t + 27 = 102$$
$$3t + 27 = 102$$
$$3t + 27 - 27 = 102 - 27$$
$$3t = 75$$
$$\frac{3t}{3} = \frac{75}{3}$$
$$t = 25$$

If $t = 25$, then $2t = 2 \cdot 25 = 50$.

Check. The $50 tie costs twice as much as the $25 tie, and the total cost of the ties is $\$25 + \$50 + \$27$, or $102. The answer checks.

State. One tie cost $25 and another cost $50.

37. Discussion and Writing Exercise

39.
$$-\frac{4}{5} - \frac{3}{8} = -\frac{4}{5} + \left(-\frac{3}{8}\right)$$
$$= -\frac{32}{40} + \left(-\frac{15}{40}\right)$$
$$= -\frac{47}{40}$$

41.
$$-\frac{4}{5} \cdot \frac{3}{8} = -\frac{4 \cdot 3}{5 \cdot 8}$$
$$= -\frac{4 \cdot 3}{5 \cdot 2 \cdot 4}$$
$$= -\frac{\cancel{4} \cdot 3}{5 \cdot 2 \cdot \cancel{4}}$$
$$= -\frac{3}{10}$$

43.
$$\frac{1}{10} \div \left(-\frac{1}{100}\right) = \frac{1}{10} \cdot \left(-\frac{100}{1}\right) = -\frac{1 \cdot 100}{10 \cdot 1} =$$
$$-\frac{\cancel{1} \cdot \cancel{10} \cdot 10}{\cancel{10} \cdot \cancel{1} \cdot 1} = -\frac{10}{1} = -10$$

45. $-25.6(-16) = 409.6$

47. $-25.6 + (-16) = -41.6$

49. *Familiarize.* Let $a =$ the original number of apples. Then $\frac{1}{3}a, \frac{1}{4}a, \frac{1}{8}a,$ and $\frac{1}{5}a$ are given to four people, respectively. The fifth and sixth people get 10 apples and 1 apple, respectively.

Translate. We reword the problem.

$$\underbrace{\text{The total number of apples}}_{\downarrow} \underset{\downarrow}{\text{is}} \underset{\downarrow}{a}$$

$$\frac{1}{3}a + \frac{1}{4}a + \frac{1}{8}a + \frac{1}{5}a + 10 + 1 = \quad a$$

Solve. We solve the equation.

$$\frac{1}{3}a + \frac{1}{4}a + \frac{1}{8}a + \frac{1}{5}a + 10 + 1 = a, \text{ LCD is } 120$$
$$120\left(\frac{1}{3}a + \frac{1}{4}a + \frac{1}{8}a + \frac{1}{5}a + 11\right) = 120 \cdot a$$
$$40a + 30a + 15a + 24a + 1320 = 120a$$
$$109a + 1320 = 120a$$
$$1320 = 11a$$
$$120 = a$$

Check. If the original number of apples was 120, then the first four people got $\frac{1}{3} \cdot 120, \frac{1}{4} \cdot 120, \frac{1}{8} \cdot 120,$ and $\frac{1}{5} \cdot 120$, or 40, 30, 15, and 24 apples, respectively. Adding all the apples we get $40 + 30 + 15 + 24 + 10 + 1$, or 120. The result checks.

State. There were originally 120 apples in the basket.

51. Divide the largest triangle into three triangles, each with a vertex at the center of the circle and with height x as shown.

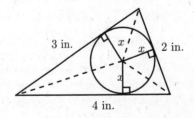

3 in.   2 in.

4 in.

Then the sum of the areas of the three smaller triangles is the area of the original triangle. We have:

$$\frac{1}{2} \cdot 3x + \frac{1}{2} \cdot 2x + \frac{1}{2} \cdot 4x = 2.9047$$

$$2\left(\frac{1}{2} \cdot 3x + \frac{1}{2} \cdot 2x + \frac{1}{2} \cdot 4x\right) = 2(2.9047)$$

$$3x + 2x + 4x = 5.8094$$

$$9x = 5.8094$$

$$x \approx 0.65$$

Thus, x is about 0.65 in.

53. *Familiarize.* Let $p =$ the price of the gasoline as registered on the pump. Then the sales tax will be 9%p.

Translate. We reword the problem.

$$\underbrace{\begin{array}{c}\text{Price} \\ \text{on pump}\end{array}}_{\downarrow \atop p} \ \text{plus} \ \underbrace{\text{sales tax}}_{\downarrow \atop 9\%p} \ \underset{\downarrow}{\text{is}} \ \underbrace{\$10}_{\downarrow \atop 10}$$

$$p \ + \ 9\%p \ = \ 10$$

Solve. We solve the equation.

$$p + 9\%p = 10$$
$$1p + 0.09p = 10$$
$$1.09p = 10$$
$$\frac{1.09p}{1.09} = \frac{10}{1.09}$$
$$p \approx 9.17$$

Check. We find 9% of $9.17 and add it to $9.17:

$$9\% \times \$9.17 = 0.09 \times \$9.17 \approx \$0.83$$

Then $9.17 + $0.83 = $10, so $9.17 checks.

State. The attendant should have filled the tank until the pump read $9.17, not $9.10.

Exercise Set 2.7

1. $x > -4$

a) Since $4 > -4$ is true, 4 is a solution.

b) Since $0 > -4$ is true, 0 is a solution.

c) Since $-4 > -4$ is false, -4 is not a solution.

d) Since $6 > -4$ is true, 6 is a solution.

e) Since $5.6 > -4$ is true, 5.6 is a solution.

3. $x \geq 6.8$

a) Since $-6 \geq 6.8$ is false, -6 is not a solution.

b) Since $0 \geq 6.8$ is false, 0 is not a solution.

c) Since $6 \geq 6.8$ is false, 6 is not a solution.

d) Since $8 \geq 6.8$ is true, 8 is a solution.

e) Since $-3\frac{1}{2} \geq 6.8$ is false, $-3\frac{1}{2}$ is not a solution.

5. The solutions of $x > 4$ are those numbers greater than 4. They are shown on the graph by shading all points to the right of 4. The open circle at 4 indicates that 4 is not part of the graph.

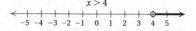

7. The solutions of $t < -3$ are those numbers less than -3. They are shown on the graph by shading all points to the left of -3. The open circle at -3 indicates that -3 is not part of the graph.

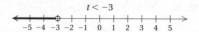

9. The solutions of $m \geq -1$ are are shown by shading the point for -1 and all points to the right of -1. The closed circle at -1 indicates that -1 is part of the graph.

11. In order to be a solution of the inequality $-3 < x \leq 4$, a number must be a solution of both $-3 < x$ and $x \leq 4$. The solution set is graphed as follows:

The open circle at -3 means that -3 is not part of the graph. The closed circle at 4 means that 4 is part of the graph.

13. In order to be a solution of the inequality $0 < x < 3$, a number must be a solution of both $0 < x$ and $x < 3$. The solution set is graphed as follows:

The open circles at 0 and at 3 mean that 0 and 3 are not part of the graph.

15.
$$x + 7 > 2$$
$$x + 7 - 7 > 2 - 7 \quad \text{Subtracting 7}$$
$$x > -5 \quad \text{Simplifying}$$

The solution set is $\{x | x > -5\}$.

The graph is as follows:

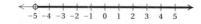

17.
$$x + 8 \leq -10$$
$$x + 8 - 8 \leq -10 - 8 \quad \text{Subtracting 8}$$
$$x \leq -18 \quad \text{Simplifying}$$

The solution set is $\{x | x \leq -18\}$.

The graph is as follows:

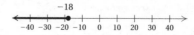

19.
$$y - 7 > -12$$
$$y - 7 + 7 > -12 + 7 \quad \text{Adding 7}$$
$$y > -5 \quad \text{Simplifying}$$

The solution set is $\{y | y > -5\}$.

21.
$$2x + 3 > x + 5$$
$$2x + 3 - 3 > x + 5 - 3 \qquad \text{Subtracting 3}$$
$$2x > x + 2 \qquad \text{Simplifying}$$
$$2x - x > x + 2 - x \qquad \text{Subtracting } x$$
$$x > 2 \qquad \text{Simplifying}$$

The solution set is $\{x | x > 2\}$.

23.
$$3x + 9 \le 2x + 6$$
$$3x + 9 - 9 \le 2x + 6 - 9 \qquad \text{Subtracting 9}$$
$$3x \le 2x - 3 \qquad \text{Simplifying}$$
$$3x - 2x \le 2x - 3 - 2x \qquad \text{Subtracting } 2x$$
$$x \le -3 \qquad \text{Simplifying}$$

The solution set is $\{x | x \le -3\}$.

25.
$$5x - 6 < 4x - 2$$
$$5x - 6 + 6 < 4x - 2 + 6$$
$$5x < 4x + 4$$
$$5x - 4x < 4x + 4 - 4x$$
$$x < 4$$

The solution set is $\{x | x < 4\}$.

27.
$$-9 + t > 5$$
$$-9 + t + 9 > 5 + 9$$
$$t > 14$$

The solution set is $\{t | t > 14\}$.

29.
$$y + \frac{1}{4} \le \frac{1}{2}$$
$$y + \frac{1}{4} - \frac{1}{4} \le \frac{1}{2} - \frac{1}{4}$$
$$y \le \frac{2}{4} - \frac{1}{4} \qquad \text{Obtaining a common denominator}$$
$$y \le \frac{1}{4}$$

The solution set is $\left\{y \middle| y \le \frac{1}{4}\right\}$.

31.
$$x - \frac{1}{3} > \frac{1}{4}$$
$$x - \frac{1}{3} + \frac{1}{3} > \frac{1}{4} + \frac{1}{3}$$
$$x > \frac{3}{12} + \frac{4}{12} \qquad \text{Obtaining a common denominator}$$
$$x > \frac{7}{12}$$

The solution set is $\left\{x \middle| x > \frac{7}{12}\right\}$.

33.
$$5x < 35$$
$$\frac{5x}{5} < \frac{35}{5} \qquad \text{Dividing by 5}$$
$$x < 7$$

The solution set is $\{x | x < 7\}$. The graph is as follows:

35.
$$-12x > -36$$
$$\frac{-12x}{-12} < \frac{-36}{-12} \qquad \text{Dividing by } -12$$
$$\llcorner\!\_\_\_\_ \text{ The symbol has to be reversed.}$$
$$x < 3 \qquad \text{Simplifying}$$

The solution set is $\{x | x < 3\}$. The graph is as follows:

$$\xleftarrow{\quad} \begin{array}{ccccccccccc} & & & & & & & & & & \\ -5 & -4 & -3 & -2 & -1 & 0 & 1 & 2 & 3 & 4 & 5 \end{array} \xrightarrow{\quad}$$

37.
$$5y \ge -2$$
$$\frac{5y}{5} \ge \frac{-2}{5} \qquad \text{Dividing by 5}$$
$$y \ge -\frac{2}{5}$$

The solution set is $\left\{y \middle| y \ge -\frac{2}{5}\right\}$.

39.
$$-2x \le 12$$
$$\frac{-2x}{-2} \ge \frac{12}{-2} \qquad \text{Dividing by } -2$$
$$\llcorner\!\_\_\_\_ \text{ The symbol has to be reversed.}$$
$$x \ge -6 \qquad \text{Simplifying}$$

The solution set is $\{x | x \ge -6\}$.

41.
$$-4y \ge -16$$
$$\frac{-4y}{-4} \le \frac{-16}{-4} \qquad \text{Dividing by } -4$$
$$\llcorner\!\_\_\_\_ \text{ The symbol has to be reversed.}$$
$$y \le 4 \qquad \text{Simplifying}$$

The solution set is $\{y | y \le 4\}$.

43.
$$-3x < -17$$
$$\frac{-3x}{-3} > \frac{-17}{-3} \qquad \text{Dividing by } -3$$
$$\llcorner\!\_\_\_\_ \text{ The symbol has to be reversed.}$$
$$x > \frac{17}{3} \qquad \text{Simplifying}$$

The solution set is $\left\{x \middle| x > \frac{17}{3}\right\}$.

45.
$$-2y > \frac{1}{7}$$
$$-\frac{1}{2} \cdot (-2y) < -\frac{1}{2} \cdot \frac{1}{7}$$
$$\llcorner\!\_\_\_\_ \text{ The symbol has to be reversed.}$$
$$y < -\frac{1}{14}$$

The solution set is $\left\{y \middle| y < -\frac{1}{14}\right\}$.

47.
$$-\frac{6}{5} \le -4x$$
$$-\frac{1}{4} \cdot \left(-\frac{6}{5}\right) \ge -\frac{1}{4} \cdot (-4x)$$
$$\frac{6}{20} \ge x$$
$$\frac{3}{10} \ge x, \text{ or } x \le \frac{3}{10}$$

The solution set is $\left\{x \middle| \frac{3}{10} \ge x\right\}$, or $\left\{x \middle| x \le \frac{3}{10}\right\}$.

49.
$$4 + 3x < 28$$
$$-4 + 4 + 3x < -4 + 28 \qquad \text{Adding } -4$$
$$3x < 24 \qquad \text{Simplifying}$$
$$\frac{3x}{3} < \frac{24}{3} \qquad \text{Dividing by 3}$$
$$x < 8$$

The solution set is $\{x | x < 8\}$.

51.
$$3x - 5 \leq 13$$
$$3x - 5 + 5 \leq 13 + 5 \quad \text{Adding 5}$$
$$3x \leq 18$$
$$\frac{3x}{3} \leq \frac{18}{3} \quad \text{Dividing by 3}$$
$$x \leq 6$$

The solution set is $\{x | x \leq 6\}$.

53.
$$13x - 7 < -46$$
$$13x - 7 + 7 < -46 + 7$$
$$13x < -39$$
$$\frac{13x}{13} < \frac{-39}{13}$$
$$x < -3$$

The solution set is $\{x | x < -3\}$.

55.
$$30 > 3 - 9x$$
$$30 - 3 > 3 - 9x - 3 \quad \text{Subtracting 3}$$
$$27 > -9x$$
$$\frac{27}{-9} < \frac{-9x}{-9} \quad \text{Dividing by } -9$$
⌐___ The symbol has to be reversed.
$$-3 < x$$

The solution set is $\{x | -3 < x\}$, or $\{x | x > -3\}$.

57.
$$4x + 2 - 3x \leq 9$$
$$x + 2 \leq 9 \quad \text{Collecting like terms}$$
$$x + 2 - 2 \leq 9 - 2$$
$$x \leq 7$$

The solution set is $\{x | x \leq 7\}$.

59.
$$-3 < 8x + 7 - 7x$$
$$-3 < x + 7 \quad \text{Collecting like terms}$$
$$-3 - 7 < x + 7 - 7$$
$$-10 < x$$

The solution set is $\{x | -10 < x\}$, or $\{x | x > -10\}$.

61.
$$6 - 4y > 4 - 3y$$
$$6 - 4y + 4y > 4 - 3y + 4y \quad \text{Adding } 4y$$
$$6 > 4 + y$$
$$-4 + 6 > -4 + 4 + y \quad \text{Adding } -4$$
$$2 > y, \text{ or } y < 2$$

The solution set is $\{y | 2 > y\}$, or $\{y | y < 2\}$.

63.
$$5 - 9y \leq 2 - 8y$$
$$5 - 9y + 9y \leq 2 - 8y + 9y$$
$$5 \leq 2 + y$$
$$-2 + 5 \leq -2 + 2 + y$$
$$3 \leq y, \text{ or } y \geq 3$$

The solution set is $\{y | 3 \leq y\}$, or $\{y | y \geq 3\}$.

65.
$$19 - 7y - 3y < 39$$
$$19 - 10y < 39 \quad \text{Collecting like terms}$$
$$-19 + 19 - 10y < -19 + 39$$
$$-10y < 20$$
$$\frac{-10y}{-10} > \frac{20}{-10}$$
⌐___ The symbol has to be reversed.
$$y > -2$$

The solution set is $\{y | y > -2\}$.

67.
$$2.1x + 45.2 > 3.2 - 8.4x$$
$$10(2.1x + 45.2) > 10(3.2 - 8.4x) \quad \begin{array}{l}\text{Multiplying by 10} \\ \text{to clear decimals}\end{array}$$
$$21x + 452 > 32 - 84x$$
$$21x + 84x > 32 - 452 \quad \begin{array}{l}\text{Adding } 84x \text{ and} \\ \text{subtracting 452}\end{array}$$
$$105x > -420$$
$$x > -4 \quad \text{Dividing by 105}$$

The solution set is $\{x | x > -4\}$.

69.
$$\frac{x}{3} - 2 \leq 1$$
$$3\left(\frac{x}{3} - 2\right) \leq 3 \cdot 1 \quad \begin{array}{l}\text{Multiplying by 3 to} \\ \text{to clear the fraction}\end{array}$$
$$x - 6 \leq 3 \quad \text{Simplifying}$$
$$x \leq 9 \quad \text{Adding 6}$$

The solution set is $\{x | x \leq 9\}$.

71.
$$\frac{y}{5} + 1 \leq \frac{2}{5}$$
$$5\left(\frac{y}{5} + 1\right) \leq 5 \cdot \frac{2}{5} \quad \text{Clearing fractions}$$
$$y + 5 \leq 2$$
$$y \leq -3 \quad \text{Subtracting 5}$$

The solution set is $\{y | y \leq -3\}$.

73.
$$3(2y - 3) < 27$$
$$6y - 9 < 27 \quad \text{Removing parentheses}$$
$$6y < 36 \quad \text{Adding 9}$$
$$y < 6 \quad \text{Dividing by 6}$$

The solution set is $\{y | y < 6\}$.

75.
$$2(3 + 4m) - 9 \geq 45$$
$$6 + 8m - 9 \geq 45 \quad \text{Removing parentheses}$$
$$8m - 3 \geq 45 \quad \text{Collecting like terms}$$
$$8m \geq 48 \quad \text{Adding 3}$$
$$m \geq 6 \quad \text{Dividing by 8}$$

The solution set is $\{m | m \geq 6\}$.

77.
$$8(2t + 1) > 4(7t + 7)$$
$$16t + 8 > 28t + 28$$
$$16t - 28t > 28 - 8$$
$$-12t > 20$$
$$t < -\frac{20}{12} \quad \begin{array}{l}\text{Dividing by } -12 \text{ and} \\ \text{reversing the symbol}\end{array}$$
$$t < -\frac{5}{3}$$

The solution set is $\left\{t \left| t < -\frac{5}{3}\right.\right\}$.

79.
$$3(r - 6) + 2 < 4(r + 2) - 21$$
$$3r - 18 + 2 < 4r + 8 - 21$$
$$3r - 16 < 4r - 13$$
$$-16 + 13 < 4r - 3r$$
$$-3 < r, \text{ or } r > -3$$

The solution set is $\{r | r > -3\}$.

81.
$$0.8(3x + 6) \geq 1.1 - (x + 2)$$
$$2.4x + 4.8 \geq 1.1 - x - 2$$
$$10(2.4x + 4.8) \geq 10(1.1 - x - 2) \quad \text{Clearing decimals}$$
$$24x + 48 \geq 11 - 10x - 20$$
$$24x + 48 \geq -10x - 9 \quad \text{Collecting like terms}$$
$$24x + 10x \geq -9 - 48$$
$$34x \geq -57$$
$$x \geq -\frac{57}{34}$$

The solution set is $\left\{x \middle| x \geq -\dfrac{57}{34}\right\}$.

83. $\dfrac{5}{3} + \dfrac{2}{3}x < \dfrac{25}{12} + \dfrac{5}{4}x + \dfrac{3}{4}$

The number 12 is the least common multiple of all the denominators. We multiply by 12 on both sides.

$$12\left(\frac{5}{3} + \frac{2}{3}x\right) < 12\left(\frac{25}{12} + \frac{5}{4}x + \frac{3}{4}\right)$$
$$12 \cdot \frac{5}{3} + 12 \cdot \frac{2}{3}x < 12 \cdot \frac{25}{12} + 12 \cdot \frac{5}{4}x + 12 \cdot \frac{3}{4}$$
$$20 + 8x < 25 + 15x + 9$$
$$20 + 8x < 34 + 15x$$
$$20 - 34 < 15x - 8x$$
$$-14 < 7x$$
$$-2 < x, \text{ or } x > -2$$

The solution set is $\{x | x > -2\}$.

85. Discussion and Writing Exercise

87. $-56 + (-18)$ Two negative numbers. Add the absolute values and make the answer negative.
$$-56 + (-18) = -74$$

89. $-\dfrac{3}{4} + \dfrac{1}{8}$ One negative and one positive number. Find the difference of the absolute values. Then make the answer negative, since the negative number has the larger absolute value.
$$-\frac{3}{4} + \frac{1}{8} = -\frac{6}{8} + \frac{1}{8} = -\frac{5}{8}$$

91. $-56 - (-18) = -56 + 18 = -38$

93. $-2.3 - 7.1 = -2.3 + (-7.1) = -9.4$

95. $5 - 3^2 + (8 - 2)^2 \cdot 4 = 5 - 3^2 + 6^2 \cdot 4$
$$= 5 - 9 + 36 \cdot 4$$
$$= 5 - 9 + 144$$
$$= -4 + 144$$
$$= 140$$

97. $5(2x - 4) - 3(4x + 1) = 10x - 20 - 12x - 3 = -2x - 23$

99. $|x| < 3$

a) Since $|0| = 0$ and $0 < 3$ is true, 0 is a solution.

b) Since $|-2| = 2$ and $2 < 3$ is true, -2 is a solution.

c) Since $|-3| = 3$ and $3 < 3$ is false, -3 is not a solution.

d) Since $|4| = 4$ and $4 < 3$ is false, 4 is not a solution.

e) Since $|3| = 3$ and $3 < 3$ is false, 3 is not a solution.

f) Since $|1.7| = 1.7$ and $1.7 < 3$ is true, 1.7 is a solution.

g) Since $|-2.8| = 2.8$ and $2.8 < 3$ is true, -2.8 is a solution.

101.
$$x + 3 \leq 3 + x$$
$$x - x \leq 3 - 3 \quad \text{Subtracting } x \text{ and } 3$$
$$0 \leq 0$$

We get an inequality that is true for all values of x, so the inequality is true for all real numbers.

Exercise Set 2.8

1. $n \geq 7$

3. $w > 2$ kg

5. 90 mph $< s <$ 110 mph

7. $a \leq 1,200,000$

9. $c \leq \$1.50$

11. $x > 8$

13. $y \leq -4$

15. $n \geq 1300$

17. $a \leq 500$ L

19. $3x + 2 < 13$, or $2 + 3x < 13$

21. *Familiarize*. Let s represent the score on the fourth test.

Translate.

$$\underbrace{\text{The average score}}_{\dfrac{82 + 76 + 78 + s}{4}} \quad \underbrace{\text{is at least}}_{\geq} \quad \underbrace{80.}_{80}$$

Solve.
$$\frac{82 + 76 + 78 + s}{4} \geq 80$$
$$4\left(\frac{82 + 76 + 78 + s}{4}\right) \geq 4 \cdot 80$$
$$82 + 76 + 78 + s \geq 320$$
$$236 + s \geq 320$$
$$s \geq 84$$

Check. As a partial check we show that the average is at least 80 when the fourth test score is 84.
$$\frac{82 + 76 + 78 + 84}{4} = \frac{320}{4} = 80$$

State. The student will get at least a B if the score on the fourth test is at least 84. The solution set is $\{s | s \geq 84\}$.

23. *Familiarize*. We use the formula for converting Celsius temperatures to Fahrenheit temperatures, $F = \dfrac{9}{5}C + 32$.

Translate.

$$\underbrace{\text{Fahrenheit temperature}}_{\dfrac{9}{5}C + 32} \quad \underbrace{\text{is less than}}_{\leq} \quad \underbrace{1945.4.}_{1945.4}$$

Solve.

$$\frac{9}{5}C + 32 < 1945.4$$

$$\frac{9}{5}C < 1913.4$$

$$\frac{5}{9} \cdot \frac{9}{5}C < \frac{5}{9}(1913.4)$$

$$C < 1063$$

Check. As a partial check we can show that the Fahrenheit temperature is less than 1945.4° for a Celsius temperature less than 1063° and is greater than 1945.4° for a Celsius temperature greater than 1063°.

$$F = \frac{9}{5} \cdot 1062 + 32 = 1943.6 < 1945.4$$

$$F = \frac{9}{5} \cdot 1064 + 32 = 1947.2 > 1945.4$$

State. Gold stays solid for temperatures less than 1063°C. The solution set is $\{C | C < 1063°\}$.

25. Familiarize. $R = -0.075t + 3.85$

In the formula R represents the world record and t represents the years since 1930. When $t = 0$ (1930), the record was $-0.075 \cdot 0 + 3.85$, or 3.85 minutes. When $t = 2$ (1932), the record was $-0.075(2) + 3.85$, or 3.7 minutes. For what values of t will $-0.075t + 3.85$ be less than 3.5?

Translate. The record is to be less than 3.5. We have the inequality

$$R < 3.5.$$

To find the t values which satisfy this condition we substitute $-0.075t + 3.85$ for R.

$$-0.075t + 3.85 < 3.5$$

Solve.

$$-0.075t + 3.85 < 3.5$$
$$-0.075t < 3.5 - 3.85$$
$$-0.075t < -0.35$$
$$t > \frac{-0.35}{-0.075}$$
$$t > 4\frac{2}{3}$$

Check. With inequalities it is impossible to check each solution. But we can check to see if the solution set we obtained seems reasonable.

When $t = 4\frac{1}{2}$, $R = -0.075(4.5) + 3.85$, or 3.5125.

When $t = 4\frac{2}{3}$, $R = -0.075\left(\frac{14}{3}\right) + 3.85$, or 3.5.

When $t = 4\frac{3}{4}$, $R = -0.075(4.75) + 3.85$, or 3.49375.

Since $r = 3.5$ when $t = 4\frac{2}{3}$ and R decreases as t increases, R will be less than 3.5 when t is greater than $4\frac{2}{3}$.

State. The world record will be less than 3.5 minutes more than $4\frac{2}{3}$ years after 1930. If we let $Y =$ the year, then the solution set is $\{Y | Y \geq 1935\}$.

27. Familiarize. As in the drawing in the text, we let $L =$ the length of the envelope. Recall that the area of a rectangle is the product of the length and the width.

Translate.

Length	times	width	is at least	$17\frac{1}{2}$ in^2
↓	↓	↓	↓	↓
L	$\cdot$	$3\frac{1}{2}$	$\geq$	$17\frac{1}{2}$

Solve.

$$L \cdot 3\frac{1}{2} \geq 17\frac{1}{2}$$

$$L \cdot \frac{7}{2} \geq \frac{35}{2}$$

$$L \cdot \frac{7}{2} \cdot \frac{2}{7} \geq \frac{35}{2} \cdot \frac{2}{7}$$

$$L \geq 5$$

The solution set is $\{L | L \geq 5\}$.

Check. We can obtain a partial check by substituting a number greater than or equal to 5 in the inequality. For example, when $L = 6$:

$$L \cdot 3\frac{1}{2} = 6 \cdot 3\frac{1}{2} = 6 \cdot \frac{7}{2} = 21 \geq 17\frac{1}{2}$$

The result appears to be correct.

State. Lengths of 5 in. or more will satisfy the constraints. The solution set is $\{L | L \geq 5 \text{ in.}\}$.

29. Familiarize. Let $c =$ the number of copies Myra has made. The total cost of the copies is the setup fee of \$5 plus \$4 times the number of copies, or $\$4 \cdot c$.

Translate.

Setup fee	plus	copying cost	cannot exceed	\$65.
↓	↓	↓	↓	↓
5	+	$4c$	$\leq$	65

Solve. We solve the inequality.

$$5 + 4c \leq 65$$
$$4c \leq 60$$
$$c \leq 15$$

Check. As a partial check, we show that Myra can have 15 copies made and not exceed her \$65 budget.

$$\$5 + \$4 \cdot 15 = 5 + 60 = \$65$$

State. Myra can have 15 or fewer copies made and stay within her budget.

31. Familiarize. Let m represent the length of a telephone call, in minutes.

Translate.

\$0.75 charge	plus	charge for time used	is at least	\$3.00.
↓	↓	↓	↓	↓
0.75	+	$0.45m$	$\geq$	3

Solve. We solve the inequality.

$$0.75 + 0.45m \geq 3$$
$$0.45m \geq 2.25$$
$$m \geq 5$$

Check. As a partial check, we can show that if a call lasts 5 minutes it costs at least $3.00:

$$\$0.75 + \$0.45(5) = \$0.75 + \$2.25 = \$3.00.$$

State. Simon's calls last at least 5 minutes each.

33. *Familiarize*. Let c = the number of courses for which Angelica registers. Her total tuition is the $35 registration fee plus $375 times the number of courses for which she registers, or $375 \cdot c$.

Translate.

Registration fee	plus	fee for courses	cannot exceed	$1000.
↓	↓	↓	↓	↓
35	+	$375 \cdot c$	≤	1000

Solve. We solve the inequality.

$$35 + 375c \leq 1000$$
$$375c \leq 965$$
$$c \leq 2.57\overline{3}$$

Check. Although the solution set of the inequality is all numbers less than or equal to $2.57\overline{3}$, since c represents the number of courses for which Angelica registers, we round down to 2. If she registers for 2 courses, her tuition is $35 + \$375 \cdot 2$, or $785 which does not exceed $1000. If she registers for 3 courses, her tuition is $35 + \$375 \cdot 3$, or $1160 which exceeds $1000.

State. Angelica can register for at most 2 courses.

35. *Familiarize*. Let s = the number of servings of fruits or vegetables Dale eats on Saturday.

Translate.

Average number of fruit or vegetable servings	is at least	5.
↓	↓	↓
$\dfrac{4+6+7+4+6+4+s}{7}$	≥	5

Solve. We first multiply by 7 to clear the fraction.

$$7\left(\frac{4+6+7+4+6+4+s}{7}\right) \geq 7 \cdot 5$$
$$4+6+7+4+6+4+s \geq 35$$
$$31 + s \geq 35$$
$$s \geq 4$$

Check. As a partial check, we show that Dale can eat 4 servings of fruits or vegetables on Saturday and average at least 5 servings per day for the week:

$$\frac{4+6+7+4+6+4+4}{7} = \frac{35}{7} = 5$$

State. Dale should eat at least 4 servings of fruits or vegetables on Saturday.

37. *Familiarize*. We first make a drawing. We let l represent the length, in feet.

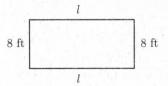

The perimeter is $P = 2l + 2w$, or $2l + 2 \cdot 8$, or $2l + 16$.

Translate. We translate to 2 inequalities.

The perimeter	is at least	200 ft.
↓	↓	↓
$2l + 16$	≥	200

The perimeter	is at most	200 ft.
↓	↓	↓
$2l + 16$	≤	200

Solve. We solve each inequality.

$$2l + 16 \geq 200 \qquad 2l + 16 \leq 200$$
$$2l \geq 184 \qquad\quad 2l \leq 184$$
$$l \geq 92 \qquad\quad\;\; l \leq 92$$

Check. We check to see if the solutions seem reasonable.

When $l = 91$ ft, $P = 2 \cdot 91 + 16$, or 198 ft.

When $l = 92$ ft, $P = 2 \cdot 92 + 16$, or 200 ft.

When $l = 93$ ft, $P = 2 \cdot 93 + 16$, or 202 ft.

From these calculations, it appears that the solutions are correct.

State. Lengths greater than or equal to 92 ft will make the perimeter at least 200 ft. Lengths less than or equal to 92 ft will make the perimeter at most 200 ft.

39. *Familiarize*. Using the label on the drawing in the text, we let L represent the length.

The area is the length times the width, or $4L$.

Translate.

Area	is less than	86 cm².
↓	↓	↓
$4L$	<	86

Solve.

$$4L < 86$$
$$L < 21.5$$

Check. We check to see if the solution seems reasonable.

When $L = 22$, the area is $22 \cdot 4$, or 88 cm².

When $L = 21.5$, the area is $21.5(4)$, or 86 cm².

When $L = 21$, the area is $21 \cdot 4$, or 84 cm².

From these calculations, it would appear that the solution is correct.

State. The area will be less than 86 cm² for lengths less than 21.5 cm.

41. *Familiarize*. Let v = the blue book value of the car. Since the car was repaired, we know that $8500 does not exceed $0.8v$ or, in other words, $0.8v$ is at least $8500.

Translate.

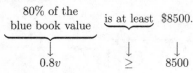

80% of the blue book value is at least $8500.

$$0.8v \qquad \geq \qquad 8500$$

Solve.

$$0.8v \geq 8500$$
$$v \geq \frac{8500}{0.8}$$
$$v \geq 10,625$$

Check. As a partial check, we show that 80% of $10,625 is at least $8500:

$$0.8(\$10,625) = \$8500$$

State. The blue book value of the car was at least $10,625.

43. Familiarize. Let r = the amount of fat in a serving of the regular peanut butter, in grams. If reduced fat peanut butter has at least 25% less fat than regular peanut butter, then it has at most 75% as much fat as the regular peanut butter.

Translate.

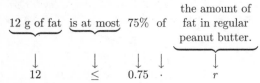

12 g of fat is at most 75% of the amount of fat in regular peanut butter.

$$12 \qquad \leq \qquad 0.75 \quad \cdot \qquad r$$

Solve.

$$12 \leq 0.75r$$
$$16 \leq r$$

Check. As a partial check, we show that 12 g of fat does not exceed 75% of 16 g of fat:

$$0.75(16) = 12$$

State. Regular peanut butter contains at least 16 g of fat per serving.

45. Familiarize. Let w = the number of weeks after July 1. After w weeks the water level has dropped $\frac{2}{3}w$ ft.

Translate.

Original depth minus drop in water level does not exceed 21 ft.

$$25 \qquad - \qquad \frac{2}{3}w \qquad \leq \qquad 21$$

Solve. We solve the inequality.

$$25 - \frac{2}{3}w \leq 21$$
$$-\frac{2}{3}w \leq -4$$
$$w \geq -\frac{3}{2}(-4)$$
$$w \geq 6$$

Check. As a partial check we show that the water level is 21 ft 6 weeks after July 1.

$$25 - \frac{2}{3} \cdot 6 = 25 - 4 = 21 \text{ ft}$$

Since the water level continues to drop during the weeks after July 1, the answer seems reasonable.

State. The water level will not exceed 21 ft for dates at least 6 weeks after July 1.

47. Familiarize. Let h = the height of the triangle, in ft. Recall that the formula for the area of a triangle with base b and height h is $A = \frac{1}{2}bh$.

Translate.

Area is at least 3 ft².

$$\frac{1}{2}\left(1\frac{1}{2}\right)h \qquad \geq \qquad 3$$

Solve. We solve the inequality.

$$\frac{1}{2}\left(1\frac{1}{2}\right)h \geq 3$$
$$\frac{1}{2} \cdot \frac{3}{2} \cdot h \geq 3$$
$$\frac{3}{4}h \geq 3$$
$$h \geq \frac{4}{3} \cdot 3$$
$$h \geq 3$$

Check. As a partial check, we show that the area of the triangle is 3 ft² when the height is 4 ft.

$$\frac{1}{2}\left(1\frac{1}{2}\right)(4) = \frac{1}{2} \cdot \frac{3}{2} \cdot \frac{4}{1} = 3$$

State. The height should be at least 4 ft.

49. Familiarize. The average number of calls per week is the sum of the calls for the three weeks divided by the number of weeks, 3. We let c represent the number of calls made during the third week.

Translate. The average of the three weeks is given by

$$\frac{17 + 22 + c}{3}.$$

Since the average must be at least 20, this means that it must be greater than or equal to 20. Thus, we can translate the problem to the inequality

$$\frac{17 + 22 + c}{3} \geq 20.$$

Solve. We first multiply by 3 to clear the fraction.

$$3\left(\frac{17 + 22 + c}{3}\right) \geq 3 \cdot 20$$
$$17 + 22 + c \geq 60$$
$$39 + c \geq 60$$
$$c \geq 21$$

Check. Suppose c is a number greater than or equal to 21. Then by adding 17 and 22 on both sides of the inequality we get

$$17 + 22 + c \geq 17 + 22 + 21$$
$$17 + 22 + c \geq 60$$

so

$$\frac{17 + 22 + c}{3} \geq \frac{60}{3}, \text{ or } 20.$$

State. 21 calls or more will maintain an average of at least 20 for the three-week period.

51. Discussion and Writing Exercise.

53. The product of an <u>even</u> number of negative numbers is always positive.

55. The <u>additive</u> inverse of a negative number is always positive.

57. Equations with the same solutions are called <u>equivalent</u> equations.

59. The <u>multiplication principle</u> for inequalities asserts that when we multiply or divide by a negative number on both sides of an inequality, the direction of the inequality symbol <u>is reversed</u>.

61. *Familiarize*. We use the formula $F = \frac{9}{5}C + 32$.

Translate. We are interested in temperatures such that $5° < F < 15°$. Substituting for F, we have:

$$5 < \frac{9}{5}C + 32 < 15$$

Solve.

$$5 < \frac{9}{5}C + 32 < 15$$

$$5 \cdot 5 < 5\left(\frac{9}{5}C + 32\right) < 5 \cdot 15$$

$$25 < 9C + 160 < 75$$

$$-135 < 9C < -85$$

$$-15 < C < -9\frac{4}{9}$$

Check. The check is left to the student.

State. Green ski wax works best for temperatures between $-15°$C and $-9\frac{4}{9}°$C.

63. *Familiarize*. Let $f =$ the fat content of a serving of regular tortilla chips, in grams. A product that contains 60% less fat than another product has 40% of the fat content of that product. If Reduced Fat Tortilla Pops cannot be labeled lowfat, then they contain at least 3 g of fat.

Translate.

40% of	the fat content of regular tortilla chips	is at least	3 grams of fat
↓ ↓	↓	↓	↓
0.4 ·	f	≥	3

Solve.

$$0.4f \geq 3$$
$$f \geq 7.5$$

Check. As a partial check, we show that 40% of 7.5 g is not less than 3 g.

$$0.4(7.5) = 3$$

State. A serving of regular tortilla chips contains at least 7.5 g of fat.

Chapter 2 Review Exercises

1.
$$x + 5 = -17$$
$$x + 5 - 5 = -17 - 5$$
$$x = -22$$
The solution is -22.

2.
$$n - 7 = -6$$
$$n - 7 + 7 = -6 + 7$$
$$n = 1$$
The solution is 1.

3.
$$x - 11 = 14$$
$$x - 11 + 11 = 14 + 11$$
$$x = 25$$
The solution is 25.

4.
$$y - 0.9 = 9.09$$
$$y - 0.9 + 0.9 = 9.09 + 0.9$$
$$y = 9.99$$
The solution is 9.99.

5.
$$-\frac{2}{3}x = -\frac{1}{6}$$
$$-\frac{3}{2} \cdot \left(-\frac{2}{3}x\right) = -\frac{3}{2} \cdot \left(-\frac{1}{6}\right)$$
$$1 \cdot x = \frac{\cancel{3} \cdot 1}{2 \cdot 2 \cdot \cancel{3}}$$
$$x = \frac{1}{4}$$
The solution is $\frac{1}{4}$.

6.
$$-8x = -56$$
$$\frac{-8x}{-8} = \frac{-56}{-8}$$
$$x = 7$$
The solution is 7.

7.
$$-\frac{x}{4} = 48$$
$$4 \cdot \frac{1}{4} \cdot (-x) = 4 \cdot 48$$
$$-x = 192$$
$$-1 \cdot (-1 \cdot x) = -1 \cdot 192$$
$$x = -192$$
The solution is -192.

8.
$$15x = -35$$
$$\frac{15x}{15} = \frac{-35}{15}$$
$$x = -\frac{\cancel{5} \cdot 7}{3 \cdot \cancel{5}}$$
$$x = -\frac{7}{3}$$
The solution is $-\frac{7}{3}$.

9.
$$\frac{4}{5}y = -\frac{3}{16}$$
$$\frac{5}{4}\cdot\frac{4}{5}y = \frac{5}{4}\cdot\left(-\frac{3}{16}\right)$$
$$y = -\frac{15}{64}$$

The solution is $-\frac{15}{64}$.

10.
$$5 - x = 13$$
$$5 - x - 5 = 13 - 5$$
$$-x = 8$$
$$-1\cdot(-1\cdot x) = -1\cdot 8$$
$$x = -8$$

The solution is -8.

11.
$$\frac{1}{4}x - \frac{5}{8} = \frac{3}{8}$$
$$\frac{1}{4}x - \frac{5}{8} + \frac{5}{8} = \frac{3}{8} + \frac{5}{8}$$
$$\frac{1}{4}x = 1$$
$$4\cdot\frac{1}{4}x = 4\cdot 1$$
$$x = 4$$

The solution is 4.

12.
$$5t + 9 = 3t - 1$$
$$5t + 9 - 3t = 3t - 1 - 3t$$
$$2t + 9 = -1$$
$$2t + 9 - 9 = -1 - 9$$
$$2t = -10$$
$$\frac{2t}{2} = \frac{-10}{2}$$
$$t = -5$$

The solution is -5.

13.
$$7x - 6 = 25x$$
$$7x - 6 - 7x = 25x - 7x$$
$$-6 = 18x$$
$$\frac{-6}{18} = \frac{18x}{18}$$
$$-\frac{\cancel{6}\cdot 1}{3\cdot\cancel{6}} = x$$
$$-\frac{1}{3} = x$$

The solution is $-\frac{1}{3}$.

14.
$$14y = 23y - 17 - 10$$
$$14y = 23y - 27 \quad\text{Collecting like terms}$$
$$14y - 23y = 23y - 27 - 23y$$
$$-9y = -27$$
$$\frac{-9y}{-9} = \frac{-27}{-9}$$
$$y = 3$$

The solution is 3.

15.
$$0.22y - 0.6 = 0.12y + 3 - 0.8y$$
$$0.22y - 0.6 = -0.68y + 3 \quad\text{Collecting like terms}$$
$$0.22y - 0.6 + 0.68y = -0.68y + 3 + 0.68y$$
$$0.9y - 0.6 = 3$$
$$0.9y - 0.6 + 0.6 = 3 + 0.6$$
$$0.9y = 3.6$$
$$\frac{0.9y}{0.9} = \frac{3.6}{0.9}$$
$$y = 4$$

The solution is 4.

16.
$$\frac{1}{4}x - \frac{1}{8}x = 3 - \frac{1}{16}x$$
$$\frac{2}{8}x - \frac{1}{8}x = 3 - \frac{1}{16}x$$
$$\frac{1}{8}x = 3 - \frac{1}{16}x$$
$$\frac{1}{8}x + \frac{1}{16}x = 3 - \frac{1}{16}x + \frac{1}{16}x$$
$$\frac{2}{16}x + \frac{1}{16}x = 3$$
$$\frac{3}{16}x = 3$$
$$\frac{16}{3}\cdot\frac{3}{16}x = \frac{16}{3}\cdot 3$$
$$x = \frac{16\cdot\cancel{3}}{\cancel{3}\cdot 1}$$
$$x = 16$$

The solution is 16.

17.
$$14y + 17 + 7y = 9 + 21y + 8$$
$$21y + 17 = 21y + 17$$
$$21y + 17 - 21y = 21y + 17 - 21y$$
$$17 = 17 \quad\text{TRUE}$$

All real numbers are solutions.

18.
$$4(x + 3) = 36$$
$$4x + 12 = 36$$
$$4x + 12 - 12 = 36 - 12$$
$$4x = 24$$
$$\frac{4x}{4} = \frac{24}{4}$$
$$x = 6$$

The solution is 6.

19.
$$3(5x - 7) = -66$$
$$15x - 21 = -66$$
$$15x - 21 + 21 = -66 + 21$$
$$15x = -45$$
$$\frac{15x}{15} = \frac{-45}{15}$$
$$x = -3$$

The solution is -3.

20. $8(x-2)-5(x+4)=20+x$

$8x-16-5x-20=20+x$

$3x-36=20+x$

$3x-36-x=20+x-x$

$2x-36=20$

$2x-36+36=20+36$

$2x=56$

$\dfrac{2x}{2}=\dfrac{56}{2}$

$x=28$

The solution is 28.

21. $-5x+3(x+8)=16$

$-5x+3x+24=16$

$-2x+24=16$

$-2x+24-24=16-24$

$-2x=-8$

$\dfrac{-2x}{-2}=\dfrac{-8}{-2}$

$x=4$

The solution is 4.

22. $6(x-2)-16=3(2x-5)+11$

$6x-12-16=6x-15+11$

$6x-28=6x-4$

$6x-28-6x=6x-4-6x$

$-28=-4 \qquad$ False

There are no solutions.

23. Since $-3 \le 4$ is true, -3 is a solution.

24. Since $7 \le 4$ is false, 7 is not a solution.

25. Since $4 \le 4$ is true, 4 is a solution.

26. $y+\dfrac{2}{3} \ge \dfrac{1}{6}$

$y+\dfrac{2}{3}-\dfrac{2}{3} \ge \dfrac{1}{6}-\dfrac{2}{3}$

$y \ge \dfrac{1}{6}-\dfrac{4}{6}$

$y \ge -\dfrac{3}{6}$

$y \ge -\dfrac{1}{2}$

The solution set is $\left\{y \middle| y \ge -\dfrac{1}{2}\right\}$.

27. $9x \ge 63$

$\dfrac{9x}{9} \ge \dfrac{63}{9}$

$x \ge 7$

The solution set is $\{x|x \ge 7\}$.

28. $2+6y > 14$

$2+6y-2 > 14-2$

$6y > 12$

$\dfrac{6y}{6} > \dfrac{12}{6}$

$y > 2$

The solution set is $\{y|y > 2\}$.

29. $7-3y \ge 27+2y$

$7-3y-2y \ge 27+2y-2y$

$7-5y \ge 27$

$7-5y-7 \ge 27-7$

$-5y \ge 20$

$\dfrac{-5y}{-5} \le \dfrac{20}{-5} \quad$ Reversing the inequality symbol

$y \le -4$

The solution set is $\{y|y \le -4\}$.

30. $3x+5 < 2x-6$

$3x+5-2x < 2x-6-2x$

$x+5 < -6$

$x+5-5 < -6-5$

$x < -11$

The solution set is $\{x|x < -11\}$.

31. $-4y < 28$

$\dfrac{-4y}{-4} > \dfrac{28}{-4} \quad$ Reversing the inequality symbol

$y > -7$

The solution set is $\{y|y > -7\}$.

32. $4-8x < 13+3x$

$4-8x-3x < 13+3x-3x$

$4-11x < 13$

$4-11x-4 < 13-4$

$-11x < 9$

$\dfrac{-11x}{-11} > \dfrac{9}{-11} \quad$ Reversing the inequality symbol

$x > -\dfrac{9}{11}$

The solution set is $\left\{x \middle| x > -\dfrac{9}{11}\right\}$.

33. $-4x \le \dfrac{1}{3}$

$-\dfrac{1}{4} \cdot (-4x) \ge -\dfrac{1}{4} \cdot \dfrac{1}{3} \quad$ Reversing the inequality symbol

$x \ge -\dfrac{1}{12}$

The solution set is $\left\{x \middle| x \ge -\dfrac{1}{12}\right\}$.

34.
$$4x - 6 < x + 3$$
$$4x - 6 - x < x + 3 - x$$
$$3x - 6 < 3$$
$$3x - 6 + 6 < 3 + 6$$
$$3x < 9$$
$$\frac{3x}{3} < \frac{9}{3}$$
$$x < 3$$

The solution set is $\{x | x < 3\}$. The graph is as follows:

35. In order to be a solution of $-2 < x \le 5$, a number must be a solution of both $-2 < x$ and $x \le 5$. The solution set is graphed as follows:

36. The solutions of $y > 0$ are those numbers greater than 0. The graph is as follows:

37.
$$C = \pi d$$
$$\frac{C}{\pi} = \frac{\pi d}{\pi}$$
$$\frac{C}{\pi} = d$$

38.
$$V = \frac{1}{3}Bh$$
$$3 \cdot V = 3 \cdot \frac{1}{3}Bh$$
$$3V = Bh$$
$$\frac{3V}{h} = \frac{Bh}{h}$$
$$\frac{3V}{h} = B$$

39.
$$A = \frac{a + b}{2}$$
$$2 \cdot A = 2 \cdot \left(\frac{a+b}{2}\right)$$
$$2A = a + b$$
$$2A - b = a + b - b$$
$$2A - b = a$$

40.
$$y = mx + b$$
$$y - b = mx + b - b$$
$$y - b = mx$$
$$\frac{y - b}{m} = \frac{mx}{m}$$
$$\frac{y - b}{m} = x$$

41. *Familiarize.* Let $w =$ the width, in miles. Then $w + 90 =$ the length. Recall that the perimeter P of a rectangle with length l and width w is given by $P = 2l + 2w$.

Translate. Substitute 1280 for P and $w + 90$ for l in the formula above.
$$P = 2l + 2w$$
$$1280 = 2(w + 90) + 2w$$

Solve. We solve the equation.
$$1280 = 2(w + 90) + 2w$$
$$1280 = 2w + 180 + 2w$$
$$1280 = 4w + 180$$
$$1280 - 180 = 4w + 180 - 180$$
$$1100 = 4w$$
$$\frac{1100}{4} = \frac{4w}{4}$$
$$275 = w$$

If $w = 275$, then $w + 90 = 275 + 90 = 365$.

Check. The length, 365 mi, is 90 mi more than the width, 275 mi. The perimeter is $2 \cdot 365 \text{ mi} + 2 \cdot 275 \text{ mi} = 730 \text{ mi} + 550 \text{ mi} = 1280 \text{ mi}$. The answer checks.

State. The length is 365 mi, and the width is 275 mi.

42. *Familiarize.* Let $x =$ the number on the first marker. Then $x + 1 =$ the number on the second marker.

Translate.

First number plus second number is 691.
$$x + (x + 1) = 691$$

Solve. We solve the equation.
$$x + (x + 1) = 691$$
$$2x + 1 = 691$$
$$2x + 1 - 1 = 691 - 1$$
$$2x = 690$$
$$\frac{2x}{2} = \frac{690}{2}$$
$$x = 345$$

If $x = 345$, then $x + 1 = 345 + 1 = 346$.

Check. 345 and 346 are consecutive integers and $345 + 346 = 691$. The answer checks.

State. The numbers on the markers are 345 and 346.

43. *Familiarize.* Let $c =$ the cost of the entertainment center in February.

Translate.

Cost in February plus \$332 is Cost in June
$$c + 332 = 2449$$

Solve. We solve the equation.
$$c + 332 = 2449$$
$$c + 332 - 332 = 2449 - 332$$
$$c = 2117$$

Check. $\$2117 + \$332 = \$2449$, so the answer checks.

State. The entertainment center cost \$2117 in February.

44. *Familiarize*. Let a = the number of appliances Ty sold.

***Translate*.**

Commission per appliance	times	number sold	is	Total commission
↓	↓	↓	↓	↓
4	×	a	=	108

***Solve*.** We solve the equation.

$$4 \cdot a = 108$$
$$\frac{4 \cdot a}{4} = \frac{108}{4}$$
$$a = 27$$

***Check*.** $\$4 \cdot 27 = \$108$, so the answer checks.

***State*.** Ty sold 27 appliances.

45. *Familiarize*. Let x = the measure of the first angle. Then $x + 50$ = the measure of the second angle, and $2x - 10$ = the measure of the third angle. Recall that the sum of measures of the angles of a triangle is $180°$.

***Translate*.**

Measure of first angle	+	measure of second angle	+	measure of third angle	is	$180°$.
↓	↓	↓	↓	↓	↓	↓
x	+	$(x+50)$	+	$(2x-10)$	=	180

***Solve*.** We solve the equation.

$$x + (x+50) + (2x-10) = 180$$
$$4x + 40 = 180$$
$$4x + 40 - 40 = 180 - 40$$
$$4x = 140$$
$$\frac{4x}{4} = \frac{140}{4}$$
$$x = 35$$

If $x = 35$, then $x + 50 = 35 + 50 = 85$ and $2x - 10 = 2 \cdot 35 - 10 = 70 - 10 = 60$.

***Check*.** The measure of the second angle is $50°$ more than the measure of the first angle, and the measure of the third angle is $10°$ less than twice the measure of the first angle. The sum of the measure is $35° + 85° + 60° = 180°$. The answer checks.

***State*.** The measures of the angles are $35°$, $85°$, and $60°$.

46. *Translate*.

What number	is	20%	of	75?
↓	↓	↓	↓	↓
a	=	20%	·	75

***Solve*.** We convert 20% to decimal notation and multiply.

$$a = 20\% \cdot 75$$
$$a = 0.2 \cdot 75$$
$$a = 15$$

Thus, 15 is 20% of 75.

47. *Translate*.

15	is	what percent	of	80?
↓	↓	↓	↓	↓
15	=	p	·	80

***Solve*.** We solve the equation.

$$15 = p \cdot 80$$
$$\frac{15}{80} = \frac{p \cdot 80}{80}$$
$$0.1875 = p$$
$$18.75\% = p$$

Thus, 15 is 18.75% of 80.

48. *Translate*.

18	is	3%	of	what number?
↓	↓	↓	↓	↓
18	=	3%	·	b

***Solve*.** We solve the equation.

$$18 = 3\% \cdot b$$
$$18 = 0.03 \cdot b$$
$$\frac{18}{0.03} = \frac{0.03 \cdot b}{0.03}$$
$$600 = b$$

Thus, 18 is 3% of 600.

49. We subtract to find the increase, in thousands.

$$1141 - 905 = 236$$

The increase is 236 thousand.

Now we find the percent of increase.

236	is	what percent	of	905?
↓	↓	↓	↓	↓
236	=	p	·	905

We divide by 905 on both sides and then convert to percent notation.

$$236 = p \cdot 905$$
$$\frac{236}{905} = \frac{p \cdot 905}{905}$$
$$0.26 \approx p$$
$$26\% \approx p$$

The percent of increase is about 26%.

50. *Familiarize*. Let p = the price before the reduction.

***Translate*.**

Price before reduction	minus	30%	of	price	is	$154.
↓	↓	↓	↓	↓		↓
p	−	30%	·	p	=	154

***Solve*.** We solve the equation.

$$p - 30\% \cdot p = 154$$
$$p - 0.3p = 154$$
$$0.7p = 154$$
$$\frac{0.7p}{0.7} = \frac{154}{0.7}$$
$$p = 220$$

***Check*.** 30% of $220 is $0.3 \cdot \$220 = \$66$ and $\$220 - \$66 = \$154$, so the answer checks.

***State*.** The price before the reduction was $220.

51. Familiarize. Let $s =$ the previous salary.

Translate.

$$\underbrace{\text{Previous salary}}_{s} \underbrace{\text{plus}}_{+} \underbrace{\text{15% of}}_{15\% \cdot} \underbrace{\text{previous salary}}_{s} \underbrace{\text{is}}_{=} \underbrace{\$61,410.}_{61,410}$$

Solve. We solve the equation.

$$s + 15\% \cdot s = 61,410$$
$$s + 0.15s = 61,410$$
$$1.15s = 61,410$$
$$\frac{1.15s}{1.15} = \frac{61,410}{1.15}$$
$$s = 53,400$$

Check. 15% of $\$53,400 = 0.15 \cdot \$53,400 = \$8010$ and $\$53,400 + \$8010 = \$61,410$, so the answer checks.

State. The previous salary was $53,400.

52. Familiarize. Let $a =$ the amount the charity actually owes. This is the price of the pump without sales tax added. Then the incorrect amount is $a + 5\%$ of a, or $a + 0.05a$, or $1.05a$.

Translate.

$$\underbrace{\text{Incorrect amount}}_{1.05a} \underbrace{\text{is}}_{=} \underbrace{\$145.90.}_{145.90}$$

Solve. We solve the equation.

$$1.05a = 145.90$$
$$\frac{1.05a}{1.05} = \frac{145.90}{1.05}$$
$$a \approx 138.95$$

Check. 5% of $\$138.95$ is $0.05 \cdot \$138.95 \approx \$6.95$, and $\$138.95 + \$6.95 = \$145.90$, so the answer checks.

State. The charity actually owes $138.95.

53. Familiarize. Let s represent the score on the next test.

Translate.

$$\underbrace{\text{The average score}}_{\frac{71+75+82+86+s}{5}} \underbrace{\text{is at least}}_{\geq} \underbrace{80.}_{80}$$

Solve.

$$\frac{71+75+82+86+s}{5} \geq 80$$
$$5\left(\frac{71+75+82+86+s}{5}\right) \geq 5 \cdot 80$$
$$71+75+82+86+s \geq 400$$
$$314 + s \geq 400$$
$$s \geq 86$$

Check. As a partial check we show that the average is at least 80 when the next test score is 86.

$$\frac{71+75+82+86+86}{5} = \frac{400}{5} = 80$$

State. The lowest grade you can get on the next test and have an average test score of 80 is 86.

54. Familiarize. Let w represent the width of the rectangle, in cm. The perimeter is given by $P = 2l+2w$, or $2\cdot43+2w$, or $86+2w$.

Translate.

$$\underbrace{\text{The perimeter}}_{86+2w} \underbrace{\text{is greater than}}_{>} \underbrace{120 \text{ cm}.}_{120}$$

Solve.

$$86 + 2w > 120$$
$$2w > 34$$
$$w > 17$$

Check. We check to see if the solution seems reasonable.

When $w = 16$ cm, $P = 2 \cdot 43 + 2 \cdot 16$, or 118 cm.

When $w = 17$ cm, $P = 2 \cdot 43 + 2 \cdot 17$, or 120 cm.

When $w = 18$ cm, $P = 2 \cdot 43 + 2 \cdot 18$, or 122 cm.

It appears that the solution is correct.

State. The solution set is $\{w | w > 17 \text{ cm}\}$.

55. Discussion and Writing Exercise. The end result is the same either way. If s is the original salary, the new salary after a 5% raise followed by an 8% raise is $1.08(1.05s)$. If the raises occur in the opposite order, the new salary is $1.05(1.08s)$. By the commutative and associate laws of multiplication, we see that these are equal. However, it would be better to receive the 8% raise first, because this increase yields a higher salary initially than a 5% raise.

56. Discussion and Writing Exercise. The inequalities are equivalent by the multiplication principle for inequalities. If we multiply both sides of one inequality by -1, the other inequality results.

57.
$$2|x| + 4 = 50$$
$$2|x| = 46$$
$$|x| = 23$$

The solutions are the numbers whose distance from 0 is 23. Those numbers are -23 and 23.

58. $|3x| = 60$

The solutions are the values of x for which the distance of $3 \cdot x$ from 0 is 60. Then we have:

$$3x = -60 \quad or \quad 3x = 60$$
$$x = -20 \quad or \quad x = 20$$

The solutions are -20 and 20.

59.
$$y = 2a - ab + 3$$
$$y - 3 = 2a - ab$$
$$y - 3 = a(2 - b)$$
$$\frac{y-3}{2-b} = a$$

Chapter 2 Test

1.
$$x + 7 = 15$$
$$x + 7 - 7 = 15 - 7$$
$$x = 8$$
The solution is 8.

2.
$$t - 9 = 17$$
$$t - 9 + 9 = 17 + 9$$
$$t = 26$$
The solution is 26.

3.
$$3x = -18$$
$$\frac{3x}{3} = \frac{-18}{3}$$
$$x = -6$$
The solution is -6.

4.
$$-\frac{4}{7}x = -28$$
$$-\frac{7}{4} \cdot \left(-\frac{4}{7}x\right) = -\frac{7}{4} \cdot (-28)$$
$$x = \frac{7 \cdot \cancel{4} \cdot 7}{\cancel{4} \cdot 1}$$
$$x = 49$$
The solution is 49.

5.
$$3t + 7 = 2t - 5$$
$$3t + 7 - 2t = 2t - 5 - 2t$$
$$t + 7 = -5$$
$$t + 7 - 7 = -5 - 7$$
$$t = -12$$
The solution is -12.

6.
$$\frac{1}{2}x - \frac{3}{5} = \frac{2}{5}$$
$$\frac{1}{2}x - \frac{3}{5} + \frac{3}{5} = \frac{2}{5} + \frac{3}{5}$$
$$\frac{1}{2}x = 1$$
$$2 \cdot \frac{1}{2}x = 2 \cdot 1$$
$$x = 2$$
The solution is 2.

7.
$$8 - y = 16$$
$$8 - y - 8 = 16 - 8$$
$$-y = 8$$
$$-1 \cdot (-1 \cdot y) = -1 \cdot 8$$
$$y = -8$$
The solution is -8.

8.
$$-\frac{2}{5} + x = -\frac{3}{4}$$
$$-\frac{2}{5} + x + \frac{2}{5} = -\frac{3}{4} + \frac{2}{5}$$
$$x = -\frac{15}{20} + \frac{8}{20}$$
$$x = -\frac{7}{20}$$
The solution is $-\frac{7}{20}$.

9.
$$3(x + 2) = 27$$
$$3x + 6 = 27$$
$$3x + 6 - 6 = 27 - 6$$
$$3x = 21$$
$$\frac{3x}{3} = \frac{21}{3}$$
$$x = 7$$
The solution is 7.

10.
$$-3x - 6(x - 4) = 9$$
$$-3x - 6x + 24 = 9$$
$$-9x + 24 = 9$$
$$-9x + 24 - 24 = 9 - 24$$
$$-9x = -15$$
$$\frac{-9x}{-9} = \frac{-15}{-9}$$
$$x = \frac{\cancel{3} \cdot 5}{\cancel{3} \cdot 3}$$
$$x = \frac{5}{3}$$
The solution is $\frac{5}{3}$.

11. We multiply by 10 to clear the decimals.
$$0.4p + 0.2 = 4.2p - 7.8 - 0.6p$$
$$10(0.4p + 0.2) = 10(4.2p - 7.8 - 0.6p)$$
$$4p + 2 = 42p - 78 - 6p$$
$$4p + 2 = 36p - 78$$
$$4p + 2 - 36p = 36p - 78 - 36p$$
$$-32p + 2 = -78$$
$$-32p + 2 - 2 = -78 - 2$$
$$-32p = -80$$
$$\frac{-32p}{-32} = \frac{-80}{-32}$$
$$p = \frac{5 \cdot \cancel{16}}{2 \cdot \cancel{16}}$$
$$p = \frac{5}{2}$$
The solution is $\frac{5}{2}$.

12. $4(3x - 1) + 11 = 2(6x + 5) - 8$

$12x - 4 + 11 = 12x + 10 - 8$

$12x + 7 = 12x + 2$

$12x + 7 - 12x = 12x + 2 - 12x$

$7 = 2$ FALSE

There are no solutions.

13. $-2 + 7x + 6 = 5x + 4 + 2x$

$7x + 4 = 7x + 4$

$7x + 4 - 7x = 7x + 4 - 7x$

$4 = 4$ TRUE

All real numbers are solutions.

14. $x + 6 \leq 2$

$x + 6 - 6 \leq 2 - 6$

$x \leq -4$

The solution set is $\{x | x \leq -4\}$.

15. $14x + 9 > 13x - 4$

$14x + 9 - 13x > 13x - 4 - 13x$

$x + 9 > -4$

$x + 9 - 9 > -4 - 9$

$x > -13$

The solution set is $\{x | x > -13\}$.

16. $12x \leq 60$

$\dfrac{12x}{12} \leq \dfrac{60}{12}$

$x \leq 5$

The solution set is $\{x | x \leq 5\}$.

17. $-2y \geq 26$

$\dfrac{-2y}{-2} \leq \dfrac{26}{-2}$ Reversing the inequality symbol

$y \leq -13$

The solution set is $\{y | y \leq -13\}$.

18. $-4y \leq -32$

$\dfrac{-4y}{-4} \geq \dfrac{-32}{-4}$ Reversing the inequality symbol

$y \geq 8$

The solution set is $\{y | y \geq 8\}$.

19. $-5x \geq \dfrac{1}{4}$

$-\dfrac{1}{5} \cdot (-5x) \leq -\dfrac{1}{5} \cdot \dfrac{1}{4}$ Reversing the inequality symbol

$x \leq -\dfrac{1}{20}$

The solution set is $\left\{ x \middle| x \leq -\dfrac{1}{20} \right\}$.

20. $4 - 6x > 40$

$4 - 6x - 4 > 40 - 4$

$-6x > 36$

$\dfrac{-6x}{-6} < \dfrac{36}{-6}$ Reversing the inequality symbol

$x < -6$

The solution set is $\{x | x < -6\}$.

21. $5 - 9x \geq 19 + 5x$

$5 - 9x - 5x \geq 19 + 5x - 5x$

$5 - 14x \geq 19$

$5 - 14x - 5 \geq 19 - 5$

$-14x \geq 14$

$\dfrac{-14x}{-14} \leq \dfrac{14}{-14}$ Reversing the inequality symbol

$x \leq -1$

The solution set is $\{x | x \leq -1\}$.

22. The solutions of $y \leq 9$ are shown by shading the point for 9 and all points to the left of 9. The closed circle at 9 indicates that 9 is part of the graph.

23. $6x - 3 < x + 2$

$6x - 3 - x < x + 2 - x$

$5x - 3 < 2$

$5x - 3 + 3 < 2 + 3$

$5x < 5$

$\dfrac{5x}{5} < \dfrac{5}{5}$

$x < 1$

The solution set is $\{x | x < 1\}$. The graph is as follows:

24. In order to be a solution of the inequality $-2 \leq x \leq 2$, a number must be a solution of both $-2 \leq x$ and $x \leq 2$. The solution set is graphed as follows:

25. *Translate.*

What number is 24% of 75?

$a = 24\% \cdot 75$

Solve. We convert 24% to decimal notation and multiply.

$a = 24\% \cdot 75$

$a = 0.24 \cdot 75$

$a = 18$

Thus, 18 is 24% of 75.

26. *Translate.*

15.84   is   what percent   of   96?

$\downarrow \quad \downarrow \qquad \downarrow \qquad \downarrow \quad \downarrow$

15.84 $=$ p $\cdot$ 96

Solve.

$$15.84 = p \cdot 96$$
$$\frac{15.84}{96} = \frac{p \cdot 96}{96}$$
$$0.165 = p$$
$$16.5\% = p$$

Thus, 15.84 is 16.5% of 96.

27. *Translate.*

800   is   2%   of   what number?

$\downarrow \quad \downarrow \quad \downarrow \quad \downarrow \qquad \downarrow$

800   $=$ 2% $\cdot$ b

Solve.

$$800 = 2\% \cdot b$$
$$800 = 0.02 \cdot b$$
$$\frac{800}{0.02} = \frac{0.02 \cdot b}{0.02}$$
$$40,000 = b$$

Thus, 800 is 2% of 40,000.

28. We subtract to find the increase.

$$89,000 - 58,000 = 31,000$$

Now we find the percent of increase.

31,000   is   what percent   of   58,000?

$\downarrow \qquad \downarrow \qquad \downarrow \qquad \downarrow \qquad \downarrow$

31,000 $=$ p $\cdot$ 58,000

We divide by 58,000 on both sides and then convert to percent notation.

$$31,000 = p \cdot 58,000$$
$$\frac{31,000}{58,000} = \frac{p \cdot 58,000}{58,000}$$
$$0.534 \approx p$$
$$53.4\% \approx p$$

The percent of increase is about 53.4%.

29. Familiarize. Let $w =$ the width of the photograph, in cm. Then $w + 4 =$ the length. Recall that the perimeter P of a rectangle with length l and width w is given by $P = 2l + 2w$.

Translate. We substitute 36 for P and $w + 4$ for l in the formula above.

$$P = 2l + 2w$$
$$36 = 2(w + 4) + 2w$$

Solve. We solve the equation.

$$36 = 2(w + 4) + 2w$$
$$36 = 2w + 8 + 2w$$
$$36 = 4w + 8$$
$$36 - 8 = 4w + 8 - 8$$
$$28 = 4w$$
$$\frac{28}{4} = \frac{4w}{4}$$
$$7 = w$$

If $w = 7$, then $w + 4 = 7 + 4 = 11$.

Check. The length, 11 cm, is 4 cm more than the width, 7 cm. The perimeter is $2 \cdot 11$ cm $+ 2 \cdot 7$ cm $= 22$ cm $+ 14$ cm $= 36$ cm. The answer checks.

State. The width is 7 cm, and the length is 11 cm.

30. Familiarize. Let $c =$ the amount that was given to charities in general in 2003, in billions of dollars.

Translate.

$86.4 billion   is   35.9%   of   what?

$\downarrow \qquad \quad \downarrow \quad \downarrow \quad \downarrow \quad \downarrow$

86.4 $=$ 35.9% $\cdot$ c

Solve. We solve the equation.

$$86.4 = 35.9\% \cdot c$$
$$86.4 = 0.359 \cdot c$$
$$\frac{86.4}{0.359} = \frac{0.359 \cdot c}{0.359}$$
$$240.7 \approx c$$

Check. 35.9% of $240.7 billion is $0.359(\$240.7$ billion$) \approx$ $86.4 billion, so the answer checks.

State. In 2003 about $240.7 billion was given to charities.

31. Familiarize. Let $x =$ the first integer. Then $x + 1 =$ the second and $x + 2 =$ the third.

Translate.

First integer   plus   second integer   plus   third integer   is 7530.

$\downarrow \qquad \downarrow \qquad \downarrow \qquad \downarrow \qquad \downarrow \qquad \downarrow \quad \downarrow$

x $+$ $(x + 1)$ $+$ $(x + 2)$ $=$ 7530

Solve.

$$x + (x + 1) + (x + 2) = 7530$$
$$3x + 3 = 7530$$
$$3x + 3 - 3 = 7530 - 3$$
$$3x = 7527$$
$$\frac{3x}{3} = \frac{7527}{3}$$
$$x = 2509$$

If $x = 2509$, then $x + 1 = 2510$ and $x + 2 = 2511$.

Check. The numbers 2509, 2510, and 2511 are consecutive integers and $2509 + 2510 + 2511 = 7530$. The answer checks.

State. The integers are 2509, 2510, and 2511.

32. *Familiarize.* Let $x =$ the amount originally invested. Using the formula for simple interest, $I = Prt$, the interest earned in one year will be $x \cdot 5\% \cdot 1$, or $5\%x$.

Translate.

$$\underbrace{\text{Amount invested}}_{x} \quad \underbrace{\text{plus}}_{+} \quad \underbrace{\text{interest}}_{5\%x} \quad \underbrace{\text{is}}_{=} \quad \underbrace{\text{amount after 1 year.}}_{924}$$

Solve. We solve the equation.
$$x + 5\%x = 924$$
$$x + 0.05x = 924$$
$$1.05x = 924$$
$$\frac{1.05x}{1.05} = \frac{924}{1.05}$$
$$x = 880$$

Check. 5% of \$880 is $0.05 \cdot \$880 = \$44$ and $\$880 + \$44 = \$924$, so the answer checks.

State. \$880 was originally invested.

33. *Familiarize.* Using the labels on the drawing in the text, we let $x =$ the length of the shorter piece, in meters, and $x + 2 =$ the length of the longer piece.

Translate.

$$\underbrace{\text{Length of shorter piece}}_{x} \quad \underbrace{\text{plus}}_{+} \quad \underbrace{\text{length of longer piece}}_{(x+2)} \quad \underbrace{\text{is}}_{=} \quad \underbrace{8\text{ m}}_{8}.$$

Solve. We solve the equation.
$$x + (x + 2) = 8$$
$$2x + 2 = 8$$
$$2x + 2 - 2 = 8 - 2$$
$$2x = 6$$
$$\frac{2x}{2} = \frac{6}{2}$$
$$x = 3$$
If $x = 3$, then $x + 2 = 3 + 2 = 5$.

Check. One piece is 2 m longer than the other and the sum of the lengths is 3 m$+5$ m, or 8 m. The answer checks.

State. The lengths of the pieces are 3 m and 5 m.

34. *Familiarize.* Let $l =$ the length of the rectangle, in yd. The perimeter is given by $P = 2l + 2w$, or $2l + 2 \cdot 96$, or $2l + 192$.

Translate.

$$\underbrace{\text{The perimeter}}_{2l + 192} \quad \underbrace{\text{is at least}}_{\geq} \quad \underbrace{540\text{ yd}}_{540}.$$

Solve.
$$2l + 192 \geq 540$$
$$2l \geq 348$$
$$l \geq 174$$

Check. We check to see if the solution seems reasonable.

When $l = 174$ yd, $P = 2 \cdot 174 + 2 \cdot 96$, or 540 yd.

When $l = 175$ yd, $P = 2 \cdot 175 + 2 \cdot 96$, or 542 yd.

It appears that the solution is correct.

State. For lengths that are at least 174 yd, the perimeter will be at least 540 yd. The solution set can be expressed as $\{l | l \geq 174 \text{ yd}\}$.

35. *Familiarize.* Let $s =$ the amount Jason spends in the sixth month.

Translate.

$$\underbrace{\text{Average spending}}_{\dfrac{98 + 89 + 110 + 85 + 83 + s}{6}} \quad \underbrace{\text{is no more than}}_{\leq} \quad \underbrace{\$95.}_{95}$$

Solve.
$$\frac{98 + 89 + 110 + 85 + 83 + s}{6} \leq 95$$
$$6\left(\frac{98 + 89 + 110 + 85 + 83 + s}{6}\right) \leq 6 \cdot 95$$
$$98 + 89 + 110 + 85 + 83 + s \leq 570$$
$$465 + s \leq 570$$
$$s \leq 105$$

Check. As a partial check we show that the average spending is \$95 when Jason spends \$105 in the sixth month.
$$\frac{98 + 89 + 110 + 85 + 83 + 105}{6} = \frac{570}{6} = 95$$

State. Jason can spend no more than \$105 in the sixth month. The solution set can be expressed as $\{s | s \leq \$105\}$.

36. *Familiarize.* Let $c =$ the number of copies made. For 3 months, the rental charge is $3 \cdot \$225$, or \$675. Expressing 1.2¢ as \$0.012, the charge for the copies is given by $\$0.012 \cdot c$.

Translate.

$$\underbrace{\text{Rental charge}}_{675} \quad \underbrace{\text{plus}}_{+} \quad \underbrace{\text{copy charge}}_{0.012c} \quad \underbrace{\text{is no more than}}_{\leq} \quad \underbrace{\$2400.}_{2400}$$

Solve.
$$675 + 0.012c \leq 2400$$
$$0.012c \leq 1725$$
$$c \leq 143,750$$

Check. We check to see if the solution seems reasonable.

When $c = 143,749$, the total cost is $\$675 + \$0.012(143,749)$, or about \$2399.99.

When $c = 143,750$, the total cost is $\$675 + \$0.012(143,750)$, or about \$2400.

It appears that the solution is correct.

State. No more than 143,750 copies can be made. The solution set can be expressed as $\{c | c \leq 143,750\}$.

37. $A = 2\pi rh$

$$\frac{A}{2\pi h} = \frac{2\pi rh}{2\pi h}$$

$$\frac{A}{2\pi h} = r$$

38. $y = 8x + b$

$y - b = 8x + b - b$

$y - b = 8x$

$$\frac{y-b}{8} = \frac{8x}{8}$$

$$\frac{y-b}{8} = x$$

39. $$c = \frac{1}{a-d}$$

$$(a-d)\cdot c = a - d \cdot \left(\frac{1}{a-d}\right)$$

$$ac - dc = 1$$

$$ac - dc - ac = 1 - ac$$

$$-dc = 1 - ac$$

$$\frac{-dc}{-c} = \frac{1-ac}{-c}$$

$$d = \frac{1-ac}{-c}$$

Since $\dfrac{1-ac}{-c} = \dfrac{-1}{-1}\cdot\dfrac{1-ac}{-c} = \dfrac{-1(1-ac)}{-1(-c)} = \dfrac{-1+ac}{c}$, or $\dfrac{ac-1}{c}$, we can also express the result as $d = \dfrac{ac-1}{c}$.

40. $3|w| - 8 = 37$

$3|w| = 45$

$|w| = 15$

The solutions are the numbers whose distance from 0 is 15. They are -15 and 15.

41. Familiarize. Let $t =$ the number of tickets given away.

Translate. We add the number of tickets given to the five people.

$$\frac{1}{3}t + \frac{1}{4}t + \frac{1}{5}t + 8 + 5 = t$$

Solve.

$$\frac{1}{3}t + \frac{1}{4}t + \frac{1}{5}t + 8 + 5 = t$$

$$\frac{20}{60}t + \frac{15}{60}t + \frac{12}{60}t + 8 + 5 = t$$

$$\frac{47}{60}t + 13 = t$$

$$13 = t - \frac{47}{60}t$$

$$13 = \frac{60}{60}t - \frac{47}{60}t$$

$$13 = \frac{13}{60}t$$

$$\frac{60}{13}\cdot 13 = \frac{60}{13}\cdot\frac{13}{60}t$$

$$60 = t$$

Check. $\dfrac{1}{3}\cdot 60 = 20,\ \dfrac{1}{4}\cdot 60 = 15,\ \dfrac{1}{5}\cdot 60 = 12$; then $20 + 15 + 12 + 8 + 5 = 60$. The answer checks.

State. 60 tickets were given away.

Cumulative Review Chapters 1 - 2

1. $\dfrac{y-x}{4} = \dfrac{12-6}{4} = \dfrac{6}{4} = \dfrac{\cancel{2}\cdot 3}{\cancel{2}\cdot 2} = \dfrac{3}{2}$

2. $\dfrac{3x}{y} = \dfrac{3\cdot 5}{4} = \dfrac{15}{4}$

3. $x - 3 = 3 - 3 = 0$

4. $2w - 4$

5. Since -4 is to the right of -6, we have $-4 > -6$.

6. Since 0 is to the right of -5, we have $0 > -5$.

7. Since -8 is to the left of 7, we have $-8 < 7$.

8. The opposite of $\dfrac{2}{5}$ is $-\dfrac{2}{5}$ because $\dfrac{2}{5} + \left(-\dfrac{2}{5}\right) = 0$.

The reciprocal of $\dfrac{2}{5}$ is $\dfrac{5}{2}$ because $\dfrac{2}{5}\cdot\dfrac{5}{2} = 1$.

9. The distance of 3 from 0 is 3, so $|3| = 3$.

10. The distance of $-\dfrac{3}{4}$ from 0 is $\dfrac{3}{4}$, so $\left|-\dfrac{3}{4}\right| = \dfrac{3}{4}$.

11. The distance of 0 from 0 is 0, so $|0| = 0$.

12. $-6.7 + 2.3$

One negative number and one positive number. The absolute values are 6.7 and 2.3. The difference of the absolute values is $6.7 - 2.3 = 4.4$. The negative number has the large absolute value, so the sum is negative.

$-6.7 + 2.3 = -4.4$

13. $-\dfrac{1}{6} - \dfrac{7}{3} = -\dfrac{1}{6} + \left(-\dfrac{7}{3}\right) = -\dfrac{1}{6} + \left(-\dfrac{14}{6}\right) = -\dfrac{15}{6} = -\dfrac{\cancel{3}\cdot 5}{2\cdot\cancel{3}} = -\dfrac{5}{2}$

14. $-\dfrac{5}{8}\left(-\dfrac{4}{3}\right) = \dfrac{5\cdot 4}{8\cdot 3} = \dfrac{5\cdot\cancel{4}}{2\cdot\cancel{4}\cdot 3} = \dfrac{5}{6}$

15. $(-7)(5)(-6)(-0.5) = -35(3) = -105$

16. $81 \div (-9) = -9$

17. $-10.8 \div 3.6 = -3$

18. $-\dfrac{4}{5} \div -\dfrac{25}{8} = -\dfrac{4}{5}\cdot -\dfrac{8}{25} = \dfrac{4\cdot 8}{5\cdot 25} = \dfrac{32}{125}$

19. $5(3x + 5y + 2z) = 5\cdot 3x + 5\cdot 5y + 5\cdot 2z = 15x + 25y + 10z$

20. $4(-3x - 2) = 4(-3x) - 4\cdot 2 = -12x - 8$

21. $-6(2y - 4x) = -6\cdot 2y - (-6)(4x) = -12y - (-24x) = -12y + 24x$

22. $64 + 18x + 24y = 2 \cdot 32 + 2 \cdot 9x + 2 \cdot 12y = 2(32 + 9x + 12y)$

23. $16y - 56 = 8 \cdot 2y - 8 \cdot 7 = 8(2y - 7)$

24. $5a - 15b + 25 = 5 \cdot a - 5 \cdot 3b + 5 \cdot 5 = 5(a - 3b + 5)$

25. $\begin{aligned} 9b + 18y + 6b + 4y &= 9b + 6b + 18y + 4y \\ &= (9 + 6)b + (18 + 4)y \\ &= 15b + 22y \end{aligned}$

26. $\begin{aligned} 3y + 4 + 6z + 6y &= 3y + 6y + 4 + 6z \\ &= (3 + 6)y + 4 + 6z \\ &= 9y + 4 + 6z \end{aligned}$

27. $\begin{aligned} -4d - 6a + 3a - 5d + 1 &= -4d - 5d - 6a + 3a + 1 \\ &= (-4 - 5)d + (-6 + 3)a + 1 \\ &= -9d - 3a + 1 \end{aligned}$

28. $\begin{aligned} 3.2x + 2.9y - 5.8x - 8.1y &= 3.2x - 5.8x + 2.9y - 8.1y \\ &= (3.2 - 5.8)x + (2.9 - 8.1)y \\ &= -2.6x - 5.2y \end{aligned}$

29. $7 - 2x - (-5x) - 8 = 7 - 2x + 5x - 8 = -1 + 3x$

30. $-3x - (-x + y) = -3x + x - y = -2x - y$

31. $-3(x - 2) - 4x = -3x + 6 - 4x = -7x + 6$

32. $10 - 2(5 - 4x) = 10 - 10 + 8x = 8x$

33. $\begin{aligned} &[3(x + 6) - 10] - [5 - 2(x - 8)] \\ &= [3x + 18 - 10] - [5 - 2x + 16] \\ &= [3x + 8] - [21 - 2x] \\ &= 3x + 8 - 21 + 2x \\ &= 5x - 13 \end{aligned}$

34. $\begin{aligned} x + 1.75 &= 6.25 \\ x + 1.75 - 1.75 &= 6.25 - 1.75 \\ x &= 4.5 \end{aligned}$

The solution is 4.5.

35. $\begin{aligned} \frac{5}{2}y &= \frac{2}{5} \\ \frac{2}{5} \cdot \frac{5}{2}y &= \frac{2}{5} \cdot \frac{2}{5} \\ y &= \frac{4}{25} \end{aligned}$

The solution is $\frac{4}{25}$.

36. $\begin{aligned} -2.6 + x &= 8.3 \\ -2.6 + x + 2.6 &= 8.3 + 2.6 \\ x &= 10.9 \end{aligned}$

The solution is 10.9.

37. $\begin{aligned} 4\frac{1}{2} + y &= 8\frac{1}{3} \\ 4\frac{1}{2} + y - 4\frac{1}{2} &= 8\frac{1}{3} - 4\frac{1}{2} \\ y &= 8\frac{2}{6} - 4\frac{3}{6} \\ y &= 7\frac{8}{6} - 4\frac{3}{6} \quad \left(8\frac{2}{6} = 7 + 1\frac{2}{6} = 7 + \frac{8}{6} = 7\frac{8}{6}\right) \\ y &= 3\frac{5}{6} \end{aligned}$

The solution is $3\frac{5}{6}$.

38. $\begin{aligned} -\frac{3}{4}x &= 36 \\ -\frac{4}{3}\left(-\frac{3}{4}x\right) &= -\frac{4}{3} \cdot 36 \\ x &= -\frac{4 \cdot 36}{3} = -\frac{4 \cdot \cancel{3} \cdot 12}{\cancel{3} \cdot 1} \\ x &= -48 \end{aligned}$

The solution is -48.

39. $\begin{aligned} -2.2y &= -26.4 \\ \frac{-2.2y}{-2.2} &= \frac{-26.4}{-2.2} \\ y &= 12 \end{aligned}$

The solution is 12.

40. $\begin{aligned} 5.8x &= -35.96 \\ \frac{5.8x}{5.8} &= \frac{-35.96}{5.8} \\ x &= -6.2 \end{aligned}$

The solution is -6.2.

41. $\begin{aligned} -4x + 3 &= 15 \\ -4x + 3 - 3 &= 15 - 3 \\ -4x &= 12 \\ \frac{-4x}{-4} &= \frac{12}{-4} \\ x &= -3 \end{aligned}$

The solution is -3.

42. $\begin{aligned} -3x + 5 &= -8x - 7 \\ -3x + 5 + 8x &= -8x - 7 + 8x \\ 5x + 5 &= -7 \\ 5x + 5 - 5 &= -7 - 5 \\ 5x &= -12 \\ \frac{5x}{5} &= \frac{-12}{5} \\ x &= -\frac{12}{5} \end{aligned}$

The solution is $-\frac{12}{5}$.

43.
$$4y - 4 + y = 6y + 20 - 4y$$
$$5y - 4 = 2y + 20$$
$$5y - 4 - 2y = 2y + 20 - 2y$$
$$3y - 4 = 20$$
$$3y - 4 + 4 = 20 + 4$$
$$3y = 24$$
$$\frac{3y}{3} = \frac{24}{3}$$
$$y = 8$$

The solution is 8.

44.
$$-3(x - 2) = -15$$
$$-3x + 6 = -15$$
$$-3x + 6 - 6 = -15 - 6$$
$$-3x = -21$$
$$\frac{-3x}{-3} = \frac{-21}{-3}$$
$$x = 7$$

The solution is 7.

45. First we will multiply by the least common multiple of all the denominators to clear the fractions.
$$\frac{1}{3}x - \frac{5}{6} = \frac{1}{2} + 2x$$
$$6\left(\frac{1}{3}x - \frac{5}{6}\right) = 6\left(\frac{1}{2} + 2x\right)$$
$$6 \cdot \frac{1}{3}x - 6 \cdot \frac{5}{6} = 6 \cdot \frac{1}{2} + 6 \cdot 2x$$
$$2x - 5 = 3 + 12x$$
$$2x - 5 - 12x = 3 + 12x - 12x$$
$$-10x - 5 = 3$$
$$-10x - 5 + 5 = 3 + 5$$
$$-10x = 8$$
$$\frac{-10x}{-10} = \frac{8}{-10}$$
$$x = -\frac{8}{10} = -\frac{2 \cdot 4}{2 \cdot 5}$$
$$x = -\frac{4}{5}$$

The solution is $-\frac{4}{5}$.

46. First we will multiply by 10 to clear the decimals.
$$-3.7x + 6.2 = -7.3x - 5.8$$
$$10(-3.7x + 6.2) = 10(-7.3x - 5.8)$$
$$-37x + 62 = -73x - 58$$
$$-37x + 62 + 73x = -73x - 58 + 73x$$
$$36x + 62 = -58$$
$$36x + 62 - 62 = -58 - 62$$
$$36x = -120$$
$$\frac{36x}{36} = \frac{-120}{36}$$
$$x = -\frac{10 \cdot \cancel{12}}{3 \cdot \cancel{12}}$$
$$x = -\frac{10}{3}$$

The solution is $-\frac{10}{3}$.

47.
$$4(x + 2) = 4(x - 2) + 16$$
$$4x + 8 = 4x - 8 + 16$$
$$4x + 8 = 4x + 8$$
$$4x + 8 - 4x = 4x + 8 - 4x$$
$$8 = 8 \qquad \text{TRUE}$$

All real numbers are solutions.

48.
$$0(x + 3) + 4 = 0$$
$$0 + 4 = 0$$
$$4 = 0 \qquad \text{FALSE}$$

There are no solutions.

49.
$$5(7 + x) = (x + 7)5$$
$$35 + 5x = 5x + 35$$
$$35 + 5x - 5x = 5x + 35 - 5x$$
$$35 = 35 \qquad \text{TRUE}$$

All real numbers are solutions.

50.
$$3x - 1 < 2x + 1$$
$$3x - 1 - 2x < 2x + 1 - 2x$$
$$x - 1 < 1$$
$$x - 1 + 1 < 1 + 1$$
$$x < 2$$

The solution set is $\{x | x < 2\}$.

51.
$$5 - y \le 2y - 7$$
$$5 - y - 2y \le 2y - 7 - 2y$$
$$5 - 3y \le -7$$
$$5 - 3y - 5 \le -7 - 5$$
$$-3y \le -12$$
$$\frac{-3y}{-3} \ge \frac{-12}{-3} \qquad \text{Reversing the inequality symbol}$$
$$y \ge 4$$

The solution set is $\{y | y \ge 4\}$.

52.
$$3y + 7 > 5y + 13$$
$$3y + 7 - 5y > 5y + 13 - 5y$$
$$-2y + 7 > 13$$
$$-2y + 7 - 7 > 13 - 7$$
$$-2y > 6$$
$$\frac{-2y}{-2} < \frac{6}{-2} \quad \text{Reversing the inequality symbol}$$
$$y < -3$$

The solution set is $\{y | y < -3\}$.

53.
$$H = 65 - m$$
$$H - 65 = 65 - m - 65$$
$$H - 65 = -m$$
$$-1(H - 65) = -1 \cdot (-1 \cdot m)$$
$$-H + 65 = m, \text{ or}$$
$$65 - H = m$$

54.
$$I = Prt$$
$$\frac{I}{rt} = \frac{Prt}{rt}$$
$$\frac{I}{rt} = P$$

55. *Translate.*

$$\underbrace{\text{What number}}_{\downarrow \atop a} \text{ is } \underset{= 24\%}{\underset{\downarrow}{} } \underset{\cdot}{\underset{\downarrow}{} } \underset{105}{\underset{\downarrow}{} }$$

Solve. We convert 24% to decimal notation and multiply.
$$a = 24\% \cdot 105$$
$$a = 0.24 \cdot 105$$
$$a = 25.2$$

Thus, 25.2 is 24% of 105.

56. *Translate.*

$$39.6 \text{ is } \underbrace{\text{what percent}}_{} \text{ of } 88?$$
$$\underset{39.6}{\underset{\downarrow}{}} \underset{=}{\underset{\downarrow}{}} \qquad \underset{p}{\underset{\downarrow}{}} \qquad \underset{\cdot}{\underset{\downarrow}{}} \underset{88}{\underset{\downarrow}{}}$$

Solve. We solve the equation.
$$39.6 = p \cdot 88$$
$$\frac{39.6}{88} = \frac{p \cdot 88}{88}$$
$$0.45 = p$$
$$45\% = p$$

Thus, 39.6 is 45% of 88.

57. *Translate.*

$$\$163.60 \text{ is } 45\% \text{ of } \underbrace{\text{what number}}_{}?$$
$$\underset{163.60}{\underset{\downarrow}{}} \underset{=}{\underset{\downarrow}{}} \underset{45\%}{\underset{\downarrow}{}} \underset{\cdot}{\underset{\downarrow}{}} \qquad \underset{b}{\underset{\downarrow}{}}$$

Solve.
$$163.60 = 45\% \cdot b$$
$$163.60 = 0.45 \cdot b$$
$$\frac{163.60}{0.45} = \frac{0.45 \cdot b}{0.45}$$
$$363.56 \approx b$$

Thus, \$163.60 is 45% of \$363.56.

58. *Translate.*

$$\text{What is } 60\% \text{ of } \underbrace{291 \text{ million}}_{}?$$
$$\underset{a}{\underset{\downarrow}{}} \underset{=}{\underset{\downarrow}{}} \underset{60\%}{\underset{\downarrow}{}} \underset{\cdot}{\underset{\downarrow}{}} \qquad \underset{291}{\underset{\downarrow}{}}$$

Solve.
$$a = 60\% \cdot 291$$
$$a = 0.6 \cdot 291$$
$$a = 174.6$$

In 2004, 174.6 million people were overweight.

59. *Familiarize.* Let s = Nadia's score on the fourth test.

Translate.

$$\underbrace{\text{The average score}}_{} \underbrace{\text{is at least}}_{} 80.$$
$$\underset{\frac{82 + 76 + 78 + s}{4}}{\underset{\downarrow}{}} \qquad \underset{\geq}{\underset{\downarrow}{}} \qquad \underset{80}{\underset{\downarrow}{}}$$

Solve.
$$\frac{82 + 76 + 78 + s}{4} \geq 80$$
$$4 \left(\frac{82 + 76 + 78 + s}{4} \right) \geq 4 \cdot 80$$
$$82 + 76 + 78 + s \geq 320$$
$$236 + s \geq 320$$
$$s \geq 84$$

Check. As a partial check we show that the average is at least 80 when the fourth test score is 84.
$$\frac{82 + 76 + 78 + 84}{4} = \frac{320}{4} = 80$$

State. Scores greater than or equal to 84 will earn Nadia at least a B. The solution set is $\{s | s \geq 84\}$.

60. *Familiarize.* Let m = the amount Melinda paid for her rollerblades. Then $m + 17$ = the amount Susan paid for hers.

Translate.

$$\underbrace{\begin{array}{c}\text{Amount} \\ \text{Melinda paid}\end{array}}_{} \text{ plus } \underbrace{\begin{array}{c}\text{amount} \\ \text{Susan paid}\end{array}}_{} \text{ is } \$107.$$
$$\underset{m}{\underset{\downarrow}{}} \qquad \underset{+}{\underset{\downarrow}{}} \qquad \underset{(m+17)}{\underset{\downarrow}{}} \qquad \underset{=}{\underset{\downarrow}{}} \underset{107}{\underset{\downarrow}{}}$$

Solve.
$$m + (m + 17) = 107$$
$$2m + 17 = 107$$
$$2m + 17 - 17 = 107 - 17$$
$$2m = 90$$
$$\frac{2m}{2} = \frac{90}{2}$$
$$m = 45$$

The exercise asks only for the amount Melinda paid, but we also find the amount Susan paid so that we can check the answer.

If $m = 45$, then $m + 17 = 45 + 17 = 62$.

Check. $62 is $17 more than $45, and $45 + $62 = $107. The answer checks.

State. Melinda paid $45 for her rollerblades.

61. Familiarize. Let $x =$ the amount originally invested. Using the formula for simple interest, $I = Prt$, the interest earned in one year will be $x \cdot 8\% \cdot 1$, or $8\%x$.

Translate.

Amount invested	plus	interest	is	amount after 1 year.
↓	↓	↓	↓	↓
x	$+$	$8\%x$	$=$	1134

Solve.

$$x + 8\%x = 1134$$
$$x + 0.08x = 1134$$
$$1.08x = 1134$$
$$\frac{1.08x}{1.08} = \frac{1134}{1.08}$$
$$x = 1050$$

Check. 8% of $1050 is $0.08 \cdot \$1050 = \$84$ and $\$1050 + \$84 = \$1134$, so the answer checks.

State. $1050 was originally invested.

62. Familiarize. Let $l =$ the length of the first piece of wire, in meters. Then $l + 3 =$ the length of the second piece and $\frac{4}{5}l =$ the length of the third piece.

Translate.

Length of first piece	plus	length of second piece	plus	length of third piece	is	143 m.
↓	↓	↓	↓	↓	↓	↓
l	$+$	$(l+3)$	$+$	$\frac{4}{5}l$	$=$	143

Solve.

$$l + (l+3) + \frac{4}{5}l = 143$$
$$\frac{5}{5}l + \frac{5}{5}l + 3 + \frac{4}{5}l = 143$$
$$\frac{14}{5}l + 3 = 143$$
$$\frac{14}{5}l + 3 - 3 = 143 - 3$$
$$\frac{14}{5}l = 140$$
$$\frac{5}{14} \cdot \frac{14}{5}l = \frac{5}{14} \cdot 140$$
$$l = \frac{5 \cdot 10 \cdot \cancel{14}}{\cancel{14} \cdot 1}$$
$$l = 50$$

If $l = 50$, then $l + 3 = 50 + 3 = 53$ and $\frac{4}{5}l = \frac{4}{5} \cdot 50 = 40$.

Check. The second piece is 3 m longer than the first piece and the third piece is $\frac{4}{5}$ as long as the first. Also, $50 \text{ m} + 53 \text{ m} + 40 \text{ m} = 143 \text{ m}$. The answer checks.

State. The lengths of the pieces are 50 m, 53 m, and 40 m.

63. Familiarize. Let $m =$ the number of miles traveled. We will express 39¢ as $0.39. The cost of the rental is the daily rate plus $0.39 times the number of miles traveled, or $49.95 + \$0.39 \cdot m$.

Translate.

The cost of the rental	is no more than	$100.
↓	↓	↓
$49.95 + 0.39 \cdot m$	$\leq$	100

Solve. First we will multiply by 100 to clear the decimals.

$$49.95 + 0.39 \cdot m \leq 100$$
$$100(49.95 + 0.39 \cdot m) \leq 100 \cdot 100$$
$$4995 + 39m \leq 10,000$$
$$39m \leq 5005$$
$$m \leq 128\frac{1}{3}$$

Check. We check to see if the solution seems reasonable. When $m = 128$, the cost is $\$49.95 + \$0.39 \cdot 128$, or $99.87. When $m = 128\frac{1}{3}$, the cost is $\$49.95 + \$0.39 \cdot 128\frac{1}{3}$, or $100.

It appears that the solution is correct.

State. At most $128\frac{1}{3}$ mi can be traveled. The solution set is $\left\{ m \middle| m \leq 128\frac{1}{3} \text{ mi} \right\}$.

64. Familiarize. Let $p =$ the price before the reduction.

Translate.

Price before reduction	minus	25% of price	is	$18.45.
↓	↓	↓ ↓ ↓	↓	↓
p	$-$	$25\% \cdot p$	$=$	18.45

Solve. We solve the equation.

$$p - 25\% \cdot p = 18.45$$
$$p - 0.25p = 18.45$$
$$0.75p = 18.45$$
$$\frac{0.75p}{0.75} = \frac{18.45}{0.75}$$
$$p = 24.6$$

Check. 25% of $24.60 is $0.25 \cdot \$24.60 = \$6.15$ and $\$24.60 - \$6.15 = \$18.45$, so the answer checks.

State. The price before the reduction was $24.60.

65.
$$-125 \div 25 \cdot 625 \div 5 = -5 \cdot 625 \div 5$$
$$= -3125 \div 5$$
$$= -625$$

Answer (c) is correct.

66. $[5(2x + 6) - 7] - [2(x + 4) + 5]$

$= [10x + 30 - 7] - [2x + 8 + 5]$

$= [10x + 23] - [2x + 13]$

$= 10x + 23 - 2x - 13$

$= 8x + 10$

Answer (d) is correct.

67. $V = IR$

$\dfrac{V}{R} = \dfrac{IR}{R}$

$\dfrac{V}{R} = I$

Answer (b) is correct.

68. ***Familiarize.*** Let $s =$ the salary at the beginning of the year. After a 4% increase the new salary is $s + 4\%s$, or $s + 0.04s$, or $1.04s$. Then after a 3% cost-of-living adjustment the final salary is $1.04s + 3\% \cdot 1.04s$, or $1.04s + 0.03 \cdot 1.04s$, or $1.04s + 0.0312s$, or $1.0712s$.

Translate.

$\underbrace{\text{Final salary}}$ is $\$48,418.24.$

$\quad\;\downarrow\qquad\quad\downarrow\qquad\;\downarrow$

$\quad 1.0712s\quad = \quad 48,418.24$

Solve.

$1.0712s = 48,418.24$

$\dfrac{1.0712s}{1.0712} = \dfrac{48,418.24}{1.0712}$

$s = 45,200$

Check. 4% of $45,200 is $0.04 \cdot \$45,200 = \$1808$ and $\$45,200 + \$1808 = \$47,008$. Then 3% of $47,008 is $0.03 \cdot \$47,008 = \$1410.24$ and $\$47,008 + \$1410.24 = \$48,418.24$. The answer checks.

State. At the beginning of the year the salary was $45,200.

69. First we subtract to find the amount of the reduction.

9 in. $-$ 6.3 in. $=$ 2.7 in.

Translate.

$\underbrace{\text{2.7 in.}}$ is $\underbrace{\text{what percent}}$ of $\underbrace{\text{9 in.?}}$

$\quad\downarrow\qquad\downarrow\qquad\quad\downarrow\qquad\quad\downarrow\qquad\downarrow$

$\quad 2.7\quad =\qquad\quad p\qquad\quad\cdot\quad 9$

Solve.

$2.7 = p \cdot 9$

$\dfrac{2.7}{9} = \dfrac{p \cdot 9}{9}$

$0.3 = p$

$30\% = p$

The drawing should be reduced 30%.

70. $4|x| - 13 = 3$

$4|x| = 16$

$|x| = 4$

The solutions are the numbers whose distance from 0 is 4. They are -4 and 4.

71. First we multiply by 28 to clear the fractions.

$\dfrac{2 + 5x}{4} = \dfrac{11}{28} + \dfrac{8x + 3}{7}$

$28\left(\dfrac{2 + 5x}{4}\right) = 28\left(\dfrac{11}{28} + \dfrac{8x + 3}{7}\right)$

$\dfrac{28(2 + 5x)}{4} = 28 \cdot \dfrac{11}{28} + \dfrac{28(8x + 3)}{7}$

$7(2 + 5x) = 11 + 4(8x + 3)$

$14 + 35x = 11 + 32x + 12$

$14 + 35x = 32x + 23$

$14 + 3x = 23$

$3x = 9$

$x = 3$

The solution is 3.

72. $p = \dfrac{2}{m + Q}$

$(m + Q) \cdot p = (m + Q) \cdot \dfrac{2}{m + Q}$

$mp + Qp = 2$

$Qp = 2 - mp$

$Q = \dfrac{2 - mp}{p}$

Chapter 3

Graphs of Linear Equations

Exercise Set 3.1

1. $(2, 5)$ is 2 units right and 5 units up.

   $(-1, 3)$ is 1 unit left and 3 units up.

   $(3, -2)$ is 3 units right and 2 units down.

   $(-2, -4)$ is 2 units left and 4 units down.

   $(0, 4)$ is 0 units left or right and 4 units up.

   $(0, -5)$ is 0 units left or right and 5 units down.

   $(5, 0)$ is 5 units right and 0 units up or down.

   $(-5, 0)$ is 5 units left and 0 units up or down.

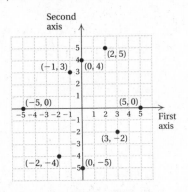

3. Since the first coordinate is negative and the second coordinate positive, the point $(-5, 3)$ is located in quadrant II.

5. Since the first coordinate is positive and the second coordinate negative, the point $(100, -1)$ is in quadrant IV.

7. Since both coordinates are negative, the point $(-6, -29)$ is in quadrant III.

9. Since one of the coordinates is 0, the point $(3.8, 0)$ lies on an axis.

11. Since the first coordinate is negative and the second coordinate is positive, the point $\left(-\frac{1}{3}, \frac{15}{7}\right)$ is in quadrant II.

13. Since the first coordinate is positive and the second coordinate is negative, the point $\left(12\frac{7}{8}, -1\frac{1}{2}\right)$ is in quadrant IV.

15.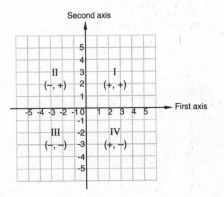

If the first coordinate is negative and the second coordinate is positive, the point is in quadrant II.

17. See the figure in Exercise 15.

If the first coordinate is positive, then the point must be in either quadrant I or quadrant IV.

19. If the first and second coordinates are equal, they must either be both positive or both negative. The point must be in either quadrant I (both positive) or quadrant III (both negative).

21.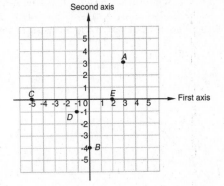

Point A is 3 units right and 3 units up. The coordinates of A are $(3, 3)$.

Point B is 0 units left or right and 4 units down. The coordinates of B are $(0, -4)$.

Point C is 5 units left and 0 units up or down. The coordinates of C are $(-5, 0)$.

Point D is 1 unit left and 1 unit down. The coordinates of D are $(-1, -1)$.

Point E is 2 units right and 0 units up or down. The coordinates of E are $(2, 0)$.

23. We substitute 2 for x and 9 for y (alphabetical order of variables).

$$
\begin{array}{c|l}
\multicolumn{2}{l}{y = 3x - 1} \\
\hline
9 \; ? \; 3 \cdot 2 - 1 & \\
6 - 1 & \\
5 & \text{FALSE}
\end{array}
$$

Since $9 = 5$ is false, the pair $(2, 9)$ is not a solution.

25. We substitute 4 for x and 2 for y.

$$
\begin{array}{c|l}
\multicolumn{2}{l}{2x + 3y = 12} \\
\hline
2 \cdot 4 + 3 \cdot 2 \; ? \; 12 & \\
8 + 6 & \\
14 & \text{FALSE}
\end{array}
$$

Since $14 = 12$ is false, the pair $(4, 2)$ is not a solution.

27. We substitute 3 for a and -1 for b.

$$
\begin{array}{c|l}
\multicolumn{2}{l}{3a - 4b = 13} \\
\hline
3 \cdot 3 - 4(-1) \; ? \; 13 & \\
9 + 4 & \\
13 & \text{TRUE}
\end{array}
$$

Since $13 = 13$ is true, the pair $(3, -1)$ is a solution.

29. To show that a pair is a solution, we substitute, replacing x with the first coordinate and y with the second coordinate in each pair.

$$
\begin{array}{c|l}
\multicolumn{2}{l}{y = x - 5} \\
\hline
-1 \; ? \; 4 - 5 & \\
-1 & \text{TRUE}
\end{array}
\qquad
\begin{array}{c|l}
\multicolumn{2}{l}{y = x - 5} \\
\hline
-4 \; ? \; 1 - 5 & \\
-4 & \text{TRUE}
\end{array}
$$

In each case the substitution results in a true equation. Thus, $(4, -1)$ and $(1, -4)$ are both solutions of $y = x - 5$. We graph these points and sketch the line passing through them.

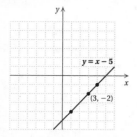

The line appears to pass through $(3, -2)$ also. We check to determine if $(3, -2)$ is a solution of $y = x - 5$.

$$
\begin{array}{c|l}
\multicolumn{2}{l}{y = x - 5} \\
\hline
-2 \; ? \; 3 - 5 & \\
-2 & \text{TRUE}
\end{array}
$$

Thus, $(3, -2)$ is another solution. There are other correct answers, including $(-1, -6)$, $(2, -3)$, $(0, -5)$, $(5, 0)$, and $(6, 1)$.

31. To show that a pair is a solution, we substitute, replacing x with the first coordinate and y with the second coordinate in each pair.

$$
\begin{array}{c|l}
\multicolumn{2}{l}{y = \dfrac{1}{2}x + 3} \\
\hline
5 \; ? \; \dfrac{1}{2} \cdot 4 + 3 & \\
2 + 3 & \\
5 & \text{TRUE}
\end{array}
\qquad
\begin{array}{c|l}
\multicolumn{2}{l}{y = \dfrac{1}{2}x + 3} \\
\hline
2 \; ? \; \dfrac{1}{2}(-2) + 3 & \\
-1 + 3 & \\
2 & \text{TRUE}
\end{array}
$$

In each case the substitution results in a true equation. Thus, $(4, 5)$ and $(-2, 2)$ are both solutions of $y = \dfrac{1}{2}x + 3$. We graph these points and sketch the line passing through them.

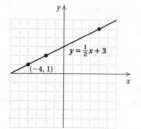

The line appears to pass through $(-4, 1)$ also. We check to determine if $(-4, 1)$ is a solution of $y = \dfrac{1}{2}x + 3$.

$$
\begin{array}{c|l}
\multicolumn{2}{l}{y = \dfrac{1}{2}x + 3} \\
\hline
1 \; ? \; \dfrac{1}{2}(-4) + 3 & \\
-2 + 3 & \\
1 & \text{TRUE}
\end{array}
$$

Thus, $(-4, 1)$ is another solution. There are other correct answers, including $(-6, 0)$, $(0, 3)$, $(2, 4)$, and $(6, 6)$.

33. To show that a pair is a solution, we substitute, replacing x with the first coordinate and y with the second coordinate in each pair.

$$
\begin{array}{c|l}
\multicolumn{2}{l}{4x - 2y = 10} \\
\hline
4 \cdot 0 - 2(-5) \; ? \; 10 & \\
10 & \text{TRUE}
\end{array}
$$

$$
\begin{array}{c|l}
\multicolumn{2}{l}{4x - 2y = 10} \\
\hline
4 \cdot 4 - 2 \cdot 3 \; ? \; 10 & \\
16 - 6 & \\
10 & \text{TRUE}
\end{array}
$$

In each case the substitution results in a true equation. Thus, $(0, -5)$ and $(4, 3)$ are both solutions of $4x - 2y = 10$. We graph these points and sketch the line passing through them.

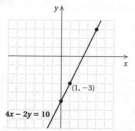

The line appears to pass through $(1, -3)$ also. We check to determine if $(1, -3)$ is a solution of $4x - 2y = 10$.

$$
\begin{array}{c|c}
\multicolumn{2}{c}{4x - 2y = 10} \\
\hline
4 \cdot 1 - 2(-3) \ ? \ 10 & \\
4 + 6 & \\
10 & \text{TRUE}
\end{array}
$$

Thus, $(1, -3)$ is another solution. There are other correct answers, including $(2, -1)$, $(3, 1)$, and $(5, 5)$.

35. $y = x + 1$

The equation is in the form $y = mx + b$. The y-intercept is $(0, 1)$. We find five other pairs.

When $x = -2$, $y = -2 + 1 = -1$.

When $x = -1$, $y = -1 + 1 = 0$.

When $x = 1$, $y = 1 + 1 = 2$.

When $x = 2$, $y = 2 + 1 = 3$.

When $x = 3$, $y = 3 + 1 = 4$.

x	y
-2	-1
-1	0
0	1
1	2
2	3
3	4

Plot these points, draw the line they determine, and label the graph $y = x + 1$.

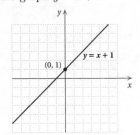

37. $y = x$

The equation is equivalent to $y = x + 0$. The y-intercept is $(0, 0)$. We find five other points.

When $x = -2$, $y = -2$.

When $x = -1$, $y = -1$.

When $x = 1$, $y = 1$.

When $x = 2$, $y = 2$.

When $x = 3$, $y = 3$.

x	y
-2	-2
-1	-1
0	0
1	1
2	2
3	3

Plot these points, draw the line they determine, and label the graph $y = x$.

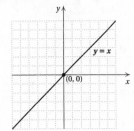

39. $y = \frac{1}{2}x$

The equation is equivalent to $y = \frac{1}{2}x + 0$. The y-intercept is $(0, 0)$. We find two other points.

When $x = -2$, $y = \frac{1}{2}(-2) = -1$.

When $x = 4$, $y = \frac{1}{2} \cdot 4 = 2$.

x	y
-2	-1
0	0
4	2

Plot these points, draw the line they determine, and label the graph $y = \frac{1}{2}x$.

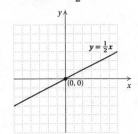

41. $y = x - 3$

The equation is equivalent to $y = x + (-3)$. The y-intercept is $(0, -3)$. We find two other points.

When $x = -2$, $y = -2 - 3 = -5$.

When $x = 4$, $y = 4 - 3 = 1$.

x	y
-2	-5
0	-3
4	1

Plot these points, draw the line they determine, and label the graph $y = x - 3$.

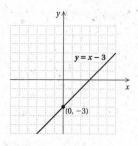

43. $y = 3x - 2 = 3x + (-2)$

The y-intercept is $(0, -2)$. We find two other points.

When $x = -2$, $y = 3(-2) + 2 = -6 + 2 = -4$.

When $x = 1$, $y = 3 \cdot 1 + 2 = 3 + 2 = 5$.

x	y
-2	-4
0	-2
1	5

Plot these points, draw the line they determine, and label the graph $y = 3x + 2$.

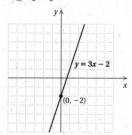

45. $y = \dfrac{1}{2}x + 1$

The y-intercept is $(0, 1)$. We find two other points using multiples of 2 for x to avoid fractions.

When $x = -4$, $y = \dfrac{1}{2}(-4) + 1 = -2 + 1 = -1$.

When $x = 4$, $y = \dfrac{1}{2} \cdot 4 + 1 = 2 + 1 = 3$.

x	y
-4	-1
0	1
4	3

Plot these points, draw the line they determine, and label the graph $y = \dfrac{1}{2}x + 1$.

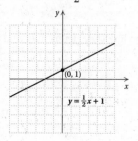

47. $x + y = -5$

$$y = -x - 5$$
$$y = -x + (-5)$$

The y-intercept is $(0, -5)$. We find two other points.

When $x = -4$, $y = -(-4) - 5 = 4 - 5 = -1$.

When $x = -1$, $y = -(-1) - 5 = 1 - 5 = -4$.

x	y
-4	-1
0	-5
-1	-4

Plot these points, draw the line they determine, and label the graph $x + y = -5$.

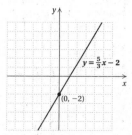

49. $y = \dfrac{5}{3}x - 2 = \dfrac{5}{3}x + (-2)$

The y-intercept is $(0, -2)$. We find two other points using multiples of 3 for x to avoid fractions.

When $x = -3$, $y = \dfrac{5}{3}(-3) - 2 = -5 - 2 = -7$.

When $x = 3$, $y = \dfrac{5}{3} \cdot 3 - 2 = 5 - 2 = 3$.

x	y
-3	-7
0	-2
3	3

Plot these points, draw the line they determine, and label the graph $y = \dfrac{5}{3}x - 2$.

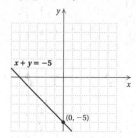

51. $x + 2y = 8$

$$2y = -x + 8$$
$$y = -\dfrac{1}{2}x + 4$$

The y-intercept is $(0, 4)$. We find two other points using multiples of 2 for x to avoid fractions.

When $x = -2$, $y = -\frac{1}{2}(-2) + 4 = 1 + 4 = 5$.

When $x = 4$, $y = -\frac{1}{2} \cdot 4 + 4 = -2 + 4 = 2$.

x	y
-2	5
0	4
4	2

Plot these points, draw the line they determine, and label the graph $x + 2y = 8$.

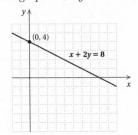

53. $y = \frac{3}{2}x + 1$

The y-intercept is $(0, 1)$. We find two other points using multiples of 2 for x to avoid fractions.

When $x = -4$, $y = \frac{3}{2}(-4) + 1 = -6 + 1 = -5$.

When $x = 2$, $y = \frac{3}{2} \cdot 2 + 1 = 3 + 1 = 4$.

x	y
-4	-5
0	1
2	4

Plot these points, draw the line they determine, and label the graph $y = \frac{3}{2}x + 1$.

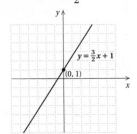

55. $8x - 2y = -10$

$-2y = -8x - 10$

$y = 4x + 5$

The y-intercept is $(0, 5)$. We find two other points.

When $x = -2$, $y = 4(-2) + 5 = -8 + 5 = -3$.

When $x = -1$, $y = 4(-1) + 5 = -4 + 5 = 1$.

x	y
-2	-3
-1	1
0	5

Plot these points, draw the line they determine, and label the graph $8x - 2y = -10$.

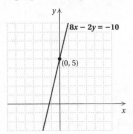

57. $8y + 2x = -4$

$8y = -2x - 4$

$y = -\frac{1}{4}x - \frac{1}{2}$

$y = -\frac{1}{4}x + \left(-\frac{1}{2}\right)$

The y-intercept is $\left(0, -\frac{1}{2}\right)$. We find two other points.

When $x = -2$, $y = -\frac{1}{4}(-2) - \frac{1}{2} = \frac{1}{2} - \frac{1}{2} = 0$.

When $x = 2$, $y = -\frac{1}{4} \cdot 2 - \frac{1}{2} = -\frac{1}{2} - \frac{1}{2} = -1$.

x	y
-2	0
0	$-\frac{1}{2}$
2	-1

Plot these points, draw the line they determine, and label the graph $8y + 2x = -4$.

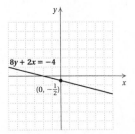

59. a) We substitute 0, 4, and 6 for t and then calculate V.

If $t = 0$, then $V = -50 \cdot 0 + 300 = \$300$.

If $t = 4$, then $V = -50 \cdot 4 + 300 = -200 + 300 = \$100$.

If $t = 6$, then $V = -50 \cdot 6 + 300 = -300 + 300 = \$0$.

b) We plot the three ordered pairs we found in part (a). Note the negative t- and V-values have no meaning in this problem.

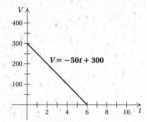

To use the graph to estimate the value of the software after 5 years we must determine which V-value is paired with $t = 5$. We locate 5 on the t-axis, go up to the graph, and then find the value on the V-axis that corresponds to that point. It appears that after 5 years the value of the software is $50.

c) Substitute 150 for V and then solve for t.
$$V = -50t + 300$$
$$150 = -50t + 300$$
$$-150 = -50t$$
$$3 = t$$

The value of the software is $150 after 3 years.

61. a) When $d = 1$, $N = 0.8(1) + 21.2 = 0.8 + 21.2 = 22$ gal.

In 2000, $d = 2000 - 1995 = 5$. When $d = 5$, $N = 0.8(5) + 21.2 = 4 + 21.2 = 25.2$ gal.

In 2006, $d = 2006 - 1995 = 11$. When $d = 11$, $N = 0.8(11) + 21.2 = 8.8 + 21.2 = 30$ gal.

In 2010, $d = 2010 - 1995 = 15$. When $d = 15$, $N = 0.8(15) + 21.2 = 12 + 21.2 = 33.2$ gal.

b) Plot the four ordered pairs we found in part (a). Note that negative d- and N-values have no meaning in this problem.

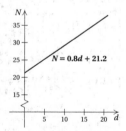

To use the graph to estimate what tea consumption was in 2002 we must determine which N-value is paired with 2002, or with $d = 7$. We locate 7 on the d-axis, go up to the graph, and then find the value on the N-axis that corresponds to that point. It appears that tea consumption was about 27 gallons in 2002.

c) Substitute 31.6 for N and then solve for d.
$$N = 0.8d + 21.2$$
$$31.6 = 0.8d + 21.2$$
$$10.4 = 0.8d$$
$$13 = d$$

Tea consumption will be about 31.6 gallons 13 years after 1995, or in 2008.

63. Discussion and Writing Exercise

65. The distance of -12 from 0 is 12, so $|-12| = 12$.

67. The distance of 0 from 0 is 0, so $|0| = 0$.

69. The distance of -3.4 from 0 is 3.4, so $|-3.4| = 3.4$.

71. The distance of $\frac{2}{3}$ from 0 is $\frac{2}{3}$, so $\left|\frac{2}{3}\right| = \frac{2}{3}$.

73. *Familiarize.* Let $p =$ the average price of a ticket to a major-league baseball game in 2000. The price in 2004 was $p + 17.9\%p$, or $p + 0.179p$, or $1.179p$.

Translate.

$$\underbrace{\text{The price in 2004}}_{1.179p} \underset{=}{\text{ was }} \underset{19.82}{\$19.82}.$$

Solve.
$$1.179p = 19.82$$
$$p = \frac{19.82}{1.179}$$
$$p \approx 16.81$$

Check. 17.9% of $16.81 is $0.179(\$16.81) \approx \$3.01$ and $\$16.81 + \$3.01 = \$19.82$. The answer checks.

State. In 2000 the average price of a major-league baseball ticket was $16.81.

75.

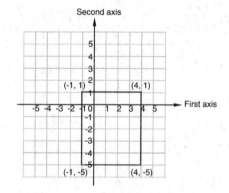

The coordinates of the fourth vertex are $(-1, -5)$.

77. Answers may vary.

We select eight points such that the sum of the coordinates for each point is 6.

$$
\begin{array}{ll}
(-1, 7) & -1 + 7 = 6 \\
(0, 6) & 0 + 6 = 6 \\
(1, 5) & 1 + 5 = 6 \\
(2, 4) & 2 + 4 = 6 \\
(3, 3) & 3 + 3 = 6 \\
(4, 2) & 4 + 2 = 6 \\
(5, 1) & 5 + 1 = 6 \\
(6, 0) & 6 + 0 = 6 \\
\end{array}
$$

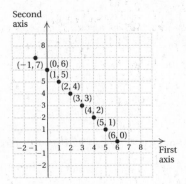

79.

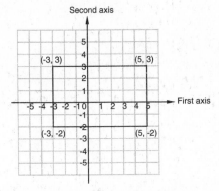

The length is 8 linear units, and the width is 5 linear units.

$P = 2l + 2w$

$P = 2 \cdot 8 + 2 \cdot 5 = 16 + 10 = 26$ linear units

81. Point A is 10 units right and 18 units up. The coordinates of A are $(10, 18)$.

Point B is 8 units right and 15 units up. The coordinates of B are $(8, 15)$.

Point C is 3 units right and 13 units up. The coordinates of C are $(3, 13)$.

Point D is 8 units right and 11 units up. The coordinates of D are $(8, 11)$.

Point E is 3 units right and 8 units up. The coordinates of E are $(3, 8)$.

Point F is 8 units right and 5 units up. The coordinates of F are $(8, 5)$.

Point G is 12 units right and 5 units up. The coordinates of G are $(12, 5)$.

Point H is 17 units right and 8 units up. The coordinates of H are $(17, 8)$.

Point I is 12 units right and 11 units up. The coordinates of I are $(12, 11)$.

Point J is 17 units right and 13 units up. The coordinates of J are $(17, 13)$.

Point K is 12 units right and 15 units up. The coordinates of K are $(12, 15)$.

83. Subtracting 3 from each of the y-coordinates found in Exercise 81 gives us the points $(10, 15)$, $(8, 12)$, $(3, 10)$, $(8, 8)$, $(3, 5)$, $(8, 2)$, $(12, 2)$, $(17, 5)$, $(12, 8)$, $(17, 10)$, and $(12, 12)$.

When we plot these points and connect them in the same order as in Exercise 81, we get a figure that has the same shape as the figure in Exercise 81 and that is translated 3 units down.

Exercise Set 3.2

1. (a) The graph crosses the y-axis at $(0, 5)$, so the y-intercept is $(0, 5)$.

(b) The graph crosses the x- axis at $(2, 0)$, so the x-intercept is $(2, 0)$.

3. (a) The graph crosses the y-axis at $(0, -4)$, so the y-intercept is $(0, -4)$.

(b) The graph crosses the x-axis at $(3, 0)$, so the x-intercept is $(3, 0)$.

5. $3x + 5y = 15$

(a) To find the y-intercept, let $x = 0$. This is the same as covering up the x-term and then solving.

$$5y = 15$$
$$y = 3$$

The y-intercept is $(0, 3)$.

(b) To find the x-intercept, let $y = 0$. This is the same as covering up the y-term and then solving.

$$3x = 15$$
$$x = 5$$

The x-intercept is $(5, 0)$.

7. $7x - 2y = 28$

(a) To find the y-intercept, let $x = 0$. This is the same as covering up the x-term and then solving.

$$-2y = 28$$
$$y = -14$$

The $y-$intercept is $(0, -14)$.

(b) To find the x-intercept, let $y = 0$. This is the same as covering up the y-term and then solving.

$$7x = 28$$
$$x = 4$$

The x-intercept is $(4, 0)$.

9. $-4x + 3y = 10$

(a) To find the y-intercept, let $x = 0$. This is the same as covering up the x-term and then solving.

$$3y = 10$$
$$y = \frac{10}{3}$$

The y-intercept is $\left(0, \frac{10}{3}\right)$.

(b) To find the x-intercept, let $y = 0$. This is the same as covering up the y-term and then solving.

$$-4x = 10$$
$$x = -\frac{5}{2}$$

The x-intercept is $\left(-\frac{5}{2}, 0\right)$.

11. $6x - 3 = 9y$

$6x - 9y = 3$ Writing the equation in the
                          form $Ax + By = C$

(a) To find the y-intercept, let $x = 0$. This is the same
    as covering up the x-term and then solving.

$$-9y = 3$$
$$y = -\frac{1}{3}$$

The y-intercept is $\left(0, -\frac{1}{3}\right)$.

(b) To find the x-intercept, let $y = 0$. This is the same
    as covering up the y-term and then solving.

$$6x = 3$$
$$x = \frac{1}{2}$$

The x-intercept is $\left(\frac{1}{2}, 0\right)$.

13. $x + 3y = 6$

To find the x-intercept, let $y = 0$. Then solve for x.

$$x + 3y = 6$$
$$x + 3 \cdot 0 = 6$$
$$x = 6$$

Thus, $(6, 0)$ is the x-intercept.

To find the y-intercept, let $x = 0$. Then solve for y.

$$x + 3y = 6$$
$$0 + 3y = 6$$
$$3y = 6$$
$$y = 2$$

Thus, $(0, 2)$ is the y-intercept.

Plot these points and draw the line.

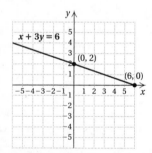

A third point should be used as a check. We substitute
any value for x and solve for y.

We let $x = 3$. Then

$$x + 3y = 6$$
$$3 + 3y = 6$$
$$3y = 3$$
$$y = 1$$

The point $(3, 1)$ is on the graph, so the graph is probably
correct.

15. $-x + 2y = 4$

To find the x-intercept, let $y = 0$. Then solve for x.

$$-x + 2y = 4$$
$$-x + 2 \cdot 0 = 4$$
$$-x = 4$$
$$x = -4$$

Thus, $(-4, 0)$ is the x-intercept.

To find the y-intercept, let $x = 0$. Then solve for y.

$$-x + 2y = 4$$
$$-0 + 2y = 4$$
$$2y = 4$$
$$y = 2$$

Thus, $(0, 2)$ is the y-intercept.

Plot these points and draw the line.

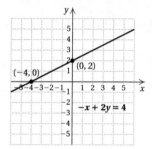

A third point should be used as a check. We substitute
any value for x and solve for y.

We let $x = 4$. Then

$$-x + 2y = 4$$
$$-4 + 2y = 4$$
$$2y = 8$$
$$y = 4$$

The point $(4, 4)$ is on the graph, so the graph is probably
correct.

17. $3x + y = 6$

To find the x-intercept, let $y = 0$. Then solve for x.

$$3x + y = 6$$
$$3x + 0 = 6$$
$$3x = 6$$
$$x = 2$$

Thus, $(2, 0)$ is the x-intercept.

To find the y-intercept, let $x = 0$. Then solve for y.

$$3x + y = 6$$
$$3 \cdot 0 + y = 6$$
$$y = 6$$

Thus, $(0, 6)$ is the y-intercept.

Plot these points and draw the line.

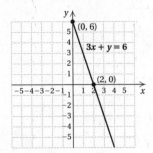

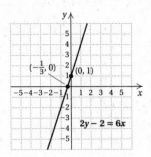

A third point should be used as a check. We substitute any value for x and solve for y.

We let $x = 1$. Then

$$3x + y = 6$$
$$3 \cdot 1 + y = 6$$
$$3 + y = 6$$
$$y = 3$$

The point $(1, 3)$ is on the graph, so the graph is probably correct.

19. $2y - 2 = 6x$

To find the x-intercept, let $y = 0$. Then solve for x.

$$2y - 2 = 6x$$
$$2 \cdot 0 - 2 = 6x$$
$$-2 = 6x$$
$$-\frac{1}{3} = x$$

Thus, $\left(-\frac{1}{3}, 0\right)$ is the x-intercept.

To find the y-intercept, let $x = 0$. Then solve for y.

$$2y - 2 = 6x$$
$$2y - 2 = 6 \cdot 0$$
$$2y - 2 = 0$$
$$2y = 2$$
$$y = 1$$

Thus, $(0, 1)$ is the y-intercept.

It is helpful to plot another point since the intercepts are so close together. This point can also serve as a check.

We let $x = 1$. Then

$$2y - 2 = 6x$$
$$2y - 2 = 6 \cdot 1$$
$$2y - 2 = 6$$
$$2y = 8$$
$$y = 4$$

Plot the point $(1, 4)$ and the intercepts and draw the line.

21. $3x - 9 = 3y$

To find the x-intercept, let $y = 0$. Then solve for x.

$$3x - 9 = 3y$$
$$3x - 9 = 3 \cdot 0$$
$$3x - 9 = 0$$
$$3x = 9$$
$$x = 3$$

Thus, $(3, 0)$ is the x-intercept.

To find the y-intercept, let $x = 0$. Then solve for y.

$$3x - 9 = 3y$$
$$3 \cdot 0 - 9 = 3y$$
$$-9 = 3y$$
$$-3 = y$$

Thus, $(0, -3)$ is the y-intercept.

Plot these points and draw the line.

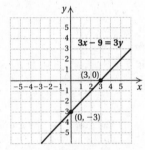

A third point should be used as a check. We substitute any value for x and solve for y.

We let $x = 1$. Then

$$3x - 9 = 3y$$
$$3 \cdot 1 - 9 = 3y$$
$$3 - 9 = 3y$$
$$-6 = 3y$$
$$-2 = y$$

The point $(1, -2)$ is on the graph, so the graph is probably correct.

23. $2x - 3y = 6$

To find the x-intercept, let $y = 0$. Then solve for x.

$$2x - 3y = 6$$
$$2x - 3 \cdot 0 = 6$$
$$2x = 6$$
$$x = 3$$

Thus, $(3, 0)$ is the x-intercept.

To find the y-intercept, let $x = 0$. Then solve for y.

$$2x - 3y = 6$$
$$2 \cdot 0 - 3y = 6$$
$$-3y = 6$$
$$y = -2$$

Thus, $(0, -2)$ is the y-intercept.

Plot these points and draw the line.

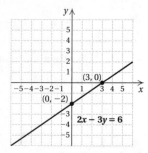

A third point should be used as a check. We substitute any value for x and solve for y.

We let $x = -3$.

$$2x - 3y = 6$$
$$2(-3) - 3y = 6$$
$$-6 - 3y = 6$$
$$-3y = 12$$
$$y = -4$$

The point $(-3, -4)$ is on the graph, so the graph is probably correct.

25. $4x + 5y = 20$

To find the x-intercept, let $y = 0$. Then solve for x.

$$4x + 5y = 20$$
$$4x + 5 \cdot 0 = 20$$
$$4x = 20$$
$$x = 5$$

Thus, $(5, 0)$ is the x-intercept.

To find the y-intercept, let $x = 0$. Then solve for y.

$$4x + 5y = 20$$
$$4 \cdot 0 + 5y = 20$$
$$5y = 20$$
$$y = 4$$

Thus, $(0, 4)$ is the y-intercept.

Plot these points and draw the graph.

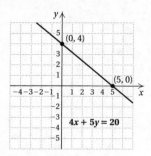

A third point should be used as a check. We substitute any value for x and solve for y.

We let $x = 4$. Then

$$4x + 5y = 20$$
$$4 \cdot 4 + 5y = 20$$
$$16 + 5y = 20$$
$$5y = 4$$
$$y = \frac{4}{5}$$

The point $\left(4, \dfrac{4}{5}\right)$ is on the graph, so the graph is probably correct.

27. $2x + 3y = 8$

To find the x-intercept, let $y = 0$. Then solve for x.

$$2x + 3y = 8$$
$$2x + 3 \cdot 0 = 8$$
$$2x = 8$$
$$x = 4$$

Thus, $(4, 0)$ is the x-intercept.

To find the y-intercept, let $x = 0$. Then solve for y.

$$2x + 3y = 8$$
$$2 \cdot 0 + 3y = 8$$
$$3y = 8$$
$$y = \frac{8}{3}$$

Thus, $\left(0, \dfrac{8}{3}\right)$ is the y-intercept.

Plot these points and draw the graph.

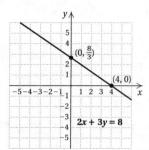

A third point should be used as a check.

We let $x = 1$. Then
$$2x + 3y = 8$$
$$2 \cdot 1 + 3y = 8$$
$$2 + 3y = 8$$
$$3y = 6$$
$$y = 2$$

The point $(1, 2)$ is on the graph, so the graph is probably correct.

29. $x - 3 = y$

To find the x-intercept, let $y = 0$. Then solve for x.
$$x - 3 = y$$
$$x - 3 = 0$$
$$x = 3$$

Thus, $(3, 0)$ is the x-intercept.

To find the y-intercept, let $x = 0$. Then solve for y.
$$x - 3 = y$$
$$0 - 3 = y$$
$$-3 = y$$

Thus, $(0, -3)$ is the y-intercept.

Plot these points and draw the line.

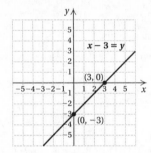

A third point should be used as a check.

We let $x = -2$. Then
$$x - 3 = y$$
$$-2 - 3 = y$$
$$-5 = y$$

The point $(-2, -5)$ is on the graph, so the graph is probably correct.

31. $3x - 2 = y$

To find the x-intercept, let $y = 0$. Then solve for x.
$$3x - 2 = y$$
$$3x - 2 = 0$$
$$3x = 2$$
$$x = \frac{2}{3}$$

Thus, $\left(\frac{2}{3}, 0\right)$ is the x-intercept.

To find the y-intercept, let $x = 0$. Then solve for y.
$$3x - 2 = y$$
$$3 \cdot 0 - 2 = y$$
$$-2 = y$$

Thus, $(0, -2)$ is the y-intercept.

Plot these points and draw the line.

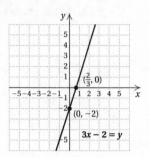

A third point should be used as a check.

We let $x = 2$. Then
$$3x - 2 = y$$
$$3 \cdot 2 - 2 = y$$
$$6 - 2 = y$$
$$4 = y$$

The point $(2, 4)$ is on the graph, so the graph is probably correct.

33. $6x - 2y = 12$

To find the x-intercept, let $y = 0$. Then solve for x.
$$6x - 2y = 12$$
$$6x - 2 \cdot 0 = 12$$
$$6x = 12$$
$$x = 2$$

Thus, $(2, 0)$ is the x-intercept.

To find the y-intercept, let $x = 0$. Then solve for y.
$$6x - 2y = 12$$
$$6 \cdot 0 - 2y = 12$$
$$-2y = 12$$
$$y = -6$$

Thus, $(0, -6)$ is the y-intercept.

Plot these points and draw the line.

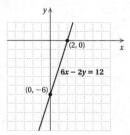

We use a third point as a check.

We let $x = 1$. Then
$$6x - 2y = 12$$
$$6 \cdot 1 - 2y = 12$$
$$6 - 2y = 12$$
$$-2y = 6$$
$$y = -3$$

The point $(1, -3)$ is on the graph, so the graph is probably correct.

35. $3x + 4y = 5$

To find the x-intercept, let $y = 0$. Then solve for x.
$$3x + 4y = 5$$
$$3x + 4 \cdot 0 = 5$$
$$3x = 5$$
$$x = \frac{5}{3}$$

Thus, $\left(\frac{5}{3}, 0\right)$ is the x-intercept.

To find the y-intercept, let $x = 0$. Then solve for y.
$$3x + 4y = 5$$
$$3 \cdot 0 + 4y = 5$$
$$4y = 5$$
$$y = \frac{5}{4}$$

Thus, $\left(0, \frac{5}{4}\right)$ is the y-intercept.

It is helpful to plot another point since the intercepts are so close together. This point can also serve as a check.

We let $x = 3$. Then
$$3x + 4y = 5$$
$$3 \cdot 3 + 4y = 5$$
$$9 + 4y = 5$$
$$4y = -4$$
$$y = -1$$

Plot the point $(3, -1)$ and the intercepts and draw the line.

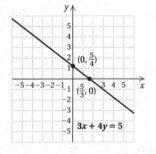

37. $y = -3 - 3x$

To find the x-intercept, let $y = 0$. Then solve for x.
$$y = -3 - 3x$$
$$0 = -3 - 3x$$
$$3x = -3$$
$$x = -1$$

Thus, $(-1, 0)$ is the x-intercept.

To find the y-intercept, let $x = 0$. Then solve for y.
$$y = -3 - 3x$$
$$y = -3 - 3 \cdot 0$$
$$y = -3$$

Thus, $(0, -3)$ is the y-intercept.

Plot these points and draw the graph.

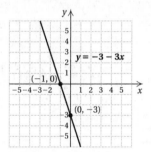

We use a third point as a check.

We let $x = -2$. Then
$$y = -3 - 3x$$
$$y = -3 - 3 \cdot (-2)$$
$$y = -3 + 6$$
$$y = 3$$

The point $(-2, 3)$ is on the graph, so the graph is probably correct.

39. $y - 3x = 0$

To find the x-intercept, let $y = 0$. Then solve for x.
$$0 - 3x = 0$$
$$-3x = 0$$
$$x = 0$$

Thus, $(0, 0)$ is the x-intercept. Note that this is also the y-intercept.

In order to graph the line, we will find a second point.

When $x = 1$, $y - 3 \cdot 1 = 0$
$$y - 3 = 0$$
$$y = 3$$

Plot the points and draw the graph.

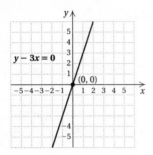

We use a third point as a check.

We let $x = -1$. Then

$$y - 3(-1) = 0$$
$$y + 3 = 0$$
$$y = -3$$

The point $(-1, -3)$ is on the graph, so the graph is probably correct.

41. $x = -2$

Any ordered pair $(-2, y)$ is a solution. The variable x must be -2, but y can be any number we choose. A few solutions are listed below. Plot these points and draw the line.

x	y
-2	-2
-2	0
-2	4

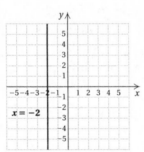

43. $y = 2$

Any ordered pair $(x, 2)$ is a solution. The variable y must be 2, but x can be any number we choose. A few solutions are listed below. Plot these points and draw the line.

x	y
-3	2
0	2
2	2

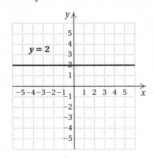

45. $x = 2$

Any ordered pair $(2, y)$ is a solution. The variable x must be 2, but y can be any number we choose. A few solutions are listed below. Plot these points and draw the line.

x	y
2	-1
2	4
2	5

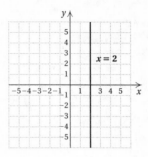

47. $y = 0$

Any ordered pair $(x, 0)$ is a solution. The variable y must be 0, but x can be any number we choose. A few solutions are listed below. Plot these points and draw the line.

x	y
-5	0
-1	0
3	0

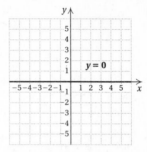

49. $x = \dfrac{3}{2}$

Any ordered pair $\left(\dfrac{3}{2}, y\right)$ is a solution. The variable x must be $\dfrac{3}{2}$, but y can be any number we choose. A few solutions are listed below. Plot these points and draw the line.

x	y
$\dfrac{3}{2}$	-2
$\dfrac{3}{2}$	0
$\dfrac{3}{2}$	4

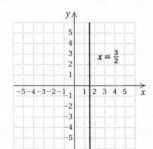

51. $3y = -5$

$$y = -\dfrac{5}{3} \qquad \text{Solving for } y$$

Any ordered pair $\left(x, -\dfrac{5}{3}\right)$ is a solution. A few solutions are listed below. Plot these points and draw the line.

x	y
-3	$-\dfrac{5}{3}$
0	$-\dfrac{5}{3}$
2	$-\dfrac{5}{3}$

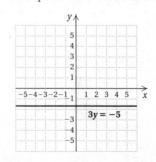

53. $4x + 3 = 0$

$$4x = -3$$

$$x = -\frac{3}{4} \quad \text{Solving for } x$$

Any ordered pair $\left(-\frac{3}{4}, y\right)$ is a solution. A few solutions are listed below. Plot these points and draw the line.

x	y
$-\dfrac{3}{4}$	-2
$-\dfrac{3}{4}$	0
$-\dfrac{3}{4}$	3

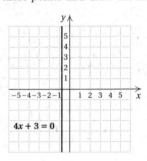

55. $48 - 3y = 0$

$$-3y = -48$$

$$y = 16 \quad \text{Solving for } y$$

Any ordered pair $(x, 16)$ is a solution. A few solutions are listed below. Plot these points and draw the line.

x	y
-4	16
0	16
2	16

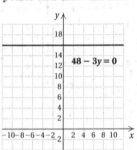

57. Note that every point on the horizontal line passing through $(0, -1)$ has -1 as the y-coordinate. Thus, the equation of the line is $y = -1$.

59. Note that every point on the vertical line passing through $(4, 0)$ has 4 as the x-coordinate. Thus, the equation of the line is $x = 4$.

61. Discussion and Writing Exercise

63. *Familiarize.* Let $p =$ the percent of desserts sold that will be pie.

Translate. We reword the problem.

40 is what percent of 250?

$\downarrow \downarrow$ $\downarrow$ $\downarrow$ $\downarrow$

$40 = \quad p \quad \cdot \ 250$

Solve. We solve the equation.

$$40 = p \cdot 250$$

$$\frac{40}{250} = \frac{p \cdot 250}{250}$$

$$0.16 = p$$

$$16\% = p$$

Check. We can find 16% of 250:

$$16\% \cdot 250 = 0.16 \cdot 250 = 40$$

The answer checks.

State. 16% of the desserts sold will be pie.

65. $-1.6x < 64$

$$\frac{-1.6x}{-1.6} > \frac{64}{-1.6} \quad \begin{array}{l}\text{Dividing by } -1.6 \text{ and reversing} \\ \text{the inequality symbol}\end{array}$$

$$x > -40$$

The solution set is $\{x | x > -40\}$.

67. $x + (x - 1) < (x + 2) - (x + 1)$

$$2x - 1 < x + 2 - x - 1$$

$$2x - 1 < 1$$

$$2x < 2$$

$$x < 1$$

The solution set is $\{x | x < 1\}$.

69. A line parallel to the x-axis has an equation of the form $y = b$. Since the y-coordinate of one point on the line is -4, then $b = -4$ and the equation is $y = -4$.

71. Substitute -4 for x and 0 for y.

$$3(-4) + k = 5 \cdot 0$$

$$-12 + k = 0$$

$$k = 12$$

Exercise Set 3.3

1. We consider (x_1, y_1) to be $(-3, 5)$ and (x_2, y_2) to be $(4, 2)$.

$$m = \frac{y_2 - y_1}{x_2 - x_1} = \frac{2 - 5}{4 - (-3)} = \frac{-3}{7} = -\frac{3}{7}$$

3. We can choose any two points. We consider (x_1, y_1) to be $(-3, -1)$ and (x_2, y_2) to be $(0, 1)$.

$$m = \frac{y_2 - y_1}{x_2 - x_1} = \frac{1 - (-1)}{0 - (-3)} = \frac{2}{3}$$

5. We can choose any two points. We consider (x_1, y_1) to be $(-4, -2)$ and (x_2, y_2) to be $(4, 4)$.

$$m = \frac{y_2 - y_1}{x_2 - x_1} = \frac{4 - (-2)}{4 - (-4)} = \frac{6}{8} = \frac{3}{4}$$

7. We consider (x_1, y_1) to be $(-4, -2)$ and (x_2, y_2) to be $(3, -2)$.

$$m = \frac{y_2 - y_1}{x_2 - x_1} = \frac{-2 - (-2)}{3 - (-4)} = \frac{0}{7} = 0$$

9. We plot $(-2, 4)$ and $(3, 0)$ and draw the line containing these points.

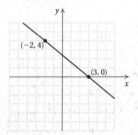

To find the slope, consider (x_1, y_1) to be $(-2, 4)$ and (x_2, y_2) to be $(3, 0)$.

$$m = \frac{y_2 - y_1}{x_2 - x_1} = \frac{0 - 4}{3 - (-2)} = \frac{-4}{5} = -\frac{4}{5}$$

11. We plot $(-4, 0)$ and $(-5, -3)$ and draw the line containing these points.

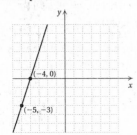

To find the slope, consider (x_1, y_1) to be $(-4, 0)$ and (x_2, y_2) to be $(-5, -3)$.

$$m = \frac{y_2 - y_1}{x_2 - x_1} = \frac{-3 - 0}{-5 - (-4)} = \frac{-3}{-1} = 3$$

13. We plot $(-4, 2)$ and $(2, -3)$ and draw the line containing these points.

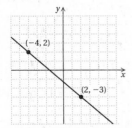

To find the slope, consider (x_1, y_1) to be $(-4, 2)$ and (x_2, y_2) to be $(2, -3)$.

$$m = \frac{y_2 - y_1}{x_2 - x_1} = \frac{-3 - 2}{2 - (-4)} = \frac{-5}{6} = -\frac{5}{6}$$

15. We plot $(5, 3)$ and $(-3, -4)$ and draw the line containing these points.

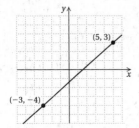

To find the slope, consider (x_1, y_1) to be $(5, 3)$ and (x_2, y_2) to be $(-3, -4)$.

$$m = \frac{y_2 - y_1}{x_2 - x_1} = \frac{-4 - 3}{-3 - 5} = \frac{-7}{-8} = \frac{7}{8}$$

17. $m = \dfrac{-\frac{1}{2} - \frac{3}{2}}{2 - 5} = \dfrac{-2}{-3} = \dfrac{2}{3}$

19. $m = \dfrac{-2 - 3}{4 - 4} = \dfrac{-5}{0}$

Since division by 0 is not defined, the slope is not defined.

21. $m = \dfrac{-3 - 7}{15 - (-11)} = \dfrac{-10}{26} = -\dfrac{5}{13}$

23. $m = \dfrac{\frac{3}{11} - \frac{3}{11}}{\frac{5}{4} - \left(-\frac{1}{2}\right)} = \dfrac{0}{\frac{7}{4}} = 0$

25. $m = \dfrac{\text{rise}}{\text{run}} = \dfrac{2.4}{8.2} = \dfrac{2.4}{8.2} \cdot \dfrac{10}{10} = \dfrac{24}{82}$

$\qquad = \dfrac{2 \cdot 12}{2 \cdot 41} = \dfrac{12}{41}$

27. $m = \dfrac{\text{rise}}{\text{run}} = \dfrac{56}{258} = \dfrac{2 \cdot 28}{2 \cdot 129} = \dfrac{28}{129}$

29. Long's Peak rises 14,255 ft − 9600 ft = 4655 ft.

Grade $= \dfrac{4655}{15,840} \approx 0.294 \approx 29.4\%$

31. The rate of change is the slope of the line. We can use any two ordered pairs to find the slope. We choose $(2, 50)$ and $(8, 200)$.

Rate of change $= \dfrac{200 \text{ mi} - 50 \text{ mi}}{8 \text{ gal} - 2 \text{ gal}} = \dfrac{150 \text{ mi}}{6 \text{ gal}} =$

25 miles per gallon

33. The rate of change is the slope of the line. We can use any two ordered pairs to find the slope. We choose $(2, 2000)$ and $(4, 1000)$. (Note that units on the vertical axis are given in thousands.)

Rate of change $= \dfrac{\$1000 - \$2000}{4 \text{ yr} - 2 \text{ yr}} = \dfrac{-\$1000}{2 \text{ yr}} = -\$500$ per year

35. The rate of change is the slope of the line. We can use any two ordered pairs to find the slope. We choose $(1990, 550,000)$ and $(2003, 649,000)$.

Rate of change $= \dfrac{649,000 - 550,000}{2003 - 1990} = \dfrac{99,000}{13} \approx$

7600 people per year

37. $y = -10x + 7$

The equation is in the form $y = mx + b$, where $m = -10$. Thus, the slope is -10.

39. $y = 3.78x - 4$

The equation is in the form $y = mx + b$, where $m = 3.78$. Thus, the slope is 3.78.

41. We solve for y, obtaining an equation of the form $y = mx + b$.

$$3x - y = 4$$
$$-y = -3x + 4$$
$$-1(-y) = -1(-3x + 4)$$
$$y = 3x - 4$$

The slope is 3.

43. We solve for y, obtaining an equation of the form $y = mx + b$.

$$x + 5y = 10$$
$$5y = -x + 10$$
$$y = \frac{1}{5}(-x + 10)$$
$$y = -\frac{1}{5}x + 2$$

The slope is $-\frac{1}{5}$.

45. We solve for y, obtaining an equation of the form $y = mx + b$.

$$3x + 2y = 6$$
$$2y = -3x + 6$$
$$y = \frac{1}{2}(-3x + 6)$$
$$y = -\frac{3}{2}x + 3$$

The slope is $-\frac{3}{2}$.

47. The graph of $x = \frac{2}{15}$ is a vertical line, so the slope is not defined.

49. $y = -2.74x$

The equation is in the form $y = mx + b$, where $m = -2.74$. Thus, the slope is -2.74.

51. We solve for y, obtaining an equation of the form $y = mx + b$.

$$9x = 3y + 5$$
$$9x - 5 = 3y$$
$$\frac{1}{3}(9x - 5) = y$$
$$3x - \frac{5}{3} = y$$

The slope is 3.

53. We solve for y, obtaining an equation of the form $y = mx + b$.

$$5x - 4y + 12 = 0$$
$$5x + 12 = 4y$$
$$\frac{1}{4}(5x + 12) = y$$
$$\frac{5}{4}x + 3 = y$$

The slope is $\frac{5}{4}$.

55. $y = 4$

The equation can be thought of as $y = 0 \cdot x + 4$, so the slope is 0.

57. Discussion and Writing Exercise

59. $16\% = \frac{16}{100} = \frac{4 \cdot 4}{4 \cdot 25} = \frac{4}{25}$

61. $37.5\% = \frac{37.5}{100} = \frac{37.5}{100} \cdot \frac{10}{10} = \frac{375}{1000} = \frac{3 \cdot 125}{8 \cdot 125} = \frac{3}{8}$

63. *Translate.*

What is 15% of \$23.80?
$$a = 15\% \cdot 23.80$$

Solve. We convert to decimal notation and multiply.

$$a = 15\% \cdot 23.80 = 0.15 \cdot 23.80 = 3.57$$

The answer is \$3.57.

65. *Familiarize.* Let p = the percent of the cost of the meal represented by the tip.

Translate. We reword the problem.

\$8.50 is what percent of \$42.50?
$$8.50 = p \cdot 42.50$$

Solve. We solve the equation.

$$8.50 = p \cdot 42.50$$
$$0.2 = p$$
$$20\% = p$$

Check. We can find 20% of 42.50.

$$20\% \cdot 42.50 = 0.2 \cdot 42.50 = 8.50$$

The answer checks.

State. The tip was 20% of the cost of the meal.

67. *Familiarize.* Let c = the cost of the meal before the tip was added. Then the tip is $15\% \cdot c$.

Translate. We reword the problem.

Cost of meal plus tip is total cost
$$c + 15\% \cdot c = 51.92$$

Solve. We solve the equation.

$$c + 15\% \cdot c = 51.92$$
$$1 \cdot c + 0.15c = 51.92$$
$$1.15c = 51.92$$
$$c \approx 45.15$$

Check. We can find 15% of 45.15 and then add this to 45.15.

$$15\% \cdot 45.15 = 0.15 \cdot 45.15 \approx 6.77 \text{ and } 45.15 + 6.77 = 51.92$$

The answer checks.

State. Before the tip the meal cost \$45.15.

69. Note that the sum of the coordinates of each point on the graph is 5. Thus, we have $x + y = 5$, or $y = -x + 5$.

71. Note that each y-coordinate is 2 more than the corresponding x-coordinate. Thus, we have $y = x + 2$.

73.

$y = 0.35x - 7$

X	Y1
-10	-10.5
-9.9	-10.47
-9.8	-10.43
-9.7	-10.4
-9.6	-10.36
-9.5	-10.33
-9.4	-10.29

X = -10

75. $y = x^3 - 5$

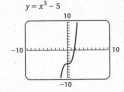

X	Y1
-10	-1005
-9.9	-975.3
-9.8	-946.2
-9.7	-917.7
-9.6	-889.7
-9.5	-862.4
-9.4	-835.6

X = -10

Exercise Set 3.4

1. $y = -4x - 9$

The equation is already in the form $y = mx + b$. The slope is -4 and the y-intercept is $(0, -9)$.

3. $y = 1.8x$

We can think of $y = 1.8x$ as $y = 1.8x + 0$. The slope is 1.8 and the y-intercept is $(0, 0)$.

5. We solve for y.

$$-8x - 7y = 21$$
$$-7y = 8x + 21$$
$$y = -\frac{1}{7}(8x + 21)$$
$$y = -\frac{8}{7}x - 3$$

The slope is $-\frac{8}{7}$ and the y-intercept is $(0, -3)$.

7. We solve for y.

$$4x = 9y + 7$$
$$4x - 7 = 9y$$
$$\frac{1}{9}(4x - 7) = y$$
$$\frac{4}{9}x - \frac{7}{9} = y$$

The slope is $\frac{4}{9}$ and the y-intercept is $\left(0, -\frac{7}{9}\right)$.

9. We solve for y.

$$-6x = 4y + 2$$
$$-6x - 2 = 4y$$
$$\frac{1}{4}(-6x - 2) = y$$
$$-\frac{3}{2}x - \frac{1}{2} = y$$

The slope is $-\frac{3}{2}$ and the y-intercept is $\left(0, -\frac{1}{2}\right)$.

11. $y = -17$

We can think of $y = -17$ as $y = 0x - 17$. The slope is 0 and the y-intercept is $(0, -17)$.

13. We substitute -7 for m and -13 for b in the equation $y = mx + b$.

$$y = -7x - 13$$

15. We substitute 1.01 for m and -2.6 for b in the equation $y = mx + b$.

$$y = 1.01x - 2.6$$

17. We know the slope is -2, so the equation is $y = -2x + b$. Using the point $(-3, 0)$, we substitute -3 for x and 0 for y in $y = -2x + b$. Then we solve for b.

$$y = -2x + b$$
$$0 = -2(-3) + b$$
$$0 = 6 + b$$
$$-6 = b$$

Thus, we have the equation $y = -2x - 6$.

19. We know the slope is $\frac{3}{4}$, so the equation is $y = \frac{3}{4}x + b$. Using the point $(2, 4)$, we substitute 2 for x and 4 for y in $y = \frac{3}{4}x + b$. Then we solve for b.

$$y = \frac{3}{4}x + b$$
$$4 = \frac{3}{4} \cdot 2 + b$$
$$4 = \frac{3}{2} + b$$
$$\frac{5}{2} = b$$

Thus, we have the equation $y = \frac{3}{4}x + \frac{5}{2}$.

21. We know the slope is 1, so the equation is $y = 1 \cdot x + b$, or $y = x + b$. Using the point $(2, -6)$, we substitute 2 for x and -6 for y in $y = x + b$. Then we solve for y.

$$y = x + b$$
$$-6 = 2 + b$$
$$-8 = b$$

Thus, we have the equation $y = x - 8$.

23. We substitute -3 for m and 3 for b in the equation $y = mx + b$.

$$y = -3x + 3$$

25. $(12, 16)$ and $(1, 5)$

First we find the slope.

$$m = \frac{16 - 5}{12 - 1} = \frac{11}{11} = 1$$

Thus, $y = 1 \cdot x + b$, or $y = x + b$. We can use either point to find b. We choose $(1, 5)$. Substitute 1 for x and 5 for y in $y = x + b$.

$$y = x + b$$
$$5 = 1 + b$$
$$4 = b$$

Thus, the equation is $y = x + 4$.

27. $(0, 4)$ and $(4, 2)$

First we find the slope.

$$m = \frac{4 - 2}{0 - 4} = \frac{2}{-4} = -\frac{1}{2}$$

Thus, $y = -\frac{1}{2}x + b$. One of the given points is the y-intercept $(0, 4)$. Thus, we substitute 4 for b in $y = -\frac{1}{2}x + b$. The equation is $y = -\frac{1}{2}x + 4$.

29. $(3, 2)$ and $(1, 5)$

First we find the slope.

$$m = \frac{2 - 5}{3 - 1} = \frac{-3}{2} = -\frac{3}{2}$$

Thus, $y = -\frac{3}{2}x + b$. We can use either point to find b. We choose $(3, 2)$. Substitute 3 for x and 2 for y in $y = -\frac{3}{2}x + b$.

$$y = -\frac{3}{2}x + b$$

$$2 = -\frac{3}{2} \cdot 3 + b$$

$$2 = -\frac{9}{2} + b$$

$$\frac{13}{2} = b$$

Thus, the equation is $y = -\frac{3}{2}x + \frac{13}{2}$.

31. $(-4, 5)$ and $(-2, -3)$

First we find the slope.

$$m = \frac{5 - (-3)}{-4 - (-2)} = \frac{8}{-2} = -4$$

Thus, $y = -4x + b$. We can use either point to find b. We choose $(-4, 5)$. Substitute -4 for x and 5 for y in $y = -4x + b$.

$$y = -4x + b$$

$$5 = -4(-4) + b$$

$$5 = 16 + b$$

$$-11 = b$$

Thus, the equation is $y = -4x - 11$.

33. a) First we find the slope.

$$m = \frac{150 - 105}{20 - 80} = \frac{45}{-60} = -0.75$$

Thus, $T = -0.75a + b$. We can use either point to find b. We choose $(20, 150)$. Substitute 20 for a and 150 for T in $T = -0.75a + b$.

$$150 = -0.75(20) + b$$

$$150 = -15 + b$$

$$165 = b$$

Thus, the equation is $T = -0.75a + 165$.

b) The rate of change is the slope, -0.75 beats per minute per year.

c) Substitute 50 for a and calculate T.

$$T = -0.75a + 165$$

$$T = -0.75(50) + 165$$

$$T = -37.5 + 165$$

$$T = 127.5$$

The target heart rate for a 50 year old person is 127.5 beats per minute.

35. Discussion and Writing Exercise

37.
$$3x - 4(9 - x) = 17$$
$$3x - 36 + 4x = 17$$
$$7x - 36 = 17$$
$$7x = 53$$
$$x = \frac{53}{7}$$

The solution is $\frac{53}{7}$.

39.
$$40(2x - 7) = 50(4 - 6x)$$
$$80x - 280 = 200 - 300x$$
$$380x - 280 = 200$$
$$380x = 480$$
$$x = \frac{480}{380}$$
$$x = \frac{24}{19}$$

The solution is $\frac{24}{19}$.

41.
$$3x - 9x + 21x - 15x = 6x - 12 - 24x + 18$$
$$0 = -18x + 6$$
$$18x = 6$$
$$x = \frac{6}{18}$$
$$x = \frac{1}{3}$$

The solution is $\frac{1}{3}$.

43.
$$3(x - 9x) + 21(x - 15x) = 6(x - 12) - 24(x + 18)$$
$$3(-8x) + 21(-14x) = 6x - 72 - 24x - 432$$
$$-24x - 294x = -18x - 504$$
$$-318x = -18x - 504$$
$$-300x = -504$$
$$x = \frac{504}{300} = \frac{\not{2} \cdot \not{6} \cdot 6 \cdot 7}{\not{2} \cdot \not{6} \cdot 5 \cdot 5}$$
$$x = \frac{42}{25}$$

The solution is $\frac{42}{25}$.

45. First find the slope of $3x - y + 4 = 0$.

$$3x - y + 4 = 0$$
$$3x + 4 = y$$

The slope is 3.

Thus, $y = 3x + b$. Using the point $(2, -3)$, we substitute 2 for x and -3 for y in $y = 3x + b$. Then we solve for b.

$$y = 3x + b$$
$$-3 = 3 \cdot 2 + b$$
$$-3 = 6 + b$$
$$-9 = b$$

Thus, the equation is $y = 3x - 9$.

47. First find the slope of $3x - 2y = 8$.

$$3x - 2y = 8$$
$$-2y = -3x + 8$$
$$y = \frac{3}{2}x - 4$$

The slope is $\frac{3}{2}$.

Then find the y-intercept of $2y + 3x = -4$.

$$2y + 3x = -4$$
$$2y = -3x - 4$$
$$y = -\frac{3}{2}x - 2$$

The y-intercept is $(0, -2)$.

Finally, write the equation of the line with slope $\frac{3}{2}$ and y-intercept $(0, -2)$.

$$y = mx + b$$
$$y = \frac{3}{2}x + (-2)$$
$$y = \frac{3}{2}x - 2$$

Exercise Set 3.5

1. Slope $\frac{2}{5}$; y-intercept $(0, 1)$

We plot $(0, 1)$ and from there move up 2 units and right 5 units. This locates the point $(5, 3)$. We plot $(5, 3)$ and draw a line passing through $(0, 1)$ and $(5, 3)$.

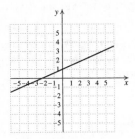

3. Slope $\frac{5}{3}$; y-intercept $(0, -2)$

We plot $(0, -2)$ and from there move up 5 units and right 3 units. This locates the point $(3, 3)$. We plot $(3, 3)$ and draw a line passing through $(0, -2)$ and $(3, 3)$.

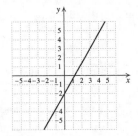

5. Slope $-\frac{3}{4}$; y-intercept $(0, 5)$

We plot $(0, 5)$. We can think of the slope as $\frac{-3}{4}$, so from $(0, 5)$ we move down 3 units and right 4 units. This locates the point $(4, 2)$. We plot $(4, 2)$ and draw a line passing through $(0, 5)$ and $(4, 2)$.

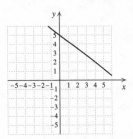

7. Slope $-\frac{1}{2}$; y-intercept $(0, 3)$

We plot $(0, 3)$. We can think of the slope as $\frac{-1}{2}$, so from $(0, 3)$ we move down 1 unit and right 2 units. This locates the point $(2, 2)$. We plot $(2, 2)$ and draw a line passing through $(0, 3)$ and $(2, 2)$

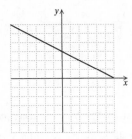

9. Slope 2; y-intercept $(0, -4)$

We plot $(0, -4)$. We can think of the slope as $\frac{2}{1}$, so from $(0, -4)$ we move up 2 units and right 1 unit. This locates the point $(1, -2)$. We plot $(1, -2)$ and draw a line passing through $(0, -4)$ and $(1, -2)$.

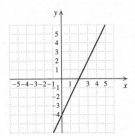

11. Slope -3; y-intercept $(0, 2)$

We plot $(0, 2)$. We can think of the slope as $\frac{-3}{1}$, so from $(0, 2)$ we move down 3 units and right 1 unit. This locates the point $(1, -1)$. We plot $(1, -1)$ and draw a line passing through $(0, 2)$ and $(1, -1)$.

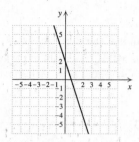

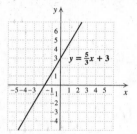

13. $y = \dfrac{3}{5}x + 2$

First we plot the y-intercept $(0, 2)$. We can start at the y-intercept and use the slope, $\dfrac{3}{5}$, to find another point. We move up 3 units and right 5 units to get a new point $(5, 5)$. Thinking of the slope as $\dfrac{-3}{-5}$ we can start at $(0, 2)$ and move down 3 units and left 5 units to get another point $(-5, -1)$.

19. $y = -\dfrac{3}{2}x - 2$

First we plot the y-intercept $(0, -2)$. We can start at the y-intercept and, thinking of the slope as $\dfrac{-3}{2}$, find another point by moving down 3 units and right 2 units to the point $(2, -5)$. Thinking of the slope as $\dfrac{3}{-2}$ we can start at $(0, -2)$ and move up 3 units and left 2 units to get another point $(-2, 1)$.

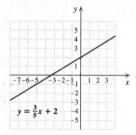

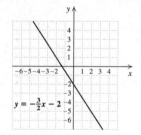

15. $y = -\dfrac{3}{5}x + 1$

First we plot the y-intercept $(0, 1)$. We can start at the y-intercept and, thinking of the slope as $\dfrac{-3}{5}$, find another point by moving down 3 units and right 5 units to the point $(5, -2)$. Thinking of the slope as $\dfrac{3}{-5}$ we can start at $(0, 1)$ and move up 3 units and left 5 units to get another point $(-5, 4)$.

21. We first rewrite the equation in slope-intercept form.
$$2x + y = 1$$
$$y = -2x + 1$$
Now we plot the y-intercept $(0, 1)$. We can start at the y-intercept and, thinking of the slope as $\dfrac{-2}{1}$, find another point by moving down 2 units and right 1 unit to the point $(1, -1)$. In a similar manner, we can move from the point $(1, -1)$ to find a third point $(2, -3)$.

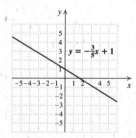

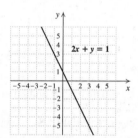

17. $y = \dfrac{5}{3}x + 3$

First we plot the y-intercept $(0, 3)$. We can start at the y-intercept and use the slope, $\dfrac{5}{3}$, to find another point. We move up 5 units and right 3 units to get a new point $(3, 8)$. Thinking of the slope as $\dfrac{-5}{-3}$ we can start at $(0, 3)$ and move down 5 units and left 3 units to get another point $(-3, -2)$.

23. We first rewrite the equation in slope-intercept form.
$$3x - y = 4$$
$$-y = -3x + 4$$
$$y = 3x - 4 \quad \text{Multiplying by } -1$$
Now we plot the y-intercept $(0, -4)$. We can start at the y-intercept and, thinking of the slope as $\dfrac{3}{1}$, find another point by moving up 3 units and right 1 unit to the point $(1, -1)$. In a similar manner, we can move from the point $(1, -1)$ to find a third point $(2, 2)$.

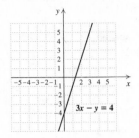

25. We first rewrite the equation in slope-intercept form.
$$2x + 3y = 9$$
$$3y = -2x + 9$$
$$y = \frac{1}{3}(-2x + 9)$$
$$y = -\frac{2}{3}x + 3$$

Now we plot the y-intercept $(0, 3)$. We can start at the y-intercept and, thinking of the slope as $\frac{-2}{3}$, find another point by moving down 2 units and right 3 units to the point $(3, 1)$. Thinking of the slope as $\frac{2}{-3}$ we can start at $(0, 3)$ and move up 2 units and left 3 units to get another point $(-3, 5)$.

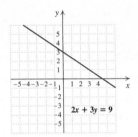

27. We first rewrite the equation in slope-intercept form.
$$x - 4y = 12$$
$$-4y = -x + 12$$
$$y = -\frac{1}{4}(-x + 12)$$
$$y = \frac{1}{4}x - 3$$

Now we plot the y-intercept $(0, -3)$. We can start at the y-intercept and use the slope, $\frac{1}{4}$, to find another point. We move up 1 unit and right 4 units to the point $(4, -2)$. Thinking of the slope as $\frac{-1}{-4}$ we can start at $(0, -3)$ and move down 1 unit and left 4 units to get another point $(-4, -4)$.

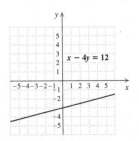

29. We first rewrite the equation in slope-intercept form.
$$x + 2y = 6$$
$$2y = -x + 6$$
$$y = \frac{1}{2}(-x + 6)$$
$$y = -\frac{1}{2}x + 3$$

Now we plot the y-intercept $(0, 3)$. We can start at the y-intercept and, thinking of the slope as $\frac{-1}{2}$, find another point by moving down 1 unit and right 2 units to the point $(2, 2)$. Thinking of the slope as $\frac{1}{-2}$ we can start at $(0, 3)$ and move up 1 unit and left 2 units to get another point $(-2, 4)$.

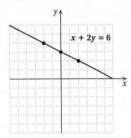

31. Discussion and Writing Exercise

33. $m = \dfrac{y_2 - y_1}{x_2 - x_1} = \dfrac{7 - (-6)}{8 - (-2)} = \dfrac{13}{10}$

35. $m = \dfrac{y_2 - y_1}{x_2 - x_1} = \dfrac{4.6 - (-2.3)}{14.5 - 4.5} = \dfrac{6.9}{10} = \dfrac{69}{100}$, or 0.69

37. $m = \dfrac{y_2 - y_1}{x_2 - x_1} = \dfrac{-6 - (-6)}{8 - (-2)} = \dfrac{0}{10} = 0$

39. $m = \dfrac{y_2 - y_1}{x_2 - x_1} = \dfrac{-4 - (-1)}{11 - 11} = \dfrac{-3}{0}$

Since division by 0 is not defined, the slope is not defined.

41. Rate of change $= \dfrac{\text{Change in number of transplants}}{\text{Change in years}} =$
$$\dfrac{15,120 - 8873}{2003 - 1988} = \dfrac{6247}{15} \approx 416$$
Kidney transplants are increasing at a rate of about 416 per year. The slope of the line is 416.

43. For residents in excess of 2, the rate of change is 1.5 ft^3 per person. For 1-2 people, the number of residents in excess of 2 is 0, so the x-intercept is $(0, 16)$. Then the equation is $y = 1.5x + 16$.

45. First we plot $(-3, 1)$. Then, thinking of the slope as $\frac{2}{1}$, from $(-3, 1)$ we move up 2 units and right 1 unit to locate the point $(-2, 3)$. We plot $(-2, 3)$ and draw a line passing through $(-3, 1)$ and $(-2, 3)$.

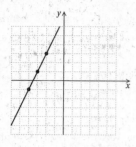

Exercise Set 3.6

1. 1. The first equation is already solved for y:
$$y = x + 4$$
2.   We solve the second equation for y:
$$y - x = -3$$
$$y = x - 3$$
The slope of each line is 1. The y-intercepts, $(0, 4)$ and $(0, -3)$, are different. The lines are parallel.

3. We solve each equation for y:

1. $y + 3 = 6x$ 2. $-6x - y = 2$
    $y = 6x - 3$ $-y = 6x + 2$
                           $y = -6x - 2$

The slope of the first line is 6 and of the second is -6. Since the slopes are different, the lines are not parallel.

5. We solve each equation for y:

1. $10y + 32x = 16.4$ 2. $y + 3.5 = 0.3125x$
    $10y = -32x + 16.4$ $y = 0.3125x - 3.5$
      $y = -3.2x + 1.64$

The slope of the first line is -3.2 and of the second is 0.3125. Since the slopes are different, the lines are not parallel.

7. 1. The first equation is already solved for y:
$$y = 2x + 7$$
2.   We solve the second equation for y:
$$5y + 10x = 20$$
$$5y = -10x + 20$$
$$y = -2x + 4$$
The slope of the first line is 2 and of the second is -2. Since the slopes are different, the lines are not parallel.

9. We solve each equation for y:

1. $3x - y = -9$ 2. $2y - 6x = -2$
  $3x + 9 = y$ $2y = 6x - 2$
                       $y = 3x - 1$

The slope of each line is 3. The y-intercepts, $(0, 9)$ and $(0, -1)$ are different. The lines are parallel.

11. $x = 3$,
$x = 4$

These are vertical lines with equations of the form $x = p$ and $x = q$, where $p \neq q$. Thus, they are parallel.

13. 1. The first equation is already solved for y:
$$y = -4x + 3$$
2.   We solve the second equation for y:
$$4y + x = -1$$
$$4y = -x - 1$$
$$y = -\frac{1}{4}x - \frac{1}{4}$$
The slopes are -4 and $-\frac{1}{4}$. Their product is $-4\left(-\frac{1}{4}\right) = 1$. Since the product of the slopes is not -1, the lines are not perpendicular.

15. We solve each equation for y:

1. $x + y = 6$ 2. $4y - 4x = 12$
    $y = -x + 6$ $4y = 4x + 12$
                       $y = x + 3$

The slopes are -1 and 1. Their product is $-1 \cdot 1 = -1$. The lines are perpendicular.

17. 1. The first equation is already solved for y:
$$y = -0.3125x + 11$$
2.   We solve the second equation for y:
$$y - 3.2x = -14$$
$$y = 3.2x - 14$$
The slopes are -0.3125 and 3.2. Their product is $-0.3125(3.2) = -1$. The lines are perpendicular.

19. 1. The first equation is already solved for y:
$$y = -x + 8$$
2.   We solve the second equation for y:
$$x - y = -1$$
$$x + 1 = y$$
The slopes are -1 and 1. Their product is $-1 \cdot 1 = -1$. The lines are perpendicular.

21. We solve each equation for y:

1. $\frac{3}{8}x - \frac{y}{2} = 1$

     $8\left(\frac{3}{8}x - \frac{y}{2}\right) = 8 \cdot 1$

    $8 \cdot \frac{3}{8}x - 8 \cdot \frac{y}{2} = 8$

          $3x - 4y = 8$

            $-4y = -3x + 8$

               $y = \frac{3}{4}x - 2$

2. $\frac{4}{3}x - y + 1 = 0$

    $\frac{4}{3}x + 1 = y$

The slopes are $\frac{3}{4}$ and $\frac{4}{3}$. Their product is $\frac{3}{4}\left(\frac{4}{3}\right) = 1$. Since

the product of the slopes is not -1, the lines are not perpendicular.

23. $x = 0$,

$y = -2$

The first line is vertical and the second is horizontal, so the lines are perpendicular.

25. We solve each equation for y:

1. $3y + 21 = 2x$ 2. $3y = 2x + 24$

   $3y = 2x - 21$ $y = \dfrac{2}{3}x + 8$

   $y = \dfrac{2}{3}x - 7$

The slope of each line is $\dfrac{2}{3}$. The y-intercepts, $(0, -7)$ and $(0, 8)$, are different. The lines are parallel.

27. We solve each equation for y:

1. $3y = 2x - 21$ 2. $2y - 16 = 3x$

   $y = \dfrac{2}{3}x - 7$ $2y = 3x + 16$

                     $y = \dfrac{3}{2}x + 8$

The slopes, $\dfrac{2}{3}$ and $\dfrac{3}{2}$, are different so the lines are not parallel. The product of the slopes is $\dfrac{2}{3} \cdot \dfrac{3}{2} = 1 \neq -1$, so the lines are not perpendicular. Thus, the lines are neither parallel nor perpendicular.

29. Discussion and Writing Exercise

31. Equations with the same solutions are called <u>equivalent equations</u>.

33. The <u>multiplication principle</u> for equations asserts that when we multiply or divide by the same non-zero number on both sides of an equation, we get equivalent equations.

35. <u>Vertical</u> lines are graphs of equations of the $x = a$.

37. The <u>x-intercept</u> of a line, if it exists, indicates where the line crosses the x-axis.

39. First we find the slope of the given line:

$y - 3x = 4$

$y = 3x + 4$

The slope is 3.

Then we use the slope-intercept equation to write the equation of a line with slope 3 and y-intercept $(0, 6)$:

$y = mx + b$

$y = 3x + 6$ Substituting 3 for m and 6 for b

41. First we find the slope of the given line:

$3y - x = 0$

$3y = x$

$y = \dfrac{1}{3}x$

The slope is $\dfrac{1}{3}$.

We can find the slope of the line perpendicular to the given line by taking the reciprocal of $\dfrac{1}{3}$ and changing the sign. We get -3.

Then we use the slope-intercept equation to write the equation of a line with slope -3 and y-intercept $(0, 2)$:

$y = mx + b$

$y = -3x + 2$ Substituting -3 for m and 2 for b

43. First we find the slope of the given line:

$4x - 8y = 12$

$-8y = -4x + 12$

$y = \dfrac{1}{2}x - \dfrac{3}{2}$

The slope is $\dfrac{1}{2}$, so the equation is $y = \dfrac{1}{2}x + b$. Substitute -2 for x and 0 for y and solve for b.

$y = \dfrac{1}{2}x + b$

$0 = \dfrac{1}{2}(-2) + b$

$0 = -1 + b$

$1 = b$

Thus, the equation is $y = \dfrac{1}{2}x + 1$.

45. We find the slope of each line:

1. $4y = kx - 6$ 2. $5x + 20y = 12$

   $y = \dfrac{k}{4}x - \dfrac{3}{2}$ $20y = -5x + 12$

                         $y = -\dfrac{1}{4}x + \dfrac{3}{5}$

The slopes are $\dfrac{k}{4}$ and $-\dfrac{1}{4}$. If the lines are perpendicular, the product of their slopes is -1.

$\dfrac{k}{4}\left(-\dfrac{1}{4}\right) = -1$

$-\dfrac{k}{16} = -1$

$k = 16$

47. First we find the equation of A, a line containing the points $(1, -1)$ and $(4, 3)$:

The slope is $\dfrac{3 - (-1)}{4 - 1} = \dfrac{4}{3}$, so the equation is $y = \dfrac{4}{3}x + b$. Use either point to find b. We choose $(1, -1)$.

$y = \dfrac{4}{3}x + b$

$-1 = \dfrac{4}{3} \cdot 1 + b$

$-1 = \dfrac{4}{3} + b$

$-\dfrac{7}{3} = b$

Thus, the equation of line A is $y = \dfrac{4}{3}x - \dfrac{7}{3}$.

The slope of A is $\frac{4}{3}$. Since A and B are perpendicular we find the slope of B by taking the reciprocal of $\frac{4}{3}$ and changing the sign. We get $-\frac{3}{4}$, so the equation is $y = -\frac{3}{4}x + b$. We use the point $(1, -1)$ to find b.

$$y = -\frac{3}{4}x + b$$

$$-1 = -\frac{3}{4} \cdot 1 + b$$

$$-1 = -\frac{3}{4} + b$$

$$-\frac{1}{4} = b$$

Thus, the equation of line B is $y = -\frac{3}{4}x - \frac{1}{4}$.

Exercise Set 3.7

1. We use alphabetical order to replace x by -3 and y by -5.

$$\frac{-x - 3y < 18}{-(-3) - 3(-5) \ ? \ 18}$$

$$\begin{array}{c|c} 3 + 15 & \\ 18 & \text{FALSE} \end{array}$$

Since $18 < 18$ is false, $(-3, -5)$ is not a solution.

3. We use alphabetical order to replace x by $\frac{1}{2}$ and y by $-\frac{1}{4}$.

$$\frac{7y - 9x \le -3}{7\left(-\frac{1}{4}\right) - 9 \cdot \frac{1}{2} \ ? \ -3}$$

$$\begin{array}{c|c} -\dfrac{7}{4} - \dfrac{9}{2} & \\ -\dfrac{7}{4} - \dfrac{18}{4} & \\ -\dfrac{25}{4} & \\ -6\dfrac{1}{4} & \text{TRUE} \end{array}$$

Since $-6\frac{1}{4} \le -3$ is true, $\left(\frac{1}{2}, -\frac{1}{4}\right)$ is a solution.

5. Graph $x > 2y$.

First graph the line $x = 2y$, or $y = \frac{1}{2}x$. Two points on the line are $(0, 0)$ and $(4, 2)$. We draw a dashed line since the inequality symbol is $>$. Then we pick a test point that is not on the line. We try $(-2, 1)$.

$$\frac{x > 2y}{-2 \ ? \ 2 \cdot 1}$$

$$\begin{array}{c|c} 2 & \text{FALSE} \end{array}$$

We see that $(-2, 1)$ is not a solution of the inequality, so we shade the points in the region that does not contain $(-2, 1)$.

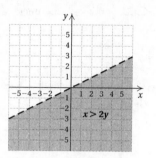

7. Graph $y \le x - 3$.

First graph the line $y = x - 3$. The intercepts are $(0, -3)$ and $(3, 0)$. We draw a solid line since the inequality symbol is $\le$. Then we pick a test point that is not on the line. We try $(0, 0)$.

$$\frac{y \le x - 3}{0 \ ? \ 0 - 3}$$

$$\begin{array}{c|c} -3 & \text{FALSE} \end{array}$$

We see that $(0, 0)$ is not a solution of the inequality, so we shade the region that does not contain $(0, 0)$.

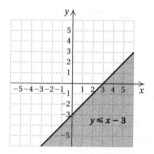

9. Graph $y < x + 1$.

First graph the line $y = x + 1$. The intercepts are $(0, 1)$ and $(-1, 0)$. We draw a dashed line since the inequality symbol is $<$. Then we pick a test point that is not on the line. We try $(0, 0)$.

$$\frac{y < x + 1}{0 \ ? \ 0 + 1}$$

$$\begin{array}{c|c} 1 & \text{TRUE} \end{array}$$

Since $(0, 0)$ is a solution of the inequality, we shade the region that contains $(0, 0)$.

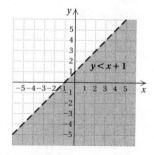

11. Graph $y \geq x - 2$.

First graph the line $y = x - 2$. The intercepts are $(0, -2)$ and $(2, 0)$. We draw a solid line since the inequality symbol is $\geq$. Then we test the point $(0, 0)$.

$$\frac{y \geq x - 2}{0 \ ? \ 0 - 2}$$
$$\begin{array}{c|c} & -2 \quad \text{TRUE} \end{array}$$

Since $(0, 0)$ is a solution of the inequality, we shade the region containing $(0, 0)$.

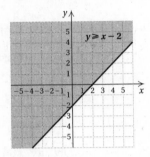

13. Graph $y \leq 2x - 1$.

First graph the line $y = 2x - 1$. The intercepts are $(0, -1)$ and $\left(\frac{1}{2}, 0\right)$. We draw a solid line since the inequality symbol is $\leq$. Then we test the point $(0, 0)$.

$$\frac{y \leq 2x - 1}{0 \ ? \ 2 \cdot 0 - 1}$$
$$\begin{array}{c|c} & -1 \quad \text{FALSE} \end{array}$$

Since $(0, 0)$ is not a solution of the inequality, we shade the region that does not contain $(0, 0)$.

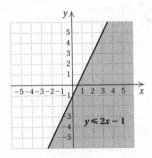

15. Graph $x + y \leq 3$.

First graph the line $x + y = 3$. The intercepts are $(0, 3)$ and $(3, 0)$. We draw a solid line since the inequality symbol is $\leq$. Then we test the point $(0, 0)$.

$$\frac{x + y \leq 3}{0 + 0 \ ? \ 3}$$
$$\begin{array}{c|c} 0 & \text{TRUE} \end{array}$$

Since $(0, 0)$ is a solution of the inequality, we shade the region that contains $(0, 0)$.

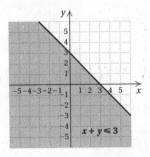

17. Graph $x - y > 7$.

First graph the line $x - y = 7$. The intercepts are $(0, -7)$ and $(7, 0)$. We draw a dashed line since the inequality symbol is $>$. Then we test the point $(0, 0)$.

$$\frac{x - y > 7}{0 - 0 \ ? \ 7}$$
$$\begin{array}{c|c} 0 & \text{FALSE} \end{array}$$

Since $(0, 0)$ is not a solution of the inequality, we shade the region that does not contain $(0, 0)$.

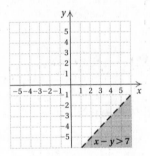

19. Graph $2x + 3y \leq 12$.

First graph the line $2x + 3y = 12$. The intercepts are $(0, 4)$ and $(6, 0)$. We draw a solid line since the inequality symbol is $\leq$. Then we test the point $(0, 0)$.

$$\frac{2x + 3y \leq 12}{2 \cdot 0 + 3 \cdot 0 \ ? \ 12}$$
$$\begin{array}{c|c} 0 & \text{TRUE} \end{array}$$

Since $(0, 0)$ is a solution of the inequality, we shade the region containing $(0, 0)$.

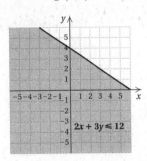

21. Graph $y \geq 1 - 2x$.

First graph the line $y = 1 - 2x$. The intercepts are $(0, 1)$ and $\left(\frac{1}{2}, 0\right)$. We draw a solid line since the inequality

symbol is $\geq$. Then we test the point $(0,0)$.

$$\frac{y \geq 1 - 2x}{0 \ ? \ 1 - 2 \cdot 0}$$
$$\begin{array}{c|c} & 1 \qquad \text{FALSE} \end{array}$$

Since $(0,0)$ is not a solution of the inequality, we shade the region that does not contain $(0,0)$.

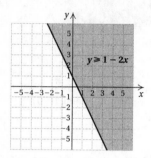

23. Graph $2x - 3y > 6$.

First graph the line $2x - 3y = 6$. The intercepts are $(0, -2)$ and $(3, 0)$. We draw a dashed line since the inequality symbol is $>$. Then we test the point $(0, 0)$.

$$\frac{2x - 3y > 6}{2 \cdot 0 - 3 \cdot 0 \ ? \ 6}$$
$$\begin{array}{c|c} 0 & \text{FALSE} \end{array}$$

Since $(0,0)$ is not a solution of the inequality, we shade the region that does not contain $(0,0)$.

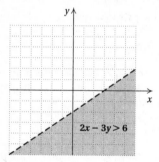

25. Graph $y \leq 3$.

First graph the line $y = 3$ using a solid line since the inequality symbol is $\leq$. Then pick a test point that is not on the line. We choose $(1, -2)$. We can write the inequality as $0x + y \leq 3$.

$$\frac{0x + y \leq 3}{0 \cdot 1 + (-2) \ ? \ 3}$$
$$\begin{array}{c|c} -2 & \text{TRUE} \end{array}$$

Since $(1, -2)$ is a solution of the inequality, we shade the region containing $(1, -2)$.

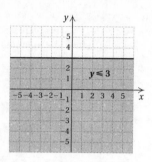

27. Graph $x \geq -1$.

Graph the line $x = 1$ using a solid line since the inequality symbol is $\geq$. Then pick a test point that is not on the line. We choose $(2, 3)$. We can write the inequality as $x + 0y \geq -1$.

$$\frac{x + 0y \geq -1}{2 + 0 \cdot 3 \ ? \ -1}$$
$$\begin{array}{c|c} 2 & \text{TRUE} \end{array}$$

Since $(2, 3)$ is a solution of the inequality, we shade the region containing $(2, 3)$.

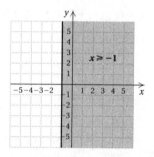

29. Discussion and Writing Exercise

31. First we solve each equation for y:

$$\begin{array}{ll} \text{1. } 5y + 50 = 4x & \qquad \text{2. } 5y = 4x + 15 \\ \quad 5y = 4x - 50 & \qquad \quad y = \frac{4}{5}x + 3 \\ \quad y = \frac{4}{5}x - 10 & \end{array}$$

The slope of each line is $\frac{4}{5}$. The y-intercepts, $(0, -10)$ and $(0, 3)$, are different. The lines are parallel.

33. First we solve each equation for y:

$$\begin{array}{ll} \text{1. } 5y + 50 = 4x & \qquad \text{2. } 4y = 5x + 12 \\ \quad 5y = 4x - 50 & \qquad \quad y = \frac{5}{4}x + 3 \\ \quad y = \frac{4}{5}x - 10 & \end{array}$$

The slope, $\frac{4}{5}$ and $\frac{5}{4}$, are different, so the lines are not parallel. The product of the slopes is $\frac{4}{5} \cdot \frac{5}{4} = 1 \neq -1$, so the lines are not perpendicular. Thus, the lines are neither parallel nor perpendicular.

35. The c children weigh $35c$ kg, and the a adults weigh $75a$ kg. Together, the children and adults weigh $35c + 75a$ kg.

When this total is more than 1000 kg the elevator is over-loaded, so we have $35c + 75a > 1000$. (Of course, c and a would also have to be nonnegative, but we will not deal with nonnegativity constraints here.)

To graph $35c + 75a > 1000$, we first graph $35c + 75a = 1000$ using a dashed line. Two points on the line are $(4, 20)$ and $(11, 5)$. (We are using alphabetical order of variables.) Then we test the point $(0, 0)$.

$$\begin{array}{c|c} 35c + 75a > 1000 \\ \hline 35 \cdot 0 + 75 \cdot 0 \ ? \ 1000 \\ 0 & \text{FALSE} \end{array}$$

Since $(0, 0)$ is not a solution of the inequality, we shade the region that does not contain $(0, 0)$.

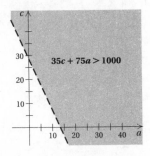

Chapter 3 Review Exercises

1. Point A is 5 units left and 1 unit down. The coordinates of A are $(-5, -1)$.

2. Point B is 2 units left and 5 units up. The coordinates of B are $(-2, 5)$.

3. Point C is 3 units right and 0 units up or down. The coordinates of C are $(3, 0)$.

4. $(2, 5)$ is 2 units right and 5 units up. See the graph following Exercise 6 below.

5. $(0, -3)$ is 0 units right or left and 3 units down. See the graph following Exercise 6 below.

6. $(-4, -2)$ is 4 units left and 2 units down. See the graph below.

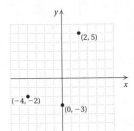

7. Since the first coordinate is positive and the second coordinate is negative, the point $(3, -8)$ is in quadrant IV.

8. Since both coordinates are negative, the point $(-20, -14)$ is in quadrant III.

9. Since both coordinates are positive, the point $(4.9, 1.3)$ is in quadrant I.

10. We substitute 2 for x and -6 for y.

$$\begin{array}{c|c} 2y - x = 10 \\ \hline 2(-6) - 2 \ ? \ 10 \\ -12 - 2 \\ -14 & \text{FALSE} \end{array}$$

Since $-14 = 10$ is false, the pair $(2, -6)$ is not a solution.

11. We substitute 0 for x and 5 for y.

$$\begin{array}{c|c} 2y - x = 10 \\ \hline 2 \cdot 5 - 0 \ ? \ 10 \\ 10 - 0 \\ 10 & \text{TRUE} \end{array}$$

Since $10 = 10$ is true, the pair $(0, 5)$ is a solution.

12. To show that a pair is a solution, we substitute, replacing x with the first coordinate and y with the second coordinate in each pair.

$$\begin{array}{c|c} 2x - y = 3 \\ \hline 2 \cdot 0 - (-3) \ ? \ 3 \\ 0 + 3 \\ 3 & \text{TRUE} \end{array}$$

$$\begin{array}{c|c} 2x - y = 3 \\ \hline 2 \cdot 2 - 1 \ ? \ 3 \\ 4 - 1 \\ 3 & \text{TRUE} \end{array}$$

In each case the substitution results in a true equation. Thus, $(0, -3)$ and $(2, 1)$ are both solutions of $2x - y = 3$. We graph these points and sketch the line passing through them.

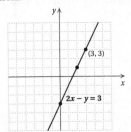

The line appears to pass through $(3, 3)$ also. We check to determine if $(3, 3)$ is a solution of $2x - y = 3$.

$$\begin{array}{c|c} 2x - y = 3 \\ \hline 2 \cdot 3 - 3 \ ? \ 3 \\ 6 - 3 \\ 3 & \text{TRUE} \end{array}$$

Thus, $(3, 3)$ is another solution. There are other correct answers, including $(-1, -5)$ and $(4, 5)$.

13. $y = 2x - 5$

The y-intercept is $(0, -5)$. We find two other points.

When $x = 2$, $y = 2 \cdot 2 - 5 = 4 - 5 = -1$.

When $x = 4$, $y = 2 \cdot 4 - 5 = 8 - 5 = 3$.

x	y
0	−5
2	−1
4	3

Plot these points, draw the line they determine, and label the graph $y = 2x - 5$.

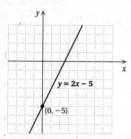

14. $y = -\dfrac{3}{4}x$

The equation is equivalent to $y = -\dfrac{3}{4}x + 0$. The y-intercept is $(0, 0)$. We find two other points.

When $x = -4$, $y = -\dfrac{3}{4}(-4) = 3$.

When $x = 4$, $y = -\dfrac{3}{4} \cdot 4 = -3$.

x	y
−4	3
0	0
4	−3

Plot these points, draw the line they determine, and label the graph $y = -\dfrac{3}{4}x$.

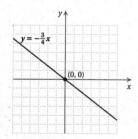

15. $y = -x + 4$

The y-intercept is $(0, 4)$. We find two other points.

When $x = -1$, $y = -(-1) + 4 = 1 + 4 = 5$.

When $x = 4$, $y = -4 + 4 = 0$.

x	y
−1	5
0	4
4	0

Plot these points, draw the line they determine, and label the graph $y = -x + 4$.

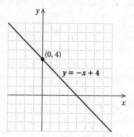

16. $y = 3 - 4x$, or $y = -4x + 3$

The y-intercept is $(0, 3)$. We find two other points.

When $x = 1$, $y = -4 \cdot 1 + 3 = -4 + 3 = -1$.

When $x = 2$, $y = -4 \cdot 2 + 3 = -8 + 3 = -5$.

x	y
0	3
1	−1
2	−5

Plot these points, draw the line they determine, and label the graph $y = 3 - 4x$.

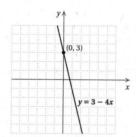

17. $y = 3$

Any ordered pair $(x, 3)$ is a solution. The variable y must be 3, but x can be any number we choose. A few solutions are listed below. Plot these points and draw the line.

x	y
−3	3
0	3
2	3

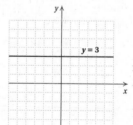

18. $5x - 4 = 0$

$5x = 4$

$x = \dfrac{4}{5}$ Solving for x

Any ordered pair $\left(\dfrac{4}{5}, y\right)$ is a solution. A few solutions are listed below. Plot these points and draw the graph.

x	y
$\frac{4}{5}$	-3
$\frac{4}{5}$	0
$\frac{4}{5}$	2

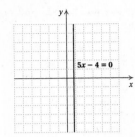

19. $x - 2y = 6$

To find the x-intercept, let $y = 0$. Then solve for x.

$$x - 2y = 6$$
$$x - 2 \cdot 0 = 6$$
$$x = 6$$

Thus, $(6, 0)$ is the x-intercept.

To find the y-intercept, let $x = 0$. Then solve for y.

$$x - 2y = 6$$
$$0 - 2y = 6$$
$$-2y = 6$$
$$y = -3$$

Thus, $(0, -3)$ is the y-intercept.

Plot these points and draw the graph.

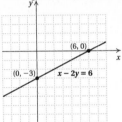

We use a third point as a check.

We let $x = -2$. Then

$$x - 2y = 6$$
$$-2 - 2y = 6$$
$$-2y = 8$$
$$y = -4.$$

The point $(-2, -4)$ is on the graph, so the graph is probably correct.

20. $5x - 2y = 10$

To find the x-intercept, let $y = 0$. Then solve for x.

$$5x - 2y = 10$$
$$5x - 2 \cdot 0 = 10$$
$$5x = 10$$
$$x = 2$$

Thus, $(2, 0)$ is the x-intercept.

To find the y-intercept, let $x = 0$. Then solve for y.

$$5x - 2y = 10$$
$$5 \cdot 0 - 2y = 10$$
$$-2y = 10$$
$$y = -5$$

Thus, $(0, -5)$ is the y-intercept.

Plot these points and draw the graph.

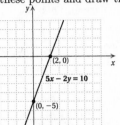

We use a third point as a check.

We let $x = 4$. Then

$$5x - 2y = 10$$
$$5 \cdot 4 - 2y = 10$$
$$20 - 2y = 10$$
$$-2y = -10$$
$$y = 5.$$

The point $(4, 5)$ is on the graph, so the graph is probably correct.

21. a) When $n = 1$, $S = \frac{3}{2} \cdot 1 + 13 = \frac{3}{2} + 13 = 1\frac{1}{2} + 13 = 14\frac{1}{2}$ ft^3.

When $n = 2$, $S = \frac{3}{2} \cdot 2 + 13 = 3 + 13 = 16$ ft^3.

When $n = 5$, $S = \frac{3}{2} \cdot 5 + 13 = \frac{15}{2} + 13 = 7\frac{1}{2} + 13 = 20\frac{1}{2}$ ft^3.

When $n = 10$, $S = \frac{3}{2} \cdot 10 + 13 = 15 + 13 = 28$ ft^3.

b) We plot the points found in part (a): $\left(1, 14\frac{1}{2}\right)$, $(2, 16)$, $\left(5, 20\frac{1}{2}\right)$ and $(10, 28)$. Then we draw the graph.

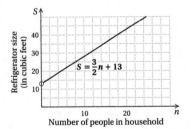

From the graph, it appears that an x-value of 3 corresponds to the S-value of $17\frac{1}{2}$, so the recommended size is $17\frac{1}{2}$ ft^3.

c) We substitute 22 for S and solve for n.

$$22 = \frac{3}{2}n + 13$$

$$9 = \frac{3}{2}n$$

$$\frac{2}{3} \cdot 9 = \frac{2}{3} \cdot \frac{3}{2}n$$

$$6 = n$$

A 22-ft^3 refrigerator is recommended for a household of 6 residents.

22. 5:30 P.M. is 2.5 hr after 3:00 P.M. In this time the number of driveways plowed was $13 - 7$, or 6.

a) Rate of change $= \dfrac{6 \text{ driveways}}{2.5 \text{ hr}} = 2.4$ driveways per hour

b) 2.5 hr $= 2.5 \times 1$ hr $= 2.5 \times 60$ min $= 150$ min

Rate of change $= \dfrac{150 \text{ min}}{6 \text{ driveways}} = 25$ minutes per driveway

23. We will use the points (11:00 A.M., 6 manicures) and (1:00 P.M., 14 manicures) to find the rate of change. Note that 1:00 P.M. is 2 hr after 11:00 A.M.

Rate of change $= \dfrac{14 \text{ manicures} - 6 \text{ manicures}}{2 \text{ hr}} =$

$\dfrac{8 \text{ manicures}}{2 \text{ hr}} = 4$ manicures per hour

24. We can choose any two points. We consider (x_1, y_1) to be $(-3, 1)$ and (x_2, y_2) to be $(3, 3)$.

$$m = \frac{y_2 - y_1}{x_2 - x_1} = \frac{3 - 1}{3 - (-3)} = \frac{2}{6} = \frac{1}{3}$$

25. We can choose any two points. We consider (x_1, y_1) to be $(3, 1)$ and (x_2, y_2) to be $(-3, 3)$.

$$m = \frac{3 - 1}{-3 - 3} = \frac{2}{-6} = -\frac{1}{3}$$

26. We plot $(-5, -2)$ and $(5, 4)$ and draw the line containing those points.

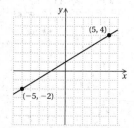

To find the slope, consider (x_1, y_1) to be $(-5, -2)$ and (x_2, y_2) to be $(5, 4)$.

$$m = \frac{y_2 - y_1}{x_2 - x_1} = \frac{4 - (-2)}{5 - (-5)} = \frac{6}{10} = \frac{3}{5}$$

27. We plot $(-5, 5)$ and $(4, -4)$ and draw the line containing those points.

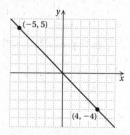

To find the slope, consider (x_1, y_1) to be $(4, -4)$ and (x_2, y_2) to be $(-5, 5)$.

$$m = \frac{y_2 - y_1}{x_2 - x_1} = \frac{5 - (-4)}{-5 - 4} = \frac{9}{-9} = -1$$

28. Grade $= \dfrac{315}{4500} = 0.07 = 7\%$

29. $y = -\dfrac{5}{8}x - 3$

The equation is in the form $y = mx + b$, where $m = -\dfrac{5}{8}$.

Thus, the slope is $-\dfrac{5}{8}$.

30. We solve for y, obtaining an equation of the form $y = mx + b$.

$$2x - 4y = 8$$

$$-4y = -2x + 8$$

$$-\frac{1}{4}(-4y) = -\frac{1}{4}(-2x + 8)$$

$$y = \frac{1}{2}x - 2$$

The slope is $\dfrac{1}{2}$.

31. The graph of $x = -2$ is a vertical line, so the slope is not defined.

32. $y = 9$, or $y = 0 \cdot x + 9$

The slope is 0.

33. $y = -9x + 46$

The equation is in the form $y = mx + b$. The slope is -9 and the y-intercept is $(0, 46)$.

34. We solve for y.

$$x + y = 9$$

$$y = -x + 9$$

The slope is -1 and the y-intercept is $(0, 9)$.

35. We solve for y.

$$3x - 5y = 4$$

$$-5y = -3x + 4$$

$$-\frac{1}{5}(-5y) = -\frac{1}{5}(-3x + 4)$$

$$y = \frac{3}{5}x - \frac{4}{5}$$

The slope is $\dfrac{3}{5}$ and the y-intercept is $\left(0, -\dfrac{4}{5}\right)$.

36. We substitute -2.8 for m and 19 for b in the equation $y = mx + b$.

$$y = -2.8x + 19$$

37. We substitute $\dfrac{5}{8}$ for m and $-\dfrac{7}{8}$ for b in the equation $y = mx + b$.

$$y = \frac{5}{8}x - \frac{7}{8}$$

38. We know the slope is 3, so the equation is $y = 3x + b$. Using the point $(1, 2)$, we substitute 1 for x and 2 for y in $y = 3x + b$. Then we solve for b.

$$y = 3x + b$$
$$2 = 3 \cdot 1 + b$$
$$2 = 3 + b$$
$$-1 = b$$

Thus, we have the equation $y = 3x - 1$.

39. We know the slope is $\dfrac{2}{3}$, so the equation is $y = \dfrac{2}{3}x + b$. Using the point $(-2, -5)$, we substitute -2 for x and -5 for y in $y = \dfrac{2}{3}x + b$. Then we solve for b.

$$y = \frac{2}{3}x + b$$
$$-5 = \frac{2}{3}(-2) + b$$
$$-5 = -\frac{4}{3} + b$$
$$-\frac{11}{3} = b$$

Thus, we have the equation $y = \dfrac{2}{3}x - \dfrac{11}{3}$.

40. The slope is -2 and the y-intercept is $(0, -4)$, so we have the equation $y = -2x - 4$.

41. First we find the slope.

$$m = \frac{1 - 7}{-1 - 5} = \frac{-6}{-6} = 1$$

Thus, $y = 1 \cdot x + b$, or $y = x + b$. We can use either point to find b. We choose $(5, 7)$. Substitute 5 for x and 7 for y in $y = x + b$.

$$y = x + b$$
$$7 = 5 + b$$
$$2 = b$$

Thus, the equation is $y = x + 2$.

42. First we find the slope.

$$m = \frac{-3 - 0}{-4 - 2} = \frac{-3}{-6} = \frac{1}{2}$$

Thus, $y = \dfrac{1}{2}x + b$. We can use either point to find b. We choose $(2, 0)$. Substitute 2 for x and 0 for y in $y = \dfrac{1}{2}x + b$.

$$y = \frac{1}{2}x + b$$
$$0 = \frac{1}{2} \cdot 2 + b$$
$$0 = 1 + b$$
$$-1 = b$$

Thus, the equation is $y = \dfrac{1}{2}x - 1$.

43. a) First we find the slope.

$$m = \frac{59.5 - 33.9}{53 - 0} = \frac{25.6}{53} \approx 0.48$$

The y-intercept is $(0, 33.9)$, so we have the equation $y = 0.48x + 33.9$.

b) The rate of change is the slope, 0.48 percent per year.

c) In 2005, $x = 2005 - 1950 = 55$.

$$y = 0.48x + 33.9$$
$$y = 0.48(55) + 33.9$$
$$y = 26.4 + 33.9$$
$$y = 60.3$$

The percent of female workers in the labor force in 2005 will be 60.3%.

44. Slope -1, y-intercept $(0, 4)$

We plot $(0, 4)$. We can think of the slope as $\dfrac{-1}{1}$, so from $(0, 4)$ we move down 1 unit and right 1 unit. This locates the point $(1, 3)$. We plot $(1, 3)$ and draw a line passing through $(0, 4)$ and $(1, 3)$.

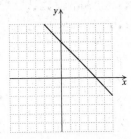

45. Slope $\dfrac{5}{3}$, y-intercept $(0, -3)$.

Plot $(0, -3)$ and from there move up 5 units and right 3 units. This locates the point $(3, 2)$. We plot $(3, 2)$ and draw a line passing through $(0, -3)$ and $(3, 2)$.

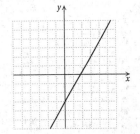

46. $y = -\dfrac{3}{5}x + 2$

First we plot the y-intercept $(0, 2)$. We can start at the y-intercept and, thinking of the slope as $\dfrac{-3}{5}$, find another point by moving down 3 units and right 5 units to the point $(5, -1)$. Thinking of the slope as $\dfrac{3}{-5}$ we can start at $(0, 2)$ and move up 3 units and left 5 units to get another point $(-5, 5)$.

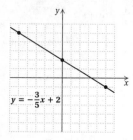

47. First we rewrite the equation in slope-intercept form.
$$2y - 3x = 6$$
$$2y = 3x + 6$$
$$y = \frac{1}{2}(3x + 6)$$
$$y = \frac{3}{2}x + 3$$

Now we plot the y-intercept $(0, 3)$. We can start at the y-intercept and use the slope, $\dfrac{3}{2}$, to find another point. We move up 3 units and right 2 units to the point $(2, 6)$. Thinking of the slope as $\dfrac{-3}{-2}$ we can start at $(0, 3)$ and move down 3 units and left 2 units to get another point, $(-2, 0)$.

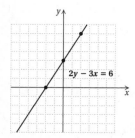

48. First we solve each equation for y:

1. $4x + y = 6$ 2. $4x + y = 8$
   $y = -4x + 6$ $y = -4x + 8$

The slope of each line is -4. The y-intercepts, $(0, 6)$ and $(0, 8)$, are different. The lines are parallel.

49. We solve the first equation for y.
$$2x + y = 10$$
$$y = -2x + 10$$

The second equation is already solved for y.
$$y = \frac{1}{2}x - 4$$

The slopes, -2 and $\dfrac{1}{2}$, are not the same so the lines are not parallel. The product of the slopes is $-2 \cdot \dfrac{1}{2} = -1$, so the lines are perpendicular.

50. First we solve each equation for y:

1. $x + 4y = 8$ 2. $x = -4y - 10$
   $4y = -x + 8$ $x + 10 = -4y$
   $y = \dfrac{1}{4}(-x + 8)$ $-\dfrac{1}{4}(x + 10) = y$
   $y = -\dfrac{1}{4}x + 2$ $-\dfrac{1}{4}x - \dfrac{5}{2} = y$

The slope of each line is $-\dfrac{1}{4}$. The y-intercepts, $(0, 2)$ and $\left(0, -\dfrac{5}{2}\right)$, are different. The lines are parallel.

51. First we solve each equation for y:

1. $3x - y = 6$ 2. $3x + y = 8$
   $-y = -3x + 6$ $y = -3x + 8$
   $y = -1(-3x + 6)$
   $y = 3x - 6$

The slopes, 3 and -3, are not the same so the lines are not parallel. The product of the slopes is $3(-3) = -9 \neq -1$, so the lines are not perpendicular. Thus, the lines are neither parallel nor perpendicular.

52.
$$\frac{x - 2y > 1}{0 - 2 \cdot 0 \; ? \; 1}$$
$$0 \;\big|\; \text{FALSE}$$

Since $0 > 1$ is false, $(0, 0)$ is not a solution.

53.
$$\frac{x - 2y > 1}{1 - 2 \cdot 3 \; ? \; 1}$$
$$1 - 6 \;\big|$$
$$-5 \;\big|\; \text{FALSE}$$

Since $-5 > 1$ is false, $(1, 3)$ is not a solution.

54.
$$\frac{x - 2y > 1}{4 - 2(-1) \; ? \; 1}$$
$$4 + 2 \;\big|$$
$$6 \;\big|\; \text{TRUE}$$

Since $6 > 1$ is true, $(4, -1)$ is a solution.

55. Graph $x < y$.

First graph the line $x = y$, or $y = x$. Two points on the line are $(0, 0)$ and $(3, 3)$. We draw a dashed line since the inequality symbol is $<$. Then we pick a test point that is not on the line. We try $(1, 2)$.

$$\frac{x < y}{1 \; ? \; 2 \;\; \text{TRUE}}$$

We see that $(1, 2)$ is a solution of the inequality, so we shade the region that contains $(1, 2)$.

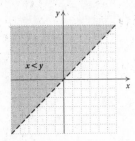

56. Graph $x + 2y \geq 4$.

First graph the line $x + 2y = 4$. The intercepts are $(0, 2)$ and $(4, 0)$. We draw a solid line since the inequality symbol is $\geq$. Then we test the point $(0, 0)$.

$$
\begin{array}{c|l}
\multicolumn{2}{c}{x + 2y \geq 4} \\ \hline
0 + 2 \cdot 0 \; ? \; 1 & \\
0 & \text{FALSE}
\end{array}
$$

Since $(0, 0)$ is not a solution of the inequality, we shade the region that does not contain $(0, 0)$.

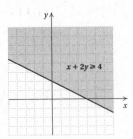

57. Graph $x > -2$.

Graph the line $x = -2$ using a dashed line since the inequality symbol is $>$. Then pick a test point that is not on the line. We choose $(0, 0)$. We can write the inequality as $x + 0y > -2$.

$$
\begin{array}{c|l}
\multicolumn{2}{c}{x + 0y > -2} \\ \hline
0 + 0 \cdot 0 \; ? \; -2 & \\
0 & \text{TRUE}
\end{array}
$$

Since $(0, 0)$ is a solution of the inequality, we shade the region containing $(0, 0)$.

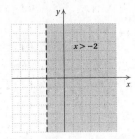

58. *Discussion and Writing Exercise.* The y-intercept is the point at which the graph crosses the y-axis. Since a point on the y-axis is neither left nor right of the origin, the first or x-coordinate of the point is 0.

59. *Discussion and Writing Exercise.* The graph of $x < 1$ on a number line consists of the points in the set $\{x | x < 1\}$.

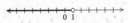

The graph of $x < 1$ on a plane consists of the points, or ordered pairs, in the set $\{(x, y) | x + 0 \cdot y < 1\}$. This is the set of ordered pairs with first coordinate less than 1.

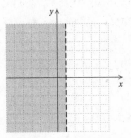

60. *Discussion and Writing Exercise.* First plot the y-intercept, $(0, 2458)$. Then, thinking of the slope as $\dfrac{37}{100}$, plot a second point on the line by moving up 37 units and right 100 units from the y-intercept and plot a third point by moving down 37 units and left 100 units. Finally, draw a line through the three points.

61. Substitute -2 for x and 5 for y and then solve for m.

$$
\begin{aligned}
y &= mx + 3 \\
5 &= m(-2) + 3 \\
5 &= -2m + 3 \\
2 &= -2m \\
-1 &= m
\end{aligned}
$$

62. We plot the given points. We see that the fourth vertex is $(-2, -3)$.

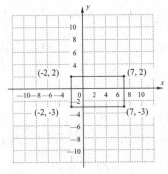

The length of the rectangle is 9 units and the width is 5 units.

$$
\begin{aligned}
A &= l \cdot w \\
A &= (9 \text{ units})(5 \text{ units}) = 45 \text{ square units}
\end{aligned}
$$

$$
\begin{aligned}
P &= 2l + 2w \\
P &= 2 \cdot 9 \text{ units} + 2 \cdot 5 \text{ units} \\
P &= 18 \text{ units} + 10 \text{ units} = 28 \text{ units}
\end{aligned}
$$

63. From 7:00 A.M. to 1:25 P.M., the elapsed time is 6 hours, 25 minutes.

$$
\begin{aligned}
6 \text{ hours} &= 6 \times 1 \text{ hour} \\
&= 6 \times 60 \text{ minutes} \\
&= 360 \text{ minutes}
\end{aligned}
$$

Then 6 hours, 25 minutes is 360 minutes + 25 minutes = 385 minutes.

The distance climbed is $29,028$ ft $- 27,600$ ft $= 1428$ ft.

a) Rate $= \dfrac{1428 \text{ ft}}{385 \text{ min}} \approx 3.709$ feet per minute

b) Rate $= \dfrac{385 \text{ min}}{1428 \text{ ft}} \approx 0.2696$ minute per foot

64. Consider a move to a, b, c, or d to be a move up; consider a move to e, f, g, or h to be a move down. Also consider a move to c, d, e, or f to be a move to the right; consider a move to a, b, g, or h to be a move to the left.

A move to a is up 1 and 2 to the left, or $\dfrac{1}{-2}$, or $-\dfrac{1}{2}$.

A move to b is up 2 and 1 to the left, or $\dfrac{2}{-1}$, or -2.

A move to c is up 2 and 1 to the right, or $\dfrac{2}{1}$, or 2.

A move to d is up 1 and 2 to the right, or $\dfrac{1}{2}$.

A move to e is down 1 and 2 to the right, or $\dfrac{-1}{2}$, or $-\dfrac{1}{2}$.

A move to f is down 2 and 1 to the right, or $\dfrac{-2}{1}$, or -2.

A move to g is down 2 and 1 to the left, or $\dfrac{-2}{-1}$, or 2.

A move to h is down 1 and 2 to the left, or $\dfrac{-1}{-2}$, or $\dfrac{1}{2}$.

Thus the slopes that are possible are $-\dfrac{1}{2}$, -2, 2, and $\dfrac{1}{2}$.

Chapter 3 Test

1. Since the first coordinate is negative and the second coordinate is positive, the point $\left(-\dfrac{1}{2}, 7\right)$ is in quadrant II.

2. Since both coordinates are negative, the point $(-5, -6)$ is in quadrant III.

3. Point A is 3 units right and 4 units up. The coordinates of A are $(3, 4)$.

4. Point B is 0 units left or right and 4 units down. The coordinates of B are $(0, -4)$.

5.
$$\begin{array}{c|c} y - 2x = 5 & \\ \hline -3 - 2(-4) \ ? \ 5 & \\ -3 + 8 & \\ 5 & \text{TRUE} \end{array}$$

$$\begin{array}{c|c} y - 2x = 5 & \\ \hline 3 - 2(-1) \ ? \ 5 & \\ 3 + 2 & \\ 5 & \text{TRUE} \end{array}$$

In each case we get a true equation, so $(-4, -3)$ and $(-1, 3)$ are solutions of $y - 2x = 5$. We plot these points and draw the line passing through them.

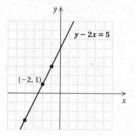

The line appears to pass through $(-2, 1)$. We check to determine if $(-2, 1)$ is a solution of $y - 2x = 5$.

$$\begin{array}{c|c} y - 2x = 5 & \\ \hline 1 - 2(-2) \ ? \ 5 & \\ 1 + 4 & \\ 5 & \text{TRUE} \end{array}$$

Thus, $(-2, 1)$ is another solution. There are other correct answers, including $(-5, -5)$, $(-3, -1)$, and $(0, 5)$.

6. $y = 2x - 1$

The y-intercept is $(0, -1)$. We find two other points.

When $x = -2$, $y = 2(-2) - 1 = -4 - 1 = -5$.

When $x = 3$, $y = 2 \cdot 3 - 1 = 6 - 1 = 5$.

x	y
-2	-5
0	-1
3	5

Plot these points, draw the line they determine, and label the graph $y = 2x - 1$.

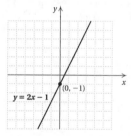

7. $y = -\dfrac{3}{2}x$, or $y = -\dfrac{3}{2}x + 0$

The y-intercept is $(0, 0)$. We find two other points.

When $x = -2$, $y = -\dfrac{3}{2}(-2) = 3$.

When $x = 2$, $y = -\dfrac{3}{2} \cdot 2 = -3$.

x	y
-2	3
0	0
2	-3

Plot these points, draw the line they determine, and label the graph $y = -\dfrac{3}{2}x$.

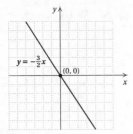

8. $2x + 8 = 0$

$$2x = -8$$

$$x = -4 \quad \text{Solving for } x$$

Any ordered pair $(-4, y)$ is a solution. A few solutions are listed below. Plot these points and draw the graph.

x	y
-4	-3
-4	0
-4	2

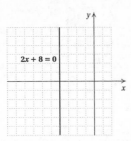

9. $y = 5$

Any ordered pair $(x, 5)$ is a solution. A few solutions are listed below. Plot these points and draw the graph.

x	y
-2	5
0	5
3	5

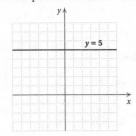

10. $2x - 4y = -8$

To find the x-intercept, let $y = 0$. Then solve for x.

$$2x - 4y = -8$$

$$2x - 4 \cdot 0 = -8$$

$$2x = -8$$

$$x = -4$$

Thus, $(-4, 0)$ is the x-intercept.

To find the y-intercept, let $x = 0$. Then solve for y.

$$2x - 4y = -8$$

$$2 \cdot 0 - 4y = -8$$

$$-4y = -8$$

$$y = 2$$

Thus, $(0, 2)$ is the y-intercept.

Plot these points and draw the graph.

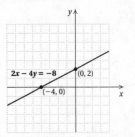

We use a third point as a check.

We let $x = 4$. Then

$$2x - 4y = -8$$

$$2 \cdot 4 - 4y = -8$$

$$8 - 4y = -8$$

$$-4y = -16$$

$$y = 4$$

The point $(4, 4)$ is on the graph, so the graph is probably correct.

11. $2x - y = 3$

To find the x-intercept, let $y = 0$. Then solve for x.

$$2x - y = 3$$

$$2x - 0 = 3$$

$$2x = 3$$

$$x = \frac{3}{2}$$

Thus, $\left(\frac{3}{2}, 0\right)$ is the x-intercept.

To find the y-intercept, let $x = 0$. Then solve for y.

$$2x - y = 3$$

$$2 \cdot 0 - y = 3$$

$$-y = 3$$

$$y = -3$$

Thus, $(0, -3)$ is the y-intercept.

Plot these points and draw the graph.

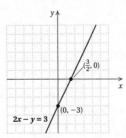

We use a third point as a check.

We let $x = 3$. Then

$$2x - y = 3$$

$$2 \cdot 3 - y = 3$$

$$6 - y = 3$$

$$-y = -3$$

$$y = 3$$

The point $(3,3)$ is on the line, so the graph is probably correct.

12. a) In 1985, $n = 0$, and $T = \dfrac{3}{5} \cdot 0 + 5 = 5$ so the cost of tuition was \$5 thousand, or \$5000.

In 1996, $n = 1996 - 1985 = 11$, and $T = \dfrac{3}{5} \cdot 11 + 5 = \dfrac{33}{5} + 5 = 6.6 + 5 = 11.6$ so the cost of tuition was \$11.6 thousand, or \$11,600.

In 2000, $n = 2000 - 1985 = 15$, and $T = \dfrac{3}{5} \cdot 15 + 5 = 9 + 5 = 14$, so the cost of tuition was \$14 thousand, or \$14,000.

In 2004, $n = 2004 - 1985 = 19$, and $T = \dfrac{3}{5} \cdot 19 + 5 = \dfrac{57}{5} + 5 = 11.4 + 5 = 16.4$, so the cost of tuition was \$16.4 thousand, or \$16,400.

b) We plot the points found in part (a), $(0,5)$, $(11, 11.6)$, $(15, 14)$, and $(19, 16.4)$. Then we draw the line passing through these points.

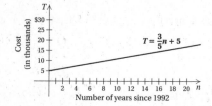

In 2005, $n = 2005 - 1985 = 20$. From the graph it appears that an n-value of 20 corresponds to the T-value of 17, so we estimate that the cost of tuition was \$17 thousand, or \$17,000, in 2005.

c) We substitute 23 for T and solve for n.
$$T = \frac{3}{5}n + 5$$
$$23 = \frac{3}{5}n + 5$$
$$18 = \frac{3}{5}n$$
$$\frac{5}{3} \cdot 18 = n$$
$$30 = n$$

A tuition cost of \$23,000 will occur 30 years after 1985, or in 2015.

13. The time that elapses from 2:38 to 2:40 is 2 minutes. In that time the elevator travels $34 - 5$, or 29 floors.

a) Rate $= \dfrac{29 \text{ floors}}{2 \text{ minutes}} = 14.5$ floors per minute

b) 2 min $= 2 \times 1$ min $= 2 \times 60$ sec $= 120$ sec

Rate $= \dfrac{120 \text{ seconds}}{29 \text{ floors}} = \dfrac{120}{29}$ seconds per floor $= 4\dfrac{4}{29}$ seconds per floor

14. The time that elapses from 1:00 P.M. to 5:00 P.M. is 4 hours.

Rate $= \dfrac{450 \text{ miles} - 100 \text{ miles}}{4 \text{ hours}} = \dfrac{350 \text{ miles}}{4 \text{ hours}} =$ 87.5 miles per hour

15. We can choose any two points. We consider (x_1, y_1) to be $(2, 4)$ and (x_2, y_2) to be $(5, -2)$.
$$m = \frac{y_2 - y_1}{x_2 - x_1} = \frac{-2 - 4}{5 - 2} = \frac{-6}{3} = -2$$

16. We plot $(-3, 1)$ and $(5, 4)$ and draw the line containing these points.

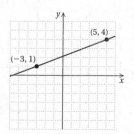

To find the slope, consider (x_1, y_1) to be $(5, 4)$ and (x_2, y_2) to be $(-3, 1)$.
$$m = \frac{y_2 - y_1}{x_2 - x_1} = \frac{1 - 4}{-3 - 5} = \frac{-3}{-8} = \frac{3}{8}$$

17. Slope $= \dfrac{-54}{1080} = -\dfrac{1}{20}$

18. a) We solve for y.
$$2x - 5y = 10$$
$$-5y = -2x + 10$$
$$y = -\frac{1}{5}(-2x + 10)$$
$$y = \frac{2}{5}x - 2$$

The slope is $\dfrac{2}{5}$.

b) The graph of $x = -2$ is a vertical line, so the slope is not defined.

19. Slope $-\dfrac{3}{2}$, y-intercept $(0, 1)$

We plot $(0, 1)$. We can think of the slope as $\dfrac{-3}{2}$, so from $(0, 1)$ we move down 3 units and right 2 units. This locates the point $(2, -2)$. We plot $(2, -2)$ and draw a line passing through $(0, 1)$ and $(2, -2)$.

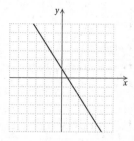

20. $y = 2x - 3$

The slope is 2 and the y-intercept is $(0, -3)$. We plot $(0, -3)$. We can start at the y-intercept and, thinking of the slope as $\dfrac{2}{1}$, find another point by moving up 2 units and right 1 unit to the point $(1, -1)$. Thinking of the slope

as $\dfrac{-2}{-1}$ we can start at $(0, -3)$ and move down 2 units and left 1 unit to the point $(-1, -5)$.

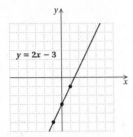

21. $y = 2x - \dfrac{1}{4}$

The equation is in the form $y = mx + b$. The slope is 2 and the y-intercept is $\left(0, -\dfrac{1}{4}\right)$.

22. We solve for y.

$$-4x + 3y = -6$$
$$3y = 4x - 6$$
$$y = \frac{1}{3}(4x - 6)$$
$$y = \frac{4}{3}x - 2$$

The slope is $\dfrac{4}{3}$ and the y-intercept is $(0, -2)$.

23. We substitute 1.8 for m and -7 for b in the equation $y = mx + b$.

$$y = 1.8x - 7$$

24. We substitute $-\dfrac{3}{8}$ for m and $-\dfrac{1}{8}$ for b in the equation $y = mx + b$.

$$y = -\frac{3}{8}x - \frac{1}{8}$$

25. We know the slope is 1, so the equation is $y = 1 \cdot x + b$, or $y = x + b$. Using the point $(3, 5)$, we substitute 3 for x and 5 for y in $y = x + b$. Then we solve for b.

$$y = x + b$$
$$5 = 3 + b$$
$$2 = b$$

Thus, the equation is $y = x + 2$.

26. We know the slope is -3, so the equation is $y = -3x + b$. Using the point $(-2, 0)$, we substitute -2 for x and 0 for y in $y = -3x + b$. Then we solve for b.

$$y = -3x + b$$
$$0 = -3(-2) + b$$
$$0 = 6 + b$$
$$-6 = b$$

Thus, the equation is $y = -3x - 6$.

27. First we find the slope.

$$m = \frac{-2 - 1}{2 - 1} = \frac{-3}{1} = -3$$

Thus, $y = -3x + b$. We can use either point to find b. We choose $(1, 1)$. Substitute 1 for both x and y in $y = -3x + b$.

$$y = -3x + b$$
$$1 = -3 \cdot 1 + b$$
$$1 = -3 + b$$
$$4 = b$$

Thus, the equation is $y = -3x + 4$.

28. First we find the slope.

$$m = \frac{-1 - (-3)}{4 - (-4)} = \frac{2}{8} = \frac{1}{4}$$

Thus, $y = \dfrac{1}{4}x + b$. We can use either point to find b. We choose $(4, -1)$. Substitute 4 for x and -1 for y in $y = \dfrac{1}{4}x + b$.

$$y = \frac{1}{4}x + b$$
$$-1 = \frac{1}{4} \cdot 4 + b$$
$$-1 = 1 + b$$
$$-2 = b$$

Thus, the equation is $y = \dfrac{1}{4}x - 2$.

29. a) First we find the slope.

$$m = \frac{956 - 203}{12 - 2} = \frac{753}{10} = 75.3$$

The slope is 75.3, so the equation is $y = 75.3x + b$.

Using the point $(2, 203)$, we substitute 2 for x and 203 for y in $y = 75.3x + b$.

$$y = 75.3x + b$$
$$203 = 75.3(2) + b$$
$$203 = 150.6 + b$$
$$52.4 = b$$

Thus, the equation is $y = 75.3x + 52.4$.

b) The rate of change is the slope, 75.3 lung transplants per year.

c) In 2005, $x = 2005 - 1988 = 17$.

$$y = 75.3x + 52.4$$
$$y = 75.3(17) + 52.4$$
$$y = 1280.1 + 52.4$$
$$y = 1332.5 \approx 1333$$

There were about 1333 lung transplants in 2005.

30. First we solve each equation for y:

   1. $2x + y = 8$ 2. $2x + y = 4$
       $y = -2x + 8$ $y = -2x + 4$

The slope of each line is -2. The y-intercepts, $(0, 8)$ and $(0, 4)$ are different. The lines are parallel.

31. We solve the first equation for y.

$$2x + 5y = 2$$
$$5y = -2x + 2$$
$$y = \frac{1}{5}(-2x + 2)$$
$$y = -\frac{2}{5}x + \frac{2}{5}$$

The second equation is in the form $y = mx + b$:

$$y = 2x + 4$$

The slopes, $-\frac{2}{5}$ and 2, are not the same, so the lines are not parallel. The product of the slopes is $-\frac{2}{5} \cdot 2 = -\frac{4}{5} \neq -1$, so the lines are not perpendicular. Thus, the lines are neither parallel nor perpendicular.

32. First we solve each equation for y:

1. $\quad x + 2y = 8$ 2. $\quad -2x + y = 8$
   $\qquad 2y = -x + 8$ $\qquad y = 2x + 8$
   $\qquad y = \frac{1}{2}(-x + 8)$
   $\qquad y = -\frac{1}{2}x + 4$

The slopes, $-\frac{1}{2}$ and 2, are not the same, so the lines are not parallel. The product of the slope is $-\frac{1}{2} \cdot 2 = -1$, so the lines are perpendicular.

33.

$$\frac{3y - 2x < -2}{3 \cdot 0 - 2 \cdot 0 \ ? \ -2}$$
$$\quad 0 \ | \qquad \text{FALSE}$$

Since $0 < -2$ is false, $(0,0)$ is not a solution.

34.

$$\frac{3y - 2x < -2}{3(-10) - 2(-4) \ ? \ -2}$$
$$\quad -30 + 8 \ |$$
$$\quad -22 \ | \qquad \text{TRUE}$$

Since $-22 < -2$ is true, $(-4, -10)$ is a solution.

35. Graph $y > x - 1$.

First graph the line $y = x - 1$. Two points on the line are $(0, -1)$ and $(4, 3)$. We draw a dashed line since the inequality symbol is >. Then we test a point that is not on the line. We try $(0,0)$.

$$\frac{y > x - 1}{0 \ ? \ 0 - 1}$$
$$\quad | \ -1 \quad \text{TRUE}$$

Since $(0,0)$ is a solution of the inequality, we shade the region containing $(0,0)$.

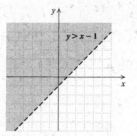

36. Graph $2x - y \leq 4$.

First graph the line $2x - y = 4$. The intercepts are $(0, -4)$ and $(2, 0)$. We draw a solid line since the inequality symbol is $\leq$. Then we test a point that is not on the line. We try $(0,0)$.

$$\frac{2x - y \leq 4}{2 \cdot 0 - 0 \ ? \ 4}$$
$$\quad 0 \ | \qquad \text{TRUE}$$

Since $(0,0)$ is a solution of the inequality, we shade the region containing $(0,0)$.

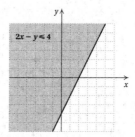

37. We plot the given points. We see that the other vertices of the square are $(-3, 4)$ and $(2, -1)$.

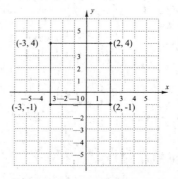

The length of the sides of the square is 5 units.

$$A = s^2 = (5 \text{ units})^2 = 25 \text{ square units}$$
$$P = 4s = 4 \cdot 5 \text{ units} = 20 \text{ units}$$

38. First solve each equation for y.

$$3x + 7y = 14$$
$$7y = -3x + 14$$
$$y = \frac{1}{7}(-3x + 14)$$
$$y = -\frac{3}{7}x + 2$$

$$ky - 7x = -3$$
$$ky = 7x - 3$$
$$y = \frac{1}{k}(7x - 3)$$
$$y = \frac{7}{k}x - \frac{3}{k}$$

If the lines are perpendicular, the product of their slopes is -1.

$$-\frac{3}{7} \cdot \frac{7}{k} = -1$$
$$-\frac{3 \cdot 7}{7 \cdot k} = -1$$
$$-\frac{3}{k} = -1$$
$$k\left(-\frac{3}{k}\right) = k(-1)$$
$$-3 = -k$$
$$3 = k$$

Chapter 4

Polynomials: Operations

Exercise Set 4.1

1. 3^4 means $3 \cdot 3 \cdot 3 \cdot 3$.

3. $(-1.1)^5$ means $(-1.1)(-1.1)(-1.1)(-1.1)(-1.1)$.

5. $\left(\dfrac{2}{3}\right)^4$ means $\left(\dfrac{2}{3}\right)\left(\dfrac{2}{3}\right)\left(\dfrac{2}{3}\right)\left(\dfrac{2}{3}\right)$.

7. $(7p)^2$ means $(7p)(7p)$.

9. $8k^3$ means $8 \cdot k \cdot k \cdot k$.

11. $-6y^4$ means $-6 \cdot y \cdot y \cdot y \cdot y$.

13. $a^0 = 1$, $a \neq 0$

15. $b^1 = b$

17. $\left(\dfrac{2}{3}\right)^0 = 1$

19. $(-7.03)^1 = -7.03$

21. $8.38^0 = 1$

23. $(ab)^1 = ab$

25. $ab^0 = a \cdot b^0 = a \cdot 1 = a$

27. $m^3 = 3^3 = 3 \cdot 3 \cdot 3 = 27$

29. $p^1 = 19^1 = 19$

31. $-x^4 = -(-3)^4 = -(-3)(-3)(-3)(-3) = -81$

33. $x^4 = 4^4 = 4 \cdot 4 \cdot 4 \cdot 4 = 256$

35. $\begin{aligned} y^2 - 7 &= 10^2 - 7 \\ &= 100 - 7 \quad \text{Evaluating the power} \\ &= 93 \qquad\quad \text{Subtracting} \end{aligned}$

37. $\begin{aligned} 161 - b^2 &= 161 - 5^2 \\ &= 161 - 25 \quad \text{Evaluating the power} \\ &= 136 \qquad\quad \text{Subtracting} \end{aligned}$

39. $\begin{aligned} x^1 + 3 &= 7^1 + 3 \\ &= 7 + 3 \quad (7^1 = 7) \\ &= 10 \end{aligned}$

$\begin{aligned} x^0 + 3 &= 7^0 + 3 \\ &= 1 + 3 \quad (7^0 = 1) \\ &= 4 \end{aligned}$

41. $\begin{aligned} A &= \pi r^2 \approx 3.14 \times (34 \text{ ft})^2 \\ &\approx 3.14 \times 1156 \text{ ft}^2 \quad \text{Evaluating the power} \\ &\approx 3629.84 \text{ ft}^2 \end{aligned}$

43. $3^{-2} = \dfrac{1}{3^2} = \dfrac{1}{9}$

45. $10^{-3} = \dfrac{1}{10^3} = \dfrac{1}{1000}$

47. $7^{-3} = \dfrac{1}{7^3} = \dfrac{1}{343}$

49. $a^{-3} = \dfrac{1}{a^3}$

51. $\dfrac{1}{8^{-2}} = 8^2 = 64$

53. $\dfrac{1}{y^{-4}} = y^4$

55. $\dfrac{1}{z^{-n}} = z^n$

57. $\dfrac{1}{4^3} = 4^{-3}$

59. $\dfrac{1}{x^3} = x^{-3}$

61. $\dfrac{1}{a^5} = a^{-5}$

63. $2^4 \cdot 2^3 = 2^{4+3} = 2^7$

65. $8^5 \cdot 8^9 = 8^{5+9} = 8^{14}$

67. $x^4 \cdot x^3 = x^{4+3} = x^7$

69. $9^{17} \cdot 9^{21} = 9^{17+21} = 9^{38}$

71. $(3y)^4 (3y)^8 = (3y)^{4+8} = (3y)^{12}$

73. $(7y)^1 (7y)^{16} = (7y)^{1+16} = (7y)^{17}$

75. $3^{-5} \cdot 3^8 = 3^{-5+8} = 3^3$

77. $x^{-2} \cdot x = x^{-2+1} = x^{-1} = \dfrac{1}{x}$

79. $x^{14} \cdot x^3 = x^{14+3} = x^{17}$

81. $x^{-7} \cdot x^{-6} = x^{-7+(-6)} = x^{-13} = \dfrac{1}{x^{13}}$

83. $a^{11} \cdot a^{-3} \cdot a^{-18} = a^{11+(-3)+(-18)} = a^{-10} = \dfrac{1}{a^{10}}$

85. $t^8 \cdot t^{-8} = t^{8+(-8)} = t^0 = 1$

87. $\dfrac{7^5}{7^2} = 7^{5-2} = 7^3$

89. $\dfrac{8^{12}}{8^6} = 8^{12-6} = 8^6$

91. $\dfrac{y^9}{y^5} = y^{9-5} = y^4$

93. $\dfrac{16^2}{16^8} = 16^{2-8} = 16^{-6} = \dfrac{1}{16^6}$

95. $\dfrac{m^6}{m^{12}} = m^{6-12} = m^{-6} = \dfrac{1}{m^6}$

97. $\dfrac{(8x)^6}{(8x)^{10}} = (8x)^{6-10} = (8x)^{-4} = \dfrac{1}{(8x)^4}$

99. $\dfrac{(2y)^9}{(2y)^9} = (2y)^{9-9} = (2y)^0 = 1$

101. $\dfrac{x}{x^{-1}} = x^{1-(-1)} = x^2$

103. $\dfrac{x^7}{x^{-2}} = x^{7-(-2)} = x^9$

105. $\dfrac{z^{-6}}{z^{-2}} = z^{-6-(-2)} = z^{-4} = \dfrac{1}{z^4}$

107. $\dfrac{x^{-5}}{x^{-8}} = x^{-5-(-8)} = x^3$

109. $\dfrac{m^{-9}}{m^{-9}} = m^{-9-(-9)} = m^0 = 1$

111. $5^2 = 5 \cdot 5 = 25$

$5^{-2} = \dfrac{1}{5^2} = \dfrac{1}{25}$

$\left(\dfrac{1}{5}\right)^2 = \dfrac{1}{5} \cdot \dfrac{1}{5} = \dfrac{1}{25}$

$\left(\dfrac{1}{5}\right)^{-2} = \dfrac{1}{\left(\dfrac{1}{5}\right)^2} = \dfrac{1}{\dfrac{1}{25}} = 1 \cdot \dfrac{25}{1} = 25$

$-5^2 = -(5)(5) = -25$

$(-5)^2 = (-5)(-5) = 25$

$-\left(-\dfrac{1}{5}\right)^2 = -\left(-\dfrac{1}{5}\right)\left(-\dfrac{1}{5}\right) = -\dfrac{1}{25}$

$\left(-\dfrac{1}{5}\right)^{-2} = \dfrac{1}{\left(-\dfrac{1}{5}\right)^2} = \dfrac{1}{\dfrac{1}{25}} = 1 \cdot \dfrac{25}{1} = 25$

113. Discussion and Writing Exercise

115. *Familiarize*. Let $x =$ the length of the shorter piece. Then $2x =$ the length of the longer piece.

***Translate*.**

$$\underbrace{\text{Length of shorter piece}}_{x} \; \underset{+}{\text{plus}} \; \underbrace{\text{length of longer piece}}_{2x} \; \underset{=}{\text{is}} \; \underset{12}{\text{12 in.}}$$

***Solve*.**

$$x + 2x = 12$$
$$3x = 12$$
$$\dfrac{3x}{3} = \dfrac{12}{3}$$
$$x = 4$$

If $x = 4$, $2x = 2 \cdot 4 = 8$.

***Check*.** The longer piece, 8 in., is twice as long as the shorter piece, 4 in. Also, 4 in. + 8 in. = 12 in., the total length of the sandwich. The answer checks.

***State*.** The lengths of the pieces are 4 in. and 8 in.

117. *Familiarize*. Let $w =$ the width. Then $w + 15 =$ the length. We draw a picture.

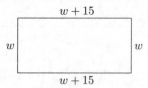

We will use the fact that the perimeter is 640 ft to find w (the width). Then we can find $w + 15$ (the length) and multiply the length and the width to find the area.

***Translate*.**

$$\underset{w}{\text{Width}} + \underset{w}{\text{Width}} + \underset{(w+15)}{\text{Length}} + \underset{(w+15)}{\text{Length}} = \underset{640}{\text{Perimeter}}$$

***Solve*.**

$$w + w + (w + 15) + (w + 15) = 640$$
$$4w + 30 = 640$$
$$4w = 610$$
$$w = 152.5$$

If the width is 152.5, then the length is $152.5 + 15$, or 167.5. The area is $(167.5)(152.5)$, or $25{,}543.75$ ft^2.

***Check*.** The length, 167.5 ft, is 15 ft greater than the width, 152.5 ft. The perimeter is $152.5 + 152.5 + 167.5 + 167.5$, or 640 ft. We should also recheck the computation we used to find the area. The answer checks.

***State*.** The area is $25{,}543.75$ ft^2.

119.

$$-6(2 - x) + 10(5x - 7) = 10$$
$$-12 + 6x + 50x - 70 = 10$$
$$56x - 82 = 10 \quad \text{Collecting like terms}$$
$$56x - 82 + 82 = 10 + 82 \quad \text{Adding 82}$$
$$56x = 92$$
$$\dfrac{56x}{56} = \dfrac{92}{56} \quad \text{Dividing by 56}$$
$$x = \dfrac{23}{14}$$

The solution is $\dfrac{23}{14}$.

121. $4x - 12 + 24y = 4 \cdot x - 4 \cdot 3 + 4 \cdot 6y = 4(x - 3 + 6y)$

123. Let $y_1 = (x+1)^2$ and $y_2 = x^2 + 1$. A graph of the equations or a table of values shows that $(x + 1)^2 = x^2 + 1$ is not correct.

125. Let $y_1 = (5x)^0$ and $y_2 = 5x^0$. A graph of the equations or a table of values shows that $(5x)^0 = 5x^0$ is not correct.

127. $(y^{2x})(y^{3x}) = y^{2x+3x} = y^{5x}$

129. $\dfrac{a^{6t}(a^{7t})}{a^{9t}} = \dfrac{a^{6t+7t}}{a^{9t}} = \dfrac{a^{13t}}{a^{9t}} = a^{13t-9t} = a^{4t}$

131. $\dfrac{(0.8)^5}{(0.8)^3(0.8)^2} = \dfrac{(0.8)^5}{(0.8)^{3+2}} = \dfrac{(0.8)^5}{(0.8)^5} = 1$

133. Since the bases are the same, the expression with the larger exponent is larger. Thus, $3^5 > 3^4$.

135. Since the exponents are the same, the expression with the larger base is larger. Thus, $4^3 < 5^3$.

137. $\dfrac{1}{-z^4} = \dfrac{1}{-(-10)^4} = \dfrac{1}{-(-10)(-10)(-10)(-10)} =$

$\dfrac{1}{-10,000} = -\dfrac{1}{10,000}$

Exercise Set 4.2

1. $(2^3)^2 = 2^{3 \cdot 2} = 2^6$

3. $(5^2)^{-3} = 5^{2(-3)} = 5^{-6} = \dfrac{1}{5^6}$

5. $(x^{-3})^{-4} = x^{(-3)(-4)} = x^{12}$

7. $(a^{-2})^9 = a^{-2 \cdot 9} = a^{-18} = \dfrac{1}{a^{18}}$

9. $(t^{-3})^{-6} = t^{(-3)(-6)} = t^{18}$

11. $(t^4)^{-3} = t^{4(-3)} = t^{-12} = \dfrac{1}{t^{12}}$

13. $(x^{-2})^{-4} = x^{-2)(-4)} = x^8$

15. $(ab)^3 = a^3b^3$ Raising each factor to the third power

17. $(ab)^{-3} = a^{-3}b^{-3} = \dfrac{1}{a^3b^3}$

19. $(mn^2)^{-3} = m^{-3}(n^2)^{-3} = m^{-3}n^{2(-3)} =$

$m^{-3}n^{-6} = \dfrac{1}{m^3n^6}$

21. $(4x^3)^2 = 4^2(x^3)^2$ Raising each factor to the second power

$= 16x^6$

23. $(3x^{-4})^2 = 3^2(x^{-4})^2 = 3^2x^{-4 \cdot 2} = 9x^{-8} = \dfrac{9}{x^8}$

25. $(x^4y^5)^{-3} = (x^4)^{-3}(y^5)^{-3} = x^{4(-3)}y^{5(-3)} =$

$x^{-12}y^{-15} = \dfrac{1}{x^{12}y^{15}}$

27. $(x^{-6}y^{-2})^{-4} = (x^{-6})^{-4}(y^{-2})^{-4} = x^{(-6)(-4)}y^{(-2)(-4)} =$

$x^{24}y^8$

29. $(a^{-2}b^7)^{-5} = (a^{-2})^{-5}(b^7)^{-5} = a^{10}b^{-35} = \dfrac{a^{10}}{b^{35}}$

31. $(5r^{-4}t^3)^2 = 5^2(r^{-4})^2(t^3)^2 = 25r^{-4 \cdot 2}t^{3 \cdot 2} =$

$25r^{-8}t^6 = \dfrac{25t^6}{r^8}$

33. $(a^{-5}b^7c^{-2})^3 = (a^{-5})^3(b^7)^3(c^{-2})^3 =$

$a^{-5 \cdot 3}b^{7 \cdot 3}c^{-2 \cdot 3} = a^{-15}b^{21}c^{-6} = \dfrac{b^{21}}{a^{15}c^6}$

35. $(3x^3y^{-8}z^{-3})^2 = 3^2(x^3)^2(y^{-8})^2(z^{-3})^2 =$

$9x^6y^{-16}z^{-6} = \dfrac{9x^6}{y^{16}z^6}$

37. $(-4x^3y^{-2})^2 = (-4)^2(x^3)^2(y^{-2})^2 = 16x^6y^{-4} = \dfrac{16x^6}{y^4}$

39. $(-a^{-3}b^{-2})^{-4} = (-1 \cdot a^{-3}b^{-2})^{-4} =$

$(-1)^{-4}(a^{-3})^{-4}(b^{-2})^{-4} = \dfrac{1}{(-1)^4} \cdot a^{12}b^8 = \dfrac{a^{12}b^8}{1} = a^{12}b^8$

41. $\left(\dfrac{y^3}{2}\right)^2 = \dfrac{(y^3)^2}{2^2} = \dfrac{y^6}{4}$

43. $\left(\dfrac{a^2}{b^3}\right)^4 = \dfrac{(a^2)^4}{(b^3)^4} = \dfrac{a^8}{b^{12}}$

45. $\left(\dfrac{y^2}{2}\right)^{-3} = \dfrac{(y^2)^{-3}}{2^{-3}} = \dfrac{y^{-6}}{2^{-3}} = \dfrac{\dfrac{1}{y^6}}{\dfrac{1}{2^3}} = \dfrac{1}{y^6} \cdot \dfrac{2^3}{1} = \dfrac{8}{y^6}$

47. $\left(\dfrac{7}{x^{-3}}\right)^2 = \dfrac{7^2}{(x^{-3})^2} = \dfrac{49}{x^{-6}} = 49x^6$

49. $\left(\dfrac{x^2y}{z}\right)^3 = \dfrac{(x^2)^3y^3}{z^3} = \dfrac{x^6y^3}{z^3}$

51. $\left(\dfrac{a^2b}{cd^3}\right)^{-2} = \dfrac{(a^2)^{-2}b^{-2}}{c^{-2}(d^3)^{-2}} = \dfrac{a^{-4}b^{-2}}{c^{-2}d^{-6}} = \dfrac{\dfrac{1}{a^4} \cdot \dfrac{1}{b^2}}{\dfrac{1}{c^2} \cdot \dfrac{1}{d^6}} = \dfrac{\dfrac{1}{a^4b^2}}{\dfrac{1}{c^2d^6}} =$

$\dfrac{1}{a^4b^2} \cdot \dfrac{c^2d^6}{1} = \dfrac{c^2d^6}{a^4b^2}$

53. $2.8,000,000,000.$

$\quad$ ⌞_____⌟ 10 places

Large number, so the exponent is positive.

$28,000,000,000 = 2.8 \times 10^{10}$

55. $9.07,000,000,000,000,000.$

$\quad$ ⌞_____⌟ 17 places

Large number, so the exponent is positive.

$907,000,000,000,000,000 = 9.07 \times 10^{17}$

57. $0.000003.04$

$\quad$ ⌞_____⌝ 6 places

Small number, so the exponent is negative.

$0.00000304 = 3.04 \times 10^{-6}$

59. $0.00000001.8$

$\quad$ ⌞_____⌝ 8 places

Small number, so the exponent is negative.

$0.000000018 = 1.8 \times 10^{-8}$

61. $1.00,000,000,000.$

$\quad$ ⌞_____⌟ 11 places

Large number, so the exponent is positive.

$100,000,000,000 = 1.0 \times 10^{11} = 10^{11}$

63. 296 million = 296,000,000

2.96,000,000.

$\underset{\text{8 places}}{\underbrace{\qquad\qquad}}$

Large number, so the exponent is positive.

296 million = 2.96×10^8

65. $\dfrac{1}{10,000,000} = 0.0000001$

0.0000001.

$\underset{\text{7 places}}{\underbrace{\qquad\qquad}}$

Small number, so the exponent is negative.

$\dfrac{1}{10,000,000} = 1 \times 10^{-7}$, or 10^{-7}

67. 8.74×10^7

Positive exponent, so the answer is a large number.

8.7400000.

$\underset{\text{7 places}}{\underbrace{\qquad\qquad}}$

$8.74 \times 10^7 = 87,400,000$

69. 5.704×10^{-8}

Negative exponent, so the answer is a small number.

0.00000005.704

$\underset{\text{8 places}}{\underbrace{\qquad\qquad}}$

$5.704 \times 10^{-8} = 0.00000005704$

71. $10^7 = 1 \times 10^7$

Positive exponent, so the answer is a large number.

1.0000000.

$\underset{\text{7 places}}{\underbrace{\qquad\qquad}}$

$10^7 = 10,000,000$

73. $10^{-5} = 1 \times 10^{-5}$

Negative exponent, so the answer is a small number.

0.00001.

$\underset{\text{5 places}}{\underbrace{\qquad\qquad}}$

$10^{-5} = 0.00001$

75. $(3 \times 10^4)(2 \times 10^5) = (3 \cdot 2) \times (10^4 \cdot 10^5)$
$= 6 \times 10^9$

77. $(5.2 \times 10^5)(6.5 \times 10^{-2}) = (5.2 \cdot 6.5) \times (10^5 \cdot 10^{-2})$
$= 33.8 \times 10^3$

The answer at this stage is 33.8×10^3 but this is not scientific notation since 33.8 is not a number between 1 and 10. We convert 33.8 to scientific notation and simplify.

$33.8 \times 10^3 = (3.38 \times 10^1) \times 10^3 = 3.38 \times (10^1 \times 10^3) = 3.38 \times 10^4$

The answer is 3.38×10^4.

79. $(9.9 \times 10^{-6})(8.23 \times 10^{-8}) = (9.9 \cdot 8.23) \times (10^{-6} \cdot 10^{-8})$
$= 81.477 \times 10^{-14}$

The answer at this stage is 81.477×10^{-14}. We convert 81.477 to scientific notation and simplify.

$81.477 \times 10^{-14} = (8.1477 \times 10^1) \times 10^{-14} =$
$8.1477 \times (10^1 \times 10^{-14}) = 8.1477 \times 10^{-13}$.

The answer is 8.1477×10^{-13}.

81. $\dfrac{8.5 \times 10^8}{3.4 \times 10^{-5}} = \dfrac{8.5}{3.4} \times \dfrac{10^8}{10^{-5}}$
$= 2.5 \times 10^{8-(-5)}$
$= 2.5 \times 10^{13}$

83. $(3.0 \times 10^6) \div (6.0 \times 10^9) = \dfrac{3.0 \times 10^6}{6.0 \times 10^9}$
$= \dfrac{3.0}{6.0} \times \dfrac{10^6}{10^9}$
$= 0.5 \times 10^{6-9}$
$= 0.5 \times 10^{-3}$

The answer at this stage is 0.5×10^{-3}. We convert 0.5 to scientific notation and simplify.

$0.5 \times 10^{-3} = (5.0 \times 10^{-1}) \times 10^{-3} =$
$5.0 \times (10^{-1} \times 10^{-3}) = 5.0 \times 10^{-4}$

85. $\dfrac{7.5 \times 10^{-9}}{2.5 \times 10^{12}} = \dfrac{7.5}{2.5} \times \dfrac{10^{-9}}{10^{12}}$
$= 3.0 \times 10^{-9-12}$
$= 3.0 \times 10^{-21}$

87. There are 60 seconds in one minute and 60 minutes in one hour, so there are 60(60), or 3600 seconds in one hour. There are 24 hours in one day and 365 days in one year, so there are 3600(24)(365), or 31,536,000 seconds in one year.

$4,200,000 \times 31,536,000$
$= (4.2 \times 10^6) \times (3.1536 \times 10^7)$
$= (4.2 \cdot 3.1536) \times (10^6 \times 10^7)$
$\approx 13.25 \times 10^{13}$
$\approx (1.325 \times 10) \times 10^{13}$
$\approx 1.325 \times (10 \times 10^{13})$
$\approx 1.325 \times 10^{14}$

About 1.325×10^{14} cubic feet of water is discharged from the Amazon River in 1 yr.

89. $\dfrac{1.908 \times 10^{24}}{6 \times 10^{21}} = \dfrac{1.908}{6} \times \dfrac{10^{24}}{10^{21}}$
$= 0.318 \times 10^3$
$= (3.18 \times 10^{-1}) \times 10^3$
$= 3.18 \times (10^{-1} \times 10^3)$
$= 3.18 \times 10^2$

The mass of Jupiter is 3.18×10^2 times the mass of Earth.

91. 10 billion trillion $= 1 \times 10 \times 10^9 \times 10^{12}$
$= 1 \times 10^{22}$

There are 1×10^{22} stars in the known universe.

93. We divide the mass of the sun by the mass of earth.

$\dfrac{1.998 \times 10^{27}}{6 \times 10^{21}} = 0.333 \times 10^6$
$= (3.33 \times 10^{-1}) \times 10^6$
$= 3.33 \times 10^5$

The mass of the sun is 3.33×10^5 times the mass of Earth.

95. First we divide the distance from the earth to the moon by 3 days to find the number of miles per day the space vehicle travels. Note that $240,000 = 2.4 \times 10^5$.

$$\frac{2.4 \times 10^5}{3} = 0.8 \times 10^5 = 8 \times 10^4$$

The space vehicle travels 8×10^4 miles per day. Now divide the distance from the earth to Mars by 8×10^4 to find how long it will take the space vehicle to reach Mars. Note that $35,000,000 = 3.5 \times 10^7$.

$$\frac{3.5 \times 10^7}{8 \times 10^4} = 0.4375 \times 10^3 = 4.375 \times 10^2$$

It takes 4.375×10^2 days for the space vehicle to travel from the earth to Mars.

97. Discussion and Writing Exercise

99. $9x - 36 = 9 \cdot x - 9 \cdot 4 = 9(x - 4)$

101. $3s + 3t + 24 = 3 \cdot s + 3 \cdot t + 3 \cdot 8 = 3(s + t + 8)$

103.
$$2x - 4 - 5x + 8 = x - 3$$
$$-3x + 4 = x - 3 \qquad \text{Collecting like terms}$$
$$-3x + 4 - 4 = x - 3 - 4 \qquad \text{Subtracting 4}$$
$$-3x = x - 7$$
$$-3x - x = x - 7 - x \qquad \text{Subtracting } x$$
$$-4x = -7$$
$$\frac{-4x}{-4} = \frac{-7}{-4} \qquad \text{Dividing by } -4$$
$$x = \frac{7}{4}$$

The solution is $\frac{7}{4}$.

105.
$$8(2x + 3) - 2(x - 5) = 10$$
$$16x + 24 - 2x + 10 = 10 \qquad \text{Removing parentheses}$$
$$14x + 34 = 10 \qquad \text{Collecting like terms}$$
$$14x + 34 - 34 = 10 - 34 \qquad \text{Subtracting 34}$$
$$14x = -24$$
$$\frac{14x}{14} = \frac{-24}{14} \qquad \text{Dividing by 14}$$
$$x = -\frac{12}{7} \qquad \text{Simplifying}$$

The solution is $-\frac{12}{7}$.

107. $y = x - 5$

The equation is equivalent to $y = x + (-5)$. The y-intercept is $(0, -5)$. We find two other points.

When $x = 2$, $y = 2 - 5 = -3$.

When $x = 4$, $y = 4 - 5 = -1$.

x	y
0	-5
2	-3
4	-1

Plot these points, draw the line they determine, and label the graph $y = x - 5$.

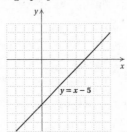

109.
$$\frac{(5.2 \times 10^6)(6.1 \times 10^{-11})}{1.28 \times 10^{-3}} = \frac{(5.2 \cdot 6.1)}{1.28} \times \frac{(10^6 \cdot 10^{-11})}{10^{-3}}$$
$$= 24.78125 \times 10^{-2}$$
$$= (2.478125 \times 10^1) \times 10^{-2}$$
$$= 2.478125 \times 10^{-1}$$

111. $\dfrac{(5^{12})^2}{5^{25}} = \dfrac{5^{24}}{5^{25}} = 5^{24-25} = 5^{-1} = \dfrac{1}{5}$

113. $\dfrac{(3^5)^4}{3^5 \cdot 3^4} = \dfrac{3^{5 \cdot 4}}{3^{5+4}} = \dfrac{3^{20}}{3^9} = 3^{20-9} = 3^{11}$

115. $\dfrac{49^{18}}{7^{35}} = \dfrac{(7^2)^{18}}{7^{35}} = \dfrac{7^{36}}{7^{35}} = 7$

117. $\dfrac{(0.4)^5}{\left((0.4)^3\right)^2} = \dfrac{(0.4)^5}{(0.4)^6} = (0.4)^{-1} = \dfrac{1}{0.4}$, or 2.5

119. False; let $x = 2$, $y = 3$, $m = 4$, and $n = 2$:
$$2^4 \cdot 3^2 = 16 \cdot 9 = 144, \text{ but}$$
$$(2 \cdot 3)^{4 \cdot 2} = 6^8 = 1,679,616$$

121. False; let $x = 5$, $y = 3$, and $m = 2$:
$$(5 - 3)^2 = 2^2 = 4, \text{ but}$$
$$5^2 - 3^2 = 25 - 9 = 16$$

123. True; $(-x)^{2m} = (-1 \cdot x)^{2m} = (-1)^{2m} \cdot x^{2m} = [(-1)^2]^m \cdot x^{2m} = 1^m \cdot x^{2m} = x^{2m}$

Exercise Set 4.3

1. $-5x + 2 = -5 \cdot 4 + 2 = -20 + 2 = -18$;

$-5x + 2 = -5(-1) + 2 = 5 + 2 = 7$

3. $2x^2 - 5x + 7 = 2 \cdot 4^2 - 5 \cdot 4 + 7 = 2 \cdot 16 - 20 + 7 = 32 - 20 + 7 = 19$;

$2x^2 - 5x + 7 = 2(-1)^2 - 5(-1) + 7 = 2 \cdot 1 + 5 + 7 = 2 + 5 + 7 = 14$

5. $x^3 - 5x^2 + x = 4^3 - 5 \cdot 4^2 + 4 = 64 - 5 \cdot 16 + 4 = 64 - 80 + 4 = -12$;

$x^3 - 5x^2 + x = (-1)^3 - 5(-1)^2 + (-1) = -1 - 5 \cdot 1 - 1 = -1 - 5 - 1 = -7$

7. $\dfrac{1}{3}x + 5 = \dfrac{1}{3}(-2) + 5 = -\dfrac{2}{3} + 5 = -\dfrac{2}{3} + \dfrac{15}{3} = \dfrac{13}{3}$;

$\dfrac{1}{3}x + 5 = \dfrac{1}{3} \cdot 0 + 5 = 0 + 5 = 5$

9. $x^2 - 2x + 1 = (-2)^2 - 2(-2) + 1 = 4 + 4 + 1 = 9$;

$x^2 - 2x + 1 = 0^2 - 2 \cdot 0 + 1 = 0 - 0 + 1 = 1$

11. $-3x^3 + 7x^2 - 3x - 2 = -3(-2)^3 + 7(-2)^2 - 3(-2) - 2 = -3(-8) + 7(4) - 3(-2) - 2 = 24 + 28 + 6 - 2 = 56$;

$-3x^3 + 7x^2 - 3x - 2 = -3 \cdot 0^3 + 7 \cdot 0^2 - 3 \cdot 0 - 2 = -3 \cdot 0 + 7 \cdot 0 - 0 - 2 = 0 + 0 - 0 - 2 = -2$

13. We evaluate the polynomial for $t = 10$:

$S = 11.12t^2 = 11.12(10)^2 = 11.12(100) = 1112$

The skydiver has fallen approximately 1112 ft.

15. a) In 2001, $t = 0$.

$E = 90.28(0) + 1138.34 = 0 + 1138.34 = 1138.34$.

The consumption of electricity in 2001 was 1138.34 billion kilowatt-hours.

In 2005, $t = 2005 - 2001 = 4$.

$E = 90.28(4) + 1138.34 = 361.12 + 1138.34 = 1499.46$

The consumption of electricity in 2005 was 1499.46 billion kilowatt-hours.

In 2010, $t = 2010 - 2001 = 9$.

$E = 90.28(9) + 1138.34 = 812.52 + 1138.34 = 1950.86$

The consumption of electricity in 2010 will be 1950.86 billion kilowatt-hours.

In 2015, $t = 2015 - 2001 = 14$.

$E = 90.28(14) + 1138.34 = 1263.92 + 1138.34 = 2402.26$

The consumption of electricity in 2015 will be 2402.26 billion kilowatt-hours.

In 2025, $t = 2025 - 2001 = 24$.

$E = 90.28(24) + 1138.34 = 2166.72 + 1138.34 = 3305.06$

The consumption of electricity in 2025 will be 3305.06 billion kilowatt-hours.

b) It appears that the points $(0, 1138.34)$, $(4, 1499.46)$, $(9, 1950.86)$, $(14, 2402.26)$, and $(24, 3305.06)$ are on the graph, so the results check.

17. We evaluate the polynomial for $x = 75$:

$$R = 280x - 0.4x^2 = 280(75) - 0.4(75)^2$$
$$= 280(75) - 0.4(5625)$$
$$= 21,000 - 2250$$
$$= 18,750$$

The total revenue from the sale of 75 TVs is $18,750.

We evaluate the polynomial for $x = 100$:

$$R = 280x - 0.4x^2 = 280(100) - 0.4(100)^2$$
$$= 280(100) - 0.4(10,000)$$
$$= 28,000 - 4000$$
$$= 24,000$$

The total revenue from the sale of 100 TVs is $24,000.

19. Locate -3 on the x-axis. Then move vertically to the graph and horizontally to the y-axis. It appears that the y-value that is paired with -3 is -4. Thus, the value of $y = 5 - x^2$ is -4 when $x = -3$.

Locate -1 on the x-axis. Then move vertically to the graph and horizontally to the y-axis. It appears that the y-value that is paired with -1 is 4. Thus, the value of $y = 5 - x^2$ is 4 when $x = -1$.

Locate 0 on the x-axis. Then move vertically to the graph. We arrive at a point on the y-axis with the y-value 5. Thus, the value of $5 - x^2$ is 5 when $x = 0$.

Locate 1.5 on the x-axis. Then move vertically to the graph and horizontally to the y-axis. It appears that the y-value that is paired with 1.5 is 2.75. Thus, the value of $y = 5 - x^2$ is 2.75 when $x = 1.5$.

Locate 2 on the x-axis. Then move vertically to the graph and horizontally to the y-axis. It appears that the y-value that is paired with 2 is 1. Thus, the value of $y = 5 - x^2$ is 1 when $x = 2$.

21. We evaluate the polynomial for $x = 20$:

$$N = -0.00006(20)^3 + 0.006(20)^2 - 0.1(20) + 1.9$$
$$= -0.00006(8000) + 0.006(400) - 0.1(20) + 1.9$$
$$= -0.48 + 2.4 - 2.0 + 1.9$$
$$= 1.82$$

There are about 1.82 million or 1,820,000 hearing-impaired Americans of age 20.

We evaluate the polynomial for $x = 40$:

$$N = -0.00006(40)^3 + 0.006(40)^2 - 0.1(40) + 1.9$$
$$= -0.00006(64,000) + 0.006(1600) - 0.1(40) + 1.9$$
$$= -3.84 + 9.6 - 4.0 + 1.9$$
$$= 3.66$$

There are about 3.66 million, or 3,660,000, hearing-impaired Americans of age 40.

23. Locate 10 on the horizontal axis. From there move vertically to the graph and then horizontally to the M-axis. This locates an M-value of about 9. Thus, about 9 words were memorized in 10 minutes.

25. Locate 8 on the horizontal axis. From there move vertically to the graph and then horizontally to the M-axis. This locates an M-value of about 6. Thus, the value of $-0.001t^3 + 0.1t^2$ for $t = 8$ is approximately 6.

27. Locate 13 on the horizontal axis. It is halfway between 12 and 14. From there move vertically to the graph and then horizontally to the M-axis. This locates an M-value of about 15. Thus, the value of $-0.001t^3 + 0.1t^2$ when t is 13 is approximately 15.

29. $2 - 3x + x^2 = 2 + (-3x) + x^2$

The terms are 2, $-3x$, and x^2.

31. $-2x^4 + \frac{1}{3}x^3 - x + 3 = -2x^4 + \frac{1}{3}x^3 + (-x) + 3$

The terms are $-2x^4$, $\frac{1}{3}x^3$, $-x$, and 3.

33. $5x^3 + 6x^2 - 3x^2$

Like terms: $6x^2$ and $-3x^2$ Same variable and exponent

35. $2x^4 + 5x - 7x - 3x^4$

Like terms: $2x^4$ and $-3x^4$ Same variable and
Like terms: $5x$ and $-7x$ exponent

37. $3x^5 - 7x + 8 + 14x^5 - 2x - 9$

Like terms: $3x^5$ and $14x^5$
Like terms: $-7x$ and $-2x$
Like terms: 8 and -9 Constant terms are like terms.

39. $-3x + 6$

The coefficient of $-3x$, the first term, is -3.

The coefficient of 6, the second term, is 6.

41. $5x^2 + \dfrac{3}{4}x + 3$

The coefficient of $5x^2$, the first term, is 5.

The coefficient of $\dfrac{3}{4}x$, the second term, is $\dfrac{3}{4}$.

The coefficient of 3, the third term, is 3.

43. $-5x^4 + 6x^3 - 2.7x^2 + 8x - 2$

The coefficient of $-5x^4$, the first term, is -5.

The coefficient of $6x^3$, the second term, is 6.

The coefficient of $-2.7x^2$, the third term, is -2.7.

The coefficient of $8x$, the fourth term, is 8.

The coefficient of -2, the fifth term, is -2.

45. $2x - 5x = (2 - 5)x = -3x$

47. $x - 9x = 1x - 9x = (1 - 9)x = -8x$

49. $5x^3 + 6x^3 + 4 = (5 + 6)x^3 + 4 = 11x^3 + 4$

51. $5x^3 + 6x - 4x^3 - 7x = (5 - 4)x^3 + (6 - 7)x =$
$1x^3 + (-1)x = x^3 - x$

53. $6b^5 + 3b^2 - 2b^5 - 3b^2 = (6 - 2)b^5 + (3 - 3)b^2 =$
$4b^5 + 0b^2 = 4b^5$

55. $\dfrac{1}{4}x^5 - 5 + \dfrac{1}{2}x^5 - 2x - 37 =$
$\left(\dfrac{1}{4} + \dfrac{1}{2}\right)x^5 - 2x + (-5 - 37) = \dfrac{3}{4}x^5 - 2x - 42$

57. $6x^2 + 2x^4 - 2x^2 - x^4 - 4x^2 =$
$6x^2 + 2x^4 - 2x^2 - 1x^4 - 4x^2 =$
$(6 - 2 - 4)x^2 + (2 - 1)x^4 = 0x^2 + 1x^4 =$
$0 + x^4 = x^4$

59. $\dfrac{1}{4}x^3 - x^2 - \dfrac{1}{6}x^2 + \dfrac{3}{8}x^3 + \dfrac{5}{16}x^3 =$
$\dfrac{1}{4}x^3 - 1x^2 - \dfrac{1}{6}x^2 + \dfrac{3}{8}x^3 + \dfrac{5}{16}x^3 =$
$\left(\dfrac{1}{4} + \dfrac{3}{8} + \dfrac{5}{16}\right)x^3 + \left(-1 - \dfrac{1}{6}\right)x^2 =$
$\left(\dfrac{4}{16} + \dfrac{6}{16} + \dfrac{5}{16}\right)x^3 + \left(-\dfrac{6}{6} - \dfrac{1}{6}\right)x^2 = \dfrac{15}{16}x^3 - \dfrac{7}{6}x^2$

61. $x^5 + x + 6x^3 + 1 + 2x^2 = x^5 + 6x^3 + 2x^2 + x + 1$

63. $5y^3 + 15y^9 + y - y^2 + 7y^8 =$
$15y^9 + 7y^8 + 5y^3 - y^2 + y$

65. $3x^4 - 5x^6 - 2x^4 + 6x^6 = x^4 + x^6 = x^6 + x^4$

67. $-2x + 4x^3 - 7x + 9x^3 + 8 = -9x + 13x^3 + 8 =$
$13x^3 - 9x + 8$

69. $3x + 3x + 3x - x^2 - 4x^2 = 9x - 5x^2 = -5x^2 + 9x$

71. $-x + \dfrac{3}{4} + 15x^4 - x - \dfrac{1}{2} - 3x^4 = -2x + \dfrac{1}{4} + 12x^4 =$
$12x^4 - 2x + \dfrac{1}{4}$

73. $2x - 4 = 2x^1 - 4x^0$

The degree of $2x$ is 1.

The degree of -4 is 0.

The degree of the polynomial is 1, the largest exponent.

75. $3x^2 - 5x + 2 = 3x^2 - 5x^1 + 2x^0$

The degree of $3x^2$ is 2.

The degree of $-5x$ is 1.

The degree of 2 is 0.

The degree of the polynomial is 2, the largest exponent.

77. $-7x^3 + 6x^2 + \dfrac{3}{5}x + 7 = -7x^3 + 6x^2 + \dfrac{3}{5}x^1 + 7x^0$

The degree of $-7x^3$ is 3.

The degree of $6x^2$ is 2.

The degree of $\dfrac{3}{5}x$ is 1.

The degree of 7 is 0.

The degree of the polynomial is 3, the largest exponent.

79. $x^2 - 3x + x^6 - 9x^4 = x^2 - 3x^1 + x^6 - 9x^4$

The degree of x^2 is 2.

The degree of $-3x$ is 1.

The degree of x^6 is 6.

The degree of $-9x^4$ is 4.

The degree of the polynomial is 6, the largest exponent.

81. See the answer section in the text.

83. In the polynomial $x^3 - 27$, there are no x^2 or x terms. The x^2 term (or second-degree term) and the x term (or first-degree term) are missing.

85. In the polynomial $x^4 - x$, there are no x^3, x^2, or x^0 terms. The x^3 term (or third-degree term), the x^2 term (or second-degree term), and the x^0 term (or zero-degree term) are missing.

87. No terms are missing in the polynomial
$2x^3 - 5x^2 + x - 3$.

89. $x^3 - 27 = x^3 + 0x^2 + 0x - 27$
$x^3 - 27 = x^3 \qquad\qquad - 27$

91. $x^4 - x = x^4 + 0x^3 + 0x^2 - x + 0x^0$

$x^4 - x = x^4 \qquad\qquad - x$

93. There are no missing terms.

95. The polynomial $x^2 - 10x + 25$ is a *trinomial* because it has just three terms.

97. The polynomial $x^3 - 7x^2 + 2x - 4$ is *none of these* because it has more than three terms.

99. The polynomial $4x^2 - 25$ is a *binomial* because it has just two terms.

101. The polynomial $40x$ is a *monomial* because it has just one term.

103. Discussion and Writing Exercise

105. ***Familiarize***. Let $a =$ the number of apples the campers had to begin with. Then the first camper ate $\frac{1}{3}a$ apples and $a - \frac{1}{3}a$, or $\frac{2}{3}a$, apples were left. The second camper ate $\frac{1}{3}\left(\frac{2}{3}a\right)$, or $\frac{2}{9}a$, apples, and $\frac{2}{3}a - \frac{2}{9}a$, or $\frac{4}{9}a$, apples were left. The third camper ate $\frac{1}{3}\left(\frac{4}{9}a\right)$, or $\frac{4}{27}a$, apples, and $\frac{4}{9}a - \frac{4}{27}a$, or $\frac{8}{27}a$, apples were left.

Translate. We write an equation for the number of apples left after the third camper eats.

Number of apples left is 8.

$$\underbrace{\qquad\qquad\qquad}$$

$$\frac{8}{27}a \qquad\qquad = 8$$

Solve. We solve the equation.

$$\frac{8}{27}a = 8$$

$$a = \frac{27}{8} \cdot 8$$

$$a = 27$$

Check. If the campers begin with 27 apples, then the first camper eats $\frac{1}{3} \cdot 27$, or 9, and $27 - 9$, or 18, are left. The second camper then eats $\frac{1}{3} \cdot 18$, or 6 apples and $18 - 6$, or 12, are left. Finally, the third camper eats $\frac{1}{3} \cdot 12$, or 4 apples and $12 - 4$, or 8, are left. The answer checks.

State. The campers had 27 apples to begin with.

107. $\frac{1}{8} - \frac{5}{6} = \frac{1}{8} + \left(-\frac{5}{6}\right)$,   LCM is 24

$$= \frac{1}{8} \cdot \frac{3}{3} + \left(-\frac{5}{6}\right)\left(\frac{4}{4}\right)$$

$$= \frac{3}{24} + \left(-\frac{20}{24}\right)$$

$$= -\frac{17}{24}$$

109. $5.6 - 8.2 = 5.6 + (-8.2) = -2.6$

111. $\qquad C = ab - r$

$\quad C + r = ab \qquad\qquad$ Adding r

$\quad \dfrac{C+r}{a} = \dfrac{ab}{a} \qquad\qquad$ Dividing by a

$\quad \dfrac{C+r}{a} = b \qquad\qquad$ Simplifying

113. $3x - 15y + 63 = 3 \cdot x - 3 \cdot 5y + 3 \cdot 21 = 3(x - 5y + 21)$

115. $(3x^2)^3 + 4x^2 \cdot 4x^4 - x^4(2x)^2 + [(2x)^2]^3 - 100x^2(x^2)^2$

$= 27x^6 + 4x^2 \cdot 4x^4 - x^4 \cdot 4x^2 + (2x)^6 - 100x^2 \cdot x^4$

$= 27x^6 + 16x^6 - 4x^6 + 64x^6 - 100x^6$

$= 3x^6$

117. $(5m^5)^2 = 5^2 m^{5 \cdot 2} = 25m^{10}$

The degree is 10.

119. Graph $y = 5 - x^2$. Then use VALUE from the CALC menu to find the y-values that correspond to $x = -3$, $x = -1$, x=0, $x = 1.5$, and $x = 2$. As before, we find that these values are -4, 4, 5, 2.75, and 1, respectively.

121. Graph $y = -0.00006x^3 + 0.006x^2 - 0.1x + 1.9$. Then use VALUE from the CALC menu to find the y-values that correspond to $x = 20$ and $x = 40$. As before, we find that these values are 1.82 and 3.66, respectively. These results represent 1,820,000 and 3,660,000 hearing-impaired Americans.

Exercise Set 4.4

1. $(3x + 2) + (-4x + 3) = (3 - 4)x + (2 + 3) = -x + 5$

3. $(-6x + 2) + \left(x^2 + \frac{1}{2}x - 3\right) = x^2 + \left(-6 + \frac{1}{2}\right)x + (2 - 3) =$

$x^2 + \left(-\frac{12}{2} + \frac{1}{2}\right)x + (2 - 3) = x^2 - \frac{11}{2}x - 1$

5. $(x^2 - 9) + (x^2 + 9) = (1 + 1)x^2 + (-9 + 9) = 2x^2$

7. $(3x^2 - 5x + 10) + (2x^2 + 8x - 40) =$

$(3 + 2)x^2 + (-5 + 8)x + (10 - 40) = 5x^2 + 3x - 30$

9. $(1.2x^3 + 4.5x^2 - 3.8x) + (-3.4x^3 - 4.7x^2 + 23) =$

$(1.2 - 3.4)x^3 + (4.5 - 4.7)x^2 - 3.8x + 23 =$

$-2.2x^3 - 0.2x^2 - 3.8x + 23$

11. $(1 + 4x + 6x^2 + 7x^3) + (5 - 4x + 6x^2 - 7x^3) =$

$(1 + 5) + (4 - 4)x + (6 + 6)x^2 + (7 - 7)x^3 =$

$6 + 0x + 12x^2 + 0x^3 = 6 + 12x^2$, or $12x^2 + 6$

13. $\left(\frac{1}{4}x^4 + \frac{2}{3}x^3 + \frac{5}{8}x^2 + 7\right) + \left(-\frac{3}{4}x^4 + \frac{3}{8}x^2 - 7\right) =$

$\left(\frac{1}{4} - \frac{3}{4}\right)x^4 + \frac{2}{3}x^3 + \left(\frac{5}{8} + \frac{3}{8}\right)x^2 + (7 - 7) =$

$-\frac{2}{4}x^4 + \frac{2}{3}x^3 + \frac{8}{8}x^2 + 0 =$

$-\frac{1}{2}x^4 + \frac{2}{3}x^3 + x^2$

15. $(0.02x^5 - 0.2x^3 + x + 0.08) + (-0.01x^5 + x^4 - 0.8x - 0.02) =$
$(0.02 - 0.01)x^5 + x^4 - 0.2x^3 + (1 - 0.8)x + (0.08 - 0.02) =$
$0.01x^5 + x^4 - 0.2x^3 + 0.2x + 0.06$

17. $9x^8 - 7x^4 + 2x^2 + 5) + (8x^7 + 4x^4 - 2x) +$
$(-3x^4 + 6x^2 + 2x - 1) = 9x^8 + 8x^7 + (-7 + 4 - 3)x^4 +$
$(2 + 6)x^2 + (-2 + 2)x + (5 - 1) =$
$9x^8 + 8x^7 - 6x^4 + 8x^2 + 4$

19. Rewrite the problem so the coefficients of like terms have the same number of decimal places.

$$
\begin{array}{r}
0.15x^4 + 0.10x^3 - 0.90x^2 \\
- 0.01x^3 + 0.01x^2 + x \\
1.25x^4 \qquad\quad + 0.11x^2 \qquad\quad + 0.01 \\
0.27x^3 \qquad\qquad\qquad\quad + 0.99 \\
-0.35x^4 \qquad\qquad + 15.00x^2 \qquad - 0.03 \\
\hline
1.05x^4 + 0.36x^3 + 14.22x^2 + x + 0.97
\end{array}
$$

21. We change the sign of the term inside the parentheses.
$-(-5x) = 5x$

23. We change the sign of every term inside the parentheses.
$-\left(-x^2 + \dfrac{3}{2}x - 2\right) = x^2 - \dfrac{3}{2}x + 2$

25. We change the sign of every term inside the parentheses.
$-(12x^4 - 3x^3 + 3) = -12x^4 + 3x^3 - 3$

27. We change the sign of every term inside parentheses.
$-(3x - 7) = -3x + 7$

29. We change the sign of every term inside parentheses.
$-(4x^2 - 3x + 2) = -4x^2 + 3x - 2$

31. We change the sign of every term inside parentheses.
$-\left(-4x^4 + 6x^2 + \dfrac{3}{4}x - 8\right) = 4x^4 - 6x^2 - \dfrac{3}{4}x + 8$

33. $(3x + 2) - (-4x + 3) = 3x + 2 + 4x - 3$
Changing the sign of every term inside parentheses
$= 7x - 1$

35. $(-6x + 2) - (x^2 + x - 3) = -6x + 2 - x^2 - x + 3$
$= -x^2 - 7x + 5$

37. $(x^2 - 9) - (x^2 + 9) = x^2 - 9 - x^2 - 9 = -18$

39. $(6x^4 + 3x^3 - 1) - (4x^2 - 3x + 3)$
$= 6x^4 + 3x^3 - 1 - 4x^2 + 3x - 3$
$= 6x^4 + 3x^3 - 4x^2 + 3x - 4$

41. $(1.2x^3 + 4.5x^2 - 3.8x) - (-3.4x^3 - 4.7x^2 + 23)$
$= 1.2x^3 + 4.5x^2 - 3.8x + 3.4x^3 + 4.7x^2 - 23$
$= 4.6x^3 + 9.2x^2 - 3.8x - 23$

43. $\dfrac{5}{8}x^3 - \dfrac{1}{4}x - \dfrac{1}{3} - \left(-\dfrac{1}{8}x^3 + \dfrac{1}{4}x - \dfrac{1}{3}\right)$
$= \dfrac{5}{8}x^3 - \dfrac{1}{4}x - \dfrac{1}{3} + \dfrac{1}{8}x^3 - \dfrac{1}{4}x + \dfrac{1}{3}$
$= \dfrac{6}{8}x^3 - \dfrac{2}{4}x$
$= \dfrac{3}{4}x^3 - \dfrac{1}{2}x$

45. $(0.08x^3 - 0.02x^2 + 0.01x) - (0.02x^3 + 0.03x^2 - 1)$
$= 0.08x^3 - 0.02x^2 + 0.01x - 0.02x^3 - 0.03x^2 + 1$
$= 0.06x^3 - 0.05x^2 + 0.01x + 1$

47.
$$
\begin{array}{l}
x^2 + 5x + 6 \\
x^2 + 2x \\
\hline
\\
x^2 + 5x + 6 \\
-x^2 - 2x \qquad \text{Changing signs} \\
\hline
3x + 6 \qquad \text{Adding}
\end{array}
$$

49.
$$
\begin{array}{l}
5x^4 + 6x^3 - 9x^2 \\
-6x^4 - 6x^3 \qquad\quad + 8x + 9 \\
\hline
\\
5x^4 + 6x^3 - 9x^2 \\
6x^4 + 6x^3 \qquad\qquad - 8x - 9 \qquad \text{Changing signs} \\
\hline
11x^4 + 12x^3 - 9x^2 - 8x - 9 \qquad \text{Adding}
\end{array}
$$

51.
$$
\begin{array}{l}
x^5 \qquad\qquad\qquad\quad - 1 \\
x^5 - x^4 + x^3 - x^2 + x - 1 \\
\hline
\\
x^5 \qquad\qquad\qquad\quad - 1 \\
-x^5 + x^4 - x^3 + x^2 - x + 1 \qquad \text{Changing signs} \\
\hline
x^4 - x^3 + x^2 - x \qquad \text{Adding}
\end{array}
$$

53. We add the lengths of the sides:
$$4a + 7 + a + \dfrac{1}{2}a + 3 + a + 2a + 3a$$
$$= \left(4 + 1 + \dfrac{1}{2} + 1 + 2 + 3\right)a + (7 + 3)$$
$$= 11\dfrac{1}{2}a + 10, \text{ or } \dfrac{23}{2}a + 10$$

55.

The area of a rectangle is the product of the length and width. The sum of the areas is found as follows:

$$
\begin{array}{ccccccccc}
& \text{Area} & & \text{Area} & & \text{Area} & & \text{Area} & \\
& \text{of } A & + & \text{of } B & + & \text{of } C & + & \text{of } D & \\
= & 3x \cdot x & + & x \cdot x & + & x \cdot x & + & 4 \cdot x & \\
= & 3x^2 & + & x^2 & + & x^2 & + & 4x & \\
= & 5x^2 & + & 4x & & & & &
\end{array}
$$

A polynomial for the sum of the areas is $5x^2 + 4x$.

57.

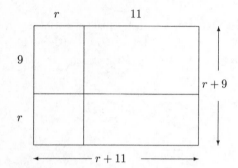

The length and width of the figure can be expressed as $r + 11$ and $r + 9$, respectively. The area of this figure (a rectangle) is the product of the length and width. An algebraic expression for the area is $(r + 11) \cdot (r + 9)$.

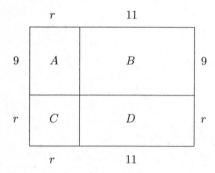

The area of the figure can also be found by adding the areas of the four rectangles A, B, C, and D. The area of a rectangle is the product of the length and the width.

$$
\begin{array}{ccccccccc}
& \text{Area} & & \text{Area} & & \text{Area} & & \text{Area} & \\
& \text{of } A & + & \text{of } B & + & \text{of } C & + & \text{of } D & \\
= & 9 \cdot r & + & 11 \cdot 9 & + & r \cdot r & + & 11 \cdot r & \\
= & 9r & + & 99 & + & r^2 & + & 11r &
\end{array}
$$

A second algebraic expression for the area of the figure is $9r + 99 + r^2 + 11r$, or $r^2 + 20r + 99$.

59.

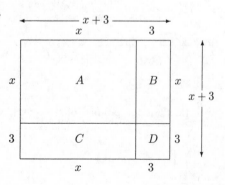

The length and width of the figure can each be expressed as $x + 3$. The area can be expressed as $(x+3) \cdot (x+3)$, or $(x+3)^2$.

Another way to express the area is to find an expression for the sum of the areas of the four rectangles A, B, C, and D. The area of each rectangle is the product of its length and width.

$$
\begin{array}{ccccccccc}
& \text{Area} & & \text{Area} & & \text{Area} & & \text{Area} & \\
& \text{of } A & + & \text{of } B & + & \text{of } C & + & \text{of } D & \\
= & x \cdot x & + & 3 \cdot x & + & 3 \cdot x & + & 3 \cdot 3 & \\
= & x^2 & + & 3x & + & 3x & + & 9 &
\end{array}
$$

Then a second algebraic expression for the area of the figure is $x^2 + 3x + 3x + 9$, or $x^2 + 6x + 9$.

61.

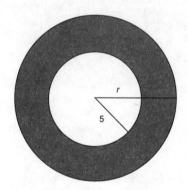

Familiarize. Recall that the area of a circle is the product of π and the square of the radius, r^2.

$$A = \pi r^2$$

Translate.

$$
\begin{array}{ccccc}
\text{Area of circle} & & \text{Area of circle} & & \text{Shaded} \\
\text{with radius } r & - & \text{with radius 5} & = & \text{area} \\
\pi \cdot r^2 & - & \pi \cdot 5^2 & = & \text{Shaded area}
\end{array}
$$

Carry out. We simplify the expression.

$$\pi \cdot r^2 - \pi \cdot 5^2 = \pi r^2 - 25\pi$$

Check. We can go over our calculations. We can also assign some value to r, say 7, and carry out the computation in two ways.

Difference of areas: $\pi \cdot 7^2 - \pi \cdot 5^2 = 49\pi - 25\pi = 24\pi$

Substituting in the polynomial: $\pi \cdot 7^2 - 25\pi = 49\pi - 25\pi = 24\pi$

Since the results are the same, our solution is probably correct.

State. A polynomial for the shaded area is $\pi r^2 - 25\pi$.

63. *Familiarize.* We label the figure with additional information.

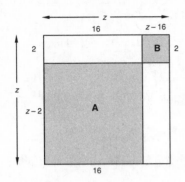

Translate.

Area of shaded sections = Area of A + Area of B

Area of shaded sections = $16(z - 2) + 2(z - 16)$

Carry out. We simplify the expression.

$16(z - 2) + 2(z - 16) = 16z - 32 + 2z - 32 = 18z - 64$

Check. We can go over the calculations. We can also assign some value to z, say 30, and carry out the computation in two ways.

Sum of areas:

$16 \cdot 28 + 2 \cdot 14 = 448 + 28 = 476$

Substituting in the polynomial:

$18 \cdot 30 - 64 = 540 - 64 = 476$

Since the results are the same, our solution is probably correct.

State. A polynomial for the shaded area is $18z - 64$.

65. Discussion and Writing Exercise

67. $8x + 3x = 66$

$\quad 11x = 66 \quad$ Collecting like terms

$\quad \dfrac{11x}{11} = \dfrac{66}{11} \quad$ Dividing by 11

$\quad\quad x = 6$

The solution is 6.

69. $\dfrac{3}{8}x + \dfrac{1}{4} - \dfrac{3}{4}x = \dfrac{11}{16} + x, \quad$ LCM is 16

$16\left(\dfrac{3}{8}x + \dfrac{1}{4} - \dfrac{3}{4}x\right) = 16\left(\dfrac{11}{16} + x\right) \quad$ Clearing fractions

$6x + 4 - 12x = 11 + 16x$

$-6x + 4 = 11 + 16x \quad$ Collecting like terms

$-6x + 4 - 4 = 11 + 16x - 4 \quad$ Subtracting 4

$-6x = 7 + 16x$

$-6x - 16x = 7 + 16x - 16x \quad$ Subtracting $16x$

$-22x = 7$

$\dfrac{-22x}{-22} = \dfrac{7}{-22} \quad\quad$ Dividing by -22

$x = -\dfrac{7}{22}$

The solution is $-\dfrac{7}{22}$.

71. $1.5x - 2.7x = 22 - 5.6x$

$10(1.5x - 2.7x) = 10(22 - 5.6x) \quad$ Clearing decimals

$15x - 27x = 220 - 56x$

$-12x = 220 - 56x \quad$ Collecting like terms

$44x = 220 \quad$ Adding $56x$

$x = \dfrac{220}{44} \quad$ Dividing by 44

$x = 5 \quad$ Simplifying

The solution is 5.

73. $6(y - 3) - 8 = 4(y + 2) + 5$

$6y - 18 - 8 = 4y + 8 + 5 \quad$ Removing parentheses

$6y - 26 = 4y + 13 \quad$ Collecting like terms

$6y - 26 + 26 = 4y + 13 + 26 \quad$ Adding 26

$6y = 4y + 39$

$6y - 4y = 4y + 39 - 4y \quad$ Subtracting $4y$

$2y = 39$

$\dfrac{2y}{2} = \dfrac{39}{2} \quad$ Dividing by 2

$y = \dfrac{39}{2}$

The solution is $\dfrac{39}{2}$.

75. $3x - 7 \leq 5x + 13$

$-2x - 7 \leq 13 \quad$ Subtracting $5x$

$-2x \leq 20 \quad$ Adding 7

$x \geq -10 \quad$ Dividing by -2 and reversing the inequality symbol

The solution set is $\{x | x \geq -10\}$.

77. *Familiarize.* The surface area is $2lw + 2lh + 2wh$, where l = length, w = width, and h = height of the rectangular solid. Here we have $l = 3$, $w = w$, and $h = 7$.

Translate. We substitute in the formula above.

$2 \cdot 3 \cdot w + 2 \cdot 3 \cdot 7 + 2 \cdot w \cdot 7$

Carry out. We simplify the expression.

$2 \cdot 3 \cdot w + 2 \cdot 3 \cdot 7 + 2 \cdot w \cdot 7$

$= 6w + 42 + 14w$

$= 20w + 42$

Check. We can go over the calculations. We can also assign some value to w, say 6, and carry out the computation in two ways.

Using the formula: $2 \cdot 3 \cdot 6 + 2 \cdot 3 \cdot 7 + 2 \cdot 6 \cdot 7 = 36 + 42 + 84 = 162$

Substituting in the polynomial: $20 \cdot 6 + 42 = 120 + 42 = 162$

Since the results are the same, our solution is probably correct.

State. A polynomial for the surface area is $20w + 42$.

79. ***Familiarize***. The surface area is $2lw + 2lh + 2wh$, where l = length, w = width, and h = height of the rectangular solid. Here we have $l = x$, $w = x$, and $h = 5$.

Translate. We substitute in the formula above.

$$2 \cdot x \cdot x + 2 \cdot x \cdot 5 + 2 \cdot x \cdot 5$$

Carry out. We simplify the expression.

$$\begin{aligned} &2 \cdot x \cdot x + 2 \cdot x \cdot 5 + 2 \cdot x \cdot 5 \\ &= 2x^2 + 10x + 10x \\ &= 2x^2 + 20x \end{aligned}$$

Check. We can go over the calculations. We can also assign some value to x, say 3, and carry out the computation in two ways.

Using the formula: $2 \cdot 3 \cdot 3 + 2 \cdot 3 \cdot 5 + 2 \cdot 3 \cdot 5 = 18 + 30 + 30 = 78$

Substituting in the polynomial: $2 \cdot 3^2 + 20 \cdot 3 = 2 \cdot 9 + 60 = 18 + 60 = 78$

Since the results are the same, our solution is probably correct.

State. A polynomial for the surface area is $2x^2 + 20x$.

81.

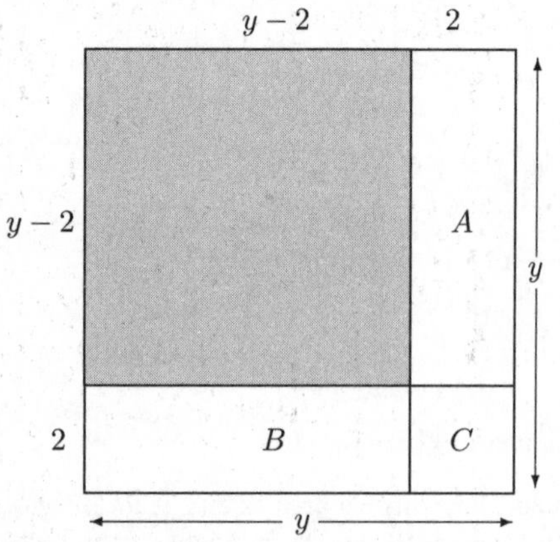

The shaded area is $(y-2)^2$. We find it as follows:

$$\begin{aligned} &\text{Shaded area} = \text{Area of square} - \text{Area of } A - \text{Area of } B - \text{Area of } C \\ &(y-2)^2 = y^2 - 2(y-2) - 2(y-2) - 2 \cdot 2 \\ &(y-2)^2 = y^2 - 2y + 4 - 2y + 4 - 4 \\ &(y-2)^2 = y^2 - 4y + 4 \end{aligned}$$

83. $(7y^2 - 5y + 6) - (3y^2 + 8y - 12) + (8y^2 - 10y + 3)$
$= 7y^2 - 5y + 6 - 3y^2 - 8y + 12 + 8y^2 - 10y + 3$
$= 12y^2 - 23y + 21$

85. $(-y^4 - 7y^3 + y^2) + (-2y^4 + 5y - 2) - (-6y^3 + y^2)$
$= -y^4 - 7y^3 + y^2 - 2y^4 + 5y - 2 + 6y^3 - y^2$
$= -3y^4 - y^3 + 5y - 2$

Exercise Set 4.5

1. $(8x^2)(5) = (8 \cdot 5)x^2 = 40x^2$

3. $(-x^2)(-x) = (-1x^2)(-1x) = (-1)(-1)(x^2 \cdot x) = x^3$

5. $(8x^5)(4x^3) = (8 \cdot 4)(x^5 \cdot x^3) = 32x^8$

7. $(0.1x^6)(0.3x^5) = (0.1)(0.3)(x^6 \cdot x^5) = 0.03x^{11}$

9. $\left(-\frac{1}{5}x^3\right)\left(-\frac{1}{3}x\right) = \left(-\frac{1}{5}\right)\left(-\frac{1}{3}\right)(x^3 \cdot x) = \frac{1}{15}x^4$

11. $(-4x^2)(0) = 0$ Any number multiplied by 0 is 0.

13. $(3x^2)(-4x^3)(2x^6) = (3)(-4)(2)(x^2 \cdot x^3 \cdot x^6) = -24x^{11}$

15. $2x(-x+5) = 2x(-x) + 2x(5)$
$= -2x^2 + 10x$

17. $-5x(x-1) = -5x(x) - 5x(-1)$
$= -5x^2 + 5x$

19. $x^2(x^3+1) = x^2(x^3) + x^2(1)$
$= x^5 + x^2$

21. $3x(2x^2 - 6x + 1) = 3x(2x^2) + 3x(-6x) + 3x(1)$
$= 6x^3 - 18x^2 + 3x$

23. $-6x^2(x^2 + x) = -6x^2(x^2) - 6x^2(x)$
$= -6x^4 - 6x^3$

25. $3y^2(6y^4 + 8y^3) = 3y^2(6y^4) + 3y^2(8y^3)$
$= 18y^6 + 24y^5$

27. $(x+6)(x+3) = (x+6)x + (x+6)3$
$= x \cdot x + 6 \cdot x + x \cdot 3 + 6 \cdot 3$
$= x^2 + 6x + 3x + 18$
$= x^2 + 9x + 18$

29. $(x+5)(x-2) = (x+5)x + (x+5)(-2)$
$= x \cdot x + 5 \cdot x + x(-2) + 5(-2)$
$= x^2 + 5x - 2x - 10$
$= x^2 + 3x - 10$

31. $(x-4)(x-3) = (x-4)x + (x-4)(-3)$
$= x \cdot x - 4 \cdot x + x(-3) - 4(-3)$
$= x^2 - 4x - 3x + 12$
$= x^2 - 7x + 12$

33. $(x+3)(x-3) = (x+3)x + (x+3)(-3)$
$= x \cdot x + 3 \cdot x + x(-3) + 3(-3)$
$= x^2 + 3x - 3x - 9$
$= x^2 - 9$

35. $(5-x)(5-2x) = (5-x)5 + (5-x)(-2x)$
$= 5 \cdot 5 - x \cdot 5 + 5(-2x) - x(-2x)$
$= 25 - 5x - 10x + 2x^2$
$= 25 - 15x + 2x^2$

37. $(2x + 5)(2x + 5) = (2x + 5)2x + (2x + 5)5$
$$= 2x \cdot 2x + 5 \cdot 2x + 2x \cdot 5 + 5 \cdot 5$$
$$= 4x^2 + 10x + 10x + 25$$
$$= 4x^2 + 20x + 25$$

39. $\left(x - \dfrac{5}{2}\right)\left(x + \dfrac{2}{5}\right) = \left(x - \dfrac{5}{2}\right)x + \left(x - \dfrac{5}{2}\right)\dfrac{2}{5}$
$$= x \cdot x - \dfrac{5}{2} \cdot x + x \cdot \dfrac{2}{5} - \dfrac{5}{2} \cdot \dfrac{2}{5}$$
$$= x^2 - \dfrac{5}{2}x + \dfrac{2}{5}x - 1$$
$$= x^2 - \dfrac{25}{10}x + \dfrac{4}{10}x - 1$$
$$= x^2 - \dfrac{21}{10}x - 1$$

41. $(x - 2.3)(x + 4.7) = (x - 2.3)x + (x - 2.3)4.7$
$$= x \cdot x - 2.3 \cdot x + x \cdot 4.7 - 2.3(4.7)$$
$$= x^2 - 2.3x + 4.7x - 10.81$$
$$= x^2 + 2.4x - 10.81$$

43. The length of the rectangle is $x+6$ and the width is $x+2$, so the area is $(x+6)(x+2)$. If we carry out the multiplication, we have $x^2 + 8x + 12$.

45. The length of the rectangle is $x+6$ and the width is $x+1$, so the area is $(x+6)(x+1)$. If we carry out the multiplication, we have $x^2 + 7x + 6$.

47. Illustrate $x(x + 5)$ as the area of a rectangle with width x and length $x + 5$.

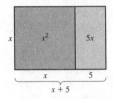

49. Illustrate $(x + 1)(x + 2)$ as the area of a rectangle with width $x + 1$ and length $x + 2$.

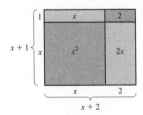

51. Illustrate $(x + 5)(x + 3)$ as the area of a rectangle with length $x + 5$ and width $x + 3$.

53. $(x^2 + x + 1)(x - 1)$
$$= (x^2 + x + 1)x + (x^2 + x + 1)(-1)$$
$$= x^2 \cdot x + x \cdot x + 1 \cdot x + x^2(-1) + x(-1) + 1(-1)$$
$$= x^3 + x^2 + x - x^2 - x - 1$$
$$= x^3 - 1$$

55. $(2x + 1)(2x^2 + 6x + 1)$
$$= 2x(2x^2 + 6x + 1) + 1(2x^2 + 6x + 1)$$
$$= 2x \cdot 2x^2 + 2x \cdot 6x + 2x \cdot 1 + 1 \cdot 2x^2 + 1 \cdot 6x + 1 \cdot 1$$
$$= 4x^3 + 12x^2 + 2x + 2x^2 + 6x + 1$$
$$= 4x^3 + 14x^2 + 8x + 1$$

57. $(y^2 - 3)(3y^2 - 6y + 2)$
$$= y^2(3y^2 - 6y + 2) - 3(3y^2 - 6y + 2)$$
$$= y^2 \cdot 3y^2 + y^2(-6y) + y^2 \cdot 2 - 3 \cdot 3y^2 - 3(-6y) - 3 \cdot 2$$
$$= 3y^4 - 6y^3 + 2y^2 - 9y^2 + 18y - 6$$
$$= 3y^4 - 6y^3 - 7y^2 + 18y - 6$$

59. $(x^3 + x^2)(x^3 + x^2 - x)$
$$= x^3(x^3 + x^2 - x) + x^2(x^3 + x^2 - x)$$
$$= x^3 \cdot x^3 + x^3 \cdot x^2 + x^3(-x) + x^2 \cdot x^3 + x^2 \cdot x^2 + x^2(-x)$$
$$= x^6 + x^5 - x^4 + x^5 + x^4 - x^3$$
$$= x^6 + 2x^5 - x^3$$

61. $(-5x^3 - 7x^2 + 1)(2x^2 - x)$
$$= (-5x^3 - 7x^2 + 1)2x^2 + (-5x^3 - 7x^2 + 1)(-x)$$
$$= -5x^3 \cdot 2x^2 - 7x^2 \cdot 2x^2 + 1 \cdot 2x^2 - 5x^3(-x) - 7x^2(-x) + 1(-x)$$
$$= -10x^5 - 14x^4 + 2x^2 + 5x^4 + 7x^3 - x$$
$$= -10x^5 - 9x^4 + 7x^3 + 2x^2 - x$$

63.

$$
\begin{array}{lll}
\quad 1 + x + x^2 & \text{Line up like terms}\\
\underline{-1 - x + x^2} & \text{in columns}\\
\quad x^2 + x^3 + x^4 & \text{Multiplying the top row by } x^2\\
-\ x - x^2 - x^3 & \text{Multiplying by } -x\\
\underline{-1 - \ x - x^2} & \text{Multiplying by } -1\\
-1 - 2x - x^2 \quad + x^4
\end{array}
$$

65.

$$
\begin{array}{ll}
\quad 2t^2 - \ t - 4\\
\quad 3t^2 + 2t - 1\\
\hline
-\ 2t^2 + \ t + 4 & \text{Multiplying by } -1\\
\ 4t^3 - \ 2t^2 - 8t & \text{Multiplying by } 2t\\
\underline{6t^4 - 3t^3 - 12t^2} & \text{Multiplying by } 3t^2\\
6t^4 + \ t^3 - 16t^2 - 7t + 4
\end{array}
$$

67.

$$
\begin{array}{ll}
\ x \quad\ - x^3 \quad\ + x^5\\
\underline{-1 + x^2 \quad\ + x^4} & \text{Rewriting in ascending order}\\
\quad\quad\quad x^5 - x^7 + x^9 & \text{Multiplying by } x^4\\
\quad x^3 - x^5 + x^7 & \text{Multiplying by } x^2\\
\underline{-x + \ x^3 - x^5} & \text{Multiplying by } -1\\
-x + 2x^3 - x^5 \quad\quad + x^9
\end{array}
$$

69.

$$
\begin{array}{l}
\quad x^3 + x^2 + x + 1\\
\quad\quad\quad\quad\quad x - 1\\
\hline
-x^3 - x^2 - x - 1\\
\underline{x^4 + x^3 + x^2 + x}\\
x^4 \quad\quad\quad\quad\quad - 1
\end{array}
$$

71. We will multiply horizontally while still aligning like terms.

$$(x+1)(x^3+7x^2+5x+4)$$

$= x^4 + 7x^3 + 5x^2 + 4x$ Multiplying by x
$\underline{\quad + x^3 + 7x^2 + 5x + 4}$ Multiplying by 1
$= x^4 + 8x^3 + 12x^2 + 9x + 4$

73. We will multiply horizontally while still aligning like terms.

$$\left(x - \frac{1}{2}\right)\left(2x^3 - 4x^2 + 3x - \frac{2}{5}\right)$$

$= 2x^4 - 4x^3 + 3x^2 - \dfrac{2}{5}x$

$\quad \underline{\; - x^3 + 2x^2 - \dfrac{3}{2}x + \dfrac{1}{5}}$

$\quad 2x^4 - 5x^3 + 5x^2 - \dfrac{19}{10}x + \dfrac{1}{5}$

75. Discussion and Writing Exercise

77. $-\dfrac{1}{4} - \dfrac{1}{2} = -\dfrac{1}{4} - \dfrac{1}{2} \cdot \dfrac{2}{2} = -\dfrac{1}{4} - \dfrac{2}{4} = -\dfrac{3}{4}$

79. $(10-2)(10+2) = 8 \cdot 12 = 96$

81. $15x - 18y + 12 = 3 \cdot 5x - 3 \cdot 6y + 3 \cdot 4 =$
$3(5x - 6y + 4)$

83. $-9x - 45y + 15 = -3 \cdot 3x - 3 \cdot 15y - 3(-5) =$
$-3(3x + 15y - 5)$

85. $y = \dfrac{1}{2}x - 3$

The equation is equivalent to $y = \dfrac{1}{2}x + (-3)$. The y-intercept is $(0, -3)$. We find two other points, using multiples of 2 for x to avoid fractions.

When $x = -2$, $y = \dfrac{1}{2}(-2) - 3 = -1 - 3 = -4$.

When $x = 4$, $y = \dfrac{1}{2} \cdot 4 - 3 = 2 - 3 = -1$.

x	y
0	-3
-2	-4
4	-1

Plot these points, draw the line they determine, and label the graph $y = \dfrac{1}{2}x - 3$.

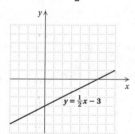

87. The shaded area is the area of the large rectangle, $6y(14y - 5)$ less the area of the unshaded rectangle, $3y(3y + 5)$. We have:

$$6y(14y - 5) - 3y(3y + 5)$$
$= 84y^2 - 30y - 9y^2 - 15y$
$= 75y^2 - 45y$

89.

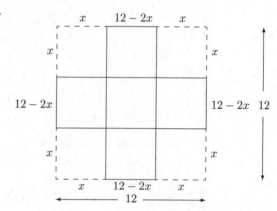

The dimensions, in inches, of the box are $12 - 2x$ by $12 - 2x$ by x. The volume is the product of the dimensions (volume = length × width × height):

Volume $= (12 - 2x)(12 - 2x)x$
$= (144 - 48x + 4x^2)x$
$= (144x - 48x^2 + 4x^3)$ in^3, or
$\quad (4x^3 - 48x^2 + 144x)$ in^3

The outside surface area is the sum of the area of the bottom and the areas of the four sides. The dimensions, in inches, of the bottom are $12 - 2x$ by $12 - 2x$, and the dimensions, in inches, of each side are x by $12 - 2x$.

$\begin{aligned} \text{Surface area} &= \text{Area of bottom} + 4 \cdot \text{Area of each side} \\ &= (12 - 2x)(12 - 2x) + 4 \cdot x(12 - 2x) \\ &= 144 - 24x - 24x + 4x^2 + 48x - 8x^2 \\ &= 144 - 48x + 4x^2 + 48x - 8x^2 \\ &= (144 - 4x^2) \text{ in}^2, \text{ or } (-4x^2 + 144) \text{ in}^2 \end{aligned}$

91. Let $n =$ the missing number.

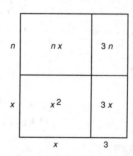

The area of the figure is $x^2 + 3x + nx + 3n$. This is equivalent to $x^2 + 8x + 15$, so we have $3x + nx = 8x$ and $3n = 15$. Solving either equation for n, we find that the missing number is 5.

93. We have a rectangular solid with dimensions x m by x m by $x+2$ m with a rectangular solid piece with dimensions 6 m by 5 m by 7 m cut out of it.

$$\text{Volume} = \frac{\text{Volume of}}{\text{large solid}} - \frac{\text{Volume of}}{\text{small solid}}$$

$$= (x \text{ m})(x \text{ m})(x+2 \text{ m}) - (6 \text{ m})(5 \text{ m})(7 \text{ m})$$
$$= x^2(x+2) \text{ m}^3 - 210 \text{ m}^3$$
$$= (x^3 + 2x^2 - 210) \text{ m}^3$$

95. $(x-2)(x-7) - (x-7)(x-2)$

First observe that, by the commutative law of multiplication, $(x-2)(x-7)$ and $(x-7)(x-2)$ are equivalent expressions. Then when we subtract $(x-7)(x-2)$ from $(x-2)(x-7)$, the result is 0.

97. $(x-a)(x-b)\cdots(x-x)(x-y)(x-z)$
$= (x-a)(x-b)\cdots 0 \cdot (x-y)(x-z)$
$= 0$

Exercise Set 4.6

1. $(x+1)(x^2+3)$
     F   O   I   L
$= x \cdot x^2 + x \cdot 3 + 1 \cdot x^2 + 1 \cdot 3$
$= x^3 + 3x + x^2 + 3$

3. $(x^3+2)(x+1)$
     F   O   I   L
$= x^3 \cdot x + x^3 \cdot 1 + 2 \cdot x + 2 \cdot 1$
$= x^4 + x^3 + 2x + 2$

5. $(y+2)(y-3)$
     F   O   I   L
$= y \cdot y + y \cdot (-3) + 2 \cdot y + 2 \cdot (-3)$
$= y^2 - 3y + 2y - 6$
$= y^2 - y - 6$

7. $(3x+2)(3x+2)$
     F   O   I   L
$= 3x \cdot 3x + 3x \cdot 2 + 2 \cdot 3x + 2 \cdot 2$
$= 9x^2 + 6x + 6x + 4$
$= 9x^2 + 12x + 4$

9. $(5x-6)(x+2)$
     F   O   I   L
$= 5x \cdot x + 5x \cdot 2 + (-6) \cdot x + (-6) \cdot 2$
$= 5x^2 + 10x - 6x - 12$
$= 5x^2 + 4x - 12$

11. $(3t-1)(3t+1)$
     F   O   I   L
$= 3t \cdot 3t + 3t \cdot 1 + (-1) \cdot 3t + (-1) \cdot 1$
$= 9t^2 + 3t - 3t - 1$
$= 9t^2 - 1$

13. $(4x-2)(x-1)$
     F   O   I   L
$= 4x \cdot x + 4x \cdot (-1) + (-2) \cdot x + (-2) \cdot (-1)$
$= 4x^2 - 4x - 2x + 2$
$= 4x^2 - 6x + 2$

15. $\left(p - \frac{1}{4}\right)\left(p + \frac{1}{4}\right)$
     F   O   I   L
$= p \cdot p + p \cdot \frac{1}{4} + \left(-\frac{1}{4}\right) \cdot p + \left(-\frac{1}{4}\right) \cdot \frac{1}{4}$
$= p^2 + \frac{1}{4}p - \frac{1}{4}p - \frac{1}{16}$
$= p^2 - \frac{1}{16}$

17. $(x-0.1)(x+0.1)$
     F   O   I   L
$= x \cdot x + x \cdot (0.1) + (-0.1) \cdot x + (-0.1)(0.1)$
$= x^2 + 0.1x - 0.1x - 0.01$
$= x^2 - 0.01$

19. $(2x^2+6)(x+1)$
     F   O   I   L
$= 2x^3 + 2x^2 + 6x + 6$

21. $(-2x+1)(x+6)$
     F   O   I   L
$= -2x^2 - 12x + x + 6$
$= -2x^2 - 11x + 6$

23. $(a+7)(a+7)$
     F   O   I   L
$= a^2 + 7a + 7a + 49$
$= a^2 + 14a + 49$

25. $(1+2x)(1-3x)$
     F   O   I   L
$= 1 - 3x + 2x - 6x^2$
$= 1 - x - 6x^2$

27. $\left(\frac{3}{8}y - \frac{5}{6}\right)\left(\frac{3}{8}y - \frac{5}{6}\right)$
     F   O   I   L
$= \frac{9}{64}y^2 - \frac{15}{48}y - \frac{15}{48}y + \frac{25}{36}$
$= \frac{9}{64}y^2 - \frac{30}{48}y + \frac{25}{36}$
$= \frac{9}{64}y^2 - \frac{5}{8}y + \frac{25}{36}$

29. $(x^2+3)(x^3-1)$
     F   O   I   L
$= x^5 - x^2 + 3x^3 - 3$

31. $(3x^2-2)(x^4-2)$
     F   O   I   L
$= 3x^6 - 6x^2 - 2x^4 + 4$

33. $(2.8x - 1.5)(4.7x + 9.3)$
$$\quad\text{F}\qquad\quad\text{O}\qquad\quad\text{I}\qquad\quad\text{L}$$
$$= 2.8x(4.7x) + 2.8x(9.3) - 1.5(4.7x) - 1.5(9.3)$$
$$= 13.16x^2 + 26.04x - 7.05x - 13.95$$
$$= 13.16x^2 + 18.99x - 13.95$$

35. $(3x^5 + 2)(2x^2 + 6)$
$$\quad\text{F}\qquad\text{O}\qquad\text{I}\qquad\text{L}$$
$$= 6x^7 + 18x^5 + 4x^2 + 12$$

37. $(8x^3 + 1)(x^3 + 8)$
$$\quad\text{F}\qquad\text{O}\qquad\text{I}\qquad\text{L}$$
$$= 8x^6 + 64x^3 + x^3 + 8$$
$$= 8x^6 + 65x^3 + 8$$

39. $(4x^2 + 3)(x - 3)$
$$\quad\text{F}\qquad\text{O}\qquad\text{I}\qquad\text{L}$$
$$= 4x^3 - 12x^2 + 3x - 9$$

41. $(4y^4 + y^2)(y^2 + y)$
$$\quad\text{F}\qquad\text{O}\qquad\text{I}\qquad\text{L}$$
$$= 4y^6 + 4y^5 + y^4 + y^3$$

43. $(x + 4)(x - 4)$ Product of sum and difference of two terms
$$= x^2 - 4^2$$
$$= x^2 - 16$$

45. $(2x + 1)(2x - 1)$ Product of sum and difference of two terms
$$= (2x)^2 - 1^2$$
$$= 4x^2 - 1$$

47. $(5m - 2)(5m + 2)$ Product of sum and difference of two terms
$$= (5m)^2 - 2^2$$
$$= 25m^2 - 4$$

49. $(2x^2 + 3)(2x^2 - 3)$ Product of sum and difference of two terms
$$= (2x^2)^2 - 3^2$$
$$= 4x^4 - 9$$

51. $(3x^4 - 4)(3x^4 + 4)$
$$= (3x^4)^2 - 4^2$$
$$= 9x^8 - 16$$

53. $(x^6 - x^2)(x^6 + x^2)$
$$= (x^6)^2 - (x^2)^2$$
$$= x^{12} - x^4$$

55. $(x^4 + 3x)(x^4 - 3x)$
$$= (x^4)^2 - (3x)^2$$
$$= x^8 - 9x^2$$

57. $(x^{12} - 3)(x^{12} + 3)$
$$= (x^{12})^2 - 3^2$$
$$= x^{24} - 9$$

59. $(2y^8 + 3)(2y^8 - 3)$
$$= (2y^8)^2 - 3^2$$
$$= 4y^{16} - 9$$

61. $\left(\dfrac{5}{8}x - 4.3\right)\left(\dfrac{5}{8}x + 4.3\right)$
$$= \left(\dfrac{5}{8}x\right)^2 - (4.3)^2$$
$$= \dfrac{25}{64}x^2 - 18.49$$

63. $(x + 2)^2 = x^2 + 2 \cdot x \cdot 2 + 2^2$ Square of a binomial sum
$$= x^2 + 4x + 4$$

65. $(3x^2 + 1)$ Square of a binomial sum
$$= (3x^2)^2 + 2 \cdot 3x^2 \cdot 1 + 1^2$$
$$= 9x^4 + 6x^2 + 1$$

67. $\left(a - \dfrac{1}{2}\right)^2$ Square of a binomial sum
$$= a^2 - 2 \cdot a \cdot \dfrac{1}{2} + \left(\dfrac{1}{2}\right)^2$$
$$= a^2 - a + \dfrac{1}{4}$$

69. $(3 + x)^2 = 3^2 + 2 \cdot 3 \cdot x + x^2$
$$= 9 + 6x + x^2$$

71. $(x^2 + 1)^2 = (x^2)^2 + 2 \cdot x^2 \cdot 1 + 1^2$
$$= x^4 + 2x^2 + 1$$

73. $(2 - 3x^4)^2 = 2^2 - 2 \cdot 2 \cdot 3x^4 + (3x^4)^2$
$$= 4 - 12x^4 + 9x^8$$

75. $(5 + 6t^2)^2 = 5^2 + 2 \cdot 5 \cdot 6t^2 + (6t^2)^2$
$$= 25 + 60t^2 + 36t^4$$

77. $\left(x - \dfrac{5}{8}\right)^2 = x^2 - 2 \cdot x \cdot \dfrac{5}{8} + \left(\dfrac{5}{8}\right)^2$
$$= x^2 - \dfrac{5}{4}x + \dfrac{25}{64}$$

79. $(3 - 2x^3)^2 = 3^2 - 2 \cdot 3 \cdot 2x^3 + (2x^3)^2$
$$= 9 - 12x^3 + 4x^6$$

81. $4x(x^2 + 6x - 3)$ Product of a monomial and a trinomial
$$= 4x \cdot x^2 + 4x \cdot 6x + 4x(-3)$$
$$= 4x^3 + 24x^2 - 12x$$

83. $\left(2x^2 - \frac{1}{2}\right)\left(2x^2 - \frac{1}{2}\right)$ Square of a binomial difference

$= (2x^2)^2 - 2 \cdot 2x^2 \cdot \frac{1}{2} + \left(\frac{1}{2}\right)^2$

$= 4x^4 - 2x^2 + \frac{1}{4}$

85. $(-1 + 3p)(1 + 3p)$

$= (3p - 1)(3p + 1)$ Product of the sum and difference of two terms

$= (3p)^2 - 1^2$

$= 9p^2 - 1$

87. $3t^2(5t^3 - t^2 + t)$ Product of a monomial and a trinomial

$= 3t^2 \cdot 5t^3 + 3t^2(-t^2) + 3t^2 \cdot t$

$= 15t^5 - 3t^4 + 3t^3$

89. $(6x^4 + 4)^2$ Square of a binomial sum

$= (6x^4)^2 + 2 \cdot 6x^4 \cdot 4 + 4^2$

$= 36x^8 + 48x^4 + 16$

91. $(3x + 2)(4x^2 + 5)$ Product of two binomials; use FOIL

$= 3x \cdot 4x^2 + 3x \cdot 5 + 2 \cdot 4x^2 + 2 \cdot 5$

$= 12x^3 + 15x + 8x^2 + 10$

93. $(8 - 6x^4)^2$ Square of a binomial difference

$= 8^2 - 2 \cdot 8 \cdot 6x^4 + (6x^4)^2$

$= 64 - 96x^4 + 36x^8$

95.

$$\begin{array}{r} t^2 + t + 1 \\ t - 1 \\ \hline -t^2 - t - 1 \\ t^3 + t^2 + t \\ \hline t^3 \qquad\qquad -1 \end{array}$$

97. $3^2 + 4^2 = 9 + 16 = 25$

$(3 + 4)^2 = 7^2 = 49$

99. $9^2 - 5^2 = 81 - 25 = 56$

$(9 - 5)^2 = 4^2 = 16$

101.

We can find the shaded area in two ways.

Method 1: The figure is a square with side $a + 1$, so the area is $(a + 1)^2 = a^2 + 2a + 1$.

Method 2: We add the areas of A, B, C, and D.

$1 \cdot a + 1 \cdot 1 + 1 \cdot a + a \cdot a = a + 1 + a + a^2 = a^2 + 2a + 1$.

Either way we find that the total shaded area is $a^2 + 2a + 1$.

103.

We can find the shaded area in two ways.

Method 1: The figure is a rectangle with dimensions $t + 6$ by $t + 4$, so the area is $(t + 6)(t + 4) = t^2 + 4t + 6t + 24 = t^2 + 10t + 24$.

Method 2: We add the areas of A, B, C, and D.

$t \cdot t + t \cdot 6 + 6 \cdot 4 + 4 \cdot t = t^2 + 6t + 24 + 4t = t^2 + 10t + 24$.

Either way, we find that the total shaded area is $t^2 + 10t + 24$.

105. Discussion and Writing Exercise

107. *Familiarize.* Let t = the number of watts used by the television set. Then $10t$ = the number of watts used by the lamps, and $40t$ = the number of watts used by the air conditioner.

Translate.

Lamp watts	+	Air conditioner watts	+	Television watts	=	Total watts
↓	↓	↓	↓	↓	↓	↓
$10t$	+	$40t$	+	t	=	2550

Solve. We solve the equation.

$10t + 40t + t = 2550$

$51t = 2550$

$t = 50$

The possible solution is:

Television, t: 50 watts

Lamps, $10t$: $10 \cdot 50$, or 500 watts

Air conditioner, $40t$: $40 \cdot 50$, or 2000 watts

Check. The number of watts used by the lamps, 500, is 10 times 50, the number used by the television. The number of watts used by the air conditioner, 2000, is 40 times 50, the number used by the television. Also, $50 + 500 + 2000 = 2550$, the total wattage used.

State. The television uses 50 watts, the lamps use 500 watts, and the air conditioner uses 2000 watts.

109. $3(x - 2) = 5(2x + 7)$

$3x - 6 = 10x + 35$ Removing parentheses

$3x - 6 + 6 = 10x + 35 + 6$ Adding 6

$3x = 10x + 41$

$3x - 10x = 10x + 41 - 10x$ Subtracting $10x$

$-7x = 41$

$\dfrac{-7x}{-7} = \dfrac{41}{-7}$ Dividing by -7

$x = -\dfrac{41}{7}$

The solution is $-\dfrac{41}{7}$.

111. $3x - 2y = 12$

$-2y = -3x + 12$ Subtracting $3x$

$\dfrac{-2y}{-2} = \dfrac{-3x + 12}{-2}$ Dividing by -2

$y = \dfrac{3x - 12}{2}$, or

$y = \dfrac{3}{2}x - 6$

113. $5x(3x - 1)(2x + 3)$

$= 5x(6x^2 + 7x - 3)$ Using FOIL

$= 30x^3 + 35x^2 - 15x$

115. $[(a - 5)(a + 5)]^2$

$= (a^2 - 25)^2$ Finding the product of a sum
                      and difference of same two terms

$= a^4 - 50a^2 + 625$ Squaring a binomial

117. $(3t^4 - 2)^2 1(3t^4 + 2)^2$

$= [(3t^4 - 2)(3t^4 + 2)]^2$

$= (9t^8 - 4)^2$

$= 81t^{16} - 72t^8 + 16$

119. $(x + 2)(x - 5) = (x + 1)(x - 3)$

$x^2 - 5x + 2x - 10 = x^2 - 3x + x - 3$

$x^2 - 3x - 10 = x^2 - 2x - 3$

$-3x - 10 = -2x - 3$ Adding $-x^2$

$-3x + 2x = 10 - 3$ Adding $2x$ and 10

$-x = 7$

$x = -7$

The solution is -7.

121. See the answer section in the text.

123. Enter $y_1 = (x - 1)^2$ and $y_2 = x^2 - 2x + 1$. Then compare the graphs or the y_1- and y_2-values in a table. It appears that the graphs are the same and that the y_1- and y_2-values are the same, so $(x - 1)^2 = x^2 - 2x + 1$ is correct.

125. Enter $y_1 = (x - 3)(x + 3)$ and $y_2 = x^2 - 6$. Then compare the graphs or the y_1- and y_2-values in a table. The graphs are not the same nor are the y_1- and y_2-values, so $(x - 3)(x + 3) = x^2 - 6$ is not correct.

Exercise Set 4.7

1. We replace x by 3 and y by -2.

$x^2 - y^2 + xy = 3^2 - (-2)^2 + 3(-2) = 9 - 4 - 6 = -1$

3. We replace x by 3 and y by -2.

$x^2 - 3y^2 + 2xy = 3^2 - 3(-2)^2 + 2 \cdot 3(-2) =$
$9 - 3 \cdot 4 + 2 \cdot 3(-2) = 9 - 12 - 12 = -15$

5. We replace x by 3, y by -2, and z by -5.

$8xyz = 8 \cdot 3 \cdot (-2) \cdot (-5) = 240$

7. We replace x by 3, y by -2, and z by -5.

$xyz^2 - z = 3(-2)(-5)^2 - (-5) = 3(-2)(25) - (-5) =$
$-150 + 5 = -145$

9. We replace h by 165 and A by 20.

$C = 0.041h - 0.018A - 2.69$
$= 0.041(165) - 0.018(20) - 2.69$
$= 6.765 - 0.36 - 2.69$
$= 6.405 - 2.69$
$= 3.715$

The lung capacity of a 20-year-old woman who is 165 cm tall is 3.715 liters.

11. Evaluate the polynomial for $h = 32$, $v = 40$, and $t = 2$.

$h = h_0 + vt - 4.9t^2$
$= 32 + 40 \cdot 2 - 4.9(2)^2$
$= 32 + 80 - 19.6$
$= 92.4$

The rocket will be 92.4 m above the ground 2 seconds after blast off.

13. Replace h by 4.7, r by 1.2, and π by 3.14.

$S = 2\pi rh + 2\pi r^2$
$\approx 2(3.14)(1.2)(4.7) + 2(3.14)(1.2)^2$
$\approx 2(3.14)(1.2)(4.7) + 2(3.14)(1.44)$
$\approx 35.4192 + 9.0432$
≈ 44.46

The surface area of the can is about 44.46 in^2.

15. Evaluate the polynomial for $h = 7\dfrac{1}{2}$, or $\dfrac{15}{2}$, $r = 1\dfrac{1}{4}$, or $\dfrac{5}{4}$, and $\pi \approx 3.14$.

$S = 2\pi rh + \pi r^2$

$\approx 2(3.14)\left(\dfrac{5}{4}\right)\left(\dfrac{15}{2}\right) + (3.14)\left(\dfrac{5}{4}\right)^2$

$\approx 2(3.14)\left(\dfrac{5}{4}\right)\left(\dfrac{15}{2}\right) + (3.14)\left(\dfrac{25}{16}\right)$

$\approx 58.875 + 4.90625$

≈ 63.78125

The surface area is about 63.78125 in^2.

17. $x^3y - 2xy + 3x^2 - 5$

Term	Coefficient	Degree	
x^3y	1	4	(Think: $x^3y = x^3y^1$)
$-2xy$	-2	2	(Think: $-2xy = -2x^1y^1$)
$3x^2$	3	2	
-5	-5	0	(Think: $-5 = -5x^0$)

The degree of the polynomial is the degree of the term of highest degree. The term of highest degree is x^3y. Its degree is 4. The degree of the polynomial is 4.

19. $17x^2y^3 - 3x^3yz - 7$

Term	Coefficient	Degree	
$17x^2y^3$	17	5	
$-3x^3yz$	-3	5	(Think: $-3x^3yz =$ $-3x^3y^1z^1$)
-7	-7	0	(Think: $-7 = -7x^0$)

The terms of highest degree are $17x^2y^3$ and $-3x^3yz$. Each has degree 5. The degree of the polynomial is 5.

21. $a + b - 2a - 3b = (1-2)a + (1-3)b = -a - 2b$

23. $3x^2y - 2xy^2 + x^2$

There are *no* like terms, so none of the terms can be collected.

25.
$$6au + 3av + 14au + 7av$$
$$= (6+14)au + (3+7)av$$
$$= 20au + 10av$$

27.
$$2u^2v - 3uv^2 + 6u^2v - 2uv^2$$
$$= (2+6)u^2v + (-3-2)uv^2$$
$$= 8u^2v - 5uv^2$$

29.
$$(2x^2 - xy + y^2) + (-x^2 - 3xy + 2y^2)$$
$$= (2-1)x^2 + (-1-3)xy + (1+2)y^2$$
$$= x^2 - 4xy + 3y^2$$

31.
$$(r - 2s + 3) + (2r + s) + (s + 4)$$
$$= (1+2)r + (-2+1+1)s + (3+4)$$
$$= 3r + 0s + 7$$
$$= 3r + 7$$

33.
$$(b^3a^2 - 2b^2a^3 + 3ba + 4) + (b^2a^3 - 4b^3a^2 + 2ba - 1)$$
$$= (1-4)b^3a^2 + (-2+1)b^2a^3 + (3+2)ba + (4-1)$$
$$= -3b^3a^2 - b^2a^3 + 5ba + 3$$

35.
$$(a^3 + b^3) - (a^2b - ab^2 + b^3 + a^3)$$
$$= a^3 + b^3 - a^2b + ab^2 - b^3 - a^3$$
$$= (1-1)a^3 - a^2b + ab^2 + (1-1)b^3$$
$$= -a^2b + ab^2, \text{ or } ab^2 - a^2b$$

37.
$$(xy - ab - 8) - (xy - 3ab - 6)$$
$$= xy - ab - 8 - xy + 3ab + 6$$
$$= (1-1)xy + (-1+3)ab + (-8+6)$$
$$= 2ab - 2$$

39.
$$(-2a + 7b - c) - (-3b + 4c - 8d)$$
$$= -2a + 7b - c + 3b - 4c + 8d$$
$$= -2a + (7+3)b + (-1-4)c + 8d$$
$$= -2a + 10b - 5c + 8d$$

41.
$$\quad\quad\quad\text{F}\quad\quad\text{O}\quad\quad\text{I}\quad\quad\text{L}$$
$$(3z - u)(2z + 3u) = 6z^2 + 9zu - 2uz - 3u^2$$
$$= 6z^2 + 7zu - 3u^2$$

43.
$$\quad\quad\quad\quad\text{F}\quad\quad\quad\text{O}\quad\quad\quad\text{I}\quad\quad\text{L}$$
$$(a^2b - 2)(a^2b - 5) = a^4b^2 - 5a^2b - 2a^2b + 10$$
$$= a^4b^2 - 7a^2b + 10$$

45.
$$(a^3 + bc)(a^3 - bc) = (a^3)^2 - (bc)^2$$
$$[(A+B)(A-B) = A^2 - B^2]$$
$$= a^6 - b^2c^2$$

47.
$$\begin{array}{r} y^4x + y^2 + 1 \\ y^2 + 1 \\ \hline y^4x + y^2 + 1 \\ y^6x + y^4 \quad\quad + y^2 \\ \hline y^6x + y^4 + y^4x + 2y^2 + 1 \end{array}$$

49. $(3xy - 1)(4xy + 2)$
$$\quad\quad\text{F}\quad\quad\text{O}\quad\quad\text{I}\quad\quad\text{L}$$
$$= 12x^2y^2 + 6xy - 4xy - 2$$
$$= 12x^2y^2 + 2xy - 2$$

51. $(3 - c^2d^2)(4 + c^2d^2)$
$$\quad\text{F}\quad\quad\text{O}\quad\quad\text{I}\quad\quad\text{L}$$
$$= 12 + 3c^2d^2 - 4c^2d^2 - c^4d^4$$
$$= 12 - c^2d^2 - c^4d^4$$

53. $(m^2 - n^2)(m + n)$
$$\quad\text{F}\quad\quad\text{O}\quad\quad\text{I}\quad\quad\text{L}$$
$$= m^3 + m^2n - mn^2 - n^3$$

55. $(xy + x^5y^5)(x^4y^4 - xy)$
$$\quad\text{F}\quad\quad\text{O}\quad\quad\text{I}\quad\quad\text{L}$$
$$= x^5y^5 - x^2y^2 + x^9y^9 - x^6y^6$$
$$= x^9y^9 - x^6y^6 + x^5y^5 - x^2y^2$$

57. $(x + h)^2$
$$= x^2 + 2xh + h^2 \quad [(A+B)^2 = A^2 + 2AB + B^2]$$

59. $(r^3t^2 - 4)^2$
$$= (r^3t^2)^2 - 2 \cdot r^3t^2 \cdot 4 + 4^2$$
$$[(A-B)^2 = A^2 - 2AB + B^2]$$
$$= r^6t^4 - 8r^3t^2 + 16$$

61. $(p^4 + m^2n^2)^2$
$$= (p^4)^2 + 2 \cdot p^4 \cdot m^2n^2 + (m^2n^2)^2$$
$$[(A+B)^2 = A^2 + 2AB + B^2]$$
$$= p^8 + 2p^4m^2n^2 + m^4n^4$$

63. $\left(2a^3 - \dfrac{1}{2}b^3\right)^2$
$$= (2a^3)^2 - 2 \cdot 2a^3 \cdot \frac{1}{2}b^3 + \left(\frac{1}{2}b^3\right)^2$$
$$[(A-B)^2 = A^2 - 2AB + B^2]$$
$$= 4a^6 - 2a^3b^3 + \frac{1}{4}b^6$$

65. $3a(a - 2b)^2 = 3a(a^2 - 4ab + 4b^2)$
$$= 3a^3 - 12a^2b + 12ab^2$$

67. $(2a - b)(2a + b) = (2a)^2 - b^2 = 4a^2 - b^2$

69. $(c^2 - d)(c^2 + d) = (c^2)^2 - d^2$
$$= c^4 - d^2$$

71. $(ab + cd^2)(ab - cd^2) = (ab)^2 - (cd^2)^2$
$$= a^2b^2 - c^2d^4$$

73. $(x + y - 3)(x + y + 3)$
$= [(x + y) - 3][(x + y) + 3]$
$= (x + y)^2 - 3^2$
$= x^2 + 2xy + y^2 - 9$

75. $[x + y + z][x - (y + z)]$
$= [x + (y + z)][x - (y + z)]$
$= x^2 - (y + z)^2$
$= x^2 - (y^2 + 2yz + z^2)$
$= x^2 - y^2 - 2yz - z^2$

77. $(a + b + c)(a - b - c)$
$= [a + (b + c)][a - (b + c)]$
$= a^2 - (b + c)^2$
$= a^2 - (b^2 + 2bc + c^2)$
$= a^2 - b^2 - 2bc - c^2$

79.
$$
\begin{array}{l}
\;\;\;x^2 \;\;\;\;\; - \; 4y \; + 2 \\
\;\;\;3x^2 \; + \; 5y \; - 3 \\
\hline
-3x^2 + 12y - 6 \\
\;\;\;\;\;\;\;+ \, 10y \;\;\;\;\;\; - 20y^2 + 5x^2y \\
\;\;\;6x^2 \;\;\;\;\;\;\;\;\;\;\;\;\;\;\;\;\;\;\; - 12x^2y + 3x^4 \\
\hline
3x^2 \; + \, 22y - 6 - 20y^2 - 7x^2y + 3x^4
\end{array}
$$

We could also write the result as
$3x^4 - 7x^2y + 3x^2 - 20y^2 + 22y - 6$.

81. Discussion and Writing Exercise

83. The first coordinate is positive and the second coordinate is negative, so $(2, -5)$ is in quadrant IV.

85. Both coordinates are positive, so $(16, 23)$ is in quadrant I.

87. $2x = -10$
$x = -5$

Any ordered pair $(-5, y)$ is a solution. The variable x must be -5, but y can be any number we choose. A few solutions are listed below. Plot these points and draw the line.

x	y
-5	-3
-5	0
-5	4

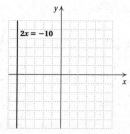

89. $8y - 16 = 0$
$8y = 16$
$y = 2$

Any ordered pair $(x, 2)$ is a solution. The variable y must be 2, but x can be any number we choose. A few solutions are listed below. Plot these points and draw the line.

x	y
-4	2
0	2
3	2

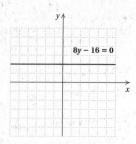

91. It is helpful to add additional labels to the figure.

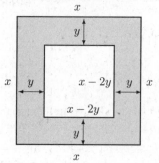

The area of the large square is $x \cdot x$, or x^2. The area of the small square is $(x - 2y)(x - 2y)$, or $(x - 2y)^2$.

Area of shaded region	$=$	Area of large square	$-$	Area of small square
Area of shaded region	$=$	x^2	$-$	$(x - 2y)^2$

$= x^2 - (x^2 - 4xy + 4y^2)$
$= x^2 - x^2 + 4xy - 4y^2$
$= 4xy - 4y^2$

93. It is helpful to add additional labels to the figure.

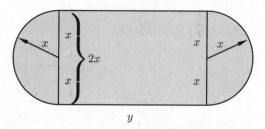

The two semicircles make a circle with radius x. The area of that circle is πx^2. The area of the rectangle is $2x \cdot y$. The sum of the two regions, $\pi x^2 + 2xy$, is the area of the shaded region.

95. The lateral surface area of the outer portion of the solid is the lateral surface area of a right circular cylinder with radius n and height h. The lateral surface area of the inner portion is the lateral surface area of a right circular cylinder with radius m and height h. Recall that the formula for the lateral surface area of a right circular cylinder with radius r and height h is $2\pi rh$.

The surface area of the top is the area of a circle with radius n less the area of a circle with radius m. The surface area of the bottom is the same as the surface area of the top.

Thus, the surface area of the solid is

$$2\pi nh + 2\pi mh + 2\pi n^2 - 2\pi m^2.$$

97. In the formula for the surface area of a silo, $S = 2\pi rh + \pi r^2$, the term πr^2 represents the area of the base. Since the base of the observatory rests on the ground, it will not need to be painted. Thus, we will subtract this term from the formula and find the remaining surface area, $2\pi rh$.

The height of the observatory is 40 ft and its radius is $30/2$, or 15 ft, so the surface area is $2\pi rh \approx$ $2(3.14)(15)(40) \approx 3768$ ft^2. Since 3768 ft^2/250 ft^2 = 15.072, 16 gallons of paint should be purchased.

99. Substitute \$10,400 for P, 8.5% or 0.085 for r, and 5 for t.
$$P(1+r)^t$$
$$= \$10,400(1 + 0.085)^5$$
$$= \$10,400(1.085)^5$$
$$\approx \$15,638.03$$

Exercise Set 4.8

1. $\dfrac{24x^4}{8} = \dfrac{24}{8} \cdot x^4 = 3x^4$

Check: We multiply.
$$3x^4 \cdot 8 = 24x^4$$

3. $\dfrac{25x^3}{5x^2} = \dfrac{25}{5} \cdot \dfrac{x^3}{x^2} = 5x^{3-2} = 5x$

Check: We multiply.
$$5x \cdot 5x^2 = 25x^3$$

5. $\dfrac{-54x^{11}}{-3x^8} = \dfrac{-54}{-3} \cdot \dfrac{x^{11}}{x^8} = 18x^{11-8} = 18x^3$

Check: We multiply.
$$18x^3(-3x^8) = -54x^{11}$$

7. $\dfrac{64a^5b^4}{16a^2b^3} = \dfrac{64}{16} \cdot \dfrac{a^5}{a^2} \cdot \dfrac{b^4}{b^3} = 4a^{5-2}b^{4-3} = 4a^3b$

Check: We multiply.
$$(4a^3b)(16a^2b^3) = 64a^5b^4$$

9. $\dfrac{24x^4 - 4x^3 + x^2 - 16}{8}$
$$= \dfrac{24x^4}{8} - \dfrac{4x^3}{8} + \dfrac{x^2}{8} - \dfrac{16}{8}$$
$$= 3x^4 - \dfrac{1}{2}x^3 + \dfrac{1}{8}x^2 - 2$$

Check: We multiply.
$$3x^4 - \dfrac{1}{2}x^3 + \dfrac{1}{8}x^2 - 2$$
$$\underline{\qquad\qquad\qquad 8}$$
$$24x^4 - 4x^3 + x^2 - 16$$

11. $\dfrac{u - 2u^2 - u^5}{u}$
$$= \dfrac{u}{u} - \dfrac{2u^2}{u} - \dfrac{u^5}{u}$$
$$= 1 - 2u - u^4$$

Check: We multiply.
$$1 - 2u - u^4$$
$$\underline{\qquad u}$$
$$u - 2u^2 - u^5$$

13. $(15t^3 + 24t^2 - 6t) \div (3t)$
$$= \dfrac{15t^3 + 24t^2 - 6t}{3t}$$
$$= \dfrac{15t^3}{3t} + \dfrac{24t^2}{3t} - \dfrac{6t}{3t}$$
$$= 5t^2 + 8t - 2$$

Check: We multiply.
$$5t^2 + 8t - 2$$
$$\underline{\qquad\qquad 3t}$$
$$15t^3 + 24t^2 - 6t$$

15. $(20x^6 - 20x^4 - 5x^2) \div (-5x^2)$
$$= \dfrac{20x^6 - 20x^4 - 5x^2}{-5x^2}$$
$$= \dfrac{20x^6}{-5x^2} - \dfrac{20x^4}{-5x^2} - \dfrac{5x^2}{-5x^2}$$
$$= -4x^4 - (-4x^2) - (-1)$$
$$= -4x^4 + 4x^2 + 1$$

Check: We multiply.
$$-4x^4 + 4x^2 + 1$$
$$\underline{\qquad\qquad -5x^2}$$
$$20x^6 - 20x^4 - 5x^2$$

17. $(24x^5 - 40x^4 + 6x^3) \div (4x^3)$
$$= \dfrac{24x^5 - 40x^4 + 6x^3}{4x^3}$$
$$= \dfrac{24x^5}{4x^3} - \dfrac{40x^4}{4x^3} + \dfrac{6x^3}{4x^3}$$
$$= 6x^2 - 10x + \dfrac{3}{2}$$

Check: We multiply.
$$6x^2 - 10x + \dfrac{3}{2}$$
$$\underline{\qquad\qquad 4x^3}$$
$$24x^5 - 40x^4 + 6x^3$$

19. $\dfrac{18x^2 - 5x + 2}{2}$
$$= \dfrac{18x^2}{2} - \dfrac{5x}{2} + \dfrac{2}{2}$$
$$= 9x^2 - \dfrac{5}{2}x + 1$$

Check: We multiply.
$$9x^2 - \dfrac{5}{2}x + 1$$
$$\underline{\qquad\qquad 2}$$
$$18x^2 - 5x + 2$$

21. $\dfrac{12x^3 + 26x^2 + 8x}{2x}$

$= \dfrac{12x^3}{2x} + \dfrac{26x^2}{2x} + \dfrac{8x}{2x}$

$= 6x^2 + 13x + 4$

Check: We multiply.

$$\begin{array}{r} 6x^2 \;+\; 13x \;+\; 4 \\ 2x \\ \hline 12x^3 + 26x^2 + 8x \end{array}$$

23. $\dfrac{9r^2s^2 + 3r^2s - 6rs^2}{3rs}$

$= \dfrac{9r^2s^2}{3rs} + \dfrac{3r^2s}{3rs} - \dfrac{6rs^2}{3rs}$

$= 3rs + r - 2s$

Check: We multiply.

$$\begin{array}{r} 3rs \;+\; r \;-\; 2s \\ 3rs \\ \hline 9r^2s^2 + 3r^2s - 6rs^2 \end{array}$$

25.
$$\begin{array}{r} x + 2 \\ x+2 \overline{\smash{\big)}\, x^2+4x+4} \\ \underline{x^2+2x} \\ 2x+4 \leftarrow (x^2+4x)-(x^2+2x)\\ \underline{2x+4} \\ 0 \leftarrow (2x+4)-(2x+4) \end{array}$$

The answer is $x + 2$.

27.
$$\begin{array}{r} x - 5 \\ x-5 \overline{\smash{\big)}\, x^2-10x-25} \\ \underline{x^2-5x} \\ -5x-25 \leftarrow (x^2-10x)-(x^2-5x)\\ \underline{-5x+25} \\ -50 \leftarrow (-5x-25)-(-5x+25) \end{array}$$

The answer is $x - 5 + \dfrac{-50}{x-5}$.

29.
$$\begin{array}{r} x - 2 \\ x+6 \overline{\smash{\big)}\, x^2+4x-14} \\ \underline{x^2+6x} \\ -2x-14 \leftarrow (x^2+4x)-(x^2+6x)\\ \underline{-2x-12} \\ -2 \leftarrow (-2x-14)-(-2x-12) \end{array}$$

The answer is $x - 2 + \dfrac{-2}{x+6}$.

31.
$$\begin{array}{r} x - 3 \\ x+3 \overline{\smash{\big)}\, x^2+0x-9} \leftarrow \text{Filling in the missing term}\\ \underline{x^2+3x} \\ -3x-9 \leftarrow x^2-(x^2+3x)\\ \underline{-3x-9} \\ 0 \leftarrow (-3x-9)-(-3x-9) \end{array}$$

The answer is $x - 3$.

33.
$$\begin{array}{r} x^4 - x^3 + x^2 - x + 1 \\ x+1 \overline{\smash{\big)}\, x^5+0x^4+0x^3+0x^2+0x+1} \leftarrow \text{Filling in missing terms}\\ \underline{x^5+ x^4} \\ -x^4 \leftarrow x^5-(x^5+x^4)\\ \underline{-x^4-x^3} \\ x^3 \leftarrow -x^4-(-x^4-x^3)\\ \underline{x^3+x^2} \\ -x^2 \leftarrow x^3-(x^3+x^2)\\ \underline{-x^2-x} \\ x+1 \leftarrow -x^2-(-x^2-x)\\ \underline{x+1} \\ 0 \leftarrow (x+1)-(x+1) \end{array}$$

The answer is $x^4 - x^3 + x^2 - x + 1$.

35.
$$\begin{array}{r} 2x^2 - 7x + 4 \\ 4x+3 \overline{\smash{\big)}\, 8x^3-22x^2-5x+12} \\ \underline{8x^3+6x^2} \\ -28x^2-5x \leftarrow (8x^3-22x^2)-(8x^3+6x^2)\\ \underline{-28x^2-21x} \\ 16x+12 \leftarrow (-28x^2-5x)-\\ (-28x^2-21x)\\ \underline{16x+12} \\ 0 \leftarrow (16x+12)-(16x+12) \end{array}$$

The answer is $2x^2 - 7x + 4$.

37.
$$\begin{array}{r} x^3 - 6 \\ x^3-7 \overline{\smash{\big)}\, x^6-13x^3+42} \\ \underline{x^6-7x^3} \\ -6x^3+42 \leftarrow (x^6-13x^3)-(x^6-7x^3)\\ \underline{-6x^3+42} \\ 0 \leftarrow (-6x^3+42)-(-6x^3+42) \end{array}$$

The answer is $x^3 - 6$.

39.
$$\begin{array}{r} 3x^2 + x + 2 \\ 5x+1 \overline{\smash{\big)}\, 15x^3+8x^2+11x+12} \\ \underline{15x^3+3x^2} \\ 5x^2+11x \\ \underline{5x^2+x} \\ 10x+12\\ \underline{10x+2} \\ 10 \end{array}$$

The answer is $3x^2 + x + 2 + \dfrac{10}{5x+1}$.

41.
$$\begin{array}{r} t^2 + 1 \\ t-1 \overline{\smash{\big)}\, t^3-t^2+t-1} \\ \underline{t^3-t^2} \\ 0+t-1 \leftarrow (t^3-t^2)-(t^3-t^2)\\ \underline{t-1} \\ 0 \leftarrow (t-1)-(t-1) \end{array}$$

The answer is $t^2 + 1$.

43. Discussion and Writing Exercise

45. The product rule asserts that when multiplying with exponential notation, if the bases are the same, keep the base and add the exponents.

47. The <u>multiplication</u> principle asserts that when we multiply or divide by the same nonzero number on each side of an equation, we get equivalent equations.

49. A <u>trinomial</u> is a polynomial with three terms, such as $5x^4 - 7x^2 + 4$.

51. The <u>absolute value</u> of a number is its distance from zero on a number line.

53.

$$
\begin{array}{r}
x^2+\ 5 \\
x^2 + 4\,\overline{)\,x^4+9x^2+20} \\
\underline{x^4+4x^2} \\
5x^2+20 \\
\underline{5x^2+20} \\
0
\end{array}
$$

The answer is $x^2 + 5$.

55.

$$
\begin{array}{r}
a +\ 3 \\
5a^2 - 7a - 2\,\overline{)\,5a^3+ 8a^2 -23a-1} \\
\underline{5a^3- 7a^2 -\ 2a} \\
15a^2-21a-1 \\
\underline{15a^2-21a-6} \\
5
\end{array}
$$

The answer is $a + 3 + \dfrac{5}{5a^2 - 7a - 2}$.

57. We rewrite the dividend in descending order.

$$
\begin{array}{r}
2x^2+\ x\ -\ 3 \\
3x^3 - 2x - 1\,\overline{)\,6x^5+3x^4 -13x^3 -4x^2+5x+3} \\
\underline{6x^5\qquad\ - 4x^3 -2x^2} \\
3x^4-\ 9x^3\ -2x^2+5x \\
\underline{3x^4\qquad\quad -2x^2-\ x} \\
-9x^3\qquad +6x+3 \\
\underline{-9x^3\qquad +6x+3} \\
0
\end{array}
$$

The answer is $2x^2 + x - 3$.

59.

$$
\begin{array}{r}
a^5+ a^4b + a^3b^2 + a^2b^3 +\ ab^4 +\ b^5 \\
a - b\,\overline{)\,a^6+0a^5b+0a^4b^2+0a^3b^3+0a^2b^4+0ab^5-b^6} \\
\underline{a^6- a^5b} \\
a^5b \\
\underline{a^5b - a^4b^2} \\
a^4b^2 \\
\underline{a^4b^2 - a^3b^3} \\
a^3b^3 \\
\underline{a^3b^3 - a^2b^4} \\
a^2b^4 \\
\underline{a^2b^4 - ab^5} \\
ab^5 -b^6 \\
\underline{ab^5 -b^6} \\
0
\end{array}
$$

The answer is $a^5 + a^4b + a^3b^2 + a^2b^3 + ab^4 + b^5$.

61.

$$
\begin{array}{r}
x + 5 \\
x - 1\,\overline{)\,x^2+4x+c} \\
\underline{x^2- x} \\
5x+c \\
\underline{5x-5} \\
c+5
\end{array}
$$

We set the remainder equal to 0.

$$c + 5 = 0$$
$$c = -5$$

Thus, c must be -5.

63.

$$
\begin{array}{r}
c^2x +(-2c+c^2) \\
x - 1\,\overline{)\,c^2x^2 -\ 2cx+1} \\
\underline{c^2x^2 -\ c^2x} \\
(-2c+ c^2)x+1 \\
\underline{(-2c+ c^2)x-(-2c+c^2)} \\
1+(-2c+c^2)
\end{array}
$$

We set the remainder equal to 0.

$$c^2 - 2c + 1 = 0$$
$$(c - 1)^2 = 0$$
$$c = 1$$

Thus, c must be 1.

Chapter 4 Review Exercises

1. $7^2 \cdot 7^{-4} = 7^{2+(-4)} = 7^{-2} = \dfrac{1}{7^2}$

2. $y^7 \cdot y^3 \cdot y = y^{7+3+1} = y^{11}$

3. $(3x)^5(3x)^9 = (3x)^{5+9} = (3x)^{14}$

4. $t^8 \cdot t^0 = t^8 \cdot 1 = t^8$, or $t^8 \cdot t^0 = t^{8+0} = t^8$

5. $\dfrac{4^5}{4^2} = 4^{5-2} = 4^3$

6. $\dfrac{a^5}{a^8} = a^{5-8} = a^{-3} = \dfrac{1}{a^3}$

7. $\dfrac{(7x)^4}{(7x)^4} = 1$

8. $(3t^4)^2 = 3^2 \cdot (t^4)^2 = 9 \cdot t^{4\cdot 2} = 9t^8$

9. $(2x^3)^2(-3x)^2 = 2^2 \cdot (x^3)^2(-3)^2x^2 = 4 \cdot x^6 \cdot 9 \cdot x^2 = 36x^8$

10. $\left(\dfrac{2x}{y}\right)^{-3} = \left(\dfrac{y}{2x}\right)^3 = \dfrac{y^3}{2^3 \cdot x^3} = \dfrac{y^3}{8x^3}$

11. $\dfrac{1}{t^5} = t^{-5}$

12. $y^{-4} = \dfrac{1}{y^4}$

13. $0.000003.\,28$

$\underline{\qquad\qquad}{\uparrow}$ 6 places

Small number, so the exponent is negative.

$0.00000328 = 3.28 \times 10^{-6}$

14. 8.3×10^6

$\quad$ 8.300000.

$\quad$ └──────┘↑ $\quad$ 6 places

Positive exponent, so the answer is a large number.

$8.3 \times 10^6 = 8,300,000$

15. $(3.8 \times 10^4)(5.5 \times 10^{-1}) = (3.8 \cdot 5.5) \times (10^4 \cdot 10^{-1})$

$\qquad\qquad\qquad\qquad = 20.9 \times 10^3$

$\qquad\qquad\qquad\qquad = (2.09 \times 10) \times 10^3$

$\qquad\qquad\qquad\qquad = 2.09 \times 10^4$

16. $\dfrac{1.28 \times 10^{-8}}{2.5 \times 10^{-4}} = \dfrac{1.28}{2.5} \times \dfrac{10^{-8}}{10^{-4}}$

$\qquad\qquad\qquad = 0.512 \times 10^{-4}$

$\qquad\qquad\qquad = (5.12 \times 10^{-1}) \times 10^{-4}$

$\qquad\qquad\qquad = 5.12 \times 10^{-5}$

17. 292 million $= 292 \times 10^6$

$\qquad\qquad\qquad = (2.92 \times 10^2) \times 10^6$

$\qquad\qquad\qquad = 2.92 \times 10^8$

Also, $15.3 = 1.53 \times 10$. Then we have

$(2.92 \times 10^8)(1.53 \times 10) = (2.92 \cdot 1.53) \times (10^8 \cdot 10)$

$\qquad\qquad\qquad\qquad\qquad = 4.4676 \times 10^9$

In 2005, 4.4676×10^9 gal of diet drinks were consumed in the United States.

18. $x^2 - 3x + 6 = (-1)^2 - 3(-1) + 6 = 1 + 3 + 6 = 10$

19. $-4y^5 + 7y^2 - 3y - 2 = -4y^5 + 7y^2 + (-3y) + (-2)$

The terms are $-4y^5$, $7y^2$, $-3y$, and -2.

20. In the polynomial $x^3 + x$ there are no x^2 or x^0 terms. Thus, the x^2 term (or second-degree term) and the x^0 term (or zero-degree term) are missing.

21. $4x^3 + 6x^2 - 5x + \dfrac{5}{3} = 4x^3 + 6x^2 - 5x^1 + \dfrac{5}{3}x^0$

The degree of $4x^3$ is 3.

The degree of $6x^2$ is 2.

The degree of $-5x$ is 1.

The degree of $\dfrac{5}{3}$ is 0.

The degree of the polynomial is 3, the largest exponent.

22. The polynomial $4x^3 - 1$ is a binomial because it has just two terms.

23. The polynomial $4 - 9t^3 - 7t^4 + 10t^2$ is none of these because it has more than three terms.

24. The polynomial $7y^2$ is a monomial because it has just one term.

25. $\quad 3x^2 - 2x + 3 - 5x^2 - 1 - x$

$= (3 - 5)x^2 + (-2 - 1)x + (3 - 1)$

$= -2x^2 - 3x + 2$

26. $\qquad -x + \dfrac{1}{2} + 14x^4 - 7x^2 - 1 - 4x^4$

$= (14 - 4)x^4 - 7x^2 - x + \left(\dfrac{1}{2} - 1\right)$

$= 10x^4 - 7x^2 - x - \dfrac{1}{2}$

27. $(3x^4 - x^3 + x - 4) + (x^5 + 7x^3 - 3x^2 - 5) + (-5x^4 + 6x^2 - x) = (3 - 5)x^4 + (-1 + 7)x^3 + (1 - 1)x + (-4 - 5) + x^5 + (-3 + 6)x^2 = -2x^4 + 6x^3 - 9 + x^5 + 3x^2$, or $x^5 - 2x^4 + 6x^3 + 3x^2 - 9$

28. $(3x^5 - 4x^4 + x^3 - 3) + (3x^4 - 5x^3 + 3x^2) + (-5x^5 - 5x^2) + (-5x^4 + 2x^3 + 5) = (3 - 5)x^5 + (-4 + 3 - 5)x^4 + (1 - 5 + 2)x^3 + (-3 + 5) + (3 - 5)x^2 = -2x^5 - 6x^4 - 2x^3 + 2 - 2x^2$, or $-2x^5 - 6x^4 - 2x^3 - 2x^2 + 2$

29. $(5x^2 - 4x + 1) - (3x^2 + 1) = 5x^2 - 4x + 1 - 3x^2 - 1$

$\qquad\qquad\qquad\qquad\qquad = 2x^2 - 4x$

30. $\quad (3x^5 - 4x^4 + 3x^2 + 3) - (2x^5 - 4x^4 + 3x^3 + 4x^2 - 5)$

$= 3x^5 - 4x^4 + 3x^2 + 3 - 2x^5 + 4x^4 - 3x^3 - 4x^2 + 5$

$= x^5 - 3x^3 - x^2 + 8$

31. $P = 2(w + 3) + 2w = 2w + 6 + 2w = 4w + 6$

$A = w(w + 3) = w^2 + 3w$

32. Regarding the figure as one large rectangle with length $t + 4$ and width $t + 3$, we have $(t + 4)(t + 3)$. We can also add the areas of the four smaller rectangles:

$3 \cdot t + 4 \cdot 3 + 4 \cdot t + t \cdot t$, or $3t + 12 + 4t + t^2$, or

$t^2 + 7t + 12$

33. $\left(x + \dfrac{2}{3}\right)\left(x + \dfrac{1}{2}\right) = x^2 + \dfrac{1}{2}x + \dfrac{2}{3}x + \dfrac{2}{6}$

$\qquad\qquad\qquad\qquad = x^2 + \dfrac{3}{6}x + \dfrac{4}{6}x + \dfrac{1}{3}$

$\qquad\qquad\qquad\qquad = x^2 + \dfrac{7}{6}x + \dfrac{1}{3}$

34. $(7x + 1)^2 = (7x)^2 + 2 \cdot 7x \cdot 1 + 1^2$

$\qquad\qquad\quad = 49x^2 + 14x + 1$

35. $\qquad\qquad\quad 4x^2 \quad - 5x \; + 1$

$\qquad\qquad\qquad\qquad 3x \; - 2$

$\qquad\qquad \overline{\qquad\qquad\qquad\qquad}$

$\qquad\qquad\quad -8x^2 + 10x - 2$

$\qquad 12x^3 - 15x^2 + 3x$

$\qquad \overline{\qquad\qquad\qquad\qquad\qquad}$

$\qquad 12x^3 - 23x^2 + 13x - 2$

36. $(3x^2 + 4)(3x^2 - 4) = (3x^2)^2 - 4^2 = 9x^4 - 16$

37. $\quad 5x^4(3x^3 - 8x^2 + 10x + 2)$

$= 5x^4 \cdot 3x^3 - 5x^4 \cdot 8x^2 + 5x^4 \cdot 10x + 5x^4 \cdot 2$

$= 15x^7 - 40x^6 + 50x^5 + 10x^4$

38. $(x + 4)(x - 7) = x^2 - 7x + 4x - 28 = x^2 - 3x - 28$

39. $(3y^2 - 2y)^2 = (3y^2)^2 - 2 \cdot 3y^2 \cdot 2y + (2y)^2 = 9y^4 - 12y^3 + 4y^2$

40. $(2t^2 + 3)(t^2 - 7) = 2t^4 - 14t^2 + 3t^2 - 21 = 4t^4 - 11t^2 - 21$

41. $\quad 2 - 5xy + y^2 - 4xy^3 + x^6$

$= 2 - 5(-1)(2) + 2^2 - 4(-1)(2)^3 + (-1)^6$

$= 2 - 5(-1)(2) + 4 - 4(-1)(8) + 1$

$= 2 + 10 + 4 + 32 + 1$

$= 49$

42. $x^5y - 7xy + 9x^2 - 8$

Term	Coefficient	Degree	
x^5y	1	6	$(x^5y = 1 \cdot x^5y^1)$
$-7xy$	-7	2	$(-7xy = -7x^1y^1)$
$9x^2$	9	2	
-8	-8	0	$(-8 = -8x^0)$

The degree of the polynomial is the degree of the term of highest degree. The term of highest degree is x^5y. Its degree is 6, so the degree of the polynomial is 6.

43. $\quad y + w - 2y + 8w - 5 = (1-2)y + (1+8)w - 5$

$\qquad\qquad\qquad\qquad\qquad = -y + 9w - 5$

44. $\quad m^6 - 2m^2n + m^2n^2 + n^2m - 6m^3 + m^2n^2 + 7n^2m$

$= m^6 - 2m^2n + (1+1)m^2n^2 + (1+7)n^2m - 6m^3$

$= m^6 - 2m^2n + 2m^2n^2 + 8n^2m - 6m^3$

45. $\quad (5x^2 - 7xy + y^2) + (-6x^2 - 3xy - y^2) + (x^2 + xy - 2y^2)$

$= (5 - 6 + 1)x^2 + (-7 - 3 + 1)xy + (1 - 1 - 2)y^2$

$= -9xy - 2y^2$

46. $\quad (6x^3y^2 - 4x^2y - 6x) - (-5x^3y^2 + 4x^2y + 6x^2 - 6)$

$= 6x^3y^2 - 4x^2y - 6x + 5x^3y^2 - 4x^2y - 6x^2 + 6$

$= (6+5)x^3y^2 + (-4-4)x^2y - 6x - 6x^2 + 6$

$= 11x^3y^2 - 8x^2y - 6x - 6x^2 + 6$

47.
$$\begin{array}{r} p^2 + pq + q^2 \\ p - q \\ \hline -p^2q - pq^2 - q^3 \\ p^3 + p^2q + pq^2 \\ \hline p^3 \qquad\qquad - q^3 \end{array}$$

48. $\quad \left(3a^4 - \dfrac{1}{3}b^3\right)^2 = (3a^4)^2 - 2 \cdot 3a^4 \cdot \dfrac{1}{3}b^3 + \left(\dfrac{1}{3}b^3\right)^2$

$\qquad\qquad\qquad\qquad = 9a^8 - 2a^4b^3 + \dfrac{1}{9}b^6$

49. $\quad \dfrac{10x^3 - x^2 + 6x}{2x} = \dfrac{10x^3}{2x} - \dfrac{x^2}{2x} + \dfrac{6x}{2x}$

$\qquad\qquad\qquad\quad = 5x^2 - \dfrac{1}{2}x + 3$

50.
$$\begin{array}{r} 3x^2 - 7x + 4 \\ 2x+3\overline{)\,6x^3 - 5x^2 - 13x + 13} \\ \underline{6x^3 + 9x^2} \\ -14x^2 - 13x \\ \underline{-14x^2 - 21x} \\ 8x + 13 \\ \underline{8x + 12} \\ 1 \end{array}$$

The answer is $3x^2 - 7x + 4 + \dfrac{1}{2x+3}$.

51. Locate -1 on the x-axis. Then move vertically to the graph and horizontally to the y-axis. It appears that the y-value that is paired with -1 is 0. Thus, the value of $y = 10x^3 - 10x$ is 0 when $x = -1$.

Locate -0.5 on the x-axis. Then move vertically to the graph and horizontally to the y-axis. It appears that the y-value that is paired with -0.5 is about 3.75. Thus, the value of $y = 10x^3 - 10x$ is 3.75 when $x = -0.5$.

Locate 0.5 on the x-axis. Then move vertically to the graph and horizontally to the y-axis. It appears that the y-value that is paired with 0.5 is about -3.75. Thus, the value of $y = 10x^3 - 10x$ is -3.75 when $x = 0.5$.

Locate 1 on the x-axis. Then move vertically to the graph and horizontally to the y-axis. It appears that the y-value that is paired with 1 is 0. Thus, the value of $y = 10x^3 - 10x$ is 0 when $x = 1$.

52. *Discussion and Writing Exercise.* 578.6×10^{-7} is not in scientific notation because 578.6 is larger than 10.

53. *Discussion and Writing Exercise.* A monomial is an expression of the type ax^n, where n is a whole number and a is a real number. A binomial is a sum of two monomials and has two terms. A trinomial is a sum of three monomials and has three terms. A general polynomial is a monomial or a sum of monomials and has one or more terms.

54. $\quad A = \dfrac{1}{2}bh$

$\quad A = \dfrac{1}{2}(x+y)(x-y) = \dfrac{1}{2}(x^2 - y^2) = \dfrac{1}{2}x^2 - \dfrac{1}{2}y^2$

55. The shaded area is the area of a square with side 20 minus the area of 4 small squares, each with side a.

$\quad A = 20^2 - 4 \cdot a^2 = 400 - 4a^2$

56. $\quad -3x^5 \cdot 3x^3 - x^6(2x)^2 + (3x^4)^2 + (2x^2)^4 - 40x^2(x^3)^2$

$= -3x^5 \cdot 3x^3 - x^6(4x^2) + 9x^8 + 16x^8 - 40x^2(x^6)$

$= -9x^8 - 4x^8 + 9x^2 + 16x^8 - 40x^8$

$= -28x^8$

57. $\qquad (x-7)(x+10) = (x-4)(x-6)$

$\quad x^2 + 10x - 7x - 70 = x^2 - 6x - 4x + 24$

$\qquad\quad x^2 + 3x - 70 = x^2 - 10x + 24$

$\qquad\qquad 3x - 70 = -10x + 24 \quad$ Subtracting x^2

$\qquad\qquad 13x - 70 = 24 \qquad\qquad$ Adding $10x$

$\qquad\qquad\quad 13x = 94 \qquad\qquad$ Adding 70

$\qquad\qquad\qquad x = \dfrac{94}{13}$

The solution is $\dfrac{94}{13}$.

58. Let P represent the other polynomial. Then we have

$(x-1)P = x^5 - 1$, or $P = \dfrac{x^5 - 1}{x - 1}$. We divide to find P.

$$
\begin{array}{r}
x^4 + x^3 + x^2 + x + 1 \\
x-1 \overline{\smash{\big)}\, x^5 + 0x^4 + 0x^3 + 0x^2 + 0x - 1} \\
\underline{x^5 - x^4} \\
x^4 \\
\underline{x^4 - x^3} \\
x^3 \\
\underline{x^3 - x^2} \\
x^2 \\
\underline{x^2 - x} \\
x - 1 \\
\underline{x - 1} \\
0
\end{array}
$$

The other polynomial is $x^4 + x^3 + x^2 + x + 1$.

59. *Familiarize*. Let w = the width of the garden, in feet. Then $2w$ = the length. From the drawing in the text we see that the width of the garden and the sidewalk together is $w + 4 + 4$, or $w + 8$, and the length of the garden and the sidewalk together is $2w + 4 + 4$, or $2w + 8$.

Translate. The area of the sidewalk is the area of the garden and sidewalk together minus the area of the garden. Recall that the formula for the area of a rectangle is $A = l \cdot w$. Thus, we have

$$256 = (2w + 8)(w + 8) - 2w \cdot w.$$

Solve. We solve the equation.

$$256 = (2w + 8)(w + 8) - 2w \cdot w$$
$$256 = 2w^2 + 16w + 8w + 64 - 2w^2$$
$$256 = 24w + 64$$
$$192 = 24w$$
$$8 = w$$

If $w = 8$, then $2w = 2 \cdot 8 = 16$.

Check. The dimensions of the garden and the sidewalk together are $2 \cdot 8 + 8$ by $8 + 8$, or 24 by 16. Then the area of the garden and sidewalk together is 24 ft $\cdot$ 16 ft, or 384 ft^2 and the area of the garden is 16 ft $\cdot$ 8 ft, or 128 ft^2. Subtracting to find the area of the sidewalk, we get 384 ft^2 − 128 ft^2, or 256 ft^2, so the answer checks.

State. The dimensions of the garden are 16 ft by 8 ft.

Chapter 4 Test

1. $6^{-2} \cdot 6^{-3} = 6^{-2+(-3)} = 6^{-5} = \dfrac{1}{6^5}$

2. $x^6 \cdot x^2 \cdot x = x^{6+2+1} = x^9$

3. $(4a)^3 \cdot (4a)^8 = (4a)^{3+8} = (4a)^{11}$

4. $\dfrac{3^5}{3^2} = 3^{5-2} = 3^3$

5. $\dfrac{x^3}{x^8} = x^{3-8} = x^{-5} = \dfrac{1}{x^5}$

6. $\dfrac{(2x)^5}{(2x)^5} = 1$

7. $(x^3)^2 = x^{3\cdot2} = x^6$

8. $(-3y^2)^3 = (-3)^3(y^2)^3 = -27y^{2\cdot3} = -27y^6$

9. $(2a^3b)^4 = 2^4(a^3)^4 \cdot b^4 = 16a^{12}b^4$

10. $\left(\dfrac{ab}{c}\right)^3 = \dfrac{(ab)^3}{c^3} = \dfrac{a^3b^3}{c^3}$

11. $(3x^2)^3(-2x^5)^3 = 3^3(x^2)^3(-2)^3(x^5)^3 = 27x^6(-8)x^{15} = -216x^{21}$

12. $3(x^2)^3(-2x^5)^3 = 3x^6(-2)^3(x^5)^3 = 3x^6(-8)x^{15} = -24x^{21}$

13. $2x^2(-3x^2)^4 = 2x^2(-3)^4(x^2)^4 = 2x^2 \cdot 81x^8 = 162x^{10}$

14. $(2x)^2(-3x^2)^4 = 2^2x^2(-3)^4(x^2)^4 = 4x^2 \cdot 81x^8 = 324x^{10}$

15. $5^{-3} = \dfrac{1}{5^3}$

16. $\dfrac{1}{y^8} = y^{-8}$

17. 3,900,000,000

    3.900,000,000.

    ⌐_____⌐ 9 places

Large number, so the exponent is positive.

$3,900,000,000 = 3.9 \times 10^9$

18. 5×10^{-8}

Negative exponent, so the answer is a small number.

    0.00000005.

    ⌐_____⌐ 8 places

$5 \times 10^{-8} = 0.00000005$

19. $\dfrac{5.6\times10^6}{3.2\times10^{-11}} = \dfrac{5.6}{3.2} \times \dfrac{10^6}{10^{-11}} = 1.75\times10^{6-(-11)} = 1.75\times10^{17}$

20. $(2.4 \times 10^5)(5.4 \times 10^{16}) = (2.4 \cdot 5.4) \times (10^5 \cdot 10^{16}) = 12.96 \times 10^{21} = (1.296 \times 10) \times 10^{21} = 1.296 \times 10^{22}$

21. 600 million = 600×1 million = $600 \times 1,000,000 = 600,000,000 = 6 \times 10^8$

$40,000 = 4 \times 10^4$

We divide:

$\dfrac{6 \times 10^8}{4 \times 10^4} = 1.5 \times 10^4$

A CD-ROM can hold 1.5×10^4 sound files.

22. $x^5 + 5x - 1 = (-2)^5 + 5(-2) - 1 = -32 - 10 - 1 = -43$

23. $\dfrac{1}{3}x^5 - x + 7$

The coefficient of $\dfrac{1}{3}x^5$ is $\dfrac{1}{3}$.

The coefficient of $-x$, or $-1 \cdot x$, is -1.

The coefficient of 7 is 7.

24. $2x^3 - 4 + 5x + 3x^6$

The degree of $2x^3$ is 3.

The degree of -4, or $-4x^0$, is 0.

The degree of $5x$, or $5x^1$, is 1.

The degree of $3x^6$ is 6.

The degree of the polynomial is 6, the largest exponent.

25. $7 - x$ is a binomial because it has just 2 terms.

26. $4a^2 - 6 + a^2 = (4+1)a^2 - 6 = 5a^2 - 6$

27. $y^2 - 3y - y + \dfrac{3}{4}y^2 = \left(1 + \dfrac{3}{4}\right)y^2 + (-3-1)y =$

$\left(\dfrac{4}{4} + \dfrac{3}{4}\right)y^2 + (-3-1)y = \dfrac{7}{4}y^2 - 4y$

28. $3 - x^2 + 2x^3 + 5x^2 - 6x - 2x + x^5$

$= 3 + (-1+5)x^2 + 2x^3 + (-6-2)x + x^5$

$= 3 + 4x^2 + 2x^3 - 8x + x^5$

$= x^5 + 2x^3 + 4x^2 - 8x + 3$

29. $(3x^5 + 5x^3 - 5x^2 - 3) + (x^5 + x^4 - 3x^3 - 3x^2 + 2x - 4)$

$= (3+1)x^5 + x^4 + (5-3)x^3 + (-5-3)x^2 + 2x + (-3-4)$

$= 4x^5 + x^4 + 2x^3 - 8x^2 + 2x - 7$

30. $\left(x^4 + \dfrac{2}{3}x + 5\right) + \left(4x^4 + 5x^2 + \dfrac{1}{3}x\right)$

$= (1+4)x^4 + 5x^2 + \left(\dfrac{2}{3} + \dfrac{1}{3}\right)x + 5$

$= 5x^4 + 5x^2 + x + 5$

31. $(2x^4 + x^3 - 8x^2 - 6x - 3) - (6x^4 - 8x^2 + 2x)$

$= 2x^4 + x^3 - 8x^2 - 6x - 3 - 6x^4 + 8x^2 - 2x$

$= (2-6)x^4 + x^3 + (-8+8)x^2 + (-6-2)x - 3$

$= -4x^4 + x^3 - 8x - 3$

32. $(x^3 - 0.4x^2 - 12) - (x^5 + 0.3x^3 + 0.4x^2 + 9)$

$= x^3 - 0.4x^2 - 12 - x^5 - 0.3x^3 - 0.4x^2 - 9$

$= -x^5 + (1 - 0.3)x^3 + (-0.4 - 0.4)x^2 + (-12 - 9)$

$= -x^5 + 0.7x^3 - 0.8x^2 - 21$

33. $-3x^2(4x^2 - 3x - 5) = -3x^2 \cdot 4x^2 - 3x^2(-3x) - 3x^2(-5) =$

$-12x^4 + 9x^3 + 15x^2$

34. $\left(x - \dfrac{1}{3}\right)^2 = x^2 - 2 \cdot x \cdot \dfrac{1}{3} + \left(\dfrac{1}{3}\right)^2 = x^2 - \dfrac{2}{3}x + \dfrac{1}{9}$

35. $(3x + 10)(3x - 10) = (3x)^2 - 10^2 = 9x^2 - 100$

36. $(3b + 5)(b - 3) = 3b^2 - 9b + 5b - 15 = 3b^2 - 4b - 15$

37. $(x^6 - 4)(x^8 + 4) = x^{14} + 4x^6 - 4x^8 - 16$, or

$x^{14} - 4x^8 + 4x^6 - 16$

38. $(8 - y)(6 + 5y) = 48 + 40y - 6y - 5y^2 = 48 + 34y - 5y^2$

39.
$$
\begin{array}{r}
3x^2 - 5x - 3 \\
2x + 1 \\
\hline
3x^2 - 5x - 3 \\
6x^3 - 10x^2 - 6x \\
\hline
6x^3 - 7x^2 - 11x - 3
\end{array}
$$

40. $(5t + 2)^2 = (5t)^2 + 2 \cdot 5t \cdot 2 + 2^2 = 25t^2 + 20t + 4$

41. $x^3y - y^3 + xy^3 + 8 - 6x^3y - x^2y^2 + 11$

$= (1 - 6)x^3y - y^3 + xy^3 + (8 + 11) - x^2y^2$

$= -5x^3y - y^3 + xy^3 + 19 - x^2y^2$

42. $(8a^2b^2 - ab + b^3) - (-6ab^2 - 7ab - ab^3 + 5b^3)$

$= 8a^2b^2 - ab + b^3 + 6ab^2 + 7ab + ab^3 - 5b^3$

$= 8a^2b^2 + (-1 + 7)ab + (1 - 5)b^3 + 6ab^2 + ab^3$

$= 8a^2b^2 + 6ab - 4b^3 + 6ab^2 + ab^3$

43. $(3x^5 - 4y^5)(3x^5 + 4y^5) = (3x^5)^2 - (4y^5)^2 =$

$9x^{10} - 16y^{10}$

44. $(12x^4 + 9x^3 - 15x^2) \div (3x^2)$

$= \dfrac{12x^4 + 9x^3 - 15x^2}{3x^2}$

$= \dfrac{12x^4}{3x^2} + \dfrac{9x^3}{3x^2} - \dfrac{15x^2}{3x^2}$

$= 4x^2 + 3x - 5$

45.
$$
\begin{array}{r}
2x^2 - 4x - 2 \\
3x + 2 \overline{\smash{\big)}\ 6x^3 - 8x^2 - 14x + 13} \\
\underline{6x^3 + 4x^2 } \\
-12x^2 - 14x \\
\underline{-12x^2 - 8x } \\
-6x + 13 \\
\underline{-6x - 4} \\
17
\end{array}
$$

The answer is $2x^2 - 4x - 2 + \dfrac{17}{3x + 2}$.

46. Locate -1 on the x-axis. Then move vertically to the graph and horizontally to the y-axis. It appears that the y-value that is paired with -1 is 3. Thus, the value of $y = x^3 - 5x - 1$ is 3 when $x = -1$.

Locate -0.5 on the x-axis. Then move vertically to the graph and horizontally to the y-axis. It appears that the y-value that is paired with -0.5 is 1.5. Thus, the value of $y = x^3 - 5x - 1$ is 1.5 when $x = -0.5$.

Locate 0.5 on the x-axis. Then move vertically to the graph and horizontally to the y-axis. It appears that the y-value that is paired with 0.5 is -3.5. Thus, the value of $y = x^3 - 5x - 1$ is -3.5 when $x = 0.5$.

Locate 1 on the x-axis. Then move vertically to the graph and horizontally to the y-axis. It appears that the y-value that is paired with 1 is -5. Thus, the value of $y = x^3 - 5x - 1$ is -5 when $x = 1$.

Locate 1.1 on the x-axis. Then move vertically to the graph and horizontally to the y-axis. It appears that the y-value that is paired with 1.1 is -5.25. Thus, the value of $y = x^3 - 5x - 1$ is -5.25 when $x = 1.1$.

47. Two sides have dimensions 9 by 5, two other sides have dimensions a by 5, and the two remaining sides have dimensions a by 9. Then the surface area is $2 \cdot 9 \cdot 5 + 2 \cdot a \cdot 5 + 2 \cdot a \cdot 9$, or $90 + 10a + 18a$, or $90 + 28a$.

48. When we regard the figure as one large rectangle with dimensions $t + 2$ by $t + 2$, we can express the area as $(t + 2)(t + 2)$.

Next we will regard the figure as the sum of four smaller rectangles with dimensions t by t, 2 by t, 2 by t, and 2 by 2. The sum of the areas of these rectangles is $t \cdot t + 2 \cdot t + 2 \cdot t + 2 \cdot 2$, or $t^2 + 2t + 2t + 4$, or $t^2 + 4t + 4$.

49. Let l = the length of the box. Then the height is $l - 1$ and the width is $l - 2$. The volume of the box is length × width × height.

$$V = l(l - 2)(l - 1)$$
$$V = (l^2 - 2l)(l - 1)$$
$$V = l^3 - l^2 - 2l^2 + 2l$$
$$V = l^3 - 3l^2 + 2l$$

50. $(x - 5)(x + 5) = (x + 6)^2$

$$x^2 - 25 = x^2 + 12x + 36$$
$$-25 = 12x + 36 \qquad \text{Subtracting } x^2$$
$$-61 = 12x$$
$$-\frac{61}{12} = x$$

The solution is $-\dfrac{61}{12}$.

Cumulative Review Chapters 1 - 4

1. For 100 ft:
$$h = -0.002(100)^2 + 0.8(100) + 6.6$$
$$= -0.002(10,000) + 0.8(100) + 6.6$$
$$= -20 + 80 + 6.6$$
$$= 66.6$$

After the arrow has traveled 100 ft horizontally, its height is 66.6 ft.

For 200 ft:
$$h = -0.002(200)^2 + 0.8(200) + 6.6$$
$$= -0.002(40,000) + 0.8(200) + 6.6$$
$$= -80 + 160 + 6.6$$
$$= 86.6$$

After the arrow has traveled 200 ft horizontally, its height is 86.6 ft.

For 300 ft:
$$h = -0.002(300)^2 + 0.8(300) + 6.6$$
$$= -0.002(90,000) + 0.8(300) + 6.6$$
$$= -180 + 240 + 6.6$$
$$= 66.6$$

After the arrow has traveled 300 ft horizontally, its height is 66.6 ft.

For 350 ft:
$$h = -0.002(350)^2 + 0.8(350) + 6.6$$
$$= -0.002(122,500) + 0.8(350) + 6.6$$
$$= -245 + 280 + 6.6$$
$$= 41.6$$

After the arrow has traveled 350 ft horizontally, its height is 41.6 ft.

2. $\dfrac{x}{2y} = \dfrac{10}{2 \cdot 2} = \dfrac{10}{4} = \dfrac{5}{2}$

3. $2x^3 + x^2 - 3 = 2(-1)^3 + (-1)^2 - 3 = 2(-1) + 1 - 3 = -2 + 1 - 3 = -4$

4. $x^3 y^2 + xy + 2xy^2 = (-1)^3(2)^2 + (-1)(2) + 2(-1)(2)^2$
$$= -1 \cdot 4 + (-1)(2) + 2(-1)(4)$$
$$= -4 - 2 - 8 = -14$$

5. The distance of -4 from 0 is 4, so $|-4| = 4$.

6. The reciprocal of 5 is $\dfrac{1}{5}$, because $5 \cdot \dfrac{1}{5} = 1$.

7. $-\dfrac{3}{5} + \dfrac{5}{12} = -\dfrac{36}{60} + \dfrac{25}{60} = -\dfrac{11}{60}$

8. $3.4 - (-0.8) = 3.4 + 0.8 = 4.2$

9. $(-2)(-1.4)(2.6) = (2.8)(2.6) = 7.28$

10. $\dfrac{3}{8} \div \left(-\dfrac{9}{10}\right) = \dfrac{3}{8} \cdot \left(-\dfrac{10}{9}\right) = -\dfrac{3 \cdot 10}{8 \cdot 9} =$
$$-\dfrac{3 \cdot 2 \cdot 5}{2 \cdot 4 \cdot 3 \cdot 3} = -\dfrac{5}{12}$$

11. $(1.1 \times 10^{10})(2 \times 10^{12}) = (1.1 \cdot 2) \times (10^{10} \cdot 10^{12}) = 2.2 \times 10^{22}$

12. $(3.2 \times 10^{-10}) \div (8 \times 10^{-6}) = \dfrac{3.2 \times 10^{-10}}{8 \times 10^{-6}} = 0.4 \times 10^{-4} = (4 \times 10^{-1}) \times 10^{-4} = 4 \times 10^{-5}$

13. $\dfrac{-9x}{3x} = \dfrac{-3 \cdot 3x}{1 \cdot 3x} = \dfrac{-3}{1} \cdot \dfrac{3x}{3x} = \dfrac{-3}{1} \cdot 1 = -3$

14. $y - (3y + 7) = y - 3y - 7 = -2y - 7$

15. $3(x-1) - 2[x - (2x + 7)] = 3(x-1) - 2[x - 2x - 7]$
$$= 3(x - 1) - 2[-x - 7]$$
$$= 3x - 3 + 2x + 14$$
$$= 5x + 11$$

16. $2 - [32 \div (4 + 2^2)] = 2 - [32 \div (4 + 4)]$
$$= 2 - [32 \div 8]$$
$$= 2 - 4$$
$$= -2$$

17. $(x^4 + 3x^3 - x + 7) + (2x^5 - 3x^4 + x - 5)$
$$= (1 - 3)x^4 + 3x^3 + (-1 + 1)x + (7 - 5) + 2x^5$$
$$= -2x^4 + 3x^3 + 2 + 2x^5, \text{ or}$$
$$2x^5 - 2x^4 + 3x^3 + 2$$

18. $(x^2 + 2xy) + (y^2 - xy) + (2x^2 - 3y^2)$
$$= (1 + 2)x^2 + (2 - 1)xy + (1 - 3)y^2$$
$$= 3x^2 + xy - 2y^2$$

19. $(x^3+3x^2-4)-(-2x^2+x+3) = x^3+3x^2-4+2x^2-x-3$
$$= x^3+(3+2)x^2-x+(-4-3)$$
$$= x^3+5x^2-x-7$$

20. $\left(\dfrac{1}{3}x^2 - \dfrac{1}{4}x - \dfrac{1}{5}\right) - \left(\dfrac{2}{3}x^2 + \dfrac{1}{2}x - \dfrac{1}{5}\right)$

$= \dfrac{1}{3}x^2 - \dfrac{1}{4}x - \dfrac{1}{5} - \dfrac{2}{3}x^2 - \dfrac{1}{2}x + \dfrac{1}{5}$

$= \left(\dfrac{1}{3} - \dfrac{2}{3}\right)x^2 + \left(-\dfrac{1}{4} - \dfrac{1}{2}\right)x + \left(-\dfrac{1}{5} + \dfrac{1}{5}\right)$

$= -\dfrac{1}{3}x^2 - \dfrac{3}{4}x$

21. $3(4x - 5y + 7) = 3 \cdot 4x - 3 \cdot 5y + 3 \cdot 7 = 12x - 15y + 21$

22. $(-2x^3)(-3x^5) = (-2)(-3)x^3 \cdot x^5 = 6x^{3+5} = 6x^8$

23. $2x^2(x^3 - 2x^2 + 4x - 5)$
$$= 2x^2 \cdot x^3 - 2x^2 \cdot 2x^2 + 2x^2 \cdot 4x - 2x^2 \cdot 5$$
$$= 2x^5 - 4x^4 + 8x^3 - 10x^2$$

24.
$$
\begin{array}{r}
3y^2 + 5y + 6 \\
y^2 - 2 \\
\hline
-6y^2 - 10y - 12 \\
3y^4 + 5y^3 + 6y^2 \\
\hline
3y^4 + 5y^3 - 10y - 12
\end{array}
$$

25.
$$
\begin{array}{r}
2p^3 + p^2q + pq^2 \\
p - pq + q \\
\hline
2p^3q + p^2q^2 + pq^3 \\
- 2p^4q - p^3q^2 - p^2q^3 \\
2p^4 + p^3q + p^2q^2 \\
\hline
2p^4 + 3p^3q + 2p^2q^2 + pq^3 - 2p^4q - p^3q^2 - p^2q^3
\end{array}
$$

26. $(2x + 3)(3x + 2) = 6x^2 + 4x + 9x + 6 = 6x^2 + 13x + 6$

27. $(3x^2 + 1)^2 = (3x^2)^2 + 2 \cdot 3x^2 \cdot 1 + 1^2 = 9x^4 + 6x^2 + 1$

28. $\left(t + \dfrac{1}{2}\right)\left(t - \dfrac{1}{2}\right) = t^2 - \left(\dfrac{1}{2}\right)^2 = t^2 - \dfrac{1}{4}$

29. $(2y^2 + 5)(2y^2 - 5) = (4y^2)^2 - 5^2 = 16y^4 - 25$

30. $(2x^4 - 3)(2x^2 + 3) = 4x^6 + 6x^4 - 6x^2 - 9$

31. $(t - 2t^2)^2 = t^2 - 2 \cdot t \cdot 2t^2 + (2t^2)^2 = t^2 - 4t^3 + 4t^4$

32. $(3p + q)(5p - 2q) = 15p^2 - 6pq + 5pq - 2q^2 =$
$15p^2 - pq - 2q^2$

33. $(18x^3 + 6x^2 - 9x) \div 3x = \dfrac{18x^3 + 6x^2 - 9x}{3x}$

$= \dfrac{18x^3}{3x} + \dfrac{6x^2}{3x} - \dfrac{9x}{3x}$

$= 6x^2 + 2x - 3$

34.
$$
\begin{array}{r}
3x^2 - 2x - 7 \\
x + 3 \overline{\smash{)}\, 3x^3 + 7x^2 - 13x - 21} \\
\underline{3x^3 + 9x^2} \\
-2x^2 - 13x \\
\underline{-2x^2 - 6x} \\
-7x - 21 \\
\underline{-7x - 21} \\
0
\end{array}
$$

The answer is $3x^2 - 2x - 7$.

35. $1.5 = 2.7 + x$
$1.5 - 2.7 = 2.7 + x - 2.7$
$-1.2 = x$

The solution is -1.2.

36. $\dfrac{2}{7}x = -6$

$\dfrac{7}{2} \cdot \dfrac{2}{7}x = \dfrac{7}{2}(-6)$

$x = -\dfrac{7 \cdot \cancel{2} \cdot 3}{\cancel{2} \cdot 1}$

$x = -21$

The solution is -21.

37. $5x - 9 = 36$
$5x - 9 + 9 = 36 + 9$
$5x = 45$
$\dfrac{5x}{5} = \dfrac{45}{5}$
$x = 9$

The solution is 9.

38. $\dfrac{2}{3} = \dfrac{-m}{10}$

$10 \cdot \dfrac{2}{3} = 10\left(\dfrac{-m}{10}\right)$

$\dfrac{20}{3} = -m$

$-1 \cdot \dfrac{20}{3} = -1 \cdot (-1 \cdot m)$

$-\dfrac{20}{3} = m$

The solution is $-\dfrac{20}{3}$.

39. $5.4 - 1.9x = 0.8x$
$5.4 - 1.9x + 1.9x = 0.8x + 1.9x$
$5.4 = 2.7x$
$\dfrac{5.4}{2.7} = \dfrac{2.7x}{2.7}$
$2 = x$

The solution is 2.

40.
$$x - \frac{7}{8} = \frac{3}{4}$$
$$x - \frac{7}{8} + \frac{7}{8} = \frac{3}{4} + \frac{7}{8}$$
$$x = \frac{6}{8} + \frac{7}{8}$$
$$x = \frac{13}{8}$$
The solution is $\frac{13}{8}$.

41.
$$2(2 - 3x) = 3(5x + 7)$$
$$4 - 6x = 15x + 21$$
$$4 - 6x + 6x = 15x + 21 + 6x$$
$$4 = 21x + 21$$
$$4 - 21 = 21x + 21 - 21$$
$$-17 = 21x$$
$$\frac{-17}{21} = \frac{21x}{21}$$
$$-\frac{17}{21} = x$$
The solution is $-\frac{17}{21}$.

42.
$$\frac{1}{4}x - \frac{2}{3} = \frac{3}{4} + \frac{1}{3}x$$
$$\frac{1}{4}x - \frac{2}{3} - \frac{1}{4}x = \frac{3}{4} + \frac{1}{3}x - \frac{1}{4}x$$
$$-\frac{2}{3} = \frac{3}{4} + \frac{4}{12}x - \frac{3}{12}x$$
$$-\frac{2}{3} = \frac{3}{4} + \frac{1}{12}x$$
$$-\frac{2}{3} - \frac{3}{4} = \frac{3}{4} + \frac{1}{12}x - \frac{3}{4}$$
$$-\frac{8}{12} - \frac{9}{12} = \frac{1}{12}x$$
$$-\frac{17}{12} = \frac{1}{12}x$$
$$12\left(-\frac{17}{12}\right) = 12\left(\frac{1}{12}x\right)$$
$$-17 = x$$
The solution is -17.

43.
$$y + 5 - 3y = 5y - 9$$
$$-2y + 5 = 5y - 9$$
$$-2y + 5 - 5y = 5y - 9 - 5y$$
$$-7y + 5 = -9$$
$$-7y + 5 - 5 = -9 - 5$$
$$-7y = -14$$
$$\frac{-7y}{-7} = \frac{-14}{-7}$$
$$y = 2$$
The solution is 2.

44.
$$\frac{1}{4}x - 7 < 5 - \frac{1}{2}x$$
$$\frac{1}{4}x - 7 + \frac{1}{2}x < 5 - \frac{1}{2}x + \frac{1}{2}x$$
$$\frac{3}{4}x - 7 < 5$$
$$\frac{3}{4}x - 7 + 7 < 5 + 7$$
$$\frac{3}{4}x < 12$$
$$\frac{4}{3} \cdot \frac{3}{4}x < \frac{4}{3} \cdot 12$$
$$x < \frac{4 \cdot \cancel{3} \cdot 4}{\cancel{3} \cdot 1}$$
$$x < 16$$
The solution set is $\{x | x < 16\}$.

45.
$$2(x + 2) \geq 5(2x + 3)$$
$$2x + 4 \geq 10x + 15$$
$$2x + 4 - 10x \geq 10x + 15 - 10x$$
$$-8x + 4 \geq 15$$
$$-8x + 4 - 4 \geq 15 - 4$$
$$-8x \geq 11$$
$$\frac{-8x}{-8} \leq \frac{11}{-8} \quad \text{Reversing the inequality symbol}$$
$$x \leq -\frac{11}{8}$$
The solution set is $\left\{x \middle| x \leq -\frac{11}{8}\right\}$.

46.
$$A = Qx - P$$
$$A + P = Qx - P + P$$
$$A + P = Qx$$
$$\frac{A + P}{Q} = \frac{Qx}{Q}$$
$$\frac{A + P}{Q} = x$$

47. The area of the surface of the pool is πr^2 ft². The area of the raft is 6 ft $\cdot$ 3 ft $= 18$ ft². We subtract to find the area of the surface of the pool not covered by the raft. We have
$$\pi r^2 \text{ ft}^2 - 18 \text{ ft}^2, \text{ or } (\pi r^2 - 18) \text{ ft}^2.$$

48. *Familiarize*. The page numbers are consecutive integers. Let $n =$ the smaller number. Then $n + 1 =$ the larger number.

Translate.

$$\underbrace{\text{The sum of the page numbers}}_{n + (n + 1)} \underset{=}{\text{ is }} \underset{37}{37.}$$

Solve.
$$n + (n + 1) = 37$$
$$2n + 1 = 37$$
$$2n = 37 - 1$$
$$2n = 36$$
$$n = 18$$

If $n = 18$, then $n + 1 = 18 + 1 = 19$.

Check. 18 and 19 are consecutive numbers and $18 + 19 = 37$. The answer checks.

State. The page numbers are 18 and 19.

49. **Familiarize**. Let l = the length of the room, in feet. Then $l - 4$ = the width. Recall that the formula for the perimeter of a rectangle is $P = 2l + 2w$.

    Translate.
    $$\underbrace{\text{Perimeter}}_{\downarrow} \quad \underset{\downarrow}{\text{is}} \quad \underset{\downarrow}{88 \text{ ft}}.$$
    $$2l + 2(l - 4) \quad = \quad 88$$

    Solve.
    $$2l + 2(l - 4) = 88$$
    $$2l + 2l - 8 = 88$$
    $$4l - 8 = 88$$
    $$4l = 88 + 8$$
    $$4l = 96$$
    $$l = \frac{96}{4}$$
    $$l = 24$$

    If $l = 24$, then $l - 4 = 24 - 4 = 20$.

    Check. The width, 20 ft, is 4 ft less than the length, 24 ft. The perimeter is $2 \cdot 24$ ft $+ 2 \cdot 20$ ft $= 48$ ft $+ 40$ ft $= 88$ ft. The answer checks.

    State. The length is 24 ft, and the width is 20 ft.

50. **Familiarize**. Let x = the measure of the first angle. Then $5x$ = the measure of the second angle and $2(x + 5x)$, or $2 \cdot 6x$, or $12x$ = the measure of the third angle. Recall that the sum of the measures of the angles of a triangle is $180°$.

    Translate.
    $$x + 5x + 12x = 180$$

    Solve.
    $$x + 5x + 12x = 180$$
    $$18x = 180$$
    $$x = \frac{180}{18}$$
    $$x = 10$$

    We are asked to find only the measure of the first angle, but we will also find the measures of the other two angles so that we can check the answer.

    If $x = 10$, then $5x = 5 \cdot 10 = 50$ and $12x = 12 \cdot 10 = 120$.

    Check. The measure of the second angle, $50°$, is five times $10°$, the measure of the first angle; the measure of the third angle, $120°$, is twice the sum of $10°$ and $50°$, or $2(10° + 50°)$, or $2 \cdot 60°$. Also, $10° + 50° + 120° = 180°$. The answer checks.

    State. The measure of the first angle is $10°$.

51. **Familiarize**. Let p = the price the store paid for the book. The store sells the book for a price of $p + 80\%p$, or $p + 0.8p$, or $1.8p$.

    Translate.
    $$\underbrace{\text{The selling price}}_{\downarrow} \quad \underset{\downarrow}{\text{is}} \quad \underset{\downarrow}{\$6.30}.$$
    $$1.8p \qquad = \quad 6.30$$

    Solve.
    $$1.8p = 6.30$$
    $$p = \frac{6.30}{1.8}$$
    $$p = 3.5$$

    Check. 80% of $3.50 is $0.8 \cdot \$3.50 = \$2.80$ and $\$3.50 + \$2.80 = \$6.30$. The answer checks.

    State. The store paid $3.50 for the book.

52. $295 \text{ million} = 295{,}000{,}000 = 2.95 \times 10^8$

    $21.1 = 2.11 \times 10$

    We multiply:
    $$(2.95 \times 10^8)(2.11 \times 10) = 6.2245 \times 10^9$$

    In 2005, 6.2245×10^9 gal of coffee will be consumed in the United States.

53. $y^2 \cdot y^{-6} \cdot y^8 = y^{2+(-6)+8} = y^4$

54. $\dfrac{x^6}{x^7} = x^{6-7} = x^{-1} = \dfrac{1}{x}$

55. $(-3x^3y^{-2})^3 = (-3)^3(x^3)^3(y^{-2})^3 = -27x^{3 \cdot 3}y^{-2 \cdot 3} =$

    $-27x^9y^{-6} = -\dfrac{27x^9}{y^6}$

56. $\dfrac{x^3x^{-4}}{x^{-5}x} = \dfrac{x^{3+(-4)}}{x^{-5+1}} = \dfrac{x^{-1}}{x^{-4}} = x^{-1-(-4)} = x^{-1+4} = x^3$

57. $\dfrac{2}{3}x^2 + 4x - 6$

    The coefficient of $\dfrac{2}{3}x^2$ is $\dfrac{2}{3}$.

    The coefficient of $4x$ is 4.

    The coefficient of -6 is -6.

58. $2x^4 + 3x^2 + 2x + 1$

    The degree of $2x^4$ is 4.

    The degree of $3x^2$ is 2.

    The degree of $2x$, or $2x^1$, is 1.

    The degree of 1, or $1 \cdot x^0$, is 0.

    The degree of the polynomial is 4, the largest exponent.

59. $2x^2 + 1$ is a binomial because it has just two terms.

60. $2x^2 + x + 1$ is a trinomial because it has just three terms.

61. $4x - 5y = 20$

    To find the x-intercept, let $y = 0$. Then solve for x.
    $$4x - 5y = 20$$
    $$4x - 5 \cdot 0 = 20$$
    $$4x = 20$$
    $$x = 5$$

    Thus, $(5, 0)$ is the x-intercept.

To find the y-intercept, let $x = 0$. Then solve for y.

$$4x - 5y = 20$$
$$4 \cdot 0 - 5y = 20$$
$$-5y = 20$$
$$y = -4$$

Thus, $(0, -4)$ is the y-intercept.

Plot these points and draw the graph.

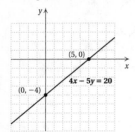

A third point should be used as a check. We substitute any value for x and solve for y.

We let $x = 4$. Then

$$4x - 5y = 20$$
$$4 \cdot 4 - 5y = 20$$
$$16 - 5y = 20$$
$$-5y = 4$$
$$y = -\frac{4}{5}$$

The point $\left(4, -\frac{4}{5}\right)$ is on the graph, so the graph is probably correct.

62. $(-2, 5)$ and $(-8, 3)$

First we find the slope.

$$m = \frac{3 - 5}{-8 - (-2)} = \frac{-2}{-6} = \frac{1}{3}$$

Thus, $y = \frac{1}{3}x + b$. We can use either point to find b. We choose $(-2, 5)$. Substitute -2 for x and 5 for y in $y = \frac{1}{3}x + b$.

$$y = \frac{1}{3}x + b$$
$$5 = \frac{1}{3}(-2) + b$$
$$5 = -\frac{2}{3} + b$$
$$\frac{15}{3} + \frac{2}{3} = b$$
$$\frac{17}{3} = b$$

Thus, the equation is $y = \frac{1}{3}x + \frac{17}{3}$.

63. Regarding the rectangle as one large rectangle with dimensions $x + 4$ by $x + 4$ we have $(x + 4)(x + 4)$. Regarding the rectangle as four smaller rectangles with dimensions x by x, 4 by x, 4 by x, and 4 by 4 we have

$$x \cdot x + 4 \cdot x + 4 \cdot x + 4 \cdot 4, \text{ or } x^2 + 8x + 16.$$

64.
$$3^2 = 9$$
$$3^{-2} = \frac{1}{3^2} = \frac{1}{9}$$
$$\left(\frac{1}{3}\right)^2 = \frac{1}{9}$$
$$\left(\frac{1}{3}\right)^{-2} = \left(\frac{3}{1}\right)^2 = 9$$
$$-3^2 = -9$$
$$(-3)^2 = 9$$
$$\left(-\frac{1}{3}\right)^2 = \frac{1}{9}$$
$$\left(-\frac{1}{3}\right)^{-2} = \left(-\frac{3}{1}\right)^2 = 9$$

65. We solve each equation for y:

1. $3y + 6 = 2x$ 2. $3y - 12 = 5x$
   $3y = 2x - 6$ $3y = 5x + 12$
   $y = \frac{2}{3}x - 2$ $y = \frac{5}{3}x + 4$

The slopes are not the same, so the lines are not parallel. The product of the slopes is $\frac{2}{3} \cdot \frac{5}{3} = \frac{10}{9} \neq -1$, so the lines are not perpendicular. Thus, the lines are neither parallel nor perpendicular.

66. We solve each equation for y:

1. $3y - 12 = 5x$ 2. $3y + 9 = 5x$
   $3y = 5x + 12$ $3y = 5x - 9$
   $y = \frac{5}{3}x + 4$ $y = \frac{5}{3}x - 3$

The slope of each line is $\frac{5}{3}$. The y-intercepts, $(0, 4)$ and $(0, -3)$ are different. The lines are parallel.

67. We solve each equation for y:

1. $5y + 3x = 4$ 2. $3y = 5x + 12$
   $5y = -3x + 4$ $y = \frac{5}{3}x + 4$
   $y = -\frac{3}{5}x + \frac{4}{5}$

The slopes are not the same, so the lines are not parallel. The product of the slopes is $-\frac{3}{5} \cdot \frac{5}{3} = -1$, so the lines are perpendicular.

68. The area of the framed picture is x in.$\cdot x$ in., or x^2 in^2. The length of a side of the picture is $(x - 2)$ in. Then the area of the picture is $(x - 2)$ in. $\cdot (x - 2)$ in., or $(x^2 - 4x + 4)$ in^2. We subtract to find the area of the frame.

$$x^2 - (x^2 - 4x + 4) = x^2 - x^2 + 4x - 4 = (4x - 4) \text{ in}^2$$

69.
$$[(2x)^2 - (3x)^3 + 2x^2x^3 + (x^2)^2] + [5x^2(2x^3) - ((2x)^2)^2]$$
$$= [4x^2 - 27x^3 + 2x^5 + x^4] + [10x^5 - (4x^2)^2]$$
$$= [4x^2 - 27x^3 + 2x^5 + x^4] + [10x^5 - 16x^4]$$
$$= 12x^5 - 15x^4 - 27x^3 + 4x^2$$

70. $(x - 3)^2 + (2x + 1)^2 = x^2 - 6x + 9 + 4x^2 + 4x + 1 = 5x^2 - 2x + 10$

71. First we find $[(3x^3 + 11x^2 + 11x + 15) \div (x + 3)]$.

$$
\begin{array}{r}
3x^2 + 2x + 5 \\
x + 3 \overline{\smash{\big)}\ 3x^3 + 11x^2 + 11x + 15} \\
\underline{3x^3 + 9x^2} \\
2x^2 + 11x \\
\underline{2x^2 + 6x} \\
5x + 15 \\
\underline{5x + 15} \\
0
\end{array}
$$

Next we find $[(2x^3 - 7x^2 + 2) \div (2x + 1)]$.

$$
\begin{array}{r}
x^2 - 4x + 2 \\
2x + 1 \overline{\smash{\big)}\ 2x^3 - 7x^2 + 0x + 2} \\
\underline{2x^3 + x^2} \\
-8x^2 + 0x \\
\underline{-8x^2 - 4x} \\
4x + 2 \\
\underline{4x + 2} \\
0
\end{array}
$$

Finally, we find the sum of the quotients.

$$(3x^2 + 2x + 5) + (x^2 - 4x + 2) = 4x^2 - 2x + 7$$

72.
$$(x+3)(2x-5)+(x-1)^2 = (3x+1)(x-3)$$
$$2x^2-5x+6x-15+x^2-2x+1 = 3x^2-9x+x-3$$
$$3x^2 - x - 14 = 3x^2 - 8x - 3$$
$$-x - 14 = -8x - 3 \qquad \text{Subtracting } 3x^2$$
$$7x - 14 = -3 \qquad \text{Adding } 8x$$
$$7x = 11 \qquad \text{Adding } 14$$
$$x = \frac{11}{7}$$

The solution is $\dfrac{11}{7}$.

73. First we find $(2x^2 + x - 6) \div (2x - 3)$.

$$
\begin{array}{r}
x + 2 \\
2x - 3 \overline{\smash{\big)}\ 2x^2 + x - 6} \\
\underline{2x^2 - 3x} \\
4x - 6 \\
\underline{4x - 6} \\
0
\end{array}
$$

Next we find $(2x^2 - 9x - 5) \div (x - 5)$.

$$
\begin{array}{r}
2x + 1 \\
x - 5 \overline{\smash{\big)}\ 2x^2 - 9x - 5} \\
\underline{2x^2 - 10x} \\
x - 5 \\
\underline{x - 5} \\
0
\end{array}
$$

Now we solve the resulting equation.
$$x + 2 = 2x + 1$$
$$-x + 2 = 1$$
$$-x = -1$$
$$x = 1$$

The solution is 1.

74.
$$20 - 3|x| = 5$$
$$-3|x| = -15$$
$$|x| = 5$$

The solutions are the numbers whose distance from 0 is 5. They are -5 and 5.

75. First we find $(x^3 - 4x^2 - 17x + 60) \div (x - 5)$.

$$
\begin{array}{r}
x^2 + x - 12 \\
x - 5 \overline{\smash{\big)}\ x^3 - 4x^2 - 17x + 60} \\
\underline{x^3 - 5x^2} \\
x^2 - 17x \\
\underline{x^2 - 5x} \\
-12x + 60 \\
\underline{-12x + 60} \\
0
\end{array}
$$

Then we solve the resulting equation.
$$(x-3)(x+4) = x^2 + x - 12$$
$$x^2 + x - 12 = x^2 + x - 12$$
$$x - 12 = x - 12 \qquad \text{Subtracting } x^2$$
$$-12 = -12 \qquad \text{TRUE}$$

$-12 = -12$ is true for all real numbers. However, we divided by $x - 5$ on the right side of the original equation and $x - 5 = 0$ when $x = 5$. Thus, all real numbers except 5 are solutions of the equation.

76.
$$V = (x + 2)(x + 2)(x + 2)$$
$$V = (x^2 + 4x + 4)(x + 2)$$
$$V = (x^2 + 4x + 4)x + (x^2 + 4x + 4)2$$
$$V = x^3 + 4x^2 + 4x + 2x^2 + 8x + 8$$
$$V = x^3 + 6x^2 + 12x + 8$$

The volume is $(x^3 + 6x^2 + 12x + 8)$ cm^3.

Chapter 5

Polynomials: Factoring

1. $x^2 = x^2$

$-6x = -1 \cdot 2 \cdot 3 \cdot x$

The coefficients have no common prime factor. The GCF of the powers of x is x because 1 is the smallest exponent of x. Thus the GCF is x.

3. $3x^4 = 3 \cdot x^4$

$x^2 = x^2$

The coefficients have no common prime factor. The GCF of the powers of x is x^2 because 2 is the smallest exponent of x. Thus the GCF is x^2.

5. $2x^2 = 2 \cdot x^2$

$2x = 2 \cdot x$

$-8 = -1 \cdot 2 \cdot 2 \cdot 2$

Each coefficient has a factor of 2. There are no other common prime factors. The GCF of the powers of x is 1 since -8 has no x-factor. Thus the GCF is 2.

7. $-17x^5y^3 = -1 \cdot 17 \cdot x^5 \cdot y^3$

$34x^3y^2 = 2 \cdot 17 \cdot x^3 \cdot y^2$

$51xy = 3 \cdot 17 \cdot x \cdot y$

Each coefficient has a factor of 17. There are no other common prime factors. The GCF of the powers of x is x because 1 is the smallest exponent of x. Similarly, the GCF of the powers of y is y because 1 is the smallest exponent of y. Thus the GCF is $17xy$.

9. $-x^2 = -1 \cdot x^2$

$-5x = -1 \cdot 5 \cdot x$

$-20x^3 = -1 \cdot 2 \cdot 2 \cdot 5 \cdot x^3$

The coefficients have no common prime factor. (Note that -1 is not a prime number.) The GCF of the powers of x is x because 1 is the smallest exponent of x. Thus the GCF is x.

11. $x^5y^5 = x^5 \cdot y^5$

$x^4y^3 = x^4 \cdot y^3$

$x^3y^3 = x^3 \cdot y^3$

$-x^2y^2 = -1 \cdot x^2 \cdot y^2$

There is no common prime factor. The GCF of the powers of x is x^2 because 2 is the smallest exponent of x. Similarly, the GCF of the powers of y is y^2 because 2 is the smallest exponent of y. Thus the GCF is x^2y^2.

13. $x^2 - 6x = x \cdot x - x \cdot 6$ Factoring each term

$= x(x - 6)$ Factoring out the common factor x

15. $2x^2 + 6x = 2x \cdot x + 2x \cdot 3$ Factoring each term

$= 2x(x + 3)$ Factoring out the common factor $2x$

17. $x^3 + 6x^2 = x^2 \cdot x + x^2 \cdot 6$ Factoring each term

$= x^2(x + 6)$ Factoring out x^2

19. $8x^4 - 24x^2 = 8x^2 \cdot x^2 - 8x^2 \cdot 3$

$= 8x^2(x^2 - 3)$ Factoring out $8x^2$

21. $2x^2 + 2x - 8 = 2 \cdot x^2 + 2 \cdot x - 2 \cdot 4$

$= 2(x^2 + x - 4)$ Factoring out 2

23. $17x^5y^3 + 34x^3y^2 + 51xy$

$= 17xy \cdot x^4y^2 + 17xy \cdot 2x^2y + 17xy \cdot 3$

$= 17xy(x^4y^2 + 2x^2y + 3)$

25. $6x^4 - 10x^3 + 3x^2 = x^2 \cdot 6x^2 - x^2 \cdot 10x + x^2 \cdot 3$

$= x^2(6x^2 - 10x + 3)$

27. $x^5y^5 + x^4y^3 + x^3y^3 - x^2y^2$

$= x^2y^2 \cdot x^3y^3 + x^2y^2 \cdot x^2y + x^2y^2 \cdot xy + x^2y^2(-1)$

$= x^2y^2(x^3y^3 + x^2y + xy - 1)$

29. $2x^7 - 2x^6 - 64x^5 + 4x^3$

$= 2x^3 \cdot x^4 - 2x^3 \cdot x^3 - 2x^3 \cdot 32x^2 + 2x^3 \cdot 2$

$= 2x^3(x^4 - x^3 - 32x^2 + 2)$

31. $1.6x^4 - 2.4x^3 + 3.2x^2 + 6.4x$

$= 0.8x(2x^3) - 0.8x(3x^2) + 0.8x(4x) + 0.8x(8)$

$= 0.8x(2x^3 - 3x^2 + 4x + 8)$

33. $\frac{5}{3}x^6 + \frac{4}{3}x^5 + \frac{1}{3}x^4 + \frac{1}{3}x^3$

$= \frac{1}{3}x^3(5x^3) + \frac{1}{3}x^3(4x^2) + \frac{1}{3}x^3(x) + \frac{1}{3}x^3(1)$

$= \frac{1}{3}x^3(5x^3 + 4x^2 + x + 1)$

35. Factor: $x^2(x + 3) + 2(x + 3)$

The binomial $x + 3$ is common to both terms:

$x^2(x + 3) + 2(x + 3) = (x^2 + 2)(x + 3)$

37. $5a^3(2a - 7) - (2a - 7)$

$= 5a^3(2a - 7) - 1(2a - 7)$

$= (5a^3 - 1)(2a - 7)$

39. $x^3 + 3x^2 + 2x + 6$

$= (x^3 + 3x^2) + (2x + 6)$

$= x^2(x + 3) + 2(x + 3)$ Factoring each binomial

$= (x^2 + 2)(x + 3)$ Factoring out the common factor $x + 3$

41. $2x^3 + 6x^2 + x + 3$

$= (2x^3 + 6x^2) + (x + 3)$

$= 2x^2(x + 3) + 1(x + 3)$ Factoring each binomial

$= (2x^2 + 1)(x + 3)$

43. $8x^3 - 12x^2 + 6x - 9 = 4x^2(2x - 3) + 3(2x - 3)$

$= (4x^2 + 3)(2x - 3)$

45. $12p^3 - 16p^2 + 3p - 4$

$= 4p^2(3p - 4) + 1(3p - 4)$ Factoring 1 out of the
second binomial

$= (4p^2 + 1)(3p - 4)$

47. $5x^3 - 5x^2 - x + 1$

$= (5x^3 - 5x^2) + (-x + 1)$

$= 5x^2(x - 1) - 1(x - 1)$ Check: $-1(x-1) = -x+1$

$= (5x^2 - 1)(x - 1)$

49. $x^3 + 8x^2 - 3x - 24 = x^2(x + 8) - 3(x + 8)$

$= (x^2 - 3)(x + 8)$

51. $2x^3 - 8x^2 - 9x + 36 = 2x^2(x - 4) - 9(x - 4)$

$= (2x^2 - 9)(x - 4)$

53. Discussion and Writing Exercise

55. $-2x < 48$

$x > -24$ Dividing by -2 and reversing the
inequality symbol

The solution set is $\{x | x > -24\}$.

57. $\dfrac{-108}{-4} = 27$ (The quotient of two negative
numbers is positive.)

59. $(y + 5)(y + 7) = y^2 + 7y + 5y + 35$ Using FOIL

$= y^2 + 12y + 35$

61. $(y + 7)(y - 7) = y^2 - 7^2 = y^2 - 49$

$[(A + B))(A - B) = A^2 - B^2]$

63. $x + y = 4$

To find the x-intercept, let $y = 0$. Then solve for x.

$x + y = 4$

$x + 0 = 4$

$x = 4$

The x-intercept is $(4, 0)$.

To find the y-intercept, let $x = 0$. Then solve for y.

$x + y = 4$

$0 + y = 4$

$y = 4$

The y-intercept is $(0, 4)$.

Plot these points and draw the line.

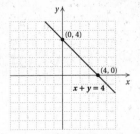

A third point should be used as a check. We substitute
any value for x and solve for y. We let $x = 2$. Then

$x + y = 4$

$2 + y = 4$

$y = 2$

The point $(2, 2)$ is on the graph, so the graph is probably
correct.

65. $5x - 3y = 15$

To find the x-intercept, let $y = 0$. Then solve for x.

$5x - 3y = 15$

$5x - 3 \cdot 0 = 15$

$5x = 15$

$x = 3$

The x-intercept is $(3, 0)$.

To find the y-intercept, let $x = 0$. Then solve for y.

$5x - 3y = 15$

$5 \cdot 0 - 3y = 15$

$-3y = 15$

$y = -5$

The y-intercept is $(0, -5)$.

Plot these points and draw the line.

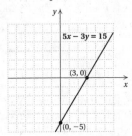

A third point should be used as a check. We substitute
any value for x and solve for y. We let $x = 6$. Then

$5x - 3y = 15$

$5 \cdot 6 - 3y = 15$

$30 - 3y = 15$

$-3y = -15$

$y = 5$

The point $(6, 5)$ is on the graph, so the graph is probably
correct.

67. $4x^5 + 6x^3 + 6x^2 + 9 = 2x^3(2x^2 + 3) + 3(2x^2 + 3)$

$= (2x^3 + 3)(2x^2 + 3)$

69. $x^{12} + x^7 + x^5 + 1 = x^7(x^5 + 1) + (x^5 + 1)$
$$= (x^7 + 1)(x^5 + 1)$$

71. $p^3 + p^2 - 3p + 10 = p^2(p + 1) - (3p - 10)$

This polynomial is not factorable using factoring by grouping.

Exercise Set 5.2

1. $x^2 + 8x + 15$

Since the constant term and coefficient of the middle term are both positive, we look for a factorization of 15 in which both factors are positive. Their sum must be 8.

Pairs of factors	Sums of factors
1, 15	16
3, 5	8

The numbers we want are 3 and 5.
$x^2 + 8x + 15 = (x + 3)(x + 5)$.

3. $x^2 + 7x + 12$

Since the constant term is positive and the coefficient of the middle term is positive, we look for a factorization of 12 in which both factors are positive. Their sum must be 7.

Pairs of factors	Sums of factors
1, 12	13
2, 6	8
3, 4	7

The numbers we want are 3 and 4.
$x^2 + 7x + 12 = (x + 3)(x + 4)$.

5. $x^2 - 6x + 9$

Since the constant term is positive and the coefficient of the middle term is negative, we look for a factorization of 9 in which both factors are negative. Their sum must be -6.

Pairs of factors	Sums of factors
$-1, -9$	-10
$-3, -3$	-6

The numbers we want are -3 and -3.
$x^2 - 6x + 9 = (x - 3)(x - 3)$, or $(x - 3)^2$.

7. $x^2 - 5x - 14$

Since the constant term is negative, we look for a factorization of -14 in which one factor is positive and one factor is negative. Their sum must be -5, the coefficient of the middle term.

Pairs of factors	Sums of factors
$-1, 14$	13
$1, -14$	-13
$-2, 7$	5
$2, -7$	-5

The numbers we want are 2 and -7.
$x^2 - 5x - 14 = (x + 2)(x - 7)$.

9. $b^2 + 5b + 4$

Since the constant term is positive and the coefficient of the middle term is positive, we look for a factorization of 4 in which both factors are positive. Their sum must be 5.

Pairs of factors	Sums of factors
1, 4	5
2, 2	4

The numbers we want are 1 and 4.
$b^2 + 5b + 4 = (b + 1)(b + 4)$.

11. $x^2 + \frac{2}{3}x + \frac{1}{9}$

Since the constant term is positive and the coefficient of the middle term is positive, we look for a factorization of $\frac{1}{9}$ in which both factors are positive. Their sum must be $\frac{2}{3}$.

Pairs of factors	Sums of factors
$1, \frac{1}{9}$	$\frac{10}{9}$
$\frac{1}{3}, \frac{1}{3}$	$\frac{2}{3}$

The numbers we want are $\frac{1}{3}$ and $\frac{1}{3}$.
$x^2 + \frac{2}{3}x + \frac{1}{9} = \left(x + \frac{1}{3}\right)\left(x + \frac{1}{3}\right)$, or $\left(x + \frac{1}{3}\right)^2$.

13. $d^2 - 7d + 10$

Since the constant term is positive and the coefficient of the middle term is negative, we look for a factorization of 10 in which both factors are negative. Their sum must be -7.

Pairs of factors	Sums of factors
$-1, -10$	-11
$-2, -5$	-7

The numbers we want are -2 and -5.
$d^2 - 7d + 10 = (d - 2)(d - 5)$.

15. $y^2 - 11y + 10$

Since the constant term is positive and the coefficient of the middle term is negative, we look for a factorization of 10 in which both factors are negative. Their sum must be -11.

Pairs of factors	Sums of factors
$-1, -10$	-11
$-2, -5$	-7

The numbers we want are -1 and -10.
$y^2 - 11y + 10 = (y - 1)(y - 10)$.

17. $x^2 + x + 1$

Since the constant term and the coefficient of the middle term are both positive, we look for a factorization of 1 in which both factors are positive. The sum must be 1. The only possible pair of factors is 1 and 1, but their sum is not 1. Thus, this polynomial is not factorable into binomials. It is prime.

19. $x^2 - 7x - 18$

Since the constant term is negative, we look for a factorization of -18 in which one factor is positive and one factor is negative. Their sum must be -7, the coefficient of the middle term.

Pairs of factors	Sums of factors
-1, 18	17
1, -18	-17
-2, 9	7
2, -9	-7
-3, 6	3
3, -6	-3

The numbers we want are 2 and -9.

$x^2 - 7x - 18 = (x + 2)(x - 9)$.

21. $x^3 - 6x^2 - 16x = x(x^2 - 6x - 16)$

After factoring out the common factor, x, we consider $x^2 - 6x - 16$. Since the constant term is negative, we look for a factorization of -16 in which one factor is positive and one factor is negative. Their sum must be -6, the coefficient of the middle term.

Pairs of factors	Sums of factors
-1, 16	15
1, -16	-15
-2, 8	6
2, -8	-6
-4, 4	0

The numbers we want are 2 and -8.

Then $x^2 - 6x - 16 = (x + 2)(x - 8)$, so $x^3 - 6x^2 - 16x = x(x + 2)(x - 8)$.

23. $y^3 - 4y^2 - 45y = y(y^2 - 4y - 45)$

After factoring out the common factor, y, we consider $y^2 - 4y - 45$. Since the constant term is negative, we look for a factorization of -45 in which one factor is positive and one factor is negative. Their sum must be -4, the coefficient of the middle term.

Pairs of factors	Sums of factors
-1, 45	44
1, -45	-44
-3, 15	12
3, -15	-12
-5, 9	4
5, -9	-4

The numbers we want are 5 and -9.

Then $y^2 - 4y - 45 = (y + 5)(y - 9)$, so $y^3 - 4y^2 - 45y = y(y + 5)(y - 9)$.

25. $-2x - 99 + x^2 = x^2 - 2x - 99$

Since the constant term is negative, we look for a factorization of -99 in which one factor is positive and one factor is negative. Their sum must be -2, the coefficient of the middle term.

Pairs of factors	Sums of factors
-1, 99	98
1, -99	-98
-3, 33	30
3, -33	-30
-9, 11	2
9, -11	-2

The numbers we want are 9 and -11.

$-2x - 99 + x^2 = (x + 9)(x - 11)$.

27. $c^4 + c^2 - 56$

Consider this trinomial as $(c^2)^2 + c^2 - 56$. We look for numbers p and q such that $c^4 + c^2 - 56 = (c^2 + p)(c^2 + q)$. Since the constant term is negative, we look for a factorization of -56 in which one factor is positive and one factor is negative. Their sum must be 1.

Pairs of factors	Sums of factors
-1, 56	55
1, -56	-55
-2, 28	26
2, -28	-26
-4, 14	12
4, -14	-12
-7, 8	1
7, -8	-1

The numbers we want are -7 and 8.

$c^4 + c^2 - 56 = (c^2 - 7)(c^2 + 8)$.

29. $a^4 + 2a^2 - 35$

Consider this trinomial as $(a^2)^2 + 2a^2 - 35$. We look for numbers p and q such that $a^4 + 2a^2 - 35 = (a^2 + p)(a^2 + q)$. Since the constant term is negative, we look for a factorization of -35 in which one factor is positive and one factor is negative. Their sum must be 2.

Pairs of factors	Sums of factors
-1, 35	34
1, -35	-34
-5, 7	2
5, -7	-2

The numbers we want are -5 and 7.

$a^4 + 2a^2 - 35 = (a^2 - 5)(a^2 + 7)$.

31. $x^2 + x - 42$

Since the constant term is negative, we look for a factorization of -42 in which one factor is positive and one factor

---

is negative. Their sum must be 1, the coefficient of the middle term.

Pairs of factors	Sums of factors
$-1,\ \ 42$	41
$1,\ -42$	-41
$-2,\ \ 21$	19
$2,\ -21$	-19
$-3,\ \ 14$	11
$3,\ -14$	-11
$-6,\ \ 7$	1
$6,\ -7$	-1

The numbers we want are -6 and 7.

$x^2 + x - 42 = (x-6)(x+7)$.

33. $7 - 2p + p^2 = p^2 - 2p + 7$

Since the constant term is positive and the coefficient of the middle term is negative, we look for a factorization of 7 in which both factors are negative. The sum must be -2. The only possible pair of factors is -1 and -7, but their sum is not -2. Thus, this polynomial is not factorable into binomials. It is prime.

35. $x^2 + 20x + 100$

We look for two factors, both positive, whose product is 100 and whose sum is 20.

They are 10 and 10. $10 \cdot 10 = 100$ and $10 + 10 = 20$.

$x^2 + 20x + 100 = (x+10)(x+10)$, or $(x+10)^2$.

37. $30 + 7x - x^2 = -x^2 + 7x + 30 = -1(x^2 - 7x - 30)$

Now we factor $x^2 - 7x - 30$. Since the constant term is negative, we look for a factorization of -30 in which one factor is positive and one factor is negative. Their sum must be -7, the coefficient of the middle term.

Pairs of factors	Sums of factors
$-1,\ \ 30$	29
$1,\ -30$	-29
$-2,\ \ 15$	13
$2,\ -15$	-13
$-3,\ \ 10$	7
$3,\ -10$	-7
$-5,\ \ 6$	1
$5,\ -6$	-1

The numbers we want are 3 and -10. Then $x^2 - 7x - 30 = (x+3)(x-10)$, so we have:

$\quad -x^2 + 7x + 30$

$= -1(x+3)(x-10)$

$= (-x-3)(x-10)$ Multiplying $x+3$ by -1

$= (x+3)(-x+10)$ Multiplying $x-10$ by -1

39. $24 - a^2 - 10a = -a^2 - 10a + 24 = -1(a^2 + 10a - 24)$

Now we factor $a^2 + 10a - 24$. Since the constant term is negative, we look for a factorization of -24 in which one factor is positive and one factor is negative. Their sum must be 10, the coefficient of the middle term.

Pairs of factors	Sums of factors
$-1,\ \ 24$	23
$1,\ -24$	-23
$-2,\ \ 12$	10
$2,\ -12$	-10
$-3,\ \ 8$	5
$3,\ -8$	-5
$-4,\ \ 6$	2
$4,\ -6$	-2

The numbers we want are -2 and 12. Then $a^2 + 10a - 24 = (a-2)(a+12)$, so we have:

$\quad -a^2 - 10a + 24$

$= -1(a-2)(a+12)$

$= (-a+2)(a+12)$ Multiplying $a-2$ by -1

$= (a-2)(-a-12)$ Multiplying $a+12$ by -1

41. $x^4 - 21x^3 - 100x^2 = x^2(x^2 - 21x - 100)$

After factoring out the common factor, x^2, we consider $x^2 - 21x - 100$. We look for two factors, one positive and one negative, whose product is -100 and whose sum is -21. They are 4 and -25. $4 \cdot (-25) = -100$ and $4 + (-25) = -21$.

Then $x^2 - 21x - 100 = (x+4)(x-25)$, so $x^4 - 21x^3 - 100x^2 = x^2(x+4)(x-25)$.

43. $x^2 - 21x - 72$

We look for two factors, one positive and one negative, whose product is -72 and whose sum is -21. They are 3 and -24.

$x^2 - 21x - 72 = (x+3)(x-24)$.

45. $x^2 - 25x + 144$

We look for two factors, both negative, whose product is 144 and whose sum is -25. They are -9 and -16.

$x^2 - 25x + 144 = (x-9)(x-16)$.

47. $a^2 + a - 132$

We look for two factors, one positive and one negative, whose product is -132 and whose sum is 1. They are -11 and 12.

$a^2 + a - 132 = (a-11)(a+12)$.

49. $120 - 23x + x^2 = x^2 - 23x + 120$

We look for two factors, both negative, whose product is 120 and whose sum is -23. They are -8 and -15.

$x^2 - 23x + 120 = (x-8)(x-15)$.

51. First write the polynomial in descending order and factor out -1.

$108 - 3x - x^2 = -x^2 - 3x + 108 = -1(x^2 + 3x - 108)$

Now we factor the polynomial $x^2 + 3x - 108$. We look for two factors, one positive and one negative, whose product is -108 and whose sum is 3. They are -9 and 12.

$x^2 + 3x - 108 = (x-9)(x+12)$

The final answer must include -1 which was factored out above.

$-x^2 - 3x + 108$

$= -1(x - 9)(x + 12)$

$= (-x + 9)(x + 12)$ Multiplying $x - 9$ by -1

$= (x - 9)(-x - 12)$ Multiplying $x + 12$ by -1

53. $y^2 - 0.2y - 0.08$

We look for two factors, one positive and one negative, whose product is -0.08 and whose sum is -0.2. They are -0.4 and 0.2.

$y^2 - 0.2y - 0.08 = (y - 0.4)(y + 0.2)$.

55. $p^2 + 3pq - 10q^2 = p^2 + 3pq - 10q^2$

Think of $3q$ as a "coefficient" of p. Then we look for factors of $-10q^2$ whose sum is $3q$. They are $5q$ and $-2q$.

$p^2 + 3pq - 10q^2 = (p + 5q)(p - 2q)$.

57. $84 - 8t - t^2 = -t^2 - 8t + 84 = -1(t^2 + 8t - 84)$

Now we factor $t^2 + 8t - 84$. We look for two factors, one positive and one negative, whose product is -84 and whose sum is 8. They are 14 and -6.

Then $t^2 + 8t - 84 = (t + 14)(t - 6)$, so we have:

$-t^2 - 8t + 84$

$= -1(t + 14)(t - 6)$

$= (-t - 14)(t - 6)$ Multiplying $t + 14$ by -1

$= (t + 14)(-t + 6)$ Multiplying $t - 6$ by -1

59. $m^2 + 5mn + 4n^2 = m^2 + 5nm + 4n^2$

We look for factors of $4n^2$ whose sum is $5n$. They are $4n$ and n.

$m^2 + 5mn + 4n^2 = (m + 4n)(m + n)$

61. $s^2 - 2st - 15t^2 = s^2 - 2ts - 15t^2$

We look for factors of $-15t^2$ whose sum is $-2t$. They are $-5t$ and $3t$.

$s^2 - 2st - 15t^2 = (s - 5t)(s + 3t)$

63. $6a^{10} - 30a^9 - 84a^8 = 6a^8(a^2 - 5a - 14)$

After factoring out the common factor, $6a^8$, we consider $a^2 - 5a - 14$. We look for two factors, one positive and one negative, whose product is -14 and whose sum is -5. They are 2 and -7.

$a^2 - 5a - 14 = (a + 2)(a - 7)$, so $6a^{10} - 30a^9 - 84a^8 = 6a^8(a + 2)(a - 7)$.

65. Discussion and Writing Exercise

67. Discussion and Writing Exercise

69. $8x(2x^2 - 6x + 1) = 8x \cdot 2x^2 - 8x \cdot 6x + 8x \cdot 1 = 16x^3 - 48x^2 + 8x$

71. $(7w + 6)^2 = (7w)^2 + 2 \cdot 7w \cdot 6 + 6^2 = 49w^2 + 84w + 36$

73. $(4w - 11)(4w + 11) = (4w)^2 - (11)^2 = 16w^2 - 121$

75. $(3x - 5y)(2x + 7y) = 3x \cdot 2x + 3x \cdot 7y - 5y \cdot 2x - 5y \cdot 7y = 6x^2 + 21xy - 10xy - 35y^2 = 6x^2 + 11xy - 35y^2$

77. $3x - 8 = 0$

$3x = 8$ Adding 8 on both sides

$x = \dfrac{8}{3}$ Dividing by 3 on both sides

The solution is $\dfrac{8}{3}$.

79. *Familiarize*. Let $n =$ the number of people arrested the year before.

Translate. We reword the problem.

Number arrested the year before	less	1.2% of	that number	is	29,200.
$\downarrow$	$\downarrow$	$\downarrow$ $\downarrow$	$\downarrow$	$\downarrow$	$\downarrow$ $\downarrow$
n	$-$	$1.2\% \cdot$	n	$=$	$29,200$

Carry out. We solve the equation.

$n - 1.2\% \cdot n = 29,200$

$1 \cdot n - 0.012n = 29,200$

$0.988n = 29,200$

$n \approx 29,555$ Rounding

Check. 1.2% of 29,555 is $0.012(29,555) \approx 355$ and $29,555 - 355 = 29,200$. The answer checks.

State. Approximately 29,555 people were arrested the year before.

81. $y^2 + my + 50$

We look for pairs of factors whose product is 50. The sum of each pair is represented by m.

Pairs of factors whose product is -50	Sums of factors
1, 50	51
$-1, -50$	-51
2, 25	27
$-2, -25$	-27
5, 10	15
$-5, -10$	-15

The polynomial $y^2 + my + 50$ can be factored if m is 51, -51, 27, -27, 15, or -15.

83. $x^2 - \dfrac{1}{2}x - \dfrac{3}{16}$

We look for two factors, one positive and one negative, whose product is $-\dfrac{3}{16}$ and whose sum is $-\dfrac{1}{2}$.

They are $-\dfrac{3}{4}$ and $\dfrac{1}{4}$.

$-\dfrac{3}{4} \cdot \dfrac{1}{4} = -\dfrac{3}{16}$ and $-\dfrac{3}{4} + \dfrac{1}{4} = -\dfrac{2}{4} = -\dfrac{1}{2}$.

$x^2 - \dfrac{1}{2}x - \dfrac{3}{16} = \left(x - \dfrac{3}{4}\right)\left(x + \dfrac{1}{4}\right)$

85. $x^2 + \dfrac{30}{7}x - \dfrac{25}{7}$

We look for two factors, one positive and one negative, whose product is $-\dfrac{25}{7}$ and whose sum is $\dfrac{30}{7}$.

They are 5 and $-\dfrac{5}{7}$.

$5 \cdot \left(-\dfrac{5}{7}\right) = -\dfrac{25}{7}$ and $5 + \left(-\dfrac{5}{7}\right) = \dfrac{35}{7} + \left(-\dfrac{5}{7}\right) = \dfrac{30}{7}$.

$x^2 + \dfrac{30}{7}x - \dfrac{25}{7} = (x + 5)\left(x - \dfrac{5}{7}\right)$

87. $b^{2n} + 7b^n + 10$

Consider this trinomial as $(b^n)^2 + 7b^n + 10$. We look for numbers p and q such that $b^{2n} + 7b^n + 10 = (b^n + p)(b^n + q)$. We find two factors, both positive, whose product is 10 and whose sum is 7. They are 5 and 2.

$b^{2n} + 7b^n + 10 = (b^n + 5)(b^n + 2)$

89. We first label the drawing with additional information.

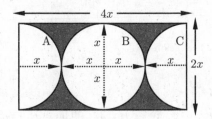

$4x$ represents the length of the rectangle and $2x$ the width. The area of the rectangle is $4x \cdot 2x$, or $8x^2$.

The area of semicircle A is $\dfrac{1}{2}\pi x^2$.

The area of circle B is πx^2.

The area of semicircle C is $\dfrac{1}{2}\pi x^2$.

$$\begin{aligned}\text{Area of shaded region} &= \text{Area of rectangle} - \begin{array}{c}\text{Area} \\ \text{of} \\ A\end{array} - \begin{array}{c}\text{Area} \\ \text{of} \\ B\end{array} - \begin{array}{c}\text{Area} \\ \text{of} \\ C\end{array}\end{aligned}$$

$$\begin{aligned}\text{Area of shaded region} &= 8x^2 - \dfrac{1}{2}\pi x^2 - \pi x^2 - \dfrac{1}{2}\pi x^2 \\ &= 8x^2 - 2\pi x^2 \\ &= 2x^2(4 - \pi)\end{aligned}$$

The shaded area can be represented by $2x^2(4 - \pi)$.

Exercise Set 5.3

1. $2x^2 - 7x - 4$

(1) Look for a common factor. There is none (other than 1 or -1).

(2) Factor the first term, $2x^2$. The only possibility is $2x$, x. The desired factorization is of the form:

$$(2x + \quad)(x + \quad)$$

(3) Factor the last term, -4, which is negative. The possibilities are -4, 1 and 4, -1 and 2, -2.

These factors can also be written as 1, -4 and -1, 4 and -2, 2.

(4) Look for combinations of factors from steps (2) and (3) such that the sum of their products is the middle term, $-7x$. We try some possibilities:

$$(2x - 4)(x + 1) = 2x^2 - 2x - 4$$
$$(2x + 4)(x - 1) = 2x^2 + 2x - 4$$
$$(2x + 2)(x - 2) = 2x^2 - 2x - 4$$
$$(2x + 1)(x - 4) = 2x^2 - 7x - 4$$

The factorization is $(2x + 1)(x - 4)$.

3. $5x^2 - x - 18$

(1) There is no common factor (other than 1 or -1).

(2) Factor the first term, $5x^2$. The only possibility is $5x$, x. The desired factorization is of the form:

$$(5x + \quad)(x + \quad)$$

(3) Factor the last term, -18. The possibilities are -18, 1 and 18, -1 and -9, 2 and 9, -2 and -6, 3 and 6, -3.

These factors can also be written as 1, -18 and -1, 18 and 2, -9 and -2, 9 and 3, -6 and -3, 6.

(4) Look for combinations of factors from steps (2) and (3) such that the sum of their products is the middle term, x. We try some possibilities:

$$(5x - 18)(x + 1) = 5x^2 - 13x - 18$$
$$(5x + 18)(x - 1) = 5x^2 + 13x - 18$$
$$(5x + 9)(x - 2) = 5x^2 - x - 18$$

The factorization is $(5x + 9)(x - 2)$.

5. $6x^2 + 23x + 7$

(1) There is no common factor (other than 1 or -1).

(2) Factor the first term, $6x^2$. The possibilities are $6x$, x and $3x$, $2x$. We have these as possibilities for factorizations:

$$(6x + \quad)(x + \quad) \text{ and } (3x + \quad)(2x + \quad)$$

(3) Factor the last term, 7. The possibilities are 7, 1 and -7, -1.

These factors can also be written as 1, 7 and -1, -7.

(4) Look for combinations of factors from steps (2) and (3) such that the sum of their products is the middle term, $23x$. Since all signs are positive, we need consider only plus signs. We try some possibilities:

$$(6x + 7)(x + 1) = 6x^2 + 13x + 7$$
$$(3x + 7)(2x + 1) = 6x^2 + 17x + 7$$
$$(6x + 1)(x + 7) = 6x^2 + 43x + 7$$
$$(3x + 1)(2x + 7) = 6x^2 + 23x + 7$$

The factorization is $(3x + 1)(2x + 7)$.

7. $3x^2 + 4x + 1$

(1) There is no common factor (other than 1 or -1).

(2) Factor the first term, $3x^2$. The only possibility is $3x$, x. The desired factorization is of the form:
$$(3x+\quad)(x+\quad)$$

(3) Factor the last term, 1. The possibilities are 1, 1 and -1, -1.

(4) Look for combinations of factors from steps (2) and (3) such that the sum of their products is the middle term, $4x$. Since all signs are positive, we need consider only plus signs. There is only one such possibility:
$$(3x+1)(x+1) = 3x^2 + 4x + 1$$

The factorization is $(3x+1)(x+1)$.

9. $4x^2 + 4x - 15$

(1) There is no common factor (other than 1 or -1).

(2) Factor the first term, $4x^2$. The possibilities are $4x$, x and $2x$, $2x$. We have these as possibilities for factorizations:
$$(4x+\quad)(x+\quad) \text{ and } (2x+\quad)(2x+\quad)$$

(3) Factor the last term, -15. The possibilities are 15, -1 and -15, 1 and 5, -3 and -5, 3.

These factors can also be written as -1, 15 and 1, -15 and -3, 5 and 3, -5.

(4) We try some possibilities:
$$(4x + 15)(x - 1) = 4x^2 + 11x - 15$$
$$(2x + 15)(2x - 1) = 4x^2 + 28x - 15$$
$$(4x - 15)(x + 1) = 4x^2 - 11x - 15$$
$$(2x - 15)(2x + 1) = 4x^2 - 28x - 15$$
$$(4x + 5)(x - 3) = 4x^2 - 7x - 15$$
$$(2x + 5)(2x - 3) = 4x^2 + 4x - 15$$

The factorization is $(2x + 5)(2x - 3)$.

11. $2x^2 - x - 1$

(1) There is no common factor (other than 1 or -1).

(2) Factor the first term, $2x^2$. The only possibility is $2x$, x. The desired factorization is of the form:
$$(2x+\quad)(x+\quad)$$

(3) Factor the last term, -1. The only possibility is -1, 1.

These factors can also be written as 1, -1.

(4) We try the possibilities:
$$(2x - 1)(x + 1) = 2x^2 + x - 1$$
$$(2x + 1)(x - 1) = 2x^2 - x - 1$$

The factorization is $(2x + 1)(x - 1)$.

13. $9x^2 + 18x - 16$

(1) There is no common factor (other than 1 or -1).

(2) Factor the first term, $9x^2$. The possibilities are $9x$, x and $3x$, $3x$. We have these as possibilities for factorizations:
$$(9x+\quad)(x+\quad) \text{ and } (3x+\quad)(3x+\quad)$$

(3) Factor the last term, -16. The possibilities are 16, -1 and -16, 1 and 8, -2 and -8, 2 and 4, -4.

These factors can also be written as -1, 16 and 1, -16 and -2, 8 and 2, -8 and -4, 4.

(4) We try some possibilities:
$$(9x + 16)(x - 1) = 9x^2 + 7x - 16$$
$$(3x + 16)(3x - 1) = 9x^2 + 45x - 16$$
$$(9x - 16)(x + 1) = 9x^2 - 7x - 16$$
$$(3x - 16)(3x + 1) = 9x^2 - 45x - 16$$
$$(9x + 8)(x - 2) = 9x^2 - 10x - 16$$
$$(3x + 8)(3x - 2) = 9x^2 + 18x - 16$$

The factorization is $(3x + 8)(3x - 2)$.

15. $3x^2 - 5x - 2$

(1) There is no common factor (other than 1 or -1).

(2) Factor the first term, $3x^2$. The only possibility is $3x$, x. The desired factorization is of the form:
$$(3x+\quad)(x+\quad)$$

(3) Factor the last term, -2. The possibilities are 2, -1 and -2 and 1.

These factors can also be written as -1, 2 and 1, -2.

(4) We try some possibilities:
$$(3x + 2)(x - 1) = 3x^2 - x - 2$$
$$(3x - 2)(x + 1) = 3x^2 + x - 2$$
$$(3x - 1)(x + 2) = 3x^2 + 5x - 2$$
$$(3x + 1)(x - 2) = 3x^2 - 5x - 2$$

The factorization is $(3x + 1)(x - 2)$.

17. $12x^2 + 31x + 20$

(1) There is no common factor (other than 1 or -1).

(2) Factor the first term, $12x^2$. The possibilities are $12x$, x and $6x$, $2x$ and $4x$, $3x$. We have these as possibilities for factorizations:
$$(12x+\quad)(x+\quad) \text{ and } (6x+\quad)(2x+\quad) \text{ and } (4x+\quad)(3x+\quad)$$

(3) Factor the last term, 20. Since all signs are positive, we need consider only positive pairs of factors. Those factor pairs are 20, 1 and 10, 2 and 5, 4.

These factors can also be written as 1, 20 and 2, 10 and 4, 5.

(4) We can immediately reject all possibilities in which either factor has a common factor, such as $(12x+20)$ or $(6x+4)$, because we determined at the outset that there are no common factors. We try some of the remaining possibilities:

$$(12x + 1)(x + 20) = 12x^2 + 241x + 20$$
$$(12x + 5)(x + 4) = 12x^2 + 53x + 20$$
$$(6x + 1)(2x + 20) = 12x^2 + 122x + 20$$
$$(4x + 5)(3x + 4) = 12x^2 + 31x + 20$$

The factorization is $(4x + 5)(3x + 4)$.

19. $14x^2 + 19x - 3$

(1) There is no common factor (other than 1 or -1).

(2) Factor the first term, $14x^2$. The possibilities are $14x, x$ and $7x, 2x$. We have these as possibilities for factorizations:

$$(14x+ \quad)(x+ \quad) \text{ and } (7x+ \quad)(2x+ \quad)$$

(3) Factor the last term, -3. The possibilities are -1, 3 and -3, 1.

These factors can also be written as 3, -1 and 1, -3.

(4) We try some possibilities:

$$(14x - 1)(x + 3) = 14x^2 + 41x - 3$$
$$(7x - 1)(2x + 3) = 7x^2 + 19x - 3$$

The factorization is $(7x - 1)(2x + 3)$.

21. $9x^2 + 18x + 8$

(1) There is no common factor (other than 1 or -1).

(2) Factor the first term, $9x^2$. The possibilities are $9x$, x and $3x$, $3x$. We have these as possibilities for factorizations:

$$(9x+ \quad)(x+ \quad) \text{ and } (3x+ \quad)(3x+ \quad)$$

(3) Factor the last term, 8. Since all signs are positive, we need consider only positive pairs of factors. Those factor pairs are 8, 1 and 4, 2.

These factors can also be written as 1, 8 and 2, 4.

(4) We try some possibilities:

$$(9x + 8)(x + 1) = 9x^2 + 17x + 8$$
$$(3x + 8)(3x + 1) = 9x^2 + 27x + 8$$
$$(9x + 4)(x + 2) = 9x^2 + 22x + 8$$
$$(3x + 4)(3x + 2) = 9x^2 + 18x + 8$$

The factorization is $(3x + 4)(3x + 2)$.

23. $49 - 42x + 9x^2 = 9x^2 - 42x + 49$

(1) There is no common factor (other than 1 or -1).

(2) Factor the first term, $9x^2$. The possibilities are $9x$, x and $3x$, $3x$. We have these as possibilities for factorizations:

$$(9x+ \quad)(x+ \quad) \text{ and } (3x+ \quad)(3x+ \quad)$$

(3) Factor 49. Since 49 is positive and the middle term is negative, we need consider only negative pairs of factors. Those factor pairs are -49, -1 and -7, -7.

The first pair of factors can also be written as -1, -49.

(4) We try some possibilities:

$$(9x - 49)(x - 1) = 9x^2 - 58x + 49$$
$$(3x - 49)(3x - 1) = 9x^2 - 150x + 49$$
$$(9x - 7)(x - 7) = 9x^2 - 70x + 49$$
$$(3x - 7)(3x - 7) = 9x^2 - 42x + 49$$

The factorization is $(3x - 7)(3x - 7)$, or $(3x - 7)^2$. This can also be expressed as follows:

$$(3x - 7)^2 = (-1)^2(3x - 7)^2 = [-1 \cdot (3x - 7)]^2 =$$
$$(-3x + 7)^2, \text{ or } (7 - 3x)^2$$

25. $24x^2 + 47x - 2$

(1) There is no common factor (other than 1 or -1).

(2) Factor the first term, $24x^2$. The possibilities are $24x$, x and $12x$, $2x$ and $6x$, $4x$ and $3x$, $8x$. We have these as possibilities for factorizations:

$$(24x+ \quad)(x+ \quad) \text{ and } (12x+ \quad)(2x+ \quad) \text{ and}$$
$$(6x+ \quad)(4x+ \quad) \text{ and } (3x+ \quad)(8x+ \quad)$$

(3) Factor the last term, -2. The possibilities are 2, -1 and -2, 1.

These factors can also be written as -1, 2 and 1, -2.

(4) We can immediately reject all possibilities in which either factor has a common factor, such as $(24x+2)$ or $(12x - 2)$, because we determined at the outset that there are no common factors. We try some of the remaining possibilities:

$$(24x - 1)(x + 2) = 24x^2 + 47x - 2$$

The factorization is $(24x - 1)(x + 2)$.

27. $35x^2 - 57x - 44$

(1) There is no common factor (other than 1 or -1).

(2) Factor the first term, $35x^2$. The possibilities are $35x$, x and $7x$, $5x$. We have these as possibilities for factorizations:

$$(35x+ \quad)(x+ \quad) \text{ and } (7x+ \quad)(5x+ \quad)$$

(3) Factor the last term, -44. The possibilities are 1, -44 and -1, 44 and 2, -22 and -2, 22 and 4, -11, and -4, 11.

These factors can also be written as -44, 1 and 44, -1 and -22, 2 and 22, -2 and -11, 4 and 11, -4.

(4) We try some possibilities:

$$(35x + 1)(x - 44) = 35x^2 - 1539x - 44$$
$$(7x + 1)(5x - 44) = 35x^2 - 303x - 44$$
$$(35x + 2)(x - 22) = 35x^2 - 768x - 44$$
$$(7x + 2)(5x - 22) = 35x^2 - 144x - 44$$

$$(35x + 4)(x - 11) = 35x^2 - 381x - 44$$
$$(7x + 4)(5x - 11) = 35x^2 - 57x - 44$$

The factorization is $(7x + 4)(5x - 11)$.

29. $20 + 6x - 2x^2 = -2x^2 + 6x + 20$

We factor out the common factor, -2. Factoring out -2 rather than 2 gives us a positive leading coefficient.

$$-2(x^2 - 3x - 10)$$

Then we factor the trinomial $x^2 - 3x - 10$. We look for a pair of factors whose product is -10 and whose sum is -3. The numbers are -5 and 2. The factorization of $x^2 - 3x - 10$ is $(x - 5)(x + 2)$. Then $20 + 6x - 2x^2 = -2(x - 5)(x + 2)$. If we think of -2 and $-1 \cdot 2$ then we can write other correct factorizations:

$$20 + 6x - 2x^2$$
$$= 2(-x + 5)(x + 2) \qquad \text{Multiplying } x - 5 \text{ by } -1$$
$$= 2(x - 5)(-x - 2) \qquad \text{Multiplying } x + 2 \text{ by } -1$$

Note that we can also express $2(-x + 5)(x + 2)$ as $2(5 - x)(x + 2)$ since $-x + 5 = 5 - x$ by the commutative law of addition.

31. $12x^2 + 28x - 24$

(1) We factor out the common factor, 4:
   $$4(3x^2 + 7x - 6)$$
   Then we factor the trinomial $3x^2 + 7x - 6$.

(2) Factor $3x^2$. The only possibility is $3x$, x. The desired factorization is of the form:
   $$(3x + \quad)(x + \quad)$$

(3) Factor -6. The possibilities are $6, -1$ and $-6, 1$ and $3, -2$ and $-3, 2$.
   These factors can also be written as $-1, 6$ and $1, -6$ and $-2, 3$ and $2, -3$.

(4) We can immediately reject all possibilities in which either factor has a common factor, such as $(3x + 6)$ or $(3x - 3)$, because we factored out the largest common factor at the outset. We try some of the remaining possibilities:
   $$(3x - 1)(x + 6) = 3x^2 + 17x - 6$$
   $$(3x - 2)(x + 3) = 3x^2 + 7x - 6$$

The factorization of $3x^2 + 7x - 6$ is $(3x - 2)(x + 3)$. We must include the common factor in order to get a factorization of the original trinomial.
$$12x^2 + 28x - 24 = 4(3x - 2)(x + 3)$$

33. $30x^2 - 24x - 54$

(1) We factor out the common factor, 6:
   $$6(5x^2 - 4x - 9)$$
   Then we factor the trinomial $5x^2 - 4x - 9$.

(2) Factor $5x^2$. The only possibility is $5x$, x. The desired factorization is of the form:
   $$(5x + \quad)(x + \quad)$$

(3) Factor -9. The possibilities are $9, -1$ and $-9, 1$ and $-3, 3$.
   These factors can also be written as $-1, 9$ and $1, -9$ and $3, -3$.

(4) We try some possibilities:
   $$(5x + 9)(x - 1) = 5x^2 + 4x - 9$$
   $$(5x - 9)(x + 1) = 5x^2 - 4x - 9$$

The factorization of $5x^2 - 4x - 9$ is $(5x - 9)(x + 1)$. We must include the common factor in order to get a factorization of the original trinomial.
$$30x^2 - 24x - 54 = 6(5x - 9)(x + 1)$$

35. $4y + 6y^2 - 10 = 6y^2 + 4y - 10$

(1) We factor out the common factor, 2:
   $$2(3y^2 + 2y - 5)$$
   Then we factor the trinomial $3y^2 + 2y - 5$.

(2) Factor $3y^2$. The only possibility is $3y$, y. The desired factorization is of the form:
   $$(3y + \quad)(y + \quad)$$

(3) Factor -5. The possibilities are $5, -1$ and $-5, 1$.
   These factors can also be written as $-1, 5$ and $1, -5$.

(4) We try some possibilities:
   $$(3y + 5)(y - 1) = 3y^2 + 2y - 5$$

Then $3y^2 + 2y - 5 = (3y + 5)(y - 1)$, so $6y^2 + 4y - 10 = 2(3y + 5)(y - 1)$.

37. $3x^2 - 4x + 1$

(1) There is no common factor (other than 1 or -1).

(2) Factor the first term, $3x^2$. The only possibility is $3x$, x. The desired factorization is of the form:
   $$(3x + \quad)(x + \quad)$$

(3) Factor the last term, 1. Since 1 is positive and the middle term is negative, we need consider only negative factor pairs. The only such pair is $-1, -1$.

(4) There is only one possibility:
   $$(3x - 1)(x - 1) = 3x^2 - 4x + 1$$

The factorization is $(3x - 1)(x - 1)$.

39. $12x^2 - 28x - 24$

(1) We factor out the common factor, 4:
   $$4(3x^2 - 7x - 6)$$
   Then we factor the trinomial $3x^2 - 7x - 6$.

(2) Factor $3x^2$. The only possibility is $3x$, x. The desired factorization is of the form:
   $$(3x + \quad)(x + \quad)$$

(3) Factor -6. The possibilities are 6, -1 and -6, 1 and 3, -2 and -3, 2.

These factors can also be written as -1, 6 and 1, -6 and -2, 3 and 2, -3.

(4) We can immediately reject all possibilities in which either factor has a common factor, such as $(3x - 6)$ or $(3x + 3)$, because we factored out the largest common factor at the outset. We try some of the remaining possibilities:

$$(3x - 1)(x + 6) = 3x^2 + 17x - 6$$
$$(3x - 2)(x + 3) = 3x^2 + 7x - 6$$
$$(3x + 2)(x - 3) = 3x^2 - 7x - 6$$

Then $3x^2 - 7x - 6 = (3x + 2)(x - 3)$, so $12x^2 - 28x - 24 = 4(3x + 2)(x - 3)$.

41. $-1 + 2x^2 - x = 2x^2 - x - 1$

(1) There is no common factor (other than 1 or -1).

(2) Factor the first term, $2x^2$. The only possibility is $2x$, x. The desired factorization is of the form:

$$(2x+ \quad)(x+ \quad)$$

(3) Factor -1. The only possibility is 1, -1.

(4) We try some possibilities:

$$(2x + 1)(x - 1) = 2x^2 - x - 1$$

The factorization is $(2x + 1)(x - 1)$.

43. $9x^2 - 18x - 16$

(1) There is no common factor (other than 1 or -1).

(2) Factor the first term, $9x^2$. The possibilities are $9x$, x and $3x$, $3x$. We have these as possibilities for factorizations:

$$(9x+ \quad)(x+ \quad) \text{ and } (3x+ \quad)(3x+ \quad)$$

(3) Factor the last term, -16. The possibilities are 16, -1 and -16, 1 and 8, -2 and -8, 2 and 4, -4.

These factors can also be written as -1, 16 and 1, -16 and -2, 8 and and 2, -8 and -4, 4.

(4) We try some possibilities:

$$(9x + 16)(x - 1) = 9x^2 + 7x - 16$$
$$(3x + 16)(3x - 1) = 9x^2 + 45x - 16$$
$$(9x + 8)(x - 2) = 9x^2 - 10x - 16$$
$$(3x + 8)(3x - 2) = 9x^2 + 18x - 16$$
$$(3x - 8)(3x + 2) = 9x^2 - 18x - 16$$

The factorization is $(3x - 8)(3x + 2)$.

45. $15x^2 - 25x - 10$

(1) Factor out the common factor, 5:

$$5(3x^2 - 5x - 2)$$

Then we factor the trinomial $3x^2 - 5x - 2$. This was done in Exercise 15. We know that $3x^2 - 5x - 2 = (3x + 1)(x - 2)$, so $15x^2 - 25x - 10 = 5(3x + 1)(x - 2)$.

47. $12p^3 + 31p^2 + 20p$

(1) We factor out the common factor, p:

$$p(12p^2 + 31p + 20)$$

Then we factor the trinomial $12p^2 + 31p + 20$. This was done in Exercise 17 although the variable is x in that exercise. We know that $12p^2 + 31p + 20 = (3p + 4)(4p + 5)$, so $12p^3 + 31p^2 + 20p = p(3p + 4)(4p + 5)$.

49.
$$16 + 18x - 9x^2 = -9x^2 + 18x + 16$$
$$= -1(9x^2 - 18x - 16)$$
$$= -1(3x - 8)(3x + 2) \quad \text{Using the}$$
$$\text{result from Exercise 43}$$

Other correct factorizations are:

$$16 + 18x - 9x^2$$
$$= (-3x + 8)(3x + 2) \quad \text{Multiplying } 3x - 8 \text{ by } -1$$
$$= (3x - 8)(-3x - 2) \quad \text{Multiplying } 3x + 2 \text{ by } -1$$

We can also express $(-3x + 8)(3x + 2)$ as $(8 - 3x)(3x + 2)$ since $-3x + 8 = 8 - 3x$ by the commutative law of addition.

51. $-15x^2 + 19x - 6 = -1(15x^2 - 19x + 6)$

Now we factor $15x^2 - 19x + 6$.

(1) There is no common factor (other than 1 or -1).

(2) Factor the first term, $15x^2$. The possibilities are $15x$, x and $5x$, $3x$. We have these as possibilities for factorizations:

$$(15x+ \quad)(x+ \quad) \text{ and } (5x+ \quad)(3x+ \quad)$$

(3) Factor the last term, 6. The possibilities are 6, 1 and -6, -1 and 3, 2 and -3, -2.

These factors can also be written as 1, 6 and -1, -6 and 2, 3 and -2, -3.

(4) We try some possibilities:

$$(15x + 1)(x + 6) = 15x^2 + 91x + 6$$
$$(5x + 3)(3x + 2) = 15x^2 - 19x + 6$$
$$(5x - 3)(3x - 2) = 15x^2 - 19x + 6$$

The factorization of $15x^2 - 19x + 6$ is $(5x - 3)(3x - 2)$.

Then $-15x^2 + 19x - 6 = -1(5x - 3)(3x - 2)$. Other correct factorizations are:

$$-15x^2 + 19x - 6$$
$$= (-5x + 3)(3x - 2) \quad \text{Multiplying } 5x - 3 \text{ by } -1$$
$$= (5x - 3)(-3x + 2) \quad \text{Multiplying } 3x - 2 \text{ by } -1$$

Note that we can also express $(-5x + 3)(3x - 2)$ as $(3 - 5x)(3x - 2)$ since $-5x + 3 = 3 - 5x$ by the commutative law of addition. Similarly, we can express $(5x - 3)(-3x + 2)$ as $(5x - 3)(2 - 3x)$.

53. $14x^4 + 19x^3 - 3x^2$

(1) Factor out the common factor, x^2: $x^2(14x^2 + 19x - 3)$

Then we factor the trinomial $14x^2 + 19x - 3$. This was done in Exercise 19. We know that $14x^2 + 19x - 3 = (7x - 1)(2x + 3)$, so $14x^4 + 19x^3 - 3x^2 = x^2(7x - 1)(2x + 3)$.

55. $168x^3 - 45x^2 + 3x$

(1) Factor out the common factor, $3x$:
$$3x(56x^2 - 15x + 1)$$
Then we factor the trinomial $56x^2 - 15x + 1$.

(2) Factor $56x^2$. The possibilities are $56x$, x and $28x$, $2x$ and $14x$, $4x$ and $7x$, $8x$. We have these as possibilities for factorizations:
$$(56x+ \quad)(x+ \quad) \text{ and } (28x+ \quad)(2x+ \quad) \text{ and}$$
$$(14x+ \quad)(4x+ \quad) \text{ and } (7x+ \quad)(8x+ \quad)$$

(3) Factor 1. Since 1 is positive and the middle term is negative we need consider only the negative factor pair $-1, -1$.

(4) We try some possibilities:
$$(56x - 1)(x - 1) = 56x^2 - 57x + 1$$
$$(28x - 1)(2x - 1) = 56x^2 - 30x + 1$$
$$(14x - 1)(4x - 1) = 56x^2 - 18x + 1$$
$$(7x - 1)(8x - 1) = 56x^2 - 15x + 1$$

Then $56x^2 - 15x + 1 = (7x - 1)(8x - 1)$, so
$168x^3 - 45x^2 + 3x = 3x(7x - 1)(8x - 1)$.

57. $15x^4 - 19x^2 + 6 = 15(x^2)^2 - 19x^2 + 6$

(1) There is no common factor (other than 1 or -1).

(2) Factor the first term, $15x^4$. The possibilities are $15x^2$, x^2 and $5x^2$, $3x^2$. We have these as possibilities for factorizations:
$$(15x^2+ \quad)(x^2+ \quad) \text{ and } (5x^2+ \quad)(3x^2+ \quad)$$

(3) Factor 6. Since 6 is positive and the middle term is negative, we need consider only negative factor pairs. Those pairs are $-6, -1$ and $-3, -2$.

These factors can also be written as $-1, -6$ and $-2, -3$.

(4) We can immediately reject all possibilities in which either factor has a common factor, such as $(15x^2 - 6)$ or $(3x^2 - 3)$, because we determined at the outset that there is no common factor. We try some of the remaining possibilities:
$$(15x^2 - 1)(x^2 - 6) = 15x^4 - 91x^2 + 6$$
$$(15x^2 - 2)(x^2 - 3) = 15x^4 - 47x^2 + 6$$
$$(5x^2 - 6)(3x^2 - 1) = 15x^4 - 23x^2 + 6$$
$$(5x^2 - 3)(3x^2 - 2) = 15x^4 - 19x^2 + 6$$

The factorization is $(5x^2 - 3)(3x^2 - 2)$.

59. $25t^2 + 80t + 64$

(1) There is no common factor (other than 1 or -1).

(2) Factor the first term, $25t^2$. The possibilities are $25t$, t and $5t$, $5t$. We have these as possibilities for factorizations:
$$(25t+ \quad)(t+ \quad) \text{ and } (5t+ \quad)(5t+ \quad)$$

(3) Factor the last term, 64. Since all signs are positive, we need consider only positive pairs of factors. Those factor pairs are 64, 1 and 32, 2 and 16, 4 and 8, 8.

These first three pairs can also be written as 1, 64 and 2, 32 and 4, 16.

(4) We try some possibilities:
$$(25t + 64)(t + 1) = 25t^2 + 89t + 64$$
$$(5t + 32)(5t + 2) = 25t^2 + 170t + 64$$
$$(25t + 16)(t + 4) = 25t^2 + 116t + 64$$
$$(5t + 8)(5t + 8) = 25t^2 + 80t + 64$$

The factorization is $(5t + 8)(5t + 8)$ or $(5t + 8)^2$.

61. $6x^3 + 4x^2 - 10x$

(1) Factor out the common factor, $2x$: $2x(3x^2 + 2x - 5)$

Then we factor the trinomial $3x^2 + 2x - 5$. We did this in Exercise 35 (after we factored 2 out of the original trinomial). We know that $3x^2 + 2x - 5 = (3x + 5)(x - 1)$, so $6x^3 + 4x^2 - 10x = 2x(3x + 5)(x - 1)$.

63. $25x^2 + 79x + 64$

We follow the same procedure as in Exercise 59. None of the possibilities works. Thus, $25x^2 + 79x + 64$ is not factorable. It is prime.

65. $6x^2 - 19x - 5$

(1) There is no common factor (other than 1 or -1).

(2) Factor the first term, $6x^2$. The possibilities are $6x$, x and $3x$, $2x$. We have these as possibilities for factorizations:
$$(6x+ \quad)(x+ \quad) \text{ and } (3x+ \quad)(2x+ \quad)$$

(3) Factor the last term, -5. The possibilities are -5, 1 and 5, -1.

These factors can also be written as 1, -5 and -1, 5.

(4) We try some possibilities:
$$(6x - 5)(x + 1) = 6x^2 + x - 5$$
$$(6x + 5)(x - 1) = 6x^2 - x - 5$$
$$(6x + 1)(x - 5) = 6x^2 - 29x - 5$$
$$(6x - 1)(x + 5) = 6x^2 + 29x - 5$$
$$(3x - 5)(2x + 1) = 6x^2 - 7x - 5$$
$$(3x + 5)(2x - 1) = 6x^2 + 7x - 5$$
$$(3x + 1)(2x - 5) = 6x^2 - 13x - 5$$
$$(3x - 1)(2x + 5) = 6x^2 + 13x - 5$$

None of the possibilities works. Thus, $6x^2 - 19x - 5$ is not factorable. It is prime.

67. $12m^2 - mn - 20n^2$

(1) There is no common factor (other than 1 or -1).

(2) Factor the first term, $12m^2$. The possibilities are $12m$, m and $6m$, $2m$ and $3m$, $4m$. We have these as possibilities for factorizations:

$(12m+\ \ )(m+\ \ )$ and $(6m+\ \ )(2m+\ \ )$

and $(3m+\ \ )(4m+\ \ )$

(3) Factor the last term, $-20n^2$. The possibilities are $20n$, $-n$ and $-20n$, n and $10n$, $-2n$ and $-10n$, $2n$ and $5n$, $-4n$ and $-5n$, $4n$.

These factors can also be written as $-n$, $20n$ and n, $-20n$ and $-2n$, $10n$ and $2n$, $-10n$ and $-4n$, $5n$ and $4n$, $-5n$.

(4) We can immediately reject all possibilities in which either factor has a common factor, such as $(12m + 20n)$ or $(4m - 2n)$, because we determined at the outset that there is no common factor. We try some of the remaining possibilities:

$(12m - n)(m + 20n) = 12m^2 + 239mn - 20n^2$
$(12m + 5n)(m - 4n) = 12m^2 - 43mn - 20n^2$
$(3m - 20n)(4m + n) = 12m^2 - 77mn - 20n^2$
$(3m - 4n)(4m + 5n) = 12m^2 - mn - 20n^2$

The factorization is $(3m - 4n)(4m + 5n)$.

69. $6a^2 - ab - 15b^2$

(1) There is no common factor (other than 1 or -1).

(2) Factor the first term, $6a^2$. The possibilities are $6a$, a and $3a$, $2a$. We have these as possibilities for factorizations:

$(6a+\ \ )(a+\ \ )$ and $(3a+\ \ )(2a+\ \ )$

(3) Factor the last term, $-15b^2$. The possibilities are $15b$, $-b$ and $-15b$, b and $5b$, $-3b$ and $-5b$, $3b$.

These factors can also be written as $-b$, $15b$ and b, $-15b$, and $-3b$, $5b$ and $3b$, $-5b$.

(4) We can immediately reject all possibilities in which either factor has a common factor, such as $(6a+15b)$ or $(3a - 3b)$, because we determined at the outset that there is no common factor. We try some of the remaining possibilities:

$(6a - b)(a + 15b) = 6a^2 + 89ab - 15b^2$
$(3a - b)(2a + 15b) = 6a^2 + 43ab - 15b^2$
$(6a + 5b)(a - 3b) = 6a^2 - 13ab - 15b^2$
$(3a + 5b)(2a - 3b) = 6a^2 + ab - 15b^2$
$(3a - 5b)(2a + 3b) = 6a^2 - ab - 15b^2$

The factorization is $(3a - 5b)(2a + 3b)$.

71. $9a^2 + 18ab + 8b^2$

(1) There is no common factor (other than 1 or -1).

(2) Factor the first term, $9a^2$. The possibilities are $9a$, a and $3a$, $3a$. We have these as possibilities for factorizations:

$(9a+\ \ )(a+\ \ )$ and $(3a+\ \ )(3a+\ \ )$

(3) Factor $8b^2$. Since all signs are positive, we need consider only pairs of factors with positive coefficients. Those factor pairs are $8b$, b and $4b$, $2b$.

These factors can also be written as b, $8b$ and $2b$, $4b$.

(4) We try some possibilities:

$(9a + 8b)(a + b) = 9a^2 + 17ab + 8b^2$
$(3a + 8b)(3a + b) = 9a^2 + 27ab + 8b^2$
$(9a + 4b)(a + 2b) = 9a^2 + 22ab + 8b^2$
$(3a + 4b)(3a + 2b) = 9a^2 + 18ab + 8b^2$

The factorization is $(3a + 4b)(3a + 2b)$.

73. $35p^2 + 34pq + 8q^2$

(1) There is no common factor (other than 1 or -1).

(2) Factor the first term, $35p^2$. The possibilities are $35p$, p and $7p$, $5p$. We have these as possibilities for factorizations:

$(35p+\ \ )(p+\ \ )$ and $(7p+\ \ )(5p+\ \ )$

(3) Factor $8q^2$. Since all signs are positive, we need consider only pairs of factors with positive coefficients. Those factor pairs are $8q$, q and $4q$, $2q$.

These factors can also be written as q, $8q$ and $2q$, $4q$.

(4) We try some possibilities:

$(35p + 8q)(p + q) = 35p^2 + 43pq + 8q^2$
$(7p + 8q)(5p + q) = 35p^2 + 47pq + 8q^2$
$(35p + 4q)(p + 2q) = 35p^2 + 74pq + 8q^2$
$(7p + 4q)(5p + 2q) = 35p^2 + 34pq + 8p^2$

The factorization is $(7p + 4q)(5p + 2q)$.

75. $18x^2 - 6xy - 24y^2$

(1) Factor out the common factor, 6:

$6(3x^2 - xy - 4y^2)$

Then we factor the trinomial $3x^2 - xy - 4y^2$.

(2) Factor $3x^2$. The only possibility is $3x$, x. The desired factorization is of the form:

$(3x+\ \ )(x+\ \ )$

(3) Factor $-4y^2$. The possibilities are $4y$, $-y$ and $-4y$, y and $2y$, $-2y$.

These factors can also be written as $-y$, $4y$ and y, $-4y$ and $-2y$, $2y$.

(4) We try some possibilities:

$(3x + 4y)(x - y) = 3x^2 + xy - 4y^2$
$(3x - 4y)(x + y) = 3x^2 - xy - 4y^2$

Then $3x^2 - xy - 4y^2 = (3x - 4y)(x + y)$, so
$18x^2 - 6xy - 24y^2 = 6(3x - 4y)(x + y)$.

77. Discussion and Writing Exercise

79.
$$A = pq - 7$$
$$A + 7 = pq \quad \text{Adding 7}$$
$$\frac{A+7}{p} = q \quad \text{Dividing by } p$$

81. $3x + 2y = 6$
$$2y = 6 - 3x \quad \text{Subtracting } 3x$$
$$y = \frac{6-3x}{2} \quad \text{Dividing by 2}$$

83. $5 - 4x < -11$
$$-4x < -16 \quad \text{Subtracting 5}$$
$$x > 4 \quad \text{Dividing by } -4 \text{ and reversing the inequality symbol}$$

The solution set is $\{x|x > 4\}$.

85. Graph: $y = \frac{2}{5}x - 1$

Because the equation is in the form $y = mx + b$, we know the y-intercept is $(0, -1)$. We find two other points on the line, substituting multiples of 5 for x to avoid fractions.

When $x = -5$, $y = \frac{2}{5}(-5) - 1 = -2 - 1 = -3$.

When $x = 5$, $y = \frac{2}{5}(5) - 1 = 2 - 1 = 1$.

x	y
0	-1
-5	-3
5	1

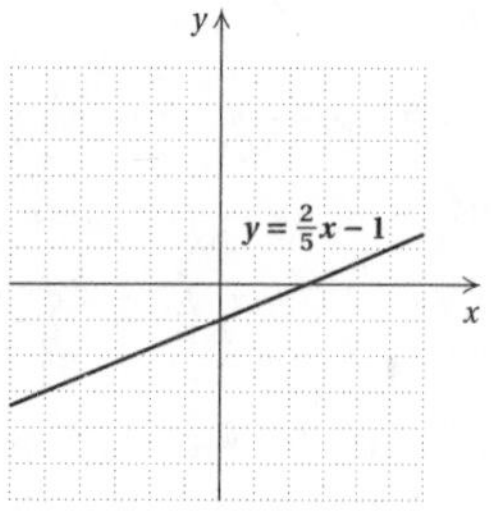

87. $4x - 16y = 64$

To find the x-intercept, let $y = 0$ and solve for x.
$$4x - 16y = 64$$
$$4x - 16 \cdot 0 = 64$$
$$4x = 64$$
$$x = 16$$

The x-intercept is $(16, 0)$.

To find the y-intercept, let $x = 0$ and solve for y.
$$4x - 16y = 64$$
$$4 \cdot 0 - 16y = 64$$
$$-16y = 64$$
$$y = -4$$

The y-intercept is $(0, -4)$.

89. $x - 1.3y = 6.5$

To find the x-intercept, let $y = 0$ and solve for x.
$$x - 1.3y = 6.5$$
$$x - 1.3(0) = 6.5$$
$$x = 6.5$$

The x-intercept is $(6.5, 0)$.

To find the y-intercept, let $x = 0$ and solve for y.
$$x - 1.3y = 6.5$$
$$0 - 1.3y = 6.5$$
$$-1.3y = 6.5$$
$$y = -5$$

The y-intercept is $(0, -5)$.

91. $y = 4 - 5x$

To find the x-intercept, let $y = 0$ and solve for x.
$$y = 4 - 5x$$
$$0 = 4 - 5x$$
$$5x = 4$$
$$x = \frac{4}{5}$$

The x-intercept is $\left(\frac{4}{5}, 0\right)$.

To find the y-intercept, let $x = 0$ and solve for y.
$$y = 4 - 5x$$
$$y = 4 - 5 \cdot 0$$
$$y = 4$$

The y-intercept is $(0, 4)$.

93. $20x^{2n} + 16x^n + 3 = 20(x^n)^2 + 16x^n + 3$

(1) There is no common factor (other than 1 and -1).

(2) Factor the first term, $20x^{2n}$. The possibilities are $20x^n$, x^n and $10x^n$, $2x^n$ and $5x^n$, $4x^n$. We have these as possibilities for factorizations:

$(20x^n + \quad)(x^n + \quad)$ and $(10x^n + \quad)(2x^n + \quad)$

and $(5x^n + \quad)(4x^n + \quad)$

(3) Factor the last term, 3. Since all signs are positive, we need consider only the positive factor pair 3, 1.

(4) We try some possibilities:
$$(20x^n + 3)(x^n + 1) = 20x^{2n} + 23x^n + 3$$
$$(10x^n + 3)(2x^n + 1) = 20x^{2n} + 16x^n + 3$$

The factorization is $(10x^n + 3)(2x^n + 1)$.

95. $3x^{6a} - 2x^{3a} - 1 = 3(x^{3a})^2 - 2x^{3a} - 1$

(1) There is no common factor (other than 1 or -1).

(2) Factor the first term, $3x^{6a}$. The only possibility is $3x^{3a}$, x^{3a}. The desired factorization is of the form:
$$(3x^{3a} + \quad)(x^{3a} + \quad)$$

(3) Factor the last term, -1. The only possibility is -1, 1.

(4) We try the possibilities:
$$(3x^{3a} - 1)(x^{3a} + 1) = 3x^{6a} + 2x^{3a} - 1$$
$$(3x^{3a} + 1)(x^{3a} - 1) = 3x^{6a} - 2x^{3a} - 1$$

The factorization is $(3x^{3a} + 1)(x^{3a} - 1)$.

97.-105. Left to the student

Exercise Set 5.4

1. $x^2 + 2x + 7x + 14 = (x^2 + 2x) + (7x + 14)$
$$= x(x + 2) + 7(x + 2)$$
$$= (x + 7)(x + 2)$$

3. $x^2 - 4x - x + 4 = (x^2 - 4x) + (-x + 4)$
$$= x(x - 4) - 1(x - 4)$$
$$= (x - 1)(x - 4)$$

5. $6x^2 + 4x + 9x + 6 = (6x^2 + 4x) + (9x + 6)$
$$= 2x(3x + 2) + 3(3x + 2)$$
$$= (2x + 3)(3x + 2)$$

7. $3x^2 - 4x - 12x + 16 = (3x^2 - 4x) + (-12x + 16)$
$$= x(3x - 4) - 4(3x - 4)$$
$$= (x - 4)(3x - 4)$$

9. $35x^2 - 40x + 21x - 24 = (35x^2 - 40x) + (21x - 24)$
$$= 5x(7x - 8) + 3(7x - 8)$$
$$= (5x + 3)(7x - 8)$$

11. $4x^2 + 6x - 6x - 9 = (4x^2 + 6x) + (-6x - 9)$
$$= 2x(2x + 3) - 3(2x + 3)$$
$$= (2x - 3)(2x + 3)$$

13. $2x^4 + 6x^2 + 5x^2 + 15 = (2x^4 + 6x^2) + (5x^2 + 15)$
$$= 2x^2(x^2 + 3) + 5(x^2 + 3)$$
$$= (2x^2 + 5)(x^2 + 3)$$

15. $2x^2 + 7x - 4$

(1) First factor out a common factor, if any. There is none (other than 1 or -1).

(2) Multiply the leading coefficient, 2 and the constant, -4: $2(-4) = -8$.

(3) Look for a factorization of -8 in which the sum of the factors is the coefficient of the middle term, 7.

Pairs of factors	Sums of factors
-1, 8	7
1, -8	-7
-2, 4	2
2, -4	-2

(4) Split the middle term: $7x = -1x + 8x$

(5) Factor by grouping:
$$2x^2 + 7x - 4 = 2x^2 - x + 8x - 4$$
$$= (2x^2 - x) + (8x - 4)$$
$$= x(2x - 1) + 4(2x - 1)$$
$$= (x + 4)(2x - 1)$$

17. $3x^2 - 4x - 15$

(1) First factor out a common factor, if any. There is none (other than 1 or -1).

(2) Multiply the leading coefficient, 3, and the constant, -15: $3(-15) = -45$.

(3) Look for a factorization of -45 in which the sum of the factors is the coefficient of the middle term, -4.

Pairs of factors	Sums of factors
-1, 45	44
1, -45	-44
-3, 15	12
3, -15	-12
-5, 9	4
5, -9	-4

(4) Split the middle term: $-4x = 5x - 9x$

(5) Factor by grouping:
$$3x^2 - 4x - 15 = 3x^2 + 5x - 9x - 15$$
$$= (3x^2 + 5x) + (-9x - 15)$$
$$= x(3x + 5) - 3(3x + 5)$$
$$= (x - 3)(3x + 5)$$

19. $6x^2 + 23x + 7$

(1) First factor out a common factor, if any. There is none (other than 1 or -1).

(2) Multiply the leading coefficient, 6, and the constant, 7: $6 \cdot 7 = 42$.

(3) Look for a factorization of 42 in which the sum of the factors is the coefficient of the middle term, 23. We only need to consider positive factors.

Pairs of factors	Sums of factors
1, 42	43
2, 21	23
3, 14	17
6, 7	13

(4) Split the middle term: $23x = 2x + 21x$

(5) Factor by grouping:
$$6x^2 + 23x + 7 = 6x^2 + 2x + 21x + 7$$
$$= (6x^2 + 2x) + (21x + 7)$$
$$= 2x(3x + 1) + 7(3x + 1)$$
$$= (2x + 7)(3x + 1)$$

21. $3x^2 - 4x + 1$

(1) First factor out a common factor, if any. There is none (other than 1 or -1).

(2) Multiply the leading coefficient, 3, and the constant, 1: $3 \cdot 1 = 3$.

(3) Look for a factorization of 3 in which the sum of the factors is the coefficient of the middle term, -4. The numbers we want are -1 and -3: $-1 \cdot (-3) = 3$ and $-1 + (-3) = -4$.

(4) Split the middle term: $-4x = -1x - 3x$

(5) Factor by grouping:
$$3x^2 - 4x + 1 = 3x^2 - x - 3x + 1$$
$$= (3x^2 - x) + (-3x + 1)$$
$$= x(3x - 1) - 1(3x - 1)$$
$$= (x - 1)(3x - 1)$$

23. $4x^2 - 4x - 15$

(1) First factor out a common factor, if any. There is none (other than 1 or -1).

(2) Multiply the leading coefficient, 4, and the constant, -15: $4(-15) = -60$.

(3) Look for a factorization of -60 in which the sum of the factors is the coefficient of the middle term, -4.

Pairs of factors	Sums of factors
$-1,\quad 60$	59
$1, -60$	-59
$-2,\quad 30$	28
$2, -30$	-28
$-3,\quad 20$	17
$3, -20$	-17
$-4,\quad 15$	11
$4, -15$	-11
$-5,\quad 12$	7
$5, -12$	-7
$-6,\quad 10$	4
$6, -10$	-4

(4) Split the middle term: $-4x = 6x - 10x$

(5) Factor by grouping:
$$4x^2 - 4x - 15 = 4x^2 + 6x - 10x - 15$$
$$= (4x^2 + 6x) + (-10x - 15)$$
$$= 2x(2x + 3) - 5(2x + 3)$$
$$= (2x - 5)(2x + 3)$$

25. $2x^2 + x - 1$

(1) First factor out a common factor, if any. There is none (other than 1 or -1).

(2) Multiply the leading coefficient, 2, and the constant, -1: $2(-1) = -2$.

(3) Look for a factorization of -2 in which the sum of the factors is the coefficient of the middle term, 1. The numbers we want are 2 and -1: $2(-1) = -2$ and $2 - 1 = 1$.

(4) Split the middle term: $x = 2x - 1x$

(5) Factor by grouping:
$$2x^2 + x - 1 = 2x^2 + 2x - x - 1$$
$$= (2x^2 + 2x) + (-x - 1)$$
$$= 2x(x + 1) - 1(x + 1)$$
$$= (2x - 1)(x + 1)$$

27. $9x^2 - 18x - 16$

(1) First factor out a common factor, if any. There is none (other than 1 or -1).

(2) Multiply the leading coefficient, 9, and the constant, -16: $9(-16) = -144$.

(3) Look for a factorization of -144, so the sum of the factors is the coefficient of the middle term, -18.

Pairs of factors	Sums of factors
$-1,\quad 144$	143
$1, -144$	-143
$-2,\quad 72$	70
$2, -72$	-70
$-3,\quad 48$	45
$3, -48$	-45
$-4,\quad 36$	32
$4, -36$	-32
$-6,\quad 24$	18
$6, -24$	-18
$-8,\quad 18$	10
$8, -18$	-10
$-9,\quad 16$	7
$9, -16$	-7
$-12,\quad 12$	0

(4) Split the middle term: $-18x = 6x - 24x$

(5) Factor by grouping:
$$9x^2 - 18x - 16 = 9x^2 + 6x - 24x - 16$$
$$= (9x^2 + 6x) + (-24x - 16)$$
$$= 3x(3x + 2) - 8(3x + 2)$$
$$= (3x - 8)(3x + 2)$$

29. $3x^2 + 5x - 2$

(1) First factor out a common factor, if any. There is none (other than 1 or -1).

(2) Multiply the leading coefficient, 3, and the constant, -2: $3(-2) = -6$.

(3) Look for a factorization of -6 in which the sum of the factors is the coefficient of the middle term, 5. The numbers we want are -1 and 6: $-1(6) = -6$ and $-1 + 6 = 5$.

(4) Split the middle term: $5x = -1x + 6x$

(5) Factor by grouping:
$$3x^2 + 5x - 2 = 3x^2 - x + 6x - 2$$
$$= (3x^2 - x) + (6x - 2)$$
$$= x(3x - 1) + 2(3x - 1)$$
$$= (x + 2)(3x - 1)$$

31. $12x^2 - 31x + 20$

(1) First factor out a common factor, if any. There is none (other than 1 or -1).

(2) Multiply the leading coefficient, 12, and the constant, 20: $12 \cdot 20 = 240$.

(3) Look for a factorization of 240 in which the sum of the factors is the coefficient of the middle term, -31. We only need to consider negative factors.

Pairs of factors	Sums of factors
$-1, -240$	-241
$-2, -120$	-122
$-3, -8$	-83
$-4, -60$	-64
$-5, -48$	-53
$-6, -40$	-46
$-8, -30$	-38
$-10, -24$	-34
$-12, -20$	-32
$-15, -16$	-31

(4) Split the middle term: $-31x = -15x - 16x$

(5) Factor by grouping:
$$12x^2 - 31x + 20 = 12x^2 - 15x - 16x + 20$$
$$= (12x^2 - 15x) + (-16x + 20)$$
$$= 3x(4x - 5) - 4(4x - 5)$$
$$= (3x - 4)(4x - 5)$$

33. $14x^2 - 19x - 3$

(1) First factor out a common factor, if any. There is none (other than 1 or -1).

(2) Multiply the leading coefficient, 14, and the constant, -3: $14(-3) = -42$.

(3) Look for a factorization of -42 so that the sum of the factors is the coefficient of the middle term, -19.

Pairs of factors	Sums of factors
$-1, 42$	41
$1, -42$	-41
$-2, 21$	19
$2, -21$	-19
$-3, 14$	11
$3, -14$	-11
$-6, 7$	1
$6, -7$	-1

(4) Split the middle term: $-19x = 2x - 21x$

(5) Factor by grouping:
$$14x^2 - 19x - 3 = 14x^2 + 2x - 21x - 3$$
$$= (14x^2 + 2x) + (-21x - 3)$$
$$= 2x(7x + 1) - 3(7x + 1)$$
$$= (2x - 3)(7x + 1)$$

35. $9x^2 + 18x + 8$

(1) First factor out a common factor, if any. There is none (other than 1 or -1).

(2) Multiply the leading coefficient, 9, and the constant, 8: $9 \cdot 8 = 72$.

(3) Look for a factorization of 72 in which the sum of the factors is the coefficient of the middle term, 18. We only need to consider positive factors.

Pairs of factors	Sums of factors
$1, 72$	73
$2, 36$	38
$3, 24$	27
$4, 18$	22
$6, 12$	18
$8, 9$	17

(4) Split the middle term: $18x = 6x + 12x$

(5) Factor by grouping:
$$9x^2 + 18x + 8 = 9x^2 + 6x + 12x + 8$$
$$= (9x^2 + 6x) + (12x + 8)$$
$$= 3x(3x + 2) + 4(3x + 2)$$
$$= (3x + 4)(3x + 2)$$

37. $49 - 42x + 9x^2 = 9x^2 - 42x + 49$

(1) First factor out a common factor, if any. There is none (other than 1 or -1).

(2) Multiply the leading coefficient, 9, and the constant, 49: $9 \cdot 49 = 441$.

(3) Look for a factorization of 441 in which the sum of the factors is the coefficient of the middle term, -42. We only need to consider negative factors.

Pairs of factors	Sums of factors
$-1, -441$	-442
$-3, -147$	-150
$-7, -63$	-70
$-9, -49$	-58
$-21, -21$	-42

(4) Split the middle term: $-42x = -21x - 21x$

(5) Factor by grouping:
$$9x^2 - 42x + 49 = 9x^2 - 21x - 21x + 49$$
$$= (9x^2 - 21x) + (-21x + 49)$$
$$= 3x(3x - 7) - 7(3x - 7)$$
$$= (3x - 7)(3x - 7), \text{ or}$$
$$(3x - 7)^2$$

39. $24x^2 - 47x - 2$

(1) First factor out a common factor, if any. There is none (other than 1 or -1).

(2) Multiply the leading coefficient, 24, and the constant, -2: $24(-2) = -48$.

(3) Look for a factorization of -48 in which the sum of the factors is the coefficient of the middle term, -47. The numbers we want are -48 and 1: $-48 \cdot 1 = -48$ and $-48 + 1 = -47$.

(4) Split the middle term: $-47x = -48x + 1x$

(5) Factor by grouping:
$$24x^2 - 47x - 2 = 24x^2 - 48x + x - 2$$
$$= (24x^2 - 48x) + (x - 2)$$
$$= 24x(x - 2) + 1(x - 2)$$
$$= (24x + 1)(x - 2)$$

41. $5 - 9a^2 - 12a = -9a^2 - 12a + 5 = -1(9a^2 + 12a - 5)$

Now we factor $9a^2 + 12a - 5$.

(1) We have already factored out the common factor, -1, to make the leading coefficient positive.

(2) Multiply the leading coefficient, 9, and the constant, -5: $9(-5) = -45$.

(3) Look for a factorization of -45 in which the sum of the factors is the coefficient of the middle term, 12. The numbers we want are 15 and -3: $15(-3) = -45$ and $15 + (-3) = 12$.

(4) Split the middle term: $12a = 15a - 3a$

(5) Factor by grouping:
$$9a^2 + 12a - 5 = 9a^2 + 15a - 3a - 5$$
$$= (9a^2 + 15a) + (-3a - 5)$$
$$= 3a(3a + 5) - (3a + 5)$$
$$= (3a - 1)(3a + 5)$$

Then we have
$$5 - 9a^2 - 12a$$
$$= -1(3a - 1)(3a + 5)$$
$$= (-3a + 1)(3a + 5) \quad \text{Multiplying } 3a - 1 \text{ by } -1$$
$$= (3a - 1)(-3a - 5) \quad \text{Multiplying } 3a + 5 \text{ by } -1$$

Note that we can also express $(-3a + 1)(3a + 5)$ as $(1 - 3a)(3a + 5)$ since $-3a + 1 = 1 - 3a$ by the commutative law of addition.

43. $20 + 6x - 2x^2 = -2x^2 + 6x + 20$

(1) Factor out the common factor -2. We factor out -2 rather than 2 in order to make the leading coefficient of the trinomial factor positive.
$$-2x^2 + 6x + 20 = -2(x^2 - 3x - 10)$$
To factor $x^2 - 3x - 10$, we look for two factors of -10 whose sum is -3. The numbers we want are -5 and 2. Then $x^2 - 3x - 10 = (x - 5)(x + 2)$, so we have:
$$20 + 6x - 2x^2$$
$$= -2(x - 5)(x + 2)$$
$$= 2(-x + 5)(x + 2) \quad \text{Multiplying } x - 5 \text{ by } -1$$
$$= 2(x - 5)(-x - 2) \quad \text{Multiplying } x + 2 \text{ by } -1$$
Note that we can also express $2(-x + 5)(x + 2)$ as $2(5 - x)(x + 2)$ since $-x + 5 = 5 - x$ by the commutative law of addition.

45. $12x^2 + 28x - 24$

(1) Factor out the common factor, 4:
$$12x^2 + 28x - 24 = 4(3x^2 + 7x - 6)$$

(2) Now we factor the trinomial $3x^2 + 7x - 6$. Multiply the leading coefficient, 3, and the constant, -6: $3(-6) = -18$.

(3) Look for a factorization of -18 in which the sum of the factors is the coefficient of the middle term, 7. The numbers we want are 9 and -2: $9(-2) = -18$ and $9 + (-2) = 7$.

(4) Split the middle term: $7x = 9x - 2x$

(5) Factor by grouping:
$$3x^2 + 7x - 6 = 3x^2 + 9x - 2x - 6$$
$$= (3x^2 + 9x) + (-2x - 6)$$
$$= 3x(x + 3) - 2(x + 3)$$
$$= (3x - 2)(x + 3)$$

We must include the common factor to get a factorization of the original trinomial.
$$12x^2 + 28x - 24 = 4(3x - 2)(x + 3)$$

47. $30x^2 - 24x - 54$

(1) Factor out the common factor, 6.
$$30x^2 - 24x - 54 = 6(5x^2 - 4x - 9)$$

(2) Now we factor the trinomial $5x^2 - 4x - 9$. Multiply the leading coefficient, 5, and the constant, -9: $5(-9) = -45$.

(3) Look for a factorization of -45 in which the sum of the factors is the coefficient of the middle term, -4. The numbers we want are -9 and 5: $-9 \cdot 5 = -45$ and $-9 + 5 = -4$.

(4) Split the middle term: $-4x = -9x + 5x$

(5) Factor by grouping:
$$5x^2 - 4x - 9 = 5x^2 - 9x + 5x - 9$$
$$= (5x^2 - 9x) + (5x - 9)$$
$$= x(5x - 9) + (5x - 9)$$
$$= (x + 1)(5x - 9)$$

We must include the common factor to get a factorization of the original trinomial.
$$30x^2 - 24x - 54 = 6(x + 1)(5x - 9)$$

49. $4y + 6y^2 - 10 = 6y^2 + 4y - 10$

(1) Factor out the common factor, 2.
$$6y^2 + 4y - 10 = 2(3y^2 + 2y - 5)$$

(2) Now we factor the trinomial $3y^2 + 2y - 5$. Multiply the leading coefficient, 3, and the constant, -5: $3(-5) = -15$.

(3) Look for a factorization of -15 in which the sum of the factors is the coefficient of the middle term, 2. The numbers we want are 5 and -3: $5(-3) = -15$ and $5 + (-3) = 2$.

(4) Split the middle term: $2y = 5y - 3y$

(5) Factor by grouping:
$$3y^2 + 2y - 5 = 3y^2 + 5y - 3y - 5$$
$$= (3y^2 + 5y) + (-3y - 5)$$
$$= y(3y + 5) - (3y + 5)$$
$$= (y - 1)(3y + 5)$$

We must include the common factor to get a factorization of the original trinomial.

$$4y + 6y^2 - 10 = 2(y - 1)(3y + 5)$$

51. $3x^2 - 4x + 1$

(1) There is no common factor (other than 1 or −1).

(2) Multiply the leading coefficient, 3, and the constant, 1: $3 \cdot 1 = 3$.

(3) Look for a factorization of 3 in which the sum of the factors is the coefficient of the middle term, −4. The numbers we want are −1 and −3: $-1(-3) = 3$ and $-1 + (-3) = -4$.

(4) Split the middle term: $-4x = -1x - 3x$

(5) Factor by grouping:
$$3x^2 - 4x + 1 = 3x^2 - x - 3x + 1$$
$$= (3x^2 - x) + (-3x + 1)$$
$$= x(3x - 1) - (3x - 1)$$
$$= (x - 1)(3x - 1)$$

53. $12x^2 - 28x - 24$

(1) Factor out the common factor, 4:
$$12x^2 - 28x - 24 = 4(3x^2 - 7x - 6)$$

(2) Now we factor the trinomial $3x^2 - 7x - 6$. Multiply the leading coefficient, 3, and the constant, −6: $3(-6) = -18$.

(3) Look for a factorization of −18 in which the sum of the factors is the coefficient of the middle term, −7. The numbers we want are −9 and 2: $-9 \cdot 2 = -18$ and $-9 + 2 = -7$.

(4) Split the middle term: $-7x = -9x + 2x$

(5) Factor by grouping:
$$3x^2 - 7x - 6 = 3x^2 - 9x + 2x - 6$$
$$= (3x^2 - 9x) + (2x - 6)$$
$$= 3x(x - 3) + 2(x - 3)$$
$$= (3x + 2)(x - 3)$$

We must include the common factor to get a factorization of the original trinomial.

$$12x^2 - 28x - 24 = 4(3x + 2)(x - 3)$$

55. $-1 + 2x^2 - x = 2x^2 - x - 1$

(1) There is no common factor (other than 1 or −1).

(2) Multiply the leading coefficient, 2, and the constant, −1: $2(-1) = -2$.

(3) Look for a factorization of −2 in which the sum of the factors is the coefficient of the middle term, −1. The numbers we want are −2 and 1: $-2 \cdot 1 = -2$ and $-2 + 1 = -1$.

(4) Split the middle term: $-x = -2x + 1x$

(5) Factor by grouping:
$$2x^2 - x - 1 = 2x^2 - 2x + x - 1$$
$$= (2x^2 - 2x) + (x - 1)$$
$$= 2x(x - 1) + (x - 1)$$
$$= (2x + 1)(x - 1)$$

57. $9x^2 + 18x - 16$

(1) There is no common factor (other than 1 or −1).

(2) Multiply the leading coefficient, 9, and the constant, −16: $9(-16) = -144$.

(3) Look for a factorization of −144 in which the sum of the factors is the coefficient of the middle term, 18. The numbers we want are 24 and −6: $24(-6) = -144$ and $24 + (-6) = 18$.

(4) Split the middle term: $18x = 24x - 6x$

(5) Factor by grouping:
$$9x^2 + 18x - 16 = 9x^2 + 24x - 6x - 16$$
$$= (9x^2 + 24x) + (-6x - 16)$$
$$= 3x(3x + 8) - 2(3x + 8)$$
$$= (3x - 2)(3x + 8)$$

59. $15x^2 - 25x - 10$

(1) Factor out the common factor, 5:
$$15x^2 - 25x - 10 = 5(3x^2 - 5x - 2)$$

(2) Now we factor the trinomial $3x^2 - 5x - 2$. Multiply the leading coefficient, 3, and the constant, −2: $3(-2) = -6$.

(3) Look for a factorization of −6 in which the sum of the factors is the coefficient of the middle term, −5. The numbers we want are −6 and 1: $-6 \cdot 1 = -6$ and $-6 + 1 = -5$.

(4) Split the middle term: $-5x = -6x + 1x$

(5) Factor by grouping:
$$3x^2 - 5x - 2 = 3x^2 - 6x + x - 2$$
$$= (3x^2 - 6x) + (x - 2)$$
$$= 3x(x - 2) + (x - 2)$$
$$= (3x + 1)(x - 2)$$

We must include the common factor to get a factorization of the original trinomial.

$$15x^2 - 25x - 10 = 5(3x + 1)(x - 2)$$

61. $12p^3 + 31p^2 + 20p$

(1) Factor out the common factor, p:
$$12p^3 + 31p^2 + 20p = p(12p^2 + 31p + 20)$$

(2) Now we factor the trinomial $12p^2 + 31p + 20$. Multiply the leading coefficient, 12, and the constant, 20: $12 \cdot 20 = 240$.

(3) Look for a factorization of 240 in which the sum of the factors is the coefficient of the middle term, 31. The numbers we want are 15 and 16: $15 \cdot 16 = 240$ and $15 + 16 = 31$.

(4) Split the middle term: $31p = 15p + 16p$

(5) Factor by grouping:
$$12p^2 + 31p + 20 = 12p^2 + 15p + 16p + 20$$
$$= (12p^2 + 15p) + (16p + 20)$$
$$= 3p(4p + 5) + 4(4p + 5)$$
$$= (3p + 4)(4p + 5)$$

We must include the common factor to get a factorization of the original trinomial.
$$12p^3 + 31p^2 + 20p = p(3p + 4)(4p + 5)$$

63. $4 - x - 5x^2 = -5x^2 - x + 4$

(1) Factor out -1 to make the leading coefficient positive:
$$-5x^2 - x + 4 = -1(5x^2 + x - 4)$$

(2) Now we factor the trinomial $5x^2 + x - 4$. Multiply the leading coefficient, 5, and the constant, -4: $5(-4) = -20$.

(3) Look for a factorization of -20 in which the sum of the factors is the coefficient of the middle term, 1. The numbers we want are 5 and -4: $5(-4) = -20$ and $5 + (-4) = 1$.

(4) Split the middle term: $x = 5x - 4x$

(5) Factor by grouping:
$$5x^2 + x - 4 = 5x^2 + 5x - 4x - 4$$
$$= (5x^2 + 5x) + (-4x - 4)$$
$$= 5x(x + 1) - 4(x + 1)$$
$$= (5x - 4)(x + 1)$$

We must include the common factor to get a factorization of the original trinomial.
$$4 - x - 5x^2$$
$$= -1(5x - 4)(x + 1)$$
$$= (-5x + 4)(x + 1) \quad \text{Multiplying } 5x - 4 \text{ by } -1$$
$$= (5x - 4)(-x - 1) \quad \text{Multiplying } x + 1 \text{ by } -1$$

Note that we can also express $(-5x + 4)(x + 1)$ as $(4 - 5x)(x + 1)$ since $-5x + 4 = 4 - 5x$ by the commutative law of addition.

65. $33t - 15 - 6t^2 = -6t^2 + 33t - 15$

(1) Factor out the common factor, -3. We factor out -3 rather than 3 in order to make the leading coefficient of the trinomial factor positive.
$$-6t^2 + 33t - 15 = -3(2t^2 - 11t + 5)$$

(2) Now we factor the trinomial $2t^2 - 11t + 5$. Multiply the leading coefficient, 2, and the constant, 5: $2 \cdot 5 = 10$.

(3) Look for a factorization of 10 in which the factors is the coefficient of the middle term, -11. The numbers we want are -1 and -10: $-1(-10) = 10$ and $-1 + (-10) = -11$.

(4) Split the middle term: $-11t = -1t - 10t$

(5) Factor by grouping:
$$2t^2 - 11t + 5 = 2t^2 - t - 10t + 5$$
$$= (2t^2 - t) + (-10t + 5)$$
$$= t(2t - 1) - 5(2t - 1)$$
$$= (t - 5)(2t - 1)$$

We must include the common factor to get a factorization of the original trinomial.
$$33t - 15 - 6t^2$$
$$= -3(t - 5)(2t - 1)$$
$$= 3(-t + 5)(2t - 1) \quad \text{Multiplying } t - 5 \text{ by } -1$$
$$= 3(t - 5)(-2t + 1) \quad \text{Multiplying } 2t - 1 \text{ by } -1$$

Note that we can also express $3(-t + 5)(2t - 1)$ as $3(5 - t)(2t - 1)$ since $-t + 5 = 5 - t$ by the commutative law of addition. Similarly, we can express $3(t - 5)(-2t + 1)$ as $3(t - 5)(1 - 2t)$.

67. $14x^4 + 19x^3 - 3x^2$

(1) Factor out the common factor, x^2:
$$14x^4 + 19x^3 - 3x^2 = x^2(14x^2 + 19x - 3)$$

(2) Now we factor the trinomial $14x^2 + 19x - 3$. Multiply the leading coefficient, 14, and the constant, -3: $14(-3) = -42$.

(3) Look for a factorization of -42 in which the sum of the factors is the coefficient of the middle term, 19. The numbers we want are 21 and -2: $21(-2) = -42$ and $21 + (-2) = 19$.

(4) Split the middle term: $19x = 21x - 2x$

(5) Factor by grouping:
$$14x^2 + 19x - 3 = 14x^2 + 21x - 2x - 3$$
$$= (14x^2 + 21x) + (-2x - 3)$$
$$= 7x(2x + 3) - (2x + 3)$$
$$= (7x - 1)(2x + 3)$$

We must include the common factor to get a factorization of the original trinomial.
$$14x^4 + 19x^3 - 3x^2 = x^2(7x - 1)(2x + 3)$$

69. $168x^3 - 45x^2 + 3x$

(1) Factor out the common factor, $3x$:
$$168x^3 - 45x^2 + 3x = 3x(56x^2 - 15x + 1)$$

(2) Now we factor the trinomial $56x^2 - 15x + 1$. Multiply the leading coefficient, 56, and the constant, 1: $56 \cdot 1 = 56$.

(3) Look for a factorization of 56 in which the sum of the factors is the coefficient of the middle term, -15. The numbers we want are -7 and -8: $-7(-8) = 56$ and $-7 + (-8) = -15$.

(4) Split the middle term: $-15x = -7x - 8x$

(5) Factor by grouping:

$$56x^2 - 15x + 1 = 56x^2 - 7x - 8x + 1$$
$$= (56x^2 - 7x) + (-8x + 1)$$
$$= 7x(8x - 1) - (8x - 1)$$
$$= (7x - 1)(8x - 1)$$

We must include the common factor to get a factorization of the original trinomial.

$$168x^3 - 45x^2 + 3x = 3x(7x - 1)(8x - 1)$$

71. $15x^4 - 19x^2 + 6$

(1) There are no common factors (other than 1 or -1).

(2) Multiply the leading coefficient, 15, and the constant, 6: $15 \cdot 6 = 90$.

(3) Look for a factorization of 90 in which the sum of the factors is the coefficient of the middle term, -19. The numbers we want are -9 and -10: $-9(-10) = 90$ and $-9 + (-10) = -19$.

(4) Split the middle term: $-19x^2 = -9x^2 - 10x^2$

(5) Factor by grouping:

$$15x^4 - 19x^2 + 6 = 15x^4 - 9x^2 - 10x^2 + 6$$
$$= (15x^4 - 9x^2) + (-10x^2 + 6)$$
$$= 3x^2(5x^2 - 3) - 2(5x^2 - 3)$$
$$= (3x^2 - 2)(5x^2 - 3)$$

73. $25t^2 + 80t + 64$

(1) There are no common factors (other than 1 or -1).

(2) Multiply the leading coefficient, 25, and the constant, 64: $25 \cdot 64 = 1600$.

(3) Look for a factorization of 1600 in which the sum of the factors is the coefficient of the middle term, 80. The numbers we want are 40 and 40: $40 \cdot 40 = 1600$ and $40 + 40 = 80$.

(4) Split the middle term: $80t = 40t + 40t$

(5) Factor by grouping:

$$25t^2 + 80t + 64 = 25t^2 + 40t + 40t + 64$$
$$= (25t^2 + 40t) + (40t + 64)$$
$$= 5t(5t + 8) + 8(5t + 8)$$
$$= (5t + 8)(5t + 8), \text{ or}$$
$$(5t + 8)^2$$

75. $6x^3 + 4x^2 - 10x$

(1) Factor out the common factor, $2x$:

$$6x^3 + 4x^2 - 10x = 2x(3x^2 + 2x - 5)$$

(2) - (5) Now we factor the trinomial $3x^2 + 2x - 5$. We did this in Exercise 49, using the variable y rather than x. We found that $3x^2 + 2x - 5 = (x - 1)(3x + 5)$. We must include the common factor to get a factorization of the original trinomial.

$$6x^3 + 4x^2 - 10x = 2x(x - 1)(3x + 5)$$

77. $25x^2 + 79x + 64$

(1) There are no common factors (other than 1 or -1).

(2) Multiply the leading coefficient, 25, and the constant, 64: $25 \cdot 64 = 1600$.

(3) Look for a factorization of 1600 in which the sum of the factors is the coefficient of the middle term, 79. It is not possible to find such a pair of numbers. Thus, $25x^2 + 79x + 64$ cannot be factored into a product of binomial factors. It is prime.

79. $6x^2 - 19x - 5$

(1) There are no common factors (other than 1 or -1).

(2) Multiply the leading coefficient, 6, and the constant, -5: $6(-5) = -30$.

(3) Look for a factorization of -30 in which the sum of the factors is the coefficient of the middle term, -19. There is no such pair of numbers. Thus, $6x^2 - 19x - 5$ cannot be factored into a product of binomial factors. It is prime.

81. $12m^2 - mn - 20n^2$

(1) There are no common factors (other than 1 or -1).

(2) Multiply the leading coefficient, 12, and the constant, -20: $12(-20) = -240$.

(3) Look for a factorization of -240 in which the sum of the factors is the coefficient of the middle term, -1. The numbers we want are 15 and -16: $15(-16) = -240$ and $15 + (-16) = -1$.

(4) Split the middle term: $-mn = 15mn - 16mn$

(5) Factor by grouping:

$$12m^2 - mn - 20n^2$$
$$= 12m^2 + 15mn - 16mn - 20n^2$$
$$= (12m^2 + 15mn) + (-16mn - 20n^2)$$
$$= 3m(4m + 5n) - 4n(4m + 5n)$$
$$= (3m - 4n)(4m + 5n)$$

83. $6a^2 - ab - 15b^2$

(1) There are no common factors (other than 1 or -1).

(2) Multiply the leading coefficient, 6, and the constant, -15: $6(-15) = -90$.

(3) Look for a factorization of -90 in which the sum of the factors is the coefficient of the middle term, -1. The numbers we want are -10 and 9: $-10 \cdot 9 = -90$ and $-10 + 9 = -1$.

(4) Split the middle term: $-ab = -10ab + 9ab$

(5) Factor by grouping:

$$6a^2 - ab - 15b^2 = 6a^2 - 10ab + 9ab - 15b^2$$
$$= (6a^2 - 10ab) + (9ab - 15b^2)$$
$$= 2a(3a - 5b) + 3b(3a - 5b)$$
$$= (2a + 3b)(3a - 5b)$$

85. $9a^2 - 18ab + 8b^2$

(1) There are no common factors (other than 1 or -1).

(2) Multiply the leading coefficient, 9, and the constant, 8: $9 \cdot 8 = 72$.

(3) Look for a factorization of 72 in which the sum of the factors is the coefficient of the middle term, -18. The numbers we want are -6 and -12: $-6(-12) = 72$ and $-6 + (-12) = -18$.

(4) Split the middle term: $-18ab = -6ab - 12ab$

(5) Factor by grouping:
$$9a^2 - 18ab + 8b^2 = 9a^2 - 6ab - 12ab + 8b^2$$
$$= (9a^2 - 6ab) + (-12ab + 8b^2)$$
$$= 3a(3a - 2b) - 4b(3a - 2b)$$
$$= (3a - 4b)(3a - 2b)$$

87. $35p^2 + 34pq + 8q^2$

(1) There are no common factors (other than 1 or -1).

(2) Multiply the leading coefficient, 35, and the constant, 8: $35 \cdot 8 = 280$.

(3) Look for a factorization of 280 in which the sum of the factors is the coefficient of the middle term, 34. The numbers we want are 14 and 20: $14 \cdot 20 = 280$ and $14 + 20 = 34$.

(4) Split the middle term: $34pq = 14pq + 20pq$

(5) Factor by grouping:
$$35p^2 + 34pq + 8q^2 = 35p^2 + 14pq + 20pq + 8q^2$$
$$= (35p^2 + 14pq) + (20pq + 8q^2)$$
$$= 7p(5p + 2q) + 4q(5p + 2q)$$
$$= (7p + 4q)(5p + 2q)$$

89. $18x^2 - 6xy - 24y^2$

(1) Factor out the common factor, 6.
$$18x^2 - 6xy - 24y^2 = 6(3x^2 - xy - 4y^2)$$

(2) Now we factor the trinomial $3x^2 - xy - 4y^2$. Multiply the leading coefficient, 3, and the constant, -4: $3(-4) = -12$.

(3) Look for a factorization of -12 in which the sum of the factors is the coefficient of the middle term, -1. The numbers we want are -4 and 3: $-4 \cdot 3 = -12$ and $-4 + 3 = -1$.

(4) Split the middle term: $-xy = -4xy + 3xy$

(5) Factor by grouping:
$$3x^2 - xy - 4y^2 = 3x^2 - 4xy + 3xy - 4y^2$$
$$= (3x^2 - 4xy) + (3xy - 4y^2)$$
$$= x(3x - 4y) + y(3x - 4y)$$
$$= (x + y)(3x - 4y)$$

We must include the common factor to get a factorization of the original trinomial.
$$18x^2 - 6xy - 24y^2 = 6(x + y)(3x - 4y)$$

91. $60x + 18x^2 - 6x^3 = -6x^3 + 18x^2 + 60x$

(1) Factor out the common factor, $-6x$. We factor out $-6x$ rather than $6x$ in order to have a positive leading coefficient in the trinomial factor.
$$-6x^3 + 18x^2 + 60x = -6x(x^2 - 3x - 10)$$

(2) - (5) We factor $x^2 - 3x - 10$ as we did in Exercise 43, getting the $(x - 5)(x + 2)$. Then we have:
$$60x + 18x^2 - 6x^3$$
$$= -6x(x - 5)(x + 2)$$
$$= 6x(-x + 5)(x + 2)$$
$$\qquad\text{Multiplying } x - 5 \text{ by } -1$$
$$= 6x(x - 5)(-x - 2)$$
$$\qquad\text{Multiplying } x + 2 \text{ by } -1$$

Note that we can express $6x(-x + 5)(x + 2)$ as $6x(5 - x)(x + 2)$ since $-x + 5 = 5 - x$ by the commutative law of addition.

93. $35x^5 - 57x^4 - 44x^3$

(1) We first factor out the common factor, x^3.
$$x^3(35x^2 - 57x - 44)$$

(2) Now we factor the trinomial $35x^2 - 57x - 44$. Multiply the leading coefficient, 35, and the constant, -44: $35(-44) = -1540$.

(3) Look for a factorization of -1540 in which the sum of the factors is the coefficient of the middle term, -57.

Pairs of factors	Sums of factors
$7, -220$	-213
$10, -154$	-144
$11, -140$	-129
$14, -110$	-96
$20, \ -77$	-57

(4) Split the middle term: $-57x = 20x - 77x$

(5) Factor by grouping:
$$35x^2 - 57x - 44 = 35x^2 + 20x - 77x - 44$$
$$= (35x^2 + 20x) + (-77x - 44)$$
$$= 5x(7x + 4) - 11(7x + 4)$$
$$= (5x - 11)(7x + 4)$$

We must include the common factor to get a factorization of the original trinomial.
$$35x^5 - 57x^4 - 44x^3 = x^3(5x - 11)(7x + 4)$$

95. Discussion and Writing Exercise

97. $-10x > 1000$
$$\frac{-10x}{-10} < \frac{1000}{-10} \qquad \begin{array}{l}\text{Dividing by } -10 \text{ and reversing} \\ \text{the inequality symbol}\end{array}$$
$$x < -100$$

The solution set is $\{x | x < -100\}$.

99. $6 - 3x \geq -18$

$\quad -3x \geq -24 \quad$ Subtracting 6

$\quad x \leq 8 \qquad$ Dividing by -3 and reversing the inequality symbol

The solution set is $\{x | x \leq 8\}$.

101. $\quad \frac{1}{2}x - 6x + 10 \leq x - 5x$

$2\left(\frac{1}{2}x - 6x + 10\right) \leq 2(x - 5x) \quad$ Multiplying by 2 to clear the fraction

$\quad x - 12x + 20 \leq 2x - 10x$

$\quad -11x + 20 \leq -8x \quad$ Collecting like terms

$\quad 20 \leq 3x \qquad$ Adding $11x$

$\quad \frac{20}{3} \leq x \qquad$ Dividing by 3

The solution set is $\left\{x | x \geq \frac{20}{3}\right\}$.

103. $3x - 6x + 2(x - 4) > 2(9 - 4x)$

$3x - 6x + 2x - 8 > 18 - 8x \quad$ Removing parentheses

$\quad -x - 8 > 18 - 8x \quad$ Collecting like terms

$\quad 7x > 26 \qquad$ Adding $8x$ and 8

$\quad x > \frac{26}{7} \qquad$ Dividing by 7

The solution set is $\left\{x | x > \frac{26}{7}\right\}$.

105. *Familiarize.* We will use the formula $C = 2\pi r$, where C is circumference and r is radius, to find the radius in kilometers. Then we will multiply that number by 0.62 to find the radius in miles.

Translate.

$$\underbrace{\text{Circumference}}_{\downarrow \quad 40,000} = \underbrace{2}_{\downarrow \quad \approx} \cdot \underbrace{\pi}_{\quad} \cdot \underbrace{\text{radius}}_{\downarrow \quad 2(3.14)r}$$

Solve. First we solve the equation.

$40,000 \approx 2(3.14)r$

$40,000 \approx 6.28r$

$6369 \approx r$

Then we multiply to find the radius in miles:

$6369(0.62) \approx 3949$

Check. If $r = 6369$, then $2\pi r = 2(3.14)(6369) \approx 40,000$. We should also recheck the multiplication we did to find the radius in miles. Both values check.

State. The radius of the earth is about 6369 km or 3949 mi. (These values may differ slightly if a different approximation is used for π.)

107. $9x^{10} - 12x^5 + 4$

(a) First factor out a common factor, if any. There is none (other than 1 or -1).

(b) Multiply the leading coefficient, 9, and the constant, 4: $9 \cdot 4 = 36$.

(c) Look for a factorization of 36 in which the sum of the factors is the coefficient of the middle term, -12. The factors we want are -6 and -6.

(d) Split the middle term: $-12x^5 = -6x^5 - 6x^5$

(e) Factor by grouping:

$9x^{10} - 12x^5 + 4 = 9x^{10} - 6x^5 - 6x^5 + 4$

$\qquad = (9x^{10} - 6x^5) + (-6x^5 + 4)$

$\qquad = 3x^5(3x^5 - 2) - 2(3x^5 - 2)$

$\qquad = (3x^5 - 2)(3x^5 - 2)$, or

$\qquad = (3x^5 - 2)^2$

109. $16x^{10} + 8x^5 + 1$

(a) First factor out a common factor, if any. There is none (other than 1 or -1).

(b) Multiply the leading coefficient, 16, and the constant, 1: $16 \cdot 1 = 16$.

(c) Look for a factorization of 16 in which the sum of the factors is the coefficient of the middle term, 8. The factors we want are 4 and 4.

(d) Split the middle term: $8x^5 = 4x^5 + 4x^5$

(e) Factor by grouping:

$16x^{10} + 8x^5 + 1 = 16x^{10} + 4x^5 + 4x^5 + 1$

$\qquad = (16x^{10} + 4x^5) + (4x^5 + 1)$

$\qquad = 4x^5(4x^5 + 1) + 1(4x^5 + 1)$

$\qquad = (4x^5 + 1)(4x^5 + 1)$, or

$\qquad = (4x^5 + 1)^2$

111.–119. Left to the student

Exercise Set 5.5

1. $x^2 - 14x + 49$

(a) We know that x^2 and 49 are squares.

(b) There is no minus sign before either x^2 or 49.

(c) If we multiply the square roots, x and 7, and double the product, we get $2 \cdot x \cdot 7 = 14x$. This is the opposite of the remaining term, $-14x$.

Thus, $x^2 - 14x + 49$ is a trinomial square.

3. $x^2 + 16x - 64$

Both x^2 and 64 are squares, but there is a minus sign before 64. Thus, $x^2 + 16x - 64$ is not a trinomial square.

5. $x^2 - 2x + 4$

(a) Both x^2 and 4 are squares.

(b) There is no minus sign before either x^2 or 4.

(c) If we multiply the square roots, x and 2, and double the product, we get $2 \cdot x \cdot 2 = 4x$. This is neither the remaining term nor its opposite.

Thus, $x^2 - 2x + 4$ is not a trinomial square.

7. $9x^2 - 36x + 24$

Only one term is a square. Thus, $9x^2 - 36x + 24$ is not a trinomial square.

9. $x^2 - 14x + 49 = x^2 - 2 \cdot x \cdot 7 + 7^2 = (x - 7)^2$
$$ \uparrow \uparrow \uparrow \uparrow \uparrow$$
$$ = A^2 - 2 A B + B^2 = (A - B)^2$$

11. $x^2 + 16x + 64 = x^2 + 2 \cdot x \cdot 8 + 8^2 = (x + 8)^2$
$$ \uparrow \uparrow \uparrow \uparrow \uparrow$$
$$ = A^2 + 2 A B + B^2 = (A + B)^2$$

13. $x^2 - 2x + 1 = x^2 - 2 \cdot x \cdot 1 + 1^2 = (x - 1)^2$

15. $4 + 4x + x^2 = x^2 + 4x + 4$ $\qquad$ Changing the order
$$= x^2 + 2 \cdot x \cdot 2 + 2^2$$
$$= (x + 2)^2$$

17. $q^4 - 6q^2 + 9 = (q^2)^2 - 2 \cdot q^2 \cdot 3 + 3^2 = (q^2 - 3)^2$

19. $49 + 56y + 16y^2 = 16y^2 + 56y + 49$
$$= (4y)^2 + 2 \cdot 4y \cdot 7 + 7^2$$
$$= (4y + 7)^2$$

21. $2x^2 - 4x + 2 = 2(x^2 - 2x + 1)$
$$= 2(x^2 - 2 \cdot x \cdot 1 + 1^2)$$
$$= 2(x - 1)^2$$

23. $x^3 - 18x^2 + 81x = x(x^2 - 18x + 81)$
$$= x(x^2 - 2 \cdot x \cdot 9 + 9^2)$$
$$= x(x - 9)^2$$

25. $12q^2 - 36q + 27 = 3(4q^2 - 12q + 9)$
$$= 3[(2q)^2 - 2 \cdot 2q \cdot 3 + 3^2]$$
$$= 3(2q - 3)^2$$

27. $49 - 42x + 9x^2 = 7^2 - 2 \cdot 7 \cdot 3x + (3x)^2$
$$= (7 - 3x)^2$$

29. $5y^4 + 10y^2 + 5 = 5(y^4 + 2y^2 + 1)$
$$= 5[(y^2)^2 + 2 \cdot y^2 \cdot 1 + 1^2]$$
$$= 5(y^2 + 1)^2$$

31. $1 + 4x^4 + 4x^2 = 1^2 + 2 \cdot 1 \cdot 2x^2 + (2x^2)^2$
$$= (1 + 2x^2)^2$$

33. $4p^2 + 12pq + 9q^2 = (2p)^2 + 2 \cdot 2p \cdot 3q + (3q)^2$
$$= (2p + 3q)^2$$

35. $a^2 - 6ab + 9b^2 = a^2 - 2 \cdot a \cdot 3b + (3b)^2$
$$= (a - 3b)^2$$

37. $81a^2 - 18ab + b^2 = (9a)^2 - 2 \cdot 9a \cdot b + b^2$
$$= (9a - b)^2$$

39. $36a^2 + 96ab + 64b^2 = 4(9a^2 + 24ab + 16b^2)$
$$= 4[(3a)^2 + 2 \cdot 3a \cdot 4b + (4b)^2]$$
$$= 4(3a + 4b)^2$$

41. $x^2 - 4$

(a) The first expression is a square: x^2

The second expression is a square: $4 = 2^2$

(b) The terms have different signs.

$x^2 - 4$ is a difference of squares.

43. $x^2 + 25$

The terms do not have different signs.

$x^2 + 25$ is not a difference of squares.

45. $x^2 - 45$

The number 45 is not a square.

$x^2 - 45$ is not a difference of squares.

47. $16x^2 - 25y^2$

(a) The first expression is a square: $16x^2 = (4x)^2$

The second expression is a square: $25y^2 = (5y)^2$

(b) The terms have different signs.

$16x^2 - 25y^2$ is a difference of squares.

49. $y^2 - 4 = y^2 - 2^2 = (y + 2)(y - 2)$

51. $p^2 - 9 = p^2 - 3^2 = (p + 3)(p - 3)$

53. $-49 + t^2 = t^2 - 49 = t^2 - 7^2 = (t + 7)(t - 7)$

55. $a^2 - b^2 = (a + b)(a - b)$

57. $25t^2 - m^2 = (5t)^2 - m^2 = (5t + m)(5t - m)$

59. $100 - k^2 = 10^2 - k^2 = (10 + k)(10 - k)$

61. $16a^2 - 9 = (4a)^2 - 3^2 = (4a + 3)(4a - 3)$

63. $4x^2 - 25y^2 = (2x)^2 - (5y)^2 = (2x + 5y)(2x - 5y)$

65. $8x^2 - 98 = 2(4x^2 - 49) = 2[(2x)^2 - 7^2] =$
$2(2x + 7)(2x - 7)$

67. $36x - 49x^3 = x(36 - 49x^2) = x[6^2 - (7x)^2] =$
$x(6 + 7x)(6 - 7x)$

69. $\dfrac{1}{16} - 49x^8 = \left(\dfrac{1}{4}\right)^2 - (7x^4)^2 = \left(\dfrac{1}{4} + 7x^4\right)\left(\dfrac{1}{4} - 7x^4\right)$

71. $0.09y^2 - 0.0004 = (0.3y)^2 - (0.02)^2 =$
$(0.3y + 0.02)(0.3y - 0.02)$

73. $49a^4 - 81 = (7a^2)^2 - 9^2 = (7a^2 + 9)(7a^2 - 9)$

75. $ a^4 - 16$
$$= (a^2)^2 - 4^2$$
$$= (a^2 + 4)(a^2 - 4) \qquad \text{Factoring a difference of squares}$$
$$= (a^2 + 4)(a + 2)(a - 2) \qquad \text{Factoring further: } a^2 - 4 \text{ is a difference of squares.}$$

77. $5x^4 - 405$

$5(x^4 - 81)$

$= 5[(x^2)^2 - 9^2]$

$= 5(x^2 + 9)(x^2 - 9)$

$= 5(x^2 + 9)(x + 3)(x - 3)$ Factoring $x^2 - 9$

79. $1 - y^8$

$= 1^2 - (y^4)^2$

$= (1 + y^4)(1 - y^4)$

$= (1 + y^4)(1 + y^2)(1 - y^2)$ Factoring $1 - y^4$

$= (1 + y^4)(1 + y^2)(1 + y)(1 - y)$ Factoring $1 - y^2$

81. $x^{12} - 16$

$= (x^6)^2 - 4^2$

$= (x^6 + 4)(x^6 - 4)$

$= (x^6 + 4)(x^3 + 2)(x^3 - 2)$ Factoring $x^6 - 4$

83. $y^2 - \dfrac{1}{16} = y^2 - \left(\dfrac{1}{4}\right)^2$

$= \left(y + \dfrac{1}{4}\right)\left(y - \dfrac{1}{4}\right)$

85. $25 - \dfrac{1}{49}x^2 = 5^2 - \left(\dfrac{1}{7}x\right)^2$

$= \left(5 + \dfrac{1}{7}x\right)\left(5 - \dfrac{1}{7}x\right)$

87. $16m^4 - t^4$

$= (4m^2)^2 - (t^2)^2$

$= (4m^2 + t^2)(4m^2 - t^2)$

$= (4m^2 + t^2)(2m + t)(2m - t)$ Factoring $4m^2 - t^2$

89. Discussion and Writing exercise

91. $-110 \div 10$ The quotient of a negative number and a positive number is negative.

$-110 \div 10 = -11$

93. $-\dfrac{2}{3} \div \dfrac{4}{5} = -\dfrac{2}{3} \cdot \dfrac{5}{4} = -\dfrac{10}{12} = -\dfrac{2 \cdot 5}{2 \cdot 6} = -\dfrac{\cancel{2} \cdot 5}{\cancel{2} \cdot 6} = -\dfrac{5}{6}$

95. $-64 \div (-32)$ The quotient of two negative numbers is a positive number.

$-64 \div (-32) = 2$

97. The shaded region is a square with sides of length $x - y - y$, or $x - 2y$. Its area is $(x - 2y)(x - 2y)$, or $(x - 2y)^2$. Multiplying, we get the polynomial $x^2 - 4xy + 4y^2$.

99. $y^5 \cdot y^7 = y^{5+7} = y^{12}$

101. $y - 6x = 6$

To find the x-intercept, let $y = 0$. Then solve for x.

$y - 6x = 6$

$0 - 6x = 6$

$-6x = 6$

$x = -1$

The x-intercept is $(-1, 0)$.

To find the y-intercept, let $x = 0$. Then solve for y.

$y - 6x = 6$

$y - 6 \cdot 0 = 6$

$y = 6$

The y-intercept is $(0, 6)$.

Plot these points and draw the line.

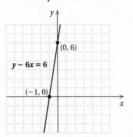

A third point should be used as a check. We substitute any value for x and solve for y. We let $x = -2$. Then

$y - 6x = 6$

$y - 6(-2) = 6$

$y + 12 = 6$

$y = -6$

The point $(-2, -6)$ is on the graph, so the graph is probably correct.

103. $49x^2 - 216$

There is no common factor. Also, $49x^2$ is a square, but 216 is not so this expression is not a difference of squares. It is not factorable. It is prime.

105. $x^2 + 22x + 121 = x^2 + 2 \cdot x \cdot 11 + 11^2$

$= (x + 11)^2$

107. $18x^3 + 12x^2 + 2x = 2x(9x^2 + 6x + 1)$

$= 2x[(3x)^2 + 2 \cdot 3x \cdot 1 + 1^2]$

$= 2x(3x + 1)^2$

109. $x^8 - 2^8$

$= (x^4 + 2^4)(x^4 - 2^4)$

$= (x^4 + 2^4)(x^2 + 2^2)(x^2 - 2^2)$

$= (x^4 + 2^4)(x^2 + 2^2)(x + 2)(x - 2)$, or

$= (x^4 + 16)(x^2 + 4)(x + 2)(x - 2)$

111. $3x^5 - 12x^3 = 3x^3(x^2 - 4) = 3x^3(x + 2)(x - 2)$

113. $18x^3 - \dfrac{8}{25}x = 2x\left(9x^2 - \dfrac{4}{25}\right) = 2x\left(3x + \dfrac{2}{5}\right)\left(3x - \dfrac{2}{5}\right)$

115. $0.49p - p^3 = p(0.49 - p^2) = p(0.7 + p)(0.7 - p)$

117. $0.64x^2 - 1.21 = (0.8x)^2 - (1.1)^2 = (0.8x + 1.1)(0.8x - 1.1)$

119. $(x + 3)^2 - 9 = [(x + 3) + 3][(x + 3) - 3] = (x + 6)x$, or $x(x + 6)$

121. $x^2 - \left(\dfrac{1}{x}\right)^2 = \left(x + \dfrac{1}{x}\right)\left(x - \dfrac{1}{x}\right)$

123. $81 - b^{4k} = 9^2 - (b^{2k})^2$

$= (9 + b^{2k})(9 - b^{2k})$

$= (9 + b^{2k})[3^2 - (b^k)^2]$

$= (9 + b^{2k})(3 + b^k)(3 - b^k)$

125. $9b^{2n} + 12b^n + 4 = (3b^n)^2 + 2 \cdot 3b^n \cdot 2 + 2^2 =$
$(3b^n + 2)^2$

127. $(y + 3)^2 + 2(y + 3) + 1$

$= (y + 3)^2 + 2 \cdot (y + 3) \cdot 1 + 1^2$

$= [(y + 3) + 1]^2$

$= (y + 4)^2$

129. If $cy^2 + 6y + 1$ is the square of a binomial, then $2 \cdot a \cdot 1 = 6$ where $a^2 = c$. Then $a = 3$, so $c = a^2 = 3^2 = 9$. (The polynomial is $9y^2 + 6y + 1$.)

131. Enter $y_1 = x^2 + 9$ and $y_2 = (x + 3)(x + 3)$ and look at a table of values. The y_1-and y_2-values are not the same, so the factorization is not correct.

133. Enter $y_1 = x^2 + 9$ and $y_2 = (x + 3)^2$ and look at a table of values. The y_1-and y_2-values are not the same, so the factorization is not correct.

Exercise Set 5.6

1. $3x^2 - 192 = 3(x^2 - 64)$ 3 is a common factor.

$= 3(x^2 - 8^2)$ Difference of squares

$= 3(x + 8)(x - 8)$

3. $a^2 + 25 - 10a = a^2 - 10a + 25$

$= a^2 - 2 \cdot a \cdot 5 + 5^2$ Trinomial square

$= (a - 5)^2$

5. $2x^2 - 11x + 12$

There is no common factor (other than 1). This polynomial has three terms, but it is not a trinomial square. Multiply the leading coefficient and the constant, 2 and 12: $2 \cdot 12 = 24$. Try to factor 24 so that the sum of the factors is -11. The numbers we want are -3 and -8: $-3(-8) = 24$ and $-3 + (-8) = -11$. Split the middle term and factor by grouping.

$2x^2 - 11x + 12 = 2x^2 - 3x - 8x + 12$

$= (2x^2 - 3x) + (-8x + 12)$

$= x(2x - 3) - 4(2x - 3)$

$= (x - 4)(2x - 3)$

7. $x^3 + 24x^2 + 144x$

$= x(x^2 + 24x + 144)$ x is a common factor.

$= x(x^2 + 2 \cdot x \cdot 12 + 12^2)$ Trinomial square

$= x(x + 12)^2$

9. $x^3 + 3x^2 - 4x - 12$

$= x^2(x + 3) - 4(x + 3)$ Factoring by grouping

$= (x^2 - 4)(x + 3)$

$= (x + 2)(x - 2)(x + 3)$ Factoring the difference
of squares

11. $48x^2 - 3 = 3(16x^2 - 1)$ 3 is a common factor.

$= 3[(4x)^2 - 1^2]$ Difference of squares

$= 3(4x + 1)(4x - 1)$

13. $9x^3 + 12x^2 - 45x$

$= 3x(3x^2 + 4x - 15)$ $3x$ is a common factor.

$= 3x(3x - 5)(x + 3)$ Factoring the trinomial

15. $x^2 + 4$ is a *sum* of squares with no common factor. It cannot be factored. It is prime.

17. $x^4 + 7x^2 - 3x^3 - 21x = x(x^3 + 7x - 3x^2 - 21)$

$= x[x(x^2 + 7) - 3(x^2 + 7)]$

$= x[(x - 3)(x^2 + 7)]$

$= x(x - 3)(x^2 + 7)$

19. $x^5 - 14x^4 + 49x^3$

$= x^3(x^2 - 14x + 49)$ x^3 is a common factor.

$= x^3(x^2 - 2 \cdot x \cdot 7 + 7^2)$ Trinomial square

$= x^3(x - 7)^2$

21. $20 - 6x - 2x^2$

$= -2(-10 + 3x + x^2)$ -2 is a common factor.

$= -2(x^2 + 3x - 10)$ Writing in descending order

$= -2(x + 5)(x - 2),$ Using trial and error

or $2(-x - 5)(x - 2),$

or $2(x + 5)(-x + 2)$

23. $x^2 - 6x + 1$

There is no common factor (other than 1 or -1). This is not a trinomial square, because $-6x \neq 2 \cdot x \cdot 1$ and $-6x \neq -2 \cdot x \cdot 1$. We try factoring using the refined trial and error procedure. We look for two factors of 1 whose sum is -6. There are none. The polynomial cannot be factored. It is prime.

25. $4x^4 - 64$

$= 4(x^4 - 16)$ 4 is a common factor.

$= 4[(x^2)^2 - 4^2]$ Difference of squares

$= 4(x^2 + 4)(x^2 - 4)$ Difference of squares

$= 4(x^2 + 4)(x + 2)(x - 2)$

27. $1 - y^8$ Difference of squares

$= (1 + y^4)(1 - y^4)$ Difference of squares

$= (1 + y^4)(1 + y^2)(1 - y^2)$ Difference of squares

$= (1 + y^4)(1 + y^2)(1 + y)(1 - y)$

29. $x^5 - 4x^4 + 3x^3$

$= x^3(x^2 - 4x + 3)$ x^3 is a common factor.

$= x^3(x - 3)(x - 1)$ Factoring the trinomial using
trial and error

31. $\dfrac{1}{81}x^6 - \dfrac{8}{27}x^3 + \dfrac{16}{9}$

$= \dfrac{1}{9}\left(\dfrac{1}{9}x^6 - \dfrac{8}{3}x^3 + 16\right)$ $\dfrac{1}{9}$ is a common factor.

$= \dfrac{1}{9}\left[\left(\dfrac{1}{3}x^3\right)^2 - 2 \cdot \dfrac{1}{3}x^3 \cdot 4 + 4^2\right]$ Trinomial square

$= \dfrac{1}{9}\left(\dfrac{1}{3}x^3 - 4\right)^2$

33. $mx^2 + my^2$

$= m(x^2 + y^2)$ m is a common factor.

The factor with more than one term cannot be factored further, so we have factored completely.

35. $9x^2y^2 - 36xy = 9xy(xy - 4)$

37. $2\pi rh + 2\pi r^2 = 2\pi r(h + r)$

39. $(a + b)(x - 3) + (a + b)(x + 4)$

$= (a + b)[(x - 3) + (x + 4)]$ $(a + b)$ is a common factor.

$= (a + b)(2x + 1)$

41. $(x - 1)(x + 1) - y(x + 1) = (x + 1)(x - 1 - y)$
$(x + 1)$ is a common factor.

43. $n^2 + 2n + np + 2p$

$= n(n + 2) + p(n + 2)$ Factoring by grouping

$= (n + p)(n + 2)$

45. $6q^2 - 3q + 2pq - p$

$= (6q^2 - 3q) + (2pq - p)$

$= 3q(2q - 1) + p(2q - 1)$ Factoring by grouping

$= (3q + p)(2q - 1)$

47. $4b^2 + a^2 - 4ab$

$= a^2 - 4ab + 4b^2$ Rearranging

$= a^2 - 2 \cdot a \cdot 2b + (2b)^2$ Trinomial square

$= (a - 2b)^2$

(Note that if we had rewritten the polynomial as $4b^2 - 4ab + a^2$, we might have written the result as $(2b-a)^2$. The two factorizations are equivalent.)

49. $16x^2 + 24xy + 9y^2$

$= (4x)^2 + 2 \cdot 4x \cdot 3y + (3y)^2$ Trinomial square

$= (4x + 3y)^2$

51. $49m^4 - 112m^2n + 64n^2$

$= (7m^2)^2 - 2 \cdot 7m^2 \cdot 8n + (8n)^2$ Trinomial square

$= (7m^2 - 8n)^2$

53. $y^4 + 10y^2z^2 + 25z^4$

$= (y^2)^2 + 2 \cdot y^2 \cdot 5z^2 + (5z^2)^2$ Trinomial square

$= (y^2 + 5z^2)^2$

55. $\dfrac{1}{4}a^2 + \dfrac{1}{3}ab + \dfrac{1}{9}b^2$

$= \left(\dfrac{1}{2}a\right)^2 + 2 \cdot \dfrac{1}{2}a \cdot \dfrac{1}{3}b + \left(\dfrac{1}{3}b\right)^2$

$= \left(\dfrac{1}{2}a + \dfrac{1}{3}b\right)^2$

57. $a^2 - ab - 2b^2 = (a - 2b)(a + b)$ Using trial and error

59. $2mn - 360n^2 + m^2$

$= m^2 + 2mn - 360n^2$ Rewriting

$= (m + 20n)(m - 18n)$ Using trial and error

61. $m^2n^2 - 4mn - 32 = (mn - 8)(mn + 4)$ Using trial and error

63. $r^5s^2 - 10r^4s + 16r^3$

$= r^3(r^2s^2 - 10rs + 16)$ r^3 is a common factor.

$= r^3(rs - 2)(rs - 8)$ Using trial and error

65. $a^5 + 4a^4b - 5a^3b^2$

$= a^3(a^2 + 4ab - 5b^2)$ a^3 is a common factor.

$= a^3(a + 5b)(a - b)$ Factoring the trinomial

67. $a^2 - \dfrac{1}{25}b^2$

$= a^2 - \left(\dfrac{1}{5}b\right)^2$ Difference of squares

$= \left(a + \dfrac{1}{5}b\right)\left(a - \dfrac{1}{5}b\right)$

69. $x^2 - y^2 = (x + y)(x - y)$ Difference of squares

71. $16 - p^4q^4$

$= 4^2 - (p^2q^2)^2$ Difference of squares

$= (4 + p^2q^2)(4 - p^2q^2)$ $4 - p^2q^2$ is a difference of squares.

$= (4 + p^2q^2)(2 + pq)(2 - pq)$

73. $1 - 16x^{12}y^{12}$

$= 1^2 - (4x^6y^6)^2$ Difference of squares

$= (1 + 4x^6y^6)(1 - 4x^6y^6)$ $1 - 4x^6y^6$ is a difference of squares.

$= (1 + 4x^6y^6)(1 + 2x^3y^3)(1 - 2x^3y^3)$

75. $q^3 + 8q^2 - q - 8$

$= q^2(q + 8) - (q + 8)$ Factoring by grouping

$= (q^2 - 1)(q + 8)$

$= (q + 1)(q - 1)(q + 8)$ Factoring the difference of squares

77. $112xy + 49x^2 + 64y^2$

$= 49x^2 + 112xy + 64y^2$ Rearranging

$= (7x)^2 + 2 \cdot 7x \cdot 8y + (8y)^2$ Trinomial square

$= (7x + 8y)^2$

79. Discussion and Writing Exercise

81. $(-4, 0)$; $m = -3$

The slope is -3, so the equation is $y = -3x + b$. Using the point $(-4, 0)$, we substitute -4 for x and 0 for y in $y = -3x + b$ and then solve for b.

$$y = -3x + b$$
$$0 = -3(-4) + b$$
$$0 = 12 + b$$
$$-12 = b$$

Then the equation is $y = -3x - 12$.

83. $(-4, 5)$; $m = -\dfrac{2}{3}$

The slope is $-\dfrac{2}{3}$, so the equation is $y = -\dfrac{2}{3}x + b$. Using the point $(-4, 5)$, we substitute -4 for x and 5 for y in $y = -\dfrac{2}{3}x + b$ and then solve for b.

$$y = -\frac{2}{3}x + b$$
$$5 = -\frac{2}{3}(-4) + b$$
$$5 = \frac{8}{3} + b$$
$$\frac{7}{3} = b$$

Then the equation is $y = -\dfrac{2}{3}x + \dfrac{7}{3}$.

85.
$$\frac{7}{5} \div \left(-\frac{11}{10}\right)$$
$$= \frac{7}{5} \cdot \left(-\frac{10}{11}\right) \quad \text{Multiplying by the reciprocal of the divisor}$$
$$= -\frac{7 \cdot 10}{5 \cdot 11}$$
$$= -\frac{7 \cdot 5 \cdot 2}{5 \cdot 11} = -\frac{7 \cdot 2}{11} \cdot \frac{5}{5}$$
$$= -\frac{14}{11}$$

87.
$$A = aX + bX - 7$$
$$A + 7 = aX + bX$$
$$A + 7 = X(a + b)$$
$$\frac{A + 7}{a + b} = X$$

89. $a^4 - 2a^2 + 1 = (a^2)^2 - 2 \cdot a^2 \cdot 1 + 1^2$
$$= (a^2 - 1)^2$$
$$= [(a + 1)(a - 1)]^2$$
$$= (a + 1)^2 (a - 1)^2$$

91. $12.25x^2 - 7x + 1 = (3.5x)^2 - 2 \cdot (3.5x) \cdot 1 + 1^2$
$$= (3.5x - 1)^2$$

93. $5x^2 + 13x + 7.2$

Multiply the leading coefficient and the constant, 5 and 7.2: $5(7.2) = 36$. Try to factor 36 so that the sum of the factors is 13. The numbers we want are 9 and 4. Split the middle term and factor by grouping:

$$5x^2 + 13x + 7.2 = 5x^2 + 9x + 4x + 7.2$$
$$= (5x^2 + 9x) + (4x + 7.2)$$
$$= 5x(x + 1.8) + 4(x + 1.8)$$
$$= (5x + 4)(x + 1.8)$$

95.
$$18 + y^3 - 9y - 2y^2$$
$$= y^3 - 2y^2 - 9y + 18$$
$$= y^2(y - 2) - 9(y - 2)$$
$$= (y^2 - 9)(y - 2)$$
$$= (y + 3)(y - 3)(y - 2)$$

97. $a^3 + 4a^2 + a + 4 = a^2(a + 4) + 1(a + 4)$
$$= (a^2 + 1)(a + 4)$$

99. $x^3 - x^2 - 4x + 4 = x^2(x - 1) - 4(x - 1)$
$$= (x^2 - 4)(x - 1)$$
$$= (x + 2)(x - 2)(x - 1)$$

101.
$$y^2(y - 1) - 2y(y - 1) + (y - 1)$$
$$= (y - 1)(y^2 - 2y + 1)$$
$$= (y - 1)(y - 1)^2$$
$$= (y - 1)^3$$

103.
$$(y + 4)^2 + 2x(y + 4) + x^2$$
$$= (y + 4)^2 + 2 \cdot (y + 4) \cdot x + x^2 \quad \text{Trinomial square}$$
$$= (y + 4 + x)^2$$

Exercise Set 5.7

1. $(x + 4)(x + 9) = 0$

$x + 4 = 0 \quad or \quad x + 9 = 0$ Using the principle of zero products

$x = -4 \quad or \quad\quad\quad x = -9$ Solving the two equations separately

Check:

For -4
$$\frac{(x + 4)(x + 9) = 0}{(-4 + 4)(-4 + 9) ? 0}$$
$$0 \cdot 5 \quad\Big|$$
$$0 \quad\Big| \quad \text{TRUE}$$

For -9
$$\frac{(x + 4)(x + 9) = 0}{(9 + 4)(-9 + 9) ? 0}$$
$$13 \cdot 0 \quad\Big|$$
$$0 \quad\Big| \quad \text{TRUE}$$

The solutions are -4 and -9.

3. $(x + 3)(x - 8) = 0$

$x + 3 = 0 \quad or \quad x - 8 = 0$ Using the principle of zero products

$x = -3 \quad or \quad\quad\quad x = 8$

Check:

For -3

$$\frac{(x+3)(x-8)=0}{}$$

$(-3+3)(-3-8)$? 0

$0(-11)$

0 | TRUE

For 8

$$\frac{(x+3)(x-8)=0}{}$$

$(8+3)(8-8)$? 0

$11 \cdot 0$

0 | TRUE

The solutions are -3 and 8.

5. $(x+12)(x-11)=0$

$x+12=0$ $\quad$ or $\quad$ $x-11=0$

$x=-12$ $\quad$ or $\quad$ $x=11$

The solutions are -12 and 11.

7. $x(x+3)=0$

$x=0$ $\quad$ or $\quad$ $x+3=0$

$x=0$ $\quad$ or $\quad$ $x=-3$

The solutions are 0 and -3.

9. $0=y(y+18)$

$y=0$ $\quad$ or $\quad$ $y+18=0$

$y=0$ $\quad$ or $\quad$ $y=-18$

The solutions are 0 and -18.

11. $(2x+5)(x+4)=0$

$2x+5=0$ $\quad$ or $\quad$ $x+4=0$

$2x=-5$ $\quad$ or $\quad$ $x=-4$

$x=-\dfrac{5}{2}$ $\quad$ or $\quad$ $x=-4$

The solutions are $-\dfrac{5}{2}$ and -4.

13. $(5x+1)(4x-12)=0$

$5x+1=0$ $\quad$ or $\quad$ $4x-12=0$

$5x=-1$ $\quad$ or $\quad$ $4x=12$

$x=-\dfrac{1}{5}$ $\quad$ or $\quad$ $x=3$

The solutions are $-\dfrac{1}{5}$ and 3.

15. $(7x-28)(28x-7)=0$

$7x-28=0$ $\quad$ or $\quad$ $28x-7=0$

$7x=28$ $\quad$ or $\quad$ $28x=7$

$x=4$ $\quad$ or $\quad$ $x=\dfrac{7}{28}=\dfrac{1}{4}$

The solutions are 4 and $\dfrac{1}{4}$.

17. $2x(3x-2)=0$

$2x=0$ $\quad$ or $\quad$ $3x-2=0$

$x=0$ $\quad$ or $\quad$ $3x=2$

$x=0$ $\quad$ or $\quad$ $x=\dfrac{2}{3}$

The solutions are 0 and $\dfrac{2}{3}$.

19. $\left(\dfrac{1}{5}+2x\right)\left(\dfrac{1}{9}-3x\right)=0$

$\dfrac{1}{5}+2x=0$ $\quad$ or $\quad$ $\dfrac{1}{9}-3x=0$

$2x=-\dfrac{1}{5}$ $\quad$ or $\quad$ $-3x=-\dfrac{1}{9}$

$x=-\dfrac{1}{10}$ $\quad$ or $\quad$ $x=\dfrac{1}{27}$

The solutions are $-\dfrac{1}{10}$ and $\dfrac{1}{27}$.

21. $(0.3x-0.1)(0.05x+1)=0$

$0.3x-0.1=0$ $\quad$ or $\quad$ $0.05x+1=0$

$0.3x=0.1$ $\quad$ or $\quad$ $0.05x=-1$

$x=\dfrac{0.1}{0.3}$ $\quad$ or $\quad$ $x=-\dfrac{1}{0.05}$

$x=\dfrac{1}{3}$ $\quad$ or $\quad$ $x=-20$

The solutions are $\dfrac{1}{3}$ and -20.

23. $9x(3x-2)(2x-1)=0$

$9x=0$ $\quad$ or $\quad$ $3x-2=0$ $\quad$ or $\quad$ $2x-1=0$

$x=0$ $\quad$ or $\quad$ $3x=2$ $\quad$ or $\quad$ $2x=1$

$x=0$ $\quad$ or $\quad$ $x=\dfrac{2}{3}$ $\quad$ or $\quad$ $x=\dfrac{1}{2}$

The solutions are 0, $\dfrac{2}{3}$, and $\dfrac{1}{2}$.

25. $x^2+6x+5=0$

$(x+5)(x+1)=0$ $\quad$ Factoring

$x+5=0$ $\quad$ or $\quad$ $x+1=0$ $\quad$ Using the principle of zero products

$x=-5$ $\quad$ or $\quad$ $x=-1$

The solutions are -5 and -1.

27. $x^2+7x-18=0$

$(x+9)(x-2)=0$ $\quad$ Factoring

$x+9=0$ $\quad$ or $\quad$ $x-2=0$ $\quad$ Using the principle of zero products

$x=-9$ $\quad$ or $\quad$ $x=2$

The solutions are -9 and 2.

29. $x^2-8x+15=0$

$(x-5)(x-3)=0$

$x-5=0$ $\quad$ or $\quad$ $x-3=0$

$x=5$ $\quad$ or $\quad$ $x=3$

The solutions are 5 and 3.

31. $x^2 - 8x = 0$

$x(x - 8) = 0$

$x = 0 \quad or \quad x - 8 = 0$

$x = 0 \quad or \qquad x = 8$

The solutions are 0 and 8.

33. $x^2 + 18x = 0$

$x(x + 18) = 0$

$x = 0 \quad or \quad x + 18 = 0$

$x = 0 \quad or \qquad x = -18$

The solutions are 0 and -18.

35. $\qquad x^2 = 16$

$\qquad x^2 - 16 = 0 \quad$ Subtracting 16

$(x - 4)(x + 4) = 0$

$x - 4 = 0 \quad or \quad x + 4 = 0$

$\quad x = 4 \quad or \qquad x = -4$

The solutions are 4 and -4.

37. $\qquad 9x^2 - 4 = 0$

$(3x - 2)(3x + 2) = 0$

$3x - 2 = 0 \quad or \quad 3x + 2 = 0$

$\quad 3x = 2 \quad or \qquad 3x = -2$

$\quad x = \dfrac{2}{3} \quad or \qquad x = -\dfrac{2}{3}$

The solutions are $\dfrac{2}{3}$ and $-\dfrac{2}{3}$.

39. $0 = 6x + x^2 + 9$

$0 = x^2 + 6x + 9 \quad$ Writing in descending order

$0 = (x + 3)(x + 3)$

$x + 3 = 0 \quad or \quad x + 3 = 0$

$\quad x = -3 \quad or \qquad x = -3$

There is only one solution, -3.

41. $\qquad x^2 + 16 = 8x$

$x^2 - 8x + 16 = 0 \quad$ Subtracting $8x$

$(x - 4)(x - 4) = 0$

$x - 4 = 0 \quad or \quad x - 4 = 0$

$\quad x = 4 \quad or \qquad x = 4$

There is only one solution, 4.

43. $\qquad 5x^2 = 6x$

$5x^2 - 6x = 0$

$x(5x - 6) = 0$

$x = 0 \quad or \quad 5x - 6 = 0$

$x = 0 \quad or \qquad 5x = 6$

$x = 0 \quad or \qquad x = \dfrac{6}{5}$

The solutions are 0 and $\dfrac{6}{5}$.

45. $\qquad 6x^2 - 4x = 10$

$6x^2 - 4x - 10 = 0$

$2(3x^2 - 2x - 5) = 0$

$2(3x - 5)(x + 1) = 0$

$3x - 5 = 0 \quad or \quad x + 1 = 0$

$\quad 3x = 5 \quad or \qquad x = -1$

$\quad x = \dfrac{5}{3} \quad or \qquad x = -1$

The solutions are $\dfrac{5}{3}$ and -1.

47. $\qquad 12y^2 - 5y = 2$

$12y^2 - 5y - 2 = 0$

$(4y + 1)(3y - 2) = 0$

$4y + 1 = 0 \quad or \quad 3y - 2 = 0$

$\quad 4y = -1 \quad or \qquad 3y = 2$

$\quad y = -\dfrac{1}{4} \quad or \qquad y = \dfrac{2}{3}$

The solutions are $-\dfrac{1}{4}$ and $\dfrac{2}{3}$.

49. $\qquad t(3t + 1) = 2$

$3t^2 + t = 2 \quad$ Multiplying on the left

$3t^2 + t - 2 = 0 \quad$ Subtracting 2

$(3t - 2)(t + 1) = 0$

$3t - 2 = 0 \quad or \quad t + 1 = 0$

$\quad 3t = 2 \quad or \qquad t = -1$

$\quad t = \dfrac{2}{3} \quad or \qquad t = -1$

The solutions are $\dfrac{2}{3}$ and -1.

51. $\qquad 100y^2 = 49$

$100y^2 - 49 = 0$

$(10y + 7)(10y - 7) = 0$

$10y + 7 = 0 \quad or \quad 10y - 7 = 0$

$\quad 10y = -7 \quad or \qquad 10y = 7$

$\quad y = -\dfrac{7}{10} \quad or \qquad y = \dfrac{7}{10}$

The solutions are $-\dfrac{7}{10}$ and $\dfrac{7}{10}$.

53. $\qquad x^2 - 5x = 18 + 2x$

$x^2 - 5x - 18 - 2x = 0 \qquad$ Subtracting 18 and $2x$

$x^2 - 7x - 18 = 0$

$(x - 9)(x + 2) = 0$

$x - 9 = 0 \quad or \quad x + 2 = 0$

$\quad x = 9 \quad or \qquad x = -2$

The solutions are 9 and -2.

55. $10x^2 - 23x + 12 = 0$

$(5x - 4)(2x - 3) = 0$

$5x - 4 = 0$ $\;or\;$ $2x - 3 = 0$

$5x = 4$ $\;or\;$ $2x = 3$

$x = \dfrac{4}{5}$ $\;or\;$ $x = \dfrac{3}{2}$

The solutions are $\dfrac{4}{5}$ and $\dfrac{3}{2}$.

57. We let $y = 0$ and solve for x.

$0 = x^2 + 3x - 4$

$0 = (x + 4)(x - 1)$

$x + 4 = 0$ $\;or\;$ $x - 1 = 0$

$x = -4$ $\;or\;$ $x = 1$

The x-intercepts are $(-4, 0)$ and $(1, 0)$.

59. We let $y = 0$ and solve for x

$0 = 2x^2 + x - 10$

$0 = (2x + 5)(x - 2)$

$2x + 5 = 0$ $\;or\;$ $x - 2 = 0$

$2x = -5$ $\;or\;$ $x = 2$

$x = -\dfrac{5}{2}$ $\;or\;$ $x = 2$

The x-intercepts are $\left(-\dfrac{5}{2}, 0\right)$ and $(2, 0)$.

61. We let $y = 0$ and solve for x.

$0 = x^2 - 2x - 15$

$0 = (x - 5)(x + 3)$

$x - 5 = 0$ $\;or\;$ $x + 3 = 0$

$x = 5$ $\;or\;$ $x = -3$

The x-intercepts are $(5, 0)$ and $(-3, 0)$.

63. The solutions of the equation are the first coordinates of the x-intercepts of the graph. From the graph we see that the x-intercepts are $(-1, 0)$ and $(4, 0)$, so the solutions of the equation are -1 and 4.

65. The solutions of the equation are the first coordinates of the x-intercepts of the graph. From the graph we see that the x-intercepts are $(-1, 0)$ and $(3, 0)$, so the solutions of the equation are -1 and 3.

67. Discussion and Writing Exercise

69. $(a + b)^2$

71. The two numbers have different signs, so their quotient is negative.

$144 \div -9 = -16$

73. $-\dfrac{5}{8} \div \dfrac{3}{16} = -\dfrac{5}{8} \cdot \dfrac{16}{3}$

$= -\dfrac{5 \cdot 16}{8 \cdot 3}$

$= -\dfrac{5 \cdot 8 \cdot 2}{8 \cdot 3}$

$= -\dfrac{10}{3}$

75. $b(b + 9) = 4(5 + 2b)$

$b^2 + 9b = 20 + 8b$

$b^2 + 9b - 8b - 20 = 0$

$b^2 + b - 20 = 0$

$(b + 5)(b - 4) = 0$

$b + 5 = 0$ $\;or\;$ $b - 4 = 0$

$b = -5$ $\;or\;$ $b = 4$

The solutions are -5 and 4.

77. $(t - 3)^2 = 36$

$t^2 - 6t + 9 = 36$

$t^2 - 6t - 27 = 0$

$(t - 9)(t + 3) = 0$

$t - 9 = 0$ $\;or\;$ $t + 3 = 0$

$t = 9$ $\;or\;$ $t = -3$

The solutions are 9 and -3.

79. $x^2 - \dfrac{1}{64} = 0$

$\left(x - \dfrac{1}{8}\right)\left(x + \dfrac{1}{8}\right) = 0$

$x - \dfrac{1}{8} = 0$ $\;or\;$ $x + \dfrac{1}{8} = 0$

$x = \dfrac{1}{8}$ $\;or\;$ $x = -\dfrac{1}{8}$

The solutions are $\dfrac{1}{8}$ and $-\dfrac{1}{8}$.

81. $\dfrac{5}{16}x^2 = 5$

$\dfrac{5}{16}x^2 - 5 = 0$

$5\left(\dfrac{1}{16}x^2 - 1\right) = 0$

$5\left(\dfrac{1}{4}x - 1\right)\left(\dfrac{1}{4}x + 1\right) = 0$

$\dfrac{1}{4}x - 1 = 0$ $\;or\;$ $\dfrac{1}{4}x + 1 = 0$

$\dfrac{1}{4}x = 1$ $\;or\;$ $\dfrac{1}{4}x = -1$

$x = 4$ $\;or\;$ $x = -4$

The solutions are 4 and -4.

83. (a) $x = -3$ $\;or\;$ $x = 4$

$x + 3 = 0$ $\;or\;$ $x - 4 = 0$

$(x + 3)(x - 4) = 0$ Principle of zero products

$x^2 - x - 12 = 0$ Multiplying

(b) $x = -3$ $\;or\;$ $x = -4$

$x + 3 = 0$ $\;or\;$ $x + 4 = 0$

$(x + 3)(x + 4) = 0$

$x^2 + 7x + 12 = 0$

(c) $x = \dfrac{1}{2}$ or $x = \dfrac{1}{2}$

$x - \dfrac{1}{2} = 0$ or $x - \dfrac{1}{2} = 0$

$\left(x - \dfrac{1}{2}\right)\left(x - \dfrac{1}{2}\right) = 0$

$x^2 - x + \dfrac{1}{4} = 0,$ or

$4x^2 - 4x + 1 = 0$ Multiplying by 4

(d) $(x - 5)(x + 5) = 0$

$x^2 - 25 = 0$

(e) $(x - 0)(x - 0.1)\left(x - \dfrac{1}{4}\right) = 0$

$x\left(x - \dfrac{1}{10}\right)\left(x - \dfrac{1}{4}\right) = 0$

$x\left(x^2 - \dfrac{7}{20}x + \dfrac{1}{40}\right) = 0$

$x^3 - \dfrac{7}{20}x^2 + \dfrac{1}{40}x = 0,$ or

$40x^3 - 14x^2 + x = 0$ Multiplying by 40

85. 2.33, 6.77

87. 0, 2.74

_____

Exercise Set 5.8

1. *Familiarize*. We make a drawing. Let w = the width, in cm. Then $w + 2$ = the length, in cm.

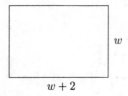

Recall that the area of a rectangle is length times width.

***Translate*.** We reword the problem.

Length times width is $\underbrace{144 \text{ cm}^2}$.

  ↓   ↓   ↓   ↓   ↓

$(w + 2)$ · w = 24

***Solve*.** We solve the equation.

$(w + 2)w = 24$

$w^2 + 2w = 24$

$w^2 + 2w - 24 = 0$

$(w + 6)(w - 4) = 0$

$w + 6 = 0$ or $w - 4 = 0$

$w = -6$ or $w = 4$

***Check*.** Since the width must be positive, -6 cannot be a solution. If the width is 4 cm, then the length is $4 + 2$, or 6 cm, and the area is $6 \cdot 4$, or 24 cm². Thus, 4 checks.

***State*.** The width is 4 cm, and the length is 6 cm.

3. *Familiarize*. Let w = the width of the table, in feet. Then $6w$ = the length, in feet. Recall that the area of a rectangle is Length · Width.

***Translate*.**

$\underbrace{\text{The area of the table}}$ is $\underbrace{24 \text{ ft}^2}$.

  ↓   ↓   ↓

$6w \cdot w$ = 24

***Solve*.** We solve the equation.

$6w \cdot w = 24$

$6w^2 = 24$

$6w^2 - 24 = 0$

$6(w^2 - 4) = 0$

$6(w + 2)(w - 2) = 0$

$w + 2 = 0$ or $w - 2 = 0$

$w = -2$ or $w = 2$

***Check*.** Since the width must be positive, -2 cannot be a solution. If the width is 2 ft, then the length is $6 \cdot 2$ ft, or 12 ft, and the area is 12 ft · 2 ft = 24 ft². These numbers check.

***State*.** The table is 12 ft long and 2 ft wide.

5. *Familiarize*. Using the labels shown on the drawing in the text, we let h = the height, in cm, and $h + 10$ = the base, in cm. Recall that the formula for the area of a triangle is $\dfrac{1}{2} \cdot$ (base) · (height).

***Translate*.**

$\dfrac{1}{2}$ times base   times height is $\underbrace{28 \text{ cm}^2}$.

↓   ↓   ↓   ↓   ↓   ↓   ↓

$\dfrac{1}{2}$ · $(h + 10)$ · h = 28

***Solve*.** We solve the equation.

$\dfrac{1}{2}(h + 10)h = 28$

$(h + 10)h = 56$ Multiplying by 2

$h^2 + 10h = 56$

$h^2 + 10h - 56 = 0$

$(h + 14)(h - 4) = 0$

$h + 14 = 0$ or $h - 4 = 0$

$h = -14$ or $h = 4$

***Check*.** Since the height of the triangle must be positive, -14 cannot be a solution. If the height is 4 cm, then the base is $4 + 10$, or 14 cm, and the area is $\dfrac{1}{2} \cdot 14 \cdot 4$, or 28 cm². Thus, 4 checks.

***State*.** The height of the triangle is 4 cm, and the base is 14 cm.

7. *Familiarize*. Using the labels shown on the drawing in the text, we let h = the height of the triangle, in meters, and $\dfrac{1}{2}h$ = the length of the base, in meters. Recall that the formula for the area of a triangle is $\dfrac{1}{2} \cdot$ (base) · (height).

Translate.

$\frac{1}{2}$ times base times height is $\underbrace{64 \text{ m}^2.}$

$\downarrow$ $\downarrow$ $\downarrow$ $\downarrow$ $\downarrow$ $\downarrow$ $\downarrow$

$\frac{1}{2}$ $\cdot$ $\frac{1}{2}h$ $\cdot$ h $=$ 64

Solve. We solve the equation.

$$\frac{1}{2} \cdot \frac{1}{2}h \cdot h = 64$$

$$\frac{1}{4}h^2 = 64$$

$$h^2 = 256 \quad \text{Multiplying by 4}$$

$$h^2 - 256 = 0$$

$$(h+16)(h-16) = 0$$

$$h + 16 = 0 \quad or \quad h - 16 = 0$$

$$h = -16 \quad or \quad h = 16$$

Check. The height of the triangle cannot be negative, so -16 cannot be a solution. If the height is 16 m, then the length of the base is $\frac{1}{2} \cdot 16$ m, or 8 m, and the area is $\frac{1}{2} \cdot 8$ m $\cdot 16$ m $= 64$ m^2. These numbers check.

State. The length of the base is 8 m, and the height is 16 m.

9. Familiarize. Reread Example 4 in Section 4.3.

Translate. Substitute 14 for n.

$$14^2 - 14 = N$$

Solve. We do the computation on the left.

$$14^2 - 14 = N$$

$$196 - 14 = N$$

$$182 = N$$

Check. We can redo the computation, or we can solve the equation $n^2 - n = 182$. The answer checks.

State. 182 games will be played.

11. Familiarize. Reread Example 4 in Section 4.3.

Translate. Substitute 132 for N.

$$n^2 - n = 132$$

Solve.

$$n^2 - n = 132$$

$$n^2 - n - 132 = 0$$

$$(n-12)(n+11) = 0$$

$$n - 12 = 0 \quad or \quad n + 11 = 0$$

$$n = 12 \quad or \quad n = -11$$

Check. The solutions of the equation are 12 and -11. Since the number of teams cannot be negative, -11 cannot be a solution. But 12 checks since $12^2 - 12 = 144 - 12 = 132$.

State. There are 12 teams in the league.

13. Familiarize. We will use the formula $N = \frac{1}{2}(n^2 - n)$.

Translate. Substitute 100 for n.

$$N = \frac{1}{2}(100^2 - 100)$$

Solve. We do the computation on the right.

$$N = \frac{1}{2}(10,000 - 100)$$

$$N = \frac{1}{2}(9900)$$

$$N = 4950$$

Check. We can redo the computation, or we can solve the equation $4950 = \frac{1}{2}(n^2 - n)$. The answer checks.

State. 4950 handshakes are possible.

15. Familiarize. We will use the formula $N = \frac{1}{2}(n^2 - n)$.

Translate. Substitute 300 for N.

$$300 = \frac{1}{2}(n^2 - n)$$

Solve. We solve the equation.

$$2 \cdot 300 = 2 \cdot \frac{1}{2}(n^2 - n) \qquad \text{Multiplying by 2}$$

$$600 = n^2 - n$$

$$0 = n^2 - n - 600$$

$$0 = (n+24)(n-25)$$

$$n + 24 = 0 \quad or \quad n - 25 = 0$$

$$n = -24 \quad or \quad n = 25$$

Check. The number of people at a meeting cannot be negative, so -24 cannot be a solution. But 25 checks since $\frac{1}{2}(25^2 - 25) = \frac{1}{2}(625 - 25) = \frac{1}{2} \cdot 600 = 300$.

State. There were 25 people at the party.

17. Familiarize. We will use the formula $N = \frac{1}{2}(n^2 - n)$, since toasts can be substituted for handshakes.

Translate. Substitute 190 for N.

$$190 = \frac{1}{2}(n^2 - n)$$

Solve.

$$190 = \frac{1}{2}(n^2 - n)$$

$$380 = n^2 - n \qquad \text{Multiplying by 2}$$

$$0 = n^2 - n - 380$$

$$0 = (n-20)(n+19)$$

$$n - 20 = 0 \quad or \quad n + 19 = 0$$

$$n = 20 \quad or \quad n = -19$$

Check. The solutions of the equation are 20 and -19. Since the number of people cannot be negative, -19 cannot be a solution. However, 20 checks since $\frac{1}{2}(20^2 - 20) = \frac{1}{2}(400 - 20) = \frac{1}{2}(380) = 190$.

State. 20 people took part in the toast.

19. Familiarize. The page numbers on facing pages are consecutive integers. Let $x =$ the smaller integer. Then $x + 1 =$ the larger integer.

Translate. We reword the problem.

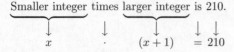

Smaller integer times larger integer is 210.

$$x \cdot (x+1) = 210$$

Solve. We solve the equation.

$$x(x+1) = 210$$
$$x^2 + x = 210$$
$$x^2 + x - 210 = 0$$
$$(x+15)(x-14) = 0$$
$$x + 15 = 0 \quad or \quad x - 14 = 0$$
$$x = -15 \quad or \qquad x = 14$$

Check. The solutions of the equation are -15 and 14. Since a page number cannot be negative, -15 cannot be a solution of the original problem. We only need to check 14. When $x = 14$, then $x+1 = 15$, and $14 \cdot 15 = 210$. This checks.

State. The page numbers are 14 and 15.

21. *Familiarize*. Let $x =$ the smaller even integer. Then $x + 2 =$ the larger even integer.

Translate. We reword the problem.

Smaller even integer times larger even integer is 168.

$$x \cdot (x+2) = 168$$

Solve.

$$x(x+2) = 168$$
$$x^2 + 2x = 168$$
$$x^2 + 2x - 168 = 0$$
$$(x+14)(x-12) = 0$$
$$x + 14 = 0 \quad or \quad x - 12 = 0$$
$$x = -14 \quad or \qquad x = 12$$

Check. The solutions of the equation are -14 and 12. When x is -14, then $x + 2$ is -12 and $-14(-12) = 168$. The numbers -14 and -12 are consecutive even integers which are solutions of the problem. When x is 12, then $x + 2$ is 14 and $12 \cdot 14 = 168$. The numbers 12 and 14 are also consecutive even integers which are solutions of the problem.

State. We have two solutions, each of which consists of a pair of numbers: -14 and -12, and 12 and 14.

23. *Familiarize*. Let $x =$ the smaller odd integer. Then $x + 2 =$ the larger odd integer.

Translate. We reword the problem.

Smaller odd integer times larger odd integer is 255.

$$x \cdot (x+2) = 255$$

Solve.

$$x(x+2) = 255$$
$$x^2 + 2x = 255$$
$$x^2 + 2x - 255 = 0$$
$$(x-15)(x+17) = 0$$
$$x - 15 = 0 \quad or \quad x + 17 = 0$$
$$x = 15 \quad or \qquad x = -17$$

Check. The solutions of the equation are 15 and -17. When x is 15, then $x + 2$ is 17 and $15 \cdot 17 = 255$. The numbers 15 and 17 are consecutive odd integers which are solutions to the problem. When x is -17, then $x + 2$ is -15 and $-17(-15) = 255$. The numbers -17 and -15 are also consecutive odd integers which are solutions to the problem.

State. We have two solutions, each of which consists of a pair of numbers: 15 and 17, and -17 and -15.

25. *Familiarize*. We make a drawing. Let $x =$ the length of the unknown leg. Then $x + 2 =$ the length of the hypotenuse.

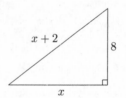

Translate. Use the Pythagorean theorem.

$$a^2 + b^2 = c^2$$
$$8^2 + x^2 = (x+2)^2$$

Solve. We solve the equation.

$$8^2 + x^2 = (x+2)^2$$
$$64 + x^2 = x^2 + 4x + 4$$
$$60 = 4x \qquad \text{Subtracting } x^2 \text{ and } 4$$
$$15 = x$$

Check. When $x = 15$, then $x+2 = 17$ and $8^2 + 15^2 = 17^2$. Thus, 15 and 17 check.

State. The lengths of the hypotenuse and the other leg are 17 ft and 15 ft, respectively.

27. *Familiarize*. Consider the drawing in the text. We let $w =$ the width of Main Street, in feet.

Translate. Use the Pythagorean theorem.

$$a^2 + b^2 = c^2$$
$$24^2 + w^2 = 40^2$$

Solve. We solve the equation.

$$24^2 + w^2 = 40^2$$
$$576 + w^2 = 1600$$
$$w^2 - 1024 = 0$$
$$(w+32)(w-32) = 0$$
$$w + 32 = 0 \quad or \quad w - 32 = 0$$
$$w = -32 \quad or \qquad w = 32$$

Check. The width of the street cannot be negative, so -32 cannot be a solution. If Main Street is 32 ft wide, we have $24^2 + 32^2 = 576 + 1024 = 1600$, which is 40^2. Thus, 32 ft checks.

State. Main Street is 32 ft wide.

29. Familiarize. Using the labels on the drawing in the text, we let $h =$ the height of a brace, in feet. Note that we have a right triangle with hypotenuse 15 ft and legs of 12 ft and h.

Translate. We use the Pythagorean theorem.

$$a^2 + b^2 = c^2$$
$$12^2 + h^2 = 15^2 \quad \text{Substituting}$$

Solve.

$$12^2 + h^2 = 15^2$$
$$144 + h^2 = 225$$
$$h^2 - 81 = 0$$
$$(h+9)(h-9) = 0$$
$$h + 9 = 0 \quad or \quad h - 9 = 0$$
$$h = -9 \quad or \quad h = 9$$

Check. The height of a brace cannot be negative, so -9 cannot be a solution. When $h = 9$, we have $12^2 + 9^2 = 144 + 81 = 225 = 15^2$. This checks.

State. The brace is 9 ft high.

31. Familiarize. We label the drawing. Let $x =$ the length of a side of the dining room, in ft. Then the dining room has dimensions x by x and the kitchen has dimensions x by 10. The entire rectangular space has dimension x by $x + 10$. Recall that we multiply these dimensions to find the area of the rectangle.

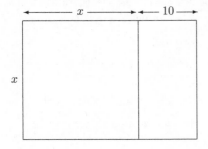

Translate.

$$\underbrace{\text{The area of the rectangular space}}_{\downarrow} \text{ is } \underbrace{264 \text{ ft}^2}.$$
$$x(x+10) \quad = \quad 264$$

Solve. We solve the equation.

$$x(x+10) = 264$$
$$x^2 + 10x = 264$$
$$x^2 + 10x - 264 = 0$$
$$(x+22)(x-12) = 0$$
$$x + 22 = 0 \quad or \quad x - 12 = 0$$
$$x = -22 \quad or \quad x = 12$$

Check. Since the length of a side of the dining room must be positive, -22 cannot be a solution. If x is 12 ft, then $x + 10$ is 22 ft, and the area of the space is $12 \cdot 22$, or 264 ft^2. The number 12 checks.

State. The dining room is 12 ft by 12 ft, and the kitchen is 12 ft by 10 ft.

33. Familiarize. We will use the formula $h = 180t - 16t^2$.

Translate. Substitute 464 for h.

$$464 = 180t - 16t^2$$

Solve. We solve the equation.

$$464 = 180t - 16t^2$$
$$16t^2 - 180t + 464 = 0$$
$$4(4t^2 - 45t + 116) = 0$$
$$4(4t - 29)(t - 4) = 0$$
$$4t - 29 = 0 \quad or \quad t - 4 = 0$$
$$4t = 29 \quad or \quad t = 4$$
$$t = \frac{29}{4} \quad or \quad t = 4$$

Check. The solutions of the equation are $\frac{29}{4}$, or $7\frac{1}{4}$, and 4. Since we want to find how many seconds it takes the rocket to *first* reach a height of 464 ft, we check the smaller number, 4. We substitute 4 for t in the formula.

$$h = 180t - 16t^2$$
$$h = 180 \cdot 4 - 16(4)^2$$
$$h = 180 \cdot 4 - 16 \cdot 16$$
$$h = 720 - 256$$
$$h = 464$$

The answer checks.

State. The rocket will first reach a height of 464 ft after 4 seconds.

35. Familiarize. Let $x =$ the smaller odd positive integer. Then $x + 2 =$ the larger odd positive integer.

Translate.

$$\underbrace{\text{Square of the smaller odd positive integer}}_{x^2} + \underbrace{\text{Square of the larger odd positive integer}}_{(x+2)^2} \text{ is } 74$$
$$x^2 \quad + \quad (x+2)^2 \quad = \quad 74$$

Solve.

$$x^2 + (x+2)^2 = 74$$
$$x^2 + x^2 + 4x + 4 = 74$$
$$2x^2 + 4x - 70 = 0$$
$$2(x^2 + 2x - 35) = 0$$
$$2(x+7)(x-5) = 0$$
$$x + 7 = 0 \quad or \quad x - 5 = 0$$
$$x = -7 \quad or \quad x = 5$$

Check. The solutions of the equation are -7 and 5. The problem asks for odd positive integers, so -7 cannot be a solution. When x is 5, $x + 2$ is 7. The numbers 5 and

7 are consecutive odd positive integers. The sum of their squares, $25 + 49$, is 74. The numbers check.

State. The integers are 5 and 7.

37. Discussion and Writing Exercise

39. To <u>factor</u> a polynomial is to express it as a product.

41. A factorization of a polynomial is an expression that names that polynomial as a <u>product</u>.

43. The expression $-5x^2 + 8x - 7$ is an example of a <u>trinomial</u>.

45. For the graph of the equation $4x - 3y = 12$, the pair $(0, -4)$ is known as the <u>y-intercept</u>.

47. **Familiarize**. First we can use the Pythagorean theorem to find x, in ft. Then the height of the telephone pole is $x + 5$.

Translate. We use the Pythagorean theorem.
$$a^2 + b^2 = c^2$$
$$\left(\frac{1}{2}x + 1\right)^2 + x^2 = 34^2$$

Solve. We solve the equation.
$$\left(\frac{1}{2}x + 1\right)^2 + x^2 = 34^2$$
$$\frac{1}{4}x^2 + x + 1 + x^2 = 1156$$
$$x^2 + 4x + 4 + 4x^2 = 4624 \quad \text{Multiplying by 4}$$
$$5x^2 + 4 + 4 = 4624$$
$$5x^2 + 4x - 4620 = 0$$
$$(5x + 154)(x - 30) = 0$$
$$5x + 154 = 0 \quad or \quad x - 30 = 0$$
$$5x = -154 \quad or \quad x = 30$$
$$x = -30.8 \quad or \quad x = 30$$

Check. Since the length x must be positive, -30.8 cannot be a solution. If x is 30 ft, then $\frac{1}{2}x + 1$ is $\frac{1}{2} \cdot 30 + 1$, or 16 ft. Since $16^2 + 30^2 = 1156 = 34^2$, the number 30 checks. When x is 30 ft, then $x + 5$ is 35 ft.

State. The height of the telephone pole is 35 ft.

49. **Familiarize**. Using the labels shown on the drawing in the text, we let $x =$ the width of the walk. Then the length and width of the rectangle formed by the pool and walk together are $40 + 2x$ and $20 + 2x$, respectively.

Translate.

Area is length times width.
$$1500 = (40 + 2x) \cdot (20 + 2x)$$

Solve. We solve the equation.
$$1500 = (40 + 2x)(20 + 2x)$$
$$1500 = 2(20 + x) \cdot 2(10 + x) \quad \text{Factoring 2 out of each factor on the right}$$
$$1500 = 4 \cdot (20 + x)(10 + x)$$
$$375 = (20 + x)(10 + x) \quad \text{Dividing by 4}$$
$$375 = 200 + 30x + x^2$$
$$0 = x^2 + 30x - 175$$
$$0 = (x + 35)(x - 5)$$
$$x + 35 = 0 \quad or \quad x - 5 = 0$$
$$x = -35 \quad or \quad x = 5$$

Check. The solutions of the equation are -35 and 5. Since the width of the walk cannot be negative, -35 is not a solution. When $x = 5$, $40 + 2x = 40 + 2 \cdot 5$, or 50 and $20 + 2x = 20 + 2 \cdot 5$, or 30. The total area of the pool and walk is $50 \cdot 30$, or 1500 ft^2. This checks.

State. The width of the walk is 5 ft.

51. **Familiarize**. We make a drawing. Let $w =$ the width of the piece of cardboard. Then $2w =$ the length.

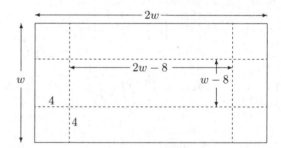

The box will have length $2w - 8$, width $w - 8$, and height 4. Recall that the formula for volume is $V = $ length $\times$ width $\times$ height.

Translate.

The volume is 616cm^3.
$$(2w - 8)(w - 8)(4) = 616$$

Solve. We solve the equation.
$$(2w - 8)(w - 8)(4) = 616$$
$$(2w^2 - 24w + 64)(4) = 616$$
$$8w^2 - 96w + 256 = 616$$
$$8w^2 - 96w - 360 = 0$$
$$8(w^2 - 12w - 45) = 0$$
$$w^2 - 12w - 45 = 0 \quad \text{Dividing by 8}$$
$$(w - 15)(w + 3) = 0$$
$$w - 15 = 0 \quad or \quad w + 3 = 0$$
$$w = 15 \quad or \quad w = -3$$

Check. The width cannot be negative, so we only need to check 15. When $w = 15$, then $2w = 30$ and the dimensions of the box are $30 - 8$ by $15 - 8$ by 4, or 22 by 7 by 4. The volume is $22 \cdot 7 \cdot 4$, or 616.

State. The cardboard is 30 cm by 15 cm.

53. *Familiarize*. Let $x =$ the length of a side of the base of the box, in feet. Then each of the four sides of the box has dimensions 9 ft by x and the top and bottom each have dimensions x by x.

Translate. We add the areas of the four sides of the box and of the top and bottom.

$$4 \cdot 9 \cdot x + 2 \cdot x^2 = 350$$

Solve. We solve the equation.

$$4 \cdot 9 \cdot x + 2 \cdot x^2 = 350$$
$$36x + 2x^2 = 350$$
$$2x^2 + 36x - 350 = 0$$
$$2(x^2 + 18x - 175) = 0$$
$$x^2 + 18x - 175 = 0 \qquad \text{Dividing by 2}$$
$$(x + 25)(x - 7) = 0$$
$$x + 25 = 0 \quad or \quad x - 7 = 0$$
$$x = -25 \quad or \qquad x = 7$$

Check. The length of a side cannot be negative, so -25 cannot be a solution. If $x = 7$, then the surface area of the box is $4 \cdot 9 \cdot 7 \text{ ft} + 2 \cdot (7 \text{ ft})^2 = 252 \text{ ft}^2 + 98 \text{ ft}^2 = 350 \text{ ft}^2$. The number 7 checks.

State. The length of a side of the base of the box is 7 ft.

Chapter 5 Review Exercises

1. $-15y^2 = -1 \cdot 3 \cdot 5 \cdot y^2$
$$25y^6 = 5 \cdot 5 \cdot y^6$$

Each coefficient has a factor of 5. There are no other common prime factors. The GCF of the powers of y is y^2 because 2 is the smallest exponent of y. Thus the GCF is $5y^2$.

2. $12x^3 = 2 \cdot 2 \cdot 3 \cdot x^3$
$$-60x^2y = -1 \cdot 2 \cdot 2 \cdot 3 \cdot 5 \cdot x^2 \cdot y$$
$$36xy = 2 \cdot 2 \cdot 3 \cdot 3 \cdot x \cdot y$$

Each coefficient has two factors of 2 and one factor of 3. There are no other common prime factors. The GCF of the powers of x is x because 1 is the smallest exponent of x. The GCF of the powers of y is 1 because $12x^3$ has no y-factor. Thus the GCF is $2 \cdot 2 \cdot 3 \cdot x \cdot 1$, or $12x$.

3. $5 - 20x^6$
$$= 5(1 - 4x^6) \qquad \text{5 is a common factor.}$$
$$= 5(1 - 2x^3)(1 + 2x^3) \quad \text{Factoring the difference of squares}$$

4. $x^2 - 3x = x(x - 3)$

5. $9x^2 - 4 = (3x + 2)(3x - 2)$ Factoring a difference of squares

6. $x^2 + 4x - 12$

We look for a pair of factors of -12 whose sum is 4. The numbers we need are 6 and -2.
$$x^2 + 4x - 12 = (x + 6)(x - 2)$$

7. $x^2 + 14x + 49 = x^2 + 2 \cdot x \cdot 7 + 7^2 = (x + 7)^2$

8. $6x^3 + 12x^2 + 3x = 3x(2x^2 + 4x + 1)$

The trinomial $2x^2 + 4x + 1$ cannot be factored, so the factorization is complete.

9. $x^3 + x^2 + 3x + 3$
$$= (x^3 + x^2) + (3x + 3)$$
$$= x^2(x + 1) + 3(x + 1) \quad \text{Factoring by grouping}$$
$$= (x^2 + 3)(x + 1)$$

10. $6x^2 - 5x + 1$

There is no common factor (other than 1). This polynomial has three terms, but it is not a trinomial square. Multiply the leading coefficient and the constant, 6 and 1: $6 \cdot 1 = 6$. Try to factor 6 so that the sum of the factors is -5. The numbers we want are -2 and -3: $-2(-3) = 6$ and $-2 + (-3) = -5$. Split the middle term and factor by grouping.

$$6x^2 - 5x + 1 = 6x^2 - 2x - 3x + 1$$
$$= (6x^2 - 2x) + (-3x + 1)$$
$$= 2x(3x - 1) - 1(3x - 1)$$
$$= (2x - 1)(3x - 1)$$

11. $x^4 - 81 = (x^2 + 9)(x^2 - 9) = (x^2 + 9)(x + 3)(x - 3)$

12. $9x^3 + 12x^2 - 45x$
$$= 3x(3x^2 + 4x - 15) \quad \text{3x is a common factor.}$$
$$= 3x(3x - 5)(x + 3) \quad \text{Using trial and error}$$

13. $2x^2 - 50 = 2(x^2 - 25) = 2(x + 5)(x - 5)$

14. $x^4 + 4x^3 - 2x - 8 = (x^4 + 4x^3) + (-2x - 8)$
$$= x^3(x + 4) - 2(x + 4)$$
$$= (x^3 - 2)(x + 4)$$

15. $16x^4 - 1 = (4x^2 + 1)(4x^2 - 1) = (4x^2 + 1)(2x + 1)(2x - 1)$

16. $8x^6 - 32x^5 + 4x^4 = 4x^4(2x^2 - 8x + 1)$

The trinomial $2x^2 - 8x + 1$ cannot be factored, so the factorization is complete.

17. $75 + 12x^2 + 60x = 12x^2 + 60x + 75$
$$= 3(4x^2 + 20x + 25)$$
$$= 3(2x + 5)^2$$

18. $x^2 + 9$ is a sum of squares with no common factor, so it is prime.

19. $x^3 - x^2 - 30x = x(x^2 - x - 30) = x(x - 6)(x + 5)$

20. $4x^2 - 25 = (2x + 5)(2x - 5)$

21. $9x^2 + 25 - 30x = 9x^2 - 30x + 25 = (3x - 5)^2$

22. $6x^2 - 28x - 48 = 2(3x^2 - 14x - 24)$
$$= 2(3x + 4)(x - 6)$$

23. $x^2 - 6x + 9 = (x - 3)^2$

24. $2x^2 - 7x - 4 = (2x + 1)(x - 4)$

25. $18x^2 - 12x + 2 = 2(9x^2 - 6x + 1) = 2(3x-1)^2$

26. $3x^2 - 27 = 3(x^2 - 9) = 3(x+3)(x-3)$

27. $15 - 8x + x^2 = x^2 - 8x + 15 = (x-3)(x-5)$

28. $25x^2 - 20x + 4 = (5x-2)^2$

29. $49b^{10} + 4a^8 - 28a^4b^5 = 49b^{10} - 28a^4b^5 + 4a^8$
$$= (7b^5)^2 - 2 \cdot 7b^5 \cdot 2a^4 + (2a^4)^2$$
$$= (7b^5 - 2a^4)^2$$

30. $x^2y^2 + xy - 12 = (xy+4)(xy-3)$

31. $12a^2 + 84ab + 147b^2 = 3(4a^2 + 28ab + 49b^2) = 3(2a+7b)^2$

32. $m^2 + 5m + mt + 5t = (m^2 + 5m) + (mt + 5t)$
$$= m(m+5) + t(m+5)$$
$$= (m+t)(m+5)$$

33. $32x^4 - 128y^4z^4 = 32(x^4 - 4y^4z^4) = 32(x^2 + 2y^2z^2)(x^2 - 2y^2z^2)$

34. $(x-1)(x+3) = 0$
$x - 1 = 0 \ or \ x + 3 = 0$
$x = 1 \ or \ \ \ \ x = -3$
The solutions are 1 and -3.

35. $x^2 + 2x - 35 = 0$
$(x+7)(x-5) = 0$
$x + 7 = 0 \ or \ x - 5 = 0$
$x = -7 \ or \ \ \ \ x = 5$
The solutions are -7 and 5.

36. $x^2 + x - 12 = 0$
$(x+4)(x-3) = 0$
$x + 4 = 0 \ or \ x - 3 = 0$
$x = -4 \ or \ \ \ \ x = 3$
The solutions are -4 and 3.

37. $3x^2 + 2 = 5x$
$3x^2 - 5x + 2 = 0$
$(3x-2)(x-1) = 0$
$3x - 2 = 0 \ or \ x - 1 = 0$
$3x = 2 \ or \ \ \ \ x = 1$
$x = \dfrac{2}{3} \ or \ \ \ \ x = 1$
The solutions are $\dfrac{2}{3}$ and 1.

38. $2x^2 + 5x = 12$
$2x^2 + 5x - 12 = 0$
$(2x-3)(x+4) = 0$
$2x - 3 = 0 \ or \ x + 4 = 0$
$2x = 3 \ or \ \ \ \ x = -4$
$x = \dfrac{3}{2} \ or \ \ \ \ x = -4$
The solutions are $\dfrac{3}{2}$ and -4.

39. $16 = x(x-6)$
$16 = x^2 - 6x$
$0 = x^2 - 6x - 16$
$0 = (x-8)(x+2)$
$x - 8 = 0 \ or \ x + 2 = 0$
$x = 8 \ or \ \ \ \ x = -2$
The solutions are 8 and -2.

40. *Familiarize.* Let $b =$ the length of the base, in cm. Then $b + 1 =$ the height.

Translate. We use the formula for the area of a triangle.
$$A = \frac{1}{2}bh$$
$$15 = \frac{1}{2}b(b+1) \quad \text{Substituting}$$

Solve.
$$15 = \frac{1}{2}b(b+1)$$
$$15 = \frac{1}{2}b^2 + \frac{1}{2}b$$
$$0 = \frac{1}{2}b^2 + \frac{1}{2}b - 15$$
$$2 \cdot 0 = 2\left(\frac{1}{2}b^2 + \frac{1}{2}b - 15\right) \quad \text{Clearing fractions}$$
$$0 = b^2 + b - 30$$
$$0 = (b+6)(b-5)$$
$$b + 6 = 0 \quad or \quad b - 5 = 0$$
$$b = -6 \quad or \quad b = 5$$

Check. The length of the base cannot be negative, so -6 cannot be a solution. If $b = 5$, then $b + 1 = 5 + 1 = 6$ and the area is $\dfrac{1}{2} \cdot 5 \cdot 6 = 15$ cm^2. The number 5 checks.

State. The base is 5 cm, and the height is 6 cm.

41. *Familiarize.* Let $x =$ the smaller integer. Then $x + 2$ is the other integer.

Translate.

$\underbrace{\text{Smaller even integer}}$	times	$\underbrace{\text{larger even integer}}$	is	288.
↓	↓	↓	↓	↓
x	$\cdot$	$(x+2)$	$=$	288

Solve.
$$x(x+2) = 288$$
$$x^2 + 2x = 288$$
$$x^2 + 2x - 288 = 0$$
$$(x+18)(x-16) = 0$$
$$x + 18 = 0 \quad or \quad x - 16 = 0$$
$$x = -18 \quad or \quad x = 16$$
If $x = -18$, then $x + 2 = -18 + 2 = -16$.
If $x = 16$, then $x + 2 = 16 + 2 = 18$.

Check. -18 and -16 are consecutive even integers and $-18(-16) = 288$. Similarly, 16 and 18 are consecutive even integers and $16 \cdot 18 = 288$. Both pairs of integers check.

State. The integers are -18 and -16 or 16 and 18.

42. *Familiarize*. Let $x =$ the smaller integer. Then $x + 2$ is the other integer.

Translate.

$$\underbrace{\text{Smaller}\atop\text{odd integer}}\quad \text{times}\quad \underbrace{\text{larger}\atop\text{odd integer}}\quad \text{is}\quad 323.$$

$$\begin{array}{ccccc} \downarrow & \downarrow & \downarrow & \downarrow & \downarrow \\ x & \cdot & (x+2) & = & 323 \end{array}$$

Solve.

$$x(x+2) = 323$$
$$x^2 + 2x = 323$$
$$x^2 + 2x - 323 = 0$$
$$(x+19)(x-17) = 0$$
$$x + 19 = 0 \quad or \quad x - 17 = 0$$
$$x = -19 \quad or \qquad x = 17$$

If $x = -19$, then $x + 2 = -19 + 2 = -17$.

If $x = 17$, then $x + 2 = 17 + 2 = 19$.

Check. -19 and -17 are consecutive odd integers and $-19(-17) = 323$. Similarly, 17 and 19 are consecutive odd integers and $17 \cdot 19 = 323$. Both pairs of integers check.

State. The integers are -19 and -17 or 17 and 19.

43. *Familiarize*. We make a drawing. Let $d =$ the height above the ground at which the cables are attached to the tree. Then $d + 2 =$ the length of the cable.

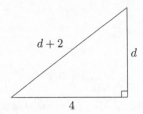

Translate. We use the Pythagorean theorem.
$$a^2 + b^2 = c^2$$
$$4^2 + d^2 = (d+2)^2 \quad \text{Substituting}$$

Solve.
$$4^2 + d^2 = (d+2)^2$$
$$16 + d^2 = d^2 + 4d + 4$$
$$16 = 4d + 4 \qquad \text{Subtracting } d^2$$
$$12 = 4d$$
$$3 = d$$

Check. If $d = 3$, then $d + 2 = 3 + 2 = 5$ and we have $4^2 + 3^2 = 16 + 9 = 25 = 5^2$. The answer checks.

State. The cables are attached to the tree 3 ft above the ground.

44. *Familiarize*. Let $s =$ the length of a side of the original square, in km. Then $s + 3 =$ the length of a side of the enlarged square.

Translate.

$$\underbrace{\text{Area of enlarged square}}\quad \text{is}\quad \underbrace{81 \text{ km}^2}.$$

$$\begin{array}{ccc} \downarrow & \downarrow & \downarrow \\ (s+3)^2 & = & 81 \end{array}$$

Solve.

$$(s+3)^2 = 81$$
$$s^2 + 6s + 9 = 81$$
$$s^2 + 6s - 72 = 0$$
$$(s+12)(s-6) = 0$$
$$s + 12 = 0 \quad or \quad s - 6 = 0$$
$$s = -12 \quad or \qquad s = 6$$

Check. The length of a side of the square cannot be negative, so -12 cannot be a solution. If the length of a side of the original square is 6 km, then the length of a side of the enlarged square is $6 + 3$, or 9 km. The area of the enlarged square is $(9 \text{ km})^2$, or 81 km^2, so the answer checks.

State. The length of a side of the original square is 6 km.

45. Let $y = 0$ and solve for x.
$$0 = x^2 + 9x + 20$$
$$0 = (x+5)(x+4)$$
$$x + 5 = 0 \quad or \quad x + 4 = 0$$
$$x = -5 \quad or \qquad x = -4$$

The x-intercepts are $(-5, 0)$ and $(-4, 0)$.

46. Let $y = 0$ and solve for x.
$$0 = 2x^2 - 7x - 15$$
$$0 = (2x+3)(x-5)$$
$$2x + 3 = 0 \quad or \quad x - 5 = 0$$
$$2x = -3 \quad or \qquad x = 5$$
$$x = -\frac{3}{2} \quad or \qquad x = 5$$

The x-intercepts are $\left(-\frac{3}{2}, 0\right)$ and $(5, 0)$.

47. *Discussion and Writing Exercise*. Answers may vary. The area of a rectangle is 90 m^2. The length is 1 m greater than the width. Find the length and the width.

48. *Discussion and Writing Exercise*. Because Sheri did not first factor out the largest common factor, 4, her factorization will not be "complete" until she removes a common factor of 2 from each binomial.

49. *Familiarize*. Let $w =$ the width of the margins, in cm. Then the printed area on each page has dimensions $20 - 2w$ by $15 - 2w$. The area of the margins constitutes one-half the area of each page, so the printed area also constitutes one-half of the area.

Translate.

$$\underbrace{\text{Printed area}}\quad \text{is}\quad \underbrace{\text{one-half}}\quad \text{of}\quad \underbrace{\text{total area}}.$$

$$\begin{array}{ccccc} \downarrow & \downarrow & \downarrow & \downarrow & \downarrow \\ (20-2w)(15-2w) & = & \dfrac{1}{2} & \cdot & 20 \cdot 15 \end{array}$$

Solve.

$$(20 - 2w)(15 - 2w) = \frac{1}{2} \cdot 20 \cdot 15$$

$$300 - 70w + 4w^2 = 150$$

$$150 - 70w + 4w^2 = 0$$

$$4w^2 - 70w + 150 = 0$$

$$2(2w^2 - 35w + 75) = 0$$

$$2w^2 - 35w + 75 = 0 \quad \text{Dividing by 2}$$

$$(2w - 5)(w - 15) = 0$$

$$2w - 5 = 0 \quad or \quad w - 15 = 0$$

$$2w = 5 \quad or \qquad w = 15$$

$$w = 2.5 \ or \qquad w = 15$$

Check. If $w = 15$, then $20 - 2w$ and $15 - 2w$ are both negative. Since the dimensions of the printed area cannot be negative, 15 cannot be a solution. If $w = 2.5$, then $20 - 2w = 20 - 2(2.5) = 20 - 5 = 15$ and $15 - 2w = 15 - 2(2.5) = 15 - 5 = 10$. Thus the printed area is $15 \cdot 10$, or 150 cm^2. This is one-half of the total area of the page, $20 \cdot 15$, or 300 cm^2. The number 2.5 checks.

State. The width of the margins is 2.5 cm.

50. Familiarize. Let $n =$ the number.

Translate.

$$\underbrace{\text{The cube of a number}}_{\substack{\downarrow \\ n^3}} \quad \underset{\substack{\downarrow \\ =}}{\text{is}} \quad \underset{\substack{\downarrow \\ 2 \cdot}}{\text{twice}} \quad \underbrace{\text{the square of the number.}}_{\substack{\downarrow \\ n^2}}$$

Solve.

$$n^3 = 2n^2$$

$$n^3 - 2n^2 = 0$$

$$n^2(n - 2) = 0$$

$$n \cdot n(n - 2) = 0$$

$$n = 0 \ or \ n = 0 \ or \ n - 2 = 0$$

$$n = 0 \ or \ n = 0 \ or \qquad n = 2$$

Check. If $n = 0$, then $n^3 = 0^3 = 0$, $n^2 = 0^2 = 0$, and $0 = 2 \cdot 0$. If $n = 2$, then $n^3 = 2^3 = 8$, $n^2 = 2^2 = 4$, and $8 = 2 \cdot 4$. Both numbers check.

State. The number is 0 or 2.

51. Familiarize. Let $w =$ the width of the original rectangle. Then $2w =$ the length. The new length and width are $2w + 20$ and $w - 1$, respectively.

Translate.

$$\underbrace{\text{The new area}}_{\substack{\downarrow \\ (2w + 20)(w - 1)}} \quad \underset{\substack{\downarrow \\ =}}{\text{is}} \quad \underset{\substack{\downarrow \\ 160}}{160.}$$

Solve.

$$(2w + 20)(w - 1) = 160$$

$$2w^2 + 18w - 20 = 160$$

$$2w^2 + 18w - 180 = 0$$

$$2(w^2 + 9w - 90) = 0$$

$$w^2 + 9w - 90 = 0$$

$$(w + 15)(w - 6) = 0$$

$$w + 15 = 0 \quad or \quad w - 6 = 0$$

$$w = -15 \quad or \qquad w = 6$$

Check. The dimensions of the rectangle cannot be negative, so -15 cannot be a solution. If $w = 6$, then $2w = 2 \cdot 6 = 12$, $2w + 20 = 12 + 20 = 32$ and $w - 1 = 6 - 1 = 5$. The area of a rectangle with dimensions 32 by 5 is $32 \cdot 5$, or 160, so the answer checks.

State. The length of the original rectangle is 12 and the width is 6.

52. $x^2 + 25 = 0$

Since $x^2 + 25$ cannot be factored, the equation has no solution.

53. $(x - 2)(x + 3)(2x - 5) = 0$

$$x - 2 = 0 \ or \ x + 3 = 0 \quad or \ 2x - 5 = 0$$

$$x = 2 \ or \qquad x = -3 \ or \qquad 2x = 5$$

$$x = 2 \ or \qquad x = -3 \ or \qquad x = \frac{5}{2}$$

The solutions are 2, -3, and $\frac{5}{2}$.

54. $(x - 3)4x^2 + 3x(x - 3) - (x - 3)10 = 0$

$$(4x^2 + 3x - 10)(x - 3) = 0 \quad \text{Factoring out } x - 3$$

$$(4x - 5)(x + 2)(x - 3) = 0 \quad \text{Factoring } 4x^2 + 3x - 10$$

$$4x - 5 = 0 \ or \ x + 2 = 0 \quad or \ x - 3 = 0$$

$$4x = 5 \ or \qquad x = -2 \ or \qquad x = 3$$

$$x = \frac{5}{4} \ or \qquad x = -2 \ or \qquad x = 3$$

The solutions are $\frac{5}{4}$, -2, and 3.

55. The shaded area is the area of a circle with radius x less the area of a square with a diagonal of length $x + x$, or $2x$. The area of the circle is πx^2. The square can be thought of as two triangles, each with base $2x$ and height x. Then the area of the square is $2 \cdot \frac{1}{2} \cdot 2x \cdot x$, or $2x^2$. We subtract to find the shaded area.

$$\pi x^2 - 2x^2 = (\pi - 2)x^2$$

Chapter 5 Test

---

1. $28x^3 = 2 \cdot 2 \cdot 7 \cdot x^3$

$48x^7 = 2 \cdot 2 \cdot 2 \cdot 2 \cdot 3 \cdot x^7$

The coefficients each have two factors of 2. There are no other common prime factors. The GCF of the powers of x is x^3 because 3 is the smallest exponent of x. Thus the GCF is $2 \cdot 2 \cdot x^3$, or $4x^3$.

2. $x^2 - 7x + 10$

We look for a pair of factors of 10 whose sum is -7. The numbers we need are -2 and -5.

$$x^2 - 7x + 10 = (x - 2)(x - 5)$$

3. $x^2 + 25 - 10x = x^2 - 10x + 25$
$\quad\quad = x^2 - 2 \cdot x \cdot 5 + 5^2$
$\quad\quad = (x-5)^2$

4. $6y^2 - 8y^3 + 4y^4 = 4y^4 - 8y^3 + 6y^2 =$
$2y^2 \cdot 2y^2 - 2y^2 \cdot 4y + 2y^3 \cdot 3 = 2y^2(2y^2 - 4y + 3)$
Since $2y^2 - 4y + 3$ cannot be factored, the factorization is complete.

5. $\quad x^3 + x^2 + 2x + 2$
$\quad = (x^3 + x^2) + (2x + 2)$
$\quad = x^2(x+1) + 2(x+1)$ Factoring by grouping
$\quad = (x^2 + 2)(x + 1)$

6. $x^2 - 5x = x \cdot x - 5 \cdot x = x(x - 5)$

7. $\quad x^3 + 2x^2 - 3x$
$\quad = x(x^2 + 2x - 3)$ x is a common factor.
$\quad = x(x+3)(x-1)$ Factoring the trinomial

8. $\quad 28x - 48 + 10x^2$
$\quad = 10x^2 + 28x - 48$
$\quad = 2(5x^2 + 14x - 24)$ 2 is a common factor.
$\quad = 2(5x - 6)(x + 4)$ Factoring the trinomial

9. $4x^2 - 9 = (2x)^2 - 3^2$ Difference of squares
$\quad\quad\quad = (2x + 3)(2x - 3)$

10. $x^2 - x - 12$
We look for a pair of factors of -12 whose sum is -1. The numbers we need are -4 and 3.
$\quad x^2 - x - 12 = (x - 4)(x + 3)$

11. $\quad 6m^3 + 9m^2 + 3m$
$\quad = 3m(2m^2 + 3m + 1)$ $3m$ is a common factor.
$\quad = 3m(2m+1)(m+1)$ Factoring the trinomial

12. $3w^2 - 75 = 3(w^2 - 25)$ 3 is a common factor.
$\quad\quad = 3(w^2 - 5^2)$ Difference of squares
$\quad\quad = 3(w+5)(w-5)$

13. $\quad 60x + 45x^2 + 20$
$\quad = 45x^2 + 60x + 20$
$\quad = 5(9x^2 + 12x + 4)$ 5 is a common factor.
$\quad = 5[(3x)^2 + 2 \cdot 3x \cdot 2 + 2^2]$ Trinomial square
$\quad = 5(3x + 2)^2$

14. $\quad 3x^4 - 48$
$\quad = 3(x^4 - 16)$ 3 is a common factor.
$\quad = 3[(x^2)^2 - 4^2]$ Difference of squares
$\quad = 3(x^2 + 4)(x^2 - 4)$
$\quad = 3(x^2 + 4)(x^2 - 2^2)$ Difference of squares
$\quad = 3(x^2 + 4)(x + 2)(x - 2)$

15. $\quad 49x^2 - 84x + 36$
$\quad = (7x)^2 - 2 \cdot 7x \cdot 6 + 6^2$ Trinomial square
$\quad = (7x - 6)^2$

16. $5x^2 - 26x + 5$
There is no common factor (other than 1). This polynomial has 3 terms, but it is not a trinomial square. Using the ac-method we first multiply the leading coefficient and the constant term: $5 \cdot 5 = 25$. Try to factor 25 so that the sum of the factors is -26. The numbers we want are -1 and -25: $-1(-25) = 25$ and $-1 + (-25) = -26$. Split the middle term and factor by grouping:
$5x^2 - 26x + 5 = 5x^2 - x - 25x + 5$
$\quad\quad\quad = (5x^2 - x) + (-25x + 5)$
$\quad\quad\quad = x(5x - 1) - 5(5x - 1)$
$\quad\quad\quad = (x - 5)(5x - 1)$

17. $\quad x^4 + 2x^3 - 3x - 6$
$\quad = (x^4 + 2x^3) + (-3x - 6)$
$\quad = x^2(x + 2) - 3(x + 2)$ Factoring by grouping
$\quad = (x^3 - 3)(x + 2)$

18. $\quad 80 - 5x^4$
$\quad = 5(16 - x^4)$ 5 is a common factor.
$\quad = 5[4^2 - (x^2)^2]$ Difference of squares
$\quad = 5(4 + x^2)(4 - x^2)$
$\quad = 5(4 + x^2)(2^2 - x^2)$ Difference of squares
$\quad = 5(4 + x^2)(2 + x)(2 - x)$

19. $4x^2 - 4x - 15$
(1) There is no common factor (other than 1 or -1).
(2) Factor the first term, $4x^2$. The possibilities are x, $4x$ and $2x$, $2x$. We have these as possibilities for factorizations:
$\quad (x+\quad)(4x+\quad)$ and $(2x+\quad)(2x+\quad)$
(3) Factor the last term, -15. The possibilities are 1, -15 and -1, 15 and 3, -5 and -3, 5. These factors can also be written as -15, 1 and 15, -1 and -5, 3 and 5, -3.
(4) We try some possibilities and find that the factorization is $(2x + 3)(2x - 5)$.

20. $\quad 6t^3 + 9t^2 - 15t$
$\quad = 3t(2t^2 + 3t - 5)$ $3t$ is a common factor.
$\quad = 3t(2t + 5)(t - 1)$ Factoring the trinomial

21. $\quad 3m^2 - 9mn - 30n^2$
$\quad = 3(m^2 - 3mn - 10n^2)$ 3 is a common factor.
$\quad = 3(m - 5n)(m + 2n)$ Factoring the trinomial

22. $\quad x^2 - x - 20 = 0$
$\quad (x - 5)(x + 4) = 0$
$\quad x - 5 = 0 \ or \ x + 4 = 0$
$\quad\quad x = 5 \ or \quad\quad x = -4$
The solutions are 5 and -4.

23.
$$2x^2 + 7x = 15$$
$$2x^2 + 7x - 15 = 0$$
$$(2x - 3)(x + 5) = 0$$
$$2x - 3 = 0 \quad or \quad x + 5 = 0$$
$$2x = 3 \quad or \qquad x = -5$$
$$x = \frac{3}{2} \quad or \qquad x = -5$$

The solutions are $\frac{3}{2}$ and -5.

24.
$$x(x - 3) = 28$$
$$x^2 - 3x = 28$$
$$x^2 - 3x - 28 = 0$$
$$(x - 7)(x + 4) = 0$$
$$x - 7 = 0 \quad or \quad x + 4 = 0$$
$$x = 7 \quad or \qquad x = -4$$

The solutions are 7 and -4.

25. **Familiarize**. Let $w =$ the width, in meters. Then $w + 2 =$ the length. Recall that the area of a rectangle is (length) $\cdot$ (width).

Translate. We use the formula for the area of a rectangle.
$$48 = (w + 2)w$$

Solve.
$$48 = (w + 2)w$$
$$48 = w^2 + 2w$$
$$0 = w^2 + 2w - 48$$
$$0 = (w + 8)(w - 6)$$
$$w + 8 = 0 \quad or \quad w - 6 = 0$$
$$w = -8 \quad or \qquad w = 6$$

Check. The width cannot be negative, so -8 cannot be a solution. If $w = 6$, then $w + 2 = 8$ and the area is $(8\text{ m}) \cdot (6\text{ m})$, or 48 m^2. The number 6 checks.

State. The length is 8 m and the width is 6 m.

26. **Familiarize**. Using the labels on the drawing in the text, we let $h =$ the height of the triangle, in cm, and $2h + 6 =$ the base. Recall that the area of a triangle is $\frac{1}{2} \cdot$ (base) $\cdot$ (height).

Translate. We use the formula for the area of a triangle.
$$28 = \frac{1}{2} \cdot (2h + 6) \cdot h$$

Solve.
$$28 = \frac{1}{2}(2h + 6)h$$
$$28 = h^2 + 3h$$
$$0 = h^2 + 3h - 28$$
$$0 = (h + 7)(h - 4)$$
$$h + 7 = 0 \quad or \quad h - 4 = 0$$
$$h = -7 \quad or \qquad h = 4$$

Check. The height cannot be negative, so -7 cannot be a solution. If $h = 4$, then $2h + 6 = 2 \cdot 4 + 6 = 8 + 6 = 14$ and

the area is $\frac{1}{2} \cdot (14\text{ cm}) \cdot (4\text{ cm})$, or 28 cm^2. The number 4 checks.

State. The height is 4 cm and the base is 14 cm.

27. **Familiarize**. Using the labels on the drawing in the text, we let $x =$ the distance between the two marked points, in feet. If the corner is a right angle, the lengths 3 ft, 4 ft, and x will satisfy the Pythagorean equation.

Translate.
$$a^2 + b^2 = c^2$$
$$3^2 + 4^2 = x^2 \quad \text{Substituting}$$

Solve.
$$3^2 + 4^2 = x^2$$
$$9 + 16 = x^2$$
$$25 = x^2$$
$$0 = x^2 - 25$$
$$0 = (x + 5)(x - 5)$$
$$x + 5 = 0 \quad or \quad x - 5 = 0$$
$$x = -5 \quad or \qquad x = 5$$

Check. The distance cannot be negative, so -5 cannot be a solution. If $x = 5$, then we have $3^2 + 4^2 = 9 + 16 = 25 = 5^2$, so the answer checks.

State. The distance between the marked points should be 5 ft.

28. We let $y = 0$ and solve for x.
$$0 = x^2 - 2x - 35$$
$$0 = (x + 5)(x - 7)$$
$$x + 5 = 0 \quad or \quad x - 7 = 0$$
$$x = -5 \quad or \qquad x = 7$$

The x-intercepts are $(-5, 0)$ and $(7, 0)$.

29. We let $y = 0$ and solve for x.
$$0 = 3x^2 - 5x + 2$$
$$0 = (3x - 2)(x - 1)$$
$$3x - 2 = 0 \quad or \quad x - 1 = 0$$
$$3x = 2 \quad or \qquad x = 1$$
$$x = \frac{2}{3} \quad or \qquad x = 1$$

The x-intercepts are $\left(\frac{2}{3}, 0\right)$ and $(1, 0)$.

30. **Familiarize**. Let $w =$ the width of the original rectangle. Then $5w =$ the length. The new width and length are $w + 2$ and $5w - 3$, respectively.

Translate. We will use the formula for the area of a rectangle, Area = (length) $\cdot$ (width).
$$60 = (5w - 3)(w + 2)$$

Solve.
$$60 = (5w - 3)(w + 2)$$
$$60 = 5w^2 + 7w - 6$$
$$0 = 5w^2 + 7w - 66$$
$$0 = (5w + 22)(w - 3)$$

$$5w + 22 = 0 \quad or \quad w - 3 = 0$$
$$5w = -22 \quad or \quad w = 3$$
$$w = -\frac{22}{5} \quad or \quad w = 3$$

Check. The width cannot be negative, so $-\frac{22}{5}$ cannot be a solution. If $w = 3$, then $5w = 5 \cdot 3 = 15$. The new dimensions are $w + 2$, or $3 + 2$, or 5 and $5w - 3$, or $15 - 3$, or 12, and the new area is $12 \cdot 5$, or 60. The number 3 checks.

State. The original length is 15 and the width is 3.

31. $(a + 3)^2 - 2(a + 3) - 35$

We can think of $a + 3$ as the variable in this expression. Then we find a pair of factors of -35 whose sum is -2. The numbers we want are -7 and 5.

$(a + 3)^2 - 2(a + 3) - 35 = [(a + 3) - 7][(a + 3) + 5] = (a - 4)(a + 8)$

We could also do this exercise as follows:
$$(a + 3)^2 - 2(a + 3) - 35 = a^2 + 6a + 9 - 2a - 6 - 35$$
$$= a^2 + 4a - 32$$
$$= (a - 4)(a + 8)$$

32.
$$20x(x + 2)(x - 1) = 5x^3 - 24x - 14x^2$$
$$(20x^2 + 40x)(x - 1) = 5x^3 - 24x - 14x^2$$
$$20x^3 + 20x^2 - 40x = 5x^3 - 24x - 14x^2$$
$$15x^3 + 34x^2 - 16x = 0$$
$$x(15x^2 + 34x - 16) = 0$$
$$x(3x + 8)(5x - 2) = 0$$
$$x = 0 \quad or \quad 3x + 8 = 0 \quad or \quad 5x - 2 = 0$$
$$x = 0 \quad or \quad 3x = -8 \quad or \quad 5x = 2$$
$$x = 0 \quad or \quad x = -\frac{8}{3} \quad or \quad x = \frac{2}{5}$$

33. $x^2 - y^2 = (x + y)(x - y) = 4 \cdot 6 = 24$, so choice (d) is correct.

Chapter 6

Rational Expressions and Equations

Exercise Set 6.1

1. $\dfrac{-3}{2x}$

To determine the numbers for which the rational expression is not defined, we set the denominator equal to 0 and solve:

$$2x = 0$$
$$x = 0$$

The expression is not defined for the replacement number 0.

3. $\dfrac{5}{x-8}$

To determine the numbers for which the rational expression is not defined, we set the denominator equal to 0 and solve:

$$x - 8 = 0$$
$$x = 8$$

The expression is not defined for the replacement number 8.

5. $\dfrac{3}{2y+5}$

Set the denominator equal to 0 and solve:

$$2y + 5 = 0$$
$$2y = -5$$
$$y = -\frac{5}{2}$$

The expression is not defined for the replacement number $-\dfrac{5}{2}$.

7. $\dfrac{x^2 + 11}{x^2 - 3x - 28}$

Set the denominator equal to 0 and solve:

$$x^2 - 3x - 28 = 0$$
$$(x - 7)(x + 4) = 0$$
$$x - 7 = 0 \quad \text{or} \quad x + 4 = 0$$
$$x = 7 \quad \text{or} \qquad x = -4$$

The expression is not defined for the replacement numbers 7 and -4.

9. $\dfrac{m^3 - 2m}{m^2 - 25}$

Set the denominator equal to 0 and solve:

$$m^2 - 25 = 0$$
$$(m + 5)(m - 5) = 0$$
$$m + 5 = 0 \quad \text{or} \quad m - 5 = 0$$
$$m = -5 \quad \text{or} \qquad m = 5$$

The expression is not defined for the replacement numbers -5 and 5.

11. $\dfrac{x-4}{3}$

Since the denominator is the constant 3, there are no replacement numbers for which the expression is not defined.

13. $\dfrac{4x}{4x} \cdot \dfrac{3x^2}{5y} = \dfrac{(4x)(3x^2)}{(4x)(5y)}$ Multiplying the numerators and the denominators

15. $\dfrac{2x}{2x} \cdot \dfrac{x-1}{x+4} = \dfrac{2x(x-1)}{2x(x+4)}$ Multiplying the numerators and the denominators

17. $\dfrac{3-x}{4-x} \cdot \dfrac{-1}{-1} = \dfrac{(3-x)(-1)}{(4-x)(-1)}$, or $\dfrac{-1(3-x)}{-1(4-x)}$

19. $\dfrac{y+6}{y+6} \cdot \dfrac{y-7}{y+2} = \dfrac{(y+6)(y-7)}{(y+6)(y+2)}$

21. $\dfrac{8x^3}{32x} = \dfrac{8 \cdot x \cdot x^2}{8 \cdot 4 \cdot x}$ Factoring numerator and denominator

$\qquad = \dfrac{8x}{8x} \cdot \dfrac{x^2}{4}$ Factoring the rational expression

$\qquad = 1 \cdot \dfrac{x^2}{4} \qquad \left(\dfrac{8x}{8x} = 1\right)$

$\qquad = \dfrac{x^2}{4}$ We removed a factor of 1.

23. $\dfrac{48p^7q^5}{18p^5q^4} = \dfrac{8 \cdot 6 \cdot p^5 \cdot p^2 \cdot q^4 \cdot q}{6 \cdot 3 \cdot p^5 \cdot q^4}$ Factoring numerator and denominator

$\qquad = \dfrac{6p^5q^4}{6p^5q^4} \cdot \dfrac{8p^2q}{3}$ Factoring the rational expression

$\qquad = 1 \cdot \dfrac{8p^2q}{3} \qquad \left(\dfrac{6p^5q^4}{6p^5q^4} = 1\right)$

$\qquad = \dfrac{8p^2q}{3}$ Removing a factor of 1

25. $\dfrac{4x-12}{4x} = \dfrac{4(x-3)}{4 \cdot x}$

$\qquad = \dfrac{4}{4} \cdot \dfrac{x-3}{x}$

$\qquad = 1 \cdot \dfrac{x-3}{x}$

$\qquad = \dfrac{x-3}{x}$

27. $\dfrac{3m^2 + 3m}{6m^2 + 9m} = \dfrac{3m(m+1)}{3m(2m+3)}$

$\qquad = \dfrac{3m}{3m} \cdot \dfrac{m+1}{2m+3}$

$\qquad = 1 \cdot \dfrac{m+1}{2m+3}$

$\qquad = \dfrac{m+1}{2m+3}$

29. $\dfrac{a^2-9}{a^2+5a+6} = \dfrac{(a-3)(a+3)}{(a+2)(a+3)}$

$= \dfrac{a-3}{a+2} \cdot \dfrac{a+3}{a+3}$

$= \dfrac{a-3}{a+2} \cdot 1$

$= \dfrac{a-3}{a+2}$

31. $\dfrac{a^2-10a+21}{a^2-11a+28} = \dfrac{(a-7)(a-3)}{(a-7)(a-4)}$

$= \dfrac{a-7}{a-7} \cdot \dfrac{a-3}{a-4}$

$= 1 \cdot \dfrac{a-3}{a-4}$

$= \dfrac{a-3}{a-4}$

33. $\dfrac{x^2-25}{x^2-10x+25} = \dfrac{(x-5)(x+5)}{(x-5)(x-5)}$

$= \dfrac{x-5}{x-5} \cdot \dfrac{x+5}{x-5}$

$= 1 \cdot \dfrac{x+5}{x-5}$

$= \dfrac{x+5}{x-5}$

35. $\dfrac{a^2-1}{a-1} = \dfrac{(a-1)(a+1)}{a-1}$

$= \dfrac{a-1}{a-1} \cdot \dfrac{a+1}{1}$

$= 1 \cdot \dfrac{a+1}{1}$

$= a+1$

37. $\dfrac{x^2+1}{x+1}$ cannot be simplified.

Neither the numerator nor the denominator can be factored.

39. $\dfrac{6x^2-54}{4x^2-36} = \dfrac{2 \cdot 3(x^2-9)}{2 \cdot 2(x^2-9)}$

$= \dfrac{2(x^2-9)}{2(x^2-9)} \cdot \dfrac{3}{2}$

$= 1 \cdot \dfrac{3}{2}$

$= \dfrac{3}{2}$

41. $\dfrac{6t+12}{t^2-t-6} = \dfrac{6(t+2)}{(t-3)(t+2)}$

$= \dfrac{6}{t-3} \cdot \dfrac{t+2}{t+2}$

$= \dfrac{6}{t-3} \cdot 1$

$= \dfrac{6}{t-3}$

43. $\dfrac{2t^2+6t+4}{4t^2-12t-16} = \dfrac{2(t^2+3t+2}{4(t^2-3t-4)}$

$= \dfrac{2(t+2)(t+1)}{2 \cdot 2(t-4)(t+1)}$

$= \dfrac{2(t+1)}{2(t+1)} \cdot \dfrac{t+2}{2(t-4)}$

$= 1 \cdot \dfrac{t+2}{2(t-4)}$

$= \dfrac{t+2}{2(t-4)}$

45. $\dfrac{t^2-4}{(t+2)^2} = \dfrac{(t-2)(t+2)}{(t+2)(t+2)}$

$= \dfrac{t-2}{t+2} \cdot \dfrac{t+2}{t+2}$

$= \dfrac{t-2}{t+2} \cdot 1$

$= \dfrac{t-2}{t+2}$

47. $\dfrac{6-x}{x-6} = \dfrac{-(-6+x)}{x-6}$

$= \dfrac{-1(x-6)}{x-6}$

$= -1 \cdot \dfrac{x-6}{x-6}$

$= -1 \cdot 1$

$= -1$

49. $\dfrac{a-b}{b-a} = \dfrac{-(-a+b)}{b-a}$

$= \dfrac{-1(b-a)}{b-a}$

$= -1 \cdot \dfrac{b-a}{b-a}$

$= -1 \cdot 1$

$= -1$

51. $\dfrac{6t-12}{2-t} = \dfrac{-6(-t+2)}{2-t}$

$= \dfrac{-6(2-t)}{2-t}$

$= \dfrac{-6(2-t)}{2-t}$

$= -6$

53. $\dfrac{x^2-1}{1-x} = \dfrac{(x+1)(x-1)}{-1(-1+x)}$

$= \dfrac{(x+1)(x-1)}{-1(x-1)}$

$= \dfrac{(x+1)(x-1)}{-1(x-1)}$

$= -(x+1)$

$= -x-1$

55. $\dfrac{4x^3}{3x} \cdot \dfrac{14}{x} = \dfrac{4x^3 \cdot 14}{3x \cdot x}$ Multiplying the numerators and the denominators

$\quad = \dfrac{4 \cdot x \cdot x \cdot x \cdot 14}{3 \cdot x \cdot x}$ Factoring the numerator and the denominator

$\quad = \dfrac{4 \cdot \cancel{x} \cdot \cancel{x} \cdot x \cdot 14}{3 \cdot \cancel{x} \cdot \cancel{x}}$ Removing a factor of 1

$\quad = \dfrac{56x}{3}$ Simplifying

57. $\dfrac{3c}{d^2} \cdot \dfrac{4d}{6c^3} = \dfrac{3c \cdot 4d}{d^2 \cdot 6c^3}$ Multiplying the numerators and the denominators

$\quad = \dfrac{3 \cdot c \cdot 2 \cdot 2 \cdot d}{d \cdot d \cdot 3 \cdot 2 \cdot c \cdot c \cdot c}$ Factoring the numerator and the denominator

$\quad = \dfrac{\cancel{3} \cdot \cancel{c} \cdot \cancel{2} \cdot 2 \cdot \cancel{d}}{d \cdot d \cdot \cancel{3} \cdot \cancel{2} \cdot \cancel{c} \cdot c \cdot c}$

$\quad = \dfrac{2}{dc^2}$

59. $\dfrac{x^2 - 3x - 10}{(x-2)^2} \cdot \dfrac{x-2}{x-5} = \dfrac{(x^2 - 3x - 10)(x-2)}{(x-2)^2(x-5)}$

$\quad = \dfrac{(x-5)(x+2)(x-2)}{(x-2)(x-2)(x-5)}$

$\quad = \dfrac{(x\cancel{-5})(x+2)(x\cancel{-2})}{(x\cancel{-2})(x-2)(x\cancel{-5})}$

$\quad = \dfrac{x+2}{x-2}$

61. $\dfrac{a^2 - 9}{a^2} \cdot \dfrac{a^2 - 3a}{a^2 + a - 12} = \dfrac{(a-3)(a+3)(a)(a-3)}{a \cdot a(a+4)(a-3)}$

$\quad = \dfrac{(a\cancel{-3})(a+3)(\cancel{a})(a-3)}{\cancel{a} \cdot a(a+4)(a\cancel{-3})}$

$\quad = \dfrac{(a-3)(a+3)}{a(a+4)}$

63. $\dfrac{4a^2}{3a^2 - 12a + 12} \cdot \dfrac{3a-6}{2a} = \dfrac{4a^2(3a-6)}{(3a^2 - 12a + 12)2a}$

$\quad = \dfrac{2 \cdot 2 \cdot a \cdot a \cdot 3 \cdot (a-2)}{3 \cdot (a-2) \cdot (a-2) \cdot 2 \cdot a}$

$\quad = \dfrac{\cancel{2} \cdot 2 \cdot \cancel{a} \cdot a \cdot \cancel{3} \cdot (a\cancel{-2})}{\cancel{3} \cdot (a\cancel{-2}) \cdot (a-2) \cdot \cancel{2} \cdot \cancel{a}}$

$\quad = \dfrac{2a}{a-2}$

65. $\dfrac{t^4 - 16}{t^4 - 1} \cdot \dfrac{t^2 + 1}{t^2 + 4}$

$\quad = \dfrac{(t^4 - 16)(t^2 + 1)}{(t^4 - 1)(t^2 + 4)}$

$\quad = \dfrac{(t^2 + 4)(t+2)(t-2)(t^2 + 1)}{(t^2 + 1)(t+1)(t-1)(t^2 + 4)}$

$\quad = \dfrac{(t^2\cancel{+4})(t+2)(t-2)(t^2\cancel{+1})}{(t^2\cancel{+1})(t+1)(t-1)(t^2\cancel{+4})}$

$\quad = \dfrac{(t+2)(t-2)}{(t+1)(t-1)}$

67. $\dfrac{(x+4)^3}{(x+2)^3} \cdot \dfrac{x^2 + 4x + 4}{x^2 + 8x + 16}$

$\quad = \dfrac{(x+4)^3(x^2 + 4x + 4)}{(x+2)^3(x^2 + 8x + 16)}$

$\quad = \dfrac{(x+4)(x+4)(x+4)(x+2)(x+2)}{(x+2)(x+2)(x+2)(x+4)(x+4)}$

$\quad = \dfrac{(x\cancel{+4})(x\cancel{+4})(x+4)(x\cancel{+2})(x\cancel{+2})}{(x\cancel{+2})(x\cancel{+2})(x+2)(x\cancel{+4})(x\cancel{+4})}$

$\quad = \dfrac{x+4}{x+2}$

69. $\dfrac{5a^2 - 180}{10a^2 - 10} \cdot \dfrac{20a + 20}{2a - 12} = \dfrac{(5a^2 - 180)(20a + 20)}{(10a^2 - 10)(2a - 12)}$

$\quad = \dfrac{5(a+6)(a-6)(2)(10)(a+1)}{10(a+1)(a-1)(2)(a-6)}$

$\quad = \dfrac{5(a+6)(a\cancel{-6})\cancel{(2)}\cancel{(10)}(a\cancel{+1})}{\cancel{10}(a\cancel{+1})(a-1)\cancel{(2)}(a\cancel{-6})}$

$\quad = \dfrac{5(a+6)}{a-1}$

71. Discussion and Writing Exercise

73. *Familiarize.* Let $x =$ the smaller even integer. Then $x + 2 =$ the larger even integer.

Translate. We reword the problem.

$$\underbrace{\text{Smaller even integer}}_{x} \cdot \underbrace{\text{larger even integer}}_{(x+2)} \; \text{is } 360.$$
$$x \cdot (x+2) = 360$$

Solve.
$$x(x+2) = 360$$
$$x^2 + 2x = 360$$
$$x^2 + 2x - 360 = 0$$
$$(x+20)(x-18) = 0$$
$$x + 20 = 0 \quad \text{or} \quad x - 18 = 0$$
$$x = -20 \quad \text{or} \quad x = 18$$

Check. The solutions of the equation are -20 and 18. When $x = -20$, then $x + 2 = -18$ and $-20(-18) = 360$. The numbers -20 and -18 are consecutive even integers which are solutions to the problem. When $x = 18$, then $x + 2 = 20$ and $18 \cdot 20 = 360$. The numbers 18 and 20 are also consecutive even integers which are solutions to the problem.

State. We have two solutions, each of which consists of a pair of numbers: -20 and -18, and 18 and 20.

75. $x^2 - x - 56$

We look for a pair of numbers whose product is -56 and whose sum is -1. The numbers are -8 and 7.
$$x^2 - x - 56 = (x-8)(x+7)$$

77. $x^5 - 2x^4 - 35x^3 = x^3(x^2 - 2x - 35) = x^3(x-7)(x+5)$

79. $16 - t^4 = 4^2 - (t^2)^2$ Difference of squares

$\quad = (4 + t^2)(4 - t^2)$

$\quad = (4 + t^2)(2^2 - t^2)$ Difference of squares

$\quad = (4 + t^2)(2 + t)(2 - t)$

81. $x^2 - 9x + 14$

We look for a pair of numbers whose product is 14 and whose sum is -9. The numbers are -2 and -7.

$$x^2 - 9x + 14 = (x - 2)(x - 7)$$

83. $\quad 16x^2 - 40xy + 25y^2$

$= (4x)^2 - 2 \cdot 4x \cdot 5y + (5y)^2 \quad$ Trinomial square

$= (4x - 5y)^2$

85. $\quad \dfrac{x^4 - 16y^2}{(x^2 + 4y^2)(x - 2y)}$

$= \dfrac{(x^2 + 4y^2)(x + 2y)(x - 2y)}{(x^2 + 4y^2)(x - 2y)}$

$= \dfrac{(x^2 + 4y^2)\,(x + 2y)(x - 2y)}{(x^2 + 4y^2)\,(x - 2y)(1)}$

$= x + 2y$

87. $\quad \dfrac{t^4 - 1}{t^4 - 81} \cdot \dfrac{t^2 - 9}{t^2 + 1} \cdot \dfrac{(t - 9)^2}{(t + 1)^2}$

$= \dfrac{(t^2 + 1)(t + 1)(t - 1)(t + 3)(t - 3)(t - 9)(t - 9)}{(t^2 + 9)(t + 3)(t - 3)(t^2 + 1)(t + 1)(t + 1)}$

$= \dfrac{(t^2 + 1)(t + 1)(t - 1)(t + 3)(t - 3)(t - 9)(t - 9)}{(t^2 + 9)(t + 3)(t - 3)(t^2 + 1)(t + 1)(t + 1)}$

$= \dfrac{(t - 1)(t - 9)(t - 9)}{(t^2 + 9)(t + 1)}, \text{ or } \dfrac{(t - 1)(t - 9)^2}{(t^2 + 9)(t + 1)}$

89. $\quad \dfrac{x^2 - y^2}{(x - y)^2} \cdot \dfrac{x^2 - 2xy + y^2}{x^2 - 4xy - 5y^2}$

$= \dfrac{(x + y)(x - y)(x - y)(x - y)}{(x - y)(x - y)(x - 5y)(x + y)}$

$= \dfrac{(x + y)(x - y)(x - y)(x - y)}{(x - y)(x - y)(x - 5y)(x + y)}$

$= \dfrac{x - y}{x - 5y}$

91. $\dfrac{5(2x + 5) - 25}{10} = \dfrac{10x + 25 - 25}{10}$

$= \dfrac{10x}{10}$

$= x$

You get the same number you selected.

To do a number trick, ask someone to select a number and then perform these operations. The person will probably be surprised that the result is the original number.

Exercise Set 6.2

1. The reciprocal of $\dfrac{4}{x}$ is $\dfrac{x}{4}$ because $\dfrac{4}{x} \cdot \dfrac{x}{4} = 1$.

3. The reciprocal of $x^2 - y^2$ is $\dfrac{1}{x^2 - y^2}$ because

$\dfrac{x^2 - y^2}{1} \cdot \dfrac{1}{x^2 - y^2} = 1$.

5. The reciprocal of $\dfrac{1}{a + b}$ is $a + b$ because $\dfrac{1}{a + b} \cdot (a + b) = 1$.

7. The reciprocal of $\dfrac{x^2 + 2x - 5}{x^2 - 4x + 7}$ is $\dfrac{x^2 - 4x + 7}{x^2 + 2x - 5}$ because

$\dfrac{x^2 + 2x - 5}{x^2 - 4x + 7} \cdot \dfrac{x^2 - 4x + 7}{x^2 + 2x - 5} = 1$.

9. $\dfrac{2}{5} \div \dfrac{4}{3} = \dfrac{2}{3} \cdot \dfrac{3}{4} \quad$ Multiplying by the reciprocal of the divisor

$= \dfrac{2 \cdot 3}{5 \cdot 4}$

$= \dfrac{2 \cdot 3}{5 \cdot 2 \cdot 2} \quad$ Factoring the denominator

$= \dfrac{2 \cdot 3}{5 \cdot 2 \cdot 2} \quad$ Removing a factor of 1

$= \dfrac{3}{10} \quad$ Simplifying

11. $\dfrac{2}{x} \div \dfrac{8}{x} = \dfrac{2}{x} \cdot \dfrac{x}{8} \quad$ Multiplying by the reciprocal of the divisor

$= \dfrac{2 \cdot x}{x \cdot 8}$

$= \dfrac{2 \cdot x \cdot 1}{x \cdot 2 \cdot 4} \quad$ Factoring the numerator and the denominator

$= \dfrac{2 \cdot x \cdot 1}{x \cdot 2 \cdot 4} \quad$ Removing a factor of 1

$= \dfrac{1}{4} \quad$ Simplifying

13. $\dfrac{a}{b^2} \div \dfrac{a^2}{b^3} = \dfrac{a}{b^2} \cdot \dfrac{b^3}{a^2} \quad$ Multiplying by the reciprocal of the divisor

$= \dfrac{a \cdot b^3}{b^2 \cdot a^2}$

$= \dfrac{a \cdot b^2 \cdot b}{b^2 \cdot a \cdot a}$

$= \dfrac{a \cdot b^2 \cdot b}{b^2 \cdot a \cdot a}$

$= \dfrac{b}{a}$

15. $\dfrac{a + 2}{a - 3} \div \dfrac{a - 1}{a + 3} = \dfrac{a + 2}{a - 3} \cdot \dfrac{a + 3}{a - 1}$

$= \dfrac{(a + 2)(a + 3)}{(a - 3)(a - 1)}$

17. $\dfrac{x^2 - 1}{x} \div \dfrac{x + 1}{x - 1} = \dfrac{x^2 - 1}{x} \cdot \dfrac{x - 1}{x + 1}$

$= \dfrac{(x^2 - 1)(x - 1)}{x(x + 1)}$

$= \dfrac{(x - 1)(x + 1)(x - 1)}{x(x + 1)}$

$= \dfrac{(x - 1)(x + 1)(x - 1)}{x(x + 1)}$

$= \dfrac{(x - 1)^2}{x}$

19. $\dfrac{x+1}{6} \div \dfrac{x+1}{3} = \dfrac{x+1}{6} \cdot \dfrac{3}{x+1}$

$\qquad = \dfrac{(x+1)\cdot 3}{6(x+1)}$

$\qquad = \dfrac{3(x+1)}{2\cdot 3(x+1)}$

$\qquad = \dfrac{1\cdot \cancel{3}(\cancel{x+1})}{2\cdot \cancel{3}(\cancel{x+1})}$

$\qquad = \dfrac{1}{2}$

21. $\dfrac{5x-5}{16} \div \dfrac{x-1}{6} = \dfrac{5x-5}{16} \cdot \dfrac{6}{x-1}$

$\qquad = \dfrac{(5x-5)\cdot 6}{16(x-1)}$

$\qquad = \dfrac{5(x-1)\cdot 2\cdot 3}{2\cdot 8(x-1)}$

$\qquad = \dfrac{5(\cancel{x-1})\cdot \cancel{2}\cdot 3}{\cancel{2}\cdot 8(\cancel{x-1})}$

$\qquad = \dfrac{15}{8}$

23. $\dfrac{-6+3x}{5} \div \dfrac{4x-8}{25} = \dfrac{-6+3x}{5} \cdot \dfrac{25}{4x-8}$

$\qquad = \dfrac{(-6+3x)\cdot 25}{5(4x-8)}$

$\qquad = \dfrac{3(x-2)\cdot 5\cdot 5}{5\cdot 4(x-2)}$

$\qquad = \dfrac{3(\cancel{x-2})\cdot \cancel{5}\cdot 5}{\cancel{5}\cdot 4(\cancel{x-2})}$

$\qquad = \dfrac{15}{4}$

25. $\dfrac{a+2}{a-1} \div \dfrac{3a+6}{a-5} = \dfrac{a+2}{a-1} \cdot \dfrac{a-5}{3a+6}$

$\qquad = \dfrac{(a+2)(a-5)}{(a-1)(3a+6)}$

$\qquad = \dfrac{(a+2)(a-5)}{(a-1)\cdot 3\cdot (a+2)}$

$\qquad = \dfrac{(\cancel{a+2})(a-5)}{(a-1)\cdot 3\cdot (\cancel{a+2})}$

$\qquad = \dfrac{a-5}{3(a-1)}$

27. $\dfrac{x^2-4}{x} \div \dfrac{x-2}{x+2} = \dfrac{x^2-4}{x} \cdot \dfrac{x+2}{x-2}$

$\qquad = \dfrac{(x^2-4)(x+2)}{x(x-2)}$

$\qquad = \dfrac{(x-2)(x+2)(x+2)}{x(x-2)}$

$\qquad = \dfrac{(\cancel{x-2})(x+2)(x+2)}{x(\cancel{x-2})}$

$\qquad = \dfrac{(x+2)^2}{x}$

29. $\dfrac{x^2-9}{4x+12} \div \dfrac{x-3}{6} = \dfrac{x^2-9}{4x+12} \cdot \dfrac{6}{x-3}$

$\qquad = \dfrac{(x^2-9)\cdot 6}{(4x+12)(x-3)}$

$\qquad = \dfrac{(x-3)(x+3)\cdot 3\cdot 2}{2\cdot 2(x+3)(x-3)}$

$\qquad = \dfrac{(\cancel{x-3})(\cancel{x+3})\cdot 3\cdot \cancel{2}}{\cancel{2}\cdot 2(\cancel{x+3})(\cancel{x-3})}$

$\qquad = \dfrac{3}{2}$

31. $\dfrac{c^2+3c}{c^2+2c-3} \div \dfrac{c}{c+1} = \dfrac{c^2+3c}{c^2+2c-3} \cdot \dfrac{c+1}{c}$

$\qquad = \dfrac{(c^2+3c)(c+1)}{(c^2+2c-3)c}$

$\qquad = \dfrac{c(c+3)(c+1)}{(c+3)(c-1)c}$

$\qquad = \dfrac{\cancel{c}(\cancel{c+3})(c+1)}{(\cancel{c+3})(c-1)\cancel{c}}$

$\qquad = \dfrac{c+1}{c-1}$

33. $\dfrac{2y^2-7y+3}{2y^2+3y-2} \div \dfrac{6y^2-5y+1}{3y^2+5y-2}$

$= \dfrac{2y^2-7y+3}{2y^2+3y-2} \cdot \dfrac{3y^2+5y-2}{6y^2-5y+1}$

$= \dfrac{(2y^2-7y+3)(3y^2+5y-2)}{(2y^2+3y-2)(6y^2-5y+1)}$

$= \dfrac{(2y-1)(y-3)(3y-1)(y+2)}{(2y-1)(y+2)(3y-1)(2y-1)}$

$= \dfrac{(\cancel{2y-1})(y-3)(\cancel{3y-1})(\cancel{y+2})}{(\cancel{2y-1})(\cancel{y+2})(\cancel{3y-1})(2y-1)}$

$= \dfrac{y-3}{2y-1}$

35. $\dfrac{x^2-1}{4x+4} \div \dfrac{2x^2-4x+2}{8x+8} = \dfrac{x^2-1}{4x+4} \cdot \dfrac{8x+8}{2x^2-4x+2}$

$\qquad = \dfrac{(x^2-1)(8x+8)}{(4x+4)(2x^2-4x+2)}$

$\qquad = \dfrac{(x+1)(x-1)(2)(4)(x+1)}{4(x+1)(2)(x-1)(x-1)}$

$\qquad = \dfrac{(\cancel{x+1})(\cancel{x-1})\cancel{(2)}\cancel{(4)}(x+1)}{\cancel{4}(\cancel{x+1})\cancel{(2)}(x-1)(\cancel{x-1})}$

$\qquad = \dfrac{x+1}{x-1}$

37. Discussion and Writing Exercise

39. *Familiarize*. Let $s =$ Bonnie's score on the last test.

Translate. The average of the four scores must be at least 90. This means it must be greater than or equal to 90. We translate.

$$\dfrac{96+98+89+s}{4} \geq 90$$

Solve. We solve the inequality. First we multiply by 4 to clear the fraction.

$$4\left(\frac{96 + 98 + 89 + s}{4}\right) \geq 4 \cdot 90$$

$$96 + 98 + 89 + s \geq 360$$

$$283 + s \geq 360$$

$$s \geq 77 \qquad \text{Subtracting 283}$$

Check. We can do a partial check by substituting a value for s less than 77 and a value for s greater than 77.

For $s = 76$: $\dfrac{96 + 98 + 89 + 76}{4} = 89.75 < 90$

For $s = 78$: $\dfrac{96 + 98 + 89 + 78}{4} = 90.25 \leq 90$

Since the average is less than 90 for a value of s less than 77 and greater than or equal to 90 for a value greater than or equal to 77, the answer is probably correct.

State. The scores on the last test that will earn Bonnie an A are $\{s \mid s \geq 77\}$.

41. $(8x^3 - 3x^2 + 7) - (8x^2 + 3x - 5) =$
$8x^3 - 3x^2 + 7 - 8x^2 - 3x + 5 =$
$8x^3 - 11x^2 - 3x + 12$

43. $(2x^{-3}y^4)^2 = 2^2(x^{-3})^2(y^4)^2$

$\qquad = 2^2 x^{-6} y^8 \qquad$ Multiplying exponents

$\qquad = 4x^{-6}y^8 \qquad (2^2 = 4)$

$\qquad = \dfrac{4y^8}{x^6} \qquad \left(x^{-6} = \dfrac{1}{x^6}\right)$

45. $\left(\dfrac{2x^3}{y^5}\right)^2 = \dfrac{2^2(x^3)^2}{(y^5)^2}$

$\qquad = \dfrac{2^2 x^6}{y^{10}} \qquad$ Multiplying exponents

$\qquad = \dfrac{4x^6}{y^{10}} \qquad (2^2 = 4)$

47. $\dfrac{3a^2 - 5ab - 12b^2}{3ab + 4b^2} \div (3b^2 - ab)$

$= \dfrac{3a^2 - 5ab - 12b^2}{3ab + 4b^2} \cdot \dfrac{1}{3b^2 - ab}$

$= \dfrac{(3a + 4b)(a - 3b)}{b(3a + 4b) \cdot b(3b - a)}$

$= \dfrac{(3a + 4b)(-1)(3b - a)}{b(3a + 4b) \cdot b(3b - a)}$

$= \dfrac{(3a + 4b)(-1)(3b - a)}{b(3a + 4b) \cdot b(3b - a)}$

$= -\dfrac{1}{b^2}$

49. $\dfrac{a^2b^2 + 3ab^2 + 2b^2}{a^2b^4 + 4b^4} \div (5a^2 + 10a)$

$= \dfrac{a^2b^2 + 3ab^2 + 2b^2}{a^2b^4 + 4b^4} \cdot \dfrac{1}{5a^2 + 10a}$

$= \dfrac{a^2b^2 + 3ab^2 + 2b^2}{(a^2b^4 + 4b^4)(5a^2 + 10a)}$

$= \dfrac{b^2(a^2 + 3a + 2)}{b^4(a^2 + 4)(5a)(a + 2)}$

$= \dfrac{b^2(a + 1)(a + 2)}{b^2 \cdot b^2(a^2 + 4)(5a)(a + 2)}$

$= \dfrac{b^2(a + 1)(a + 2)}{b^2 \cdot b^2(a^2 + 4)(5a)(a + 2)}$

$= \dfrac{a + 1}{5ab^2(a^2 + 4)}$

Exercise Set 6.3

1. $12 = 2 \cdot 2 \cdot 3$
$27 = 3 \cdot 3 \cdot 3$
LCM $= 2 \cdot 2 \cdot 3 \cdot 3 \cdot 3$, or 108

3. $8 = 2 \cdot 2 \cdot 2$
$9 = 3 \cdot 3$
LCM $= 2 \cdot 2 \cdot 2 \cdot 3 \cdot 3$, or 72

5. $6 = 2 \cdot 3$
$9 = 3 \cdot 3$
$21 = 3 \cdot 7$
LCM $= 2 \cdot 3 \cdot 3 \cdot 7$, or 126

7. $24 = 2 \cdot 2 \cdot 2 \cdot 3$
$36 = 2 \cdot 2 \cdot 3 \cdot 3$
$40 = 2 \cdot 2 \cdot 2 \cdot 5$
LCM $= 2 \cdot 2 \cdot 2 \cdot 3 \cdot 3 \cdot 5$, or 360

9. $10 = 2 \cdot 5$
$100 = 2 \cdot 2 \cdot 5 \cdot 5$
$500 = 2 \cdot 2 \cdot 5 \cdot 5 \cdot 5$
LCM $= 2 \cdot 2 \cdot 5 \cdot 5 \cdot 5$, or 500

(We might have observed at the outset that both 10 and 100 are factors of 500, so the LCM is 500.)

11. $24 = 2 \cdot 2 \cdot 2 \cdot 3$
$18 = 2 \cdot 3 \cdot 3$
LCD $= 2 \cdot 2 \cdot 2 \cdot 3 \cdot 3$, or 72

$\dfrac{7}{24} + \dfrac{11}{18} = \dfrac{7}{2 \cdot 2 \cdot 2 \cdot 3} \cdot \dfrac{3}{3} + \dfrac{11}{2 \cdot 3 \cdot 3} \cdot \dfrac{2 \cdot 2}{2 \cdot 2}$

$\qquad = \dfrac{21}{2 \cdot 2 \cdot 2 \cdot 3 \cdot 3} + \dfrac{44}{2 \cdot 2 \cdot 2 \cdot 3 \cdot 3}$

$\qquad = \dfrac{65}{72}$

13. $\dfrac{1}{6} + \dfrac{3}{40}$

$= \dfrac{1}{2 \cdot 3} + \dfrac{3}{2 \cdot 2 \cdot 2 \cdot 5}$

$\qquad$ LCD is $2 \cdot 2 \cdot 2 \cdot 3 \cdot 5$, or 120

$= \dfrac{1}{2 \cdot 3} \cdot \dfrac{2 \cdot 2 \cdot 5}{2 \cdot 2 \cdot 5} + \dfrac{3}{2 \cdot 2 \cdot 2 \cdot 5} \cdot \dfrac{3}{3}$

$= \dfrac{20 + 9}{2 \cdot 2 \cdot 2 \cdot 3 \cdot 5}$

$= \dfrac{29}{120}$

15. $\dfrac{1}{20} + \dfrac{1}{30} + \dfrac{2}{45}$

$= \dfrac{1}{2 \cdot 2 \cdot 5} + \dfrac{1}{2 \cdot 3 \cdot 5} + \dfrac{2}{3 \cdot 3 \cdot 5}$

$\qquad$ LCD is $2 \cdot 2 \cdot 3 \cdot 3 \cdot 5$, or 180

$= \dfrac{1}{2 \cdot 2 \cdot 5} \cdot \dfrac{3 \cdot 3}{3 \cdot 3} + \dfrac{1}{2 \cdot 3 \cdot 5} \cdot \dfrac{2 \cdot 3}{2 \cdot 3} + \dfrac{2}{3 \cdot 3 \cdot 5} \cdot \dfrac{2 \cdot 2}{2 \cdot 2}$

$= \dfrac{9 + 6 + 8}{2 \cdot 2 \cdot 3 \cdot 3 \cdot 5}$

$= \dfrac{23}{180}$

17. $6x^2 = 2 \cdot 3 \cdot x \cdot x$

$12x^3 = 2 \cdot 2 \cdot 3 \cdot x \cdot x \cdot x$

LCM $= 2 \cdot 2 \cdot 3 \cdot x \cdot x \cdot x$, or $12x^3$

19. $2x^2 = 2 \cdot x \cdot x$

$6xy = 2 \cdot 3 \cdot x \cdot y$

$18y^2 = 2 \cdot 3 \cdot 3 \cdot y \cdot y$

LCM $= 2 \cdot 3 \cdot 3 \cdot x \cdot x \cdot y \cdot y$, or $18x^2y^2$

21. $2(y-3) = 2 \cdot (y-3)$

$6(y-3) = 2 \cdot 3 \cdot (y-3)$

LCM $= 2 \cdot 3 \cdot (y-3)$, or $6(y-3)$

23. $t, t+2, t-2$

The expressions are not factorable, so the LCM is their product:

LCM $= t(t+2)(t-2)$

25. $x^2 - 4 = (x+2)(x-2)$

$x^2 + 5x + 6 = (x+3)(x+2)$

LCM $= (x+2)(x-2)(x+3)$

27. $t^3 + 4t^2 + 4t = t(t^2 + 4t + 4) = t(t+2)(t+2)$

$t^2 - 4t = t(t-4)$

LCM $= t(t+2)(t+2)(t-4) = t(t+2)^2(t-4)$

29. $a + 1 = a + 1$

$(a-1)^2 = (a-1)(a-1)$

$a^2 - 1 = (a+1)(a-1)$

LCM $= (a+1)(a-1)(a-1) = (a+1)(a-1)^2$

31. $m^2 - 5m + 6 = (m-3)(m-2)$

$m^2 - 4m + 4 = (m-2)(m-2)$

LCM $= (m-3)(m-2)(m-2) = (m-3)(m-2)^2$

33. $2 + 3x = 2 + 3x$

$4 - 9x^2 = (2+3x)(2-3x)$

$2 - 3x = 2 - 3x$

LCM $= (2+3x)(2-3x)$

35. $10v^2 + 30v = 10v(v+3) = 2 \cdot 5 \cdot v(v+3)$

$5v^2 + 35v + 60 = 5(v^2 + 7v + 12)$

$\qquad = 5(v+4)(v+3)$

LCM $= 2 \cdot 5 \cdot v(v+3)(v+4) = 10v(v+3)(v+4)$

37. $9x^3 - 9x^2 - 18x = 9x(x^2 - x - 2)$

$\qquad = 3 \cdot 3 \cdot x(x-2)(x+1)$

$6x^5 - 24x^4 + 24x^3 = 6x^3(x^2 - 4x + 4)$

$\qquad = 2 \cdot 3 \cdot x \cdot x \cdot x(x-2)(x-2)$

LCM $= 2 \cdot 3 \cdot 3 \cdot x \cdot x \cdot x(x-2)(x-2)(x+1) =$
$18x^3(x-2)^2(x+1)$

39. $x^5 + 4x^4 + 4x^3 = x^3(x^2 + 4x + 4)$

$\qquad = x \cdot x \cdot x(x+2)(x+2)$

$3x^2 - 12 = 3(x^2 - 4) = 3(x+2)(x-2)$

$2x + 4 = 2(x+2)$

LCM $= 2 \cdot 3 \cdot x \cdot x \cdot x(x+2)(x+2)(x-2)$

$\qquad = 6x^3(x+2)^2(x-2)$

41. Discussion and Writing Exercise

43. $x^2 - 6x + 9 = x^2 - 2 \cdot x \cdot 3 + 3^2$ $\qquad$ Trinomial square

$\qquad = (x-3)^2$

45. $x^2 - 9 = x^2 - 3^2$ $\qquad$ Difference of squares

$\qquad = (x+3)(x-3)$

47. $x^2 + 6x + 9 = x^2 + 2 \cdot x \cdot 3 + 3^2$ $\qquad$ Trinomial square

$\qquad = (x+3)^2$

49. $40x^3 = 2 \cdot 2 \cdot 2 \cdot 5 \cdot x \cdot x \cdot x$

$24x^4 = 2 \cdot 2 \cdot 2 \cdot 3 \cdot x \cdot x \cdot x \cdot x$

LCM $= 2 \cdot 2 \cdot 2 \cdot 3 \cdot 5 \cdot x \cdot x \cdot x \cdot x = 120x^4$

GCF $= 2 \cdot 2 \cdot 2 \cdot x \cdot x \cdot x = 8x^3$

$120x^4(8x^3) = 960x^7$

51. $20x^2 = 2 \cdot 2 \cdot 5 \cdot x \cdot x$

$10x = 2 \cdot 5 \cdot x$

LCM $= 2 \cdot 2 \cdot 5 \cdot x \cdot x = 20x^2$

GCF $= 2 \cdot 5 \cdot x = 10x$

$20x^2(10x) = 200x^3$

53. $10x^2 = 2 \cdot 5 \cdot x \cdot x$

$24x^3 = 2 \cdot 2 \cdot 2 \cdot 3 \cdot x \cdot x \cdot x$

LCM $= 2 \cdot 2 \cdot 2 \cdot 3 \cdot 5 \cdot x \cdot x \cdot x = 120x^3$

GCF $= 2 \cdot x \cdot x = 2x^2$

$120x^3(2x^2) = 240x^5$

55. The time it takes Pedro and Maria to meet again at the starting place is the LCM of the times it takes them to complete one round of the course.

$6 = 2 \cdot 3$

$8 = 2 \cdot 2 \cdot 2$

$\text{LCM} = 2 \cdot 2 \cdot 2 \cdot 3$, or 24

It takes 24 min.

Exercise Set 6.4

1. $\dfrac{5}{8} + \dfrac{3}{8} + \dfrac{5+3}{8} = \dfrac{8}{8} = 1$

3. $\dfrac{1}{3+x} + \dfrac{5}{3+x} = \dfrac{1+5}{3+x} = \dfrac{6}{3+x}$

5. $\dfrac{x^2+7x}{x^2-5x} + \dfrac{x^2-4x}{x^2-5x} = \dfrac{(x^2+7x)+(x^2-4x)}{x^2-5x}$

$= \dfrac{2x^2+3x}{x^2-5x}$

$= \dfrac{x(2x+3)}{x(x-5)}$

$= \dfrac{\cancel{x}(2x+3)}{\cancel{x}(x-5)}$

$= \dfrac{2x+3}{x-5}$

7. $\dfrac{2}{x} + \dfrac{5}{x^2} = \dfrac{2}{x} + \dfrac{5}{x \cdot x}$ $\text{LCD} = x \cdot x$, or x^2

$= \dfrac{2}{x} \cdot \dfrac{x}{x} + \dfrac{5}{x \cdot x}$

$= \dfrac{2x+5}{x^2}$

9. $\left.\begin{array}{l} 6r = 2 \cdot 3 \cdot r \\ 8r = 2 \cdot 2 \cdot 2 \cdot r \end{array}\right\}\text{LCD} = 2 \cdot 2 \cdot 2 \cdot 3 \cdot r$, or $24r$

$\dfrac{5}{6r} + \dfrac{7}{8r} = \dfrac{5}{6r} \cdot \dfrac{4}{4} + \dfrac{7}{8r} \cdot \dfrac{3}{3}$

$= \dfrac{20+21}{24r}$

$= \dfrac{41}{24r}$

11. $\left.\begin{array}{l} xy^2 = x \cdot y \cdot y \\ x^2y = x \cdot x \cdot y \end{array}\right\}\text{LCD} = x \cdot x \cdot y \cdot y$, or x^2y^2

$\dfrac{4}{xy^2} + \dfrac{6}{x^2y} = \dfrac{4}{xy^2} \cdot \dfrac{x}{x} + \dfrac{6}{x^2y} \cdot \dfrac{y}{y}$

$= \dfrac{4x+6y}{x^2y^2}$

13. $\left.\begin{array}{l} 9t^3 = 3 \cdot 3 \cdot t \cdot t \cdot t \\ 6t^2 = 2 \cdot 3 \cdot t \cdot t \end{array}\right\}\text{LCD} = 2 \cdot 3 \cdot 3 \cdot t \cdot t \cdot t$, or $18t^3$

$\dfrac{2}{9t^3} + \dfrac{1}{6t^2} = \dfrac{2}{9t^3} \cdot \dfrac{2}{2} + \dfrac{1}{6t^2} \cdot \dfrac{3t}{3t}$

$= \dfrac{4+3t}{18t^3}$

15. $\text{LCD} = x^2y^2$ (See Exercise 11.)

$\dfrac{x+y}{xy^2} + \dfrac{3x+y}{x^2y} = \dfrac{x+y}{xy^2} \cdot \dfrac{x}{x} + \dfrac{3x+y}{x^2y} \cdot \dfrac{y}{y}$

$= \dfrac{x(x+y)+y(3x+y)}{x^2y^2}$

$= \dfrac{x^2+xy+3xy+y^2}{x^2y^2}$

$= \dfrac{x^2+4xy+y^2}{x^2y^2}$

17. The denominators do not factor, so the LCD is their product, $(x-2)(x+2)$.

$\dfrac{3}{x-2} + \dfrac{3}{x+2} = \dfrac{3}{x-2} \cdot \dfrac{x+2}{x+2} + \dfrac{3}{x+2} \cdot \dfrac{x-2}{x-2}$

$= \dfrac{3(x+2)+3(x-2)}{(x-2)(x+2)}$

$= \dfrac{3x+6+3x-6}{(x-2)(x+2)}$

$= \dfrac{6x}{(x-2)(x+2)}$

19. $\left.\begin{array}{l} 3x = 3 \cdot x \\ x+1 = x+1 \end{array}\right\}\text{LCD} = 3x(x+1)$

$\dfrac{3}{x+1} + \dfrac{2}{3x} = \dfrac{3}{x+1} \cdot \dfrac{3x}{3x} + \dfrac{2}{3x} \cdot \dfrac{x+1}{x+1}$

$= \dfrac{9x+2(x+1)}{3x(x+1)}$

$= \dfrac{9x+2x+2}{3x(x+1)}$

$= \dfrac{11x+2}{3x(x+1)}$

21. $\left.\begin{array}{l} x^2-16 = (x+4)(x-4) \\ x-4 = x-4 \end{array}\right\}\text{LCD} = (x+4)(x-4)$

$\dfrac{2x}{x^2-16} + \dfrac{x}{x-4} = \dfrac{2x}{(x+4)(x-4)} + \dfrac{x}{x-4} \cdot \dfrac{x+4}{x+4}$

$= \dfrac{2x+x(x+4)}{(x+4)(x-4)}$

$= \dfrac{2x+x^2+4x}{(x+4)(x-4)}$

$= \dfrac{x^2+6x}{(x+4)(x-4)}$

23. $\dfrac{5}{z+4} + \dfrac{3}{3z+12} = \dfrac{5}{z+4} + \dfrac{3}{3(z+4)}$ $\text{LCD} = 3(z+4)$

$= \dfrac{5}{z+4} \cdot \dfrac{3}{3} + \dfrac{3}{3(z+4)}$

$= \dfrac{15+3}{3(z+4)} = \dfrac{18}{3(z+4)}$

$= \dfrac{3 \cdot 6}{3(z+4)} = \dfrac{\cancel{3} \cdot 6}{\cancel{3}(z+4)}$

$= \dfrac{6}{z+4}$

25. $\dfrac{3}{x-1} + \dfrac{2}{(x-1)^2}$ $\quad$ LCD $= (x-1)^2$

$= \dfrac{3}{x-1} \cdot \dfrac{x-1}{x-1} + \dfrac{2}{(x-1)^2}$

$= \dfrac{3(x-1)+2}{(x-1)^2}$

$= \dfrac{3x-3+2}{(x-1)^2}$

$= \dfrac{3x-1}{(x-1)^2}$

27. $\dfrac{4a}{5a-10} + \dfrac{3a}{10a-20} = \dfrac{4a}{5(a-2)} + \dfrac{3a}{2 \cdot 5(a-2)}$

$\qquad\qquad\qquad\qquad$ LCD $= 2 \cdot 5(a-2)$

$\qquad\qquad = \dfrac{4a}{5(a-2)} \cdot \dfrac{2}{2} + \dfrac{3a}{2 \cdot 5(a-2)}$

$\qquad\qquad = \dfrac{8a+3a}{10(a-2)}$

$\qquad\qquad = \dfrac{11a}{10(a-2)}$

29. $\dfrac{x+4}{x} + \dfrac{x}{x+4}$ $\quad$ LCD $= x(x+4)$

$= \dfrac{x+4}{x} \cdot \dfrac{x+4}{x+4} + \dfrac{x}{x+4} \cdot \dfrac{x}{x}$

$= \dfrac{(x+4)^2 + x^2}{x(x+4)}$

$= \dfrac{x^2+8x+16+x^2}{x(x+4)}$

$= \dfrac{2x^2+8x+16}{x(x+4)}$

31. $\dfrac{4}{a^2-a-2} + \dfrac{3}{a^2+4a+3}$

$= \dfrac{4}{(a-2)(a+1)} + \dfrac{3}{(a+3)(a+1)}$

$\qquad\qquad$ LCD $= (a-2)(a+1)(a+3)$

$= \dfrac{4}{(a-2)(a+1)} \cdot \dfrac{a+3}{a+3} + \dfrac{3}{(a+3)(a+1)} \cdot \dfrac{a-2}{a-2}$

$= \dfrac{4(a+3)+3(a-2)}{(a-2)(a+1)(a+3)}$

$= \dfrac{4a+12+3a-6}{(a-2)(a+1)(a+3)}$

$= \dfrac{7a+6}{(a-2)(a+1)(a+3)}$

33. $\dfrac{x+3}{x-5} + \dfrac{x-5}{x+3}$ $\quad$ LCD $= (x-5)(x+3)$

$= \dfrac{x+3}{x-5} \cdot \dfrac{x+3}{x+3} + \dfrac{x-5}{x+3} \cdot \dfrac{x-5}{x-5}$

$= \dfrac{(x+3)^2 + (x-5)^2}{(x-5)(x+3)}$

$= \dfrac{x^2+6x+9+x^2-10x+25}{(x-5)(x+3)}$

$= \dfrac{2x^2-4x+34}{(x-5)(x+3)}$

35. $\dfrac{a}{a^2-1} + \dfrac{2a}{a^2-a}$

$= \dfrac{a}{(a+1)(a-1)} + \dfrac{2a}{a(a-1)}$

$\qquad\qquad\qquad$ LCD $= a(a+1)(a-1)$

$= \dfrac{a}{(a+1)(a-1)} \cdot \dfrac{a}{a} + \dfrac{2a}{a(a-1)} \cdot \dfrac{a+1}{a+1}$

$= \dfrac{a^2+2a(a+1)}{a(a+1)(a-1)} = \dfrac{a^2+2a^2+2a}{a(a+1)(a-1)}$

$= \dfrac{3a^2+2a}{a(a+1)(a-1)} = \dfrac{a(3a+2)}{a(a+1)(a-1)}$

$= \dfrac{\cancel{a}(3a+2)}{\cancel{a}(a+1)(a-1)} = \dfrac{3a+2}{(a+1)(a-1)}$

37. $\dfrac{7}{8} + \dfrac{5}{-8} = \dfrac{7}{8} + \dfrac{5}{-8} \cdot \dfrac{-1}{-1}$

$\qquad\quad = \dfrac{7}{8} + \dfrac{-5}{8}$

$\qquad\quad = \dfrac{7+(-5)}{8}$

$\qquad\quad = \dfrac{2}{8} = \dfrac{\cancel{2} \cdot 1}{4 \cdot \cancel{2}}$

$\qquad\quad = \dfrac{1}{4}$

39. $\dfrac{3}{t} + \dfrac{4}{-t} = \dfrac{3}{t} + \dfrac{4}{-t} \cdot \dfrac{-1}{-1}$

$\qquad\quad = \dfrac{3}{t} + \dfrac{-4}{t}$

$\qquad\quad = \dfrac{3+(-4)}{t}$

$\qquad\quad = \dfrac{-1}{t}$

$\qquad\quad = -\dfrac{1}{t}$

41. $\dfrac{2x+7}{x-6} + \dfrac{3x}{6-x} = \dfrac{2x+7}{x-6} + \dfrac{3x}{6-x} \cdot \dfrac{-1}{-1}$

$\qquad\qquad\qquad = \dfrac{2x+7}{x-6} + \dfrac{-3x}{x-6}$

$\qquad\qquad\qquad = \dfrac{(2x+7)+(-3x)}{x-6}$

$\qquad\qquad\qquad = \dfrac{-x+7}{x-6}$

43.
$$\frac{y^2}{y-3} + \frac{9}{3-y} = \frac{y^2}{y-3} + \frac{9}{3-y} \cdot \frac{-1}{-1}$$
$$= \frac{y^2}{y-3} + \frac{-9}{y-3}$$
$$= \frac{y^2 + (-9)}{y-3}$$
$$= \frac{y^2 - 9}{y-3}$$
$$= \frac{(y+3)(y-3)}{y-3}$$
$$= \frac{(y+3)(\cancel{y-3})}{1(\cancel{y-3})}$$
$$= y+3$$

45.
$$\frac{b-7}{b^2-16} + \frac{7-b}{16-b^2} = \frac{b-7}{b^2-16} + \frac{7-b}{16-b^2} \cdot \frac{-1}{-1}$$
$$= \frac{b-7}{b^2-16} + \frac{b-7}{b^2-16}$$
$$= \frac{(b-7) + (b-7)}{b^2-16}$$
$$= \frac{2b-14}{b^2-16}$$

47.
$$\frac{a^2}{a-b} + \frac{b^2}{b-a} = \frac{a^2}{a-b} + \frac{b^2}{b-a} \cdot \frac{-1}{-1}$$
$$= \frac{a^2}{a-b} + \frac{-b^2}{a-b}$$
$$= \frac{a^2 + (-b^2)}{a-b}$$
$$= \frac{a^2 - b^2}{a-b}$$
$$= \frac{(a+b)(a-b)}{a-b}$$
$$= \frac{(a+b)(\cancel{a-b})}{1(\cancel{a-b})}$$
$$= a+b$$

49.
$$\frac{x+3}{x-5} + \frac{2x-1}{5-x} + \frac{2(3x-1)}{x-5}$$
$$= \frac{x+3}{x-5} + \frac{2x-1}{5-x} \cdot \frac{-1}{-1} + \frac{2(3x-1)}{x-5}$$
$$= \frac{x+3}{x-5} + \frac{1-2x}{x-5} + \frac{2(3x-1)}{x-5}$$
$$= \frac{(x+3) + (1-2x) + (6x-2)}{x-5}$$
$$= \frac{5x+2}{x-5}$$

51.
$$\frac{2(4x+1)}{5x-7} + \frac{3(x-2)}{7-5x} + \frac{-10x-1}{5x-7}$$
$$= \frac{2(4x+1)}{5x-7} + \frac{3(x-2)}{7-5x} \cdot \frac{-1}{-1} + \frac{-10x-1}{5x-7}$$
$$= \frac{2(4x+1)}{5x-7} + \frac{-3(x-2)}{5x-7} + \frac{-10x-1}{5x-7}$$
$$= \frac{(8x+2) + (-3x+6) + (-10x-1)}{5x-7}$$
$$= \frac{-5x+7}{5x-7}$$
$$= \frac{-1(5x-7)}{5x-7}$$
$$= \frac{-1(\cancel{5x-7})}{\cancel{5x-7}}$$
$$= -1$$

53.
$$\frac{x+1}{(x+3)(x-3)} + \frac{4(x-3)}{(x-3)(x+3)} + \frac{(x-1)(x-3)}{(3-x)(x+3)}$$
$$= \frac{x+1}{(x+3)(x-3)} + \frac{4(x-3)}{(x-3)(x+3)} + \frac{(x-1)(x-3)}{(3-x)(x+3)} \cdot \frac{-1}{-1}$$
$$= \frac{x+1}{(x+3)(x-3)} + \frac{4(x-3)}{(x-3)(x+3)} + \frac{-1(x^2-4x+3)}{(x-3)(x+3)}$$
$$= \frac{(x+1) + (4x-12) + (-x^2+4x-3)}{(x+3)(x-3)}$$
$$= \frac{-x^2+9x-14}{(x+3)(x-3)}$$

55.
$$\frac{6}{x-y} + \frac{4x}{y^2-x^2}$$
$$= \frac{6}{x-y} + \frac{4x}{(y-x)(y+x)}$$
$$= \frac{6}{x-y} + \frac{4x}{(y-x)(y+x)} \cdot \frac{-1}{-1}$$
$$= \frac{6}{x-y} + \frac{-4x}{(x-y)(x+y)}$$
$$[-1(y-x) = x-y; y+x = x+y]$$
$$\text{LCD} = (x-y)(x+y)$$
$$= \frac{6}{x-y} \cdot \frac{x+y}{x+y} + \frac{-4x}{(x-y)(x+y)}$$
$$= \frac{6(x+y) - 4x}{(x-y)(x+y)}$$
$$= \frac{6x + 6y - 4x}{(x-y)(x+y)}$$
$$= \frac{2x+6y}{(x-y)(x+y)}$$

57.
$$\frac{4-a}{25-a^2} + \frac{a+1}{a-5}$$

$$= \frac{4-a}{25-a^2} \cdot \frac{-1}{-1} + \frac{a+1}{a-5}$$

$$= \frac{a-4}{a^2-25} + \frac{a+1}{a-5}$$

$$= \frac{a-4}{(a+5)(a-5)} + \frac{a+1}{a-5}$$

$$\text{LCD} = (a+5)(a-5)$$

$$= \frac{a-4}{(a+5)(a-5)} + \frac{a+1}{a-5} \cdot \frac{a+5}{a+5}$$

$$= \frac{a-4}{(a+5)(a-5)} + \frac{(a+1)(a+5)}{(a+5)(a-5)}$$

$$= \frac{(a-4)+(a+1)(a+5)}{(a+5)(a-5)}$$

$$= \frac{a-4+a^2+6a+5}{(a+5)(a-5)}$$

$$= \frac{a^2+7a+1}{(a+5)(a-5)}$$

59.
$$\frac{2}{t^2+t-6} + \frac{3}{t^2-9}$$

$$= \frac{2}{(t+3)(t-2)} + \frac{3}{(t+3)(t-3)}$$

$$\text{LCD} = (t+3)(t-2)(t-3)$$

$$= \frac{2}{(t+3)(t-2)} \cdot \frac{t-3}{t-3} + \frac{3}{(t+3)(t-3)} \cdot \frac{t-2}{t-2}$$

$$= \frac{2(t-3)+3(t-2)}{(t+3)(t-2)(t-3)}$$

$$= \frac{2t-6+3t-6}{(t+3)(t-2)(t-3)}$$

$$= \frac{5t-12}{(t+3)(t-2)(t-3)}$$

61. Discussion and Writing Exercise

63. $(x^2+x)-(x+1) = x^2+x-x-1 = x^2-1$

65. $(2x^4y^3)^{-3} = \frac{1}{(2x^4y^3)^3} = \frac{1}{2^3(x^4)^3(y^3)^3} = \frac{1}{8x^{12}y^9}$

67. $\left(\dfrac{x^{-4}}{y^7}\right)^3 = \dfrac{(x^{-4})^3}{(y^7)^3} = \dfrac{x^{-12}}{y^{21}} = \dfrac{1}{x^{12}y^{21}}$

69. $y = \frac{1}{2}x - 5 = \frac{1}{2}x + (-5)$

The y-intercept is $(0,-5)$. We find two other pairs.

When $x = 2$, $y = \frac{1}{2} \cdot 2 - 5 = 1 - 5 = -4$.

When $x = 4$, $y = \frac{1}{2} \cdot 4 - 5 = 2 - 5 = -3$.

x	y
0	-5
2	-4
4	-3

Plot these points, draw the line they determine, and label the graph $y = \frac{1}{2}x - 5$.

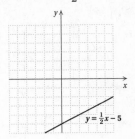

71. $y = 3$

Any ordered pair $(x, 3)$ is a solution. The variable y must be 3, but x can be any number we choose. A few solutions are listed below. Plot these points and draw the line.

x	y
-4	3
0	3
3	3

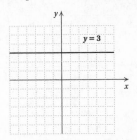

73.
$$3x - 7 = 5x + 9$$
$$-2x - 7 = 9 \qquad \text{Subtracting } 5x$$
$$-2x = 16 \qquad \text{Adding } 7$$
$$x = -8 \qquad \text{Dividing by } -2$$

The solution is -8.

75.
$$x^2 - 8x + 15 = 0$$
$$(x-3)(x-5) = 0$$
$$x - 3 = 0 \text{ or } x - 5 = 0 \quad \text{Principle of zero products}$$
$$x = 3 \text{ or } \qquad x = 5$$

The solutions are 3 and 5.

77. To find the perimeter we add the lengths of the sides:

$$\frac{y+4}{3} + \frac{y+4}{3} + \frac{y-2}{5} + \frac{y-2}{5} \quad \text{LCD} = 3 \cdot 5$$

$$= \frac{y+4}{3} \cdot \frac{5}{5} + \frac{y+4}{3} \cdot \frac{5}{5} + \frac{y-2}{5} \cdot \frac{3}{3} + \frac{y-2}{5} \cdot \frac{3}{3}$$

$$= \frac{5y+20+5y+20+3y-6+3y-6}{3 \cdot 5}$$

$$= \frac{16y+28}{15}$$

To find the area we multiply the length and the width:

$$\left(\frac{y+4}{3}\right)\left(\frac{y-2}{5}\right) = \frac{(y+4)(y-2)}{3 \cdot 5} = \frac{y^2+2y-8}{15}$$

79. $\dfrac{5}{z+2} + \dfrac{4z}{z^2-4} + 2$

$= \dfrac{5}{z+2} + \dfrac{4z}{(z+2)(z-2)} + \dfrac{2}{1}$ LCD $= (z+2)(z-2)$

$= \dfrac{5}{z+2} \cdot \dfrac{z-2}{z-2} + \dfrac{4z}{(z+2)(z-2)} + \dfrac{2}{1} \cdot \dfrac{(z+2)(z-2)}{(z+2)(z-2)}$

$= \dfrac{5z-10+4z+2(z^2-4)}{(z+2)(z-2)}$

$= \dfrac{5z-10+4z+2z^2-8}{(z+2)(z-2)} = \dfrac{2z^2+9z-18}{(z+2)(z-2)}$

$= \dfrac{(2z-3)(z+6)}{(z+2)(z-2)}$

81. $\dfrac{3z^2}{z^4-4} + \dfrac{5z^2-3}{2z^4+z^2-6}$

$= \dfrac{3z^2}{(z^2+2)(z^2-2)} + \dfrac{5z^2-3}{(2z^2-3)(z^2+2)}$

LCD $= (z^2+2)(z^2-2)(2z^2-3)$

$= \dfrac{3z^2}{(z^2+2)(z^2-2)} \cdot \dfrac{2z^2-3}{2z^2-3} +$

$\qquad \dfrac{5z^2-3}{(2z^2-3)(z^2+2)} \cdot \dfrac{z^2-2}{z^2-2}$

$= \dfrac{6z^4-9z^2+5z^4-13z^2+6}{(z^2+2)(z^2-2)(2z^2-3)}$

$= \dfrac{11z^4-22z^2+6}{(z^2+2)(z^2-2)(2z^2-3)}$

83.-85. Left to the student

Exercise Set 6.5

1. $\dfrac{7}{x} - \dfrac{3}{x} = \dfrac{7-3}{x} = \dfrac{4}{x}$

3. $\dfrac{y}{y-4} - \dfrac{4}{y-4} = \dfrac{y-4}{y-4} = 1$

5. $\dfrac{2x-3}{x^2+3x-4} - \dfrac{x-7}{x^2+3x-4}$

$= \dfrac{2x-3-(x-7)}{x^2+3x-4}$

$= \dfrac{2x-3-x+7}{x^2+3x-4}$

$= \dfrac{x+4}{x^2+3x-4}$

$= \dfrac{x+4}{(x+4)(x-1)}$

$= \dfrac{(x+4) \cdot 1}{(x+4)(x-1)}$

$= \dfrac{1}{x-1}$

7. $\dfrac{a-2}{10} - \dfrac{a+1}{5} = \dfrac{a-2}{10} - \dfrac{a+1}{5} \cdot \dfrac{2}{2}$ LCD $= 10$

$= \dfrac{a-2}{10} - \dfrac{2(a+1)}{10}$

$= \dfrac{(a-2)-2(a+1)}{10}$

$= \dfrac{a-2-2a-2}{10}$

$= \dfrac{-a-4}{10}$

9. $\dfrac{4z-9}{3z} - \dfrac{3z-8}{4z} = \dfrac{4z-9}{3z} \cdot \dfrac{4}{4} - \dfrac{3z-8}{4z} \cdot \dfrac{3}{3}$

LCD $= 3 \cdot 4 \cdot z$, or $12z$

$= \dfrac{16z-36}{12z} - \dfrac{9z-24}{12z}$

$= \dfrac{16z-36-(9z-24)}{12z}$

$= \dfrac{16z-36-9z+24}{12z}$

$= \dfrac{7z-12}{12z}$

11. $\dfrac{4x+2t}{3xt^2} - \dfrac{5x-3t}{x^2t}$ LCD $= 3x^2t^2$

$= \dfrac{4x+2t}{3xt^2} \cdot \dfrac{x}{x} - \dfrac{5x-3t}{x^2t} \cdot \dfrac{3t}{3t}$

$= \dfrac{4x^2+2tx}{3x^2t^2} - \dfrac{15xt-9t^2}{3x^2t^2}$

$= \dfrac{4x^2+2tx-(15xt-9t^2)}{3x^2t^2}$

$= \dfrac{4x^2+2tx-15xt+9t^2}{3x^2t^2}$

$= \dfrac{4x^2-13xt+9t^2}{3x^2t^2}$

13. $\dfrac{5}{x+5} - \dfrac{3}{x-5}$ LCD $= (x+5)(x-5)$

$= \dfrac{5}{x+5} \cdot \dfrac{x-5}{x-5} - \dfrac{3}{x-5} \cdot \dfrac{x+5}{x+5}$

$= \dfrac{5x-25}{(x+5)(x-5)} - \dfrac{3x+15}{(x+5)(x-5)}$

$= \dfrac{5x-25-(3x+15)}{(x+5)(x-5)}$

$= \dfrac{5x-25-3x-15}{(x+5)(x-5)}$

$= \dfrac{2x-40}{(x+5)(x-5)}$

15. $\dfrac{3}{2t^2 - 2t} - \dfrac{5}{2t-2}$

$= \dfrac{3}{2t(t-1)} - \dfrac{5}{2(t-1)}$ $\text{LCD} = 2t(t-1)$

$= \dfrac{3}{2t(t-1)} - \dfrac{5}{2(t-1)} \cdot \dfrac{t}{t}$

$= \dfrac{3}{2t(t-1)} - \dfrac{5t}{2t(t-1)}$

$= \dfrac{3 - 5t}{2t(t-1)}$

17. $\dfrac{2s}{t^2 - s^2} - \dfrac{s}{t-s}$ $\text{LCD} = (t-s)(t+s)$

$= \dfrac{2s}{(t-s)(t+s)} - \dfrac{s}{t-s} \cdot \dfrac{t+s}{t+s}$

$= \dfrac{2s}{(t-s)(t+s)} - \dfrac{st + s^2}{(t-s)(t+s)}$

$= \dfrac{2s - (st + s^2)}{(t-s)(t+s)}$

$= \dfrac{2s - st - s^2}{(t-s)(t+s)}$

19. $\dfrac{y-5}{y} - \dfrac{3y-1}{4y} = \dfrac{y-5}{y} \cdot \dfrac{4}{4} - \dfrac{3y-1}{4y}$ $\text{LCD} = 4y$

$= \dfrac{4y - 20}{4y} - \dfrac{3y-1}{4y}$

$= \dfrac{4y - 20 - (3y-1)}{4y}$

$= \dfrac{4y - 20 - 3y + 1}{4y}$

$= \dfrac{y - 19}{4y}$

21. $\dfrac{a}{x+a} - \dfrac{a}{x-a}$ $\text{LCD} = (x+a)(x-a)$

$= \dfrac{a}{x+a} \cdot \dfrac{x-a}{x-a} - \dfrac{a}{x-a} \cdot \dfrac{x+a}{x+a}$

$= \dfrac{ax - a^2}{(x+a)(x-a)} - \dfrac{ax + a^2}{(x+a)(x-a)}$

$= \dfrac{ax - a^2 - (ax + a^2)}{(x+a)(x-a)}$

$= \dfrac{ax - a^2 - ax - a^2}{(x+a)(x-a)}$

$= \dfrac{-2a^2}{(x+a)(x-a)}$

23. $\dfrac{11}{6} - \dfrac{5}{-6} = \dfrac{11}{6} - \dfrac{5}{-6} \cdot \dfrac{-1}{-1}$

$= \dfrac{11}{6} - \dfrac{-5}{6}$

$= \dfrac{11 - (-5)}{6}$

$= \dfrac{11 + 5}{6}$

$= \dfrac{16}{6}$

$= \dfrac{8}{3}$

25. $\dfrac{5}{a} - \dfrac{8}{-a} = \dfrac{5}{a} - \dfrac{8}{-a} \cdot \dfrac{-1}{-1}$

$= \dfrac{5}{a} - \dfrac{-8}{a}$

$= \dfrac{5 - (-8)}{a}$

$= \dfrac{5 + 8}{a}$

$= \dfrac{13}{a}$

27. $\dfrac{4}{y-1} - \dfrac{4}{1-y} = \dfrac{4}{y-1} - \dfrac{4}{1-y} \cdot \dfrac{-1}{-1}$

$= \dfrac{4}{y-1} - \dfrac{4(-1)}{(1-y)(-1)}$

$= \dfrac{4}{y-1} - \dfrac{-4}{y-1}$

$= \dfrac{4 - (-4)}{y-1}$

$= \dfrac{4 + 4}{y-1}$

$= \dfrac{8}{y-1}$

29. $\dfrac{3-x}{x-7} - \dfrac{2x-5}{7-x} = \dfrac{3-x}{x-7} - \dfrac{2x-5}{7-x} \cdot \dfrac{-1}{-1}$

$= \dfrac{3-x}{x-7} - \dfrac{(2x-5)(-1)}{(7-x)(-1)}$

$= \dfrac{3-x}{x-7} - \dfrac{5-2x}{x-7}$

$= \dfrac{(3-x) - (5-2x)}{x-7}$

$= \dfrac{3 - x - 5 + 2x}{x-7}$

$= \dfrac{x-2}{x-7}$

31.
$$\frac{a-2}{a^2-25} - \frac{6-a}{25-a^2} = \frac{a-2}{a^2-25} - \frac{6-a}{25-a^2} \cdot \frac{-1}{-1}$$
$$= \frac{a-2}{a^2-25} - \frac{(6-a)(-1)}{(25-a^2)(-1)}$$
$$= \frac{a-2}{a^2-25} - \frac{a-6}{a^2-25}$$
$$= \frac{(a-2)-(a-6)}{a^2-25}$$
$$= \frac{a-2-a+6}{a^2-25}$$
$$= \frac{4}{a^2-25}$$

33.
$$\frac{4-x}{x-9} - \frac{3x-8}{9-x} = \frac{4-x}{x-9} - \frac{3x-8}{9-x} \cdot \frac{-1}{-1}$$
$$= \frac{4-x}{x-9} - \frac{8-3x}{x-9}$$
$$= \frac{(4-x)-(8-3x)}{x-9}$$
$$= \frac{4-x-8+3x}{x-9}$$
$$= \frac{2x-4}{x-9}$$

35.
$$\frac{5x}{x^2-9} - \frac{4}{3-x}$$
$$= \frac{5x}{(x+3)(x-3)} - \frac{4}{3-x} \qquad \begin{array}{l} x-3 \text{ and } 3-x \\ \text{are opposites} \end{array}$$
$$= \frac{5x}{(x+3)(x-3)} - \frac{4}{3-x} \cdot \frac{-1}{-1}$$
$$= \frac{5x}{(x+3)(x-3)} - \frac{-4}{x-3} \qquad \text{LCD} = (x+3)(x-3)$$
$$= \frac{5x}{(x+3)(x-3)} - \frac{-4}{x-3} \cdot \frac{x+3}{x+3}$$
$$= \frac{5x}{(x+3)(x-3)} - \frac{-4x-12}{(x+3)(x-3)}$$
$$= \frac{5x-(-4x-12)}{(x+3)(x-3)}$$
$$= \frac{5x+4x+12}{(x+3)(x-3)}$$
$$= \frac{9x+12}{(x+3)(x-3)}$$

37.
$$\frac{t^2}{2t^2-2t} - \frac{1}{2t-2}$$
$$= \frac{t^2}{2t(t-1)} - \frac{1}{2(t-1)} \qquad \text{LCD} = 2t(t-1)$$
$$= \frac{t^2}{2t(t-1)} - \frac{1}{2(t-1)} \cdot \frac{t}{t}$$
$$= \frac{t^2}{2t(t-1)} - \frac{t}{2t(t-1)}$$
$$= \frac{t^2-t}{2t(t-1)}$$
$$= \frac{t(t-1)}{2t(t-1)}$$
$$= \frac{t(t-1)(1)}{2t(t-1)}$$
$$= \frac{1}{2}$$

39.
$$\frac{x}{x^2+5x+6} - \frac{2}{x^2+3x+2}$$
$$= \frac{x}{(x+3)(x+2)} - \frac{2}{(x+2)(x+1)}$$
$$\qquad\qquad \text{LCD} = (x+3)(x+2)(x+1)$$
$$= \frac{x}{(x+3)(x+2)} \cdot \frac{x+1}{x+1} - \frac{2}{(x+2)(x+1)} \cdot \frac{x+3}{x+3}$$
$$= \frac{x^2+x}{(x+3)(x+2)(x+1)} - \frac{2x+6}{(x+3)(x+2)(x+1)}$$
$$= \frac{x^2+x-(2x+6)}{(x+3)(x+2)(x+1)}$$
$$= \frac{x^2+x-2x-6}{(x+3)(x+2)(x+1)}$$
$$= \frac{x^2-x-6}{(x+3)(x+2)(x+1)}$$
$$= \frac{(x-3)(x+2)}{(x+3)(x+2)(x+1)}$$
$$= \frac{(x-3)(x+2)}{(x+3)(x+2)(x+1)}$$
$$= \frac{x-3}{(x+3)(x+1)}$$

41.
$$\frac{3(2x+5)}{x-1} - \frac{3(2x-3)}{1-x} + \frac{6x+1}{x-1}$$
$$= \frac{3(2x+5)}{x-1} - \frac{3(2x-3)}{1-x} \cdot \frac{-1}{-1} + \frac{6x-1}{x-1}$$
$$= \frac{3(2x+5)}{x-1} - \frac{-3(2x-3)}{x-1} + \frac{6x-1}{x-1}$$
$$= \frac{(6x+15)-(-6x+9)+(6x-1)}{x-1}$$
$$= \frac{6x+15+6x-9+6x-1}{x-1}$$
$$= \frac{18x+5}{x-1}$$

43. $\dfrac{x-y}{x^2-y^2} + \dfrac{x+y}{x^2-y^2} - \dfrac{2x}{x^2-y^2}$

$= \dfrac{x-y+x+y-2x}{x^2-y^2}$

$= \dfrac{0}{x^2-y^2}$

$= 0$

45. $\dfrac{2(x-1)}{2x-3} - \dfrac{3(x+2)}{2x-3} - \dfrac{x-1}{3-2x}$

$= \dfrac{2(x-1)}{2x-3} - \dfrac{3(x+2)}{2x-3} - \dfrac{x-1}{3-2x} \cdot \dfrac{-1}{-1}$

$= \dfrac{2(x-1)}{2x-3} - \dfrac{3(x+2)}{2x-3} - \dfrac{1-x}{2x-3}$

$= \dfrac{(2x-2)-(3x+6)-(1-x)}{2x-3}$

$= \dfrac{2x-2-3x-6-1+x}{2x-3}$

$= \dfrac{-9}{2x-3}$

47. $\dfrac{10}{2y-1} - \dfrac{6}{1-2y} + \dfrac{y}{2y-1} + \dfrac{y-4}{1-2y}$

$= \dfrac{10}{2y-1} - \dfrac{6}{1-2y} \cdot \dfrac{-1}{-1} + \dfrac{y}{2y-1} + \dfrac{y-4}{1-2y} \cdot \dfrac{-1}{-1}$

$= \dfrac{10}{2y-1} - \dfrac{-6}{2y-1} + \dfrac{y}{2y-1} + \dfrac{4-y}{2y-1}$

$= \dfrac{10-(-6)+y+4-y}{2y-1}$

$= \dfrac{10+6+y+4-y}{2y-1}$

$= \dfrac{20}{2y-1}$

49. $\dfrac{a+6}{4-a^2} - \dfrac{a+3}{a+2} + \dfrac{a-3}{2-a}$

$= \dfrac{a+6}{(2+a)(2-a)} - \dfrac{a+3}{2+a} + \dfrac{a-3}{2-a}$

$\qquad a+2 = 2+a; \text{ LCD} = (2+a)(2-a)$

$= \dfrac{a+6}{(2+a)(2-a)} - \dfrac{a+3}{2+a} \cdot \dfrac{2-a}{2-a} + \dfrac{a-3}{2-a} \cdot \dfrac{2+a}{2+a}$

$= \dfrac{(a+6)-(a+3)(2-a)+(a-3)(2+a)}{(2+a)(2-a)}$

$= \dfrac{a+6-(-a^2-a+6)+(a^2-a-6)}{(2+a)(2-a)}$

$= \dfrac{a+6+a^2+a-6+a^2-a-6}{(2+a)(2-a)}$

$= \dfrac{2a^2+a-6}{(2+a)(2-a)}$

$= \dfrac{(2a-3)(a+2)}{(2+a)(2-a)}$

$= \dfrac{(2a-3)\cancel{(2+a)}}{\cancel{(2+a)}(2-a)}$

$= \dfrac{2a-3}{2-a}$

51. $\dfrac{2z}{1-2z} + \dfrac{3z}{2z+1} - \dfrac{3}{4z^2-1}$

$= \dfrac{2z}{1-2z} \cdot \dfrac{-1}{-1} + \dfrac{3z}{2z+1} - \dfrac{3}{4z^2-1}$

$= \dfrac{-2z}{2z-1} + \dfrac{3z}{2z+1} - \dfrac{3}{(2z-1)(2z+1)}$

$\qquad\qquad \text{LCD} = (2z-1)(2z+1)$

$= \dfrac{-2z}{2z-1} \cdot \dfrac{2z+1}{2z+1} + \dfrac{3z}{2z+1} \cdot \dfrac{2z-1}{2z-1} -$

$\qquad\qquad\qquad \dfrac{3}{(2z-1)(2z+1)}$

$= \dfrac{(-4z^2-2z)+(6z^2-3z)-3}{(2z-1)(2z+1)}$

$= \dfrac{2z^2-5z-3}{(2z-1)(2z+1)}$

$= \dfrac{(z-3)(2z+1)}{(2z-1)(2z+1)}$

$= \dfrac{(z-3)\cancel{(2z+1)}}{(2z-1)\cancel{(2z+1)}}$

$= \dfrac{z-3}{2z-1}$

53.

$$\frac{1}{x+y} - \frac{1}{x-y} + \frac{2x}{x^2 - y^2}$$

$$= \frac{1}{x+y} - \frac{1}{x-y} + \frac{2x}{(x+y)(x-y)}$$

$$\text{LCD} = (x+y)(x-y)$$

$$= \frac{1}{x+y} \cdot \frac{x-y}{x-y} - \frac{1}{x-y} \cdot \frac{x+y}{x+y} \cdot \frac{x+y}{x+y} + \frac{2x}{(x+y)(x-y)}$$

$$= \frac{x - y - (x+y) + 2x}{(x+y)(x-y)}$$

$$= \frac{x - y - x - y + 2x}{(x+y)(x-y)}$$

$$= \frac{2x - 2y}{(x+y)(x-y)}$$

$$= \frac{2(x-y)}{(x+y)(x-y)}$$

$$= \frac{2\cancel{(x-y)}}{(x+y)\cancel{(x-y)}}$$

$$= \frac{2}{x+y}$$

55. Discussion and Writing Exercise

57. $\dfrac{x^8}{x^3} = x^{8-3} = x^5$

59. $(a^2 b^{-5})^{-4} = a^{2(-4)} b^{-5(-4)} = a^{-8} b^{20} = \dfrac{b^{20}}{a^8}$

61. $\dfrac{66x^2}{11x^5} = \dfrac{6 \cdot \cancel{11} \cdot \cancel{x^2}}{\cancel{11} \cdot \cancel{x^2} \cdot x^3} = \dfrac{6}{x^3}$

63. The shaded area has dimensions $x - 6$ by $x - 3$. Then the area is $(x-6)(x-3)$, or $x^2 - 9x + 18$.

65.

$$\frac{2x+11}{x-3} \cdot \frac{3}{x+4} + \frac{2x+1}{4+x} \cdot \frac{3}{3-x}$$

$$= \frac{6x+33}{(x-3)(x+4)} + \frac{6x+3}{(4+x)(3-x)}$$

$$= \frac{6x+33}{(x-3)(x+4)} + \frac{6x+3}{(4+x)(3-x)} \cdot \frac{-1}{-1}$$

$$= \frac{6x+33}{(x-3)(x+4)} + \frac{-6x-3}{(x+4)(x-3)}$$

$$= \frac{6x+33-6x-3}{(x-3)(x+4)}$$

$$= \frac{30}{(x-3)(x+4)}$$

67.

$$\frac{x}{x^4 - y^4} - \left(\frac{1}{x+y}\right)^2$$

$$= \frac{x}{(x^2+y^2)(x+y)(x-y)} - \frac{1}{(x+y)^2}$$

$$\text{LCD} = (x^2+y^2)(x+y)^2(x-y)$$

$$= \frac{x}{(x^2+y^2)(x+y)(x-y)} \cdot \frac{x+y}{x+y} - \frac{1}{(x+y)^2} \cdot \frac{(x^2+y^2)(x-y)}{(x^2+y^2)(x-y)}$$

$$= \frac{x(x+y) - (x^2+y^2)(x-y)}{(x^2+y^2)(x+y)^2(x-y)}$$

$$= \frac{x^2 + xy - (x^3 - x^2y + xy^2 - y^3)}{(x^2+y^2)(x+y)^2(x-y)}$$

$$= \frac{x^2 + xy - x^3 + x^2y - xy^2 + y^3}{(x^2+y^2)(x+y)^2(x-y)}$$

69. Let $l = $ the length of the missing side.

$$\frac{a^2 - 5a - 9}{a-6} + \frac{a^2 - 6}{a-6} + l = 2a + 5$$

$$\frac{2a^2 - 5a - 15}{a-6} + l = 2a + 5$$

$$l = 2a + 5 - \frac{2a^2 - 5a - 15}{a-6}$$

$$l = (2a+5) \cdot \frac{a-6}{a-6} - \frac{2a^2 - 5a - 15}{a-6}$$

$$l = \frac{2a^2 - 7a - 30}{a-6} - \frac{2a^2 - 5a - 15}{a-6}$$

$$l = \frac{2a^2 - 7a - 30 - (2a^2 - 5a - 15)}{a-6}$$

$$l = \frac{2a^2 - 7a - 30 - 2a^2 + 5a + 15}{a-6}$$

$$l = \frac{-2a - 15}{a-6}$$

The length of the missing side is $\dfrac{-2a-15}{a-6}$.

Now find the area.

$$A = \frac{1}{2} \cdot b \cdot h$$

$$A = \frac{1}{2}\left(\frac{-2a-15}{a-6}\right)\left(\frac{a^2-6}{a-6}\right)$$

$$A = \frac{(-2a-15)(a^2-6)}{2(a-6)^2}, \text{ or}$$

$$A = \frac{-2a^3 - 15a^2 + 12a + 90}{2a^2 - 24a + 72}$$

71.–73. Left to the student

Exercise Set 6.6

1. $\quad \dfrac{4}{5} - \dfrac{2}{3} = \dfrac{x}{9}$, LCM = 45

$$45\left(\dfrac{4}{5} - \dfrac{2}{3}\right) = 45 \cdot \dfrac{x}{9}$$

$$45 \cdot \dfrac{4}{5} - 45 \cdot \dfrac{2}{3} = 45 \cdot \dfrac{x}{9}$$

$$36 - 30 = 5x$$

$$6 = 5x$$

$$\dfrac{6}{5} = x$$

Check:

$$\dfrac{4}{5} - \dfrac{2}{3} = \dfrac{x}{9}$$

$$\begin{array}{c|c} & \dfrac{6}{9} \\ \hline \dfrac{4}{5} - \dfrac{2}{3} \ ? \ \dfrac{5}{9} & \\ \dfrac{12}{15} - \dfrac{10}{15} & \dfrac{6}{5} \cdot \dfrac{1}{9} \\ \dfrac{2}{15} & \dfrac{2}{15} \quad \text{TRUE} \end{array}$$

This checks, so the solution is $\dfrac{6}{5}$.

3. $\quad \dfrac{3}{5} + \dfrac{1}{8} = \dfrac{1}{x}$, LCM = 40x

$$40x\left(\dfrac{3}{5} + \dfrac{1}{8}\right) = 40x \cdot \dfrac{1}{x}$$

$$40x \cdot \dfrac{3}{5} + 40x \cdot \dfrac{1}{8} = 40x \cdot \dfrac{1}{x}$$

$$24x + 5x = 40$$

$$29x = 40$$

$$x = \dfrac{40}{29}$$

Check:

$$\dfrac{3}{5} + \dfrac{1}{8} = \dfrac{1}{x}$$

$$\begin{array}{c|c} \dfrac{3}{5} + \dfrac{1}{8} \ ? \ \dfrac{1}{\dfrac{40}{29}} & \\ \dfrac{24}{40} + \dfrac{5}{40} & 1 \cdot \dfrac{29}{40} \\ \dfrac{29}{40} & \dfrac{29}{40} \quad \text{TRUE} \end{array}$$

This checks, so the solution is $\dfrac{40}{29}$.

5. $\quad \dfrac{3}{8} + \dfrac{4}{5} = \dfrac{x}{20}$, LCM = 40

$$40\left(\dfrac{3}{8} + \dfrac{4}{5}\right) = 40 \cdot \dfrac{x}{20}$$

$$40 \cdot \dfrac{3}{8} + 40 \cdot \dfrac{4}{5} = 40 \cdot \dfrac{x}{20}$$

$$15 + 32 = 2x$$

$$47 = 2x$$

$$\dfrac{47}{2} = x$$

Check:

$$\dfrac{3}{8} + \dfrac{4}{5} = \dfrac{x}{20}$$

$$\begin{array}{c|c} & \dfrac{\dfrac{47}{2}}{20} \\ \hline \dfrac{3}{8} + \dfrac{4}{5} \ ? & \\ \dfrac{15}{40} + \dfrac{32}{40} & \dfrac{47}{2} \cdot \dfrac{1}{20} \\ \dfrac{47}{40} & \dfrac{47}{40} \quad \text{TRUE} \end{array}$$

This checks, so the solution is $\dfrac{47}{2}$.

7. $\quad \dfrac{1}{x} = \dfrac{2}{3} - \dfrac{5}{6}$, LCM = 6x

$$6x \cdot \dfrac{1}{x} = 6x\left(\dfrac{2}{3} - \dfrac{5}{6}\right)$$

$$6x \cdot \dfrac{1}{x} = 6x \cdot \dfrac{2}{3} - 6x \cdot \dfrac{5}{6}$$

$$6 = 4x - 5x$$

$$6 = -x$$

$$-6 = x$$

Check:

$$\dfrac{1}{x} = \dfrac{2}{3} - \dfrac{5}{6}$$

$$\begin{array}{c|c} \dfrac{1}{-6} \ ? \ \dfrac{2}{3} - \dfrac{5}{6} & \\ -\dfrac{1}{6} & \dfrac{4}{6} - \dfrac{5}{6} \\ & -\dfrac{1}{6} \quad \text{TRUE} \end{array}$$

This checks, so the solution is -6.

9. $\quad \dfrac{1}{6} + \dfrac{1}{8} = \dfrac{1}{t}$, LCM = 24t

$$24t\left(\dfrac{1}{6} + \dfrac{1}{8}\right) = 24t \cdot \dfrac{1}{t}$$

$$24t \cdot \dfrac{1}{6} + 24t \cdot \dfrac{1}{8} = 24t \cdot \dfrac{1}{t}$$

$$4t + 3t = 24$$

$$7t = 24$$

$$t = \dfrac{24}{7}$$

Check:

$$\frac{\dfrac{1}{6} + \dfrac{1}{8} = \dfrac{1}{t}}{\begin{array}{c|c} \dfrac{1}{6} + \dfrac{1}{8} \;?\; \dfrac{1}{24/7} \\[2mm] \dfrac{4}{24} + \dfrac{3}{24} & 1 \cdot \dfrac{7}{24} \\[3mm] \dfrac{7}{24} & \dfrac{7}{24} \qquad \text{TRUE} \end{array}}$$

This checks, so the solution is $\dfrac{24}{7}$.

11. $\qquad x + \dfrac{4}{x} = -5, \ \text{LCM} = x$

$$x\left(x + \frac{4}{x}\right) = x(-5)$$

$$x \cdot x + x \cdot \frac{4}{x} = x(-5)$$

$$x^2 + 4 = -5x$$

$$x^2 + 5x + 4 = 0$$

$$(x+4)(x+1) = 0$$

$$x + 4 = 0 \quad \text{or} \quad x + 1 = 0$$

$$x = -4 \quad \text{or} \qquad x = -1$$

Check:

$$\frac{x + \dfrac{4}{x} = -5}{\begin{array}{c|c} -4 + \dfrac{4}{-4} \;?\; -5 \\[3mm] -4 - 1 & \\[1mm] -5 & \text{TRUE} \end{array}} \qquad \frac{x + \dfrac{4}{x} = -5}{\begin{array}{c|c} -1 + \dfrac{4}{-1} \;?\; -5 \\[3mm] -1 - 4 & \\[1mm] -5 & \text{TRUE} \end{array}}$$

Both of these check, so the two solutions are -4 and -1.

13. $\qquad \dfrac{x}{4} - \dfrac{4}{x} = 0, \ \text{LCM} = 4x$

$$4x\left(\frac{x}{4} - \frac{4}{x}\right) = 4x \cdot 0$$

$$4x \cdot \frac{x}{4} - 4x \cdot \frac{4}{x} = 4x \cdot 0$$

$$x^2 - 16 = 0$$

$$(x+4)(x-4) = 0$$

$$x + 4 = 0 \quad \text{or} \quad x - 4 = 0$$

$$x = -4 \quad \text{or} \qquad x = 4$$

Check:

$$\frac{\dfrac{x}{4} - \dfrac{4}{x} = 0}{\begin{array}{c|c} \dfrac{-4}{4} - \dfrac{4}{-4} \;?\; 0 \\[3mm] -1 - (-1) & \\[1mm] -1 + 1 & \\[1mm] 0 & \text{TRUE} \end{array}} \qquad \frac{\dfrac{x}{4} - \dfrac{4}{x} = 0}{\begin{array}{c|c} \dfrac{4}{4} - \dfrac{4}{4} \;?\; 0 \\[3mm] 1 - 1 & \\[1mm] 0 & \text{TRUE} \end{array}}$$

Both of these check, so the two solutions are -4 and 4.

15. $\qquad \dfrac{5}{x} = \dfrac{6}{x} - \dfrac{1}{3}, \ \text{LCM} = 3x$

$$3x \cdot \frac{5}{x} = 3x\left(\frac{6}{x} - \frac{1}{3}\right)$$

$$3x \cdot \frac{5}{x} = 3x \cdot \frac{6}{x} - 3x \cdot \frac{1}{3}$$

$$15 = 18 - x$$

$$-3 = -x$$

$$3 = x$$

Check:

$$\frac{\dfrac{5}{x} = \dfrac{6}{x} - \dfrac{1}{3}}{\begin{array}{c|c} \dfrac{5}{3} \;?\; \dfrac{6}{3} - \dfrac{1}{3} \\[3mm] & \dfrac{5}{3} \qquad \text{TRUE} \end{array}}$$

This checks, so the solution is 3.

17. $\qquad \dfrac{5}{3x} + \dfrac{3}{x} = 1, \ \text{LCM} = 3x$

$$3x\left(\frac{5}{3x} + \frac{3}{x}\right) = 3x \cdot 1$$

$$3x \cdot \frac{5}{3x} + 3x \cdot \frac{3}{x} = 3x \cdot 1$$

$$5 + 9 = 3x$$

$$14 = 3x$$

$$\frac{14}{3} = x$$

Check:

$$\frac{\dfrac{5}{3x} + \dfrac{3}{x} = 1}{\begin{array}{c|c} \dfrac{5}{3 \cdot (14/3)} + \dfrac{3}{(14/3)} \;?\; 1 \\[3mm] \dfrac{5}{14} + \dfrac{9}{14} & \\[3mm] \dfrac{14}{14} & \\[2mm] 1 & \text{TRUE} \end{array}}$$

This checks, so the solution is $\dfrac{14}{3}$.

19. $\qquad \dfrac{t-2}{t+3} = \dfrac{3}{8}, \ \text{LCM} = 8(t+3)$

$$8(t+3)\left(\frac{t-2}{t+3}\right) = 8(t+3)\left(\frac{3}{8}\right)$$

$$8(t-2) = 3(t+3)$$

$$8t - 16 = 3t + 9$$

$$5t = 25$$

$$t = 5$$

Check:

$$\frac{t-2}{t+3} = \frac{3}{8}$$

$$\frac{5-2}{5+3} \;?\; \frac{3}{8}$$

$$\frac{3}{8} \;\Big|\; \text{TRUE}$$

This checks, so the solution is 5.

21. $$\frac{2}{x+1} = \frac{1}{x-2}, \text{ LCM} = (x+1)(x-2)$$

$$(x+1)(x-2) \cdot \frac{2}{x+1} = (x+1)(x-2) \cdot \frac{1}{x-2}$$

$$2(x-2) = x+1$$

$$2x - 4 = x + 1$$

$$x = 5$$

This checks, so the solution is 5.

23. $$\frac{x}{6} - \frac{x}{10} = \frac{1}{6}, \text{ LCM} = 30$$

$$30\Big(\frac{x}{6} - \frac{x}{10}\Big) = 30 \cdot \frac{1}{6}$$

$$30 \cdot \frac{x}{6} - 30 \cdot \frac{x}{10} = 30 \cdot \frac{1}{6}$$

$$5x - 3x = 5$$

$$2x = 5$$

$$x = \frac{5}{2}$$

This checks, so the solution is $\frac{5}{2}$.

25. $$\frac{t+2}{5} - \frac{t-2}{4} = 1, \text{ LCM} = 20$$

$$20\Big(\frac{t+2}{5} - \frac{t-2}{4}\Big) = 20 \cdot 1$$

$$20\Big(\frac{t+2}{5}\Big) - 20\Big(\frac{t-2}{4}\Big) = 20 \cdot 1$$

$$4(t+2) - 5(t-2) = 20$$

$$4t + 8 - 5t + 10 = 20$$

$$-t + 18 = 20$$

$$-t = 2$$

$$t = -2$$

This checks, so the solution is -2.

27. $$\frac{5}{x-1} = \frac{3}{x+2},$$

$$\text{LCD} = (x-1)(x+2)$$

$$(x-1)(x+2) \cdot \frac{5}{x-1} = (x-1)(x+2) \cdot \frac{3}{x+2}$$

$$5(x+2) = 3(x-1)$$

$$5x + 10 = 3x - 3$$

$$2x = -13$$

$$x = -\frac{13}{2}$$

This checks, so the solution is $-\frac{13}{2}$.

29. $$\frac{a-3}{3a+2} = \frac{1}{5}, \text{ LCM} = 5(3a+2)$$

$$5(3a+2) \cdot \frac{a-3}{3a+2} = 5(3a+2) \cdot \frac{1}{5}$$

$$5(a-3) = 3a+2$$

$$5a - 15 = 3a + 2$$

$$2a = 17$$

$$a = \frac{17}{2}$$

This checks, so the solution is $\frac{17}{2}$.

31. $$\frac{x-1}{x-5} = \frac{4}{x-5}, \text{ LCM} = x-5$$

$$(x-5) \cdot \frac{x-1}{x-5} = (x-5) \cdot \frac{4}{x-5}$$

$$x - 1 = 4$$

$$x = 5$$

The number 5 is not a solution because it makes a denominator zero. Thus, there is no solution.

33. $$\frac{2}{x+3} = \frac{5}{x}, \text{ LCM} = x(x+3)$$

$$x(x+3) \cdot \frac{2}{x+3} = x(x+3) \cdot \frac{5}{x}$$

$$2x = 5(x+3)$$

$$2x = 5x + 15$$

$$-15 = 3x$$

$$-5 = x$$

This checks, so the solution is -5.

35.
$$\frac{x-2}{x-3} = \frac{x-1}{x+1}, \text{ LCM} = (x-3)(x+1)$$

$$(x-3)(x+1) \cdot \frac{x-2}{x-3} = (x-3)(x+1) \cdot \frac{x-1}{x+1}$$

$$(x+1)(x-2) = (x-3)(x-1)$$

$$x^2 - x - 2 = x^2 - 4x + 3$$

$$-x - 2 = -4x + 3$$

$$3x = 5$$

$$x = \frac{5}{3}$$

This checks, so the solution is $\frac{5}{3}$.

37.
$$\frac{1}{x+3} + \frac{1}{x-3} = \frac{1}{x^2-9},$$
$$\text{LCM} = (x+3)(x-3)$$

$$(x+3)(x-3)\left(\frac{1}{x+3} + \frac{1}{x-3}\right) = (x+3)(x-3) \cdot \frac{1}{(x+3)(x-3)}$$

$$(x-3) + (x+3) = 1$$

$$2x = 1$$

$$x = \frac{1}{2}$$

This checks, so the solution is $\frac{1}{2}$.

39.
$$\frac{x}{x+4} - \frac{4}{x-4} = \frac{x^2+16}{x^2-16},$$
$$\text{LCM} = (x+4)(x-4)$$

$$(x+4)(x-4)\left(\frac{x}{x+4} - \frac{x}{x-4}\right) = (x+4)(x-4) \cdot \frac{x^2+16}{(x+4)(x-4)}$$

$$x(x-4) - 4(x+4) = x^2 + 16$$

$$x^2 - 4x - 4x - 16 = x^2 + 16$$

$$x^2 - 8x - 16 = x^2 + 16$$

$$-8x - 16 = 16$$

$$-8x = 32$$

$$x = -4$$

The number -4 is not a solution because it makes a denominator zero. Thus, there is no solution.

41.
$$\frac{4-a}{8-a} = \frac{4}{a-8} \qquad \begin{array}{l}8-a \text{ and } a-8 \\ \text{are opposites}\end{array}$$

$$\frac{4-a}{8-a} \cdot \frac{-1}{-1} = \frac{4}{a-8}$$

$$\frac{a-4}{a-8} = \frac{4}{a-8}, \text{ LCM} = a-8$$

$$(a-8)\left(\frac{a-4}{a-8}\right) = (a-8)\left(\frac{4}{a-8}\right)$$

$$a - 4 = 4$$

$$a = 8$$

The number 8 is not a solution because it makes a denominator zero. Thus, there is no solution.

43.
$$2 - \frac{a-2}{a+3} = \frac{a^2-4}{a+3}, \text{ LCM} = a+3$$

$$(a+3)\left(2 - \frac{a-2}{a+3}\right) = (a+3) \cdot \frac{a^2-4}{a+3}$$

$$2(a+3) - (a-2) = a^2 - 4$$

$$2a + 6 - a + 2 = a^2 - 4$$

$$0 = a^2 - a - 12$$

$$0 = (a-4)(a+3)$$

$$a - 4 = 0 \text{ or } a + 3 = 0$$

$$a = 4 \text{ or } \qquad a = -3$$

Only 4 checks, so the solution is 4.

45.
$$\frac{x+1}{x+2} = \frac{x+3}{x+4},$$
$$\text{LCM} = (x+2)(x+4)$$

$$(x+2)(x+4)\left(\frac{x+1}{x+2}\right) = (x+2)(x+4)\left(\frac{x+3}{x+4}\right)$$

$$(x+4)(x+1) = (x+2)(x+3)$$

$$x^2 + 5x + 4 = x^2 + 5x + 6$$

$$4 = 6 \qquad \text{Subtracting } x^2 \text{ and } 5x$$

We get a false equation, so the original equation has no solution.

47.
$$4a - 3 = \frac{a+13}{a+1}, \text{ LCM} = a+1$$

$$(a+1)(4a-3) = (a+1) \cdot \frac{a+13}{a+1}$$

$$4a^2 + a - 3 = a + 13$$

$$4a^2 - 16 = 0$$

$$4(a+2)(a-2) = 0$$

$$a + 2 = 0 \quad \text{ or } \quad a - 2 = 0$$

$$a = -2 \quad \text{ or } \qquad a = 2$$

Both of these check, so the two solutions are -2 and 2.

49.
$$\frac{4}{y-2} - \frac{2y-3}{y^2-4} = \frac{5}{y+2},$$
$$\text{LCM} = (y+2)(y-2)$$

$$(y+2)(y-2)\left(\frac{4}{y-2} - \frac{2y-3}{(y+2)(y-2)}\right) =$$
$$(y+2)(y-2) \cdot \frac{5}{y+2}$$

$$4(y+2) - (2y-3) = 5(y-2)$$

$$4y + 8 - 2y + 3 = 5y - 10$$

$$2y + 11 = 5y - 10$$

$$21 = 3y$$

$$7 = y$$

This checks, so the solution is 7.

51. Discussion and Writing Exercise

53. A rational expression is a quotient of two polynomials.

55. Two expressions are reciprocals of each other if their product is 1.

57. To find the LCM, use each factor the greatest number of times that it appears in any one factorization.

59. The quotient rule asserts that when dividing with exponential notation, if the bases are the same, keep the base and subtract the exponent of the denominator from the exponent of the numerator.

61.
$$\frac{x}{x^2 + 3x - 4} + \frac{x + 1}{x^2 + 6x + 8} = \frac{2x}{x^2 + x - 2}$$
$$\frac{x}{(x+4)(x-1)} + \frac{x+1}{(x+4)(x+2)} = \frac{2x}{(x+2)(x-1)}$$
$$x(x+2) + (x+1)(x-1) = 2x(x+4)$$

Multiplying by the LCM, $(x+4)(x-1)(x+2)$
$$x^2 + 2x + x^2 - 1 = 2x^2 + 8x$$
$$2x^2 + 2x - 1 = 2x^2 + 8x$$
$$2x - 1 = 8x$$
$$-1 = 6x$$
$$-\frac{1}{6} = x$$

This checks, so the solution is $-\frac{1}{6}$.

63. Left to the student

Exercise Set 6.7

1. Familiarize. The job takes Mandy 4 hours working alone and Omar 5 hours working alone. Then in 1 hour Mandy does $\frac{1}{4}$ of the job and Omar does $\frac{1}{5}$ of the job. Working together, they can do $\frac{1}{4} + \frac{1}{5}$, or $\frac{9}{20}$ of the job in 1 hour. In two hours, Mandy does $2\left(\frac{1}{4}\right)$ of the job and Omar does $2\left(\frac{1}{5}\right)$ of the job. Working together they can do $2\left(\frac{1}{4}\right) + 2\left(\frac{1}{5}\right)$, or $\frac{9}{10}$ of the job in 2 hours. In 3 hours they can do $3\left(\frac{1}{4}\right) + 3\left(\frac{1}{5}\right)$, or $1\frac{7}{20}$ of the job which is more of the job then needs to be done. The answer is somewhere between 2 hr and 3 hr.

Translate. If they work together t hours, then Mandy does $t\left(\frac{1}{4}\right)$ of the job and Omar does $t\left(\frac{1}{5}\right)$ of the job. We want some number t such that
$$t\left(\frac{1}{4}\right) + t\left(\frac{1}{5}\right) = 1, \text{ or } \frac{t}{4} + \frac{t}{5} = 1.$$

Solve. We solve the equation.
$$\frac{t}{4} + \frac{t}{5} = 1, \text{ LCM} = 20$$
$$20\left(\frac{t}{4} + \frac{t}{5}\right) = 20 \cdot 1$$
$$20 \cdot \frac{t}{4} + 20 \cdot \frac{t}{5} = 20$$
$$5t + 4t = 20$$
$$9t = 20$$
$$t = \frac{20}{9}, \text{ or } 2\frac{2}{9}$$

Check. The check can be done by repeating the computations. We also have a partial check in that we expected from our familiarization step that the answer would be between 2 hr and 3 hr.

State. Working together, it takes them $2\frac{2}{9}$ hr to complete the job.

3. Familiarize. The job takes Vern 45 min working alone and Nina 60 min working alone. Then in 1 minute Vern does $\frac{1}{45}$ of the job and Nina does $\frac{1}{60}$ of the job. Working together, they can do $\frac{1}{45} + \frac{1}{60}$, or $\frac{7}{180}$ of the job in 1 minute. In 20 minutes, Vern does $\frac{20}{45}$ of the job and Nina does $\frac{20}{60}$ of the job. Working together, they can do $\frac{20}{45} + \frac{20}{60}$, or $\frac{7}{9}$ of the job. In 30 minutes, they can do $\frac{30}{45} + \frac{30}{60}$, or $\frac{7}{6}$ of the job which is more of the job than needs to be done. The answer is somewhere between 20 minutes and 30 minutes.

Translate. If they work together t minutes, then Vern does $t\left(\frac{1}{45}\right)$ of the job and Nina does $t\left(\frac{1}{60}\right)$ of the job. We want some number t such that
$$t\left(\frac{1}{45}\right) + t\left(\frac{1}{60}\right) = 1, \text{ or } \frac{t}{45} + \frac{t}{60} = 1.$$

Solve. We solve the equation.
$$\frac{t}{45} + \frac{t}{60} = 1, \text{ LCM} = 180$$
$$180\left(\frac{t}{45} + \frac{t}{60}\right) = 180 \cdot 1$$
$$180 \cdot \frac{t}{45} + 180 \cdot \frac{t}{60} = 180$$
$$4t + 3t = 180$$
$$7t = 180$$
$$t = \frac{180}{7}, \text{ or } 25\frac{5}{7}$$

Check. The check can be done by repeating the computations. We also have a partial check in that we expected from our familiarization step that the answer would be between 20 minutes and 30 minutes.

State. It would take them $25\frac{5}{7}$ minutes to complete the job working together.

5. Familiarize. The job takes Kenny Dewitt 9 hours working alone and Betty Wohat 7 hours working alone. Then in 1 hour Kenny does $\frac{1}{9}$ of the job and Betty does $\frac{1}{7}$ of the job. Working together they can do $\frac{1}{9} + \frac{1}{7}$, or $\frac{16}{63}$ of the job in 1 hour. In two hours, Kenny does $2\left(\frac{1}{9}\right)$ of the job and Betty does $2\left(\frac{1}{7}\right)$ of the job. Working together they can do $2\left(\frac{1}{9}\right) + 2\left(\frac{1}{7}\right)$, or $\frac{32}{63}$ of the job in two hours. In five hours they can do $5\left(\frac{1}{9}\right) + 5\left(\frac{1}{7}\right)$, or $\frac{80}{63}$, or $1\frac{17}{63}$ of the job which is more of the job than needs to be done. The answer is somewhere between 2 hr and 5 hr.

Translate. If they work together t hours, Kenny does $t\left(\frac{1}{9}\right)$ of the job and Betty does $t\left(\frac{1}{7}\right)$ of the job. We want some number t such that

$$t\left(\frac{1}{9}\right) + t\left(\frac{1}{7}\right) = 1, \text{ or } \frac{t}{9} + \frac{t}{7} = 1.$$

Solve. We solve the equation.

$$\frac{t}{9} + \frac{t}{7} = 1, \text{ LCM} = 63$$

$$63\left(\frac{t}{9} + \frac{t}{7}\right) = 63 \cdot 1$$

$$63 \cdot \frac{t}{9} + 63 \cdot \frac{t}{7} = 63$$

$$7t + 9t = 63$$

$$16t = 63$$

$$t = \frac{63}{16}, \text{ or } 3\frac{15}{16}$$

Check. The check can be done by repeating the computations. We also have a partial check in that we expected from our familiarization step that the answer would be between 2 hr and 5 hr.

State. Working together, it takes them $3\frac{15}{16}$ hr to complete the job.

7. Familiarize. Let t = the number of minutes it takes Nicole and Glen to weed the garden, working together.

Translate. We use the work principle.

$$t\left(\frac{1}{50}\right) + t\left(\frac{1}{40}\right) = 1, \text{ or } \frac{t}{50} + \frac{t}{40} = 1$$

Solve. We solve the equation.

$$\frac{t}{50} + \frac{t}{40} = 1, \text{ LCM} = 200$$

$$200\left(\frac{t}{50} + \frac{t}{40}\right) = 200 \cdot 1$$

$$200 \cdot \frac{t}{50} + 200 \cdot \frac{t}{40} = 200$$

$$4t + 5t = 200$$

$$9t = 200$$

$$t = \frac{200}{9}, \text{ or } 22\frac{2}{9}$$

Check. In $\frac{200}{9}$ min, the portion of the job done is

$$\frac{1}{50} \cdot \frac{200}{9} + \frac{1}{40} \cdot \frac{200}{9} = \frac{4}{9} + \frac{5}{9} = 1. \text{ The answer checks.}$$

State. It would take $22\frac{2}{9}$ min to weed the garden if Nicole and Glen worked together.

9. Familiarize. Let t = the number of minutes it would take the two machines to make one copy of the report, working together.

Translate. We use the work principle.

$$t\left(\frac{1}{10}\right) + t\left(\frac{1}{6}\right) = 1, \text{ or } \frac{t}{10} + \frac{t}{6} = 1$$

Solve. We solve the equation.

$$\frac{t}{10} + \frac{t}{6} = 1, \text{ LCM} = 30$$

$$30\left(\frac{t}{10} + \frac{t}{6}\right) = 30 \cdot 1$$

$$30 \cdot \frac{t}{10} + 30 \cdot \frac{t}{6} = 30$$

$$3t + 5t = 30$$

$$8t = 30$$

$$t = \frac{15}{4}, \text{ or } 3\frac{3}{4}$$

Check. In $\frac{15}{4}$ min, the portion of the job done is

$$\frac{1}{10} \cdot \frac{15}{4} + \frac{1}{6} \cdot \frac{15}{4} = \frac{3}{8} + \frac{5}{8} = 1. \text{ The answer checks.}$$

State. It would take the two machines $3\frac{3}{4}$ min to make one copy of the report, working together.

11. Familiarize. We complete the table shown in the text.

$$d = r \cdot t$$

	Distance	Speed	Time
Car	150	r	t
Truck	350	$r + 40$	t

$\rightarrow 150 = r(t)$
$\rightarrow 350 = (r + 40)t$

Translate. We apply the formula $d = rt$ along the rows of the table to obtain two equations:

$$150 = rt,$$

$$350 = (r + 40)t$$

Then we solve each equation for t and set the results equal:

Solving $150 = rt$ for t: $t = \dfrac{150}{r}$

Solving $350 = (r + 40)t$ for t: $t = \dfrac{350}{r + 40}$

Thus, we have

$$\frac{150}{r} = \frac{350}{r + 40}.$$

Solve. We multiply by the LCM, $r(r+40)$.

$$r(r+40) \cdot \frac{150}{r} = r(r+40) \cdot \frac{350}{r+40}$$

$$150(r+40) = 350r$$

$$150r + 6000 = 350r$$

$$6000 = 200r$$

$$30 = r$$

Check. If r is 30 km/h, then $r+40$ is 70 km/h. The time for the car is 150/30, or 5 hr. The time for the truck is 350/70, or 5 hr. The times are the same. The values check.

State. The speed of Sarah's car is 30 km/h, and the speed of Rick's truck is 70 km/h.

13. *Familiarize*. We complete the table shown in the text.

$$d = r \cdot t$$

	Distance	Speed	Time
Freight	330	$r - 14$	t
Passenger	400	r	t

Translate. From the rows of the table we have two equations:

$$330 = (r-14)t,$$
$$400 = rt$$

We solve each equation for t and set the results equal:

Solving $330 = (r-14)t$ for t: $t = \dfrac{330}{r-14}$

Solving $400 = rt$ for t: $t = \dfrac{400}{r}$

Thus, we have

$$\frac{330}{r-14} = \frac{400}{r}.$$

Solve. We multiply by the LCM, $r(r-14)$.

$$r(r-14) \cdot \frac{330}{r-14} = r(r-14) \cdot \frac{400}{r}$$

$$330r = 400(r-14)$$

$$330r = 400r - 5600$$

$$-70r = -5600$$

$$r = 80$$

Then substitute 80 for r in either equation to find t:

$$t = \frac{400}{r}$$

$$t = \frac{400}{80} \qquad \text{Substituting 80 for } r$$

$$t = 5$$

Check. If $r = 80$, then $r - 14 = 66$. In 5 hr the freight train travels $66 \cdot 5$, or 330 mi, and the passenger train travels $80 \cdot 5$, or 400 mi. The values check.

State. The speed of the passenger train is 80 mph. The speed of the freight train is 66 mph.

15. *Familiarize*. We let r represent the speed going. Then $2r$ is the speed returning. We let t represent the time going. Then $t - 3$ represents the time returning. We organize the information in a table.

$$d = r \cdot t$$

	Distance	Speed	Time
Going	120	r	t
Returning	120	$2r$	$t - 3$

Translate. The rows of the table give us two equations:

$$120 = rt,$$
$$120 = 2r(t-3)$$

We can solve each equation for r and set the results equal:

Solving $120 = rt$ for r: $r = \dfrac{120}{t}$

Solving $120 = 2r(t-3)$ for r: $r = \dfrac{120}{2(t-3)}$, or

$$r = \frac{60}{t-3}$$

Then $\dfrac{120}{t} = \dfrac{60}{t-3}$.

Solve. We multiply on both sides by the LCM, $t(t-3)$.

$$t(t-3) \cdot \frac{120}{t} = t(t-3) \cdot \frac{60}{t-3}$$

$$120(t-3) = 60t$$

$$120t - 360 = 60t$$

$$-360 = -60t$$

$$6 = t$$

Then substitute 6 for t in either equation to find r, the speed going:

$$r = \frac{120}{t}$$

$$r = \frac{120}{6} \qquad \text{Substituting 6 for } t$$

$$r = 20$$

Check. If $r = 20$ and $t = 6$, then $2r = 2 \cdot 20$, or 40 mph and $t - 3 = 6 - 3$, or 3 hr. The distance going is $6 \cdot 20$, or 120 mi. The distance returning is $40 \cdot 3$, or 120 mi. The numbers check.

State. The speed going is 20 mph.

17. *Familiarize*. Let $r =$ Kelly's speed, in km/h, and $t =$ the time the bicyclists travel, in hours. Organize the information in a table.

	Distance	Speed	Time
Hank	42	$r - 5$	t
Kelly	57	r	t

Translate. We can replace the t's in the table above using the formula $t = d/r$.

	Distance	Speed	Time
Hank	42	$r-5$	$\dfrac{42}{r-5}$
Kelly	57	r	$\dfrac{57}{r}$

Since the times are the same for both bicyclists, we have the equation

$$\frac{42}{r-5} = \frac{57}{r}.$$

Solve. We first multiply by the LCD, $r(r-5)$.

$$r(r-5)\cdot\frac{42}{r-5} = r(r-5)\cdot\frac{57}{r}$$
$$42r = 57(r-5)$$
$$42r = 57r - 285$$
$$-15r = -285$$
$$r = 19$$

If $r = 19$, then $r - 5 = 14$.

Check. If Hank's speed is 14 km/h and Kelly's speed is 19 km/h, then Hank bicycles 5 km/h slower than Kelly. Hank's time is 42/14, or 3 hr. Kelly's time is 57/19, or 3 hr. Since the times are the same, the answer checks.

State. Hank travels at 14 km/h, and Kelly travels at 19 km/h.

19. **Familiarize.** Let r = Ralph's speed, in km/h. Then Bonnie's speed is $r + 3$. Also set t = the time, in hours, that Ralph and Bonnie walk. We organize the information in a table.

	Distance	Speed	Time
Ralph	7.5	r	t
Bonnie	12	$r+3$	t

Translate. We can replace the t's in the table shown above using the formula $t = d/r$.

	Distance	Speed	Time
Ralph	7.5	r	$\dfrac{7.5}{r}$
Bonnie	12	$r+3$	$\dfrac{12}{r+3}$

Since the times are the same for both walkers, we have the equation

$$\frac{7.5}{r} = \frac{12}{r+3}.$$

Solve. We first multiply by the LCD, $r(r+3)$.

$$r(r+3)\cdot\frac{7.5}{r} = r(r+3)\cdot\frac{12}{r+3}$$
$$7.5(r+3) = 12r$$
$$7.5r + 22.5 = 12r$$
$$22.5 = 4.5r$$
$$5 = r$$

If $r = 5$, then $r + 3 = 8$.

Check. If Ralph's speed is 5 km/h and Bonnie's speed is 8 km/h, then Bonnie walks 3 km/h faster than Ralph. Ralph's time is 7.5/5, or 1.5 hr. Bonnie's time is 12/8, or 1.5 hr. Since the times are the same, the answer checks.

State. Ralph's speed is 5 km/h, and Bonnie's speed is 8 km/h.

21. **Familiarize.** Let t = the time it takes Caledonia to drive to town and organize the given information in a table.

	Distance	Speed	Time
Caledonia	15	r	t
Manley	20	r	$t+1$

Translate. We can replace the r's in the table above using the formula $r = d/t$.

	Distance	Speed	Time
Caledonia	15	$\dfrac{15}{t}$	t
Manley	20	$\dfrac{20}{t+1}$	$t+1$

Since the speeds are the same for both riders, we have the equation

$$\frac{15}{t} = \frac{20}{t+1}.$$

Solve. We multiply by the LCD, $t(t+1)$.

$$t(t+1)\cdot\frac{15}{t} = t(t+1)\cdot\frac{20}{t+1}$$
$$15(t+1) = 20t$$
$$15t + 15 = 20t$$
$$15 = 5t$$
$$3 = t$$

If $t = 3$, then $t + 1 = 3 + 1$, or 4.

Check. If Caledonia's time is 3 hr and Manley's time is 4 hr, then Manley's time is 1 hr more than Caledonia's. Caledonia's speed is 15/3, or 5 mph. Manley's speed is 20/4, or 5 mph. Since the speeds are the same, the answer checks.

State. It takes Caledonia 3 hr to drive to town.

23. $\dfrac{10 \text{ divorces}}{18 \text{ marriages}} = \dfrac{10}{18}$ divorce/marriage $=$ $\dfrac{5}{9}$ divorce/marriage

25. $\dfrac{4.6 \text{ km}}{2 \text{ hr}} = 2.3$ km/h

27. **Familiarize.** A 120-lb person should eat at least 44 g of protein each day, and we wish to find the minimum protein required for a 180-lb person. We can set up ratios. We let p = the minimum number of grams of protein a 180-lb person should eat each day.

Translate. If we assume the rates of protein intake are the same, the ratios are the same and we have an equation.

$$\text{Protein} \rightarrow \frac{44}{120} = \frac{p}{180} \leftarrow \text{Protein}$$
$$\text{Weight} \rightarrow \phantom{\frac{44}{120}} \phantom{\frac{p}{180}} \leftarrow \text{Weight}$$

Solve. We solve the proportion.

$$360 \cdot \frac{44}{120} = 360 \cdot \frac{p}{180} \quad \text{Multiplying by the LCM, 360}$$

$$3 \cdot 44 = 2 \cdot p$$

$$132 = 2p$$

$$66 = p$$

Check. $\frac{44}{120} = \frac{4 \cdot 11}{4 \cdot 30} = \frac{\cancel{4} \cdot 11}{\cancel{4} \cdot 30} = \frac{11}{30}$ and

$\frac{66}{180} = \frac{6 \cdot 11}{6 \cdot 30} = \frac{\cancel{6} \cdot 11}{\cancel{6} \cdot 30} = \frac{11}{30}$. The ratios are the same.

State. A 180-lb person should eat a minimum of 66 g of protein each day.

29. *Familiarize.* 10 cc of human blood contains 1.2 grams of hemoglobin, and we wish to find how many grams of hemoglobin are contained in 16 cc of the same blood. We can set up ratios. Let $H =$ the amount of hemoglobin in 16 cc of the same blood.

Translate. Assuming the two ratios are the same, we can translate to a proportion.

$$\text{Grams} \rightarrow \frac{H}{16} = \frac{1.2}{10} \leftarrow \text{Grams}$$
$$\text{cm}^3 \rightarrow \phantom{\frac{H}{16}} \phantom{\frac{1.2}{10}} \leftarrow \text{cm}^3$$

Solve. We solve the proportion.

We multiply by 16 to get H alone.

$$16 \cdot \frac{H}{16} = 16 \cdot \frac{1.2}{10}$$

$$H = \frac{19.2}{10}$$

$$H = 1.92$$

Check.
$$\frac{1.92}{16} = 0.12 \qquad \frac{1.2}{10} = 0.12$$
The ratios are the same.

State. 16 cc of the same blood would contain 1.92 grams of hemoglobin.

31. *Familiarize.* Let $h =$ the amount of honey, in pounds, that 35,000 trips to flowers would produce.

Translate. We translate to a proportion.

$$\text{Honey} \rightarrow \frac{1}{20,000} = \frac{h}{35,000} \leftarrow \text{Honey}$$
$$\text{Trips} \rightarrow \phantom{\frac{1}{20,000}} \phantom{\frac{h}{35,000}} \leftarrow \text{Trips}$$

Solve. We solve the proportion.

$$35,000 \cdot \frac{1}{20,000} = 35,000 \cdot \frac{h}{35,000}$$

$$1.75 = h$$

Check. $\frac{1}{20,000} = 0.00005$ and $\frac{1.75}{35,000} = 0.00005$.

The ratios are the same.

State. 35,000 trips to gather nectar will produce 1.75 lb of honey.

33. *Familiarize.* The ratio of the weight of copper to the weight of zinc in a U.S. penny is $\frac{1}{39}$, and we wish to find how much copper is needed if 50 kg of zinc is being turned into pennies. We can set up a second ratio to go with the one we already have. Let $C =$ the amount of copper needed, in kg, if 50 kg of zinc is being turned into pennies.

Translate. We translate to a proportion.

$$\frac{1}{39} = \frac{C}{50}$$

Solve. We solve the proportion.

$$50 \cdot \frac{1}{39} = 50 \cdot \frac{C}{50}$$

$$\frac{50}{39} = C, \text{ or}$$

$$1\frac{11}{39} = C$$

Check. $\frac{50/39}{50} = \frac{1}{39}$, so the ratios are the same.

State. $1\frac{11}{39}$ kg of copper is needed if 50 kg of zinc is turned into pennies.

35. (a) $\frac{72}{217} \approx 0.332$

Suzuki's batting average was 0.332.

(b) Let $h =$ the number of hits Suzuki would get in the 162-game season. We translate to a proportion and solve it.

$$\frac{72}{48} = \frac{h}{162}$$

$$162 \cdot \frac{72}{48} = 162 \cdot \frac{h}{162}$$

$$243 \approx h$$

Suzuki would get 243 hits in the 162-game season.

(c) Let $h =$ the number of hits Suzuki would get if he batted 700 times. We translate to a proportion and solve it.

$$\frac{72}{217} = \frac{h}{700}$$

$$700 \cdot \frac{72}{217} = 700 \cdot \frac{h}{700}$$

$$232 \approx h$$

Suzuki would get 232 hits if he batted 700 times.

37. Let $h =$ the head circumference, in inches. We translate to a proportion and solve it.

$$\frac{6\frac{3}{4}}{21\frac{1}{5}} = \frac{7}{h}$$

$$6\frac{3}{4} \cdot h = 21\frac{1}{5} \cdot 7$$

$$\frac{27}{4} \cdot h = \frac{106}{5} \cdot 7$$

$$h = \frac{4}{27} \cdot \frac{106}{5} \cdot 7$$

$$h \approx 22$$

The head circumference is 22 in.

Now let c = the head circumference, in centimeters. We translate to a proportion and solve it.

$$\frac{6\frac{3}{4}}{53.8} = \frac{7}{c}$$

$$\frac{6.75}{53.8} = \frac{7}{c} \quad \left(6\frac{3}{4} = 6.75\right)$$

$$6.75 \cdot c = 53.8 \cdot 7$$

$$c = \frac{53.8 \cdot 7}{6.75}$$

$$c \approx 55.8$$

The head circumference is 55.8 cm.

39. Let h = the hat size. We translate to a proportion and solve it.

$$\frac{6\frac{3}{4}}{21\frac{1}{5}} = \frac{h}{22\frac{4}{5}}$$

$$6\frac{3}{4} \cdot 22\frac{4}{5} = 21\frac{1}{5} \cdot h$$

$$\frac{27}{4} \cdot \frac{114}{5} = \frac{106}{5} \cdot h$$

$$\frac{5}{106} \cdot \frac{27}{4} \cdot \frac{114}{5} = h$$

$$7.26 \approx h$$

$$7\frac{1}{4} \approx h$$

The hat size is $7\frac{1}{4}$.

Now let c = the head circumference, in centimeters. We translate to a proportion and solve it. We use the hat size found above in the translation.

$$\frac{6\frac{3}{4}}{53.8} = \frac{7\frac{1}{4}}{c}$$

$$\frac{6.75}{53.8} = \frac{7.25}{c}$$

$$6.75 \cdot c = 53.8 \cdot 7.25$$

$$c = \frac{53.8 \cdot 7.25}{6.75}$$

$$c \approx 57.8$$

The head circumference is 57.8 cm. (Answers may vary slightly depending on when rounding occurs.)

41. Let h = the hat size. We translate to a proportion and solve it.

$$\frac{6\frac{3}{4}}{53.8} = \frac{h}{59.8}$$

$$\frac{6.75}{53.8} = \frac{h}{59.8}$$

$$59.8 \cdot \frac{6.75}{53.8} = h$$

$$7.5 \approx h, \text{ or}$$

$$7\frac{1}{2} \approx h$$

The hat size is $7\frac{1}{2}$.

Now let c = the head circumference, in inches. We translate to a proportion and solve it. We use the hat size found above in the translation.

$$\frac{6\frac{3}{4}}{21\frac{1}{5}} = \frac{7\frac{1}{2}}{c}$$

$$6\frac{3}{4} \cdot c = 21\frac{1}{5} \cdot 7\frac{1}{2}$$

$$\frac{27}{4} \cdot c = \frac{106}{5} \cdot \frac{15}{2}$$

$$c = \frac{4}{27} \cdot \frac{106}{5} \cdot \frac{15}{2}$$

$$c \approx 23.6, \text{ or}$$

$$c \approx 23\frac{3}{5}$$

The head circumference is $23\frac{3}{5}$ in.

43. *Familiarize.* The ratio of trout tagged to the total trout population, P, is $\frac{112}{P}$. Of the 82 trout checked later, 32 were tagged. The ratio of trout tagged to trout checked is $\frac{32}{82}$.

Translate. Assuming the two ratios are the same, we can translate to a proportion.

$$\begin{array}{l} \text{Trout tagged} \\ \text{originally} \\ \text{Trout} \\ \text{population} \end{array} \begin{array}{l} \longrightarrow \\ \\ \longrightarrow \end{array} \frac{112}{P} = \frac{32}{82} \begin{array}{l} \longleftarrow \text{Tagged trout} \\ \quad\ \text{caught later} \\ \longleftarrow \text{Trout caught} \\ \quad\ \text{later} \end{array}$$

Solve. We solve the equation.

$$82P \cdot \frac{112}{P} = 82P \cdot \frac{32}{82} \quad \begin{array}{l} \text{Multiplying by the LCM,} \\ 82P \end{array}$$

$$82 \cdot 112 = P \cdot 32$$

$$9184 = 32P$$

$$287 = P$$

Check.

$$\frac{112}{287} \approx 0.390 \text{ and } \frac{32}{82} \approx 0.390.$$

The ratios are the same.

State. The trout population is 287.

45. *Familiarize*. A sample of 144 firecrackers contained 9 duds, and we wish to find how many duds could be expected in a sample of 3200 firecrackers. We can set up ratios, letting d = the number of duds expected in a sample of 3200 firecrackers.

Translate. Assuming the rates of occurrence of duds are the same, we can translate to a proportion.

$$\text{Duds} \rightarrow \frac{9}{144} = \frac{d}{3200} \leftarrow \text{Duds}$$
$$\text{Sample size} \rightarrow \hspace{1.5em} \hspace{1.5em} \leftarrow \text{Sample size}$$

Solve. We solve the equation. We multiply by 3200 to get d alone.

$$3200 \cdot \frac{9}{144} = 3200 \cdot \frac{d}{3200}$$
$$\frac{28,800}{144} = d$$
$$200 = d$$

Check.
$$\frac{9}{144} = 0.0625 \text{ and } \frac{200}{3200} = 0.0625$$
The ratios are the same.

State. You would expect 200 duds in a sample of 3200 firecrackers.

47. *Familiarize*. The ratio of the weight of an object on Mars to the weight of an object on earth is 0.4 to 1.

a) We wish to find how much a 12-ton rocket would weigh on Mars.

b) We wish to find how much a 120-lb astronaut would weigh on Mars.

We can set up ratios. We let r = the weight of a 12-ton rocket and a = the weight of a 120-lb astronaut on Mars.

Translate. Assuming the ratios are the same, we can translate to proportions.

a) $\begin{array}{l}\text{Weight} \\ \text{on Mars} \rightarrow \\ \text{Weight} \rightarrow \\ \text{on earth}\end{array} \dfrac{0.4}{1} = \dfrac{r}{12} \begin{array}{l}\leftarrow \text{Weight} \\ \text{on Mars} \\ \leftarrow \text{Weight} \\ \text{on earth}\end{array}$

b) $\begin{array}{l}\text{Weight} \\ \text{on Mars} \rightarrow \\ \text{Weight} \rightarrow \\ \text{on earth}\end{array} \dfrac{0.4}{1} = \dfrac{a}{120} \begin{array}{l}\leftarrow \text{Weight} \\ \text{on Mars} \\ \leftarrow \text{Weight} \\ \text{on earth}\end{array}$

Solve. We solve each proportion.

a) $\dfrac{0.4}{1} = \dfrac{r}{12}$ b) $\dfrac{0.4}{1} = \dfrac{1}{120}$

$\quad 12(0.4) = r \hspace{3em} 120(0.4) = a$

$\quad\quad 4.8 = r \hspace{4em} 48 = a$

Check. $\dfrac{0.4}{1} = 0.4$, $\dfrac{4.8}{12} = 0.4$, and $\dfrac{48}{120} = 0.4$.
The ratios are the same.

State. a) A 12-ton rocket would weigh 4.8 tons on Mars.

b) A 120-lb astronaut would weigh 48 lb on Mars.

49. We write a proportion and then solve it.

$$\frac{b}{6} = \frac{7}{4}$$
$$b = \frac{7}{4} \cdot 6 \hspace{2em} \text{Multiplying by 6}$$
$$b = \frac{42}{4}$$
$$b = \frac{21}{2}, \text{ or } 10.5$$

$\left(\text{Note that the proportions } \dfrac{6}{b} = \dfrac{4}{7}, \dfrac{b}{7} = \dfrac{6}{4}, \text{ or } \dfrac{7}{b} = \dfrac{4}{6} \text{ could also be used.}\right)$

51. We write a proportion and then solve it.

$$\frac{4}{f} = \frac{6}{4}$$
$$4f \cdot \frac{4}{f} = 4f \cdot \frac{6}{4}$$
$$16 = 6f$$
$$\frac{8}{3} = f \hspace{2em} \text{Simplifying}$$

$\left(\text{One of the following proportions could also be used: } \dfrac{f}{4} = \dfrac{4}{6}, \dfrac{4}{f} = \dfrac{9}{6}, \dfrac{f}{4} = \dfrac{6}{9}, \dfrac{4}{9} = \dfrac{f}{6}, \dfrac{9}{4} = \dfrac{6}{f}\right)$

53. We write a proportion and then solve it.

$$\frac{h}{7} = \frac{10}{6}$$
$$h = \frac{10}{6} \cdot 7 \hspace{1em} \text{Multiplying by 7}$$
$$h = \frac{70}{6}$$
$$h = \frac{35}{3} \hspace{2em} \text{Simplifying}$$

$\left(\text{Note that the proportions } \dfrac{7}{h} = \dfrac{6}{10}, \dfrac{h}{10} = \dfrac{7}{6}, \text{ or } \dfrac{10}{h} = \dfrac{6}{7} \text{ could also be used.}\right)$

55. We write a proportion and then solve it.

$$\frac{4}{10} = \frac{6}{l}$$
$$10l \cdot \frac{4}{10} = 10l \cdot \frac{6}{l}$$
$$4l = 60$$
$$l = 15 \text{ ft}$$

$\left(\text{One of the following proportions could also be used: } \dfrac{4}{6} = \dfrac{10}{l}, \dfrac{10}{4} = \dfrac{l}{6}, \text{ or } \dfrac{6}{4} = \dfrac{l}{10}\right)$

57. Discussion and Writing Exercise

59. $x^5 \cdot x^6 = x^{5+6} = x^{11}$

61. $x^{-5} \cdot x^{-6} = x^{-5+(-6)} = x^{-11} = \dfrac{1}{x^{11}}$

63. Graph: $y = 2x - 6$.

We select some x-values and compute y-values.

If $x = 1$, then $y = 2 \cdot 1 - 6 = -4$.

If $x = 3$, then $y = 2 \cdot 3 - 6 = 0$.

If $x = 5$, then $y = 2 \cdot 5 - 6 = 4$.

x	y	(x,y)
1	-4	$(1,-4)$
3	0	$(3,0)$
5	4	$(5,4)$

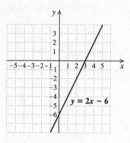

65. Graph: $3x + 2y = 12$.

We can replace either variable with a number and then calculate the other coordinate. We will find the intercepts and one other point.

If $y = 0$, we have:

$$3x + 2 \cdot 0 = 12$$
$$3x = 12$$
$$x = 4$$

The x-intercept is $(4,0)$.

If $x = 0$, we have:

$$3 \cdot 0 + 2y = 12$$
$$2y = 12$$
$$y = 6$$

The y-intercept is $(0,6)$.

If $y = -3$, we have:

$$3x + 2(-3) = 12$$
$$3x - 6 = 12$$
$$3x = 18$$
$$x = 6$$

The point $(6,-3)$ is on the graph.

We plot these points and draw a line through them.

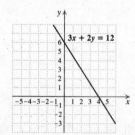

67. Graph: $y = -\dfrac{3}{4}x + 2$.

We select some x-values and compute y-values. We use multiples of 4 to avoid fractions.

If $x = -4$, then $y = -\dfrac{3}{4}(-4) + 2 = 5$.

If $x = 0$, then $y = -\dfrac{3}{4} \cdot 0 + 2 = 2$.

If $x = 4$, then $y = -\dfrac{3}{4} \cdot 4 + 2 = -1$.

x	y	(x,y)
-4	5	$(-4,5)$
0	2	$(0,2)$
4	-1	$(4,-1)$

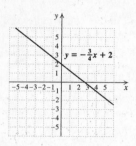

69. Familiarize. Let $t =$ the time it would take for Ann to complete the report working alone. Then $t + 6 =$ the time it would take Betty to complete the report working alone. In 1 hour they would complete $\dfrac{1}{t} + \dfrac{1}{t+6}$ of the report and in 4 hours they would complete $4\left(\dfrac{1}{t} + \dfrac{1}{t+6}\right)$, or $\dfrac{4}{t} + \dfrac{4}{t+6}$ of the report.

Translate. In 4 hours one entire job is done, so we have

$$\frac{4}{t} + \frac{4}{t+6} = 1.$$

Solve. We solve the equation.

$$\frac{4}{t} + \frac{4}{t+6} = 1, \text{ LCM} = t(t+6)$$

$$t(t+6)\left(\frac{4}{t} + \frac{4}{t+6}\right) = t(t+6) \cdot 1$$

$$t(t+6) \cdot \frac{4}{t} + t(t+6) \cdot \frac{4}{t+6} = t^2 + 6t$$

$$4(t+6) + 4t = t^2 + 6t$$

$$4t + 24 + 4t = t^2 + 6t$$

$$0 = t^2 - 2t - 24$$

$$0 = (t-6)(t+4)$$

$$t - 6 = 0 \ or \ t + 4 = 0$$

$$t = 6 \ or \ \quad t = -4$$

Check. The time cannot be negative, so we check only 6. If it takes Ann 6 hr to complete the report, then it would take Betty $6 + 6$, or 12 hr, to complete the report. In 4 hr Ann does $4 \cdot \dfrac{1}{6}$, or $\dfrac{2}{3}$, of the report, Betty does $4 \cdot \dfrac{1}{12}$, or $\dfrac{1}{3}$, of the report, and together they do $\dfrac{2}{3} + \dfrac{1}{3}$, or 1 entire job. The answer checks.

State. It would take Ann 6 hr and Betty 12 hr to complete the report working alone.

71. Familiarize. Let $t =$ the number of minutes after 5:00 at which the hands of the clock will first be together. While the minute hand moves through t minutes, the hour hand moves through $t/12$ minutes. At 5:00 the hour hand is on the 25-minute mark. We wish to find when a move of the minute hand through t minutes is equal to $25 + t/12$ minutes.

Translate. We use the last sentence of the familiarization step to write an equation.

$$t = 25 + \frac{t}{12}$$

Solve. We solve the equation.

$$t = 25 + \frac{t}{12}$$

$$12 \cdot t = 12\left(25 + \frac{t}{12}\right)$$

$$12t = 300 + t \qquad \text{Multiplying by 12}$$

$$11t = 300$$

$$t = \frac{300}{11} \text{ or } 27\frac{3}{11}$$

Check. At $27\frac{3}{11}$ minutes after 5:00, the minute hand is at the $27\frac{3}{11}$-minutes mark and the hour hand is at the $25 + \dfrac{27\frac{3}{11}}{12}$-minute mark. Simplifying $25 + \dfrac{27\frac{3}{11}}{12}$, we get

$$25 + \frac{\frac{300}{11}}{12} = 25 + \frac{300}{11} \cdot \frac{1}{12} = 25 + \frac{25}{11} = 25 + 2\frac{3}{11} = 27\frac{3}{11}.$$

Thus, the hands are together.

State. The hands are first together $27\frac{3}{11}$ minutes after 5:00.

73. $\qquad \dfrac{t}{a} + \dfrac{t}{b} = 1, \text{ LCM} = ab$

$$ab\left(\frac{t}{a} + \frac{t}{b}\right) = ab \cdot 1$$

$$ab \cdot \frac{t}{a} + ab \cdot \frac{t}{b} = ab$$

$$bt + at = ab$$

$$t(b + a) = ab$$

$$t = \frac{ab}{b + a}$$

Exercise Set 6.8

1. $\qquad \dfrac{1 + \dfrac{9}{16}}{1 - \dfrac{3}{4}} \qquad$ LCM of the denominators is 16.

$$= \frac{1 + \frac{9}{16}}{1 - \frac{3}{4}} \cdot \frac{16}{16} \qquad \text{Multiplying by 1 using } \frac{16}{16}$$

$$= \frac{\left(1 + \frac{9}{16}\right)16}{\left(1 - \frac{3}{4}\right)16} \qquad \begin{array}{l}\text{Multiplying numerator and denom-}\\ \text{inator by 16}\end{array}$$

$$= \frac{1(16) + \frac{9}{16}(16)}{1(16) - \frac{3}{4}(16)}$$

$$= \frac{16 + 9}{16 - 12}$$

$$= \frac{25}{4}$$

3. $\qquad \dfrac{1 - \dfrac{3}{5}}{1 + \dfrac{1}{5}}$

$$= \frac{1 \cdot \frac{5}{5} - \frac{3}{5}}{1 \cdot \frac{5}{5} + \frac{1}{5}} \qquad \begin{array}{l}\text{Getting a common denominator in}\\ \text{numerator and in denominator}\end{array}$$

$$= \frac{\frac{5}{5} - \frac{3}{5}}{\frac{5}{5} + \frac{1}{5}}$$

$$= \frac{\frac{2}{5}}{\frac{6}{5}} \qquad \begin{array}{l}\text{Subtracting in numerator; adding}\\ \text{in denominator}\end{array}$$

$$= \frac{2}{5} \cdot \frac{5}{6} \qquad \begin{array}{l}\text{Multiplying by the reciprocal of the}\\ \text{divisor}\end{array}$$

$$= \frac{2 \cdot 5}{5 \cdot 2 \cdot 3}$$

$$= \frac{\cancel{2} \cdot \cancel{5} \cdot 1}{\cancel{5} \cdot \cancel{2} \cdot 3}$$

$$= \frac{1}{3}$$

5. $\qquad \dfrac{\dfrac{1}{2} + \dfrac{3}{4}}{\dfrac{5}{8} - \dfrac{5}{6}} = \dfrac{\dfrac{1}{2} \cdot \dfrac{2}{2} + \dfrac{3}{4}}{\dfrac{5}{8} \cdot \dfrac{3}{3} - \dfrac{5}{6} \cdot \dfrac{4}{4}} \qquad \begin{array}{l}\text{Getting a common denomina-}\\ \text{tor in numerator and in de-}\\ \text{nominator}\end{array}$

$$= \frac{\frac{2}{4} + \frac{3}{4}}{\frac{15}{24} - \frac{20}{24}}$$

$$= \frac{\frac{5}{4}}{\frac{-5}{24}} \qquad \begin{array}{l}\text{Adding in numerator;}\\ \text{subtracting in denominator}\end{array}$$

$$= \frac{5}{4} \cdot \frac{24}{-5} \qquad \begin{array}{l}\text{Multiplying by the reciprocal}\\ \text{of the divisor}\end{array}$$

$$= \frac{5 \cdot 4 \cdot 6}{4 \cdot (-1) \cdot 5}$$

$$= \frac{\cancel{5} \cdot \cancel{4} \cdot 6}{\cancel{4} \cdot (-1) \cdot \cancel{5}}$$

$$= -6$$

7.

$$\dfrac{\dfrac{1}{x}+3}{\dfrac{1}{x}-5} \qquad \text{LCM of the denominators is } x.$$

$$= \dfrac{\dfrac{1}{x}+3}{\dfrac{1}{x}-5} \cdot \dfrac{x}{x} \qquad \text{Multiplying by 1 using } \dfrac{x}{x}$$

$$= \dfrac{\left(\dfrac{1}{x}+3\right)x}{\left(\dfrac{1}{x}-5\right)x}$$

$$= \dfrac{\dfrac{1}{x}\cdot x + 3\cdot x}{\dfrac{1}{x}\cdot x - 5\cdot x}$$

$$= \dfrac{1+3x}{1-5x}$$

9.

$$\dfrac{4-\dfrac{1}{x^2}}{2-\dfrac{1}{x}} \qquad \text{LCM of the denominators is } x^2.$$

$$= \dfrac{4-\dfrac{1}{x^2}}{2-\dfrac{1}{x}} \cdot \dfrac{x^2}{x^2}$$

$$= \dfrac{\left(4-\dfrac{1}{x^2}\right)x^2}{\left(2-\dfrac{1}{x}\right)x^2}$$

$$= \dfrac{4\cdot x^2 - \dfrac{1}{x^2}\cdot x^2}{2\cdot x^2 - \dfrac{1}{x}\cdot x^2}$$

$$= \dfrac{4x^2-1}{2x^2-x}$$

$$= \dfrac{(2x+1)(2x-1)}{x(2x-1)} \qquad \begin{array}{l}\text{Factoring numerator and}\\ \text{denominator}\end{array}$$

$$= \dfrac{(2x+1)(2x\!-\!\cancel{1})}{x(2x\!-\!\cancel{1})}$$

$$= \dfrac{2x+1}{x}$$

11.

$$\dfrac{8+\dfrac{8}{d}}{1+\dfrac{1}{d}} = \dfrac{8\cdot\dfrac{d}{d}+\dfrac{8}{d}}{1\cdot\dfrac{d}{d}+\dfrac{1}{d}}$$

$$= \dfrac{\dfrac{8d+8}{d}}{\dfrac{d+1}{d}}$$

$$= \dfrac{8d+8}{d}\cdot\dfrac{d}{d+1}$$

$$= \dfrac{8(d+1)(d)}{d(d+1)}$$

$$= \dfrac{8\cancel{(d+1)}\cancel{(d)}}{\cancel{d}\cancel{(d+1)}(1)}$$

$$= 8$$

13.

$$\dfrac{\dfrac{x}{8}-\dfrac{8}{x}}{\dfrac{1}{8}+\dfrac{1}{x}} \qquad \text{LCM of the denominators is } 8x.$$

$$= \dfrac{\dfrac{x}{8}-\dfrac{8}{x}}{\dfrac{1}{8}+\dfrac{1}{x}}\cdot\dfrac{8x}{8x}$$

$$= \dfrac{\left(\dfrac{x}{8}-\dfrac{8}{x}\right)8x}{\left(\dfrac{1}{8}+\dfrac{1}{x}\right)8x}$$

$$= \dfrac{\dfrac{x}{8}(8x)-\dfrac{8}{x}(8x)}{\dfrac{1}{8}(8x)+\dfrac{1}{x}(8x)}$$

$$= \dfrac{x^2-64}{x+8}$$

$$= \dfrac{(x+8)(x-8)}{x+8}$$

$$= \dfrac{\cancel{(x+8)}(x-8)}{1\cancel{(x+8)}}$$

$$= x-8$$

15.

$$\dfrac{1+\dfrac{1}{y}}{1-\dfrac{1}{y^2}} = \dfrac{1\cdot\dfrac{y}{y}+\dfrac{1}{y}}{1\cdot\dfrac{y^2}{y^2}-\dfrac{1}{y^2}}$$

$$= \dfrac{\dfrac{y+1}{y}}{\dfrac{y^2-1}{y^2}}$$

$$= \dfrac{y+1}{y}\cdot\dfrac{y^2}{y^2-1}$$

$$= \dfrac{(y+1)y\cdot y}{y(y+1)(y-1)}$$

$$= \dfrac{\cancel{(y+1)}\cancel{y}\cdot y}{\cancel{y}\cancel{(y+1)}(y-1)}$$

$$= \dfrac{y}{y-1}$$

17. $\dfrac{\dfrac{1}{5}-\dfrac{1}{a}}{\dfrac{5-a}{5}}$ LCM of the denominators is $5a$.

$=\dfrac{\dfrac{1}{5}-\dfrac{1}{a}}{\dfrac{5-a}{5}}\cdot\dfrac{5a}{5a}$

$=\dfrac{\left(\dfrac{1}{5}-\dfrac{1}{a}\right)5a}{\left(\dfrac{5-a}{5}\right)5a}$

$=\dfrac{\dfrac{1}{5}(5a)-\dfrac{1}{a}(5a)}{a(5-a)}$

$=\dfrac{a-5}{5a-a^2}$

$=\dfrac{a-5}{-a(-5+a)}$

$=\dfrac{1(a-5)}{-a(a-5)}$

$=-\dfrac{1}{a}$

19. $\dfrac{\dfrac{1}{a}+\dfrac{1}{b}}{\dfrac{1}{a^2}-\dfrac{1}{b^2}}$ LCM of the denominators is a^2b^2.

$=\dfrac{\dfrac{1}{a}+\dfrac{1}{b}}{\dfrac{1}{a^2}-\dfrac{1}{b^2}}\cdot\dfrac{a^2b^2}{a^2b^2}$

$=\dfrac{\left(\dfrac{1}{a}+\dfrac{1}{b}\right)\cdot a^2b^2}{\left(\dfrac{1}{a^2}-\dfrac{1}{b^2}\right)\cdot a^2b^2}$

$=\dfrac{\dfrac{1}{a}\cdot a^2b^2+\dfrac{1}{b}\cdot a^2b^2}{\dfrac{1}{a^2}\cdot a^2b^2-\dfrac{1}{b^2}\cdot a^2b^2}$

$=\dfrac{ab^2+a^2b}{b^2-a^2}$

$=\dfrac{ab(b+a)}{(b+a)(b-a)}$

$=\dfrac{ab(b+a)}{(b+a)(b-a)}$

$=\dfrac{ab}{b-a}$

21. $\dfrac{\dfrac{p}{q}+\dfrac{q}{p}}{\dfrac{1}{p}+\dfrac{1}{q}}$ LCM of the denominators is pq.

$=\dfrac{\left(\dfrac{p}{q}+\dfrac{q}{p}\right)\cdot pq}{\left(\dfrac{1}{p}+\dfrac{1}{q}\right)\cdot pq}$

$=\dfrac{\dfrac{p}{q}\cdot pq+\dfrac{q}{p}\cdot pq}{\dfrac{1}{p}\cdot pq+\dfrac{1}{q}\cdot pq}$

$=\dfrac{p^2+q^2}{q+p}$

23. $\dfrac{\dfrac{2}{a}+\dfrac{4}{a^2}}{\dfrac{5}{a^3}-\dfrac{3}{a}}$ LCD is a^3

$=\dfrac{\dfrac{2}{a}+\dfrac{4}{a^2}}{\dfrac{5}{a^3}-\dfrac{3}{a}}\cdot\dfrac{a^3}{a^3}$

$=\dfrac{\dfrac{2}{a}\cdot a^3+\dfrac{4}{a^2}\cdot a^3}{\dfrac{5}{a^3}\cdot a^3-\dfrac{3}{a}\cdot a^3}$

$=\dfrac{2a^2+4a}{5-3a^2}$

(Although the numerator can be factored, doing so will not enable us to simplify further.)

25. $\dfrac{\dfrac{2}{7a^4}-\dfrac{1}{14a}}{\dfrac{3}{5a^2}+\dfrac{2}{15a}}=\dfrac{\dfrac{2}{7a^4}\cdot\dfrac{2}{2}-\dfrac{1}{14a}\cdot\dfrac{a^3}{a^3}}{\dfrac{3}{5a^2}\cdot\dfrac{3}{3}+\dfrac{2}{15a}\cdot\dfrac{a}{a}}$

$=\dfrac{\dfrac{4-a^3}{14a^4}}{\dfrac{9+2a}{15a^2}}$

$=\dfrac{4-a^3}{14a^4}\cdot\dfrac{15a^2}{9+2a}$

$=\dfrac{15\cdot a^2(4-a^3)}{14a^2\cdot a^2(9+2a)}$

$=\dfrac{15(4-a^3)}{14a^2(9+2a)},\ \text{or}\ \dfrac{60-15a^3}{126a^2+28a^3}$

27.
$$\dfrac{\dfrac{a}{b}+\dfrac{c}{d}}{\dfrac{b}{a}+\dfrac{d}{c}} = \dfrac{\dfrac{a}{b}\cdot\dfrac{d}{d}+\dfrac{c}{d}\cdot\dfrac{b}{b}}{\dfrac{b}{a}\cdot\dfrac{c}{c}+\dfrac{d}{c}\cdot\dfrac{a}{a}}$$

$$= \dfrac{\dfrac{ad+bc}{bd}}{\dfrac{bc+ad}{ac}}$$

$$= \dfrac{ad+bc}{bd}\cdot\dfrac{ac}{bc+ad}$$

$$= \dfrac{ac(ad+bc)}{bd(bc+ad)}$$

$$= \dfrac{ac}{bd}\cdot\dfrac{ad+bc}{bc+ad}$$

$$= \dfrac{ac}{bd}\cdot 1$$

$$= \dfrac{ac}{bd}$$

29.
$$\dfrac{\dfrac{x}{5y^3}+\dfrac{3}{10y}}{\dfrac{3}{10y}+\dfrac{x}{5y^3}}$$

Observe that, by the commutative law of addition, the numerator and denominator are equivalent, so the result is 1. We could also simplify this expression as follows:

$$\dfrac{\dfrac{x}{5y^3}+\dfrac{3}{10y}}{\dfrac{3}{10y}+\dfrac{x}{5y^3}} = \dfrac{\dfrac{x}{5y^3}+\dfrac{3}{10y}}{\dfrac{3}{10y}+\dfrac{x}{5y^3}}\cdot\dfrac{10y^3}{10y^3}$$

$$= \dfrac{\dfrac{x}{5y^3}\cdot 10y^3+\dfrac{3}{10y}\cdot 10y^3}{\dfrac{3}{10y}\cdot 10y^3+\dfrac{x}{5y^3}\cdot 10y^3}$$

$$= \dfrac{2x+3y^2}{3y^2+2x}$$

$$= 1$$

31.
$$\dfrac{\dfrac{3}{x+1}+\dfrac{1}{x}}{\dfrac{2}{x+1}+\dfrac{3}{x}} = \dfrac{\dfrac{3}{x+1}+\dfrac{1}{x}}{\dfrac{2}{x+1}+\dfrac{3}{x}}\cdot\dfrac{x(x+1)}{x(x+1)}$$

$$= \dfrac{\dfrac{3}{x+1}\cdot x(x+1)+\dfrac{1}{x}\cdot x(x+1)}{\dfrac{2}{x+1}\cdot x(x+1)+\dfrac{3}{x}\cdot x(x+1)}$$

$$= \dfrac{3x+x+1}{2x+3(x+1)}$$

$$= \dfrac{4x+1}{2x+3x+3}$$

$$= \dfrac{4x+1}{5x+3}$$

33. Discussion and Writing Exercise

35.
$$(2x^3-4x^2+x-7)+(4x^4+x^3+4x^2+x)$$
$$= 4x^4+3x^3+2x-7$$

37. $p^2-10p+25 = p^2-2\cdot p\cdot 5+5^2$ Trinomial square
$$= (p-5)^2$$

39. $50p^2-100 = 50(p^2-2)$ Factoring out the common factor

Since p^2-2 cannot be factored, we have factored completely.

41. *Familiarize.* Let w = the width of the rectangle. Then $w+3$ = the length. Recall that the formula for the area of a rectangle is $A = lw$ and the formula for the perimeter of a rectangle is $P = 2l+2w$.

Translate. We substitute in the formula for area.
$$10 = lw$$
$$10 = (w+3)w$$

Solve.
$$10 = (w+3)w$$
$$10 = w^2+3w$$
$$0 = w^2+3w-10$$
$$0 = (w+5)(w-2)$$
$$w+5 = 0 \quad \text{or} \quad w-2 = 0$$
$$w = -5 \quad \text{or} \quad w = 2$$

Check. Since the width cannot be negative, we only check 2. If $w=2$, then $w+3 = 2+3$, or 5. Since $2\cdot 5 = 10$, the given area, the answer checks. Now we find the perimeter:
$$P = 2l+2w$$
$$P = 2\cdot 5+2\cdot 2$$
$$P = 10+4$$
$$P = 14$$

We can check this by repeating the calculation.

State. The perimeter is 14 yd.

43.
$$\dfrac{1}{\dfrac{2}{x-1}-\dfrac{1}{3x-2}}$$

$$= \dfrac{1}{\dfrac{2}{x-1}-\dfrac{1}{3x-2}}\cdot\dfrac{(x-1)(3x-2)}{(x-1)(3x-2)}$$

$$= \dfrac{(x-1)(3x-2)}{\left(\dfrac{2}{x-1}-\dfrac{1}{3x-2}\right)(x-1)(3x-2)}$$

$$= \dfrac{(x-1)(3x-2)}{\dfrac{2}{x-1}(x-1)(3x-2)-\dfrac{1}{3x-2}(x-1)(3x-2)}$$

$$= \dfrac{(x-1)(3x-2)}{2(3x-2)-(x-1)}$$

$$= \dfrac{(x-1)(3x-2)}{6x-4-x+1}$$

$$= \dfrac{(x-1)(3x-2)}{5x-3}$$

45. $1 + \dfrac{1}{1 + \dfrac{1}{1 + \dfrac{1}{1 + \dfrac{1}{x}}}} = 1 + \dfrac{1}{1 + \dfrac{1}{1 + \dfrac{1}{x + 1}}}$

$= 1 + \dfrac{1}{1 + \dfrac{1}{1 + \dfrac{x}{x + 1}}}$

$= 1 + \dfrac{1}{1 + \dfrac{1}{\dfrac{x + 1 + x}{x + 1}}}$

$= 1 + \dfrac{1}{1 + \dfrac{1}{\dfrac{2x + 1}{x + 1}}}$

$= 1 + \dfrac{1}{1 + \dfrac{x + 1}{2x + 1}}$

$= 1 + \dfrac{1}{\dfrac{2x + 1 + x + 1}{2x + 1}}$

$= 1 + \dfrac{1}{\dfrac{3x + 2}{2x + 1}}$

$= 1 + \dfrac{2x + 1}{3x + 2}$

$= \dfrac{3x + 2 + 2x + 1}{3x + 2}$

$= \dfrac{5x + 3}{3x + 2}$

Exercise Set 6.9

1. We substitute to find k.

$y = kx$

$36 = k \cdot 9$ Substituting 36 for y and 9 for x

$\dfrac{36}{9} = k$

$4 = k$ k is the variation constant.

The equation of the variation is $y = 4x$.

To find the value of y when $x = 20$ we substitute 20 for x in the equation of variation.

$y = 4x$

$y = 4 \cdot 20$

$y = 80$

The value of y is 80 when $x = 20$.

3. We substitute to find k.

$y = kx$

$0.8 = k \cdot 0.5$ Substituting 0.8 for y and 0.5 for x

$\dfrac{0.8}{0.5} = k$

$\dfrac{8}{5} = k$ k is the variation constant.

The equation of the variation is $y = \dfrac{8}{5}x$.

To find the value of y when $x = 20$ we substitute 20 for x in the equation of variation.

$y = \dfrac{8}{5}x$

$y = \dfrac{8}{5} \cdot 20$

$y = \dfrac{160}{5}$

$y = 32$

The value of y is 32 when $x = 20$.

5. We substitute to find k.

$y = kx$

$630 = k \cdot 175$ Substituting 630 for y and 175 for x

$\dfrac{630}{175} = k$

$3.6 = k$ k is the variation constant.

The equation of the variation is $y = 3.6x$.

To find the value of y when $x = 20$ we substitute 20 for x in the equation of variation.

$y = 3.6x$

$y = 3.6(20)$

$y = 72$

The value of y is 72 when $x = 20$.

7. We substitute to find k.

$y = kx$

$500 = k \cdot 60$ Substituting 500 for y and 60 for x

$\dfrac{500}{60} = k$

$\dfrac{25}{3} = k$ k is the variation constant.

The equation of the variation is $y = \dfrac{25}{3}x$.

To find the value of y when $x = 20$ we substitute 20 for x in the equation of variation.

$y = \dfrac{25}{3}x$

$y = \dfrac{25}{3} \cdot 20$

$y = \dfrac{500}{3}$

The value of y is $\dfrac{500}{3}$ when $x = 20$.

9. Familiarize and Translate. The problem states that we have direct variation between the variables P and H. Thus, an equation $P = kH$, $k > 0$, applies. As the number of hours increases, the paycheck increases.

Solve.

a) First find an equation of variation.

$$P = kH$$

$$84 = k \cdot 15 \quad \text{Substituting 78.75 for } P \text{ and 15 for } H$$

$$\frac{84}{15} = k$$

$$5.6 = k$$

The equation of variation is $P = 5.6H$.

b) Use the equation to find the pay for 35 hours work.

$$P = 5.6H$$

$$P = 5.6(35) \quad \text{Substituting 35 for } H$$

$$P = 196$$

Check. This check might be done by repeating the computations. We might also do some reasoning about the answer. The paycheck increased from \$84 to \$196. Similarly, the hours increased from 15 to 35.

State. a) The equation of variation is $P = 5.6H$.

b) For 35 hours work, the paycheck is \$196.

11. Familiarize and Translate. The problem states that we have direct variation between the variables C and S. Thus, an equation $C = kS$, $k > 0$, applies. As the depth increases, the cost increases.

Solve.

a) First find an equation of variation.

$$C = kS$$

$$67.5 = k \cdot 6 \quad \text{Substituting 75 for } C \text{ and 6 for } S$$

$$\frac{67.5}{6} = k$$

$$11.25 = k$$

The equation of variation is $C = 11.25S$.

b) Use the equation to find the cost of filling the sandbox to a depth of 9 inches.

$$C = 11.25S$$

$$C = 11.25(9) \quad \text{Substituting 9 for } S$$

$$C = 101.25$$

Check. In addition to repeating the computations, we can also do some reasoning. The depth increased from 6 inches to 9 inches. Similarly, the cost increased from \$67.50 to \$101.25.

State. a) The equation of variation is $C = 11.25S$.

b) The sand will cost \$101.25.

13. Familiarize and Translate. The problem states that we have direct variation between the variables M and E. Thus, an equation $M = kE$, $k > 0$, applies. As the weight on earth increases, the weight on the moon increases.

Solve.

a) First find an equation of variation.

$$M = kE$$

$$32 = k \cdot 192 \quad \text{Substituting 32 for } M \text{ and 192 for } E$$

$$\frac{32}{192} = k$$

$$\frac{1}{6} = k$$

The equation of variation is $M = \frac{1}{6}E$.

b) Use the equation to find how much a 110-lb person would weigh on the moon.

$$M = \frac{1}{6}E$$

$$M = \frac{1}{6} \cdot 110 \quad \text{Substituting 110 for } E$$

$$M = \frac{110}{6}, \text{ or } 18.\overline{3}$$

c) Use the equation to find how much a person who weighs 5 lb on the moon would weigh on Earth.

$$M = \frac{1}{6}E$$

$$5 = \frac{1}{6}E$$

$$30 = E \quad \text{Multiplying by 6}$$

Check. In addition to repeating the computations we can do some reasoning. When the weight on Earth decreased from 192 lb to 110 lb, the weight on the moon decreased from 32 lb to $18.\overline{3}$ lb. Similarly, when the weight on the moon decreased from 32 lb to 5 lb, the weight on Earth decreased from 192 lb to 30 lb.

State. a) The equation of variation is $M = \frac{1}{6}E$.

b) A person who weighs 110 lb on Earth would weigh $18.\overline{3}$ lb on the moon.

c) A person who weighs 5 lb on the moon would weigh 30 lb on Earth.

15. Familiarize and Translate. The problem states that we have direct variation between the variables N and S. Thus, an equation $N = kS$, $k > 0$, applies. As the speed of the internal processor increases, the number of instructions increases.

Solve.

a) First find an equation of variation.

$$N = kS$$

$$2,000,000 = k \cdot 25 \quad \text{Substituting 2,000,000 for } N \text{ and 25 for } S$$

$$\frac{2,000,000}{25} = k$$

$$80,000 = k$$

The equation of variation is $N = 80,000S$.

b) Use the equation to find how many instructions the processor will perform at a speed of 200 megahertz.

$$N = 80,000S$$

$$N = 80,000 \cdot 200 \quad \text{Substituting 200 for } S$$

$$N = 16,000,000$$

Check. In addition to repeating the computations we can do some reasoning. The speed of the processor increased from 25 to 200 megahertz. Similarly, the number of instructions performed per second increased from 2,000,000 to 16,000,000.

State. a) The equation of variation is $N = 80,000S$.

b) The processor will perform 16,000,000 instructions per second running at a speed of 200 megahertz.

17. Familiarize and Translate. This problem states that we have direct variation between the variables S and W. Thus, an equation $S = kW$, $k > 0$, applies. As the weight increases, the number of servings increases.

Solve.

a) First find an equation of variation.

$$S = kW$$

$$70 = k \cdot 9 \quad \text{Substituting 70 for } S \text{ and 9 for } W$$

$$\frac{70}{9} = k$$

The equation of variation is $S = \frac{70}{9}W$.

b) Use the equation to find the number of servings from 12 kg of round steak.

$$S = \frac{70}{9}W$$

$$S = \frac{70}{9} \cdot 12 \qquad \text{Substituting 12 for } W$$

$$S = \frac{840}{9}$$

$$S = \frac{280}{3}, \text{ or } 93\frac{1}{3}$$

Check. A check can always be done by repeating the computations. We can also do some reasoning about the answer. When the weight increased from 9 kg to 12 kg, the number of servings increased from 70 to $93\frac{1}{3}$.

State. $93\frac{1}{3}$ servings can be obtained from 12 kg of round steak.

19. We substitute to find k.

$$y = \frac{k}{x}$$

$$3 = \frac{k}{25} \quad \text{Substituting 3 for } y \text{ and 25 for } x$$

$$25 \cdot 3 = k$$

$$75 = k$$

The equation of variation is $y = \frac{75}{x}$.

To find the value of y when $x = 10$ we substitute 10 for x in the equation of variation.

$$y = \frac{75}{x}$$

$$y = \frac{75}{10}$$

$$y = \frac{15}{2}, \text{ or } 7.5$$

The value of y is $\frac{15}{2}$, or 7.5, when $x = 10$.

21. We substitute to find k.

$$y = \frac{k}{x}$$

$$10 = \frac{k}{8} \quad \text{Substituting 10 for } y \text{ and 8 for } x$$

$$8 \cdot 10 = k$$

$$80 = k$$

The equation of variation is $y = \frac{80}{x}$.

To find the value of y when $x = 10$ we substitute 10 for x in the equation of variation.

$$y = \frac{80}{x}$$

$$y = \frac{80}{10}$$

$$y = 8$$

The value of y is 8 when $x = 10$.

23. We substitute to find k.

$$y = \frac{k}{x}$$

$$6.25 = \frac{k}{0.16} \quad \text{Substituting 6.25 for } y \text{ and 0.16 for } x$$

$$0.16(6.25) = k$$

$$1 = k$$

The equation of variation is $y = \frac{1}{x}$.

To find the value of y when $x = 10$ we substitute 10 for x in the equation of variation.

$$y = \frac{1}{x}$$

$$y = \frac{1}{10}$$

The value of y is $\frac{1}{10}$ when $x = 10$.

25. We substitute to find k.

$$y = \frac{k}{x}$$

$$50 = \frac{k}{42} \quad \text{Substituting 50 for } y \text{ and 42 for } x$$

$$42 \cdot 50 = k$$

$$2100 = k$$

The equation of variation is $y = \dfrac{2100}{x}$.

To find the value of y when $x = 10$ we substitute 10 for x in the equation of variation.

$$y = \frac{2100}{x}$$

$$y = \frac{2100}{10}$$

$$y = 210$$

The value of y is 210 when $x = 10$.

27. We substitute to find k.

$$y = \frac{k}{x}$$

$$0.2 = \frac{k}{0.3} \quad \text{Substituting 0.2 for } y \text{ and 0.3 for } x$$

$$0.06 = k$$

The equation of variation is $y = \dfrac{0.06}{x}$.

To find the value of y when $x = 10$ we substitute 10 for x in the equation of variation.

$$y = \frac{0.06}{x}$$

$$y = \frac{0.06}{10}$$

$$y = \frac{6}{1000} \quad \text{Multiplying } \frac{0.06}{10} \text{ by } \frac{100}{100}$$

$$y = \frac{3}{500}, \text{ or } 0.006$$

The value of y is $\dfrac{3}{500}$, or 0.006, when $x = 10$.

29. a) It seems reasonable that, as the number of hours of production increases, the number of compact-disc players produced will increase, so direct variation might apply.

   b) **Familiarize.** Let $H =$ the number of hours the production line is working, and let $P =$ the number of compact-disc players produced. An equation $P = kH$, $k > 0$, applies. (See part (a)).

   Translate. We write an equation of variation.

   Number of players produced varies directly as hours of production. This translates to $P = kH$.

   Solve.

   a) First we find an equation of variation.

   $$P = kH$$

   $$15 = k \cdot 8 \quad \text{Substituting 8 for } H \text{ and 15 for } P$$

   $$\frac{15}{8} = k$$

   The equation of variation is $P = \dfrac{15}{8}H$.

   b) Use the equation to find the number of players produced in 37 hr.

   $$P = \frac{15}{8}H$$

   $$P = \frac{15}{8} \cdot 37 \quad \text{Substituting 37 for } H$$

   $$P = \frac{555}{8} = 69\frac{3}{8}$$

   Check. In addition to repeating the computations, we can do some reasoning. The number of hours increased from 8 to 37. Similarly, the number of compact disc players produced increased from 15 to $69\frac{3}{8}$.

   State. About $69\frac{3}{8}$ compact-disc players can be produced in 37 hr.

31. a) It seems reasonable that, as the number of workers increases, the number of hours required to do the job decreases, so inverse variation might apply.

   b) **Familiarize.** Let $T =$ the time required to cook the meal and $N =$ the number of cooks. An equation $T = k/N$, $k > 0$, applies. (See part (a)).

   Translate. We write an equation of variation. Time varies inversely as the number of cooks. This translates to $T = \dfrac{k}{N}$.

   Solve.

   a) First find the equation of variation.

   $$T = \frac{k}{N}$$

   $$4 = \frac{k}{9} \quad \text{Substituting 4 for } T \text{ and 9 for } N$$

   $$36 = k$$

   The equation of variation is $T = \dfrac{36}{N}$.

   b) Use the equation to find the amount of time it takes 8 cooks to prepare the dinner.

   $$T = \frac{36}{N}$$

   $$T = \frac{36}{8} \quad \text{Substituting 8 for } N$$

   $$T = 4.5$$

   Check. The check might be done by repeating the computation. We might also analyze the results. The number of cooks decreased from 9 to 8, and the time increased from 4 hr to 4.5 hr. This is what we would expect with inverse variation.

   State. It will take 8 cooks 4.5 hr to prepare the dinner.

33. **Familiarize.** The problem states that we have inverse variation between the variables N and P. Thus, an equation $N = k/P$, $k > 0$, applies. As the miles per gallon rating increases, the number of gallons required to travel the fixed distance decreases.

Translate. We write an equation of variation. Number of gallons varies inversely as miles per gallon rating. This translates to $N = \dfrac{k}{P}$.

Solve.

a) First find an equation of variation.

$$N = \frac{k}{P}$$

$$20 = \frac{k}{14} \quad \text{Substituting 20 for } N \text{ and 14 for } P$$

$$280 = k$$

The equation is $N = \dfrac{280}{P}$.

b) Use the equation to find the number of gallons of gasoline needed for a car that gets 28 mpg.

$$N = \frac{k}{P}$$

$$N = \frac{280}{28} \quad \text{Substituting 28 for } P$$

$$N = 10$$

Check. In addition to repeating the computations, we can analyze the results. The number of miles per gallon increased from 14 to 28, and the number of gallons required decreased from 20 to 10. This is what we would expect with inverse variation.

State. a) The equation of variation is $N = \dfrac{280}{P}$.

b) A car that gets 28 mpg will need 10 gallons of gasoline to travel the fixed distance.

35. *Familiarize*. The problem states that we have inverse variation between the variables I and R. Thus, an equation $I = k/R$, $k > 0$, applies. As the resistance increases, the current decreases.

Translate. We write an equation of variation. Current varies inversely as resistance. This translates to $I = \dfrac{k}{R}$.

Solve.

a) First find an equation of variation.

$$I = \frac{k}{R}$$

$$96 = \frac{k}{20} \quad \text{Substituting 96 for } I \text{ and 20 for } R$$

$$1920 = k$$

The equation of variation is $I = \dfrac{1920}{R}$.

b) Use the equation to find the current when the resistance is 60 ohms.

$$I = \frac{1920}{R}$$

$$I = \frac{1920}{60} \quad \text{Substituting 60 for } R$$

$$I = 32$$

Check. The check might be done by repeating the computations. We might also analyze the results. The resistance increased from 20 ohms to 60 ohms, and the current decreased from 96 amperes to 32 amperes. This is what we would expect with inverse variation.

State. a) The equation of variation is $I = \dfrac{1920}{R}$.

b) The current is 32 amperes when the resistance is 60 ohms.

37. *Familiarize*. The problem states that we have inverse variation between the variables m and n. Thus, an equation $m = k/n$, $k > 0$, applies. As the number of questions increases, the number of minutes allowed for each question decreases.

Translate. We write an equation of variation. Time allowed per question varies inversely as the number of questions.

Solve.

a) First find an equation of variation.

$$m = \frac{k}{n}$$

$$2.5 = \frac{k}{16} \quad \begin{array}{l} \text{Substituting 2.5 for } m \text{ and} \\ \text{16 for } n \end{array}$$

$$40 = k$$

The equation of variation is $m = \dfrac{40}{n}$.

b) Use the equation to find the number of questions on a quiz when students have 4 min per question.

$$m = \frac{40}{n}$$

$$4 = \frac{40}{n} \quad \text{Substituting 4 for } m$$

$$4n = 40 \quad \text{Multiplying by } n$$

$$n = 10 \quad \text{Dividing by 4}$$

Check. The check might be done by repeating the computations. We might also analyze the results. The time allowed for each question increased from 2.5 min to 4 min, and the number of questions decreased from 16 to 10. This is what we would expect with inverse variation.

State. a) The equation of variation is $m = \dfrac{40}{n}$.

b) There would be 10 questions on a quiz for which students have 4 min per question.

39. *Familiarize*. The problem states that we have inverse variation between the variables A and d. Thus, an equation $A = k/d$, $k > 0$, applies. As the distance increases, the apparent size decreases.

Translate. We write an equation of variation. Apparent size varies inversely as the distance. This translates to $A = \dfrac{k}{d}$.

Solve.

a) First find an equation of variation.

$$A = \frac{k}{d}$$

$$27.5 = \frac{k}{30} \quad \text{Substituting 27.5 for } A \text{ and 30 for } d$$

$$825 = k$$

The equation of variation is $A = \dfrac{825}{d}$.

b) Use the equation to find the apparent size when the distance is 100 ft.

$$A = \frac{825}{d}$$

$$A = \frac{825}{100} \quad \text{Substituting 100 for } d$$

$$A = 8.25$$

Check. The check might be done by repeating the computations. We might also analyze the results. The distance increased from 30 ft to 100 ft, and the apparent size decreased from 27.5 ft to 8.25 ft. This is what we would expect with inverse variation.

State. The flagpole will appear to be 8.25 ft tall when it is 100 ft from the observer.

41. Discussion and Writing Exercise

43. Discussion and Writing Exercise

45.
$$\frac{x+2}{x+5} = \frac{x-4}{x-6}, \quad \text{LCM is } (x+5)(x-6)$$

$$(x+5)(x-6) \cdot \frac{x+2}{x+5} = (x+5)(x-6) \cdot \frac{x-4}{x-6}$$

$$(x-6)(x+2) = (x+5)(x-4)$$

$$x^2 - 4x - 12 = x^2 + x - 20$$

$$-4x - 12 = x - 20 \quad \text{Subtracting } x^2$$

$$-5x = -8 \quad \begin{array}{l}\text{Subtracting } x \text{ and} \\ \text{adding 12}\end{array}$$

$$x = \frac{8}{5}$$

The number $\dfrac{8}{5}$ checks and is the solution.

47. $x^2 - 25x + 144 = 0$

$(x-9)(x-16) = 0$

$x - 9 = 0 \quad or \quad x - 16 = 0$

$x = 9 \quad or \qquad x = 16$

The solutions are 9 and 16.

49.
$$35x^2 + 8 = 34x$$

$$35x^2 - 34x + 8 = 0$$

$$(7x - 4)(5x - 2) = 0$$

$$7x - 4 = 0 \quad or \quad 5x - 2 = 0$$

$$7x = 4 \quad or \qquad 5x = 2$$

$$x = \frac{4}{7} \quad or \qquad x = \frac{2}{5}$$

The solutions are $\dfrac{4}{7}$ and $\dfrac{2}{5}$.

51. We do the divisions in order from left to right.

$$3^7 \div 3^4 \div 3^3 \div 3 = 3^3 \div 3^3 \div 3$$

$$= 1 \div 3$$

$$= \frac{1}{3}$$

53. $-5^2 + 4 \cdot 6 = -25 + 4 \cdot 6$

$$= -25 + 24$$

$$= -1$$

55.

X	Y1	
1	62.5	
2	125	
3	187.5	
4	250	
5	312.5	
6	375	
7	437.5	
X=1		

The y-values become larger.

57. $P^2 = kt$

59. $P = kV^3$

Chapter 6 Review Exercises

1. $\dfrac{3}{x}$

The denominator is 0 when $x = 0$, so the expression is not defined for the replacement number 0.

2. $\dfrac{4}{x-6}$

To determine the numbers for which the rational expression is not defined, we see the denominator equal to 0 and solve:

$$x - 6 = 0$$

$$x = 6$$

The expression is not defined for the replacement number 6.

3. $\dfrac{x+5}{x^2 - 36}$

To determine the numbers for which the rational expression is not defined, we see the denominator equal to 0 and solve:

$$x^2 - 36 = 0$$

$$(x+6)(x-6) = 0$$

$$x + 6 = 0 \quad or \quad x - 6 = 0$$

$$x = -6 \quad or \qquad x = 6$$

The expression is not defined for the replacement numbers -6 and 6.

4. $\dfrac{x^2 - 3x + 2}{x^2 + x - 30}$

To determine the numbers for which the rational expression is not defined, we see the denominator equal to 0 and solve:

$$x^2 + x - 30 = 0$$

$$(x+6)(x-5) = 0$$

$$x + 6 = 0 \quad or \quad x - 5 = 0$$
$$x = -6 \quad or \qquad x = 5$$

The expression is not defined for the replacement numbers -6 and 5.

5. $\dfrac{-4}{(x+2)^2}$

To determine the numbers for which the rational expression is not defined, we see the denominator equal to 0 and solve:

$$(x+2)^2 = 0$$
$$(x+2)(x+2) = 0$$
$$x + 2 = 0 \quad or \quad x + 2 = 0$$
$$x = -2 \quad or \qquad x = -2$$

The expression is not defined for the replacement number -2.

6. $\dfrac{x-5}{5}$

Since the denominator is the constant 5, there are no replacement numbers for which the expression is not defined.

7.
$$\frac{4x^2 - 8x}{4x^2 + 4x} = \frac{4x(x-2)}{4x(x+1)}$$
$$= \frac{4x}{4x} \cdot \frac{x-2}{x+1}$$
$$= 1 \cdot \frac{x-2}{x+1}$$
$$= \frac{x-2}{x+1}$$

8.
$$\frac{14x^2 - x - 3}{2x^2 - 7x + 3} = \frac{(2x-1)(7x+3)}{(2x-1)(x-3)}$$
$$= \frac{2x-1}{2x-1} \cdot \frac{7x+3}{x-3}$$
$$= 1 \cdot \frac{7x+3}{x-3}$$
$$= \frac{7x+3}{x-3}$$

9.
$$\frac{(y-5)^2}{y^2 - 25} = \frac{(y-5)(y-5)}{(y+5)(y-5)}$$
$$= \frac{y-5}{y+5} \cdot \frac{y-5}{y-5}$$
$$= \frac{y-5}{y+5} \cdot 1$$
$$= \frac{y-5}{y+5}$$

10.
$$\frac{a^2 - 36}{10a} \cdot \frac{2a}{a+6} = \frac{(a^2-36)(2a)}{10a(a+6)}$$
$$= \frac{(a+6)(a-6) \cdot 2 \cdot a}{2 \cdot 5 \cdot a \cdot (a+6)}$$
$$= \frac{(a+6)(a-6) \cdot 2 \cdot a}{2 \cdot 5 \cdot a \cdot (a+6)}$$
$$= \frac{a-6}{5}$$

11.
$$\frac{6t-6}{2t^2 + t - 1} \cdot \frac{t^2 - 1}{t^2 - 2t + 1}$$
$$= \frac{(6t-6)(t^2-1)}{(2t^2 + t - 1)(t^2 - 2t + 1)}$$
$$= \frac{6(t-1)(t+1)(t-1)}{(2t-1)(t+1)(t-1)(t-1)}$$
$$= \frac{6(t-1)(t+1)(t-1)}{(2t-1)(t+1)(t-1)(t-1)}$$
$$= \frac{6}{2t-1}$$

12.
$$\frac{10-5t}{3} \div \frac{t-2}{12t} = \frac{10-5t}{3} \cdot \frac{12t}{t-2}$$
$$= \frac{(10-5t)(12t)}{3(t-2)}$$
$$= \frac{5(2-t) \cdot 3 \cdot 4t}{3(t-2)}$$
$$= \frac{5(-1)(t-2) \cdot 3 \cdot 4t}{3(t-2)} \quad 2 - t = -1(t-2)$$
$$= \frac{5(-1)(t-2) \cdot 3 \cdot 4t}{3(t-2) \cdot 1}$$
$$= -20t$$

13.
$$\frac{4x^4}{x^2-1} \div \frac{2x^3}{x^2 - 2x + 1} = \frac{4x^4}{x^2-1} \cdot \frac{x^2 - 2x + 1}{2x^3}$$
$$= \frac{4x^4(x^2 - 2x + 1)}{(x^2-1)(2x^3)}$$
$$= \frac{2 \cdot 2 \cdot x \cdot x^3(x-1)(x-1)}{(x+1)(x-1) \cdot 2 \cdot x^3}$$
$$= \frac{2 \cdot 2 \cdot x \cdot x^3(x-1)(x-1)}{(x+1)(x-1) \cdot 2 \cdot x^3}$$
$$= \frac{2x(x-1)}{x+1}, \text{ or}$$
$$\frac{2x^2 - 2x}{x+1}$$

14. $3x^2 = 3 \cdot x \cdot x$
$$10xy = 2 \cdot 5 \cdot x \cdot y$$
$$15y^2 = 3 \cdot 5 \cdot y \cdot y$$
$$\text{LCM} = 2 \cdot 3 \cdot 5 \cdot x \cdot x \cdot y \cdot y, \text{ or } 30x^2y^2$$

15. $a - 2 = a - 2$
$$4a - 8 = 4(a-2)$$
$$\text{LCM} = 4(a-2)$$

16. $y^2 - y - 2 = (y-2)(y+1)$
$$y^2 - 4 = (y+2)(y-2)$$
$$\text{LCM} = (y-2)(y+1)(y+2)$$

17. $\dfrac{x+8}{x+7} + \dfrac{10-4x}{x+7} = \dfrac{x+8+10-4x}{x+7} = \dfrac{-3x+18}{x+7}$

18.
$$\frac{3}{3x-9} + \frac{x-2}{3-x} = \frac{3}{3(x-3)} + \frac{x-2}{3-x}$$
$$= \frac{3}{3(x-3)} + \frac{x-2}{3-x} \cdot \frac{-1}{-1}$$
$$= \frac{3}{3(x-3)} + \frac{-1(x-2)}{-1(3-x)}$$
$$= \frac{3}{3(x-3)} + \frac{-x+2}{x-3}$$
$$= \frac{3}{3(x-3)} + \frac{-x+2}{x-3} \cdot \frac{3}{3}$$
$$= \frac{3}{3(x-3)} + \frac{-3x+6}{3(x-3)}$$
$$= \frac{3-3x+6}{3(x-3)}$$
$$= \frac{-3x+9}{3(x-3)}$$
$$= \frac{-3(x-3)}{3(x-3)}$$
$$= \frac{-1 \cdot 3(x-3)}{1 \cdot 3(x-3)}$$
$$= -1$$

19.
$$\frac{2a}{a+1} + \frac{4a}{a^2-1}$$
$$= \frac{2a}{a+1} + \frac{4a}{(a+1)(a-1)}, \text{ LCM is } (a+1)(a-1)$$
$$= \frac{2a}{a+1} \cdot \frac{a-1}{a-1} + \frac{4a}{(a+1)(a-1)}$$
$$= \frac{2a(a-1)+4a}{(a+1)(a-1)}$$
$$= \frac{2a^2-2a+4a}{(a+1)(a-1)}$$
$$= \frac{2a^2+2a}{(a+1)(a-1)}$$
$$= \frac{2a(a+1)}{(a+1)(a-1)}$$
$$= \frac{2a(a+1)}{(a+1)(a-1)}$$
$$= \frac{2a}{a-1}$$

20.
$$\frac{d^2}{d-c} + \frac{c^2}{c-d} = \frac{d^2}{d-c} + \frac{c^2}{c-d} \cdot \frac{-1}{-1}$$
$$= \frac{d^2}{d-c} + \frac{-c^2}{d-c}$$
$$= \frac{d^2-c^2}{d-c}$$
$$= \frac{(d+c)(d-c)}{d-c}$$
$$= \frac{(d+c)(d-c)}{(d-c) \cdot 1}$$
$$= d+c$$

21.
$$\frac{6x-3}{x^2-x-12} - \frac{2x-15}{x^2-x-12} = \frac{6x-3-(2x-15)}{x^2-x-12}$$
$$= \frac{6x-3-2x+15}{x^2-x-12}$$
$$= \frac{4x+12}{x^2-x-12}$$
$$= \frac{4(x+3)}{(x-4)(x+3)}$$
$$= \frac{4(x+3)}{(x-4)(x+3)}$$
$$= \frac{4}{x-4}$$

22. $\frac{3x-1}{2x} - \frac{x-3}{x}$, LCM is $2x$
$$= \frac{3x-1}{2x} - \frac{x-3}{x} \cdot \frac{2}{2}$$
$$= \frac{3x-1}{2x} - \frac{2(x-3)}{2x}$$
$$= \frac{3x-1-2(x-3)}{2x}$$
$$= \frac{3x-1-2x+6}{2x}$$
$$= \frac{x+5}{2x}$$

23. $\frac{x+3}{x-2} - \frac{x}{2-x} = \frac{x+3}{x-2} - \frac{x}{2-x} \cdot \frac{-1}{-1}$
$$= \frac{x+3}{x-2} - \frac{-x}{x-2}$$
$$= \frac{x+3-(-x)}{x-2}$$
$$= \frac{x+3+x}{x-2}$$
$$= \frac{2x+3}{x-2}$$

24. $\frac{1}{x^2-25} - \frac{x-5}{x^2-4x-5}$
$$= \frac{1}{(x+5)(x-5)} - \frac{x-5}{(x-5)(x+1)},$$
$$\text{LCM is } (x+5)(x-5)(x+1)$$
$$= \frac{1}{(x+5)(x-5)} \cdot \frac{x+1}{x+1} - \frac{x-5}{(x-5)(x+1)} \cdot \frac{x+5}{x+5}$$
$$= \frac{x+1}{(x+5)(x-5)(x+1)} - \frac{(x-5)(x+5)}{(x+5)(x-5)(x+1)}$$
$$= \frac{x+1-(x^2-25)}{(x+5)(x-5)(x+1)}$$
$$= \frac{x+1-x^2+25}{(x+5)(x-5)(x+1)}$$
$$= \frac{-x^2+x+26}{(x+5)(x-5)(x+1)}$$

25.
$$\frac{3x}{x+2} - \frac{x}{x-2} + \frac{8}{x^2-4}$$
$$= \frac{3x}{x+2} - \frac{x}{x-2} + \frac{8}{(x+2)(x-2)}, \text{ LCM is } (x+2)(x-2)$$
$$= \frac{3x}{x+2} \cdot \frac{x-2}{x-2} - \frac{x}{x-2} \cdot \frac{x+2}{x+2} + \frac{8}{(x+2)(x-2)}$$
$$= \frac{3x(x-2)}{(x+2)(x-2)} - \frac{x(x+2)}{(x+2)(x-2)} + \frac{8}{(x+2)(x-2)}$$
$$= \frac{3x(x-2) - x(x+2) + 8}{(x+2)(x-2)}$$
$$= \frac{3x^2 - 6x - x^2 - 2x + 8}{(x+2)(x-2)}$$
$$= \frac{2x^2 - 8x + 8}{(x+2)(x-2)}$$
$$= \frac{2(x^2 - 4x + 4)}{(x+2)(x-2)}$$
$$= \frac{2(x-2)(x-2)}{(x+2)(x-2)}$$
$$= \frac{2(x-2)(x-2)}{(x+2)(x-2)}$$
$$= \frac{2(x-2)}{x+2}$$

26.
$$\frac{\frac{1}{z}+1}{\frac{1}{z^2}-1} \text{ LCM of the denominators is } z^2.$$
$$= \frac{\frac{1}{z}+1}{\frac{1}{z^2}-1} \cdot \frac{z^2}{z^2}$$
$$= \frac{\left(\frac{1}{z}+1\right)z^2}{\left(\frac{1}{z^2}-1\right)z^2}$$
$$= \frac{\frac{1}{z} \cdot z^2 + 1 \cdot z^2}{\frac{1}{z^2} \cdot z^2 - 1 \cdot z^2}$$
$$= \frac{z + z^2}{1 - z^2}$$
$$= \frac{z(1+z)}{(1+z)(1-z)}$$
$$= \frac{z(1+z)}{(1+z)(1-z)}$$
$$= \frac{z}{1-z}$$

27.
$$\frac{\frac{c}{d}-\frac{d}{c}}{\frac{1}{c}+\frac{1}{d}}$$
$$= \frac{\frac{c}{d}\cdot\frac{c}{c}-\frac{d}{c}\cdot\frac{d}{d}}{\frac{1}{c}\cdot\frac{d}{d}+\frac{1}{d}\cdot\frac{c}{c}} \quad \begin{array}{l}\text{Getting a common denominator in}\\\text{numerator and in denominator}\end{array}$$
$$= \frac{\frac{c^2}{cd}-\frac{d^2}{cd}}{\frac{d}{cd}+\frac{c}{cd}}$$
$$= \frac{\frac{c^2-d^2}{cd}}{\frac{d+c}{cd}}$$
$$= \frac{c^2-d^2}{cd} \cdot \frac{cd}{d+c}$$
$$= \frac{(c^2-d^2)cd}{cd(d+c)}$$
$$= \frac{(c+d)(c-d)cd}{cd(d+c)}$$
$$= \frac{(c+d)(c-d)cd}{cd(d+c)\cdot 1}$$
$$= c - d$$

28.
$$\frac{3}{y} - \frac{1}{4} = \frac{1}{y}, \text{ LCM} = 4y$$
$$4y\left(\frac{3}{y} - \frac{1}{4}\right) = 4y \cdot \frac{1}{y}$$
$$4y \cdot \frac{3}{y} - 4y \cdot \frac{1}{4} = 4y \cdot \frac{1}{y}$$
$$12 - y = 4$$
$$-y = -8$$
$$y = 8$$
This checks, so the solution is 8.

29.
$$\frac{15}{x} - \frac{15}{x+2} = 2, \text{ LCM} = x(x+2)$$
$$x(x+2)\left(\frac{15}{x} - \frac{15}{x+2}\right) = x(x+2) \cdot 2$$
$$x(x+2)\cdot\frac{15}{x} - x(x+2)\cdot\frac{15}{x+2} = 2x(x+2)$$
$$15(x+2) - 15x = 2x(x+2)$$
$$15x + 30 - 15x = 2x^2 + 4x$$
$$30 = 2x^2 + 4x$$
$$0 = 2x^2 + 4x - 30$$
$$0 = 2(x^2 + 2x - 15)$$
$$0 = x^2 + 2x - 15 \text{ Dividing by 2}$$
$$0 = (x+5)(x-3)$$
$$x + 5 = 0 \quad or \quad x - 3 = 0$$
$$x = -5 \quad or \quad x = 3$$
Both numbers check. The solutions are -5 and 3.

30. Familiarize. Let $t =$ the time the job would take if the crews worked together.

Translate. We use the work principle, substituting 9 for a and 12 for b.

$$\frac{t}{a} + \frac{t}{b} = 1$$

$$\frac{t}{9} + \frac{t}{12} = 1 \quad \text{Substituting}$$

Solve.

$$\frac{t}{9} + \frac{t}{12} = 1, \quad \text{LCM} = 36$$

$$36\left(\frac{t}{9} + \frac{t}{12}\right) = 36 \cdot 1$$

$$36 \cdot \frac{t}{9} + 36 \cdot \frac{t}{12} = 36$$

$$4t + 3t = 36$$

$$7t = 36$$

$$t = \frac{36}{7}, \text{ or } 5\frac{1}{7}$$

Check. In $\frac{36}{7}$ hr, the portion of the job done is $\frac{36}{7} \cdot \frac{1}{9} + \frac{36}{7} \cdot \frac{1}{12} = \frac{4}{7} + \frac{3}{7} = 1$. The answer checks.

State. The job would take $5\frac{1}{7}$ hr if the crews worked together.

31. Familiarize. Let $r =$ the speed of the slower train, in km/h. Then $r + 40 =$ the speed of the faster train. We organize the information in a table.

	Distance	Speed	Time
Slower train	60	r	t
Faster train	70	$r + 40$	t

Translate. We use the formula $d = rt$ in each row of the table to obtain two equations.

$$60 = rt,$$
$$70 = (r + 40)t$$

Since the times are the same, we solve each equation for t and set the results equal to each other.

$$60 = rt, \text{ so } t = \frac{60}{r}.$$

$$70 = (r + 40)t, \text{ so } t = \frac{70}{r + 40}.$$

Then we have

$$\frac{60}{r} = \frac{70}{r + 40}.$$

Solve.

$$\frac{60}{r} = \frac{70}{r + 40}, \quad \text{LCM} = r(r + 40)$$

$$r(r + 40) \cdot \frac{60}{r} = r(r + 40) \cdot \frac{70}{r + 40}$$

$$60(r + 40) = 70r$$

$$60r + 2400 = 70r$$

$$2400 = 10r$$

$$240 = r$$

If $r = 240$, then $r + 40 = 240 + 40 = 280$.

Check. If the speeds are 240 km/h and 280 km/h, then the speed of the faster train is 40 km/h faster than the speed of the slower train. At 240 km/h, the slower train travels 60 km in 60/240, or 1/4 hr. At 280 km/h, the faster train travels 70 km in 70/280, or 1/4 hr. Since the times are the same, the answer checks.

State. The speed of the slower train is 240 km/h; the speed of the faster train is 280 km/h.

32. Familiarize. Let $r =$ the speed of the slower plane, in mph. Then $r + 80 =$ the speed of the faster plane. We organize the information in a table.

	Distance	Speed	Time
Slower plane	950	r	t
Faster plane	1750	$r + 80$	t

Translate. We use the formula $d = rt$ in each row of the table to obtain two equations.

$$950 = rt,$$
$$1750 = (r + 80)t$$

Since the times are the same, we solve each equation for t and set the results equal to each other.

$$950 = rt, \text{ so } t = \frac{950}{r}.$$

$$1750 = (r + 80)t, \text{ so } t = \frac{1750}{r + 80}.$$

Then we have

$$\frac{950}{r} = \frac{1750}{r + 80}.$$

Solve.

$$\frac{950}{r} = \frac{1750}{r + 80}, \quad \text{LCM} = r(r + 80)$$

$$r(r + 80) \cdot \frac{950}{r} = r(r + 80) \cdot \frac{1750}{r + 80}$$

$$950(r + 80) = 1750r$$

$$950r + 76{,}000 = 1750r$$

$$76{,}000 = 800r$$

$$95 = r$$

If $95 = r$, then $r + 80 = 95 + 80 = 175$.

Check. If the speeds are 95 mph and 175 mph, then the speed of the faster plane is 80 mph faster than the speed of the slower plane. At 95 mph, the slower plane travels 950 mi in 950/95, or 10 hr. At 175 mph, the faster plane

travels 1750 mi in 1750/175, or 10 hr. The times are the same, so the answer checks.

State. The speed of the slower plane is 95 mph; the speed of the faster plane is 175 mph.

33. **Familiarize**. We can translate to a proportion, letting $d =$ the number of defective calculators that can be expected in a sample of 5000.

Translate.

$$\text{Number defective} \rightarrow \frac{8}{250} = \frac{d}{5000} \leftarrow \text{Number defective}$$
$$\text{Sample size} \rightarrow \qquad \qquad \leftarrow \text{Sample size}$$

Solve. We solve the proportion.

$$\frac{8}{250} = \frac{d}{5000}$$
$$5000 \cdot \frac{8}{250} = 5000 \cdot \frac{d}{5000}$$
$$160 = d$$

Check.

$$\frac{8}{250} = 0.032 \quad \text{and} \quad \frac{160}{5000} = 0.032.$$

The ratios are the same, so the answer checks.

State. You would expect to find 160 defective calculators in a sample of 5000.

34. a) Let $x =$ the number of cups of onion that would be used. Then we can write and solve a proportion.

$$\frac{6}{13} = \frac{x}{2}$$
$$2 \cdot \frac{6}{13} = 2 \cdot \frac{x}{2}$$
$$\frac{12}{13} = x$$

Thus, $\frac{12}{13}$ cup of onion would be used.

b) Let $c =$ the number of cups of cheese that would be used. We can write and solve a proportion.

$$\frac{5}{7} = \frac{3}{c}$$
$$7c \cdot \frac{5}{7} = 7c \cdot \frac{3}{c}$$
$$5c = 21$$
$$c = \frac{21}{5}, \text{ or } 4\frac{1}{5}$$

Thus, $4\frac{1}{5}$ cups of cheese would be used.

c) Let $c =$ the number of cups of cheese that would be used. We can write and solve a proportion.

$$\frac{9}{14} = \frac{6}{c}$$
$$14c \cdot \frac{9}{14} = 14c \cdot \frac{6}{c}$$
$$9c = 84$$
$$c = \frac{84}{9} = \frac{28}{3}, \text{ or } 9\frac{1}{3}$$

Thus, $9\frac{1}{3}$ cups of cheese would be used.

35. **Familiarize**. The ratio of blue whales tagged to the total blue whale population, P, is $\frac{500}{P}$. Of the 400 blue whales checked later, 20 were tagged. The ratio of blue whales tagged to blue whales checked is $\frac{20}{400}$.

Translate. Assuming the two ratios are the same, we can translate to a proportion.

$$\begin{array}{c}\text{Whales tagged} \\ \text{originally}\end{array} \longrightarrow \frac{500}{P} = \frac{20}{400} \begin{array}{c}\leftarrow \text{Tagged whales} \\ \text{caught later}\end{array}$$
$$\begin{array}{c}\text{Whale} \\ \text{population}\end{array} \longrightarrow \qquad\qquad \begin{array}{c}\leftarrow \text{Whales caught} \\ \text{later}\end{array}$$

Solve. We solve the proportion.

$$400P \cdot \frac{500}{P} = 400P \cdot \frac{20}{400} \quad \begin{array}{l}\text{Multiplying by the LCM,} \\ 400P\end{array}$$
$$400 \cdot 500 = P \cdot 20$$
$$200,000 = 20P$$
$$10,000 = P$$

Check.

$$\frac{500}{10,000} = \frac{1}{20} \quad \text{and} \quad \frac{20}{400} = \frac{1}{20}.$$

The ratios are the same.

State. The blue whale population is about 10,000.

36. We write a proportion and solve it.

$$\frac{3.4}{8.5} = \frac{2.4}{x}$$
$$8.5x \cdot \frac{3.4}{8.5} = 8.5x \cdot \frac{2.4}{x}$$
$$3.4x = 20.4$$
$$x = 6$$

(Note that the proportions $\frac{8.5}{3.4} = \frac{x}{2.4}$, $\frac{8.5}{x} = \frac{3.4}{2.4}$, and $\frac{x}{8.5} = \frac{2.4}{3.4}$ could also be used.)

37. We substitute to find k.

$$y = kx$$
$$12 = k \cdot 4$$
$$3 = k$$

The equation of variation is $y = 3x$.

To find the value of y when $x = 20$ we substitute 20 for x in the equation of variation.

$$y = 3x$$
$$y = 3 \cdot 20$$
$$y = 60$$

The value of y is 60 when $x = 20$.

38. We substitute to find k.

$$y = kx$$
$$4 = k \cdot 8$$
$$\frac{1}{2} = k$$

The equation of variation is $y = \frac{1}{2}x$.

To find the value of y when $x = 20$ we substitute 20 for x in the equation of variation.

$$y = \frac{1}{2}x$$
$$y = \frac{1}{2} \cdot 20$$
$$y = 10$$

The value of y is 10 when $x = 20$.

39. We substitute to find k.

$$y = kx$$
$$0.4 = k \cdot 0.5$$
$$\frac{0.4}{0.5} = k$$
$$\frac{4}{5} = k \quad \left(\frac{0.4}{0.5} = \frac{0.4}{0.5} \cdot \frac{10}{10} = \frac{4}{5}\right)$$

The equation of variation is $y = \frac{4}{5}x$.

To find the value of y when $x = 20$ we substitute 20 for x in the equation of variation.

$$y = \frac{4}{5}x$$
$$y = \frac{4}{5} \cdot 20$$
$$y = 16$$

The value of y is 16 when $x = 20$.

40. We substitute to find k.

$$y = \frac{k}{x}$$
$$5 = \frac{k}{6}$$
$$30 = k$$

The equation of variation is $y = \frac{30}{x}$.

To find the value of y when $x = 5$ we substitute 5 for x in the equation of variation.

$$y = \frac{30}{x}$$
$$y = \frac{30}{5}$$
$$y = 6$$

The value of y is 6 when $x = 5$.

41. We substitute to find k.

$$y = \frac{k}{x}$$
$$0.5 = \frac{k}{2}$$
$$1 = k$$

The equation of variation is $y = \frac{1}{x}$.

To find the value of y when $x = 5$ we substitute 5 for x in the equation of variation.

$$y = \frac{1}{x}$$
$$y = \frac{1}{5}$$

The value of y is $\frac{1}{5}$ when $x = 5$.

42. We substitute to find k.

$$y = \frac{k}{x}$$
$$1.3 = \frac{k}{0.5}$$
$$0.65 = k$$

The equation of variation is $y = \frac{0.65}{x}$.

To find the value of y when $x = 5$ we substitute 5 for x in the equation of variation.

$$y = \frac{0.65}{x}$$
$$y = \frac{0.65}{5}$$
$$y = 0.13$$

The value of y is 0.13 when $x = 5$.

43. *Familiarize and Translate.* The problem states that we have direct variation between the variables P and H. Thus, an equation $P = kH$, $k > 0$, applies.

Solve.

a) First we find an equation of variation.

$$P = kH$$
$$165 = k \cdot 20$$
$$8.25 = k$$

The equation of variation is $P = 8.25H$.

b) Use the equation to find the pay for 35 hr of work.

$$P = 8.25H$$
$$P = 8.25(35)$$
$$P = 288.75$$

Check. We can repeat the computations. Also note that when the number of hours worked increased from 20 to 35, the pay increased from $165.00 to $288.75 so the answer seems reasonable.

State. The pay for 35 hr of work is $288.75.

44. *Familiarize and Translate.* Let $M =$ the number of washing machines used and $T =$ the time required to do the laundry. The problem states that we have inverse variation between T and M, so an equation $T = k/M$, $k > 0$, applies.

Solve.

a) First we find an equation of variation.

$$T = \frac{k}{M}$$

$$5 = \frac{k}{2}$$

$$10 = k$$

The equation of variation is $T = \frac{10}{M}$.

b) Use the equation to find the time required to do the laundry if 10 washing machines are used.

$$T = \frac{10}{M}$$

$$T = \frac{10}{10}$$

$$T = 1$$

Check. We can repeat the computations. Also note that as the number of washing machines increased from 5 to 10, the time required to do the laundry decreased from 2 hr to 1 hr, so the answer seems reasonable.

State. It would take 1 hr for 10 washing machines to do the laundry.

45. *Discussion and Writing Exercise.* $\dfrac{5x + 6}{(x + 2)(x - 2)}$; used to find an equivalent expression for each rational expression with the LCM as the least common denominator

46. *Discussion and Writing Exercise.* $\dfrac{3x + 10}{(x - 2)(x + 2)}$; used to find an equivalent expression for each rational expression with the LCM as the least common denominator

47. *Discussion and Writing Exercise.* 4; used to clear fractions

48. *Discussion and Writing Exercise.* $\dfrac{4(x - 2)}{x(x + 4)}$; method 1: used to multiply by 1 using LCM/LCM; method 2: LCM of the denominators in the numerator used to subtract in the numerator and LCM of the denominators in the denominator used to add in the denominator

49.
$$\frac{2a^2 + 5a - 3}{a^2} \cdot \frac{5a^3 + 30a^2}{2a^2 + 7a - 4} \div \frac{a^2 + 6a}{a^2 + 7a + 12}$$

$$= \frac{2a^2 + 5a - 3}{a^2} \cdot \frac{5a^3 + 30a^2}{2a^2 + 7a - 4} \cdot \frac{a^2 + 7a + 12}{a^2 + 6a}$$

$$= \frac{(2a - 1)(a + 3)}{a^2} \cdot \frac{5a^2(a + 6)}{(2a - 1)(a + 4)} \cdot \frac{(a + 3)(a + 4)}{a(a + 6)}$$

$$= \frac{(2a - 1)(a + 3)(5a^2)(a + 6)(a + 3)(a + 4)}{a^2(2a - 1)(a + 4)(a)(a + 6)}$$

$$= \frac{(2a - 1)(a + 3) \cdot 5 \cdot a^2 \,(a + 6)(a + 3)(a + 4)}{a^2 \,(2a - 1)(a + 4)(a)(a + 6)}$$

$$= \frac{5(a + 3)^2}{a}$$

50.
$$\frac{12a}{(a - b)(b - c)} - \frac{2a}{(b - a)(c - b)}$$

$$= \frac{12a}{(a - b)(b - c)} - \frac{2a}{-1(a - b)(-1)(b - c)}$$ Factoring -1 out of $b - a$ and $c - b$

$$= \frac{12a}{(a - b)(b - c)} - \frac{2a}{(a - b)(b - c)}$$

$$= \frac{10a}{(a - b)(b - c)}$$

51.
$$\frac{A + B}{B} = \frac{C + D}{D}$$

$$\frac{A}{B} + \frac{B}{B} = \frac{C}{D} + \frac{D}{D}$$

$$\frac{A}{B} + 1 = \frac{C}{D} + 1$$

$$\frac{A}{B} = \frac{C}{D}$$

The two given proportions are equivalent.

Chapter 6 Test

1. $\dfrac{8}{2x}$

To determine the numbers for which the rational expression is not defined, we set the denominator equal to 0 and solve:

$$2x = 0$$

$$x = 0$$

The expression is not defined for the replacement number 0.

2. $\dfrac{5}{x + 8}$

To determine the numbers for which the rational expression is not defined, we set the denominator equal to 0 and solve:

$$x + 8 = 0$$

$$x = -8$$

The expression is not defined for the replacement number -8.

3. $\dfrac{x - 7}{x^2 - 49}$

To determine the numbers for which the rational expression is not defined, we set the denominator equal to 0 and solve:

$$x^2 - 49 = 0$$

$$(x + 7)(x - 7) = 0$$

$$x + 7 = 0 \quad or \quad x - 7 = 0$$

$$x = -7 \quad or \qquad x = 7$$

The expression is not defined for the replacement numbers -7 and 7.

4. $\dfrac{x^2 + x - 30}{x^2 - 3x + 2}$

To determine the numbers for which the rational expression is not defined, we set the denominator equal to 0 and solve:

$$x^2 - 3x + 2 = 0$$
$$(x - 1)(x - 2) = 0$$
$$x - 1 = 0 \ \ or \ \ x - 2 = 0$$
$$x = 1 \ \ or \ \ \ \ \ \ x = 2$$

The expression is not defined for the replacement numbers 1 and 2.

5. $\dfrac{11}{(x - 1)^2}$

To determine the numbers for which the rational expression is not defined, we set the denominator equal to 0 and solve:

$$(x - 1)^2 = 0$$
$$(x - 1)(x - 1) = 0$$
$$x - 1 = 0 \ \ or \ \ x - 1 = 0$$
$$x = 1 \ \ or \ \ \ \ \ \ x = 1$$

The expression is not defined for the replacement number 1.

6. $\dfrac{x + 2}{2}$

Since the denominator is the constant 2, there are no replacement numbers for which the expression is not defined.

7. $\dfrac{6x^2 + 17x + 7}{2x^2 + 7x + 3} = \dfrac{(2x + 1)(3x + 7)}{(2x + 1)(x + 3)}$

$$= \dfrac{2x + 1}{2x + 1} \cdot \dfrac{3x + 7}{x + 3}$$

$$= 1 \cdot \dfrac{3x + 7}{x + 3}$$

$$= \dfrac{3x + 7}{x + 3}$$

8. $\dfrac{a^2 - 25}{6a} \cdot \dfrac{3a}{a - 5} = \dfrac{(a^2 - 25)(3a)}{6a(a - 5)}$

$$= \dfrac{(a + 5)(a - 5) \cdot 3 \cdot a}{2 \cdot 3 \cdot a \cdot (a - 5)}$$

$$= \dfrac{(a + 5)(a \!\!\!\!\diagdown 5) \cdot \cancel{3} \cdot \cancel{a}}{2 \cdot \cancel{3} \cdot \cancel{a} \cdot (a \!\!\!\!\diagdown 5)}$$

$$= \dfrac{a + 5}{2}$$

9. $\dfrac{25x^2 - 1}{9x^2 - 6x} \div \dfrac{5x^2 + 9x - 2}{3x^2 + x - 2}$

$$= \dfrac{25x^2 - 1}{9x^2 - 6x} \cdot \dfrac{3x^2 + x - 2}{5x^2 + 9x - 2}$$

$$= \dfrac{(25x^2 - 1)(3x^2 + x - 2)}{(9x^2 - 6x)(5x^2 + 9x - 2)}$$

$$= \dfrac{(5x + 1)(5x - 1)(3x - 2)(x + 1)}{3x(3x - 2)(5x - 1)(x + 2)}$$

$$= \dfrac{(5x + 1)(5x \!\!\!\!\diagup 1)(3x \!\!\!\!\diagup 2)(x + 1)}{3x(3x \!\!\!\!\diagup 2)(5x \!\!\!\!\diagup 1)(x + 2)}$$

$$= \dfrac{(5x + 1)(x + 1)}{3x(x + 2)}$$

10. $y^2 - 9 = (y + 3)(y - 3)$

$$y^2 + 10y + 21 = (y + 3)(y + 7)$$
$$y^2 + 4y - 21 = (y - 3)(y + 7)$$
$$\text{LCM} = (y + 3)(y - 3)(y + 7)$$

11. $\dfrac{16 + x}{x^3} + \dfrac{7 - 4x}{x^3} = \dfrac{16 + x + 7 - 4x}{x^3} = \dfrac{23 - 3x}{x^3}$

12. $\dfrac{5 - t}{t^2 + 1} - \dfrac{t - 3}{t^2 + 1} = \dfrac{5 - t - (t - 3)}{t^2 + 1}$

$$= \dfrac{5 - t - t + 3}{t^2 + 1}$$

$$= \dfrac{8 - 2t}{t^2 + 1}$$

13. $\dfrac{x - 4}{x - 3} + \dfrac{x - 1}{3 - x} = \dfrac{x - 4}{x - 3} + \dfrac{x - 1}{3 - x} \cdot \dfrac{-1}{-1}$

$$= \dfrac{x - 4}{x - 3} + \dfrac{-x + 1}{x - 3}$$

$$= \dfrac{x - 4 - x + 1}{x - 3}$$

$$= \dfrac{-3}{x - 3}$$

14. $\dfrac{x - 4}{x - 3} - \dfrac{x - 1}{3 - x} = \dfrac{x - 4}{x - 3} - \dfrac{x - 1}{3 - x} \cdot \dfrac{-1}{-1}$

$$= \dfrac{x - 4}{x - 3} - \dfrac{-x + 1}{x - 3}$$

$$= \dfrac{x - 4 - (-x + 1)}{x - 3}$$

$$= \dfrac{x - 4 + x - 1}{x - 3}$$

$$= \dfrac{2x - 5}{x - 3}$$

15. $\dfrac{5}{t-1} + \dfrac{3}{t}$, LCD is $t(t-1)$.

$= \dfrac{5}{t-1} \cdot \dfrac{t}{t} + \dfrac{3}{t} \cdot \dfrac{t-1}{t-1}$

$= \dfrac{5t}{t(t-1)} + \dfrac{3(t-1)}{t(t-1)}$

$= \dfrac{5t}{t(t-1)} + \dfrac{3t-3}{t(t-1)}$

$= \dfrac{5t+3t-3}{t(t-1)}$

$= \dfrac{8t-3}{t(t-1)}$

16. $\dfrac{1}{x^2-16} - \dfrac{x+4}{x^2-3x-4}$

$= \dfrac{1}{(x+4)(x-4)} - \dfrac{x+4}{(x-4)(x+1)}$, LCD is $(x+4)(x-4)(x+1)$.

$= \dfrac{1}{(x+4)(x-4)} \cdot \dfrac{x+1}{x+1} - \dfrac{x+4}{(x-4)(x+1)} \cdot \dfrac{x+4}{x+4}$

$= \dfrac{x+1-(x+4)(x+4)}{(x+4)(x-4)(x+1)}$

$= \dfrac{x+1-(x^2+8x+16)}{(x+4)(x-4)(x+1)}$

$= \dfrac{x+1-x^2-8x-16}{(x+4)(x-4)(x+1)}$

$= \dfrac{-x^2-7x-15}{(x+4)(x-4)(x+1)}$

17. $\dfrac{1}{x-1} + \dfrac{4}{x^2-1} - \dfrac{2}{x^2-2x+1}$

$= \dfrac{1}{x-1} + \dfrac{4}{(x+1)(x-1)} - \dfrac{2}{(x-1)(x-1)}$,

$\qquad$ LCD is $(x+1)(x-1)(x-1)$

$= \dfrac{1}{x-1} \cdot \dfrac{(x+1)(x-1)}{(x+1)(x-1)} + \dfrac{4}{(x+1)(x-1)} \cdot \dfrac{x-1}{x-1} -$

$\qquad \dfrac{2}{(x-1)(x-1)} \cdot \dfrac{x+1}{x+1}$

$= \dfrac{(x+1)(x-1) + 4(x-1) - 2(x+1)}{(x+1)(x-1)(x-1)}$

$= \dfrac{x^2-1+4x-4-2x-2}{(x+1)(x-1)(x-1)}$

$= \dfrac{x^2+2x-7}{(x+1)(x-1)^2}$

18. We multiply the numerator and the denominator by the LCM of the denominators, y^2.

$\dfrac{9 - \dfrac{1}{y^2}}{3 - \dfrac{1}{y}} = \dfrac{9 - \dfrac{1}{y^2}}{3 - \dfrac{1}{y}} \cdot \dfrac{y^2}{y^2}$

$= \dfrac{\left(9 - \dfrac{1}{y^2}\right) \cdot y^2}{\left(3 - \dfrac{1}{y}\right) \cdot y^2}$

$= \dfrac{9 \cdot y^2 - \dfrac{1}{y^2} \cdot y^2}{3 \cdot y^2 - \dfrac{1}{y} \cdot y^2}$

$= \dfrac{9y^2 - 1}{3y^2 - y}$

$= \dfrac{(3y+1)(3y-1)}{y(3y-1)}$

$= \dfrac{(3y+1)(3y-1)}{y(3y-1)}$

$= \dfrac{3y+1}{y}$

19. $\dfrac{7}{y} - \dfrac{1}{3} = \dfrac{1}{4}$, LCM is $12y$

$12y\left(\dfrac{7}{y} - \dfrac{1}{3}\right) = 12y \cdot \dfrac{1}{4}$

$12y \cdot \dfrac{7}{y} - 12y \cdot \dfrac{1}{3} = 12y \cdot \dfrac{1}{4}$

$84 - 4y = 3y$

$84 = 7y$

$12 = y$

The number 12 checks, so it is the solution.

20. $\dfrac{15}{x} - \dfrac{15}{x-2} = -2$, LCM is $x(x-2)$

$x(x-2)\left(\dfrac{15}{x} - \dfrac{15}{x-2}\right) = x(x-2)(-2)$

$x(x-2) \cdot \dfrac{15}{x} - x(x-2) \cdot \dfrac{15}{x-2} = -2x(x-2)$

$15(x-2) - 15x = -2x(x-2)$

$15x - 30 - 15x = -2x^2 + 4x$

$-30 = -2x^2 + 4x$

$2x^2 - 4x - 30 = 0$

$2(x^2 - 2x - 15) = 0$

$x^2 - 2x - 15 = 0 \quad$ Dividing by 2

$(x-5)(x+3) = 0$

$x - 5 = 0 \quad or \quad x + 3 = 0$

$x = 5 \quad or \qquad x = -3$

Both numbers check. The solutions are 5 and -3.

21. We substitute to find k.

$$y = kx$$
$$6 = k \cdot 3$$
$$2 = k$$

The equation of variation is $y = 2x$.

To find the value of y when $x = 25$ we substitute 25 for x in the equation of variation.

$$y = 2x$$
$$y = 2 \cdot 25$$
$$y = 50$$

The value of y is 50 when $x = 25$.

22. We substitute to find k.

$$y = kx$$
$$1.5 = k \cdot 3$$
$$0.5 = k$$

The equation of variation is $y = 0.5x$.

To find the value of y when $x = 25$ we substitute 25 for x in the equation of variation.

$$y = 0.5x$$
$$y = 0.5(25)$$
$$y = 12.5$$

The value of y is 12.5 when $x = 25$.

23. We substitute to find k.

$$y = \frac{k}{x}$$
$$6 = \frac{k}{3}$$
$$18 = k$$

The equation of variation is $y = \frac{18}{x}$.

To find the value of y when $x = 100$ we substitute 100 for x in the equation of variation.

$$y = \frac{18}{x}$$
$$y = \frac{18}{100}$$
$$y = \frac{9}{50}$$

The value of y is $\frac{9}{50}$ when $x = 100$.

24. We substitute to find k.

$$y = \frac{k}{x}$$
$$11 = \frac{k}{2}$$
$$22 = k$$

The equation of variation is $y = \frac{22}{x}$.

To find the value of y when $x = 100$ we substitute 100 for x in the equation of variation.

$$y = \frac{22}{x}$$
$$y = \frac{22}{100}$$
$$y = \frac{11}{50}$$

The value of y is $\frac{11}{50}$ when $x = 100$.

25. *Familiarize and Translate.* The problem states that we have direct variation between the variables d and t. Thus, an equation $d = kt$, $k > 0$, applies.

Solve.

a) First find an equation of variation.

$$d = kt$$
$$60 = k \cdot \frac{1}{2}$$
$$120 = k$$

The equation of variation is $d = 120t$.

b) Use the equation to find the distance the train will travel in 2 hr.

$$d = 120t$$
$$d = 120 \cdot 2$$
$$d = 240$$

Check. We can repeat the computations. Also note that when the time increases from $\frac{1}{2}$ hr to 2 hr, the distance traveled increases from 60 km to 240 km so the answer seems reasonable.

State. The train will travel 240 km in 2 hr.

26. *Familiarize and Translate.* Let $T =$ the time required to do the job and let $M =$ the number of concrete mixers used. We have inverse variation between T and M so an equation $T = \frac{k}{M}$, $k > 0$, applies.

Solve.

a) First we find an equation of variation.

$$T = \frac{k}{M}$$
$$3 = \frac{k}{2}$$
$$6 = k$$

The equation of variation is $T = \frac{6}{M}$.

b) Use the equation to find the time required to do the job when 5 mixers are used.

$$T = \frac{6}{M}$$
$$T = \frac{6}{5}$$
$$T = 1\frac{1}{5}$$

Check. We can repeat the computations. Also note that as the number of mixers increased from 2 to 5, the time required to mix the concrete decreased from 3 hr to $1\frac{1}{5}$ hr, so the answer seems reasonable.

State. It would take $1\frac{1}{5}$ hr to do the job if 5 concrete mixers are used.

27. *Familiarize*. We can translate to a proportion, letting d = the number of defective spark plugs that would be expected in a sample of 500.

    Translate.

    Number defective $\rightarrow$ $\dfrac{4}{125} = \dfrac{d}{500}$ $\leftarrow$ Number defective
    Sample size $\rightarrow$ $\leftarrow$ Sample size

    Solve. We solve the proportion.

    $$\frac{4}{125} = \frac{d}{500}$$

    $$500 \cdot \frac{4}{125} = 500 \cdot \frac{d}{500}$$

    $$16 = d$$

    Check.

    $$\frac{4}{125} = 0.032 \quad \text{and} \quad \frac{16}{500} = 0.032.$$

    The ratios are the same, so the answer checks.

    State. You would expect to find 16 defective spark plugs in a sample of 500.

28. *Familiarize*. The ratio of zebras tagged to the total zebra population, P, is $\dfrac{15}{P}$. Of the 20 zebras checked later, 6 were tagged. The ratio of zebras tagged to zebras checked is $\dfrac{6}{20}$.

    Translate. Assuming the two ratios are the same, we can translate to a proportion.

    Zebras tagged
    originally $\longrightarrow$ $\dfrac{15}{P} = \dfrac{6}{20}$ $\longleftarrow$ Tagged zebras caught later
    Zebra $\longrightarrow$ $\longleftarrow$ Zebras caught
    population   later

    Solve. We solve the proportion.

    $$\frac{15}{P} = \frac{6}{20}$$

    $$20P \cdot \frac{15}{P} = 20P \cdot \frac{6}{20}$$

    $$300 = 6P$$

    $$50 = P$$

    Check.

    $$\frac{15}{50} = 0.3 \quad \text{and} \quad \frac{6}{20} = 0.3.$$

    The ratios are the same, so the answer checks.

    State. The zebra population is 50.

29. *Familiarize*. Let t = the time, in minutes, required to copy the report using both copy machines working together.

    Translate. We use the work principle, substituting 20 for a and 30 for b.

    $$\frac{t}{a} + \frac{t}{b} = 1$$

    $$\frac{t}{20} + \frac{t}{30} = 1$$

Solve.

$$\frac{t}{20} + \frac{t}{30} = 1, \text{ LCM is } 60$$

$$60\left(\frac{t}{20} + \frac{t}{30}\right) = 60 \cdot 1$$

$$60 \cdot \frac{t}{20} + 60 \cdot \frac{t}{30} = 60$$

$$3t + 2t = 60$$

$$5t = 60$$

$$t = 12$$

Check. In 12 min, the portion of the job done is

$$12 \cdot \frac{1}{20} + 12 \cdot \frac{1}{30} = \frac{3}{5} + \frac{2}{5} = 1.$$

The answer checks.

State. It would take 12 min to copy the report using both machines working together.

30. *Familiarize*. Let r = Marilyn's speed, in km/h. Then $r + 20$ = Craig's speed. We organize the information in a table.

	Distance	Speed	Time
Marilyn	225	r	t
Craig	325	$r + 20$	t

Translate. We use the formula $d = rt$ in each row of the table to obtain two equations.

$$225 = rt,$$
$$325 = (r + 20)t$$

Since the times are the same, we solve each equation for t and set the results equal to each other.

$$225 = rt, \text{ so } t = \frac{225}{r}.$$

$$325 = (r + 20)t, \text{ so } t = \frac{325}{r + 20}.$$

Then we have

$$\frac{225}{r} = \frac{325}{r + 20}.$$

Solve.

$$\frac{225}{r} = \frac{325}{r + 20}, \text{ LCM is } r(r + 20)$$

$$r(r + 20) \cdot \frac{225}{r} = r(r + 20) \cdot \frac{325}{r + 20}$$

$$225(r + 20) = 325r$$

$$225r + 4500 = 325r$$

$$4500 = 100r$$

$$45 = r$$

If $r = 45$, then $r + 20 = 45 + 20 = 65$.

Check. If Marilyn's speed is 45 km/h and Craig's speed is 65 km/h, then Craig's speed is 20 km/h faster than Marilyn's. At 45 km/h, Marilyn travels 225 km in 225/45, or 5 hr. At 65 km/h, Craig travels 325 km in 325/65, or 5 hr. Since the times are the same, the answer checks.

State. The speed of Marilyn's car is 45 km/h and the speed of Craig's car is 65 km/h.

31. We write a proportion and solve it.

$$\frac{12}{9} = \frac{20}{x}$$

$$9x \cdot \frac{12}{9} = 9x \cdot \frac{20}{x}$$

$$12x = 180$$

$$x = 15$$

(Note that the proportions $\frac{9}{12} = \frac{x}{20}$, $\frac{12}{20} = \frac{9}{x}$, and $\frac{20}{12} = \frac{x}{9}$ could also be used.)

32. *Familiarize*. Let $r =$ the number of hours it would take Rema to do the job working alone. Then $r + 6 =$ the number of hours it would take Reggie to do the job working alone.

Translate. We use the work principle, substituting $2\frac{6}{7}$, or $\frac{20}{7}$, for t, r for a, and $r + 6$ for b.

$$\frac{t}{a} + \frac{t}{b} = 1$$

$$\frac{\frac{20}{7}}{r} + \frac{\frac{20}{7}}{r + 6} = 1$$

Solve.

$$\frac{\frac{20}{7}}{r} + \frac{\frac{20}{7}}{r + 6} = 1$$

$$\frac{20}{7} \cdot \frac{1}{r} + \frac{20}{7} \cdot \frac{1}{r + 6} = 1$$

$$\frac{20}{7r} + \frac{20}{7(r+6)} = 1, \text{ LCM is } 7r(r+6)$$

$$7r(r+6)\left(\frac{20}{7r} + \frac{20}{7(r+6)}\right) = 7r(r+6) \cdot 1$$

$$7r(r+6) \cdot \frac{20}{7r} + 7r(r+6) \cdot \frac{20}{7(r+6)} = 7r(r+6)$$

$$20(r+6) + 20r = 7r^2 + 42r$$

$$20r + 120 + 20r = 7r^2 + 42r$$

$$40r + 120 = 7r^2 + 42r$$

$$0 = 7r^2 + 2r - 120$$

$$0 = (7r + 30)(r - 4)$$

$$7r + 30 = 0 \quad or \quad r - 4 = 0$$

$$7r = -30 \quad or \quad r = 4$$

$$r = -\frac{30}{7} \quad or \quad r = 4$$

Check. Since the time cannot be negative, $-\frac{30}{7}$ cannot be a solution. If $r = 4$, then $r + 6 = 4 + 6 = 10$. In $2\frac{6}{7}$ hr, or $\frac{20}{7}$ hr, then the portion of the job done is

$$\frac{\frac{20}{7}}{4} + \frac{\frac{20}{7}}{10} = \frac{20}{7} \cdot \frac{1}{4} + \frac{20}{7} \cdot \frac{1}{10} = \frac{5}{7} + \frac{2}{7} = 1.$$

The answer checks.

State. Working alone it would take Rema 4 hr to do the job and it would take Reggie 10 hr.

33.

$$1 + \cfrac{1}{1 + \cfrac{1}{1 + \cfrac{1}{a}}} = 1 + \cfrac{1}{1 + \cfrac{1}{\frac{a}{a} + \frac{1}{a}}}$$

$$= 1 + \cfrac{1}{1 + \cfrac{1}{\frac{a+1}{a}}}$$

$$= 1 + \cfrac{1}{1 + 1 \cdot \cfrac{a}{a+1}}$$

$$= 1 + \cfrac{1}{1 + \cfrac{a}{a+1}}$$

$$= 1 + \cfrac{1}{\frac{a+1}{a+1} + \frac{a}{a+1}}$$

$$= 1 + \cfrac{1}{\frac{a+1+a}{a+1}}$$

$$= 1 + \cfrac{1}{\frac{2a+1}{a+1}}$$

$$= 1 + 1 \cdot \frac{a+1}{2a+1}$$

$$= 1 + \frac{a+1}{2a+1}$$

$$= \frac{2a+1}{2a+1} + \frac{a+1}{2a+1}$$

$$= \frac{2a+1+a+1}{2a+1}$$

$$= \frac{3a+2}{2a+1}$$

Cumulative Review Chapters 1 - 6

1. The distance of 3.5 from 0 is 3.5, so $|3.5| = 3.5$.

2. $x^3 - 2x^2 + x - 1$

The coefficient of x^3, or $1 \cdot x^3$, is 1.

The coefficient of $-2x^2$ is -2.

The coefficient of x, or $1 \cdot x$, is 1.

The coefficient of 1 is 1.

3. $x^3 - 2x^2 + x - 1$

The degree of x^3 is 3.

The degree of $-2x^2$ is 2.

The degree of x, or x^1, is 1.

The degree of -1, or $-1 \cdot x^0$, is 0.

The degree of the polynomial is 3, the largest exponent.

4. The polynomial $x^3 - 2x^2 + x - 1$ is none of these because it has more than three terms.

5. Let p = the average price of a ticket in 2000. Then the amount of the increase in 2001 was $12.9\%p$, or $0.129p$, and the ticket price in 2001 was $p + 12.9\%p$, or $112.9\%p$, or $1.129p$. Now we reword the problem and translate to an equation.

$$\underset{\downarrow}{\$18.90} \;\; \underset{\downarrow}{\text{is}} \;\; \underbrace{\text{2001 ticket price}}_{\downarrow}.$$

$$18.9 \;\; = \;\; 1.129p$$

To solve the equation we divide both sides by 1.129.

$$18.9 = 1.129p$$
$$\frac{18.9}{1.129} = p$$
$$16.74 \approx p$$

Thus, the average ticket price was about \$16.74 in 2000.

6. _Familiarize_. Let w = the number of pound the driver's head would seem to weigh at a speed over 200 mph.

Translate. We translate to a proportion.

$$\begin{array}{l} \text{Actual weight} \rightarrow \\ \text{Perceived weight} \rightarrow \end{array} \frac{7.5}{22.5} = \frac{9}{w} \begin{array}{l} \leftarrow \text{Actual weight} \\ \leftarrow \text{Perceived weight} \end{array}$$

Solve. We solve the proportion.

$$\frac{7.5}{22.5} = \frac{9}{w}$$
$$22.5w \cdot \frac{7.5}{22.5} = 22.5w \cdot \frac{9}{w}$$
$$7.5w = 202.5$$
$$w = 27$$

Check.

$$\frac{7.5}{22.5} = 0.\overline{3} \text{ and } \frac{9}{27} = 0.\overline{3}.$$

The ratios are the same, so the answer checks.

State. At a speed over 200 mph the driver's head would seem to weigh 27 lb.

7. _Familiarize_. Let t = the number of minutes required to shovel the snow if Dina and Neil work together.

Translate. We use the work principle, substituting 50 for a and 75 for b.

$$\frac{t}{a} + \frac{t}{b} = 1$$
$$\frac{t}{50} + \frac{t}{75} = 1$$

Solve.

$$\frac{t}{50} + \frac{t}{75} = 1, \text{ LCD is } 150$$
$$150\left(\frac{t}{50} + \frac{t}{75}\right) = 150 \cdot 1$$
$$150 \cdot \frac{t}{50} + 150 \cdot \frac{t}{75} = 150$$
$$3t + 2t = 150$$
$$5t = 150$$
$$t = 30$$

Check. The portion of the job done in 30 min is

$$30 \cdot \frac{1}{50} + 30 \cdot \frac{1}{75} = \frac{3}{5} + \frac{2}{5} = 1.$$

The answer checks.

State. Working together it would take Dina and Neil 30 min to shovel the snow.

8. a) Let M = the muscle weight, in pounds, and B = the body weight, in pounds. We find an equation of direct variation.

$$M = kB$$
$$70 = k \cdot 175$$
$$0.4 = k$$

The equation of variation is $M = 0.4B$.

b) We use the equation to find the value of M when $B = 192$ lb.

$$M = 0.4B$$
$$M = 0.4(192)$$
$$M = 76.8$$

Mike's muscle weight is 76.8 lb.

9. Let a = the amount originally borrowed. At a 6% rate, the amount of simple interest owed after 1 yr is $6\%a$, or $0.06a$. We reword the problem and translate to an equation.

$$\underbrace{\text{Amount}}_{\text{borrowed}} \;\; \text{plus interest is } \$2650.$$

$$\underset{a}{\downarrow} \;\; \underset{+}{\downarrow} \;\; \underset{0.06a}{\downarrow} \;\; \underset{=}{\downarrow} \;\; \underset{2650}{\downarrow}$$

Now we solve the equation.

$$a + 0.06a = 2650$$
$$1.06a = 2650$$
$$a = \frac{2650}{1.06}$$
$$a = 2500$$

The amount originally borrowed was \$2500.

10. _Familiarize_. Let r = the speed of the faster car, in mph. Then $r - 10$ = the speed of the slower car. We organize the information in a table.

	Distance	Speed	Time
Faster car	105	r	t
Slower car	75	$r - 10$	t

Translate. We use the formula $d = rt$ in each row of the table to obtain two equations.

$$105 = rt,$$
$$75 = (r - 10)t$$

Since the times are the same, we solve each equation for t and set the results equal to each other.

$$105 = rt, \text{ so } t = \frac{105}{r}.$$
$$75 = (r - 10)t, \text{ so } t = \frac{75}{r - 10}.$$

Then we have
$$\frac{105}{r} = \frac{75}{r-10}.$$

Solve.
$$\frac{105}{r} = \frac{75}{r-10}, \text{ LCM is } r(r-10)$$
$$r(r-10) \cdot \frac{105}{r} = r(r-10) \cdot \frac{75}{r-10}$$
$$105(r-10) = 75r$$
$$105r - 1050 = 75r$$
$$-1050 = -30r$$
$$35 = r$$

If $r = 35$, then $r - 10 = 35 - 10 = 25$.

Check. If the speeds are 35 mph and 25 mph, then one car travels 10 mph slower than the other. At 35 mph, the faster car travels 105 mi in 105/35, or 3 hr. At 25 mph, the slower car travels 75 mi in 75/25, or 3 hr. The times are the same, so the answer checks.

State. The speed of the faster car is 35 mph, and the slower car's speed is 25 mph.

11. **Familiarize**. Let s = the length of a side of the original square, in ft. Then $s + 2$ = the length of a side of the enlarged square. Recall that the area of a square with side s is given by $s \cdot s$, or s^2.

   Translate.

   $$\underbrace{\text{Area of}}_{\begin{array}{c}s^2\end{array}} \begin{array}{c}\text{original}\\\text{square}\end{array} \quad \text{plus} \quad \underbrace{\begin{array}{c}\text{Area of}\\\text{enlarged}\\\text{square}\end{array}}_{(s+2)^2} \quad \text{is} \quad \underbrace{452 \text{ ft}^2}_{452}$$

   $$s^2 \quad + \quad (s+2)^2 \quad = \quad 452$$

   Solve.
   $$s^2 + (s+2)^2 = 452$$
   $$s^2 + s^2 + 4s + 4 = 452$$
   $$2s^2 + 4s + 4 = 452$$
   $$2s^2 + 4s - 448 = 0$$
   $$2(s^2 + 2s - 224) = 0$$
   $$s^2 + 2s - 224 = 0$$
   $$(s-14)(s+16) = 0$$
   $$s - 14 = 0 \quad or \quad s + 16 = 0$$
   $$s = 14 \quad or \qquad s = -16$$

Check. Since the length of a side cannot be negative, -16 cannot be a solution. If $s = 14$, then $s + 2 = 14 + 2 = 16$. The area of square with a side of 14 ft is $(14 \text{ ft})^2$, or 196 ft^2, and the area of a square with a side of 16 ft is $(16 \text{ ft})^2$, or 256 ft^2. The sum of the area is $196 \text{ ft}^2 + 256 \text{ ft}^2$, or 452 ft^2, so the answer checks.

State. The length of a side of the original square is 14 ft.

12. **Familiarize**. The page numbers are consecutive integers. Let n = the smaller number. Then $n + 1$ = the larger number.

   Translate.

   $$\underbrace{\text{Smaller number}}_{n} \quad \text{plus} \quad \underbrace{\text{larger number}}_{(n+1)} \quad \underbrace{\text{is}}_{=} \quad \underbrace{69.}_{69}$$

   $$n \qquad + \qquad (n+1) \qquad = \qquad 69$$

   Solve.
   $$n + (n+1) = 69$$
   $$2n + 1 = 69$$
   $$2n = 68$$
   $$n = 34$$

If $n = 34$, then $n + 1 = 34 + 1 = 35$.

Check. 34 and 35 are consecutive integers and $34 + 35 = 69$, so the answer checks.

State. The page numbers are 34 and 35.

13. $$x^2 - 3x^3 - 4x^2 + 5x^3 - 2$$
   $$= -3x^3 + 5x^3 + x^2 - 4x^2 - 2$$
   $$= (-3+5)x^3 + (1-4)x^2 - 2$$
   $$= 2x^3 - 3x^2 - 2$$

14. $$\frac{1}{2}x - \left[\frac{3}{8}x - \left(\frac{2}{3} + \frac{1}{4}x\right) - \frac{1}{3}\right]$$
   $$= \frac{1}{2}x - \left[\frac{3}{8}x - \frac{2}{3} - \frac{1}{4}x - \frac{1}{3}\right]$$
   $$= \frac{1}{2}x - \left[\frac{1}{8}x - 1\right]$$
   $$= \frac{1}{2}x - \frac{1}{8}x + 1$$
   $$= \frac{3}{8}x + 1$$

15. $$\left(\frac{2x^3}{3x^{-1}}\right)^{-2} = \left(\frac{2x^{3-(-1)}}{3}\right)^{-2} = \left(\frac{2x^4}{3}\right)^{-2} =$$
   $$\left(\frac{3}{2x^4}\right)^{2} = \frac{3^2}{(2x^4)^2} = \frac{9}{2^2(x^4)^2} = \frac{9}{4x^8}$$

16. We begin by getting a common denominator in the numerator and in the denominator.

$$\frac{\frac{4}{x}-\frac{6}{x^2}}{\frac{5}{x}+\frac{7}{2x}}=\frac{\frac{4}{x}\cdot\frac{x}{x}-\frac{6}{x^2}}{\frac{5}{x}\cdot\frac{2}{2}+\frac{7}{2x}}$$

$$=\frac{\frac{4x}{x^2}-\frac{6}{x^2}}{\frac{10}{2x}+\frac{7}{2x}}$$

$$=\frac{\frac{4x-6}{x^2}}{\frac{17}{2x}}$$

$$=\frac{4x-6}{x^2}\cdot\frac{2x}{17}$$

$$=\frac{(4x-6)(2x)}{17x^2}$$

$$=\frac{(4x-6)\cdot2\cdot\cancel{x}}{17\cdot\cancel{x}\cdot x}$$

$$=\frac{8x-12}{17x}$$

17. $(5xy^2-6x^2y^2-3xy^3)-(-4xy^3+7xy^2-2x^2y^2)$
$=5xy^2-6x^2y^2-3xy^3+4xy^3-7xy^2+2x^2y^2$
$=(5-7)xy^2+(-6+2)x^2y^2+(-3+4)xy^3$
$=-2xy^2-4x^2y^2+xy^3$

18. $(4x^4+6x^3-6x^2-4)+(2x^5+2x^4-4x^3-4x^2+3x-5)$
$=2x^5+(4+2)x^4+(6-4)x^3+(-6-4)x^2+3x+(-4-5)$
$=2x^5+6x^4+2x^3-10x^2+3x-9$

19. $\frac{2y+4}{21}\cdot\frac{7}{y^2+4y+4}=\frac{(2y+4)(7)}{21(y^2+4y+4)}$

$$=\frac{2(y+2)(7)}{3\cdot7(y+2)(y+2)}$$

$$=\frac{2(\cancel{y+2})(\cancel{7})}{3\cdot\cancel{7}(\cancel{y+2})(y+2)}$$

$$=\frac{2}{3(y+2)}$$

20. $\frac{x^2-9}{x^2+8x+15}\div\frac{x-3}{2x+10}=\frac{x^2-9}{x^2+8x+15}\cdot\frac{2x+10}{x-3}$

$$=\frac{(x^2-9)(2x+10)}{(x^2+8x+15)(x-3)}$$

$$=\frac{(x+3)(x-3)(2)(x+5)}{(x+3)(x+5)(x-3)}$$

$$=\frac{(\cancel{x+3})(\cancel{x-3})(2)(\cancel{x+5})}{(\cancel{x+3})(\cancel{x+5})(\cancel{x-3})\cdot1}$$

$$=2$$

21. $\frac{x^2}{x-4}+\frac{16}{4-x}=\frac{x^2}{x-4}+\frac{16}{4-x}\cdot\frac{-1}{-1}$

$$=\frac{x^2}{x-4}+\frac{-16}{x-4}$$

$$=\frac{x^2-16}{x-4}$$

$$=\frac{(x+4)(x-4)}{x-4}$$

$$=\frac{(x+4)(\cancel{x-4})}{(\cancel{x-4})\cdot1}$$

$$=x+4$$

22. $\frac{5x}{x^2-4}-\frac{-3}{2-x}=\frac{5x}{(x+2)(x-2)}-\frac{-3}{2-x}$

$$=\frac{5x}{(x+2)(x-2)}-\frac{-3}{2-x}\cdot\frac{-1}{-1}$$

$$=\frac{5x}{(x+2)(x-2)}-\frac{3}{x-2}$$

$$=\frac{5x}{(x+2)(x-2)}-\frac{3}{x-2}\cdot\frac{x+2}{x+2}$$

$$=\frac{5x}{(x+2)(x-2)}-\frac{3(x+2)}{(x+2)(x-2)}$$

$$=\frac{5x-3(x+2)}{(x+2)(x-2)}$$

$$=\frac{5x-3x-6}{(x+2)(x-2)}$$

$$=\frac{2x-6}{(x+2)(x-2)}$$

23. We use FOIL.

$(2.5a+7.5)(0.4a-1.2)=a^2-3a+3a-9=a^2-9$

24. $(6x-5)^2=(6x)^2-2\cdot6x\cdot5+5^2=36x^2-60x+25$

25. $(2x^3+1)(2x^3-1)=(2x^3)^2-1^2=4x^6-1$

26. $9a^2+52a-12$

There is no common factor (other than 1). This polynomial has three terms but it is not a trinomial square. Using the ac-method, we multiply the leading coefficient and the constant term: $9(-12)=-108$. Now try to factor -108 so that the sum of the factors is 52. The numbers we want are 54 and -2: $54(-2)=-108$ and $54+(-2)=52$. Use these factors to split the middle term and then factor by grouping.

$$9a^2+52a-12=9a^2+54a-2a-12$$
$$=(9a^2+54a)+(-2a-12)$$
$$=9a(a+6)-2(a+6)$$
$$=(9a-2)(a+6)$$

27. $9x^2-30xy+25y^2=(3x)^2-2\cdot3x\cdot5y+(5y)^2=(3x-5y)^2$

28. $49x^2-1=(7x)^2-1^2=(7x+1)(7x-1)$

29. $x - [x - (x - 1)] = 2$

$\quad\quad x - [x - x + 1] = 2$

$\quad\quad\quad\quad x - [1] = 2$

$\quad\quad\quad\quad\quad x - 1 = 2$

$\quad\quad\quad\quad\quad\quad x = 3$

The solution is 3.

30. $\quad\quad 2x^2 + 7x = 4$

$\quad\quad 2x^2 + 7x - 4 = 0$

$\quad (2x - 1)(x + 4) = 0$

$2x - 1 = 0 \quad or \quad x + 4 = 0$

$\quad 2x = 1 \quad or \quad\quad x = -4$

$\quad\quad x = \dfrac{1}{2} \quad or \quad\quad x = -4$

The solutions are $\dfrac{1}{2}$ and -4.

31. $\quad x^2 + x - 20 = 0$

$\quad (x + 5)(x - 4) = 0$

$x + 5 = 0 \quad or \quad x - 4 = 0$

$\quad x = -5 \quad or \quad\quad x = 4$

The solutions are -5 and 4.

32. $3(x - 2) \le 4(x + 5)$

$\quad 3x - 6 \le 4x + 20$

$\quad\quad -6 \le x + 20$

$\quad\quad -26 \le x$

The solution set is $\{x | x \ge -26\}$.

33. $x(x - 4) = 0$

$\quad x = 0 \quad or \quad x - 4 = 0$

$\quad x = 0 \quad or \quad\quad x = 4$

The solutions are 0 and 4.

34. $\quad\quad x^2 = 10x$

$\quad x^2 - 10x = 0$

$\quad x(x - 10) = 0$

$x = 0 \quad or \quad x - 10 = 0$

$x = 0 \quad or \quad\quad x = 10$

The solutions are 0 and 10.

35. $\quad\quad\quad 2x^2 = 800$

$\quad\quad 2x^2 - 800 = 0$

$\quad 2(x^2 - 400) = 0$

$2(x + 20)(x - 20) = 0$

$(x + 20)(x - 20) = 0 \quad$ Dividing by 2

$x + 20 = 0 \quad or \quad x - 20 = 0$

$\quad x = -20 \quad or \quad\quad x = 20$

The solutions are -20 and 20.

36. $\quad\quad t = ax + ay$

$\quad\quad t = a(x + y)$

$\quad \dfrac{t}{x + y} = a$

37. $\quad \dfrac{5x - 2}{4} - \dfrac{4x - 5}{3} = 1, \text{ LCM is } 12$

$12\left(\dfrac{5x - 2}{4} - \dfrac{4x - 5}{3}\right) = 12 \cdot 1$

$12 \cdot \dfrac{5x - 2}{4} - 12 \cdot \dfrac{4x - 5}{3} = 12$

$3(5x - 2) - 4(4x - 5) = 12$

$15x - 6 - 16x + 20 = 12$

$-x + 14 = 12$

$-x = -2$

$x = 2$

The solution is 2.

38. $\quad\quad \dfrac{2x}{x - 3} - \dfrac{6}{x} = \dfrac{18}{x^2 - 3x}$

$\quad\quad \dfrac{2x}{x - 3} - \dfrac{6}{x} = \dfrac{18}{x(x - 3)}$

$x(x - 3)\left(\dfrac{2x}{x - 3} - \dfrac{6}{x}\right) = x(x - 3) \cdot \dfrac{18}{x(x - 3)}$

$x(x - 3) \cdot \dfrac{2x}{x - 3} - x(x - 3) \cdot \dfrac{6}{x} = 18$

$x \cdot 2x - 6(x - 3) = 18$

$2x^2 - 6x + 18 = 18$

$2x^2 - 6x = 0$

$2x(x - 3) = 0$

$2x = 0 \quad or \quad x - 3 = 0$

$x = 0 \quad or \quad\quad x = 3$

We look at the original equation and see that 0 and 3 each make a denominator 0, so the equation has no solution.

39. $y = \dfrac{1}{2}x$

First we find some ordered pairs that are solutions of the equation.

When $x = -4$, $y = \dfrac{1}{2}(-4) = -2$.

When $x = 0$, $y = \dfrac{1}{2} \cdot 0 = 0$.

When $x = 2$, $y = \dfrac{1}{2} \cdot 2 = 1$.

x	y
-4	-2
0	0
2	1

We plot these points and draw the line containing them

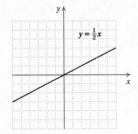

40. $3x - 5y = 15$

First we will find the intercepts. To find the y-intercept we let $x = 0$ and solve for y.

$$3 \cdot 0 - 5y = 15$$
$$-5y = 15$$
$$y = -3$$

The y-intercept is $(-3, 0)$.

To find the x-intercept we let $y = 0$ and solve for x.

$$3x - 5 \cdot 0 = 15$$
$$3x = 15$$
$$x = 5$$

The x-intercept is $(5, 0)$.

Plot these points and draw the line containing them.

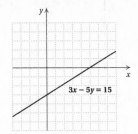

We can find a third point as a check. When $x = 3$, we have

$$3 \cdot 3 - 5y = 15$$
$$9 - 5y = 15$$
$$-5y = 6$$
$$y = -\frac{6}{5}$$

The point $\left(3, -\frac{6}{5}\right)$ appears to be on the graph, so the graph is probably correct.

41. $y = 1$

We can think of this equation as $y = 0 \cdot x + 1$. No matter what number we choose for x, the value of y will be 1. Some possible solutions are shown below. The graph is a horizontal line.

x	y
-4	1
-1	1
3	1

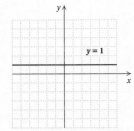

42. $x = -3$

We can think of this equation as $x + 0 \cdot y = -3$. No matter what number we choose for y, the value of x will be -3. Some solutions are shown below. The graph is a vertical line.

x	y
-3	3
-3	0
-3	4

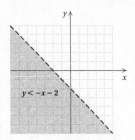

43. $y < -x - 2$

First we graph the line $y = -x - 2$. We draw a dashed line because the inequality symbol is $<$. Now we test a point that is not on the graph. We use $(0, 0)$.

$$\begin{array}{c|c} y < -x - 2 \\ \hline 0 \; ? \; -0 - 2 \\ | \; -2 \quad \text{FALSE} \end{array}$$

We see that $(0, 0)$ is not a solution, so we shade the half-plane that does not contain $(0, 0)$.

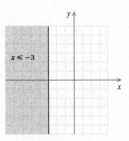

44. $x \leq -3$

First we graph the line $x = -3$. We draw a solid line since the inequality symbol is $\leq$. Then we test a point that is not on the line. We use $(0, 0)$.

$$\begin{array}{c|c} x \leq -3 \\ \hline 0 \; ? \; -3 \quad \text{FALSE} \\ | \end{array}$$

We see that $(0, 0)$ is not a solution so we shade the half-plane that does not contain $(0, 0)$.

45. We substitute to find k.

$$y = kx$$
$$8 = k \cdot 12$$
$$\frac{2}{3} = k$$

The equation of variation is $y = \frac{2}{3}x$.

To find the value of y when $x = 125$ we substitute 125 for x in the equation of variation.

$$y = \frac{2}{3}x$$
$$y = \frac{2}{3} \cdot 125$$
$$y = \frac{250}{3}$$

The value of y is $\frac{250}{3}$ when $x = 125$.

46. We substitute to find k.

$$y = \frac{k}{x}$$
$$20 = \frac{k}{0.5}$$
$$10 = k$$

The equation of variation is $y = \frac{10}{x}$.

To find the value of y when $x = 125$ we substitute 125 for x in the equation of variation.

$$y = \frac{10}{x}$$
$$y = \frac{10}{125}$$
$$y = \frac{2}{25}$$

The value of y is $\frac{2}{25}$ when $x = 125$.

47. $m = \frac{y_2 - y_1}{x_2 - x_1} = \frac{-1 - 6}{-2 - (-2)} = \frac{-7}{0}$

The slope is not defined.

48. $m = \frac{y_2 - y_1}{x_2 - x_1} = \frac{1 - (-2)}{-4 - 3} = \frac{3}{-7} = -\frac{3}{7}$

49. First we solve for y.

$$4x - 3y = 6$$
$$-3y = -4x + 6$$
$$y = -\frac{1}{3}(-4x + 6)$$
$$y = \frac{4}{3}x - 2$$

The slope is $\frac{4}{3}$ and the y-intercept is $(0, -2)$.

50. The slope is -4, so the equation is $y = -4x + b$. Now we substitute 2 for x and -3 for y in $y = -4x + b$ and solve for b.

$$y = -4x + b$$
$$-3 = -4 \cdot 2 + b$$
$$-3 = -8 + b$$
$$5 = b$$

Then the equation is $y = -4x + 5$.

51. First we find the slope.

$$m = \frac{-2 - (-3)}{5 - (-1)} = \frac{1}{6}$$

Then the equation is $y = \frac{1}{6}x + b$. Now we use one of the given points to find b. We use $(5, -2)$.

$$y = \frac{1}{6}x + b$$
$$-2 = \frac{1}{6} \cdot 5 + b$$
$$-2 = \frac{5}{6} + b$$
$$-\frac{17}{6} = b$$

The equation is $y = \frac{1}{6}x - \frac{17}{6}$.

52. First we solve each equation for y.

$$\begin{array}{ll} 1. \quad 2x = 7 - 3y & 2. \quad 7 + 2x = 3y \\ \quad\;\; 2x - 7 = -3y & \quad\;\; \frac{7}{3} + \frac{2}{2}x = y \\ \quad -\frac{2}{3}x + \frac{7}{3} = y & \end{array}$$

The slopes are not the same, so the lines are not parallel. The product of the slopes is $-\frac{2}{3} \cdot \frac{2}{3} = -\frac{4}{9} \neq -1$, so the lines are not perpendicular. Thus, the graphs are neither parallel nor perpendicular.

53. We solve the first equation for y.

$$x - y = 4$$
$$-y = -x + 4$$
$$y = x - 4$$

The second equation is already solved for y: $y = x + 5$

The slopes are both 1. The y-intercepts, $(0, -4)$ and $(0, 5)$ are different, so the graphs are parallel.

54. $m = \frac{938,750,000 - 968,845,000}{10 - 0} = \frac{-30,095,000}{10} = -3,009,500$

55. From the graph we see that the y-intercept is $(0, 968,845,000)$.

56. From Exercise 54, we know that the rate of change is $-3,009,500$ acres per year.

57. The slope is $-3,009,500$ and the y-intercept is $(0, 968,845,000)$, so the equation is $y = -3,009,500x + 968,845,000$.

58. In 2010, $x = 2010 - 1993 = 17$. Using the equation found in Exercise 57 we have:

$y = -3,009,500(17) + 968,845,000 = 917,683,500$ acres

59. We substitute $850,000,000$ for y in the equation found in Exercise 57 and solve for x.

$$850,000,000 = -3,009,500x + 968,845,000$$
$$-118,845,000 = -3,009,500x$$
$$39.5 \approx x$$

The number of acres will be 850,000,000 about 39.5 yr after 1993, or in 2032.

60. $C = \pi d \approx 3.14(1 \text{ ft}) \approx 3.14 \text{ ft}$, so answer (b) is correct.

61.
$$(x + 7)(x - 4) - (x + 8)(x - 5)$$
$$= x^2 + 3x - 28 - (x^2 + 3x - 40)$$
$$= x^2 + 3x - 28 - x^2 - 3x + 40$$
$$= 12$$

62.
$$[4y^3 - (y^2 - 3)][4y^3 + (y^2 - 3)]$$
$$= (4y^3)^2 - (y^2 - 3)^2$$
$$= 16y^6 - (y^4 - 6y^2 + 9)$$
$$= 16y^6 - y^4 + 6y^2 - 9$$

63.
$$2a^{32} - 13,122b^{40}$$
$$= 2(a^{32} - 6561b^{40})$$
$$= 2[(a^{16})^2 - (81b^{20})^2]$$
$$= 2(a^{16} + 81b^{20})(a^{16} - 81b^{20})$$
$$= 2(a^{16} + 81b^{20})[(a^8)^2 - (9b^{10})^2]$$
$$= 2(a^{16} + 81b^{20})(a^8 + 9b^{10})(a^8 - 9b^{10})$$
$$= 2(a^{16} + 81b^{20})(a^8 + 9b^{10})[(a^4)^2 - (3b^5)^2]$$
$$= 2(a^{16} + 81b^{20})(a^8 + 9b^{10})(a^4 + 3b^5)(a^4 - 3b^5)$$

64. $(x - 4)(x + 7)(x - 12) = 0$
$$x - 4 = 0 \quad or \quad x + 7 = 0 \quad or \quad x - 12 = 0$$
$$x = 4 \quad or \quad x = -7 \quad or \quad x = 12$$
The solutions are 4, -7, and 12.

65. First we find the slope of the given line.
$$2x - 3y = -12$$
$$-3y = -2x - 12$$
$$y = -\frac{1}{3}(-2x - 12)$$
$$y = \frac{2}{3}x + 4$$

The slope is $\frac{2}{3}$, so the desired equation is $y = \frac{2}{3}x + b$. Use the point $(-3, -2)$ to find b.
$$y = \frac{2}{3}x + b$$
$$-2 = \frac{2}{3}(-3) + b$$
$$-2 = -2 + b$$
$$0 = b$$

The desired equation is $y = \frac{2}{3}x + 0$, or $y = \frac{2}{3}x$.

66. $\dfrac{\dfrac{1}{x} + x}{2 + \dfrac{1}{x - 3}}$

$\frac{1}{x}$ is not defined when $x = 0$; $x - 3$ is not defined when $x - 3 = 0$, or $x = 3$. In addition, the expression is not defined when $2 + \dfrac{1}{x - 3}$ is 0. We find the value(s) of x for which this is the case.

$$2 + \frac{1}{x - 3} = 0$$
$$(x - 3)\left(2 + \frac{1}{x - 3}\right) = (x - 3) \cdot 0$$
$$2(x - 3) + (x - 3) \cdot \frac{1}{x - 3} = 0$$
$$2x - 6 + 1 = 0$$
$$2x - 5 = 0$$
$$2x = 5$$
$$x = \frac{5}{2}$$

Thus, the expression is not defined for 0, 3, and $\frac{5}{2}$.

Chapter 7

Systems of Equations

1. We check by substituting alphabetically 1 for x and 5 for y.

$$
\begin{array}{c|c}
5x - 2y = -5 \\
\hline
5 \cdot 1 - 2 \cdot 5 \; ? \; -5 \\
5 - 10 \\
-5 & \text{TRUE}
\end{array}
\qquad
\begin{array}{c|c}
3x - 7y = -32 \\
\hline
3 \cdot 1 - 7 \cdot 5 \; ? \; -32 \\
3 - 35 \\
-32 & \text{TRUE}
\end{array}
$$

The ordered pair $(1, 5)$ is a solution of both equations, so it is a solution of the system of equations.

3. We check by substituting alphabetically 4 for a and 2 for b.

$$
\begin{array}{c|c}
3b - 2a = -2 \\
\hline
3 \cdot 2 - 2 \cdot 4 \; ? \; -2 \\
6 - 8 \\
-2 & \text{TRUE}
\end{array}
\qquad
\begin{array}{c|c}
b + 2a = 8 \\
\hline
2 + 2 \cdot 4 \; ? \; 8 \\
2 + 8 \\
10 & \text{FALSE}
\end{array}
$$

The ordered pair $(4, 2)$ is not a solution of $b + 2a = 8$, so it is not a solution of the system of equations.

5. We check by substituting alphabetically 15 for x and 20 for y.

$$
\begin{array}{c|c}
3x - 2y = 5 \\
\hline
3 \cdot 15 - 2 \cdot 20 \; ? \; 5 \\
45 - 40 \\
5 & \text{TRUE}
\end{array}
\qquad
\begin{array}{c|c}
6x - 5y = -10 \\
\hline
6 \cdot 15 - 5 \cdot 20 \; ? \; -10 \\
90 - 100 \\
-10 & \text{TRUE}
\end{array}
$$

The ordered pair $(15, 20)$ is a solution of both equations, so it is a solution of the system of equations.

7. We check by substituting alphabetically -1 for x and 1 for y.

$$
\begin{array}{c}
x = -1 \\
\hline
-1 \; ? \; -1 \quad \text{TRUE} \\

\end{array}
\qquad
\begin{array}{c|c}
x - y = -2 \\
\hline
-1 - 1 \; ? \; -2 \\
-2 & \text{TRUE}
\end{array}
$$

The ordered pair $(-1, 1)$ is a solution of both equations, so it is a solution of the system of equations.

9. We check by substituting alphabetically 18 for x and 3 for y.

$$
\begin{array}{c|c}
y = \dfrac{1}{6}x \\
\hline
3 \; ? \; \dfrac{1}{6} \cdot 18 \\
3 & \text{TRUE}
\end{array}
\qquad
\begin{array}{c|c}
2x - y = 33 \\
\hline
2 \cdot 18 \; ? \; 33 \\
36 - 3 \\
33 & \text{TRUE}
\end{array}
$$

The ordered pair $(18, 3)$ is a solution of both equations, so it is a solution of the system of equations.

11. We graph the equations.

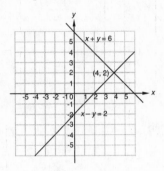

The point of intersection looks as if it has coordinates $(4, 2)$.

Check:

$$
\begin{array}{c|c}
x - y = 2 \\
\hline
4 - 2 \; ? \; 2 \\
2 & \text{TRUE}
\end{array}
\qquad
\begin{array}{c|c}
x + y = 6 \\
\hline
4 + 2 \; ? \; 6 \\
6 & \text{TRUE}
\end{array}
$$

The solution is $(4, 2)$.

13. We graph the equations.

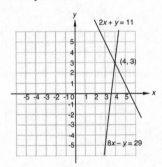

The point of intersection looks as if it has coordinates $(4, 3)$.

Check:

$$
\begin{array}{c|c}
8x - y = 29 \\
\hline
8 \cdot 4 - 3 \; ? \; 29 \\
32 - 3 \\
29 & \text{TRUE}
\end{array}
\qquad
\begin{array}{c|c}
2x + y = 11 \\
\hline
2 \cdot 4 + 3 \; ? \; 11 \\
8 + 3 \\
11 & \text{TRUE}
\end{array}
$$

The solution is $(4, 3)$.

15. We graph the equations.

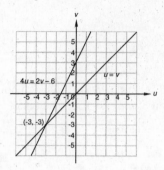

The point of intersection looks as if it has coordinates $(-3, -3)$.

Check:

$u = v$			$4u = 2v - 6$		
-3 ? -3	TRUE		$4(-3)$? $2(-3) - 6$		
			-12	$-6 - 6$	
				-12	TRUE

The solution is $(-3, -3)$.

17. We graph the equations.

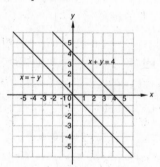

The lines are parallel. There is no solution.

19. We graph the equations.

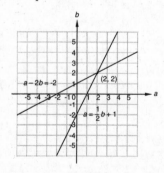

The point of intersection looks as if it has coordinates $(2, 2)$.

Check:

$a = \dfrac{1}{2}b + 1$		$a - 2b = -2$		
2 ? $\dfrac{1}{2} \cdot 2 + 1$		$2 - 2 \cdot 2$? -2		
$1 + 1$		$2 - 4$		
2	TRUE	-2	TRUE

The solution is $(2, 2)$.

21. We graph the equations.

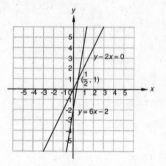

The point of intersection looks as if it has coordinates $\left(\dfrac{1}{2}, 1\right)$.

Check:

$y - 2x = 0$		$y = 6x - 2$		
$1 - 2 \cdot \dfrac{1}{2}$? 0		1 ? $6 \cdot \dfrac{1}{2} - 2$		
$1 - 1$		$3 - 2$		
0	TRUE	1	TRUE

The solution is $\left(\dfrac{1}{2}, 1\right)$.

23. We graph the equations.

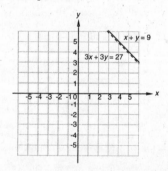

The lines coincide. The system has an infinite number of solutions.

25. We graph the equations.

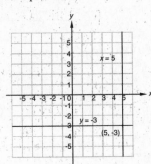

The point of intersection looks as if it has coordinates $(5, -3)$.

Check:

$$\frac{x = 5}{5 \ ? \ 5 \ \ \text{TRUE}} \qquad \frac{y = -3}{-3 \ ? \ -3 \ \ \text{TRUE}}$$

The solution is $(5, -3)$.

27. Discussion and Writing Exercise

29. $\dfrac{1}{x} - \dfrac{1}{x^2} + \dfrac{1}{x+1}$, LCM is $x^2(x+1)$

$= \dfrac{1}{x} \cdot \dfrac{x(x+1)}{x(x+1)} - \dfrac{1}{x^2} \cdot \dfrac{x+1}{x+1} + \dfrac{1}{x+1} \cdot \dfrac{x^2}{x^2}$

$= \dfrac{x(x+1) - (x+1) + x^2}{x^2(x+1)}$

$= \dfrac{x^2 + x - x - 1 + x^2}{x^2(x+1)}$

$= \dfrac{2x^2 - 1}{x^2(x+1)}$

31. $\dfrac{x+2}{x-4} - \dfrac{x+1}{x+4}$, LCM is $(x-4)(x+4)$

$= \dfrac{x+2}{x-4} \cdot \dfrac{x+4}{x+4} - \dfrac{x+1}{x+4} \cdot \dfrac{x-4}{x-4}$

$= \dfrac{(x+2)(x+4) - (x+1)(x-4)}{(x-4)(x+4)}$

$= \dfrac{x^2 + 6x + 8 - (x^2 - 3x - 4)}{(x-4)(x+4)}$

$= \dfrac{x^2 + 6x + 8 - x^2 + 3x + 4}{(x-4)(x+4)}$

$= \dfrac{9x + 12}{(x-4)(x+4)}$

33. The polynomial has exactly three terms, so it is a trinomial.

35. The polynomial has exactly one term, so it is a monomial.

37. $(2, -3)$ is a solution of $Ax - 3y = 13$. Substitute 2 for x and -3 for y and solve for A.

$$Ax - 3y = 13$$
$$A \cdot 2 - 3(-3) = 13$$
$$2A + 9 = 13$$
$$2A = 4$$
$$A = 2$$

$(2, -3)$ is a solution of $x - By = 8$. Substitute 2 for x and -3 for y and solve for B.

$$x - By = 8$$
$$2 - B(-3) = 8$$
$$2 + 3B = 8$$
$$3B = 6$$
$$B = 2$$

39. Answers may vary. Any two equations with a solution of $(6, -2)$ will do. One possibility is

$$x + y = 4,$$
$$x - y = 8.$$

41.-47. Left to the student

Exercise Set 7.2

1. $x + y = 10$, (1)

$y = x + 8$ (2)

We substitute $x + 8$ for y in Equation (1) and solve for x.

$$x + y = 10 \qquad \text{Equation (1)}$$
$$x + (x+8) = 10 \qquad \text{Substituting}$$
$$2x + 8 = 10 \qquad \text{Collecting like terms}$$
$$2x = 2 \qquad \text{Subtracting 8}$$
$$x = 1 \qquad \text{Dividing by 2}$$

Next we substitute 1 for x in either equation of the original system and solve for y. We choose Equation (2) since it has y alone on one side.

$$y = x + 8 \qquad \text{Equation (2)}$$
$$y = 1 + 8 \qquad \text{Substituting}$$
$$y = 9$$

We check the ordered pair $(1, 9)$.

$$\frac{x + y = 10}{1 + 9 \ ? \ 10} \qquad \frac{y = x + 8}{9 \ ? \ 1 + 8}$$
$$ 10 \ \bigm| \ \text{TRUE} \qquad 9 \ \bigm| \ \text{TRUE}$$

Since $(1, 9)$ checks in both equations, it is the solution.

3. $y = x - 6$, (1)

$x + y = -2$ (2)

We substitute $x - 6$ for y in Equation (2) and solve for x.

$$x + y = -2 \qquad \text{Equation (2)}$$
$$x + (x-6) = -2 \qquad \text{Substituting}$$
$$2x - 6 = -2 \qquad \text{Collecting like terms}$$
$$2x = 4 \qquad \text{Adding 6}$$
$$x = 2 \qquad \text{Dividing by 2}$$

Next we substitute 2 for x in either equation of the original system and solve for y. We choose Equation (1) since it has y alone on one side.

$$y = x - 6 \quad \text{Equation (1)}$$
$$y = 2 - 6 \quad \text{Substituting}$$
$$y = -4$$

We check the ordered pair $(2, -4)$.

$y = x - 6$	$x + y = -2$
$-4 \ ? \ 2 - 6$	$2 + (-4) \ ? \ -2$
-4 TRUE	-2 TRUE

Since $(2, -4)$ checks in both equations, it is the solution.

5. $\ y = 2x - 5, \quad (1)$
$\quad 3y - x = 5 \quad (2)$

We substitute $2x - 5$ for y in Equation (2) and solve for x.

$$3y - x = 5 \quad \text{Equation (2)}$$
$$3(2x - 5) - x = 5 \quad \text{Substituting}$$
$$6x - 15 - x = 5 \quad \text{Removing parentheses}$$
$$5x - 15 = 5 \quad \text{Collecting like terms}$$
$$5x = 20 \quad \text{Adding 15}$$
$$x = 4 \quad \text{Dividing by 5}$$

Next we substitute 4 for x in either equation of the original system and solve for y.

$$y = 2x - 5 \quad \text{Equation (1)}$$
$$y = 2 \cdot 4 - 5 \quad \text{Substituting}$$
$$y = 8 - 5$$
$$y = 3$$

We check the ordered pair $(4, 3)$.

$y = 2x - 5$	$3y - x = 5$
$3 \ ? \ 2 \cdot 4 - 5$	$3 \cdot 3 - 4 \ ? \ 5$
$8 - 5$	$9 - 4$
3 TRUE	5 TRUE

Since $(4, 3)$ checks in both equations, it is the solution.

7. $\ x = -2y, \quad (1)$
$\quad x + 4y = 2 \quad (2)$

We substitute $-2y$ for x in Equation (2) and solve for y.

$$x + 4y = 2 \quad \text{Equation (2)}$$
$$-2y + 4y = 2 \quad \text{Substituting}$$
$$2y = 2 \quad \text{Collecting like terms}$$
$$y = 1 \quad \text{Dividing by 2}$$

Next we substitute 1 for y in either equation of the original system and solve for x.

$$x = -2y \quad \text{Equation (1)}$$
$$x = -2 \cdot 1$$
$$x = -2$$

We check the ordered pair $(-2, 1)$.

$x = -2y$	$3y - x = 5$
$-2 \ ? \ -2 \cdot 1$	$3 \cdot 1 - (-2) \ ? \ 5$
-2 TRUE	$3 + 2$
	5 TRUE

Since $(-2, 1)$ checks in both equations, it is the solution.

9. $\ x - y = 6, \quad (1)$
$\quad x + y = -2 \quad (2)$

We solve Equation (1) for x.

$$x - y = 6 \quad \text{Equation (1)}$$
$$x = y + 6 \quad \text{Adding } y \quad (3)$$

We substitute $y + 6$ for x in Equation (2) and solve for y.

$$x + y = -2 \quad \text{Equation (2)}$$
$$(y + 6) + y = -2 \quad \text{Substituting}$$
$$2y + 6 = -2 \quad \text{Collecting like terms}$$
$$2y = -8 \quad \text{Subtracting 6}$$
$$y = -4 \quad \text{Dividing by 2}$$

Now we substitute -4 for y in Equation (3) and compute x.

$$x = y + 6 = -4 + 6 = 2$$

The ordered pair $(2, -4)$ checks in both equations. It is the solution.

11. $\ y - 2x = -6, \quad (1)$
$\quad 2y - x = 5 \quad (2)$

We solve Equation (1) for y.

$$y - 2x = -6 \quad \text{Equation (1)}$$
$$y = 2x - 6 \quad (3)$$

We substitute $2x - 6$ for y in Equation (2) and solve for x.

$$2y - x = 5 \quad \text{Equation (2)}$$
$$2(2x - 6) - x = 5 \quad \text{Substituting}$$
$$4x - 12 - x = 5 \quad \text{Removing parentheses}$$
$$3x - 12 = 5 \quad \text{Collecting like terms}$$
$$3x = 17 \quad \text{Adding 12}$$
$$x = \frac{17}{3} \quad \text{Dividing by 3}$$

We substitute $\frac{17}{3}$ for x in Equation (3) and compute y.

$$y = 2x - 6 = 2\left(\frac{17}{3}\right) - 6 = \frac{34}{3} - \frac{18}{3} = \frac{16}{3}$$

The ordered pair $\left(\frac{17}{3}, \frac{16}{3}\right)$ checks in both equations. It is the solution.

13. $\ 2x + 3y = -2, \quad (1)$
$\quad 2x - y = 9 \quad (2)$

We solve Equation (2) for y.

$$2x - y = 9 \quad \text{Equation (2)}$$
$$2x = 9 + y \quad \text{Adding } y$$
$$2x - 9 = y \quad \text{Subtracting 9} \quad (3)$$

We substitute $2x - 9$ for y in Equation (1) and solve for x.

$$2x + 3y = -2 \quad \text{Equation (1)}$$
$$2x + 3(2x - 9) = -2 \quad \text{Substituting}$$
$$2x + 6x - 27 = -2 \quad \text{Removing parentheses}$$
$$8x - 27 = -2 \quad \text{Collecting like terms}$$
$$8x = 25 \quad \text{Adding 27}$$
$$x = \frac{25}{8} \quad \text{Dividing by 8}$$

Now we substitute $\frac{25}{8}$ for x in Equation (3) and compute y.

$$y = 2x - 9 = 2\left(\frac{25}{8}\right) - 9 = \frac{25}{4} - \frac{36}{4} = -\frac{11}{4}$$

The ordered pair $\left(\frac{25}{8}, -\frac{11}{4}\right)$ checks in both equations. It is the solution.

15. $\quad x - y = -3, \quad (1)$
$\qquad 2x + 3y = -6 \quad (2)$

We solve Equation (1) for x.

$$x - y = -3 \quad \text{Equation (1)}$$
$$x = y - 3 \qquad\qquad (3)$$

We substitute $y - 3$ for x in Equation (2) and solve for y.

$$2x + 3y = -6 \quad \text{Equation (2)}$$
$$2(y - 3) + 3y = -6 \quad \text{Substituting}$$
$$2y - 6 + 3y = -6 \quad \text{Removing parentheses}$$
$$5y - 6 = -6 \quad \text{Collecting like terms}$$
$$5y = 0 \quad \text{Adding 6}$$
$$y = 0 \quad \text{Dividing by 5}$$

Now we substitute 0 for y in Equation (3) and compute x.

$$x = y - 3 = 0 - 3 = -3$$

The ordered pair $(-3, 0)$ checks in both equations. It is the solution.

17. $\quad r - 2s = 0, \quad (1)$
$\qquad 4r - 3s = 15 \quad (2)$

We solve Equation (1) for r.

$$r - 2s = 0 \quad \text{Equation (1)}$$
$$r = 2s \qquad\qquad (3)$$

We substitute $2s$ for r in Equation (2) and solve for s.

$$4r - 3s = 15 \quad \text{Equation (2)}$$
$$4(2s) - 3s = 15 \quad \text{Substituting}$$
$$8s - 3s = 15 \quad \text{Removing parentheses}$$
$$5s = 15 \quad \text{Collecting like terms}$$
$$s = 3 \quad \text{Dividing by 5}$$

Now we substitute 3 for s in Equation (3) and compute r.

$$r = 2s = 2 \cdot 3 = 6$$

The ordered pair $(6, 3)$ checks in both equations. It is the solution.

19. *Familiarize*. We let w = the width of the court, in ft, and l = the length, in ft. Recall that the perimeter of a rectangle with length l and width w is given by $2l + 2w$.

***Translate*.**

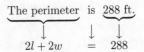

The resulting system is

$$2l + 2w = 288, \quad (1)$$
$$l = 44 + w. \qquad (2)$$

***Solve*.** We solve the system. Substitute $44 + w$ for l in the first equation and solve for w.

$$2(44 + w) + 2w = 288$$
$$88 + 2w + 2w = 288$$
$$88 + 4w = 288$$
$$4w = 200$$
$$w = 50$$

Now substitute 50 for w in Equation (2).

$$l = 44 + w = 44 + 50 = 94$$

***Check*.** If the length is 94 ft and the width is 50 ft, then the length is 44 ft more than the width and the perimeter is $2 \cdot 94 + 2 \cdot 50$, or $188 + 100$, or 288 ft. The answer checks.

***State*.** The length of the court is 94 ft, and the width is 50 ft.

21. *Familiarize*. We make a drawing. We let l = the length and w = the width, in inches.

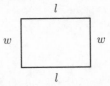

***Translate*.** The perimeter is $2l + 2w$. We translate the first statement.

The resulting system is

$$2l + 2w = 10, \quad (1)$$
$$l = w + 2. \qquad (2)$$

Solve. We solve the system. We substitute $w + 2$ for l in Equation (1) and solve for w.

$$2l + 2w = 10 \qquad \text{Equation (1)}$$
$$2(w + 2) + 2w = 10 \qquad \text{Substituting}$$
$$2w + 4 + 2w = 10 \qquad \text{Removing parentheses}$$
$$4w + 4 = 10$$
$$4w = 6 \qquad \text{Collecting like terms}$$
$$w = \frac{3}{2}, \text{ or } 1\frac{1}{2}$$

Now we substitute $\frac{3}{2}$ for w in Equation (2) and solve for l.

$$l = w + 2 \qquad \text{Equation (2)}$$
$$l = \frac{3}{2} + 2 \qquad \text{Substituting}$$
$$l = \frac{3}{2} + \frac{4}{2}$$
$$l = \frac{7}{2}, \text{ or } 3\frac{1}{2}$$

Check. A possible solution is a length of $\frac{7}{2}$, or $3\frac{1}{2}$ in. and a width of $\frac{3}{2}$, or $1\frac{1}{2}$ in. The perimeter would be $2 \cdot \frac{7}{2} + 2 \cdot \frac{3}{2}$, or $7 + 3$, or 10 in. Also, the width plus 2 is $\frac{3}{2} + 2$, or $\frac{7}{2}$, which is the length. These numbers check.

State. The length is $3\frac{1}{2}$ in., and the width is $1\frac{1}{2}$ in.

23. **Familiarize**. Let l = the length, in mi, and w = the width, in mi. Recall that the perimeter of a rectangle with length l and width w is given by $2l + 2w$.

Translate.

$$\underbrace{\text{The perimeter}}_{2l + 2w} \underbrace{\text{is}}_{=} \underbrace{\text{1280 mi.}}_{1280}$$

$$\underbrace{\text{The width}}_{w} \underbrace{\text{is}}_{=} \underbrace{\text{the length}}_{l} \underbrace{\text{less}}_{-} \underbrace{\text{90 mi.}}_{90}$$

The resulting system is

$$2l + 2w = 1280, \quad (1)$$
$$w = l - 90. \qquad (2)$$

Solve. We solve the system. Substitute $l - 90$ for w in the first equation and solve for l.

$$2l + 2(l - 90) = 1280$$
$$2l + 2l - 180 = 1280$$
$$4l - 180 = 1280$$
$$4l = 1460$$
$$l = 365$$

Now substitute 365 for l in Equation (2).

$$w = l - 90 = 365 - 90 = 275$$

Check. If the length is 365 mi and the width is 275 mi, then the width is 90 mi less than the length and the perimeter is $2 \cdot 365 + 2 \cdot 275$, or $730 + 550$, or 1280. The answer checks.

State. The length is 365 mi, and the width is 275 mi.

25. **Familiarize**. Let l = the length in ft, and w = the width, in ft. Recall that the perimeter of a rectangle with length l and width w is given by $2l + 2w$.

Translate.

$$\underbrace{\text{The perimeter}}_{2l + 2w} \underbrace{\text{is}}_{=} \underbrace{\text{120 ft.}}_{120}$$

$$\underbrace{\text{The length}}_{l} \underbrace{\text{is}}_{=} \underbrace{\text{twice the width.}}_{2w}$$

The resulting system is

$$2l + 2w = 120, \quad (1)$$
$$l = 2w. \qquad (2)$$

Solve. We solve the system.

Substitute $2w$ for l in Equation (1) and solve for w.

$$2 \cdot 2w + 2w = 120 \quad (1)$$
$$4w + 2w = 120$$
$$6w = 120$$
$$w = 20$$

Now substitute 20 for w in Equation (2).

$$l = 2w \qquad (2)$$
$$l = 2 \cdot 20 \quad \text{Substituting}$$
$$l = 40$$

Check. If the length is 40 ft and the width is 20 ft, the perimeter would be $2 \cdot 40 + 2 \cdot 20$, or $80 + 40$, or 120 ft. Also, the length is twice the width. These numbers check.

State. The length is 40 ft, and the width is 20 ft.

27. **Familiarize**. Let l = the length and w = the width, in yards. The perimeter is $l + l + w + w$, or $2l + 2w$.

Translate.

$$\underbrace{\text{The perimeter}}_{2l + 2w} \underbrace{\text{is}}_{=} \underbrace{\text{340 yd.}}_{340}$$

$$\underbrace{\text{The length}}_{l} \underbrace{\text{is}}_{=} \underbrace{\text{10 yd less than twice the width.}}_{2w - 10}$$

The resulting system is

$$2l + 2w = 340, \quad (1)$$
$$l = 2w - 10. \qquad (2)$$

Solve. We solve the system. We substitute $2w - 10$ for l in Equation (1) and solve for w.

$$2l + 2w = 340 \quad (1)$$
$$2(w - 10) + 2w = 340$$
$$4w - 20 + 2w = 340$$
$$6w - 20 = 340$$
$$6w = 360$$
$$w = 60$$

Next we substitute 60 for w in Equation (2) and solve for l.

$$l = 2w - 10 = 2 \cdot 60 - 10 = 120 - 10 = 110$$

Check. The perimeter is $2 \cdot 110 + 2 \cdot 60$, or 340 yd. Also 10 yd less than twice the width is $2 \cdot 60 - 10 = 120 - 10 = 110$. The answer checks.

State. The length is 110 yd, and the width is 60 yd.

29. Familiarize. We let $x =$ the larger number and $y =$ the smaller number.

Translate. We translate the first statement.

The sum of two numbers is 37.

$$x + y \qquad = 37$$

Now we translate the second statement.

One number is 5 more than the other.

$$x \qquad = 5 \quad + \quad y$$

The resulting system is
$$x + y = 37, \quad (1)$$
$$x = 5 + y. \quad (2)$$

Solve. We solve the system of equations. We substitute $5 + y$ for x in Equation (1) and solve for y.

$$x + y = 37 \qquad \text{Equation (1)}$$
$$(5 + y) + y = 37 \qquad \text{Substituting}$$
$$5 + 2y = 37 \qquad \text{Collecting like terms}$$
$$2y = 32 \qquad \text{Subtracting 5}$$
$$y = 16 \qquad \text{Dividing by 2}$$

We go back to the original equations and substitute 16 for y. We use Equation (2).

$$x = 5 + y \qquad \text{Equation (2)}$$
$$x = 5 + 16 \qquad \text{Substituting}$$
$$x = 21$$

Check. The sum of 21 and 16 is 37. The number 21 is 5 more than the number 16. These numbers check.

State. The numbers are 21 and 16.

31. Familiarize. Let $x =$ one number and $y =$ the other.

Translate. We reword and translate.

The sum of two numbers is 52.

$$x + y \qquad = 52$$

The difference of two numbers is 28.

$$x - y \qquad = 28$$

(The second statement could also be translated as $y - x = 28$.)

The resulting system is
$$x + y = 52, \quad (1)$$
$$x - y = 28. \quad (2)$$

Solve. We solve the system. First we solve Equation (2) for x.

$$x - y = 28 \qquad \text{Equation (2)}$$
$$x = y + 28 \qquad \text{Adding } y \qquad (3)$$

We substitute $y + 28$ for x in Equation (1) and solve for y.

$$x + y = 52 \qquad \text{Equation (1)}$$
$$(y + 28) + y = 52 \qquad \text{Substituting}$$
$$2y + 28 = 52 \qquad \text{Collecting like terms}$$
$$2y = 24 \qquad \text{Subtracting 28}$$
$$y = 12 \qquad \text{Dividing by 2}$$

Now we substitute 12 for y in Equation (3) and compute x.

$$x = y + 28 = 12 + 28 = 40$$

Check. The sum of 40 and 12 is 52, and their difference is 28. These numbers check.

State. The numbers are 40 and 12.

33. Familiarize. We let $x =$ the larger number and $y =$ the smaller number.

Translate. We translate the first statement.

The difference between two numbers is 12.

$$x - y \qquad = 12$$

Now we translate the second statement.

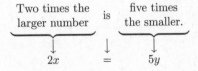

Two times the larger number is five times the smaller.

$$2x \qquad = \qquad 5y$$

The resulting system is
$$x - y = 12, \quad (1)$$
$$2x = 5y. \quad (2)$$

Solve. We solve the system. First we solve Equation (1) for x.

$$x - y = 12 \qquad \text{Equation (1)}$$
$$x = y + 12 \qquad \text{Adding } y \qquad (3)$$

We substitute $y + 12$ for x in Equation (2) and solve for y.

$2x = 5y$ Equation (2)

$2(y + 12) = 5y$ Substituting

$2y + 24 = 5y$ Removing parentheses

$24 = 3y$ Subtracting 2y

$8 = y$ Dividing by 3

Now we substitute 8 for y in Equation (3) and compute x.

$x = y + 12 = 8 + 12 = 20$

Check. The difference between 20 and 8 is 12. Two times 20, or 40, is five times 8. These numbers check.

State. The numbers are 20 and 8.

35. Discussion and Writing exercise

37. Graph: $2x - 3y = 6$

To find the x-intercept, let $y = 0$. Then solve for x.

$2x - 3 \cdot 0 = 6$

$2x = 6$

$x = 3$

The x-intercept is $(3, 0)$.

To find the y-intercept, let $x = 0$. Then solve for y.

$2 \cdot 0 - 3y = 6$

$-3y = 6$

$y = -2$

The y-intercept is $(0, -2)$.

We plot these points and draw the line.

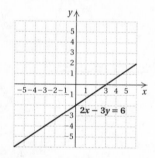

A third point should be used as a check. We let $x = -3$:

$2(-3) - 3y = 6$

$-6 - 3y = 6$

$-3y = 12$

$y = -4$

The point $(-3, -4)$ is on the graph, so our graph is probably correct.

39. Graph: $y = 2x - 5$

We select several values for x and compute the corresponding y-values.

When $x = 0$, $y = 2 \cdot 0 - 5 = 0 - 5 = -5$.

When $x = 2$, $y = 2 \cdot 2 - 5 = 4 - 5 = -1$.

When $x = 4$, $y = 2 \cdot 4 - 5 = 8 - 5 = 3$.

x	y	(x, y)
0	-5	$(0, -5)$
2	-1	$(2, -1)$
4	3	$(4, 3)$

We plot these points and draw the line connecting them.

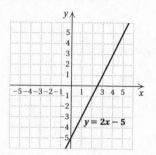

41. $6x^2 - 13x + 6$

The possibilities are $(x+\ \ )(6x+\ \ )$ and $(2x+\ \ )(3x+\ \ )$. We look for a pair of factors of the last term, 6, which produces the correct middle term. Since the last term is positive and the middle term is negative, we need only consider negative pairs. The factorization is $(2x - 3)(3x - 2)$.

43. $4x^2 + 3x + 2$

The possibilities are $(x+\ \ )(4x+\ \ )$ and $(2x+\ \ )(2x+\ \ )$. We look for a pair of factors of the last term, 2, which produce the correct middle term. Since the last term and the middle term are both positive, we need only consider positive pairs. We find that there is no possibility that works. The trinomial cannot be factored.

45. $\dfrac{x^{-2}}{x^{-5}} = x^{-2-(-5)} = x^3$

47. $x^{-2} \cdot x^{-5} = x^{-2+(-5)} = x^{-7} = \dfrac{1}{x^7}$

49. First put the equations in "$y =$" form by solving for y. We get

$y_1 = x - 5$,

$y_2 = (-1/2)x + 7/2$.

Then graph these equations in the standard window and use the INTERSECT feature from the CALC menu to find the coordinates of the point of intersection of the graphs. The solution is $(5.\overline{6}, 0.\overline{6})$.

51. First put the equations in "$y =$" form by solving for y. We get

$y_1 = 2.35x - 5.97$,

$y_2 = (1/2.14)x + (4.88/2.14)$.

Then graph these equations in the standard window and use the INTERSECT feature from the CALC menu to find the coordinates of the point of intersection of the graphs. The solution is approximately $(4.38, 4.33)$.

53. *Familiarize.* Let s = the perimeter of a softball diamond, in yards, and b = the perimeter of a baseball diamond, in yards.

Translate.

Perimeter of a softball diamond is $\frac{2}{3}$ of perimeter of a baseball diamond.

$$s = \frac{2}{3} \cdot b$$

The sum of the perimeters is 200 yd.

$$s + b = 200$$

The resulting system is

$$s = \frac{2}{3}b, \quad (1)$$

$$s + b = 200. \quad (2)$$

Solve. We solve the system of equations. We substitute $\frac{2}{3}b$ for s in Equation (2) and solve for b.

$$s + b = 200$$
$$\frac{2}{3}b + b = 200$$
$$\frac{5}{3}b = 200$$
$$\frac{3}{5} \cdot \frac{5}{3}b = \frac{3}{5} \cdot 200$$
$$b = 120$$

Next we substitute 120 for b in Equation (1) and solve for s.

$$s = \frac{2}{3}b = \frac{2}{3} \cdot 120 = 80$$

Each diamond has four sides of equal length, so we divide each perimeter by 4 to find the distance between bases in each sport. For the softball diamond the distance is $80/4$, or 20 yd. For the baseball diamond it is $120/4$, or 30 yd.

Check. The perimeter of the softball diamond, 80 yd, is $\frac{2}{3}$ of 120 yd, the perimeter of the baseball diamond. The sum of the perimeters is $80 + 120$, or 200 yd. We can also recheck the calculations of the distances between the bases. The answer checks.

State. The distance between bases on a softball diamond is 20 yd and the distance between bases on a baseball diamond is 30 yd.

Exercise Set 7.3

1.
$$x - y = 7 \quad (1)$$
$$\underline{x + y = 5} \quad (2)$$
$$2x \quad\quad = 12 \quad \text{Adding}$$
$$x = 6 \quad \text{Dividing by 2}$$

Substitute 6 for x in either of the original equations and solve for y.

$$x + y = 5 \quad \text{Equation (2)}$$
$$6 + y = 5 \quad \text{Substituting}$$
$$y = -1 \quad \text{Subtracting 6}$$

Check:

$x - y = 7$		$x + y = 5$	
$6 - (-1) ? 7$		$6 + (-1) ? 5$	
$6 + 1$		5	TRUE
7	TRUE		

Since $(6, -1)$ checks, it is the solution.

3.
$$x + y = 8 \quad (1)$$
$$\underline{-x + 2y = 7} \quad (2)$$
$$3y = 15 \quad \text{Adding}$$
$$y = 5 \quad \text{Dividing by 3}$$

Substitute 5 for y in either of the original equations and solve for x.

$$x + y = 8 \quad \text{Equation (1)}$$
$$x + 5 = 8 \quad \text{Substituting}$$
$$x = 3$$

Check:

$x + y = 8$		$-x + 2y = 7$	
$3 + 5 ? 8$		$-3 + 2 \cdot 5 ? 7$	
8	TRUE	$-3 + 10$	
		7	TRUE

Since $(3, 5)$ checks, it is the solution.

5.
$$5x - y = 5 \quad (1)$$
$$\underline{3x + y = 11} \quad (2)$$
$$8x \quad\quad = 16 \quad \text{Adding}$$
$$x = 2 \quad \text{Dividing by 8}$$

Substitute 2 for x in either of the original equations and solve for y.

$$3x + y = 11 \quad \text{Equation (2)}$$
$$3 \cdot 2 + y = 11 \quad \text{Substituting}$$
$$6 + y = 11$$
$$y = 5$$

Check:

$5x - y = 5$		$3x + y = 11$	
$5 \cdot 2 - 5 ? 5$		$3 \cdot 2 + 5 ? 11$	
$10 - 5$		$6 + 5$	
5	TRUE	11	TRUE

Since $(2, 5)$ checks, it is the solution.

7.
$$4a + 3b = 7 \quad (1)$$
$$\underline{-4a + b = 5} \quad (2)$$
$$4b = 12 \quad \text{Adding}$$
$$b = 3$$

Substitute 3 for b in either of the original equations and solve for a.

$$4a + 3b = 7 \qquad \text{Equation (1)}$$
$$4a + 3 \cdot 3 = 7 \qquad \text{Substituting}$$
$$4a + 9 = 7$$
$$4a = -2$$
$$a = -\frac{1}{2}$$

Check:

$$\frac{4a + 3b = 7}{}$$
$$4\left(-\frac{1}{2}\right) + 3 \cdot 3 \ ? \ 7$$
$$-2 + 9 \ \Big|$$
$$7 \ \Big| \ \text{TRUE}$$

$$\frac{-4a + b = 5}{}$$
$$-4\left(-\frac{1}{2}\right) + 3 \ ? \ 5$$
$$2 + 3 \ \Big|$$
$$5 \ \Big| \ \text{TRUE}$$

Since $\left(-\frac{1}{2}, 3\right)$ checks, it is the solution.

9. $8x - 5y = -9 \qquad (1)$

$\underline{3x + 5y = -2 \qquad (2)}$

$11x = -11 \qquad \text{Adding}$

$x = -1$

Substitute -1 for x in either of the original equations and solve for y.

$$3x + 5y = -2 \qquad \text{Equation (2)}$$
$$3(-1) + 5y = -2 \qquad \text{Substituting}$$
$$-3 + 5y = -2$$
$$5y = 1$$
$$y = \frac{1}{5}$$

Check:

$$\frac{8x - 5y = -9}{}$$
$$8(-1) - 5\left(\frac{1}{5}\right) \ ? \ -9$$
$$-8 - 1 \ \Big|$$
$$-9 \ \Big| \ \text{TRUE}$$

$$\frac{3x + 5y = -2}{}$$
$$3(-1) + 5\left(\frac{1}{5}\right) \ ? \ -2$$
$$-3 + 1 \ \Big|$$
$$-2 \ \Big| \ \text{TRUE}$$

Since $\left(-1, \frac{1}{5}\right)$ checks, it is the solution.

11. $4x - 5y = 7$

$\underline{-4x + 5y = 7}$

$ 0 = 14 \qquad \text{Adding}$

We obtain a false equation, $0 = 14$, so there is no solution.

13. $x + y = -7, \qquad (1)$

$3x + y = -9 \qquad (2)$

We multiply on both sides of Equation (1) by -1 and then add.

$-x - y = 7 \qquad \text{Multiplying by } -1$

$\underline{3x + y = -9 \qquad \text{Equation (2)}}$

$2x = -2 \qquad \text{Adding}$

$x = -1$

Substitute -1 for x in one of the original equations and solve for y.

$$x + y = -7 \qquad \text{Equation (1)}$$
$$-1 + y = -7 \qquad \text{Substituting}$$
$$y = -6$$

Check:

$$\frac{x + y = -7}{}$$
$$-1 + (-6) \ ? \ -7$$
$$-7 \ \Big| \ \text{TRUE}$$

$$\frac{3x + y = -9}{}$$
$$3(-1) + (-6) \ ? \ -9$$
$$-3 - 6 \ \Big|$$
$$-9 \ \Big| \ \text{TRUE}$$

Since $(-1, -6)$ checks, it is the solution.

15. $3x - y = 8, \qquad (1)$

$x + 2y = 5 \qquad (2)$

We multiply on both sides of Equation (1) by 2 and then add.

$6x - 2y = 16 \qquad \text{Multiplying by } 2$

$\underline{x + 2y = 5 \qquad \text{Equation (2)}}$

$7x = 21 \qquad \text{Adding}$

$x = 3$

Substitute 3 for x in one of the original equations and solve for y.

$$x + 2y = 5 \qquad \text{Equation (2)}$$
$$3 + 2y = 5 \qquad \text{Substituting}$$
$$2y = 2$$
$$y = 1$$

Check:

$$\frac{3x - y = 8}{}$$
$$3 \cdot 3 - 1 \ ? \ 8$$
$$9 - 1 \ \Big|$$
$$8 \ \Big| \ \text{TRUE}$$

$$\frac{x + 2y = 5}{}$$
$$3 + 2 \cdot 1 \ ? \ 5$$
$$3 + 2 \ \Big|$$
$$5 \ \Big| \ \text{TRUE}$$

Since $(3, 1)$ checks, it is the solution.

17. $x - y = 5, \qquad (1)$

$4x - 5y = 17 \qquad (2)$

We multiply on both sides of Equation (1) by -4 and then add.

$-4x + 4y = -20 \qquad \text{Multiplying by } -4$

$\underline{4x - 5y = 17 \qquad \text{Equation (2)}}$

$ -y = -3 \qquad \text{Adding}$

$ y = 3$

Substitute 3 for y in one of the original equations and solve for x.

$$x - y = 5 \qquad \text{Equation (1)}$$
$$x - 3 = 5 \qquad \text{Substituting}$$
$$x = 8$$

Check:

$$\begin{array}{c|c} x - y = 5 \\ \hline 8 - 3 \ ? \ 5 \\ 5 & \text{TRUE} \end{array} \qquad \begin{array}{c|c} 4x - 5y = 17 \\ \hline 4 \cdot 8 - 5 \cdot 3 \ ? \ 17 \\ 32 - 15 \\ 17 & \text{TRUE} \end{array}$$

Since $(8, 3)$ checks, it is the solution.

19. $2w - 3z = -1,$ (1)

     $3w + 4z = 24$ (2)

We use the multiplication principle with both equations and then add.

$$\begin{array}{ll} 8w - 12z = -4 & \text{Multiplying (1) by 4} \\ \underline{9w + 12z = 72} & \text{Multiplying (2) by 3} \\ 17w \qquad\quad = 68 & \text{Adding} \\ \qquad\quad w = 4 \end{array}$$

Substitute 4 for w in one of the original equations and solve for z.

$$\begin{array}{ll} 3w + 4z = 24 & \text{Equation (2)} \\ 3 \cdot 4 + 4z = 24 & \text{Substituting} \\ 12 + 4z = 24 \\ 4z = 12 \\ z = 3 \end{array}$$

Check:

$$\begin{array}{c|c} 2w - 3z = -1 \\ \hline 2 \cdot 4 - 3 \cdot 3 \ ? \ -1 \\ 8 - 9 \\ -1 & \text{TRUE} \end{array} \qquad \begin{array}{c|c} 3w + 4z = 24 \\ \hline 3 \cdot 4 + 4 \cdot 3 \ ? \ 24 \\ 12 + 12 \\ 24 & \text{TRUE} \end{array}$$

Since $(4, 3)$ checks, it is the solution.

21. $2a + 3b = -1,$ (1)

     $3a + 5b = -2$ (2)

We use the multiplication principle with both equations and then add.

$$\begin{array}{ll} -10a - 15b = 5 & \text{Multiplying (1) by } -5 \\ \underline{9a + 15b = -6} & \text{Multiplying (2) by 3} \\ -a \qquad\quad = -1 & \text{Adding} \\ \quad a = 1 \end{array}$$

Substitute 1 for a in one of the original equations and solve for b.

$$\begin{array}{ll} 2a + 3b = -1 & \text{Equation (1)} \\ 2 \cdot 1 + 3b = -1 & \text{Substituting} \\ 2 + 3b = -1 \\ 3b = -3 \\ b = -1 \end{array}$$

Check:

$$\begin{array}{c|c} 2a + 3b = -1 \\ \hline 2 \cdot 1 + 3(-1) \ ? \ -1 \\ 2 - 3 \\ -1 & \text{TRUE} \end{array} \qquad \begin{array}{c|c} 3a + 5b = -2 \\ \hline 3 \cdot 1 + 5(-1) \ ? \ -2 \\ 3 - 5 \\ -2 & \text{TRUE} \end{array}$$

Since $(1, -1)$ checks, it is the solution.

23. $x = 3y,$ (1)

     $5x + 14 = y$ (2)

We first get each equation in the form $Ax + By = C$.

$$\begin{array}{lll} x - 3y = 0, & \text{(1a)} & \text{Adding } -3y \\ 5x - y = -14 & \text{(2a)} & \text{Adding } -y - 14 \end{array}$$

We multiply by -5 on both sides of Equation (1a) and add.

$$\begin{array}{ll} -5x + 15y = 0 & \text{Multiplying by } -5 \\ \underline{5x - \quad y = -14} \\ \quad\quad 14y = -14 & \text{Adding} \\ \quad\quad\quad y = -1 \end{array}$$

Substitute -1 for y in Equation (1) and solve for x.

$$\begin{array}{ll} x - 3y = 0 \\ x - 3(-1) = 0 & \text{Substituting} \\ x + 3 = 0 \\ x = -3 \end{array}$$

Check:

$$\begin{array}{c|c} x - 3y = 0 \\ \hline -3 - 3(-1) \ ? \ 0 \\ -3 + 3 \\ 0 & \text{TRUE} \end{array} \qquad \begin{array}{c|c} 5x - y = -14 \\ \hline 5(-3) - (-1) \ ? \ -14 \\ -15 + 1 \\ -14 & \text{TRUE} \end{array}$$

Since $(-3, -1)$ checks, it is the solution.

25. $2x + 5y = 16,$ (1)

     $3x - 2y = 5$ (2)

We use the multiplication principle with both equations and then add.

$$\begin{array}{ll} 4x + 10y = 32 & \text{Multiplying (1) by 2} \\ \underline{15x - 10y = 25} & \text{Multiplying (2) by 5} \\ 19x \qquad\quad = 57 \\ \qquad\quad x = 3 \end{array}$$

Substitute 3 for x in one of the original equations and solve for y.

$$\begin{array}{ll} 2x + 5y = 16 & \text{Equation (1)} \\ 2 \cdot 3 + 5y = 16 & \text{Substituting} \\ 6 + 5y = 16 \\ 5y = 10 \\ y = 2 \end{array}$$

Check:

$2x + 5y = 16$		$3x - 2y = 5$	
$2 \cdot 3 + 5 \cdot 2 \ ? \ 16$		$3 \cdot 3 - 2 \cdot 2 \ ? \ 5$	
$6 + 10$		$9 - 4$	
16	TRUE	5	TRUE

Since $(3, 2)$ checks, it is the solution.

27. $p = 32 + q,$ (1)

$\quad 3p = 8q + 6$ (2)

First we write each equation in the form $Ap + Bq = C$.

$\quad p - q = 32,$ (1a)   Subtracting q

$\quad 3p - 8q = 6$ (2a)   Subtracting $8q$

Now we multiply both sides of Equation (1a) by -3 and then add.

$\quad -3p + \ 3q = -96$ Multiplying by -3

$\quad \underline{3p - \ 8q = 6}$ Equation (2a)

$\qquad\quad -5q = -90$ Adding

$\qquad\qquad\ \ q = 18$

Substitute 18 for q in Equation (1) and solve for p.

$\quad p = 32 + q$

$\quad p = 32 + 18$ Substituting

$\quad p = 50$

Check:

$p - q = 32$		$3p - 8q = 6$	
$50 - 18 \ ? \ 32$		$3 \cdot 50 - 8 \cdot 18 \ ? \ 6$	
32	TRUE	$150 - 144$	
		6	TRUE

Since $(50, 18)$ checks, it is the solution.

29. $3x - 2y = 10,$ (1)

$\quad -6x + 4y = -20$ (2)

We multiply by 2 on both sides of Equation (1) and add.

$\quad 6x - 4y = 20$

$\quad \underline{-6x + 4y = -20}$

$\qquad\quad 0 = 0$

We get an obviously true equation, so the system has an infinite number of solutions.

31. $0.06x + 0.05y = 0.07,$

$\quad 0.04x - 0.03y = 0.11$

We first multiply each equation by 100 to clear the decimals.

$\quad 6x + 5y = 7,$ (1)

$\quad 4x - 3y = 11$ (2)

We use the multiplication principle with both equations of the resulting system.

$18x + 15y = 21$ Multiplying (1) by 3

$\underline{20x - 15y = 55}$ Multiplying (2) by 5

$38x \qquad\ \ = 76$ Adding

$\quad\ x = 2$

Substitute 2 for x in Equation (1) and solve for y.

$\quad 6x + 5y = 7$

$\quad 6 \cdot 2 + 5y = 7$

$\quad 12 + 5y = 7$

$\qquad\ \ 5y = -5$

$\qquad\quad y = -1$

Check:

$0.06x + 0.05y = 0.07$	
$0.06(2) + 0.05(-1) \ ? \ 0.07$	
$0.12 - 0.05$	
0.07	TRUE

$0.04x - 0.03y = 0.11$	
$0.04(2) - 0.03(-1) \ ? \ 0.11$	
$0.08 + 0.03$	
0.11	TRUE

Since $(2, -1)$ checks, it is the solution.

33. $\dfrac{1}{3}x + \dfrac{3}{2}y = \dfrac{5}{4},$

$\quad \dfrac{3}{4}x - \dfrac{5}{6}y = \dfrac{3}{8}$

First we clear the fractions. We multiply on both sides of the first equation by 12 and on both sides of the second equation by 24.

$$12\left(\frac{1}{3}x + \frac{3}{2}y\right) = 12 \cdot \frac{5}{4}$$

$$12 \cdot \frac{1}{3}x + 12 \cdot \frac{3}{2}y = 15$$

$$4x + 18y = 15$$

$$24\left(\frac{3}{4}x - \frac{5}{6}y\right) = 24 \cdot \frac{3}{8}$$

$$24 \cdot \frac{3}{4}x - 24 \cdot \frac{5}{6}y = 9$$

$$18x - 20y = 9$$

The resulting system is

$\quad 4x + 18y = 15,$ (1)

$\quad 18x - 20y = 9.$ (2)

We use the multiplication principle with both equations.

$72x + 324y = 270$ Multiplying (1) by 18

$\underline{-72x + \ 80y = -36}$ Multiplying (2) by -4

$404y = 234$

$\qquad y = \dfrac{234}{404}, \text{ or } \dfrac{117}{202}$

Substitute $\frac{117}{202}$ for y in Equation (1) and solve for x.

$$4x + 18\left(\frac{117}{202}\right) = 15$$

$$4x + \frac{1053}{101} = 15$$

$$4x = \frac{462}{101}$$

$$x = \frac{1}{4} \cdot \frac{462}{101}$$

$$x = \frac{231}{202}$$

The ordered pair $\left(\frac{231}{202}, \frac{117}{202}\right)$ checks in both equations. It is the solution.

35. $\quad -4.5x + 7.5y = 6,$

$\quad\quad -x + 1.5y = 5$

First we clear the decimals by multiplying by 10 on both sides of each equation.

$$10(-4.5x + 7.5y) = 10 \cdot 6$$

$$-45x + 75y = 60$$

$$10(-x + 1.5y) = 10 \cdot 5$$

$$-10x + 15y = 50$$

The resulting system is

$\quad -45x + 75y = 60,\quad$ (1)

$\quad -10x + 15y = 50.\quad$ (2)

We multiply both sides of Equation (2) by -5 and then add.

$$\begin{array}{ll} -45x + 75y = 60 & \text{Equation (1)} \\ \underline{50x - 75y = -250} & \text{Multiplying by } -5 \\ 5x \qquad\quad = -190 & \text{Adding} \\ \quad x = -38 \end{array}$$

Substitute -38 for x in Equation (2) and solve for y.

$$-10x + 15y = 50$$

$$-10(-38) + 15y = 50$$

$$380 + 15y = 50$$

$$15y = -330$$

$$y = -22$$

The ordered pair $(-38, -22)$ checks in both equations. It is the solution.

37. Discussion and Writing Exercise

39. Parallel lines have the same <u>slope</u> and different <u>y-intercepts</u>.

41. A <u>solution</u> of a system of two equations is an ordered pair that makes both equations true.

43. The graph of $y = b$ is a <u>horizontal</u> line.

45. The equation $y = mx + b$ is called the <u>slope-intercept</u> equation.

47.-55. Left to the student

57.-65. Left to the student

67. $\quad 3(x - y) = 9,$

$\quad\quad x + y = 7$

First we remove parentheses in the first equation.

$$3x - 3y = 9, \quad (1)$$

$$x + y = 7 \quad (2)$$

Then we multiply Equation (2) by 3 and add.

$$\begin{array}{l} 3x - 3y = 9 \\ \underline{3x + 3y = 21} \\ 6x \qquad\quad = 30 \\ \quad x = 5 \end{array}$$

Now we substitute 5 for x in Equation (2) and solve for y.

$$x + y = 7$$

$$5 + y = 7$$

$$y = 2$$

The ordered pair $(5, 2)$ checks and is the solution.

69. $\quad 2(5a - 5b) = 10,$

$\quad\quad -5(6a + 2b) = 10$

First we remove parentheses.

$$10a - 10b = 10, \quad (1)$$

$$-30a - 10b = 10 \quad (2)$$

Then we multiply Equation (2) by -1 and add.

$$\begin{array}{l} 10a - 10b = 10 \\ \underline{30a + 10b = -10} \\ 40a \qquad\quad = 0 \\ \quad a = 0 \end{array}$$

Substitute 0 for a in Equation (1) and solve for b.

$$10 \cdot 0 - 10b = 10$$

$$-10b = 10$$

$$b = -1$$

The ordered pair $(0, -1)$ checks and is the solution.

71. $\quad y = -\dfrac{2}{7}x + 3, \quad (1)$

$\quad\quad y = \dfrac{4}{5}x + 3 \quad (2)$

Observe that these equations represent lines with different slopes and the same y-intercept. Thus, their point of intersection is the y-intercept, $(0, 3)$ and this is the solution of the system of equations.

We could also solve this system of equations algebraically. First substitute $\dfrac{4}{5}x + 3$ for y in Equation (1) and solve for x.

$$\frac{4}{5}x + 3 = -\frac{2}{7}x + 3$$

$$35\left(\frac{4}{5}x + 3\right) = 35\left(-\frac{2}{7}x + 3\right) \quad \text{Clearing fractions}$$

$$35 \cdot \frac{4}{5}x + 35 \cdot 3 = 35\left(-\frac{2}{7}x\right) + 35 \cdot 3$$

$$28x + 105 = -10x + 105$$

$$28x = -10x$$

$$38x = 0$$

$$x = 0$$

Now substitute 0 for x in one of the original equations and find y. We will use Equation (1).

$$y = -\frac{2}{7}x + 3 = -\frac{2}{7} \cdot 0 + 3 = 0 + 3 = 3$$

The ordered pair $(0, 3)$ checks and is the solution.

73. $y = ax + b$, (1)
   $y = x + c$ (2)

Substitute $x + c$ for y in Equation (1) and solve for x.

$$y = ax + b$$

$$x + c = ax + b \quad \text{Substituting}$$

$$x - ax = b - c$$

$$(1 - a)x = b - c$$

$$x = \frac{b - c}{1 - a}$$

Substitute $\frac{b-c}{1-a}$ for x in Equation (2) and simplify to find y.

$$y = x + c$$

$$y = \frac{b - c}{1 - a} + c$$

$$y = \frac{b - c}{1 - a} + c \cdot \frac{1 - a}{1 - a}$$

$$y = \frac{b - c + c - ac}{1 - a}$$

$$y = \frac{b - ac}{1 - a}$$

The ordered pair $\left(\frac{b-c}{1-a}, \frac{b-ac}{1-a}\right)$ checks and is the solution. This ordered pair could also be expressed as $\left(\frac{c-b}{a-1}, \frac{ac-b}{a-1}\right)$.

Exercise Set 7.4

1. Familiarize. Let $x =$ the number of two-point baskets made and $y =$ the number of three-point baskets made. Then $2x$ points were scored from two-point baskets and $3y$ points were scored from three-point baskets.

Translate. Since a total of 39 baskets were made we have one equation: $x + y = 39$. Since a total of 85 points was scored, we have a second equation: $2x + 3y = 85$.

The resulting system is

$$x + y = 39, \quad (1)$$
$$2x + 3y = 85. \quad (2)$$

Solve. We use the elimination method. First we multiply both sides of Equation (1) by -2 and add.

$$-2x - 2y = -78$$
$$\underline{2x + 3y = 85}$$
$$y = 7$$

Now we substitute 7 for y in Equation (1) and solve for x.

$$x + y = 39$$
$$x + 7 = 39$$
$$x = 32$$

Check. If 32 two-point baskets and 7 three-point baskets are made, then a total of $32 + 7$, or 39, baskets are made. The points scored are $2 \cdot 32 + 3 \cdot 7$, or $64 + 21$, or 85. The answer checks.

State. The Spurs made 32 two-point shots and 7 three-point shots.

3. Familiarize. Let $x =$ the number of 24-exposure rolls that were shot and $y =$ the number of 36-exposure rolls that were processed. It costs $9x$ dollars to process the 24-exposure rolls and $12.60y$ dollars to process the 36-exposure rolls.

Translate. Since 17 rolls of film were processed, we have one equation: $x + y = 17$.

The total cost of processing the film was $171, so we have a second equation: $9x + 12.6y = 171$. The resulting system is

$$x + y = 17, \quad (1)$$
$$9x + 12.6y = 171. \quad (2)$$

Solve. We use the elimination method. First we multiply both sides of Equation (1) by -9 and add.

$$-9x - 9y = -153$$
$$\underline{9x + 12.6y = 171}$$
$$3.6y = 18$$
$$y = 5$$

Now we substitute 5 for y in Equation (1) and solve for x.

$$x + y = 17$$
$$x + 5 = 17$$
$$x = 12$$

Check. If 12 rolls of 24-exposure film and 5 rolls of 36-exposure film are processed, then $12 + 5$, or 17 rolls are shot. The processing cost is $\$9(12) + \$12.60(5)$, or $\$108 + \$63$, or $171. The answer checks.

State. 12 rolls of 24-exposure film and 5 rolls of 36-exposure film were processed.

5. Familiarize. We let $x =$ the number of pounds of hay and $y =$ the number of pounds of grain that should be fed to the horse each day. We arrange the information in a table.

Type of feed	Hay	Grain	Mixture
Amount of feed	x	y	15
Percent of protein	6%	12%	8%
Amount of protein in mixture	6%x	12%y	8% × 15, or 1.2 lb

Translate. The first and last rows of the table give us two equations. The total amount of feed is 15 lb, so we have

$x + y = 15$.

The amount of protein in the mixture is to be 8% of 15 lb, or 1.2 lb. The amounts of protein from the two feeds are 6%x and 12%y. Thus

$6\%x + 12\%y = 1.2$, or

$0.06x + 0.12y = 1.2$, or

$6x + 12y = 120$ Clearing decimals

The resulting system is

$x + y = 15$,   (1)

$6x + 12y = 120$.   (2)

Solve. We use the elimination method. Multiply on both sides of Equation (1) by -6 and then add.

$-6x - 6y = -90$

$\underline{6x + 12y = 120}$

$6y = 30$

$y = 5$

We go back to Equation (1) and substitute 5 for y.

$x + y = 15$

$x + 5 = 15$

$x = 10$

Check. The sum of 10 and 5 is 15. Also, 6% of 10 is 0.6 and 12% of 5 is 0.6, and $0.6 + 0.6 = 1.2$. These numbers check.

State. Brianna should feed her horse 10 lb of hay and 5 lb of grain each day.

7. **Familiarize.** Let $x =$ the number of $50 bonds and $y =$ the number of $100 bonds. Then the total value of the $50 bonds is $50x$ and the total value of the $100 bonds is $100y$.

Translate.

Total value of bonds is $1250.

$\downarrow$

$50x + 100y \quad = \quad 1250$

Number of $50 bonds is 7 more than number of $100 bonds.

$\downarrow \qquad = 7 \quad + \qquad \downarrow$

$x \qquad = 7 \quad + \qquad y$

The resulting system is

$50x + 100y = 1250$,   (1)

$x = 7 + y$.   (2)

Solve. We use the substitution method, substituting $7 + y$ for x in Equation (1).

$50x + 100y = 1250$ (1)

$50(7 + y) + 100y = 1250$

$350 + 50y + 100y = 1250$

$350 + 150y = 1250$

$150y = 900$

$y = 6$

Now we substitute 6 for y in Equation (2) to find x.

$x = 7 + y$ (2)

$x = 7 + 6 = 13$

Check. If there are 13 $50 bonds and 6 $100 bonds, there are 7 more $50 bonds than $100 bonds. The total value of the bonds is $50 \cdot 13 + \$100 \cdot 6$, or $650 + \$600$, or $1250. The answer checks.

State. Cassandra has 13 $50 bonds and 6 $100 bonds.

9. **Familiarize.** Let $x =$ the number of cardholders tickets that were sold and $y =$ the number of non-cardholders tickets. We arrange the information in a table.

	Card-holders	Non-card-holders	Total
Price	$2.25	$3	
Number sold	x	y	203
Money taken in	2.25x	3y	$513

Translate. The last two rows of the table give us two equations. The total number of tickets sold was 203, so we have

$x + y = 203$.

The total amount of money collected was $513, so we have

$2.25x + 3y = 513$.

We can multiply the second equation on both sides by 100 to clear decimals. The resulting system is

$x + y = 203$,   (1)

$225x + 300y = 51,300$.   (2)

Solve. We use the elimination method. We multiply on both sides of Equation (1) by -225 and then add.

$-225x - 225y = -46,675$ Multiplying by -225

$\underline{225x + 300y = 51,300}$

$75y = 5625$

$y = 75$

We go back to Equation (1) and substitute 75 for y.

$x + y = 203$

$x + 75 = 203$

$x = 128$

Check. The number of tickets sold was $128 + 75$, or 203. The money collected was $2.25(128) + \$3(75)$, or $288 + \$225$, or $513. These numbers check.

State. 128 cardholders tickets and 75 non-cardholders tickets were sold.

11. *Familiarize*. Let a = the number of adults and c = the number of children who visited the exhibit. We organize the information in a table.

	Adults	Children	Total
Price	$7	$6	
Number bought	a	c	1630
Receipts	$7a$	$6c$	$11,080

Translate. The last two rows of the table give us two equations. The total number of admissions was 1630, so we have

$$a + c = 1630.$$

The total receipts were $11,080, so we have

$$7a + 6c = 11,080.$$

The resulting system is

$$a + c = 1630, \qquad (1)$$
$$7a + 6c = 11,080. \qquad (2)$$

Solve. We use the elimination method. We multiply Equation (1) by -7 and add.

$$-7a - 7c = -11,410, \qquad \text{Multiplying by } -7$$
$$\underline{7a + 6c = 11,080}$$
$$-c = -330$$
$$c = 330$$

We go back to Equation (1) and substitute 330 for c.

$$a + c = 1630$$
$$a + 330 = 1630$$
$$a = 1300$$

Check. The total admissions were $1300 + 330$, or 1630. The total receipts were $\$7 \cdot 1300 + \$6 \cdot 330$, or $\$9100 + \$1980$, or $11,080. The answer checks.

State. 1300 adults, and 330 children visited the exhibit.

13. *Familiarize*. We complete the table in the text. Note that x represents the number of liters of solution A to be used and y represents the number of liters of solution B.

Type of solution	A	B	Mixture
Amount of solution	x	y	100 L
Percent of acid	50%	80%	68%
Amount of acid in solution	50%x	80%y	68% × 100, or 68 L

Equation from first row: $x + y = 100$

Equation from second row: $50\%x + 80\%y = 68$

Translate. The first and third rows of the table give us two equations. Since the total amount of solution is 100 liters, we have

$$x + y = 100.$$

The amount of acid in the mixture is to be 68% of 100, or 68 liters. The amounts of acid from the two solutions are $50\%x$ and $80\%y$. Thus

$$50\%x + 80\%y = 68,$$
$$\text{or} \qquad 0.5x + 0.8y = 68,$$
$$\text{or} \qquad 5x + 8y = 680 \quad \text{Clearing decimals}$$

The resulting system is

$$x + y = 100, \quad (1)$$
$$5x + 8y = 680. \quad (2)$$

Solve. We use the elimination method. We multiply on both sides of Equation (1) by -5 and then add.

$$-5x - 5y = -500 \qquad \text{Multiplying by } -5$$
$$\underline{5x + 8y = 680}$$
$$3y = 180$$
$$y = 60$$

We go back to Equation (1) and substitute 60 for y.

$$x + y = 100$$
$$x + 60 = 100$$
$$x = 40$$

Check. We consider $x = 40$ and $y = 60$. The sum is 100. Now 50% of 40 is 20 and 80% of 60 is 48. These add up to 68. The numbers check.

State. 40 liters of solution A and 60 liters of solution B should be used.

15. *Familiarize*. Let d represent the number of dimes and q the number of quarters. Then, $10d$ represents the value of the dimes in cents, and $25q$ represents the value of the quarters in cents. The total value is $15.25, or 1525¢. The total number of coins is 103.

Translate.

Number of dimes	plus	number of quarters	is	103.
d	$+$	q	$=$	103

Value of dimes	plus	value of quarters	is	$15.25.
$10d$	$+$	$25q$	$=$	1525

The resulting system is

$$d + q = 103, \qquad (1)$$
$$10d + 25q = 1525. \qquad (2)$$

Solve. We use the addition method. We multiply Equation (1) by -10 and then add.

$$-10d - 10q = -1030 \quad \text{Multiplying by } -10$$
$$\underline{10d + 25q = 1525}$$
$$15q = 495 \quad \text{Adding}$$
$$q = 33$$

Now we substitute 33 for q in one of the original equations and solve for d.

$$d + q = 103 \quad (1)$$
$$d + 33 = 103 \quad \text{Substituting}$$
$$d = 70$$

Check. The number of dimes plus the number of quarters is $70 + 33$, or 103. The total value in cents is $10 \cdot 70 + 25 \cdot 33$, or $700 + 825$, or 1525. This is equal to $15.25. This checks.

State. There are 70 dimes and 33 quarters.

17. Familiarize. We complete the table in the text. Note that x represents the number of pounds of Brazilian coffee to be used and y represents the number of pounds of Turkish coffee.

Type of coffee	Brazilian	Turkish	Mixture
Cost of coffee	$19	$22	$20
Amount (in pounds)	x	y	300
Mixture	$19x$	$22y$	$20(300)$, or $6000

Equation from second row: $x + y = 300$

Equation from third row: $19x + 22y = 6000$

Translate. The second and third rows of the table give us two equations. Since the total amount of the mixture is 300 lb, we have

$$x + y = 300.$$

The value of the Brazilian coffee is $19x$ (x lb at $19 per pound), the value of the Turkish coffee is $22y$ (y lb at $22 per pound), and the value of the mixture is $20(300)$ or $6000. Thus we have

$$19x + 22y = 6000.$$

The resulting system is

$$x + y = 300, \quad (1)$$
$$19x + 22y = 6000. \quad (2)$$

Solve. We use the elimination method. We multiply on both sides of Equation (1) by -19 and then add.

$$-19x - 19y = -5700 \quad \text{Multiplying by } -19$$
$$\underline{19x + 22y = 6000}$$
$$3y = 300$$
$$y = 100$$

We go back to Equation (1) and substitute 100 for y.

$$x + y = 300$$
$$x + 100 = 300$$
$$x = 200$$

Check. The sum of 200 and 100 is 300. The value of the mixture is $19(200) + $22(100), or $3800 + $2200, or $6000. These values check.

State. 200 lb of Brazilian coffee and 100 lb of Turkish coffee should be used.

19. Familiarize. Let x and y represent the number of liters of 28%-fungicide solution and 40%-fungicide solution to be used in the mixture, respectively.

Translate. We organize the given information in a table.

Type of solution	28%	40%	36%
Amount of solution	x	y	300
Percent fungicide	28%	40%	36%
Amount of fungicide in solution	$0.28x$	$0.4y$	$0.36(300)$, or 108

We get a system of equations from the first and third rows of the table.

$$x + \quad y = 300,$$
$$0.28x + 0.4y = 108$$

Clearing decimals we have

$$x + \quad y = 300, \quad (1)$$
$$28x + 40y = 10,800 \quad (2)$$

Solve. We use the elimination method. Multiply Equation (1) by -28 and add.

$$-28x - 28y = -8400$$
$$\underline{28x + 40y = 10,800}$$
$$12y = 2400$$
$$y = 200$$

Now substitute 200 for y in Equation (1) and solve for x.

$$x + y = 300$$
$$x + 200 = 300$$
$$x = 100$$

Check. The sum of 100 and 200 is 300. The amount of fungicide in the mixture is $0.28(100) + 0.4(200)$, or $28 + 80$, or 108 L. These numbers check.

State. 100 L of the 28%-fungicide solution and 200 L of the 40%-fungicide solution should be used in the mixture.

21. Familiarize. We let $x =$ the number of pages in large type and $y =$ the number of pages in small type. We arrange the information in a table.

Size of type	Large	Small	Mixture (Book)
Words per page	830	1050	
Number of pages	x	y	12
Number of words	$830x$	$1050y$	11,720

Translate. The last two rows of the table give us two equations. The total number of pages in the document is 12, so we have

$$x + y = 12.$$

The number of words on the pages with large type is $830x$ (x pages with 830 words per page), and the number of words on the pages with small type is $1050y$ (y pages with 1050 words per page). The total number of words is 11,720, so we have

$$830x + 1050y = 11,720.$$

The resulting system is

$$x + y = 12, \qquad (1)$$
$$830x + 1050y = 11,720. \quad (2)$$

Solve. We use the elimination method. We multiply on both sides of Equation (1) by -830 and then add.

$$-830x - 830y = -9,960 \quad \text{Multiplying by } -830$$
$$\underline{830x + 1050y = 11,720}$$
$$220y = 1760$$
$$y = 8$$

We go back to Equation (1) and substitute 8 for y.

$$x + y = 12$$
$$x + 8 = 12$$
$$x = 4$$

Check. The sum of 4 and 8 is 12. The number of words in large type is $830 \cdot 4$, or 3320, and the number of words in small type is $1050 \cdot 8$, or 8400. Then the total number of words is $3320 + 8400$, or 11,720. These numbers check.

State. There were 4 pages in large type and 8 pages in small type.

23. Familiarize. Let x = the number of pounds of the 70% mixture and y = the number of pounds of the 45% mixture to be used. We organize the information in the table.

Percent of cashews	70%	45%	
Amount	x	y	60
Mixture	70%x, or 0.7x	45%x, or 0.45x	60(60%) or 60(0.6) or 36

Translate. The last two rows of the table give us two equations. The total weight of the mixture is 60 lb, so we have

$$x + y = 60.$$

The amount of cashews in the mixture is 36 lb, so we have

$$0.7x + 0.45y = 36, \text{ or}$$
$$70x + 45y = 3600 \quad \text{Clearing decimals}$$

The resulting system is

$$x + y = 60, \qquad (1)$$
$$70x + 45y = 3600. \quad (2)$$

Solve. We use the elimination method. We multiply Equation (1) by -45 and then add.

$$-45x - 45y = -2700$$
$$\underline{70x + 45y = 3600}$$
$$25x \qquad = 900$$
$$x = 36$$

Next we substitute 36 for x in one of the original equations and solve for y.

$$x + y = 60 \quad (1)$$
$$36 + y = 60$$
$$y = 24$$

Check. The total weight of the mixture is 36 lb + 24 lb, or 60 lb. The amount of cashews in the mixture is 0.7(36 lb)+ 0.45(24 lb), or 36 lb. Since 36 lb is 60% of 60 lb, the answer checks.

State. The new mixture should contain 36 lb of the 70% cashew mixture and 24 lb of the 45% cashew mixture.

25. Familiarize. We arrange the information in a table. Let a = the number of type A questions and b = the number of type B questions.

Type of question	A	B	Mixture (Test)
Number	a	b	16
Time	3 min	6 min	
Value	10 points	15 points	
Mixture (Test)	3a min, 10a points	6b min, 15b points	60 min, 180 points

Translate. The table actually gives us three equations. Since the total number of questions is 16, we have

$$a + b = 16.$$

The total time is 60 min, so we have

$$3a + 6b = 60.$$

The total number of points is 180, so we have

$$10a + 15b = 180.$$

The resulting system is

$$a + b = 16, \qquad (1)$$
$$3a + 6b = 60, \qquad (2)$$
$$10a + 15b = 180. \quad (3)$$

Solve. We will solve the system composed of Equations (1) and (2) and then check to see that this solution also satisfies Equation (3). We multiply equation (1) by -3 and add.

$$-3a - 3b = -48$$
$$\underline{3a + 6b = 60}$$
$$3b = 12$$
$$b = 4$$

Now we substitute 4 for b in Equation (1) and solve for a.

$$a + b = 16$$
$$a + 4 = 16$$
$$a = 12$$

Check. We consider $a = 12$ questions and $b = 4$ questions. The total number of questions is 16. The time required is $3 \cdot 12 + 6 \cdot 4$, or $36 + 24$, or 60 min. The total points are $10 \cdot 12 + 15 \cdot 4$, or $120 + 60$, or 180. These values check.

State. 12 questions of type A and 4 questions of type B were answered. Assuming all the answers were correct, the score was 180 points.

27. **Familiarize**. Let $k =$ the age of the Kuyatt's house now and $m =$ the age of the Marconi's house now. Eight years ago the houses' ages were $k - 8$ and $m - 8$.

Translate. We reword and translate.

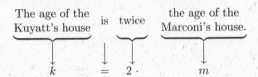

The resulting system is

$$k = 2m, \qquad (1)$$
$$k - 8 = 3(m - 8). \qquad (2)$$

Solve. We use the substitution method. We substitute $2m$ for k in Equation (2) and solve for m.

$$k - 8 = 3(m - 8)$$
$$2m - 8 = 3(m - 8)$$
$$2m - 8 = 3m - 24$$
$$-8 = m - 24$$
$$16 = m$$

We find k by substituting 16 for m in Equation (1).

$$k = 2m$$
$$k = 2 \cdot 16$$
$$k = 32$$

Check. The age of the Kuyatt's house, 32 years, is twice the age of the Marconi's house, 16 years. Eight years ago, when the Kuyatt's house was 24 years old and the Marconi's house was 8 years old, the Kuyatt's house was three times as old as the Marconi's house. These numbers check.

State. The Kuyatt's house is 32 years old, and the Marconi's house is 16 years old.

29. **Familiarize**. Let $R =$ Randy's age now and $M =$ Mandy's age now. In twelve years their ages will be $R + 12$ and $M + 12$.

Translate. We reword and translate.

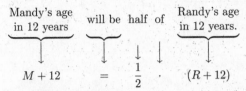

The resulting system is

$$R = 4M, \qquad (1)$$
$$M + 12 = \frac{1}{2}(R + 12). \qquad (2)$$

Solve. We use the substitution method. We substitute $4M$ for R in Equation (2) and solve for M.

$$M + 12 = \frac{1}{2}(R + 12)$$
$$M + 12 = \frac{1}{2}(4M + 12)$$
$$M + 12 = 2M + 6$$
$$12 = M + 6$$
$$6 = M$$

We find R by substituting 6 for M in Equation (1).

$$R = 4M$$
$$R = 4 \cdot 6$$
$$R = 24$$

Check. Randy's age now, 24, is 4 times 6, Mandy's age. In 12 yr, when Randy will be 36 and Mandy 18, Mandy's age will be half of Randy's age. These numbers check.

State. Randy is 24 years old now, and Mandy is 6.

31. **Familiarize**. Let $x =$ the smaller angle and $y =$ the larger angle.

Translate. We reword the problem.

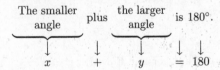

The resulting system is

$$x + y = 180,$$
$$y = 30 + 2x.$$

Solve. We solve the system. We will use the elimination method although we could also easily use the substitution method. First we get the second equation in the form $Ax + By = C$.

$$x + y = 180 \quad (1)$$
$$-2x + y = 30 \quad (2) \text{ Adding } -2x$$

Now we multiply Equation (2) by -1 and add.

$$\begin{array}{l} x + y = 180 \\ \underline{2x - y = -30} \\ 3x = 150 \\ x = 50 \end{array}$$

Then we substitute 50 for x in Equation (1) and solve for y.

$$x + y = 180 \quad \text{Equation (1)}$$
$$50 + y = 180 \quad \text{Substituting}$$
$$y = 130$$

Check. The sum of the angles is $50° + 130°$, or $180°$, so the angles are supplementary. Also, $30°$ more than two times the $50°$ angle is $30° + 2 \cdot 50°$, or $30° + 100°$, or $130°$, the other angle. These numbers check.

State. The angles are $50°$ and $130°$.

33. Familiarize. We let $x =$ the larger angle and $y =$ the smaller angle.

Translate. We reword and translate the first statement.

The sum of two angles is $90°$.
$$\underbrace{}$$
$$x + y = 90$$

We reword and translate the second statement.

The difference of two angles is $34°$.
$$\underbrace{}$$
$$x - y = 34$$

The resulting system is

$$x + y = 90,$$
$$x - y = 34.$$

Solve. We solve the system.

$$\begin{array}{l} x + y = 90, \quad (1) \\ \underline{x - y = 34} \quad (2) \\ 2x = 124 \quad \text{Adding} \\ x = 62 \end{array}$$

Now we substitute 62 for x in Equation (1) and solve for y.

$$x + y = 90 \quad \text{Equation (1)}$$
$$62 + y = 90 \quad \text{Substituting}$$
$$y = 28$$

Check. The sum of the angles is $62° + 28°$, or $90°$, so the angles are complementary. The difference of the angles is $62° - 28°$, or $34°$. These numbers check.

State. The angles are $62°$ and $28°$.

35. Familiarize. Let $x =$ the number of gallons of 87-octane gas and $y =$ the number of gallons of 93-octane gas that should be used. We arrange the information in a table.

Type of gas	87-octane	93-octane	Mixture
Amount of gas	x	y	18
Octane rating	87	93	89
Mixture	$87x$	$93y$	$18 \cdot 89$, or 1602

Translate. The first and last rows of the table give us a system of equations.

$$x + y = 18, \quad (1)$$
$$87x + 93y = 1602 \quad (2)$$

Solve. We multiply Equation (1) by -87 and then add.

$$\begin{array}{l} -87x - 87y = -1566 \\ \underline{87x + 93y = 1602} \\ 6y = 36 \\ y = 6 \end{array}$$

Then substitute 6 for y in Equation (1) and solve for x.

$$x + y = 18$$
$$x + 6 = 18$$
$$x = 12$$

Check. The total amount of gas is 12 gal $+$ 6 gal, or 18 gal. Also $87(12) + 93(6) = 1044 + 558 = 1602$. The answer checks.

State. 12 gal of 87-octane gas and 6 gal of 93-octane gas should be blended.

37. Familiarize. Let $x =$ the number of ounces of Dr. Zeke's cough syrup and $y =$ the number of ounces of Vitabrite cough syrup that should be used. We organize the information in a table.

	Dr. Zeke's	Vitabrite	Mixture
Percent of alcohol	2%	5%	3%
Amount	x	y	80
Mixture	2%x, or 0.02x	5%x, or 0.05y	3% $\cdot$ 80, or 0.03 $\cdot$ 80, or 2.4

Translate. The last two rows of the table give us a system of equations.

$$x + y = 80,$$
$$0.02x + 0.05y = 2.4.$$

Clearing decimals, we have

$$x + y = 80, \quad (1)$$
$$2x + 5y = 240. \quad (2)$$

Solve. We use the elimination method. First we multiply Equation (1) by -2 and then add.

$$\begin{array}{l} -2x - 2y = -160 \\ \underline{2x + 5y = 240} \\ 3y = 80 \\ y = 26\frac{2}{3} \end{array}$$

Substitute $26\frac{2}{3}$ for y in one of the original equations and solve for x.

$$x + y = 80$$
$$x + 26\frac{2}{3} = 80$$
$$x = 53\frac{1}{3}$$

Check. The number of ounces in the mixture is $53\frac{1}{3} + 26\frac{2}{3}$, or 80. The amount of alcohol in the mixture is $0.02\left(53\frac{1}{3}\right) + 0.05\left(26\frac{2}{3}\right)$, or 2.4 oz. Since 2.4 oz is 3% of 80 oz, the answer checks.

State. The mixture should contain $53\frac{1}{3}$ oz of Dr. Zeke's cough syrup and $26\frac{2}{3}$ oz of Vitabrite cough syrup.

39. Discussion and Writing Exercise

41. $25x^2 - 81 = (5x)^2 - 9^2$
$$= (5x + 9)(5x - 9)$$

43. $4x^2 + 100 = 4(x^2 + 25)$

45. $y = -2x - 3$

The equation is in the form $y = mx + b$, so the y-intercept is $(0, -3)$.

To find the x-intercept, we let $y = 0$ and solve for x.

$$0 = -2x - 3$$
$$2x = -3$$
$$x = -\frac{3}{2}$$

The x-intercept is $\left(-\frac{3}{2}, 0\right)$.

We plot the intercepts and draw the line.

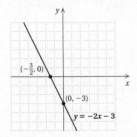

A third point should be used as a check.

For example, let $x = -3$. Then

$$y = -2(-3) - 3 = 6 - 3 = 3.$$

It appears that the point $(-3, 3)$ is on the graph, so the graph is probably correct.

47. $5x - 2y = -10$

To find the y-intercept, let $x = 0$ and solve for y.

$$5 \cdot 0 - 2y = -10$$
$$-2y = -10$$
$$y = 5$$

The y-intercept is $(0, 5)$.

To find the x-intercept, let $y = 0$ and solve for x.

$$5x - 2 \cdot 0 = -10$$
$$5x = -10$$
$$x = -2$$

The x-intercept is $(-2, 0)$.

We plot the intercepts and draw the line.

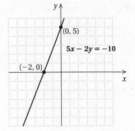

A third point should be used as a check. For example, let $x = -4$. then

$$5(-4) - 2y = -10$$
$$-20 - 2y = -10$$
$$-2y = 10$$
$$y = -5$$

It appears that the point $(-4, -5)$ is on the graph, so the graph is probably correct.

49. $\dfrac{x^2 - 5x + 6}{x^2 - 4} = \dfrac{(x - 3)(x - 2)}{(x + 2)(x - 2)}$
$$= \dfrac{(x - 3)(x - 2)}{(x + 2)(x - 2)}$$
$$= \dfrac{x - 3}{x + 2}$$

51. $\dfrac{x - 2}{x + 3} - \dfrac{2x - 5}{x - 4}$ LCD is $(x + 3)(x - 4)$

$$= \dfrac{x - 2}{x + 3} \cdot \dfrac{x - 4}{x - 4} - \dfrac{2x - 5}{x - 4} \cdot \dfrac{x + 3}{x + 3}$$
$$= \dfrac{(x - 2)(x - 4)}{(x + 3)(x - 4)} - \dfrac{(2x - 5)(x + 3)}{(x - 4)(x + 3)}$$
$$= \dfrac{x^2 - 6x + 8}{(x + 3)(x - 4)} - \dfrac{2x^2 + x - 15}{(x - 4)(x + 3)}$$
$$= \dfrac{x^2 - 6x + 8 - (2x^2 + x - 15)}{(x + 3)(x - 4)}$$
$$= \dfrac{x^2 - 6x + 8 - 2x^2 - x + 15}{(x + 3)(x - 4)}$$
$$= \dfrac{-x^2 - 7x + 23}{(x + 3)(x - 4)}$$

53. *Familiarize*. We arrange the information in a table. Let x = the number of liters of skim milk and y = the number of liters of 3.2% milk.

Type of milk	4.6%	Skim	3.2% (Mixture)
Amount of milk	100 L	x	y
Percent of butterfat	4.6%	0%	3.2%
Amount of butterfat in milk	4.6% × 100, or 4.6 L	0% · x, or 0 L	3.2%y

Translate. The first and third rows of the table give us two equations.

Amount of milk: $100 + x = y$

Amount of butterfat: $4.6 + 0 = 3.2\%y$, or $4.6 = 0.032y$.

The resulting system is

$$100 + x = y,$$
$$4.6 = 0.032y.$$

Solve. We solve the second equation for y.

$$4.6 = 0.032y$$
$$\frac{4.6}{0.032} = y$$
$$143.75 = y$$

We substitute 143.75 for y in the first equation and solve for x.

$$100 + x = y$$
$$100 + x = 143.75$$
$$x = 43.75$$

Check. We consider $x = 43.75$ L and $y = 143.75$ L. The difference between 143.75 L and 43.75 L is 100 L. There is no butterfat in the skim milk. There are 4.6 liters of butterfat in the 100 liters of the 4.6% milk. Thus there are 4.6 liters of butterfat in the mixture. This checks because 3.2% of 143.75 is 4.6.

State. 43.75 L of skim milk should be used.

55. *Familiarize*. In a table we arrange the information regarding the solution <u>after</u> some of the 30% solution is drained and replaced with pure antifreeze. We let x represent the amount of the original (30%) solution remaining, and we let y represent the amount of the 30% mixture that is drained and replaced with pure antifreeze.

Type of solution	Original (30%)	Pure antifreeze	Mixture
Amount of solution	x	y	16
Percent of antifreeze	30%	100%	50%
Amount of antifreeze in solution	0.3x	1 · y, or y	0.5(16), or 8

Translate. The table gives us two equations.

Amount of solution: $x + y = 16$

Amount of antifreeze in solution: $0.3x + y = 8$, or $3x + 10y = 80$

The resulting system is

$$x + y = 16, \quad (1)$$
$$3x + 10y = 80. \quad (2)$$

Solve. We multiply Equation (1) by -3 and then add.

$$-3x - 3y = -48$$
$$\underline{3x + 10y = 80}$$
$$7y = 32$$
$$y = \frac{32}{7}, \text{ or } 4\frac{4}{7}$$

Then we substitute $4\frac{4}{7}$ for y in Equation (1) and solve for x.

$$x + y = 16$$
$$x + 4\frac{4}{7} = 16$$
$$x = 11\frac{3}{7}$$

Check. When $x = 11\frac{3}{7}$ L and $y = 4\frac{4}{7}$ L, the total is 16 L. The amount of antifreeze in the mixture is $0.3\left(11\frac{3}{7}\right) + 4\frac{4}{7}$, or $\frac{3}{10} \cdot \frac{80}{7} + \frac{32}{7}$, or $\frac{24}{7} + \frac{32}{7} = \frac{56}{7}$, or 8 L. This is 50% of 16 L, so the numbers check.

State. $4\frac{4}{7}$ of the original mixture should be drained and replaced with pure antifreeze.

57. *Familiarize*. Let x = the tens digit and y = the units digit. Then the number is $10x + y$.

Translate. The number is six times the sum of its digit, so we have

$$10x + y = 6(x + y)$$
$$10x + y = 6x + 6y$$
$$4x - 5y = 0.$$

The tens digit is 1 more than the units digit so we have

$$x = y + 1.$$

The resulting system is

$$4x - 5y = 0, \quad (1)$$
$$x = y + 1. \quad (2)$$

Solve. First substitute $y + 1$ for x in Equation (1) and solve for y.

$$4x - 5y = 0 \quad (1)$$
$$4(y + 1) - 5y = 0$$
$$4y + 4 - 5y = 0$$
$$-y + 4 = 0$$
$$4 = y$$

Now substitute 4 for y in one of the original equations and solve for x.

$$x = y + 1 \quad (2)$$
$$x = 4 + 1$$
$$x = 5$$

Check. If the number is 54, then the sum of the digits is $5 + 4$, or 9, and $54 = 6 \cdot 9$. Also, the tens digit, 5, is one more than the units digit, 4. The answer checks.

State. The number is 54.

Exercise Set 7.5

1. Familiarize. We first make a drawing.

$$\begin{array}{c} \text{30 mph} \\ \rule{4cm}{0.4pt}\!\!\!\rightarrow \end{array}$$
Slow car t hours d miles

$$\begin{array}{c} \text{46 mph} \\ \rule{5cm}{0.4pt}\!\!\!\rightarrow \end{array}$$
Fast car t hours $d + 72$ miles

We let $d =$ the distance the slow car travels. Then $d + 72 =$ the distance the fast car travels. We call the time t. We complete the table in the text, filling in the distances as well as the other information.

$$d = r \cdot t$$

	Distance	Speed	Time
Slow car	d	30	t
Fast car	$d + 72$	46	t

Translate. We get an equation $d = rt$ from each row of the table. Thus we have

$$d = 30t, \qquad (1)$$
$$d + 72 = 46t. \quad (2)$$

Solve. We use the substitution method. We substitute $30t$ for d in Equation (2).

$$d + 72 = 46t$$
$$30t + 72 = 46t \quad \text{Substituting}$$
$$72 = 16t \quad \text{Subtracting } 30t$$
$$4.5 = t \quad \text{Dividing by 16}$$

Check. In 4.5 hr the slow car travels $30(4.5)$, or 135 mi, and the fast car travels $46(4.5)$, or 207 mi. Since 207 is 72 more than 135, our result checks.

State. The trains will be 72 mi apart in 4.5 hr.

3. Familiarize. First make a drawing.

Station   72 mph
$$\rule{4cm}{0.4pt}\!\!\!\rightarrow$$
Slow train $t + 3$ hours d miles

Station   120 mph
$$\rule{4cm}{0.4pt}\!\!\!\rightarrow$$
Fast train t hours d miles

Trains meet here.

From the drawing we see that the distances are the same. Let's call the distance d. Let t represent the time for the faster train and $t+3$ represent the time for the slower train. We complete the table in the text.

$$d = r \cdot t$$

	Distance	Speed	Time
Slow train	d	72	$t+3$
Fast train	d	120	t

Equation from first row: $d = 72(t + 3)$

Equation from second row: $d = 120t$

Translate. Using $d = rt$ in each row of the table, we get the following system of equations:

$$d = 72(t + 3), \quad (1)$$
$$d = 120t. \qquad (2)$$

Solve. Substitute $120t$ for d in Equation (1) and solve for t.

$$d = 72(t + 3)$$
$$120t = 72(t + 3) \quad \text{Substituting}$$
$$120t = 72t + 216$$
$$48t = 216$$
$$t = \frac{216}{48}$$
$$t = 4.5$$

Check. When $t = 4.5$ hours, the faster train will travel $120(4.5)$, or 540 mi, and the slower train will travel $72(7.5)$, or 540 mi. In both cases we get the distance 540 mi.

State. In 4.5 hours after the second train leaves, the second train will overtake the first train. We can also state the answer as 7.5 hours after the first train leaves.

5. Familiarize. We first make a drawing.

With the current $r + 6$
$$\rule{4cm}{0.4pt}\!\!\!\rightarrow$$
4 hours d kilometers

Against the current $r - 6$
$$\rule{5cm}{0.4pt}\!\!\!\rightarrow$$
10 hours d kilometers

From the drawing we see that the distances are the same. Let d represent the distance. Let r represent the speed of the canoe in still water. Then, when the canoe is traveling with the current, its speed is $r + 6$. When it is traveling against the current, its speed is $r - 6$. We complete the table in the text.

$$d = r \cdot t$$

	Distance	Speed	Time
With current	d	$r + 6$	4
Against current	d	$r - 6$	10

Equation from first row: $d = (r + 6)4$

Equation from second row: $d = (r - 6)10$

Translate. Using $d = rt$ in each row of the table, we get the following system of equations:

$$d = (r + 6)4, \quad (1)$$
$$d = (r - 6)10 \quad (2)$$

Solve. Substitute $(r + 6)4$ for d in Equation (2) and solve for r.

$$d = (r - 6)10$$
$$(r + 6)4 = (r - 6)10 \quad \text{Substituting}$$
$$4r + 24 = 10r - 60$$
$$84 = 6r$$
$$14 = r$$

Check. When $r = 14$, $r + 6 = 20$ and $20 \cdot 4 = 80$, the distance. When $r = 14$, $r - 6 = 8$ and $8 \cdot 10 = 80$. In both cases, we get the same distance.

State. The speed of the canoe in still water is 14 km/h.

7. Familiarize. First make a drawing.

Passenger   96 km/h

$t - 2$ hours d kilometers

Freight   64 km/h

t hours d kilometers

Central City   Clear Creek

From the drawing we see that the distances are the same. Let d represent the distance. Let t represent the time for the freight train. Then the time for the passenger train is $t - 2$. We organize the information in a table.

$$d = r \cdot t$$

	Distance	Speed	Time
Passenger	d	96	$t - 2$
Freight	d	64	t

Translate. From each row of the table we get an equation.

$$d = 96(t - 2), \quad (1)$$
$$d = 64t \quad (2)$$

Solve. Substitute $64t$ for d in Equation (1) and solve for t.

$$d = 96(t - 2)$$
$$64t = 96(t - 2) \quad \text{Substituting}$$
$$64t = 96t - 192$$
$$192 = 32t$$
$$6 = t$$

Next we substitute 6 for t in one of the original equations and solve for d.

$$d = 64t \quad \text{Equation (2)}$$
$$d = 64 \cdot 6 \quad \text{Substituting}$$
$$d = 384$$

Check. If the time is 6 hr, then the distance the passenger train travels is $96(6 - 2)$, or 384 km. The freight train travels $64(6)$, or 384 km. The distances are the same.

State. It is 384 km from Central City to Clear Creek.

9. Familiarize. We first make a drawing.

Downstream $r + 6$

3 hours d miles

Upstream $r - 6$

5 hours d miles

We let r represent the speed of the boat in still water and d represent the distance Antoine traveled downstream before he turned back. We organize the information in a table.

$$d = r \cdot t$$

	Distance	Speed	Time
Downstream	d	$r + 6$	3
Upstream	d	$r - 6$	5

Translate. Using $d = rt$ in each row of the table, we get the following system of equations:

$$d = (r + 6)3, \quad (1)$$
$$d = (r - 6)5 \quad (2)$$

Solve. Substitute $(r + 6)3$ for d in Equation (2) and solve for r.

$$d = (r - 6)5$$
$$(r + 6)3 = (r - 6)5 \quad \text{Substituting}$$
$$3r + 18 = 5r - 30$$
$$48 = 2r$$
$$24 = r$$

If $r = 24$, then $d = (r + 6)3 = (24 + 6)3 = 30 \cdot 3 = 90$.

Check. If $r = 24$, then $r + 6 = 24 + 6 = 30$ and $r - 6 = 24 - 6 = 18$. If Antoine travels for 3 hr at 30 mph, then he travels $3 \cdot 30$, or 90 mi, downstream. If he travels for 5 hr at 18 mph, then he also travels $5 \cdot 18$, or 90 mi, upstream. Since the distances are the same, the answer checks.

State. (a) Antoine must travel at a speed of 24 mph.

(b) Antoine traveled 90 mi downstream before he turned back.

11. Familiarize. We first make a drawing.

230 ft/min

Toddler $t + 1$ min d ft

660 ft/min

Mother t min d ft

They meet here.

From the drawing we see that the distances are the same. Let's call the distance d. Let $t =$ the time the mother runs.

Then $t + 1 =$ the time the toddler runs. We arrange the information in a table.

d	$=$	r	$\cdot$	t

	Distance	Speed	Time
Toddler	d	230	$t + 1$
Mother	d	660	t

Translate. Using $d = rt$ in each row of the table we get two equations.

$$d = 230(t + 1), \quad (1)$$
$$d = 660t \quad (2)$$

Solve. Substitute $660t$ for d in Equation (1) and solve for t.

$$d = 230(t + 1)$$
$$660t = 230(t + 1) \quad \text{Substituting}$$
$$660t = 230t + 230$$
$$430t = 230$$
$$t = \frac{230}{430}, \text{ or } \frac{23}{43}$$

Check. When $t = \frac{23}{43}$ the toddler will travel $230\left(1\frac{23}{43}\right)$, or $230 \cdot \frac{66}{43}$, or $\frac{15,180}{43}$ ft and the mother will travel $660 \cdot \frac{23}{43}$, or $\frac{15,180}{43}$ ft. Since the distances are the same, our result checks.

State. The mother will overtake the toddler $\frac{23}{43}$ min after she starts running. We can also state the answer as $1\frac{23}{43}$ min after the toddler starts running.

13. *Familiarize*. First make a drawing.

Home t hr 45 mph $|$ $(2 - t)$ hr 6 mph Work
Motorcycle distance $|$ Walking distance

$\longleftarrow$———— 25 miles ————$\longrightarrow$

Let t represent the time the motorcycle was driven. Then $2 - t$ represents the time the rider walked. We organize the information in a table.

d	$=$	r	$\cdot$	t

	Distance	Speed	Time
Motorcycling	Motorcycle distance	45	t
Walking	Walking distance	6	$2 - t$
Total		25	

Translate. From the drawing we see that

Motorcycle distance + Walking distance $= 25$

Then using $d = rt$ in each row of the table we get

$$45t + 6(2 - t) = 25$$

Solve. We solve this equation for t.

$$45t + 12 - 6t = 25$$
$$39t + 12 = 25$$
$$39t = 13$$
$$t = \frac{13}{39}$$
$$t = \frac{1}{3}$$

Check. The problem asks us to find how far the motorcycle went before it broke down. If $t = \frac{1}{3}$, then $45t$ (the distance the motorcycle traveled) $= 45 \cdot \frac{1}{3}$, or 15 and $6(2 - t)$ (the distance walked) $= 6\left(2 - \frac{1}{3}\right) = 6 \cdot \frac{5}{3}$, or 10. The total of these distances is 25, so $\frac{1}{3}$ checks.

State. The motorcycle went 15 miles before it broke down.

15. Discussion and Writing Exercise

17. $\dfrac{8x^2}{24x} = \dfrac{8}{24} \cdot \dfrac{x^2}{x} = \dfrac{1}{3} \cdot x^{2-1} = \dfrac{x}{3}$

19. $\dfrac{5a + 15}{10} = \dfrac{5(a + 3)}{5 \cdot 2}$

$= \dfrac{\cancel{5}(a + 3)}{\cancel{5} \cdot 2}$

$= \dfrac{a + 3}{2}$

21. $\dfrac{2x^2 - 50}{x^2 - 25} = \dfrac{2(x^2 - 25)}{x^2 - 25} = \dfrac{2}{1} \cdot \dfrac{x^2 - 25}{x^2 - 25} = 2$

23. $\dfrac{x^2 - 3x - 10}{x^2 - 2x - 15} = \dfrac{(x - 5)(x + 2)}{(x - 5)(x + 3)}$

$= \dfrac{\cancel{(x - 5)}(x + 2)}{\cancel{(x - 5)}(x + 3)}$

$= \dfrac{x + 2}{x + 3}$

25. $\dfrac{(x^2 + 6x + 9)(x - 2)}{(x^2 - 4)(x + 3)} = \dfrac{(x + 3)(x + 3)(x - 2)}{(x + 2)(x - 2)(x + 3)}$

$= \dfrac{(x+3)(x + 3)\cancel{(x-2)}}{(x + 2)\cancel{(x-2)}(x+3)}$

$= \dfrac{x + 3}{x + 2}$

27. $\dfrac{6x^2 + 18x + 12}{6x^2 - 6} = \dfrac{6(x^2 + 3x + 2)}{6(x^2 - 1)}$

$= \dfrac{6(x + 1)(x + 2)}{6(x + 1)(x - 1)}$

$= \dfrac{\cancel{6}\cancel{(x+1)}(x + 2)}{\cancel{6}\cancel{(x+1)}(x - 1)}$

$= \dfrac{x + 2}{x - 1}$

29. Familiarize. We arrange the information in a table. Let d = the length of the route and t = Lindbergh's time. Note that 16 hr and 57 min = $16\frac{57}{60}$ hr = 16.95 hr.

$$d = r \cdot t$$

	Distance	Speed	Time
Lindbergh	d	107.4	t
Hughes	d	217.1	$t - 16.95$

Translate. From the rows of the table we get two equations.

$$d = 107.4t, \qquad (1)$$
$$d = 217.1(t - 16.95) \quad (2)$$

Solve. We substitute $107.4t$ for d in Equation (2) and solve for t.

$$d = 217.1(t - 16.95)$$
$$107.4t = 217.1(t - 16.95)$$
$$107.4t = 217.1t - 3679.845$$
$$-109.7t = -3679.845$$
$$t \approx 33.54$$

Now we go back to Equation (1) and substitute 33.54 for t.

$$d = 107.4t$$
$$d = 107.4(33.54)$$
$$d \approx 3602$$

Check. When $t \approx 33.54$, Lindbergh traveled $107.4(33.54) \approx 3602$ mi, and Hughes traveled $217.1(16.59) \approx 3602$ mi. Since the distances are the same, our result checks.

State. The route was 3602 mi long. (Answers may vary slightly due to rounding differences.)

31. Familiarize. We arrange the information in a table. Let's call the distance d. When the riverboat is traveling upstream its speed is $12 - 4$, or 8 mph. Its speed traveling downstream is $12 + 4$, or 16 mph.

$$d = r \cdot t$$

	Distance	Speed	Time
Upstream	d	8	Time upstream
Downstream	d	16	Time downstream
Total			1

Translate. From the table we see that (Time upstream) + (Time downstream) = 1. Then using $d = rt$, in the form $\frac{d}{r} = t$, in each row of the table we get

$$\frac{d}{8} + \frac{d}{16} = 1.$$

Solve. We solve the equation. The LCM is 16.

$$\frac{d}{8} + \frac{d}{16} = 1$$
$$16\left(\frac{d}{8} + \frac{d}{16}\right) = 16 \cdot 1$$
$$16 \cdot \frac{d}{8} + 16 \cdot \frac{d}{16} = 16$$
$$2d + d = 16$$
$$3d = 16$$
$$d = \frac{16}{3}, \text{or } 5\frac{1}{3}$$

Check. When $d = \frac{16}{3}$,

(Time upstream) + (Time downstream)

$$= \frac{\frac{16}{3}}{8} + \frac{\frac{16}{3}}{16}$$
$$= \frac{16}{3} \cdot \frac{1}{8} + \frac{16}{3} \cdot \frac{1}{16}$$
$$= \frac{2}{3} + \frac{1}{3}$$
$$= 1 \text{ hr}$$

Thus the distance of $\frac{16}{3}$ mi, or $5\frac{1}{3}$ mi checks.

State. The pilot should travel $5\frac{1}{3}$ mi upstream before turning around.

Chapter 7 Review Exercises

1. We check by substituting alphabetically 6 for x and -1 for y.

$$\begin{array}{c|c} x - y = 3 \\ \hline 6 - (-1) \; ? \; 3 \\ 6 + 1 \\ 7 & \text{FALSE} \end{array}$$

Since $(6, -1)$ is not a solution of the first equation, it is not a solution of the system of equations.

2. We check by substituting alphabetically 2 for x and -3 for y.

$$\begin{array}{c|c} 2x + y = 1 \\ \hline 2 \cdot 2 + (-3) \; ? \; 1 \\ 4 - 3 \\ 1 & \text{TRUE} \end{array} \qquad \begin{array}{c|c} x - y = 5 \\ \hline 2 - (-3) \; ? \; 5 \\ 2 + 3 \\ 5 & \text{TRUE} \end{array}$$

The ordered pair $(2, -3)$ is a solution of both equations, so it is a solution of the system of equations.

3. We check by substituting alphabetically -2 for x and 1 for y.

$$\begin{array}{c|c} x + 3y = 1 \\ \hline -2 + 3 \cdot 1 \; ? \; 1 \\ -2 + 3 \\ 1 & \text{TRUE} \end{array} \qquad \begin{array}{c|c} 2x - y = -5 \\ \hline 2(-2) - 1 \; ? \; -5 \\ -4 - 1 \\ -5 & \text{TRUE} \end{array}$$

The ordered pair $(-2, 1)$ is a solution of both equations, so it is a solution of the system of equations.

4. We check by substituting alphabetically -4 for x and -1 for y.

$$\frac{x - y = 3}{\begin{array}{c|c} -4 - (-1) \ ? \ 3 & \\ -4 + 1 & \\ -3 & \text{FALSE} \end{array}}$$

Since $(-4, -1)$ is not a solution of the first equation, it is not a solution of the system of equations.

5. We graph the equations.

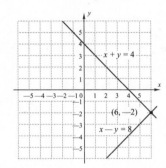

The point of intersection looks as if it has coordinates $(6, -2)$.

Check:

$$\frac{x + y = 4}{\begin{array}{c|c} 6 + (-2) \ ? \ 4 & \\ 4 & \text{TRUE} \end{array}} \qquad \frac{x - y = 8}{\begin{array}{c|c} 6 - (-2) \ ? \ 8 & \\ 6 + 2 & \\ 8 & \text{TRUE} \end{array}}$$

The solution is $(6, -2)$.

6. We graph the equations.

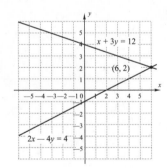

The point of intersection looks as if it has coordinates $(6, 2)$.

Check:

$$\frac{x + 3y = 12}{\begin{array}{c|c} 6 + 3 \cdot 2 \ ? \ 12 & \\ 6 + 6 & \\ 12 & \text{TRUE} \end{array}} \qquad \frac{2x - 4y = 4}{\begin{array}{c|c} 2 \cdot 6 - 4 \cdot 2 \ ? \ 8 & \\ 12 - 8 & \\ 4 & \text{TRUE} \end{array}}$$

The solution is $(6, 2)$.

7. We graph the equations.

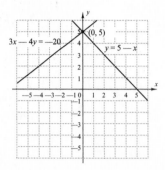

The point of intersection looks as if it has coordinates $(0, 5)$.

Check:

$$\frac{y = 5 - x}{\begin{array}{c|c} 5 \ ? \ 5 - 0 & \\ 5 & \text{TRUE} \end{array}} \qquad \frac{3x - 4y = -20}{\begin{array}{c|c} 3 \cdot 0 - 4 \cdot 5 \ ? \ -20 & \\ -20 & \text{TRUE} \end{array}}$$

The solution is $(0, 5)$.

8. We graph the equations.

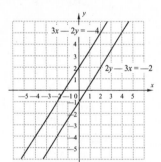

The lines are parallel. There is no solution.

9. $\quad y = 5 - x, \qquad (1)$
$\quad 3x - 4y = -20 \quad (2)$

We substitute $5 - x$ for y in Equation (2) and solve for x.

$$\begin{aligned} 3x - 4y &= -20 \quad (2) \\ 3x - 4(5 - x) &= -20 \\ 3x - 20 + 4x &= -20 \\ 7x - 20 &= -20 \\ 7x &= 0 \\ x &= 0 \end{aligned}$$

Next we substitute 0 for x in one of the original equations and solve for y.

$$\begin{aligned} y &= 5 - x \quad (1) \\ y &= 5 - 0 \\ y &= 5 \end{aligned}$$

The ordered pair $(0, 5)$ checks in both equations. It is the solution.

10. $x + y = 6,$ (1)

$\quad\ y = 3 - 2x$ (2)

We substitute $3 - 2x$ for y in Equation (1) and solve for x.

$$x + y = 6 \qquad (1)$$
$$x + (3 - 2x) = 6$$
$$-x + 3 = 6$$
$$-x = 3$$
$$x = -3$$

Now we substitute -3 for x in one of the original equations and solve for y.

$$y = 3 - 2x \qquad (2)$$
$$y = 3 - 2(-3)$$
$$y = 3 + 6$$
$$y = 9$$

The ordered pair $(-3, 9)$ checks in both equations. It is the solution.

11. $x - y = 4,$ (1)

$\quad\ y = 2 - x$ (2)

We substitute $2 - x$ for y in Equation (1) and solve for x.

$$x - y = 4 \quad (1)$$
$$x - (2 - x) = 4$$
$$x - 2 + x = 4$$
$$2x - 2 = 4$$
$$2x = 6$$
$$x = 3$$

Now substitute 3 for x in one of the original equations and solve for y.

$$y = 2 - x \quad (2)$$
$$y = 2 - 3$$
$$y = -1$$

The ordered pair $(3, -1)$ checks in both equations. It is the solution.

12. $s + t = 5,$ (1)

$\quad\ s = 13 - 3t$ (2)

We substitute $13 - 3t$ for s in Equation (1) and solve for t.

$$s + t = 5 \qquad (1)$$
$$(13 - 3t) + t = 5$$
$$13 - 2t = 5$$
$$-2t = -8$$
$$t = 4$$

Now substitute 4 for t in one of the original equations and solve for s.

$$s = 13 - 3t \qquad (2)$$
$$s = 13 - 3 \cdot 4$$
$$s = 13 - 12$$
$$s = 1$$

The ordered pair $(1, 4)$ checks in both equations. It is the solution.

13. $x + 2y = 6,$ (1)

$\quad\ 2x + 3y = 8$ (2)

We solve Equation (1) for x.

$$x + 2y = 6 \qquad (1)$$
$$x = -2y + 6 \quad (3)$$

We substitute $-2y + 6$ for x in Equation (2) and solve for y.

$$2x + 3y = 8 \qquad (2)$$
$$2(-2y + 6) + 3y = 8$$
$$-4y + 12 + 3y = 8$$
$$-y + 12 = 8$$
$$-y = -4$$
$$y = 4$$

Now substitute 4 for y in Equation (3) and compute x.

$$x = -2y + 6 = -2 \cdot 4 + 6 = -8 + 6 = -2$$

The ordered pair $(-2, 4)$ checks in both equations. It is the solution.

14. $3x + y = 1,$ (1)

$\quad\ x - 2y = 5$ (2)

We solve Equation (2) for x.

$$x - 2y = 5 \qquad (1)$$
$$x = 2y + 5 \quad (3)$$

We substitute $2y + 5$ for x in Equation (1) and solve for y.

$$3x + y = 1 \qquad (1)$$
$$3(2y + 5) + y = 1$$
$$6y + 15 + y = 1$$
$$7y + 15 = 1$$
$$7y = -14$$
$$y = -2$$

Now substitute -2 for y in Equation (3) and compute x.

$$x = 2y + 5 = 2(-2) + 5 = -4 + 5 = 1$$

The ordered pair $(1, -2)$ checks in both equations. It is the solution.

15. $x + y = 4,$ (1)

$\quad\underline{\ 2x - y = 5\ }$ (2)

$\quad\ 3x \quad\ = 9$

$\qquad\ x = 3$

Substitute 3 for x in either of the original equations and solve for y.

$$x + y = 4 \quad (1)$$
$$3 + y = 4$$
$$y = 1$$

The ordered pair $(3, 1)$ checks in both equations. It is the solution.

16. $x + 2y = 9,$ (1)

$\quad\underline{\ 3x - 2y = -5\ }$ (2)

$\quad\ 4x \qquad = 4$

$\qquad\ x = 1$

Substitute 1 for x in either of the original equations and solve for y.

$$x + 2y = 9 \quad (1)$$
$$1 + 2y = 9 \quad (2)$$
$$2y = 8$$
$$y = 4$$

The ordered pair $(1, 4)$ checks in both equations. It is the solution.

17.
$$x - y = 8, \quad (1)$$
$$\underline{2x + y = 7} \quad (2)$$
$$3x \quad\;\; = 15$$
$$x = 5$$

Substitute 5 for x in either of the original equations and solve for y.

$$2x + y = 7 \quad (2)$$
$$2 \cdot 5 + y = 7$$
$$10 + y = 7$$
$$y = -3$$

The ordered pair $(5, -3)$ checks in both equations. It is the solution.

18.
$$2x + 3y = 8, \quad (1)$$
$$5x + 2y = -2 \quad (2)$$

We use the multiplication principle with both equations and then add.

$$4x + 6y = 16 \quad \text{Multiplying (1) by 2}$$
$$\underline{-15x - 6y = 6} \quad \text{Multiplying (2) by } -3$$
$$-11x \quad\quad = 22$$
$$x = -2$$

Substitute -2 for x in one of the original equations and solve for y.

$$2x + 3y = 8 \quad (1)$$
$$2(-2) + 3y = 8$$
$$-4 + 3y = 8$$
$$3y = 12$$
$$y = 4$$

The ordered pair $(-2, 4)$ checks in both equations. It is the solution.

19.
$$5x - 2y = 2, \quad (1)$$
$$3x - 7y = 36 \quad (2)$$

We use the multiplication principle with both equations and then add.

$$35x - 14y = 14 \quad \text{Multiplying (1) by 7}$$
$$\underline{-6x + 14y = -72} \quad \text{Multiplying (2) by } -2$$
$$29x \quad\quad = -58$$
$$x = -2$$

Substitute -2 for x in one of the original equations and solve for y.

$$5x - 2y = 2 \quad (1)$$
$$5(-2) - 2y = 2$$
$$-10 - 2y = 2$$
$$-2y = 12$$
$$y = -6$$

The ordered pair $(-2, -6)$ checks in both equations. It is the solution.

20.
$$-x - y = -5, \quad (1)$$
$$2x - y = 4 \quad\;\; (2)$$

We multiply Equation (1) by -1 and then add.

$$x + y = 5$$
$$\underline{2x - y = 4}$$
$$3x \quad\;\; = 9$$
$$x = 3$$

Substitute 3 for x in one of the original equations and solve for y.

$$-x - y = -5 \quad (1)$$
$$-3 - y = -5$$
$$-y = -2$$
$$y = 2$$

The ordered pair $(3, 2)$ checks in both equations. It is the solution.

21.
$$6x + 2y = 4, \quad (1)$$
$$10x + 7y = -8 \quad (2)$$

We use the multiplication principle with both equations and then add.

$$42x + 14y = 28 \quad \text{Multiplying (1) by 7}$$
$$\underline{-20x - 14y = 16} \quad \text{Multiplying (2) by } -2$$
$$22x \quad\quad = 44$$
$$x = 2$$

Substitute 2 for x in one of the original equations and solve for y.

$$6x + 2y = 4 \quad (1)$$
$$6 \cdot 2 + 2y = 4$$
$$12 + 2y = 4$$
$$2y = -8$$
$$y = -4$$

The ordered pair $(2, -4)$ checks in both equations. It is the solution.

22.
$$-6x - 2y = 5, \quad (1)$$
$$12x + 4y = -10 \quad (2)$$

We multiply Equation (1) by 2 and then add.

$$-12x - 4y = 10$$
$$\underline{12x + 4y = -10}$$
$$0 = 0$$

We get an obviously true equation, so the system has an infinite number of solutions.

23. $\dfrac{2}{3}x + y = -\dfrac{5}{3}$

$x - \dfrac{1}{3}y = -\dfrac{13}{3}$

First we multiply both sides of each equation to clear the fractions.

$$3\left(\dfrac{2}{3}x + y\right) = 3\left(-\dfrac{5}{3}\right)$$

$$3 \cdot \dfrac{2}{3}x + 3y = -5$$

$$2x + 3y = -5$$

$$3\left(x - \dfrac{1}{3}y\right) = 3\left(-\dfrac{13}{3}\right)$$

$$3x - 3 \cdot \dfrac{1}{3}y = -13$$

$$3x - y = -13$$

The resulting system is

$$2x + 3y = -5, \quad (1)$$

$$3x - y = -13. \quad (2)$$

Now we multiply Equation (2) by 3 and then add.

$$2x + 3y = -5$$
$$\underline{9x - 3y = -39}$$
$$11x = -44$$
$$x = -4$$

Substitute -4 for x in Equation (1) and solve for y.

$$2x + 3y = -5$$
$$2(-4) + 3y = -5$$
$$-8 + 3y = -5$$
$$3y = 3$$
$$y = 1$$

The ordered pair $(-4, 1)$ checks in both equations. It is the solution.

24. Familiarize. We make a drawing. We let l = the length and w = the width, in cm.

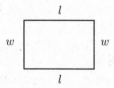

Translate. The perimeter is $2l + 2w$. We translate the first statement.

$$\underbrace{\text{The perimeter}}_{2l + 2w} \underbrace{\text{is}}_{=} \underbrace{\text{96 cm.}}_{96}$$

We translate the second statement.

$$\underbrace{\text{The length}}_{l} \underbrace{\text{is}}_{=} \underbrace{\text{27 cm}}_{27} \underbrace{\text{more than}}_{+} \underbrace{\text{the width.}}_{w}$$

The resulting system is

$$2l + 2w = 96, \quad (1)$$

$$l = 27 + w. \quad (2)$$

Solve. First we substitute $27 + w$ for l in Equation (1) and solve for w.

$$2l + 2w = 96 \quad (1)$$
$$2(27 + w) + 2w = 96$$
$$54 + 2w + 2w = 96$$
$$54 + 4w = 96$$
$$4w = 42$$
$$w = 10.5$$

Now we substitute 10.5 for w in Equation (2) and find l.

$$l = 27 + w = 27 + 10.5 = 37.5$$

Check. If the length is 37.5 cm and the width is 10.5 cm, then the perimeter is $2(37.5) + 2(10.5)$, or $75 + 21$, or 96 cm. Also, the length is 27 cm more than the width. The answer checks.

State. The length of the rectangle is 37.5 cm, and the width is 10.5 cm.

25. Familiarize. Let x = the number of orchestra seats sold and y = the number of balcony seats sold. We organize the information in a table.

	Orchestra	Balcony	Total
Price	$25	$18	
Number bought	x	y	508
Receipts	$25x$	$18y$	$11,223

Translate. The last two rows of the table give us a system of equations.

$$x + y = 508, \quad (1)$$

$$25x + 18y = 11,223. \quad (2)$$

Solve. First we multiply Equation (1) by -18 and then add.

$$-18x - 18y = -9144$$
$$\underline{25x + 18y = 11,223}$$
$$7x = 2079$$
$$x = 297$$

Now we substitute 297 for x in Equation (1) and solve for y.

$$x + y = 508$$
$$297 + y = 508$$
$$y = 211$$

Check. The total number of tickets sold was $297 + 211$, or 508. The total receipts were $\$25 \cdot 297 + \$18 \cdot 211$, or $\$7425 + \$3798$, or $11,223. The answer checks.

State. 297 orchestra seats and 211 balcony seats were sold.

26. *Familiarize.* Let c = the number of liters of Clear Shine and s = the number of liters of Sunstream window cleaner to be used in the mixture. We organize the information in a table.

Type of cleaner	Clear Shine	Sunstream	Mixture
Amount used	c	s	80
Percent of alcohol	30%	60%	45%
Amount of alcohol in solution	$0.3c$	$0.6s$	$45\% \times 80$, or 36 L

Translate. The first and third rows of the table give us a system of equations.

$$c + s = 80,$$
$$0.3c + 0.6s = 36$$

After we clear decimals we have

$$c + s = 80, \quad (1)$$
$$3c + 6s = 360. \quad (2)$$

Solve. First we multiply Equation (1) by -3 and then add.

$$-3c - 3s = -240$$
$$\underline{3c + 6s = 360}$$
$$3s = 120$$
$$s = 40$$

Now we substitute 40 for s in Equation (1) and solve for c.

$$c + s = 80$$
$$c + 40 = 80$$
$$c = 40$$

Check. If 40 L of Clear Shine and 40 L of Sunstream are used, then there is 40 L + 40 L, or 80 L, of solution. The amount of alcohol in the solution is $0.3(40) + 0.6(40)$, or $12 + 24$, or 36 L. The answer checks.

State. 40 L of each window cleaner should be used.

27. *Familiarize.* Let x = the weight of the Asian elephant and y = the weight of the African elephant, in kg.

Translate.

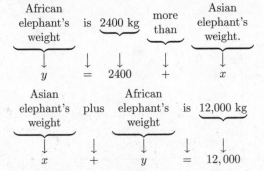

The resulting system is

$$y = 2400 + x, \quad (1)$$
$$x + y = 12,000. \quad (2)$$

Solve. First we substitute $2400 + x$ for y in Equation (2) and solve for x.

$$x + y = 12,000$$
$$x + (2400 + x) = 12,000$$
$$2x + 2400 = 12,000$$
$$2x = 9600$$
$$x = 4800$$

Now substitute 4800 for x in Equation (1) and find y.

$$y = 2400 + x = 2400 + 4800 = 7200$$

Check. If the Asian elephant weighs 7200 kg and the African elephant weighs 4800 kg, then the African elephant weighs 2400 kg more than the Asian elephant and their total weight is $4800 + 7200$, or 12,000 kg. The answer checks.

State. The Asian elephant weighs 4800 kg, and the African elephant weighs 7200 kg.

28. *Familiarize.* Let x = the number of pounds of peanuts and y = the number of pounds of fancy nuts to be used. We organize the information in a table.

Type of nuts	Peanuts	Fancy	Mixture
Cost per pound	$4.50	$7.00	
Amount	x	y	13
Mixture	$4.5x$	$7y$	$71

Translate. The last two rows of the table give us a system of equations.

$$x + y = 13,$$
$$4.5x + 7y = 71$$

After clearing decimals, we have

$$x + y = 13, \quad (1)$$
$$45x + 70y = 710. \quad (2)$$

Solve. First we multiply Equation (1) by -45 and then add.

$$-45x - 45y = -585$$
$$\underline{45x + 70y = 710}$$
$$25y = 125$$
$$y = 5$$

Now substitute 5 for y in one of the original equations and solve for x.

$$x + y = 13 \quad (1)$$
$$x + 5 = 13$$
$$x = 8$$

Check. If 8 lb of peanuts and 5 lb of fancy nuts are used, the mixture weighs 13 lb. The value of the mixture is $\$4.50(8) + \$7.00(5) = \$36 + \$35 = \$71$. The answer checks.

State. 8 lb of peanuts and 5 lb of fancy nuts should be used.

29. *Familiarize*. Let x = the number of minutes used in the $29.95 plan and y = the number of minutes used in the 7¢ a minute plan. Expressing 7¢ as $0.07, the 7¢ plan costs $0.07y + $3.95 per month.

Translate.

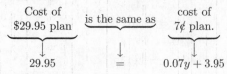

$$\underbrace{\text{Cost of}}_{29.95}\ \underbrace{\text{is the same as}}_{=}\ \underbrace{\text{cost of}}_{0.07y + 3.95}$$

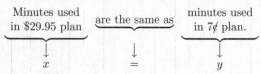

$$\underbrace{\text{Minutes used}}_{x}\ \underbrace{\text{are the same as}}_{=}\ \underbrace{\text{minutes used}}_{y}$$

After clearing decimals we have the following system of equations.

$$2995 = 7y + 395, \quad (1)$$
$$x = y \quad\quad\quad\quad (2)$$

Solve. First we solve Equation (1) for y.

$$2995 = 7y + 395$$
$$2600 = 7y$$
$$371 \approx y$$

Now substitute 371 for y in Equation (2) and find x.

$$x = y$$
$$x = 371$$

Check. For 371 min the 7¢ plan costs $0.07(371) + $3.95, or $29.92. We rounded the value of y and $29.92 \approx $29.95, so the answer checks.

State. The two plans cost the same for about 371 min.

30. *Familiarize*. Let x = the number of gallons of 87-octane gas and y = the number of gallons of 95-octane gas to be used. We arrange the information in a table.

Type of gas	87-octane	95-octane	Mixture
Amount	x	y	10
Octane rating	87	95	93
Mixture	$87x$	$95y$	$10 \cdot 93$, or 930

Translate. The first and last rows of the table give us a system of equations.

$$x + \quad y = 10, \quad (1)$$
$$87x + 95y = 930 \quad (2)$$

Solve. We multiply Equation (1) by -87 and then add.

$$-87x - 87y = -870$$
$$\underline{87x + 95y = 930}$$
$$\quad\quad 8y = 60$$
$$\quad\quad\ y = 7.5$$

Now substitute 7.5 for y in one of the original equations and solve for x.

$$x + y = 10 \quad (1)$$
$$x + 7.5 = 10$$
$$x = 2.5$$

Check. If 2.5 gal of 87-octane gas and 7.5 gal of 95-octane gas are used, then there are $2.5 + 7.5$, or 10 gal, of gas in the mixture. Also, $87(2.5) + 95(7.5) = 217.5 + 712.5 = 930$, so the answer checks.

State. 2.5 gal of 87-octane gas and 7.5 gal of 95-octane gas should be used.

31. *Familiarize*. Let x = Jeff's age now and y = his son's age now. In 13 yr their ages will be $x + 13$ and $y + 13$.

Translate.

$$\underbrace{\text{Jeff's age}}_{x}\ \underbrace{\text{is}}_{=}\ \underbrace{\text{three}}_{3}\ \underbrace{\text{times}}_{\cdot}\ \underbrace{\text{his son's age.}}_{y}$$

In 13 yr,

$$\underbrace{\text{Jeff's age}}_{x + 13}\ \underbrace{\text{will be}}_{=}\ \underbrace{\text{two}}_{2}\ \underbrace{\text{times}}_{\cdot}\ \underbrace{\text{his son's age.}}_{(y + 13)}$$

The resulting system is

$$x = 3y, \quad\quad\quad\quad\quad (1)$$
$$x + 13 = 2(y + 13). \quad (2)$$

Solve. First we substitute $3y$ for x in Equation (2) and solve for y.

$$x + 13 = 2(y + 13)$$
$$3y + 13 = 2(y + 13)$$
$$3y + 13 = 2y + 26$$
$$y + 13 = 26$$
$$y = 13$$

Now substitute 13 for y in Equation (1) and find x.

$$x = 3y = 3 \cdot 13 = 39$$

Check. If Jeff is 39 years old and his son is 13 years old, then Jeff is three times as old as his son. In 13 yr Jeff's age will be $39 + 13$, or 52, his son's age will be $13 + 13$, or 26, and $52 = 2 \cdot 26$. The answer checks.

State. Jeff is 39 years old now, and his son is 13 years old.

32. *Familiarize*. Let x = the measure of the larger angle and y = the measure of the smaller angle. Recall that the sum of the measures of complementary angles is 90°.

Translate.

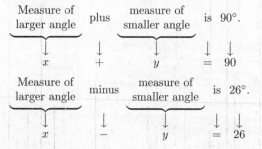

$$\underbrace{\text{Measure of}}_{x}\ \underbrace{\text{plus}}_{+}\ \underbrace{\text{measure of}}_{y}\ \underbrace{\text{is}}_{=}\ \underbrace{90°.}_{90}$$

$$\underbrace{\text{Measure of}}_{x}\ \underbrace{\text{minus}}_{-}\ \underbrace{\text{measure of}}_{y}\ \underbrace{\text{is}}_{=}\ \underbrace{26°.}_{26}$$

The resulting system is

$$x + y = 90, \quad (1)$$
$$x - y = 26. \quad (2)$$

Solve. We add.

$$x + y = 90$$
$$\underline{x - y = 26}$$
$$2x \quad\;\; = 116$$
$$x = 58$$

Substitute 58 for x in one of the original equations and solve for y.

$$x + y = 90 \quad (1)$$
$$58 + y = 90$$
$$y = 32$$

Check. $58° + 32° = 90°$ and $58° - 32° = 26°$, so the answer checks.

State. The measures of the angles are 58°and 32°.

33. Familiarize. Let $x =$ the measure of the larger angle and $y =$ the measure of the smaller angle. Recall that the sum of the measures of supplementary angles is 180°.

Translate.

Measure of larger angle plus measure of smaller angle is 180°.

$$x + y = 180$$

Measure of larger angle minus measure of smaller angle is 26°.

$$x - y = 26$$

The resulting system is

$$x + y = 180, \quad (1)$$
$$x - y = 26. \quad (2)$$

Solve. We add.

$$x + y = 180$$
$$\underline{x - y = 26}$$
$$2x \quad\;\; = 206$$
$$x = 103$$

Now substitute 103 for x in one of the original equations and solve for y.

$$x + y = 180 \quad (1)$$
$$103 + y = 180$$
$$y = 77$$

Check. $103° + 77° = 180°$ and $103° - 77° = 26°$; so the answer checks.

State. The measures of the angles are 103°and 77°.

34. Familiarize. Let $r =$ the speed of the airplane in still air, in km/h. Then $r + 15 =$ the speed with a 15 km/h tail wind and $r - 15 =$ the speed against a 15 km/h wind. We fill in the table in the text.

	Distance	Speed	Time
Going	d	$r + 15$	4
Returning	d	$r - 15$	5

Translate. We get an equation $d = rt$ from each row of the table.

$$d = (r + 15)4, \quad (1)$$
$$d = (r - 15)5 \quad (2)$$

Solve. We substitute $(r + 15)4$ for d in Equation (2).

$$d = (r - 15)5$$
$$(r + 15)4 = (r - 15)5$$
$$4r + 60 = 5r - 75$$
$$60 = r - 75$$
$$135 = r$$

Check. The plane's speed with the tail wind is $135 + 15$, or 150 km/h. At that speed, in 4 hr it will travel $150 \cdot 4$, or 600 km. The plane's speed against the wind is $135 - 15$, or 120 km/h. At that speed, in 5 hr it will travel $120 \cdot 5$, or 600 km. Since the distances are the same, the answer checks.

State. The speed of the airplane in still air is 135 km/h.

35. Familiarize. Let $t =$ the number of hours the slower car travels before the second car catches up to it. Then $t - 2 =$ the number of hours the faster car travels. We fill in the table in the text.

	Distance	Speed	Time
Slow car	d	55	t
Fast car	d	75	$t - 2$

Translate. We get an equation $d = rt$ from each row of the table.

$$d = 55t, \quad (1)$$
$$d = 75(t - 2) \quad (2)$$

Solve. We substitute $55t$ for d in Equation (2).

$$d = 75(t - 2)$$
$$55t = 75(t - 2)$$
$$55t = 75t - 150$$
$$-20t = -150$$
$$t = 7.5$$

Now substitute 7.5 for t in Equation (1) and find d.

$$d = 55t = 55(7.5) = 412.5$$

Check. From the calculation of d above, we see that the slow car travels 412.5 mi in 7.5 hr. The fast car travels $7.5 - 2$, or 5.5 hr. At a speed of 75 mph, it travels $75(5.5)$, or 412.5 mi. Since the distances are the same, the answer checks.

State. The second car catches up to the first car 412.5 mi from Phoenix.

36. Discussion and Writing Exercise. The equations have the same slope but different y-intercepts, so they represent parallel lines. Thus the system of equations has no solution.

37. *Discussion and Writing Exercise.* Answers will vary.

38. Familiarize. Let c = the compensation agreed upon for 12 months of work and let h = the value of the horse. After 7 months, Stephanie would be owed $\frac{7}{12}$ of the compensation agreed upon for 12 months of work, or $\frac{7}{12}c$.

Translate.

$$\underbrace{\text{Compensation for 12 months}} \;\; \text{is} \;\; \underbrace{\text{a horse}} \;\; \text{plus} \;\; \$2400.$$

$$\downarrow \qquad\qquad \downarrow \quad\; \downarrow \quad\; \downarrow \qquad\quad \downarrow$$
$$c \qquad\qquad = \quad\; h \quad\; + \qquad 2400$$

$$\underbrace{\text{Compensation for 7 months}} \;\; \text{is} \;\; \underbrace{\text{a horse}} \;\; \text{plus} \;\; \$1000.$$

$$\downarrow \qquad\qquad \downarrow \quad\; \downarrow \quad\; \downarrow \qquad\quad \downarrow$$
$$\frac{7}{12}c \qquad\qquad = \quad\; h \quad\; + \qquad 1000$$

After clearing the fractions we have the following system of equations.

$$c = h + 2400, \qquad (1)$$
$$7c = 12h + 12,000 \qquad (2)$$

Solve. We substitute $h + 2400$ for c in Equation (2) and solve for h.

$$7c = 12h + 12,000$$
$$7(h + 2400) = 12h + 12,000$$
$$7h + 16,800 = 12h + 12,000$$
$$16,800 = 5h + 12,000$$
$$4800 = 5h$$
$$960 = h$$

Check. If the value of the horse is $960, then the compensation for 12 months of work is $960 + $2400, or $3360; $\frac{7}{12}$ of $3360 is $1960. This is the value of the horse, $960, plus $1000, so the answer checks.

State. The value of the horse was $960.

39. We substitute 6 for x and 2 for y in each equation.

$$2x - Dy = 6 \qquad\qquad Cx + 4y = 14$$
$$2 \cdot 6 - D \cdot 2 = 6 \qquad\qquad C \cdot 6 + 4 \cdot 2 = 14$$
$$12 - 2D = 6 \qquad\qquad 6C + 8 = 14$$
$$-2D = -6 \qquad\qquad 6C = 6$$
$$D = 3 \qquad\qquad C = 1$$

40. $\quad 3(x - y) = 4 + x, \quad (1)$
$\qquad x = 5y + 2 \qquad\qquad (2)$

Substitute $5y + 2$ for x in Equation (1) and solve for y.

$$3(x - y) = 4 + x$$
$$3(5y + 2 - y) = 4 + 5y + 2$$
$$3(4y + 2) = 5y + 6$$
$$12y + 6 = 5y + 6$$
$$7y + 6 = 6$$
$$7y = 0$$
$$y = 0$$

Now substitute 0 for y in Equation (2) and find x.

$$x = 5y + 2 = 5 \cdot 0 + 2 = 0 + 2 = 2$$

The solution is $(2, 0)$.

41. The line graphed in red contains the points $(0, 0)$ and $(3, 2)$. We find the slope:

$$m = \frac{2 - 0}{3 - 0} = \frac{2}{3}$$

The y-intercept is $(0, 0)$, so the equation of the line is

$$y = \frac{2}{3}x + 0, \text{ or } y = \frac{2}{3}x.$$

The line graphed in blue contains the points $(0, 5)$ and $(3, 2)$. We find the slope:

$$m = \frac{2 - 5}{3 - 0} = \frac{-3}{3} = -1$$

The y-intercept is $(0, 5)$, so the equation of the line is $y = -1 \cdot x + 5$, or $y = -x + 5$.

42. The line graphed in red contains the points $(-3, 0)$ and $(0, -3)$. We find the slope:

$$m = \frac{-3 - 0}{0 - (-3)} = \frac{-3}{3} = -1$$

The y-intercept is $(0, -3)$, so the equation of the line is $y = -1 \cdot x - 3$, or $y = -x - 3$, or $x + y = -3$.

The line graphed in blue contains the points $(0, 4)$ and $(4, 0)$. We find the slope:

$$m = \frac{0 - 4}{4 - 0} = \frac{-4}{4} = -1$$

The y-intercept is $(0, 4)$, so the equation of the line is $y = -1 \cdot x + 4$, or $y = -x + 4$, or $x + y = 4$.

43. Familiarize. Let x = the number of rabbits and y = the number of pheasants. Then the rabbits have a total of x heads and $4x$ feet; the pheasants have a total of y heads and $2y$ feet.

Translate.

$$\underbrace{\text{Rabbit heads}} \;\; \text{plus} \;\; \underbrace{\text{pheasant heads}} \;\; \text{is} \;\; \underbrace{\text{35 heads.}}$$
$$\downarrow \qquad\quad \downarrow \qquad\quad \downarrow \qquad\quad \downarrow \quad\; \downarrow$$
$$x \qquad\quad + \qquad\quad y \qquad\quad = \quad\; 35$$

$$\underbrace{\text{Rabbit feet}} \;\; \text{plus} \;\; \underbrace{\text{pheasant feet}} \;\; \text{is} \;\; \underbrace{\text{94 feet.}}$$
$$\downarrow \qquad\quad \downarrow \qquad\quad \downarrow \qquad\quad \downarrow \quad\; \downarrow$$
$$4x \qquad\quad + \qquad\quad 2y \qquad\quad = \quad\; 94$$

The resulting system is

$$x + y = 35, \qquad (1)$$
$$4x + 2y = 94. \qquad (2)$$

Solve. First we multiply Equation (1) by -2 and then add.

$$-2x - 2y = -70$$
$$\underline{4x + 2y = 94}$$
$$2x \qquad\quad = 24$$
$$x = 12$$

Now substitute 12 for x in one of the original equations and solve for y.

$$x + y = 35 \quad (1)$$
$$12 + y = 35$$
$$y = 23$$

Check. If there are 12 rabbits and 23 pheasants, then there are $12 + 23$, or 35, heads and $4 \cdot 12 + 2 \cdot 23$, or $48 + 46$, or 94, feet. The answer checks.

State. There are 12 rabbits and 23 pheasants.

Chapter 7 Test

1. We check by substituting alphabetically -2 for x and -1 for y.

$$\begin{array}{c|c} x = 4 + 2y \\ \hline -2 \ ? \ 4 + 2(-1) \\ \ \big| \ 4 - 2 \\ \ \big| \ 2 \qquad \text{FALSE} \end{array}$$

Since $(-2, -1)$ is not a solution of the first equation, it is not a solution of the system of equations.

2. We graph the equations.

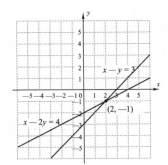

The point of intersection looks as if it has coordinates $(2, -1)$.

Check:

$$\begin{array}{c|c} x - y = 3 \\ \hline 2 - (-1) \ ? \ 3 \\ 2 + 1 \ \big| \\ 3 \ \big| \ \text{TRUE} \end{array} \qquad \begin{array}{c|c} x - 2y = 4 \\ \hline 2 - 2(-1) \ ? \ 4 \\ 2 + 2 \ \big| \\ 4 \ \big| \ \text{TRUE} \end{array}$$

The solution is $(2, -1)$.

3. $y = 6 - x, \quad (1)$
   $2x - 3y = 22 \quad (2)$

We substitute $6 - x$ for y in Equation (2) and solve for x.

$$2x - 3y = 22 \quad (2)$$
$$2x - 3(6 - x) = 22$$
$$2x - 18 + 3x = 22$$
$$5x - 18 = 22$$
$$5x = 40$$
$$x = 8$$

Next we substitute 8 for x in one of the original equations and solve for y.

$$y = 6 - x \quad (1)$$
$$y = 6 - 8$$
$$y = -2$$

The ordered pair $(8, -2)$ checks in both equations. It is the solution.

4. $x + 2y = 5, \quad (1)$
   $x + y = 2 \quad (2)$

We solve Equation (1) for x.

$$x + 2y = 5 \qquad (1)$$
$$x = -2y + 5 \quad (3)$$

We substitute $-2y + 5$ for x in Equation (2) and solve for y.

$$x + y = 2 \qquad (2)$$
$$-2y + 5 + y = 2$$
$$-y + 5 = 2$$
$$-y = -3$$
$$y = 3$$

Now substitute 3 for y in Equation (3) and compute x.

$$x = -2y + 5 = -2 \cdot 3 + 5 = -6 + 5 = -1.$$

The ordered pair $(-1, 3)$ checks in both equations. It is the solution.

5. $y = 5x - 2, \quad (1)$
   $y - 2 = 5x \quad (2)$

Substitute $5x - 2$ for y in Equation (2) and solve for x.

$$y - 2 = 5x \quad (2)$$
$$5x - 2 - 2 = 5x$$
$$5x - 4 = 5x$$
$$-4 = 0$$

We obtain a false equation, so there is no solution.

6. $x - y = 6 \qquad (1)$
   $\underline{3x + y = -2} \quad (2)$
   $4x = 4$
   $x = 1$

Substitute 1 for x in either of the original equations and solve for y.

$$3x + y = -2 \quad (2)$$
$$3 \cdot 1 + y = -2$$
$$3 + y = -2$$
$$y = -5$$

The ordered pair $(1, -5)$ checks in both equations. It is the solution.

7. $\dfrac{1}{2}x - \dfrac{1}{3}y = 8,$
   $\dfrac{2}{3}x + \dfrac{1}{2}y = 5$

First we multiply each equation by 6 to clear the fractions.

$$6\left(\frac{1}{2}x - \frac{1}{3} + y\right) = 6 \cdot 8$$

$$6 \cdot \frac{1}{2}x - 6 \cdot \frac{1}{3}y = 48$$

$$3x - 2y = 48$$

$$6\left(\frac{2}{3}x + \frac{1}{2}y\right) = 6 \cdot 5$$

$$6 \cdot \frac{2}{3}x + 6 \cdot \frac{1}{2}y = 30$$

$$4x + 3y = 30$$

The resulting system is

$$3x - 2y = 48, \quad (1)$$
$$4x + 3y = 30. \quad (2)$$

Now we multiply Equation (1) by 3 and Equation (2) by 2 and then add.

$$9x - 6y = 144$$
$$\underline{8x + 6y = 60}$$
$$17x \qquad = 204$$
$$x = 12$$

Next we substitute 12 for x in Equation (2) and solve for y.

$$4x + 3y = 30 \qquad (2)$$
$$4 \cdot 12 + 3y = 30$$
$$48 + 3y = 30$$
$$3y = -18$$
$$y = -6$$

The ordered pair $(12, -6)$ checks in both equations. It is the solution.

8. $-4x - 9y = 4, \quad (1)$
$6x + 3y = 1 \quad (2)$

Multiply Equation (2) by 3 and then add.

$$-4x - 9y = 4$$
$$\underline{18x + 9y = 3}$$
$$14x \qquad = 7$$
$$x = \frac{1}{2}$$

Now substitute $\frac{1}{2}$ for x in one of the original equations and solve for y.

$$6x + 3y = 1 \qquad (2)$$
$$6 \cdot \frac{1}{2} + 3y = 1$$
$$3 + 3y = 1$$
$$3y = -2$$
$$y = -\frac{2}{3}$$

The ordered pair $\left(\frac{1}{2}, -\frac{2}{3}\right)$ checks in both equations. It is the solution.

9. $2x + 3y = 13, \quad (1)$
$3x - 5y = 10 \quad (2)$

Multiply Equation (1) by 5 and Equation (2) by 3 and then add.

$$10x + 15y = 65$$
$$\underline{9x - 15y = 30}$$
$$19x \qquad = 95$$
$$x = 5$$

Now substitute 5 for x in one of the original equations and solve for y.

$$2x + 3y = 13 \quad (1)$$
$$2 \cdot 5 + 3y = 13$$
$$10 + 3y = 13$$
$$3y = 3$$
$$y = 1$$

The ordered pair $(5, 1)$ checks in both equations. It is the solution.

10. *Familiarize*. Let l = the length and w = the width, in yd. Recall that the perimeter of a rectangle is given by the formula $P = 2l + 2w$.

***Translate*.**

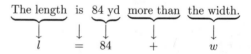

The resulting system is

$$2l + 2w = 8266, \quad (1)$$
$$l = 84 + w. \qquad (2)$$

***Solve*.** First we substitute $84 + w$ for l in Equation (1) and solve for w.

$$2l + 2w = 8266$$
$$2(84 + w) + 2w = 8266$$
$$168 + 2w + 2w = 8266$$
$$168 + 4w = 8266$$
$$4w = 8098$$
$$w = 2024.5$$

Now substitute 2024.5 for w in Equation (2) and find l.

$$l = 84 + w = 84 + 2024.5 = 2108.5$$

***Check*.** If the length is 2108.5 yd and the width is 2024.5 yd, the length is 84 yd more than the width and the perimeter is $2(2108.5) + 2(2024.5)$, or $4217 + 4049$, or 8266 yd. The answer checks.

***State*.** The length is 2108.5 yd and the width is 2024.5 yd.

11. Familiarize. Let a = the amount of solution A and b = the amount of solution B in the mixture, in liters. We organize the information in a table.

Solution	A	B	Mixture
Amount	a	b	60 L
Percent of acid	25%	40%	30%
Amount of acid	$0.25a$	$0.4b$	$0.3(60)$ or 18 L

Translate. The first and third rows of the table give us a system of equations.

$$a + b = 60,$$
$$0.25a + 0.4b = 18$$

After clearing decimals, we have

$$a + b = 60, \quad (1)$$
$$25a + 40b = 1800. \quad (2)$$

Solve. First we multiply Equation (1) by -25 and then add.

$$-25a - 25b = -1500$$
$$\underline{25a + 40b = 1800}$$
$$15b = 300$$
$$b = 20$$

Now we substitute 20 for b in one of the original equations and solve for a.

$$a + b = 60 \quad (1)$$
$$a + 20 = 60$$
$$a = 40$$

Check. If 40 L of solution A and 20 L of solution B are used, then there are $40 + 20$, or 60 L, in the mixture. The amount of acid in the mixture is $0.25(40) + 0.4(20)$, or $10 + 8$, or 18 L. The answer checks.

State. 40 L of solution A and 20 L of solution B should be used.

12. Familiarize. Let r = the speed of the motorboat in still water, in km/h. Then the speed of the boat with the current is $r + 8$ and the speed against the current is $r - 8$. We organize the information in a table.

	Distance	Speed	Time
With current	d	$r + 8$	2
Against current	d	$r - 8$	3

Translate. We get an equation from each row of the table.

$$d = (r + 8)2 \quad (1)$$
$$d = (r - 8)3 \quad (2)$$

Solve. We substitute $(r + 8)2$ for d in Equation (2) and solve for r.

$$d = (r - 8)3 \quad (2)$$
$$(r + 8)2 = (r - 8)3$$
$$2r + 16 = 3r - 24$$
$$16 = r - 24$$
$$40 = r$$

Check. When $r = 40$, then $r + 8 = 40 + 8 = 48$ and in 2 hr at this speed the boat travels $48 \cdot 2$, or 96 km. Also, $r - 8 = 40 - 8 = 32$, and in 3 hr at this speed the boat travels $32 \cdot 3$, or 96 km. The distances are the same, so the answer checks.

State. The speed of the motorboat in still water is 40 km/h.

13. Familiarize. Let c = the receipts from concessions and r = the receipts from the rides.

Translate.

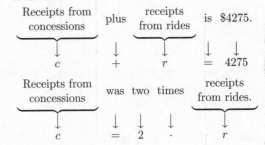

The resulting system is

$$c + r = 4275, \quad (1)$$
$$c = 2r. \quad (2)$$

Solve. First we substitute $2r$ for c in Equation (1) and solve for r.

$$c + r = 4275 \quad (1)$$
$$2r + r = 4275$$
$$3r = 4275$$
$$r = 1425$$

Now we substitute 1425 for r in one of the original equations and find c.

$$c = 2r = 2 \cdot 1425 = 2850$$

Check. If concessions brought in \$2850 and rides brought in \$1425, then the total receipts were \$2850 + \$1425, or \$4275. Also, $2 \cdot \$1425 = \$2850$, so concessions brought in twice as much as rides. The answer checks.

State. Concessions brought in \$2850; rides brought in \$1425.

14. Familiarize. Let x = the number of acres of hay planted and y = the number of acres of oats planted.

Translate.

Acres of hay plus acres of oats is 650 acres.
$$x \qquad + \qquad y \qquad = \qquad 650$$

Acres of hay is 180 more than acres of oats.
$$x \qquad = 180 \quad + \qquad y$$

The resulting system is

$$x + y = 650, \quad (1)$$
$$x = 180 + y. \quad (2)$$

Solve. First we substitute $180 + y$ for x in Equation (1) and solve for y.

$$\begin{aligned} x + y &= 650 \quad (1) \\ (180 + y) + y &= 650 \\ 180 + 2y &= 650 \\ 2y &= 470 \\ y &= 235 \end{aligned}$$

Now substitute 235 for y in Equation (2) and find x.

$$x = 180 + y = 180 + 235 = 415$$

Check. If 415 acres of hay and 235 acres of oats are planted, a total of 415+235, or 650 acres, is planted. Also, $235 + 180 = 415$, so 180 acres more of hay than of oats are planted. The answer checks.

State. 415 acres of hay and 235 acres of oats should be planted.

15. ***Familiarize.*** Let $x =$ the measure of the larger angle and $y =$ the measure of the smaller angle. Recall that the sum of the measures of supplementary angles is 180°.

Translate.

Measure of larger angle | plus | measure of smaller angle | is | 180°.

$x \quad + \quad y \quad = \quad 180$

Measure of larger angle | is | 45° | more than | twice | measure of smaller angle.

$x \quad = \quad 45 \quad + \quad 2\cdot \quad y$

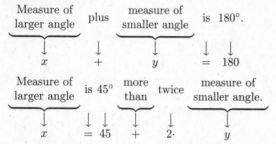

The resulting system is

$$\begin{aligned} x + y &= 180, \quad (1) \\ x &= 45 + 2y. \quad (2) \end{aligned}$$

Solve. First we substitute $45 + 2y$ for x in Equation (1) and solve for y.

$$\begin{aligned} x + y &= 180 \quad (1) \\ (45 + 2y) + y &= 180 \\ 45 + 3y &= 180 \\ 3y &= 135 \\ y &= 45 \end{aligned}$$

Now substitute 45 for y in Equation (2) and find x.

$$x = 45 + 2y = 45 + 2 \cdot 45 = 45 + 90 = 135$$

Check. $135° + 45° = 180°$; and $135° = 45° + 2 \cdot 45°$, so the answer checks.

State. The measures of the angles are 135°and 45°.

16. ***Familiarize.*** Let $x =$ the number of gallons of 87-octane gas and $y =$ the number of gallons of 93-octane gas that should be used. We organize the information in a table.

Type of gas	87-octane	93-octane	Mixture
Amount	x	y	12
Octane rating	87	93	91
Mixture	$87x$	$93y$	$12 \cdot 91$, or 1092

Translate. The first and last rows of the table give us a system of equations.

$$\begin{aligned} x + \quad y &= \quad 12, \quad (1) \\ 87x + 93y &= 1092 \quad (2) \end{aligned}$$

Solve. We multiply Equation (1) by -87 and then add.

$$\begin{aligned} -87x - 87y &= -1044 \\ 87x + 93y &= 1092 \\ \hline 6y &= 48 \\ y &= 8 \end{aligned}$$

Then substitute 8 for y in one of the original equations and solve for x.

$$\begin{aligned} x + y &= 12 \\ x + 8 &= 12 \\ x &= 4 \end{aligned}$$

Check. If 4 gal of 87-octane gas and 8 gal of 93-octane gas are used, then there are $4 + 8$, or 12 gal, of gas in the mixture. Also $87 \cdot 4 + 93 \cdot 8 = 348 + 744 = 1092$, so the answer checks.

State. 4 gal of 87-octane gas and 8 gal of 93-octane gas should be used.

17. ***Familiarize.*** Let $x =$ the number of minutes used in the \$2.95 plan and $y =$ the number of minutes used in the \$1.95 plan. Expressing 10¢ as \$0.10 and 15¢ as \$0.15, the \$2.95 plan costs $\$2.95 + \$0.10x$ per month and the \$1.95 plan costs $\$1.95 + \$0.15y$ per month.

Translate.

Cost of \$2.95 plan | is the same as | cost of \$1.95 plan.

$2.95 + 0.10x \quad = \quad 1.95 + 0.15y$

Minutes used in \$2.95 plan | are the same as | minutes used in \$1.95 plan.

$x \quad = \quad y$

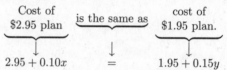

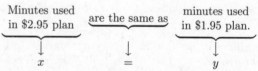

After clearing decimals we have the following system of equations.

$$\begin{aligned} 295 + 10x &= 195 + 15y, \quad (1) \\ x &= y \quad (2) \end{aligned}$$

Solve. We substitute x for y in Equation (1) and solve for x.

$$\begin{aligned} 295 + 10x &= 195 + 15y \quad (1) \\ 295 + 10x &= 195 + 15x \\ 295 &= 195 + 5x \\ 100 &= 5x \\ 20 &= x \end{aligned}$$

Substitute 20 for x in Equation (2) to find y.

$$\begin{aligned} x &= y \\ 20 &= y \end{aligned}$$

Check. For 20 min the \$2.95 plan costs \$2.95+\$0.10(20), or \$4.95 per month and the \$1.95 plan costs

$1.95 + $0.15(20)$, or \$4.95 per month. The costs are the same, so the answer checks.

State. The two plans cost the same for 20 min.

18. ***Familiarize***. Let $t = $ the number of hours it will take the SUV to catch up with the car. Then $t + 2 = $ the number of hours the car travels. We organize the information in a table.

	Distance	Speed	Time
Car	d	55	$t + 2$
SUV	d	65	t

Translate. We get an equation $d = rt$ from each row of the table.
$$d = 55(t + 2), \quad (1)$$
$$d = 65t \qquad (2)$$

Solve. Substitute $65t$ for d in Equation (1) and solve for t.
$$d = 55(t + 2) \quad (1)$$
$$65t = 55(t + 2)$$
$$65t = 55t + 110$$
$$10t = 110$$
$$t = 11$$

Check. At a speed of 55 mph for $11 + 2$, or 13 hr, the car travels $55 \cdot 13$, or 715 mi; at a speed of 65 mph for 11 hr, the SUV travels $65 \cdot 11$, or 715 mi. Since the distances are the same, the answer checks.

State. It will take the SUV 11 hr to catch up to the car.

19. Substitute -2 for x and 3 for y in each equation.

$$\begin{array}{ll} Cx - 4y = 7 & 3x + Dy = 8 \\ C(-2) - 4 \cdot 3 = 7 & 3(-2) + D \cdot 3 = 8 \\ -2C - 12 = 7 & -6 + 3D = 8 \\ -2C = 19 & 3D = 14 \\ C = -\dfrac{19}{2} & D = \dfrac{14}{3} \end{array}$$

20. ***Familiarize***. Let $a = $ the number of people ahead of you and $b = $ the number of people behind you. Then in the entire line there are the people ahead of you, the people behind you, and you yourself, or $a + b + 1$.

Translate.

$$\underbrace{\text{Number ahead of you}}_{a} \underset{=}{\text{is}} \underset{2}{\text{two}} \underset{+}{\overset{\text{more}}{\text{than}}} \underbrace{\text{number behind you.}}_{b}$$

$$\underbrace{\text{Number in entire line}}_{a+b+1} \underset{=}{\text{is}} \underset{3}{\text{three}} \underset{\cdot}{\text{times}} \underbrace{\text{number behind you.}}_{b}$$

The resulting system is
$$a = 2 + b, \qquad (1)$$
$$a + b + 1 = 3b. \quad (2)$$

Solve. We substitute $2 + b$ for a in Equation (2) and solve for b.
$$a + b + 1 = 3b \quad (2)$$
$$(2 + b) + b + 1 = 3b$$
$$2b + 3 = 3b$$
$$3 = b$$

Now substitute 3 for b in Equation (1) and find a.
$$a = 2 + b = 2 + 3 = 5$$

Check. If there are 5 people ahead of you and 3 people behind you, then the number of people ahead of you is two more than the number behind you. Also, in the entire line there are $5 + 1 + 3$, or 9 people, and $3 \cdot 3 = 9$. The answer checks.

State. There are 5 people ahead of you.

21. The line graphed in red contains the points $(-2, 3)$ and $(3, 0)$.
$$m = \frac{3 - 0}{-2 - 3} = \frac{3}{-5} = -\frac{3}{5}$$

Then the equation of the line is $y = -\dfrac{3}{5}x + b$. To find b we substitute the coordinates of either point for x and y in the equation. We use $(3, 0)$.
$$y = -\frac{3}{5}x + b$$
$$0 = -\frac{3}{5} \cdot 3 + b$$
$$0 = -\frac{9}{5} + b$$
$$\frac{9}{5} = b$$

The equation of the line is $y = -\dfrac{3}{5}x + \dfrac{9}{5}$.

The line graphed in blue contains the points $(-2, 3)$ and $(3, 4)$.
$$m = \frac{3 - 4}{-2 - 3} = \frac{-1}{-5} = \frac{1}{5}$$

Then the equation is $y - \dfrac{1}{5}x + b$. To find b we substitute the coordinates of either point for x and y in the equation. We use $(3, 4)$.
$$y = \frac{1}{5}x + b$$
$$4 = \frac{1}{5} \cdot 3 + b$$
$$4 = \frac{3}{5} + b$$
$$\frac{17}{5} = b$$

The equation of the line is $y = \dfrac{1}{5}x + \dfrac{17}{5}$.

22. The line graphed in red is a vertical line, so its equation is of the form $x = a$. The line contains the point $(3, -2)$, so the equation is $x = 3$.

The line graphed in blue is a horizontal line, so its equation is of the form $y = b$. The line contains the point $(3, -2)$, so the equation is $y = -2$.

Chapter 8

Radical Expressions and Equations

Exercise Set 8.1

1. The square roots of 4 are 2 and -2, because $2^2 = 4$ and $(-2)^2 = 4$.

3. The square roots of 9 are 3 and -3, because $3^2 = 9$ and $(-3)^2 = 9$.

5. The square roots of 100 are 10 and -10, because $10^2 = 100$ and $(-10)^2 = 100$.

7. The square roots of 169 are 13 and -13, because $13^2 = 169$ and $(-13)^2 = 169$.

9. The square roots of 256 are 16 and -16, because $16^2 = 256$ and $(-16)^2 = 256$.

11. $\sqrt{4} = 2$, taking the principal square root.

13. $\sqrt{9} = 3$, so $-\sqrt{9} = -3$.

15. $\sqrt{36} = 6$, so $-\sqrt{36} = -6$.

17. $\sqrt{225} = 15$, so $-\sqrt{225} = -15$.

19. $\sqrt{361} = 19$, taking the principal square root.

21. 2.236

23. 20.785

25. $\sqrt{347.7} \approx 18.647$, so $-\sqrt{347.7} \approx -18.647$.

27. 2.779

29. $\sqrt{8 \cdot 9 \cdot 200} = 120$, so $-\sqrt{8 \cdot 9 \cdot 200} = -120$.

31. a) We substitute 25 into the formula:
$$N = 2.5\sqrt{25} = 2.5(5) = 12.5 \approx 13 \text{ spaces}$$

    b) We substitute 89 into the formula and use a calculator to find an approximation.
$$N = 2.5\sqrt{89} \approx 2.5(9.434) = 23.585 \approx 24 \text{ spaces}$$

33. Substitute 33.5 in the formula.
$$T = 0.144\sqrt{33.5} \approx 0.833 \text{ sec}$$

35. Substitute 40 in the formula.
$$T = 0.144\sqrt{40} \approx 0.911 \text{ sec}$$

37. The radicand is the expression under the radical, 200.

39. The radicand is the expression under the radical, $a - 4$.

41. The radicand is the expression under the radical, $t^2 + 1$.

43. The radicand is the expression under the radical, $\dfrac{3}{x + 2}$.

45. No, because the radicand is negative

47. Yes, because the radicand is nonnegative

49. No, because the radicand is negative

51. $\sqrt{c^2} = c$ Since c is assumed to be nonnegative

53. $\sqrt{9x^2} = \sqrt{(3x)^2} = 3x$ Since $3x$ is assumed to be nonnegative

55. $\sqrt{(8p)^2} = 8p$ Since $8p$ is assumed to be nonnegative

57. $\sqrt{(ab)^2} = ab$

59. $\sqrt{(34d)^2} = 34d$

61. $\sqrt{(x+3)^2} = x + 3$

63. $\sqrt{a^2 - 10a + 25} = \sqrt{(a-5)^2} = a - 5$

65. $\sqrt{4a^2 - 20a + 25} = \sqrt{(2a-5)^2} = 2a - 5$

67. Discussion and Writing Exercise

69. *Familiarize*. Let x and y represent the angles. Recall that supplementary angles are angles whose sum is 180°.

Translate. We reword the problem.

The resulting system is
$$x + y = 180, \quad (1)$$
$$x = 2y - 3. \quad (2)$$

Solve. We use substitution. We substitute $2y - 3$ for x in Equation (1) and solve for y.
$$(2y - 3) + y = 180$$
$$3y - 3 = 180$$
$$3y = 183$$
$$y = 61$$

We substitute 61 for y in Equation (2) to find x.
$$x = 2(61) - 3 = 122 - 3 = 119$$

Check. $61° + 119° = 180°$, so the angles are supplementary. Also, 3° less than twice 61° is $2 \cdot 61° - 3°$, or $122° - 3°$, or 119°. The numbers check.

State. The angles are 61° and 119°.

71. *Familiarize*. This problem states that we have direct variation between F and I. Thus, an equation $F = kI$, $k > 0$, applies. As the income increases, the amount spent on food increases.

Translate. We write an equation of variation.

Amount spent on food varies directly as the income.

This translates to $F = kI$.

Solve.

a) First find an equation of variation.

$$F = kI$$

$$10,192 = k \cdot 39,200 \quad \text{Substituting 10,192 for } F$$
$$\text{and 39,200 for } I$$

$$\frac{10,192}{39,200} = k$$

$$0.26 = k$$

The equation of variation is $F = 0.26I$.

b) We use the equation to find how much a family spends on food when their income is \$41,000.

$$F = 0.26I$$

$$F = 0.26(\$41,000) \quad \text{Substituting \$41,000 for } I$$

$$F = \$10,660$$

Check. Let us do some reasoning about the answer. The income increased from \$39,200 to \$41,000. Similarly, the amount spend on food increased from \$10,192 to \$10,660. This is what we would expect with direct variation.

State. The amount spent on food is \$10,660.

73. $$\frac{x^2 + 10x - 11}{x^2 - 1} \div \frac{x + 11}{x + 1}$$

$$= \frac{x^2 + 10x - 11}{x^2 - 1} \cdot \frac{x + 1}{x + 11}$$

$$= \frac{(x^2 + 10x - 11)(x + 1)}{(x^2 - 1)(x + 11)}$$

$$= \frac{(x + 11)(x - 1)(x + 1)}{(x + 1)(x - 1)(x + 11)}$$

$$= 1$$

75. To approximate $\sqrt{3}$, locate 3 on the x-axis, move up vertically to the graph, and then move left horizontally to the y-axis to read the approximation.

$$\sqrt{3} \approx 1.7 \quad \text{(Answers may vary.)}$$

To approximate $\sqrt{5}$, locate 5 on the x-axis, move up vertically to the graph, and then move left horizontally to the y-axis to read the approximation.

$$\sqrt{5} \approx 2.2 \quad \text{(Answers may vary.)}$$

To approximate $\sqrt{7}$, locate 7 on the x-axis, move up vertically to the graph, and then move left horizontally to the y-axis to read the approximation.

$$\sqrt{7} \approx 2.6 \quad \text{(Answers may vary.)}$$

77. If $\sqrt{x^2} = 16$, then $x^2 = 256$ since $\sqrt{256} = 16$. Thus $x = 16$ or $x = -16$.

79. If $t^2 = 49$ then the values of t are the square roots of 49, 7 and -7.

Exercise Set 8.2

1. $\sqrt{12} = \sqrt{4 \cdot 3}$ 4 is a perfect square.

       $= \sqrt{4}\,\sqrt{3}$ Factoring into a product of radicals

       $= 2\sqrt{3}$ Taking the square root

3. $\sqrt{75} = \sqrt{25 \cdot 3}$ 25 is a perfect square.

       $= \sqrt{25}\,\sqrt{3}$ Factoring into a product of radicals

       $= 5\sqrt{3}$ Taking the square root

5. $\sqrt{20} = \sqrt{4 \cdot 5}$ 4 is a perfect square.

       $= \sqrt{4}\,\sqrt{5}$ Factoring into a product of radicals

       $= 2\sqrt{5}$ Taking the square root

7. $\sqrt{600} = \sqrt{100 \cdot 6}$ 100 is a perfect square.

       $= \sqrt{100} \cdot \sqrt{6}$ Factoring into a product of radicals

       $= 10\sqrt{6}$ Taking the square root

9. $\sqrt{486} = \sqrt{81 \cdot 6}$ 81 is a perfect square.

       $= \sqrt{81} \cdot \sqrt{6}$ Factoring into a product of radicals

       $= 9\sqrt{6}$ Taking the square root

11. $\sqrt{9x} = \sqrt{9 \cdot x} = \sqrt{9}\,\sqrt{x} = 3\sqrt{x}$

13. $\sqrt{48x} = \sqrt{16 \cdot 3 \cdot x} = \sqrt{16}\,\sqrt{3x} = 4\sqrt{3x}$

15. $\sqrt{16a} = \sqrt{16 \cdot a} = \sqrt{16}\,\sqrt{a} = 4\sqrt{a}$

17. $\sqrt{64y^2} = \sqrt{64}\,\sqrt{y^2} = 8y$, or

      $\sqrt{64y^2} = \sqrt{(8y)^2} = 8y$

19. $\sqrt{13x^2} = \sqrt{13}\,\sqrt{x^2} = \sqrt{13} \cdot x$, or $x\sqrt{13}$

21. $\sqrt{8t^2} = \sqrt{2 \cdot 4 \cdot t^2} = \sqrt{4}\,\sqrt{t^2}\,\sqrt{2} = 2t\sqrt{2}$

23. $\sqrt{180} = \sqrt{36 \cdot 5} = 6\sqrt{5}$

25. $\sqrt{288y} = \sqrt{144 \cdot 2 \cdot y} = \sqrt{144}\,\sqrt{2y} = 12\sqrt{2y}$

27. $\sqrt{28x^2} = \sqrt{4 \cdot 7 \cdot x^2} = \sqrt{4}\,\sqrt{x^2}\,\sqrt{7} = 2x\sqrt{7}$

29. $\sqrt{x^2 - 6x + 9} = \sqrt{(x - 3)^2} = x - 3$

31. $\sqrt{8x^2 + 8x + 2} = \sqrt{2(4x^2 + 4x + 1)} =$

      $\sqrt{2(2x + 1)^2} = \sqrt{2}\,\sqrt{(2x + 1)^2} = \sqrt{2}\,(2x + 1)$

33. $\sqrt{36y + 12y^2 + y^3} = \sqrt{y(36 + 12y + y^2)} =$

      $\sqrt{y(6 + y)^2} = \sqrt{y}\,\sqrt{(6 + y)^2} = \sqrt{y}\,(6 + y)$

35. $\sqrt{t^6} = \sqrt{(t^3)^2} = t^3$

37. $\sqrt{x^{12}} = \sqrt{(x^6)^2} = x^6$

39. $\sqrt{x^5} = \sqrt{x^4 \cdot x}$ One factor is a perfect square

$\qquad = \sqrt{x^4}\,\sqrt{x}$

$\qquad = \sqrt{(x^2)^2}\,\sqrt{x}$

$\qquad = x^2\sqrt{x}$

41. $\sqrt{t^{19}} = \sqrt{t^{18} \cdot t} = \sqrt{t^{18}}\,\sqrt{t} = \sqrt{(t^9)^2}\,\sqrt{t} = t^9\sqrt{t}$

43. $\sqrt{(y-2)^8} = \sqrt{[(y-2)^4]^2} = (y-2)^4$

45. $\sqrt{4(x+5)^{10}} = \sqrt{4[(x+5)^5]^2} = \sqrt{4}\,\sqrt{[(x+5)^5]^2} =$
$2(x+5)^5$

47. $\sqrt{36m^3} = \sqrt{36 \cdot m^2 \cdot m} = \sqrt{36}\,\sqrt{m^2}\,\sqrt{m} = 6m\sqrt{m}$

49. $\sqrt{8a^5} = \sqrt{2 \cdot 4 \cdot a^4 \cdot a} = \sqrt{2 \cdot 4 \cdot (a^2)^2 \cdot a} =$
$\sqrt{4}\,\sqrt{(a^2)^2}\,\sqrt{2a} = 2a^2\sqrt{2a}$

51. $\sqrt{104p^{17}} = \sqrt{4 \cdot 26 \cdot p^{16} \cdot p} = \sqrt{4 \cdot 26 \cdot (p^8)^2 \cdot p} =$
$\sqrt{4}\,\sqrt{(p^8)^2}\,\sqrt{26p} = 2p^8\sqrt{26p}$

53. $\sqrt{448x^6y^3} = \sqrt{64 \cdot 7 \cdot x^6 \cdot y^2 \cdot y} =$
$\sqrt{64 \cdot 7 \cdot (x^3)^2 \cdot y^2 \cdot y} =$
$\sqrt{64}\,\sqrt{(x^3)^2}\,\sqrt{y^2}\,\sqrt{7y} = 8x^3y\,\sqrt{7y}$

55. $\sqrt{3}\,\sqrt{18} = \sqrt{3 \cdot 18}$ Multiplying

$\qquad = \sqrt{3 \cdot 3 \cdot 6}$ Looking for perfect-square factors or pairs of factors

$\qquad = \sqrt{3 \cdot 3}\,\sqrt{6}$

$\qquad = 3\sqrt{6}$

57. $\sqrt{15}\,\sqrt{6} = \sqrt{15 \cdot 6}$ Multiplying

$\qquad = \sqrt{5 \cdot 3 \cdot 3 \cdot 2}$ Looking for perfect-square factors or pairs of factors

$\qquad = \sqrt{3 \cdot 3}\,\sqrt{5 \cdot 2}$

$\qquad = 3\sqrt{10}$

59. $\sqrt{18}\,\sqrt{14x} = \sqrt{18 \cdot 14x} = \sqrt{3 \cdot 3 \cdot 2 \cdot 2 \cdot 7 \cdot x} =$
$\sqrt{3 \cdot 3}\,\sqrt{2 \cdot 2}\,\sqrt{7x} = 3 \cdot 2\sqrt{7x} = 6\sqrt{7x}$

61. $\sqrt{3x}\,\sqrt{12y} = \sqrt{3x \cdot 12y} = \sqrt{3 \cdot x \cdot 3 \cdot 4 \cdot y} =$
$\sqrt{3 \cdot 3 \cdot 4 \cdot x \cdot y} = \sqrt{3 \cdot 3}\,\sqrt{4}\,\sqrt{x \cdot y} = 3 \cdot 2\sqrt{xy} = 6\sqrt{xy}$

63. $\sqrt{13}\,\sqrt{13} = \sqrt{13 \cdot 13} = 13$

65. $\sqrt{5b}\,\sqrt{15b} = \sqrt{5b \cdot 15b} = \sqrt{5 \cdot b \cdot 5 \cdot 3 \cdot b} =$
$\sqrt{5 \cdot 5 \cdot b \cdot b \cdot 3} = \sqrt{5 \cdot 5}\,\sqrt{b \cdot b}\,\sqrt{3} = 5b\sqrt{3}$

67. $\sqrt{2t}\,\sqrt{2t} = \sqrt{2t \cdot 2t} = 2t$

69. $\sqrt{ab}\,\sqrt{ac} = \sqrt{ab \cdot ac} = \sqrt{a \cdot a \cdot b \cdot c} = \sqrt{a \cdot a}\,\sqrt{b \cdot c} =$
$a\sqrt{bc}$

71. $\sqrt{2x^2y}\,\sqrt{4xy^2} = \sqrt{2x^2y \cdot 4xy^2} = \sqrt{2 \cdot x^2 \cdot y \cdot 4 \cdot x \cdot y^2} =$
$\sqrt{4}\,\sqrt{x^2}\,\sqrt{y^2}\,\sqrt{2xy} = 2xy\sqrt{2xy}$

73. $\sqrt{18}\,\sqrt{18} = \sqrt{18 \cdot 18} = 18$

75. $\sqrt{5}\,\sqrt{2x-1} = \sqrt{5(2x-1)} = \sqrt{10x-5}$

77. $\sqrt{x+2}\,\sqrt{x+2} = \sqrt{(x+2)^2} = x+2$

79. $\sqrt{18x^2y^3}\,\sqrt{6xy^4} = \sqrt{18x^2y^3 \cdot 6xy^4} =$
$\sqrt{3 \cdot 6 \cdot x^2 \cdot y^2 \cdot y \cdot 6 \cdot x \cdot y^4} = \sqrt{6 \cdot 6 \cdot x^2 \cdot y^6 \cdot 3 \cdot x \cdot y} =$
$\sqrt{6 \cdot 6}\,\sqrt{x^2}\,\sqrt{y^6}\,\sqrt{3xy} = 6xy^3\sqrt{3xy}$

81. $\sqrt{50x^4y^6}\,\sqrt{10xy} = \sqrt{50x^4y^6 \cdot 10xy} =$
$\sqrt{5 \cdot 10 \cdot x^4 \cdot y^6 \cdot 10 \cdot x \cdot y} = \sqrt{10 \cdot 10 \cdot x^4 \cdot y^6 \cdot 5 \cdot x \cdot y} =$
$\sqrt{10 \cdot 10}\,\sqrt{x^4}\,\sqrt{y^6}\,\sqrt{5xy} = 10x^2y^3\sqrt{5xy}$

83. $\sqrt{99p^4q^3}\,\sqrt{22p^5q^2} = \sqrt{99p^4q^3 \cdot 22p^5q^2} =$
$\sqrt{9 \cdot 11 \cdot p^4 \cdot q^2 \cdot q \cdot 2 \cdot 11 \cdot p^4 \cdot p \cdot q^2} =$
$\sqrt{9 \cdot 11 \cdot 11 \cdot p^4 \cdot q^2 \cdot p^4 \cdot q^2 \cdot 2 \cdot q \cdot p} =$
$3 \cdot 11 \cdot p^2 \cdot q \cdot p^2 \cdot q\sqrt{2pq} = 33p^4q^2\sqrt{2pq}$

85. $\sqrt{24a^2b^3c^4}\sqrt{32a^5b^4c^7} = \sqrt{24a^2b^3c^4 \cdot 32a^5b^4c^7} =$
$\sqrt{4 \cdot 2 \cdot 3 \cdot a^2 \cdot b^2 \cdot b \cdot c^4 \cdot 16 \cdot 2 \cdot a^4 \cdot a \cdot b^4 \cdot c^6 \cdot c} =$
$\sqrt{4 \cdot 2 \cdot 2 \cdot 16 \cdot a^2 \cdot b^2 \cdot c^4 \cdot a^4 \cdot b^4 \cdot c^6 \cdot 3 \cdot b \cdot a \cdot c} =$
$2 \cdot 2 \cdot 4 \cdot a \cdot b \cdot c^2 \cdot a^2 \cdot b^2 \cdot c^3\sqrt{3abc} =$
$16a^3b^3c^5\sqrt{3abc}$

87. Discussion and Writing Exercise

89. $\quad x - y = -6 \quad$ (1)

$\quad \underline{x + y = 2} \qquad$ (2)

$\quad 2x \qquad = -4 \quad$ Adding

$\qquad x = -2$

Now we substitute -2 for x in one of the original equations and solve for y.

$\qquad x + y = 2 \quad$ Equation (2)

$\qquad -2 + y = 2 \quad$ Substituting

$\qquad\qquad y = 4$

Since $(-2, 4)$ checks in both equations, it is the solution.

91. $\quad 3x - 2y = 4, \quad$ (1)

$\quad 2x + 5y = 9 \quad$ (2)

We will us the elimination method. We multiply on both sides of Equation (1) by 5 and on both sides of Equation (2) by 2. Then we add

$\qquad 15x - 10y = 20$

$\qquad \underline{4x + 10y = 18}$

$\qquad 19x \qquad = 38$

$\qquad\qquad x = 2$

Now we substitute 2 for x in one of the original equations and solve for y.

$\qquad 2x + 5y = 9 \quad$ Equation (2)

$\qquad 2 \cdot 2 + 5y = 9$

$\qquad 4 + 5y = 9$

$\qquad\qquad 5y = 5$

$\qquad\qquad y = 1$

Since $(2, 1)$ checks, it is the solution.

93. *Familiarize.* We let l = the length of the rectangle and w = the width. Recall that the perimeter of a rectangle is $2l + 2w$, and the area is lw.

Translate. We translate the first statement.

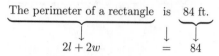

Now we translate the second statement.

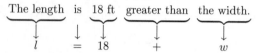

The resulting system is

$$2l + 2w = 84, \quad (1)$$
$$l = 18 + w. \quad (2)$$

Solve. We use the substitution method. We substitute $15 + w$ for l in Equation (1) and solve for w.

$$2l + 2w = 84 \quad \text{Equation (1)}$$
$$2(18 + w) + 2w = 84 \quad \text{Substituting}$$
$$36 + 2w + 2w = 84$$
$$36 + 4w = 84$$
$$4w = 48$$
$$w = 12$$

Substitute 12 for w in Equation (2) and compute l.

$$l = 18 + w = 18 + 12 = 30$$

When $l = 30$ ft and $w = 12$ ft, the area is 30 ft $\cdot$ 12 ft, or 360 ft^2.

Check. When $l = 30$ ft and $w = 12$ ft, the perimeter is $2 \cdot 30$ ft $+ 2 \cdot 12$ ft, or 60 ft $+$ 24 ft, or 84 ft. The length, 30 ft, is 18 ft greater than 12 ft, the width. We recheck the computation of the area.

State. The area of the rectangle is 360 ft^2.

95. *Familiarize.* Let x = the number of liters of 30% solution and y = the number of liters of 50% solution to be used. We organize the information in a table.

Type of solution	30% insecticide	50% insecticide	Mixture
Amount of solution	x	y	200 L
Percent of insecticide	30%	50%	42%
Amount of insecticide in solution	30%x	50%y	42% × 200, or 84 L

Translate. The first and last rows of the table give us two equations. Since the total amount of solution is 200 L, we have

$$x + y = 200.$$

The amount of insecticide in the mixture is to be 42% of 200, or 84 L. The amounts of insecticide from the two solutions are 30%x and 50%y. Thus

$$30\%x + 50\%y = 84,$$
$$\text{or} \quad 0.3x + 0.5y = 84,$$
$$\text{or} \quad 3x + 5y = 840 \quad \text{Clearing decimals}$$

The resulting system is

$$x + y = 200, \quad (1)$$
$$3x + 5y = 840. \quad (2)$$

Solve. We use the elimination method. Multiply by -3 on both sides of Equation (1) and then add.

$$-3x - 3y = -600$$
$$\underline{3x + 5y = 840}$$
$$2y = 240$$
$$y = 120$$

We go back to Equation (1) and substitute 120 for y.

$$x + y = 200$$
$$x + 120 = 200$$
$$x = 80$$

Check. The sum of 80 L and 120 L is 200 L. Also, 30% of 80 is 24 and 50% of 120 is 60 and $24 + 60 = 84$. The answer checks.

State. 80 L of 30% solution and 120 L of 50% solution should be used.

97. $\sqrt{5x - 5} = \sqrt{5(x-1)} = \sqrt{5}\sqrt{x-1}$

99. $\sqrt{x^2 - 36} = \sqrt{(x+6)(x-6)} = \sqrt{x+6}\sqrt{x-6}$

101. $\sqrt{x^3 - 2x^2} = \sqrt{x^2(x-2)} = \sqrt{x^2}\sqrt{x-2} = x\sqrt{x-2}$

103. $\sqrt{0.25} = \sqrt{(0.5)^2} = 0.5$

105. $\sqrt{2y}\sqrt{3}\sqrt{8y} = \sqrt{2y \cdot 3 \cdot 8y} = \sqrt{2 \cdot y \cdot 3 \cdot 2 \cdot 4 \cdot y} = \sqrt{2 \cdot 2 \cdot 4 \cdot y \cdot y \cdot 3} = \sqrt{2 \cdot 2}\sqrt{4}\sqrt{y \cdot y}\sqrt{3} = 2 \cdot 2 \cdot y\sqrt{3} = 4y\sqrt{3}$

107. $\sqrt{27(x+1)}\sqrt{12y(x+1)^2}$
$\sqrt{27(x+1) \cdot 12y(x+1)^2} =$
$\sqrt{9 \cdot 3 \cdot (x+1) \cdot 4 \cdot 3 \cdot y(x+1)^2} =$
$\sqrt{9 \cdot 3 \cdot 3 \cdot 4 \cdot (x+1)^2 \cdot (x+1)y} =$
$\sqrt{9}\sqrt{3 \cdot 3}\sqrt{4}\sqrt{(x+1)^2}\sqrt{(x+1)y}$
$3 \cdot 3 \cdot 2(x+1)\sqrt{(x+1)y} = 18(x+1)\sqrt{(x+1)y}$

109. $\sqrt{x}\sqrt{2x}\sqrt{10x^5} = \sqrt{x \cdot 2x \cdot 10x^5} =$
$\sqrt{x \cdot 2 \cdot x \cdot 2 \cdot 5 \cdot x^4 \cdot x} = \sqrt{x \cdot x \cdot 2 \cdot 2 \cdot x^4 \cdot 5 \cdot x} =$
$\sqrt{x \cdot x}\sqrt{2 \cdot 2}\sqrt{x^4}\sqrt{5x} = x \cdot 2 \cdot x^2\sqrt{5x} = 2x^3\sqrt{5x}$

Exercise Set 8.3

1. $\dfrac{\sqrt{18}}{\sqrt{2}} = \sqrt{\dfrac{18}{2}} = \sqrt{9} = 3$

3. $\dfrac{\sqrt{108}}{\sqrt{3}} = \sqrt{\dfrac{108}{3}} = \sqrt{36} = 6$

5. $\dfrac{\sqrt{65}}{\sqrt{13}} = \sqrt{\dfrac{65}{13}} = \sqrt{5}$

7. $\dfrac{\sqrt{3}}{\sqrt{75}} = \sqrt{\dfrac{3}{75}} = \sqrt{\dfrac{1}{25}} = \dfrac{1}{5}$

9. $\dfrac{\sqrt{12}}{\sqrt{75}} = \sqrt{\dfrac{12}{75}} = \sqrt{\dfrac{4}{25}} = \dfrac{2}{5}$

11. $\dfrac{\sqrt{8x}}{\sqrt{2x}} = \sqrt{\dfrac{8x}{2x}} = \sqrt{4} = 2$

13. $\dfrac{\sqrt{63y^3}}{\sqrt{7y}} = \sqrt{\dfrac{63y^3}{7y}} = \sqrt{9y^2} = 3y$

15. $\sqrt{\dfrac{16}{49}} = \dfrac{\sqrt{16}}{\sqrt{49}} = \dfrac{4}{7}$

17. $\sqrt{\dfrac{1}{36}} = \dfrac{\sqrt{1}}{\sqrt{36}} = \dfrac{1}{6}$

19. $-\sqrt{\dfrac{16}{81}} = -\dfrac{\sqrt{16}}{\sqrt{81}} = -\dfrac{4}{9}$

21. $\sqrt{\dfrac{64}{289}} = \dfrac{\sqrt{64}}{\sqrt{289}} = \dfrac{8}{17}$

23. $\sqrt{\dfrac{1690}{1960}} = \sqrt{\dfrac{169 \cdot 10}{196 \cdot 10}} = \sqrt{\dfrac{169}{196} \cdot \dfrac{10}{10}} = \sqrt{\dfrac{169}{196} \cdot 1} =$

$\sqrt{\dfrac{169}{196}} = \dfrac{\sqrt{169}}{\sqrt{196}} = \dfrac{13}{14}$

25. $\sqrt{\dfrac{25}{x^2}} = \dfrac{\sqrt{25}}{\sqrt{x^2}} = \dfrac{5}{x}$

27. $\sqrt{\dfrac{9a^2}{625}} = \dfrac{\sqrt{9a^2}}{\sqrt{625}} = \dfrac{3a}{25}$

29. $\dfrac{\sqrt{50y^{15}}}{\sqrt{2y^{25}}} = \sqrt{\dfrac{50y^{15}}{2y^{25}}} = \sqrt{\dfrac{25}{y^{10}}} = \dfrac{\sqrt{25}}{\sqrt{y^{10}}} = \dfrac{5}{y^5}$

31. $\dfrac{\sqrt{7x^{23}}}{\sqrt{343x^5}} = \sqrt{\dfrac{7x^{23}}{343x^5}} = \sqrt{\dfrac{x^{18}}{49}} = \dfrac{\sqrt{x^{18}}}{\sqrt{49}} = \dfrac{x^9}{7}$

33. $\sqrt{\dfrac{2}{5}} = \sqrt{\dfrac{2}{5} \cdot \dfrac{5}{5}} = \sqrt{\dfrac{10}{25}} = \dfrac{\sqrt{10}}{\sqrt{25}} = \dfrac{\sqrt{10}}{5}$

35. $\sqrt{\dfrac{7}{8}} = \sqrt{\dfrac{7}{8} \cdot \dfrac{2}{2}} = \sqrt{\dfrac{14}{16}} = \dfrac{\sqrt{14}}{\sqrt{16}} = \dfrac{\sqrt{14}}{4}$

37. $\sqrt{\dfrac{1}{12}} = \sqrt{\dfrac{1}{12} \cdot \dfrac{3}{3}} = \sqrt{\dfrac{3}{36}} = \dfrac{\sqrt{3}}{\sqrt{36}} = \dfrac{\sqrt{3}}{6}$

39. $\sqrt{\dfrac{5}{18}} = \sqrt{\dfrac{5}{18} \cdot \dfrac{2}{2}} = \sqrt{\dfrac{10}{36}} = \dfrac{\sqrt{10}}{\sqrt{36}} = \dfrac{\sqrt{10}}{6}$

41. $\dfrac{3}{\sqrt{5}} = \dfrac{3}{\sqrt{5}} \cdot \dfrac{\sqrt{5}}{\sqrt{5}} = \dfrac{3\sqrt{5}}{5}$

43. $\sqrt{\dfrac{8}{3}} = \sqrt{\dfrac{8}{3} \cdot \dfrac{3}{3}} = \sqrt{\dfrac{24}{9}} = \dfrac{\sqrt{4 \cdot 6}}{\sqrt{9}} = \dfrac{\sqrt{4}\,\sqrt{6}}{\sqrt{9}} = \dfrac{2\sqrt{6}}{3}$

45. $\sqrt{\dfrac{3}{x}} = \sqrt{\dfrac{3}{x} \cdot \dfrac{x}{x}} = \sqrt{\dfrac{3x}{x^2}} = \dfrac{\sqrt{3x}}{\sqrt{x^2}} = \dfrac{\sqrt{3x}}{x}$

47. $\sqrt{\dfrac{x}{y}} = \sqrt{\dfrac{x}{y} \cdot \dfrac{y}{y}} = \sqrt{\dfrac{xy}{y^2}} = \dfrac{\sqrt{xy}}{\sqrt{y^2}} = \dfrac{\sqrt{xy}}{y}$

49. $\sqrt{\dfrac{x^2}{20}} = \sqrt{\dfrac{x^2}{20} \cdot \dfrac{5}{5}} = \sqrt{\dfrac{5x^2}{100}} = \dfrac{\sqrt{x^2 \cdot 5}}{\sqrt{100}} = \dfrac{\sqrt{x^2}\,\sqrt{5}}{\sqrt{100}} = \dfrac{x\sqrt{5}}{10}$

51. $\dfrac{\sqrt{7}}{\sqrt{2}} = \dfrac{\sqrt{7}}{\sqrt{2}} \cdot \dfrac{\sqrt{2}}{\sqrt{2}} = \dfrac{\sqrt{14}}{2}$

53. $\dfrac{\sqrt{9}}{\sqrt{8}} = \dfrac{\sqrt{9}}{\sqrt{8}} \cdot \dfrac{\sqrt{2}}{\sqrt{2}} = \dfrac{\sqrt{9 \cdot 2}}{\sqrt{16}} = \dfrac{3\sqrt{2}}{4}$

55. $\dfrac{\sqrt{3}}{\sqrt{2}} = \dfrac{\sqrt{3}}{\sqrt{2}} \cdot \dfrac{\sqrt{2}}{\sqrt{2}} = \dfrac{\sqrt{6}}{2}$

57. $\dfrac{2}{\sqrt{2}} = \dfrac{2}{\sqrt{2}} \cdot \dfrac{\sqrt{2}}{\sqrt{2}} = \dfrac{2\sqrt{2}}{2} = \sqrt{2}$

59. $\dfrac{\sqrt{5}}{\sqrt{11}} = \dfrac{\sqrt{5}}{\sqrt{11}} \cdot \dfrac{\sqrt{11}}{\sqrt{11}} = \dfrac{\sqrt{55}}{11}$

61. $\dfrac{\sqrt{7}}{\sqrt{12}} = \dfrac{\sqrt{7}}{\sqrt{12}} \cdot \dfrac{\sqrt{3}}{\sqrt{3}} = \dfrac{\sqrt{21}}{\sqrt{36}} = \dfrac{\sqrt{21}}{6}$

63. $\dfrac{\sqrt{48}}{\sqrt{32}} = \sqrt{\dfrac{48}{32}} = \sqrt{\dfrac{3}{2}} = \sqrt{\dfrac{3}{2} \cdot \dfrac{2}{2}} = \sqrt{\dfrac{6}{4}} = \dfrac{\sqrt{6}}{\sqrt{4}} = \dfrac{\sqrt{6}}{2}$

65. $\dfrac{\sqrt{450}}{\sqrt{18}} = \sqrt{\dfrac{450}{18}} = \sqrt{25} = 5$

67. $\dfrac{\sqrt{3}}{\sqrt{x}} = \dfrac{\sqrt{3}}{\sqrt{x}} \cdot \dfrac{\sqrt{x}}{\sqrt{x}} = \dfrac{\sqrt{3x}}{x}$

69. $\dfrac{4y}{\sqrt{5}} = \dfrac{4y}{\sqrt{5}} \cdot \dfrac{\sqrt{5}}{\sqrt{5}} = \dfrac{4y\sqrt{5}}{5}$

71. $\dfrac{\sqrt{a^3}}{\sqrt{8}} = \dfrac{\sqrt{a^3}}{\sqrt{8}} \cdot \dfrac{\sqrt{2}}{\sqrt{2}} = \dfrac{\sqrt{2a^3}}{\sqrt{16}} = \dfrac{\sqrt{a^2 \cdot 2a}}{\sqrt{16}} = \dfrac{a\sqrt{2a}}{4}$

73. $\dfrac{\sqrt{56}}{\sqrt{12x}} = \sqrt{\dfrac{56}{12x}} = \sqrt{\dfrac{14}{3x}} = \sqrt{\dfrac{14}{3x} \cdot \dfrac{3x}{3x}} = \sqrt{\dfrac{42x}{3x \cdot 3x}} =$

$\dfrac{\sqrt{42x}}{3x}$

75. $\dfrac{\sqrt{27c}}{\sqrt{32c^3}} = \sqrt{\dfrac{27c}{32c^3}} = \sqrt{\dfrac{27}{32c^2}} = \sqrt{\dfrac{27}{32c^2} \cdot \dfrac{2}{2}} = \sqrt{\dfrac{54}{64c^2}} =$

$\sqrt{\dfrac{9 \cdot 6}{64c^2}} = \dfrac{3\sqrt{6}}{8c}$

77. $\dfrac{\sqrt{y^5}}{\sqrt{xy^2}} = \sqrt{\dfrac{y^5}{xy^2}} = \sqrt{\dfrac{y^3}{x}} = \sqrt{\dfrac{y^3}{x} \cdot \dfrac{x}{x}} = \sqrt{\dfrac{xy^3}{x^2}} =$

$\sqrt{\dfrac{y^2 \cdot xy}{x^2}} = \dfrac{y\sqrt{xy}}{x}$

79. $\dfrac{\sqrt{45mn^2}}{\sqrt{32m}} = \sqrt{\dfrac{45mn^2}{32m}} = \sqrt{\dfrac{45n^2}{32}} = \sqrt{\dfrac{45n^2}{32} \cdot \dfrac{2}{2}} =$

$\sqrt{\dfrac{90n^2}{64}} = \dfrac{\sqrt{90n^2}}{\sqrt{64}} = \dfrac{\sqrt{9 \cdot n^2 \cdot 10}}{8} = \dfrac{3n\sqrt{10}}{8}$

81. Discussion and Writing Exercise

83. $x = y + 2$,   (1)

$\quad x + y = 6$ (2)

We substitute $y + 2$ for x in Equation (2) and solve for y.

$$(y + 2) + y = 6$$
$$2y + 2 = 6$$
$$2y = 4$$
$$y = 2$$

Substitute 2 for y in Equation (1) to find x.

$$x = 2 + 2 = 4$$

The ordered pair $(4, 2)$ checks in both equations. It is the solution.

85. $2x - 3y = 7$ (1)

$\quad 2x - 3y = 9$ (2)

We multiply Equation (2) by -1 and add.

$$\begin{aligned} 2x - 3y &= 7 \\ -2x + 3y &= -9 \\ \hline 0 &= -2 \end{aligned}$$

We get a false equation. The system of equations has no solution.

87. $x + y = -7$ (1)

$\quad \underline{x - y = 2} \qquad (2)$

$\quad 2x \quad\quad = -5$ Adding

$$x = -\frac{5}{2}$$

Substitute $-\frac{5}{2}$ for x in Equation (1) to find y.

$$x + y = -7 \quad \text{Equation (1)}$$
$$-\frac{5}{2} + y = -7 \quad \text{Substituting}$$
$$y = -\frac{9}{2}$$

The ordered pair $\left(-\frac{5}{2}, -\frac{9}{2}\right)$ checks in both equations. It is the solution.

89.
$$\frac{x^2 - 49}{x + 8} \div \frac{x^2 - 14x + 49}{x^2 + 15x + 56}$$
$$= \frac{x^2 - 49}{x + 8} \cdot \frac{x^2 + 15x + 56}{x^2 - 14x + 49}$$
$$= \frac{(x^2 - 49)(x^2 + 15x + 56)}{(x + 8)(x^2 - 14x + 49)}$$
$$= \frac{(x + 7)(x - 7)(x + 7)(x + 8)}{(x + 8)(x - 7)(x - 7)}$$
$$= \frac{(x + 7)(x - 7)(x + 7)(x + 8)}{(x + 8)(x - 7)(x - 7)}$$
$$= \frac{(x + 7)(x + 7)}{x - 7}, \text{ or } \frac{(x + 7)^2}{x - 7}$$

91.
$$\frac{a^2 - 25}{6} \div \frac{a + 5}{3} = \frac{a^2 - 25}{6} \cdot \frac{3}{a + 5}$$
$$= \frac{(a^2 - 25) \cdot 3}{6(a + 5)}$$
$$= \frac{(a + 5)(a - 5) \cdot 3}{2 \cdot 3 \cdot (a + 5)}$$
$$= \frac{(a + 5)(a - 5) \cdot 3}{2 \cdot 3 \cdot (a + 5)}$$
$$= \frac{a - 5}{2}$$

93. $(3x - 7)(3x + 7) = (3x)^2 - 7^2 = 9x^2 - 49$

95.
$$9x - 5y + 12x - 4y$$
$$= (9x + 12x) + (-5y - 4y)$$
$$= 21x - 9y$$

97. 2 ft: $T \approx 2(3.14)\sqrt{\dfrac{2}{32}} \approx 6.28\sqrt{\dfrac{1}{16}} \approx 6.28\left(\dfrac{1}{4}\right) \approx$ 1.57 sec

8 ft: $T \approx 2(3.14)\sqrt{\dfrac{8}{32}} \approx 6.28\sqrt{\dfrac{1}{4}} \approx 6.28\left(\dfrac{1}{2}\right) \approx$ 3.14 sec

64 ft: $T \approx 2(3.14)\sqrt{\dfrac{64}{32}} \approx 6.28\sqrt{2} \approx$ $(6.28)(1.414) \approx 8.88$ sec

100 ft: $T \approx 2(3.14)\sqrt{\dfrac{100}{32}} \approx 6.28\sqrt{\dfrac{50}{16}} \approx \dfrac{6.28\sqrt{50}}{4} \approx$ $\dfrac{6.28(7.071)}{4} \approx 11.10$ sec

99. $T = 2\pi\sqrt{\dfrac{\frac{32}{\pi^2}}{32}} = 2\pi\sqrt{\dfrac{32}{\pi^2} \cdot \dfrac{1}{32}} = 2\pi\sqrt{\dfrac{1}{\pi^2}} = 2\pi\left(\dfrac{1}{\pi}\right) = 2$ sec

The time it takes the pendulum to swing from one side to the other and back is 2 sec, so it takes 1 sec to swing from one side to the other.

101. $\sqrt{\dfrac{5}{1600}} = \dfrac{\sqrt{5}}{\sqrt{1600}} = \dfrac{\sqrt{5}}{40}$

103. $\sqrt{\dfrac{1}{5x^3}} = \sqrt{\dfrac{1}{5x^3} \cdot \dfrac{5x}{5x}} = \sqrt{\dfrac{5x}{25x^4}} = \dfrac{\sqrt{5x}}{\sqrt{25x^4}} = \dfrac{\sqrt{5x}}{5x^2}$

105. $\sqrt{\dfrac{3a}{b}} = \sqrt{\dfrac{3a}{b} \cdot \dfrac{b}{b}} = \sqrt{\dfrac{3ab}{b^2}} = \dfrac{\sqrt{3ab}}{\sqrt{b^2}} = \dfrac{\sqrt{3ab}}{b}$

107. $\sqrt{0.009} = \sqrt{\dfrac{9}{1000}} = \sqrt{\dfrac{9}{1000} \cdot \dfrac{10}{10}} = \sqrt{\dfrac{90}{10,000}} =$

$\dfrac{\sqrt{90}}{\sqrt{10,000}} = \dfrac{\sqrt{9 \cdot 10}}{100} = \dfrac{\sqrt{9}\sqrt{10}}{100} = \dfrac{3\sqrt{10}}{100}$

109. $\sqrt{\dfrac{1}{x^2} - \dfrac{2}{xy} + \dfrac{1}{y^2}}$, LCD is x^2y^2

$= \sqrt{\dfrac{1}{x^2} \cdot \dfrac{y^2}{y^2} - \dfrac{2}{xy} \cdot \dfrac{xy}{xy} + \dfrac{1}{y^2} \cdot \dfrac{x^2}{x^2}}$

$= \sqrt{\dfrac{y^2 - 2xy + x^2}{x^2y^2}}$

$= \sqrt{\dfrac{(y-x)^2}{x^2y^2}}$

$= \dfrac{\sqrt{(y-x)^2}}{\sqrt{x^2y^2}}$

$= \dfrac{y-x}{xy}$

Exercise Set 8.4

1. $7\sqrt{3} + 9\sqrt{3} = (7+9)\sqrt{3}$
$= 16\sqrt{3}$

3. $7\sqrt{5} - 3\sqrt{5} = (7-3)\sqrt{5}$
$= 4\sqrt{5}$

5. $6\sqrt{x} + 7\sqrt{x} = (6+7)\sqrt{x}$
$= 13\sqrt{x}$

7. $4\sqrt{d} - 13\sqrt{d} = (4-13)\sqrt{d}$
$= -9\sqrt{d}$

9. $5\sqrt{8} + 15\sqrt{2} = 5\sqrt{4 \cdot 2} + 15\sqrt{2}$
$= 5 \cdot 2\sqrt{2} + 15\sqrt{2}$
$= 10\sqrt{2} + 15\sqrt{2}$
$= 25\sqrt{2}$

11. $\sqrt{27} - 2\sqrt{3} = \sqrt{9 \cdot 3} - 2\sqrt{3}$
$= 3\sqrt{3} - 2\sqrt{3}$
$= (3-2)\sqrt{3}$
$= 1\sqrt{3}$
$= \sqrt{3}$

13. $\sqrt{45} - \sqrt{20} = \sqrt{9 \cdot 5} - \sqrt{4 \cdot 5}$
$= 3\sqrt{5} - 2\sqrt{5}$
$= (3-2)\sqrt{5}$
$= 1\sqrt{5}$
$= \sqrt{5}$

15. $\sqrt{72} + \sqrt{98} = \sqrt{36 \cdot 2} + \sqrt{49 \cdot 2}$
$= 6\sqrt{2} + 7\sqrt{2}$
$= (6+7)\sqrt{2}$
$= 13\sqrt{2}$

17. $2\sqrt{12} + \sqrt{27} - \sqrt{48} = 2\sqrt{4 \cdot 3} + \sqrt{9 \cdot 3} - \sqrt{16 \cdot 3}$
$= 2 \cdot 2\sqrt{3} + 3\sqrt{3} - 4\sqrt{3}$
$= 4\sqrt{3} + 3\sqrt{3} - 4\sqrt{3}$
$= (4 + 3 - 4)\sqrt{3}$
$= 3\sqrt{3}$

19. $\sqrt{18} - 3\sqrt{8} + \sqrt{50} = \sqrt{9 \cdot 2} - 3\sqrt{4 \cdot 2} + \sqrt{25 \cdot 2}$
$= 3\sqrt{2} - 3 \cdot 2\sqrt{2} + 5\sqrt{2}$
$= 3\sqrt{2} - 6\sqrt{2} + 5\sqrt{2}$
$= (3 - 6 + 5)\sqrt{2}$
$= 2\sqrt{2}$

21. $2\sqrt{27} - 3\sqrt{48} + 3\sqrt{12} = 2\sqrt{9 \cdot 3} - 3\sqrt{16 \cdot 3} + 3\sqrt{4 \cdot 3}$
$= 2 \cdot 3\sqrt{3} - 3 \cdot 4\sqrt{3} + 3 \cdot 2\sqrt{3}$
$= 6\sqrt{3} - 12\sqrt{3} + 6\sqrt{3}$
$= (6 - 12 + 6)\sqrt{3}$
$= 0\sqrt{3}$
$= 0$

23. $\sqrt{4x} + \sqrt{81x^3} = \sqrt{4 \cdot x} + \sqrt{81 \cdot x^2 \cdot x}$
$= 2\sqrt{x} + 9x\sqrt{x}$
$= (2 + 9x)\sqrt{x}$

25. $\sqrt{27} - \sqrt{12x^2} = \sqrt{9 \cdot 3} - \sqrt{4 \cdot 3 \cdot x^2}$
$= 3\sqrt{3} - 2x\sqrt{3}$
$= (3 - 2x)\sqrt{3}$

27. $\sqrt{8x + 8} + \sqrt{2x + 2} = \sqrt{4(2x+2)} + \sqrt{2x+2}$
$= 2\sqrt{2x+2} + 1\sqrt{2x+2}$
$= (2+1)\sqrt{2x+2}$
$= 3\sqrt{2x+2}$

29. $\sqrt{x^5 - x^2} + \sqrt{9x^3 - 9} = \sqrt{x^2(x^3-1)} + \sqrt{9(x^3-1)}$
$= x\sqrt{x^3-1} + 3\sqrt{x^3-1}$
$= (x+3)\sqrt{x^3-1}$

31. $4a\sqrt{a^2b} + a\sqrt{a^2b^3} - 5\sqrt{b^3}$
$= 4a\sqrt{a^2 \cdot b} + a\sqrt{a^2 \cdot b^2 \cdot b} - 5\sqrt{b^2 \cdot b}$
$= 4a \cdot a\sqrt{b} + a \cdot a \cdot b\sqrt{b} - 5 \cdot b\sqrt{b}$
$= 4a^2\sqrt{b} + a^2b\sqrt{b} - 5b\sqrt{b}$
$= (4a^2 + a^2b - 5b)\sqrt{b}$

33. $\sqrt{3} - \sqrt{\dfrac{1}{3}} = \sqrt{3} - \sqrt{\dfrac{1}{3} \cdot \dfrac{3}{3}}$
$= \sqrt{3} - \dfrac{\sqrt{3}}{3}$
$= \left(1 - \dfrac{1}{3}\right)\sqrt{3}$
$= \dfrac{2}{3}\sqrt{3}$, or $\dfrac{2\sqrt{3}}{3}$

35. $5\sqrt{2} + 3\sqrt{\dfrac{1}{2}} = 5\sqrt{2} + 3\sqrt{\dfrac{1}{2} \cdot \dfrac{2}{2}}$

$$= 5\sqrt{2} + \dfrac{3}{2}\sqrt{2}$$

$$= \left(5 + \dfrac{3}{2}\right)\sqrt{2}$$

$$= \dfrac{13}{2}\sqrt{2}, \text{ or } \dfrac{13\sqrt{2}}{2}$$

37. $\sqrt{\dfrac{2}{3}} - \sqrt{\dfrac{1}{6}} = \sqrt{\dfrac{2}{3} \cdot \dfrac{3}{3}} - \sqrt{\dfrac{1}{6} \cdot \dfrac{6}{6}}$

$$= \dfrac{\sqrt{6}}{3} - \dfrac{\sqrt{6}}{6}$$

$$= \left(\dfrac{1}{3} - \dfrac{1}{6}\right)\sqrt{6}$$

$$= \dfrac{1}{6}\sqrt{6}, \text{ or } \dfrac{\sqrt{6}}{6}$$

39. $\sqrt{3}(\sqrt{5} - 1) = \sqrt{3}\,\sqrt{5} - \sqrt{3} \cdot 1$

$$= \sqrt{15} - \sqrt{3}$$

41. $(2 + \sqrt{3})(5 - \sqrt{7})$

$= 2 \cdot 5 - 2\sqrt{7} + \sqrt{3} \cdot 5 - \sqrt{3}\sqrt{7}$ Using FOIL

$= 10 - 2\sqrt{7} + 5\sqrt{3} - \sqrt{21}$

43. $(2 - \sqrt{5})^2$

$= 2^2 - 2 \cdot 2 \cdot \sqrt{5} + (\sqrt{5})^2$

          Using $(A - B)^2 = A^2 - 2AB + B^2$

$= 4 - 4\sqrt{5} + 5$

$= 9 - 4\sqrt{5}$

45. $(\sqrt{2} + 8)(\sqrt{2} - 8)$

$= (\sqrt{2})^2 - 8^2$ Using $(A + B)(A - B) = A^2 - B^2$

$= 2 - 64$

$= -62$

47. $(\sqrt{6} - \sqrt{5})(\sqrt{6} + \sqrt{5})$

$= (\sqrt{6})^2 - (\sqrt{5})^2$ Using $(A+B)(A-B) = A^2 - B^2$

$= 6 - 5$

$= 1$

49. $(3\sqrt{5} - 2)(\sqrt{5} + 1)$

$= 3\sqrt{5}\,\sqrt{5} + 3\sqrt{5} - 2\sqrt{5} - 2$ Using FOIL

$= 3 \cdot 5 + 3\sqrt{5} - 2\sqrt{5} - 2$

$= 15 + \sqrt{5} - 2$

$= 13 + \sqrt{5}$

51. $(\sqrt{x} - \sqrt{y})^2 = (\sqrt{x})^2 - 2\sqrt{x}\,\sqrt{y} + (\sqrt{y})^2$

          Using $(A - B)^2 = A^2 - 2AB + B^2$

$= x - 2\sqrt{xy} + y$

53. We multiply by 1 using the conjugate of $\sqrt{3} - \sqrt{5}$, which is $\sqrt{3} + \sqrt{5}$, as the numerator and denominator.

$$\dfrac{2}{\sqrt{3} - \sqrt{5}} = \dfrac{2}{\sqrt{3} - \sqrt{5}} \cdot \dfrac{\sqrt{3} + \sqrt{5}}{\sqrt{3} + \sqrt{5}} \quad \text{Multiplying by 1}$$

$$= \dfrac{2(\sqrt{3} + \sqrt{5})}{(\sqrt{3} - \sqrt{5})(\sqrt{3} + \sqrt{5})} \quad \text{Multiplying}$$

$$= \dfrac{2\sqrt{3} + 2\sqrt{5}}{(\sqrt{3})^2 - (\sqrt{5})^2} = \dfrac{2\sqrt{3} + 2\sqrt{5}}{3 - 5}$$

$$= \dfrac{2\sqrt{3} + 2\sqrt{5}}{-2} = \dfrac{2(\sqrt{3} + \sqrt{5})}{-2}$$

$$= -(\sqrt{3} + \sqrt{5}) = -\sqrt{3} - \sqrt{5}$$

55. We multiply by 1 using the conjugate of $\sqrt{3} + \sqrt{2}$, which is $\sqrt{3} - \sqrt{2}$, as the numerator and denominator.

$$\dfrac{\sqrt{3} - \sqrt{2}}{\sqrt{3} + \sqrt{2}} = \dfrac{\sqrt{3} - \sqrt{2}}{\sqrt{3} + \sqrt{2}} \cdot \dfrac{\sqrt{3} - \sqrt{2}}{\sqrt{3} - \sqrt{2}} \quad \text{Multiplying by 1}$$

$$= \dfrac{(\sqrt{3} - \sqrt{2})^2}{(\sqrt{3} + \sqrt{2})(\sqrt{3} - \sqrt{2})}$$

$$= \dfrac{(\sqrt{3})^2 - 2\sqrt{3}\,\sqrt{2} + (\sqrt{2})^2}{(\sqrt{3})^2 - (\sqrt{2})^2}$$

$$= \dfrac{3 - 2\sqrt{6} + 2}{3 - 2} = \dfrac{5 - 2\sqrt{6}}{1}$$

$$= 5 - 2\sqrt{6}$$

57. We multiply by 1 using the conjugate of $\sqrt{10} + 1$, which is $\sqrt{10} - 1$, as the numerator and denominator.

$$\dfrac{4}{\sqrt{10} + 1} = \dfrac{4}{\sqrt{10} + 1} \cdot \dfrac{\sqrt{10} - 1}{\sqrt{10} - 1}$$

$$= \dfrac{4(\sqrt{10} - 1)}{(\sqrt{10} + 1)(\sqrt{10} - 1)}$$

$$= \dfrac{4\sqrt{10} - 4}{(\sqrt{10})^2 - 1^2} = \dfrac{4\sqrt{10} - 4}{10 - 1}$$

$$= \dfrac{4\sqrt{10} - 4}{9}$$

59. We multiply by 1 using the conjugate of $3 + \sqrt{7}$, which is $3 - \sqrt{7}$, as the numerator and denominator.

$$\dfrac{1 - \sqrt{7}}{3 + \sqrt{7}} = \dfrac{1 - \sqrt{7}}{3 + \sqrt{7}} \cdot \dfrac{3 - \sqrt{7}}{3 - \sqrt{7}}$$

$$= \dfrac{(1 - \sqrt{7})(3 - \sqrt{7})}{(3 + \sqrt{7})(3 - \sqrt{7})}$$

$$= \dfrac{3 - \sqrt{7} - 3\sqrt{7} + \sqrt{7}\,\sqrt{7}}{3^2 - (\sqrt{7})^2}$$

$$= \dfrac{3 - \sqrt{7} - 3\sqrt{7} + 7}{9 - 7} = \dfrac{10 - 4\sqrt{7}}{2}$$

$$= \dfrac{2(5 - 2\sqrt{7})}{2} = 5 - 2\sqrt{7}$$

61. We multiply by 1 using the conjugate of $4 + \sqrt{x}$, which is $4 - \sqrt{x}$, as the numerator and denominator.

$$\frac{3}{4 + \sqrt{x}} = \frac{3}{4 + \sqrt{x}} \cdot \frac{4 - \sqrt{x}}{4 - \sqrt{x}}$$

$$= \frac{3(4 - \sqrt{x})}{(4 + \sqrt{x})(4 - \sqrt{x})}$$

$$= \frac{12 - 3\sqrt{x}}{4^2 - (\sqrt{x})^2}$$

$$= \frac{12 - 3\sqrt{x}}{16 - x}$$

63. We multiply by 1 using the conjugate of $8 - \sqrt{x}$, which is $8 + \sqrt{x}$, as the numerator and denominator.

$$\frac{3 + \sqrt{2}}{8 - \sqrt{x}} = \frac{3 + \sqrt{2}}{8 - \sqrt{x}} \cdot \frac{8 + \sqrt{x}}{8 + \sqrt{x}}$$

$$= \frac{(3 + \sqrt{2})(8 + \sqrt{x})}{(8 - \sqrt{x})(8 + \sqrt{x})}$$

$$= \frac{3 \cdot 8 + 3 \cdot \sqrt{x} + \sqrt{2} \cdot 8 + \sqrt{2} \cdot \sqrt{x}}{8^2 - (\sqrt{x})^2}$$

$$= \frac{24 + 3\sqrt{x} + 8\sqrt{2} + \sqrt{2x}}{64 - x}$$

65. Discussion and Writing Exercise

67.
$$3x + 5 + 2(x - 3) = 4 - 6x$$
$$3x + 5 + 2x - 6 = 4 - 6x$$
$$5x - 1 = 4 - 6x$$
$$11x - 1 = 4$$
$$11x = 5$$
$$x = \frac{5}{11}$$

The solution is $\frac{5}{11}$.

69.
$$x^2 - 5x = 6$$
$$x^2 - 5x - 6 = 0$$
$$(x + 1)(x - 6) = 0$$
$$x + 1 = 0 \quad or \quad x - 6 = 0$$
$$x = -1 \quad or \quad x = 6$$

The solutions are -1 and 6.

71. $\frac{7x^9}{27} \cdot \frac{9}{7x^3} = \frac{63x^9}{189x^3} = \frac{63}{189}x^{9-3} = \frac{1}{3}x^6$, or $\frac{x^6}{3}$

73. Familiarize. Let $x =$ the number of liters of Jolly Juice and $y =$ the number of liters of Real Squeeze in the mixture. We organize the given information in a table.

	Jolly Juice	Real Squeeze	Mixture
Amount	x	y	8
Percent real fruit juice	3%	6%	5.4%
Amount of real fruit juice	0.03x	0.06y	0.054(8), or 0.432

Translate. We get two equation from the first and third rows of the table.

$$x + y = 8,$$
$$0.03x + 0.06y = 0.432$$

Clearing decimals gives

$$x + y = 8, \quad (1)$$
$$30x + 60y = 432. \quad (2)$$

Carry out. We use elimination. Multiply Equation (1) by -30 and add.

$$-30x - 30y = -240$$
$$\underline{30x + 60y = 432}$$
$$30y = 192$$
$$y = 6.4$$

Now substitute 6.4 for y in Equation (1) and solve for x.

$$x + y = 8$$
$$x + 6.4 = 8$$
$$x = 1.6$$

Check. The sum of 1.6 and 6.4 is 8. The amount of real fruit juice in this mixture is $0.03(1.6) + 0.06(6.4)$, or $0.048 + 0.384$, or 0.432 L. The answer checks.

State. 1.6 L of Jolly Juice and 6.4 L of Real Squeeze should be used.

75. For $x = -1$, $y = (-1)^3 - 5(-1)^2 + (-1) - 2 = -1 - 5 - 1 - 2 = -9$.
For $x = 0$, $y = 0^3 - 5 \cdot 0^2 + 0 - 2 = 0 - 0 + 0 - 2 = -2$.
For $x = 1$, $y = 1^3 - 5 \cdot 1^2 + 1 - 2 = 1 - 5 + 1 - 2 = -5$.
For $x = 3$, $y = 3^3 - 5 \cdot 3^2 + 3 - 2 = 27 - 45 + 3 - 2 = -17$.
For $x = 4.85$, $y = (4.85)^3 - 5(4.85)^2 + 4.85 - 2 =$
$114.084125 - 117.6125 + 4.85 - 2 = -0.678375$

These values could have been estimated using the graph also.

77. Since $\sqrt{a^2 + b^2} \neq \sqrt{a^2} + \sqrt{b^2}$ for $a = 2$ and $b = 3$, the two expressions are not equivalent.

79. Enter $y_1 = \sqrt{x^2 + 4}$ and $y_2 = \sqrt{x} + 2$ and look at the graphs or a table of values. Since the graphs do not coincide and the values for y_1 and y_2 are different, the given statement is not correct.

81.
$$\frac{1}{3}\sqrt{27} + \sqrt{8} + \sqrt{300} - \sqrt{18} - \sqrt{162}$$
$$= \frac{1}{3}\sqrt{9 \cdot 3} + \sqrt{4 \cdot 2} + \sqrt{100 \cdot 3} - \sqrt{9 \cdot 2} - \sqrt{81 \cdot 2}$$
$$= \frac{1}{3} \cdot 3\sqrt{3} + 2\sqrt{2} + 10\sqrt{3} - 3\sqrt{2} - 9\sqrt{2}$$
$$= \sqrt{3} + 2\sqrt{2} + 10\sqrt{3} - 3\sqrt{2} - 9\sqrt{2}$$
$$= (1 + 10)\sqrt{3} + (2 - 3 - 9)\sqrt{2}$$
$$= 11\sqrt{3} - 10\sqrt{2}$$

83. $(3\sqrt{x+2})^2 = (3\sqrt{x+2})(3\sqrt{x+2})2 = (3 \cdot 3)(\sqrt{x+2}\sqrt{x+2}) = 9(x+2)$

The statement is true.

Exercise Set 8.5

1. $\sqrt{x} = 6$

$(\sqrt{x})^2 = 6^2$ Squaring both sides

$x = 36$ Simplifying

Check: $\sqrt{x} = 6$

$\sqrt{36}$? 6

6 | TRUE

The solution is 36.

3. $\sqrt{x} = 4.3$

$(\sqrt{x})^2 = (4.3)^2$ Squaring both sides

$x = 18.49$ Simplifying

Check: $\sqrt{x} = 4.3$

$\sqrt{18.49}$? 4.3

4.3 | TRUE

The solution is 18.49.

5. $\sqrt{y+4} = 13$

$(\sqrt{y+4})^2 = 13^2$ Squaring both sides

$y + 4 = 169$ Simplifying

$y = 165$ Subtracting 4

Check: $\sqrt{y+4} = 13$

$\sqrt{165+4}$? 13

$\sqrt{169}$

13 | TRUE

The solution is 165.

7. $\sqrt{2x+4} = 25$

$(\sqrt{2x+4})^2 = 25^2$ Squaring both sides

$2x + 4 = 625$ Simplifying

$2x = 621$ Subtracting 4

$x = \dfrac{621}{2}$ Dividing by 2

Check: $\sqrt{2x+4} = 25$

$\sqrt{2 \cdot \dfrac{621}{2} + 4}$? 25

$\sqrt{621+4}$

$\sqrt{625}$

25 | TRUE

The solution is $\dfrac{621}{2}$.

9. $3 + \sqrt{x-1} = 5$

$\sqrt{x-1} = 2$ Subtracting 3

$(\sqrt{x-1})^2 = 2^2$ Squaring both sides

$x - 1 = 4$

$x = 5$

Check: $3 + \sqrt{x-1} = 5$

$3 + \sqrt{5-1}$? 5

$3 + \sqrt{4}$

$3 + 2$

5 | TRUE

The solution is 5.

11. $6 - 2\sqrt{3n} = 0$

$6 = 2\sqrt{3n}$ Adding $2\sqrt{3n}$

$6^2 = (2\sqrt{3n})^2$ Squaring both sides

$36 = 4 \cdot 3n$

$36 = 12n$

$3 = n$

Check: $6 - 2\sqrt{3n} = 0$

$6 - 2\sqrt{3 \cdot 3}$? 0

$6 - 2 \cdot 3$

$6 - 6$

0 | TRUE

The solution is 3.

13. $\sqrt{5x-7} = \sqrt{x+10}$

$(\sqrt{5x-7})^2 = (\sqrt{x+10})^2$ Squaring both sides

$5x - 7 = x + 10$

$4x = 17$

$x = \dfrac{17}{4}$

Check: $\sqrt{5x-7} = \sqrt{x+10}$

$\sqrt{5 \cdot \dfrac{17}{4} - 7}$? $\sqrt{\dfrac{17}{4} + 10}$

$\sqrt{\dfrac{85}{4} - \dfrac{28}{4}}$ | $\sqrt{\dfrac{57}{4}}$

$\sqrt{\dfrac{57}{4}}$ | TRUE

The solution is $\dfrac{17}{4}$.

15. $\sqrt{x} = -7$

There is no solution. The principal square root of x cannot be negative.

17. $\sqrt{2y+6} = \sqrt{2y-5}$

$(\sqrt{2y+6})^2 = (\sqrt{2y-5})^2$

$2y + 6 = 2y - 5$

$6 = -5$

The equation $6 = -5$ is false; there is no solution.

19.
$$x - 7 = \sqrt{x - 5}$$
$$(x - 7)^2 = (\sqrt{x - 5})^2$$
$$x^2 - 14x + 49 = x - 5$$
$$x^2 - 15 + 54 = 0$$
$$(x - 9)(x - 6) = 0$$
$$x - 9 = 0 \quad \text{or} \quad x - 6 = 0$$
$$x = 9 \quad \text{or} \quad x = 6$$

Check:
$$\frac{x - 7 = \sqrt{x - 5}}{9 - 7 \; ? \; \sqrt{9 - 5}}$$
$$2 \;\Big|\; \sqrt{4}$$
$$\Big|\; 2 \qquad \text{TRUE}$$

$$\frac{x - 7 = \sqrt{x - 5}}{6 - 7 \;\Big|\; \sqrt{6 - 5}}$$
$$-1 \;\Big|\; \sqrt{1}$$
$$\Big|\; 1 \qquad \text{FALSE}$$

The number 9 checks, but 6 does not. The solution is 9.

21.
$$x - 9 = \sqrt{x - 3}$$
$$(x - 9)^2 = (\sqrt{x - 3})^2$$
$$x^2 - 18x + 81 = x - 3$$
$$x^2 - 19x + 84 = 0$$
$$(x - 12)(x - 7) = 0$$
$$x - 12 = 0 \quad \text{or} \quad x - 7 = 0$$
$$x = 12 \quad \text{or} \quad x = 7$$

Check:
$$\frac{x - 9 = \sqrt{x - 3}}{12 - 9 \; ? \; \sqrt{12 - 3}}$$
$$3 \;\Big|\; \sqrt{9}$$
$$\Big|\; 3 \qquad \text{TRUE}$$

$$\frac{x - 9 = \sqrt{x - 3}}{7 - 9 \; ? \; \sqrt{7 - 3}}$$
$$-2 \;\Big|\; \sqrt{4}$$
$$\Big|\; 2 \qquad \text{FALSE}$$

The number 12 checks, but 7 does not. The solution is 12.

23.
$$2\sqrt{x - 1} = x - 1$$
$$(2\sqrt{x - 1})^2 = (x - 1)^2$$
$$4(x - 1) = x^2 - 2x + 1$$
$$4x - 4 = x^2 - 2x + 1$$
$$0 = x^2 - 6x + 5$$
$$0 = (x - 5)(x - 1)$$
$$x - 5 = 0 \quad \text{or} \quad x - 1 = 0$$
$$x = 5 \quad \text{or} \quad x = 1$$

Both numbers check. The solutions are 5 and 1.

25.
$$\sqrt{5x + 21} = x + 3$$
$$(\sqrt{5x + 21})^2 = (x + 3)^2$$
$$5x + 21 = x^2 + 6x + 9$$
$$0 = x^2 + x - 12$$
$$0 = (x + 4)(x - 3)$$
$$x + 4 = 0 \quad \text{or} \quad x - 3 = 0$$
$$x = -4 \quad \text{or} \quad x = 3$$

Check:
$$\frac{\sqrt{5x + 21} = x + 3}{\sqrt{5(-4) + 21} \; ? \; -4 + 3}$$
$$\sqrt{1} \;\Big|\; -1$$
$$1 \;\Big|\; \qquad \text{FALSE}$$

$$\frac{\sqrt{5x + 21} = x + 3}{\sqrt{5 \cdot 3 + 21} \; ? \; 3 + 3}$$
$$\sqrt{36} \;\Big|\; 6$$
$$6 \;\Big|\; \qquad \text{TRUE}$$

The number 3 checks, but −4 does not. The solution is 3.

27.
$$\sqrt{2x - 1} + 2 = x$$
$$\sqrt{2x - 1} = x - 2 \qquad \text{Isolating the radical}$$
$$(\sqrt{2x - 1})^2 = (x - 2)^2$$
$$2x - 1 = x^2 - 4x + 4$$
$$0 = x^2 - 6x + 5$$
$$0 = (x - 5)(x - 1)$$
$$x - 5 = 0 \quad \text{or} \quad x - 1 = 0$$
$$x = 5 \quad \text{or} \quad x = 1$$

Check:
$$\frac{\sqrt{2x - 1} + 2 = x}{\sqrt{2 \cdot 5 - 1} + 2 \; ? \; 5}$$
$$\sqrt{10 - 1} + 2 \;\Big|\;$$
$$\sqrt{9} + 2 \;\Big|\;$$
$$3 + 2 \;\Big|\;$$
$$5 \;\Big|\; \qquad \text{TRUE}$$

$$\frac{\sqrt{2x - 1} + 2 = x}{\sqrt{2 \cdot 1 - 1} + 2 \; ? \; 1}$$
$$\sqrt{2 - 1} + 2 \;\Big|\;$$
$$\sqrt{1} + 2 \;\Big|\;$$
$$1 + 2 \;\Big|\;$$
$$3 \;\Big|\; \qquad \text{FALSE}$$

The number 5 checks, but 1 does not. The solution is 5.

29. $\sqrt{x^2+6}-x+3=0$

$\qquad\sqrt{x^2+6}=x-3$ Isolating the radical

$\qquad(\sqrt{x^2+6})^2=(x-3)^2$

$\qquad\quad x^2+6=x^2-6x+9$

$\qquad\qquad\;-3=-6x$ Adding $-x^2$ and -9

$\qquad\qquad\quad\dfrac{1}{2}=x$

Check: $\dfrac{\sqrt{x^2+6}-x+3=0}{}$

$\qquad\sqrt{\left(\dfrac{1}{2}\right)^2+6}-\dfrac{1}{2}+3\;?\;0$

$\qquad\quad\sqrt{\dfrac{25}{4}}-\dfrac{1}{2}+3$

$\qquad\qquad\dfrac{5}{2}-\dfrac{1}{2}+3$

$\qquad\qquad\qquad\quad 5\;\bigg|\quad$ FALSE

The number $\dfrac{1}{2}$ does not check. There is no solution.

31. $\sqrt{x^2-4}-x=6$

$\qquad\sqrt{x^2-4}=x+6$ Isolating the radical

$\qquad(\sqrt{x^2-4})^2=(x+6)^2$

$\qquad\quad x^2-4=x^2+12x+36$

$\qquad\qquad\;-40=12x$ Adding $-x^2$ and -36

$\qquad\quad-\dfrac{40}{12}=x$

$\qquad\quad-\dfrac{10}{3}=x$

The number $-\dfrac{10}{3}$ checks. It is the solution.

33. $\sqrt{(p+6)(p+1)}-2=p+1$

$\qquad\sqrt{(p+6)(p+1)}=p+3$ Isolating the radical

$\qquad\left(\sqrt{(p+6)(p+1)}\right)^2=(p+3)^2$

$\qquad\quad(p+6)(p+1)=p^2+6p+9$

$\qquad\quad p^2+7p+6=p^2+6p+9$

$\qquad\qquad\qquad p=3$

The number 3 checks. It is the solution.

35. $\sqrt{4x-10}=\sqrt{2-x}$

$\qquad(\sqrt{4x-10})^2=(\sqrt{2-x})^2$

$\qquad\quad 4x-10=2-x$

$\qquad\qquad 5x=12$ Adding 10 and x

$\qquad\qquad\;\; x=\dfrac{12}{5}$

Check: $\dfrac{\sqrt{4x-10}=\sqrt{2-x}}{}$

$\qquad\sqrt{4\cdot\dfrac{12}{5}-10}\;?\;\sqrt{2-\dfrac{12}{5}}$

$\qquad\quad\sqrt{\dfrac{48}{5}-10}\;\bigg|\;\sqrt{-\dfrac{2}{5}}$

Since $\sqrt{-\dfrac{2}{5}}$ does not represent a real number, there is no solution that is a real number.

37. $\sqrt{x-5}=5-\sqrt{x}$

$\qquad(\sqrt{x-5})^2=(5-\sqrt{x})^2$ Squaring both sides

$\qquad\quad x-5=25-10\sqrt{x}+x$

$\qquad\quad-30=-10\sqrt{x}$ Isolating the radical

$\qquad\qquad 3=\sqrt{x}$ Dividing by -10

$\qquad\qquad 3^2=(\sqrt{x})^2$ Squaring both sides

$\qquad\qquad 9=x$

The number 9 checks. It is the solution.

39. $\sqrt{y+8}-\sqrt{y}=2$

$\qquad\sqrt{y+8}=\sqrt{y}+2$ Isolating one radical

$\qquad(\sqrt{y+8})^2=(\sqrt{y}+2)^2$ Squaring both sides

$\qquad\quad y+8=y+4\sqrt{y}+4$

$\qquad\qquad 4=4\sqrt{x}$ Isolating the radical

$\qquad\qquad 1=\sqrt{y}$ Dividing by 4

$\qquad\qquad 1^2=(\sqrt{y})^2$

$\qquad\qquad 1=y$

The number 1 checks. It is the solution.

41. $\sqrt{x-4}+\sqrt{x+1}=5$

$\qquad\sqrt{x-4}=5-\sqrt{x+1}$ Isolating one radical

$\qquad(\sqrt{x-4})^2=(5-\sqrt{x+1})^2$

$\qquad\quad x-4=25-10\sqrt{x+1}+x+1$

$\qquad\quad-30=-10\sqrt{x+1}$ Isolating the radical

$\qquad\qquad 3=\sqrt{x+1}$ Dividing by -10

$\qquad\qquad 3^2=(\sqrt{x+1})^2$

$\qquad\qquad 9=x+1$

$\qquad\qquad 8=x$

The number 8 checks. It is the solution.

43. $\sqrt{x}-1=\sqrt{x-31}$

$\qquad(\sqrt{x}-1)^2=(\sqrt{x-31})^2$

$\qquad\quad x-2\sqrt{x}+1=x-31$

$\qquad\qquad-2\sqrt{x}=-32$

$\qquad\qquad\sqrt{x}=16$

$\qquad\qquad(\sqrt{x})^2=16^2$

$\qquad\qquad x=256$

The number 256 checks. It is the solution.

45. Substitute 27,000 for h in the equation.

$$D = \sqrt{2h}$$
$$D = \sqrt{2 \cdot 27{,}000}$$
$$D = \sqrt{54{,}000}$$
$$D \approx 232$$

You can see about 232 mi to the horizon.

47. Substitute 180 for D in the equation and solve for h.

$$D = \sqrt{2h}$$
$$180 = \sqrt{2h}$$
$$180^2 = (\sqrt{2h})^2$$
$$32{,}400 = 2h$$
$$16{,}200 = h$$

A pilot must fly 16,200 ft above sea level in order to see a horizon that is 180 mi away.

49. For 65 mph, substitute 64 for S in the equation and solve for x.

$$S = 2\sqrt{5x}$$
$$65 = 2\sqrt{5x}$$
$$65^2 = (2\sqrt{5x})^2$$
$$4225 = 4 \cdot 5x$$
$$4225 = 20x$$
$$211.25 = x$$

At 65 mph, a car will skid 211.25 ft.

For 75 mph, substitute 75 for S in the equation and solve for x.

$$S = 2\sqrt{5x}$$
$$75 = 2\sqrt{5x}$$
$$75^2 = (2\sqrt{5x})^2$$
$$5625 = 4 \cdot 5x$$
$$5625 = 20x$$
$$\frac{5625}{20} = x$$
$$281.25 = x$$

At 75 mph, a car will skid 281.25 ft.

51. Discussion and Writing Exercise

53. Parallel lines have the same slope and different y-intercepts.

55. The number c is a principal square root of a if $c^2 = a$ and c is either zero or positive.

57. The quotient rule asserts that when dividing with exponential notation, if the bases are the same, keep the base and subtract the exponent of the denominator from the exponent of the numerator.

59. The quotient rule for radicals asserts that for any nonnegative number A and any positive number B, $\dfrac{\sqrt{A}}{\sqrt{B}} = \sqrt{\dfrac{A}{B}}$.

61.
$$\sqrt{5x^2 + 5} = 5$$
$$(\sqrt{5x^2 + 5})^2 = 5^2$$
$$5x^2 + 5 = 25$$
$$5x^2 - 20 = 0$$
$$5(x^2 - 4) = 0$$
$$5(x + 2)(x - 2) = 0$$
$$x + 2 = 0 \quad or \quad x - 2 = 0$$
$$x = -2 \quad or \quad x = 2$$

Both numbers check, so the solutions are 2 and -2.

63.
$$4 + \sqrt{19 - x} = 6 + \sqrt{4 - x}$$
$$\sqrt{19 - x} = 2 + \sqrt{4 - x} \quad \text{Isolating one radical}$$
$$(\sqrt{19 - x})^2 = (2 + \sqrt{4 - x})^2$$
$$19 - x = 4 + 4\sqrt{4 - x} + (4 - x)$$
$$19 - x = 4\sqrt{4 - x} + 8 - x$$
$$11 = 4\sqrt{4 - x}$$
$$11^2 = (4\sqrt{4 - x})^2$$
$$121 = 16(4 - x)$$
$$121 = 64 - 16x$$
$$57 = -16x$$
$$-\frac{57}{16} = x$$

$-\dfrac{57}{16}$ checks, so it is the solution.

65.
$$\sqrt{x + 3} = \frac{8}{\sqrt{x - 9}}$$
$$(\sqrt{x + 3})^2 = \left(\frac{8}{\sqrt{x - 9}}\right)^2$$
$$x + 3 = \frac{64}{x - 9}$$
$$(x - 9)(x + 3) = 64 \quad \text{Multiplying by } x - 9$$
$$x^2 - 6x - 27 = 64$$
$$x^2 - 6x - 91 = 0$$
$$(x - 13)(x + 7) = 0$$
$$x - 13 = 0 \quad or \quad x + 7 = 0$$
$$x = 13 \quad or \quad x = -7$$

The number 13 checks, but -7 does not. The solution is 13.

67.–69. Left to the student

Exercise Set 8.6

1.
$$a^2 + b^2 = c^2$$
$$8^2 + 15^2 = c^2 \quad \text{Substituting}$$
$$64 + 225 = c^2$$
$$289 = c^2$$
$$\sqrt{289} = c$$
$$17 = c$$

3. $a^2 + b^2 = c^2$

$4^2 + 4^2 = c^2$ Substituting

$16 + 16 = c^2$

$32 = c^2$

$\sqrt{32} = c$ Exact answer

$5.657 \approx c$ Approximation

5. $a^2 + b^2 = c^2$

$5^2 + b^2 = 13^2$

$25 + b^2 = 169$

$b^2 = 144$

$b = 12$

7. $a^2 + b^2 = c^2$

$(4\sqrt{3})^2 + b^2 = 8^2$

$16 \cdot 3 + b^2 = 64$

$48 + b^2 = 64$

$b^2 = 16$

$b = 4$

9. $a^2 + b^2 = c^2$

$10^2 + 24^2 = c^2$

$100 + 576 = c^2$

$676 = c^2$

$26 = c$

11. $a^2 + b^2 = c^2$

$9^2 + b^2 = 15^2$

$81 + b^2 = 225$

$b^2 = 144$

$b = 12$

13. $a^2 + b^2 = c^2$

$a^2 + 1^2 = (\sqrt{5})^2$

$a^2 + 1 = 5$

$a^2 = 4$

$a = 2$

15. $a^2 + b^2 = c^2$

$1^2 + b^2 = (\sqrt{3})^2$

$1 + b^2 = 3$

$b^2 = 2$

$b = \sqrt{2}$ Exact answer

$b \approx 1.414$ Approximation

17. $a^2 + b^2 = c^2$

$a^2 + (5\sqrt{3})^2 = 10^2$

$a^2 + 25 \cdot 3 = 100$

$a^2 + 75 = 100$

$a^2 = 25$

$a = 5$

19. $a^2 + b^2 = c^2$

$(\sqrt{2})^2 + (\sqrt{7})^2 = c^2$

$2 + 7 = c^2$

$9 = c^2$

$3 = c$

21. We use the drawing in the text, labeling the horizontal distance h.

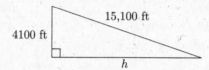

We know that $4100^2 + h^2 = 15,100^2$. We solve this equation.

$16,810,000 + h^2 = 228,010,000$

$h^2 = 211,200,000$

$h = \sqrt{211,200,000}$ ft   Exact answer

$h \approx 14,533$ ft   Approximation

23. We first make a drawing. Let d represent the distance Becky can move away from the building while using the telephone.

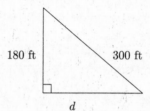

We know that $180^2 + d^2 = 300^2$.

We solve this equation.

$180^2 + d^2 = 300^2$

$32,400 + d^2 = 90,000$

$d^2 = 57,600$

$d = 240$

Becky can use her telephone 240 ft into her backyard.

25. We first make a drawing. We label the diagonal d.

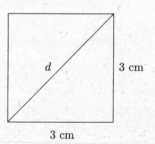

We know that $3^2 + 3^2 = d^2$. We solve this equation.

$$3^2 + 3^2 = d^2$$
$$9 + 9 = d^2$$
$$18 = d^2$$
$$\sqrt{18} \text{ cm} = d \quad \text{Exact answer}$$
$$4.243 \text{ cm} \approx d \quad \text{Approximation}$$

27. We first make a drawing. We label the length of the guy wire w.

We know that $8^2 + 12^2 = w^2$. We solve this equation.
$$8^2 + 12^2 = w^2$$
$$64 + 144 = w^2$$
$$208 = w^2$$
$$\sqrt{208} \text{ ft} = w \quad \text{Exact answer}$$
$$14.422 \text{ ft} \approx w \quad \text{Approximation}$$

29. Discussion and Writing Exercise

31. $5x + 7 = 8y,$
$3x = 8y - 4$

$5x - 8y = -7 \quad (1) \quad$ Rewriting
$3x - 8y = -4 \quad (2) \quad$ the equations

We multiply Equation (2) by -1 and add.
$$5x - 8y = -7$$
$$\underline{-3x + 8y = 4}$$
$$2x \qquad = -3$$
$$x = -\frac{3}{2}$$

Substitute $-\frac{3}{2}$ for x in Equation (1) and solve for y.
$$5x - 8y = -7$$
$$5\left(-\frac{3}{2}\right) - 8y = -7$$
$$-\frac{15}{2} - 8y = -7$$
$$-8y = \frac{1}{2}$$
$$y = -\frac{1}{16}$$

The ordered pair $\left(-\frac{3}{2}, -\frac{1}{16}\right)$ checks. It is the solution.

33. $3x - 4y = -11 \quad (1)$
$5x + 6y = 12 \quad (2)$

We multiply Equation (1) by 3 and Equation (2) by 2, and then we add.

$$9x - 12y = -33$$
$$\underline{10x + 12y = 24}$$
$$19x \qquad = -9$$
$$x = -\frac{9}{19}$$

Substitute $-\frac{9}{19}$ for x in Equation (2) and solve for y.
$$5x + 6y = 12$$
$$5\left(-\frac{9}{19}\right) + 6y = 12$$
$$-\frac{45}{19} + 6y = 12$$
$$6y = \frac{273}{19} \quad \text{Adding } \frac{45}{19}$$
$$y = \frac{273}{6 \cdot 19} \quad \text{Dividing by 6}$$
$$y = \frac{91}{38} \quad \text{Simplifying}$$

The ordered pair $\left(-\frac{9}{19}, \frac{91}{38}\right)$ checks. It is the solution.

35. Write the equation in the slope-intercept form.
$$4 - x = 3y$$
$$\frac{1}{3}(4 - x) = y$$
$$\frac{4}{3} - \frac{1}{3}x = y, \text{ or}$$
$$y = -\frac{1}{3}x + \frac{4}{3}$$

The slope is $-\frac{1}{3}$.

37.

$$a^2 + 5^2 = 7^2$$
$$a^2 + 25 = 49$$
$$a^2 = 24$$
$$a = \sqrt{24}, \text{ or } 2\sqrt{6}$$

$$(a + x)^2 + 5^2 = 13^2$$
$$(2\sqrt{6} + x)^2 + 5^2 = 13^2 \quad \text{Substituting } 2\sqrt{6} \text{ for } a$$
$$(2\sqrt{6} + x)^2 + 25 = 169$$
$$(2\sqrt{6} + x)^2 = 144$$
$$2\sqrt{6} + x = 12 \quad \text{Taking the principal square}$$
$$\qquad\qquad\qquad\qquad \text{root}$$
$$x = 12 - 2\sqrt{6}$$
$$x \approx 7.101$$

Chapter 8 Review Exercises

1. The square roots of 64 are 8 and -8, because $8^2 = 64$ and $(-8)^2 = 64$.

2. The square roots of 400 are 20 and -20, because $20^2 = 400$ and $(-20)^2 = 400$.

3. $\sqrt{36} = 6$, taking the principal square root

4. $\sqrt{169} = 13$, so $-\sqrt{169} = -13$.

5. 1.732

6. 9.950

7. $\sqrt{320.12} \approx 17.892$, so $-\sqrt{320.12} \approx -17.892$.

8. 0.742

9. $\sqrt{\dfrac{47.3}{11.2}} \approx 2.055$, so $-\sqrt{\dfrac{47.3}{11.2}} \approx -2.055$.

10. 394.648

11. The radicand is the expression under the radical, $x^2 + 4$.

12. The radicand is the expression under the radical, $5ab^3$.

13. No, because the radicand is negative

14. Yes, because the radicand is nonnegative

15. No, because the radicand is negative

16. No, because the radicand is negative

17. No, because the radicand is negative

18. No, because the radicand is negative

19. $\sqrt{m^2} = m$

20. $\sqrt{(x-4)^2} = x - 4$

21. $\sqrt{3}\sqrt{7} = \sqrt{3 \cdot 7} = \sqrt{21}$

22. $\sqrt{x-3}\sqrt{x+3} = \sqrt{(x-3)(x+3)} = \sqrt{x^2 - 9}$

23. $-\sqrt{48} = -\sqrt{16 \cdot 3} = -\sqrt{16}\sqrt{3} = -4\sqrt{3}$

24. $\sqrt{32t^2} = \sqrt{16 \cdot 2 \cdot t^2} = \sqrt{16}\sqrt{t^2}\sqrt{2} = 4t\sqrt{2}$

25. $\sqrt{t^2 - 49} = \sqrt{(t+7)(t-7)} = \sqrt{t+7}\sqrt{t-7}$

26. $\sqrt{x^2 + 16x + 64} = \sqrt{(x+8)^2} = x + 8$

27. $\sqrt{x^8} = \sqrt{(x^4)^2} = x^4$

28. $\sqrt{m^{15}} = \sqrt{m^{14} \cdot m} = \sqrt{m^{14}}\sqrt{m} = \sqrt{(m^7)^2}\sqrt{m} = m^7\sqrt{m}$

29. $\sqrt{6}\sqrt{10} = \sqrt{6 \cdot 10} = \sqrt{2 \cdot 3 \cdot 2 \cdot 5} = \sqrt{2 \cdot 2}\sqrt{3 \cdot 5} = 2\sqrt{15}$

30. $\sqrt{5x}\sqrt{8x} = \sqrt{5x \cdot 8x} = \sqrt{5 \cdot x \cdot 8 \cdot x} = \sqrt{x \cdot x}\sqrt{5 \cdot 8} = x\sqrt{40}$

31. $\sqrt{5x}\sqrt{10xy^2} = \sqrt{5x \cdot 10xy^2} = \sqrt{5 \cdot x \cdot 2 \cdot 5 \cdot x \cdot y^2} = \sqrt{5 \cdot 5}\sqrt{x \cdot x}\sqrt{y^2}\sqrt{2} = 5xy\sqrt{2}$

32. $\sqrt{20a^3b}\sqrt{5a^2b^2} = \sqrt{20a^3b \cdot 5a^2b^2} = \sqrt{4 \cdot 5 \cdot a^2 \cdot a \cdot b \cdot 5 \cdot a^2 \cdot b^2} = \sqrt{4}\sqrt{5 \cdot 5}\sqrt{a^2}\sqrt{a^2}\sqrt{b^2}\sqrt{ab} = 2 \cdot 5 \cdot a \cdot a \cdot b\sqrt{a \cdot b} = 10a^2b\sqrt{ab}$

33. $\sqrt{\dfrac{25}{64}} = \dfrac{\sqrt{25}}{\sqrt{64}} = \dfrac{5}{8}$

34. $\sqrt{\dfrac{20}{45}} = \sqrt{\dfrac{4}{9}} = \dfrac{\sqrt{4}}{\sqrt{9}} = \dfrac{2}{3}$

35. $\sqrt{\dfrac{49}{t^2}} = \dfrac{\sqrt{49}}{\sqrt{t^2}} = \dfrac{7}{t}$

36. $\sqrt{\dfrac{1}{2}} = \sqrt{\dfrac{1}{2} \cdot \dfrac{2}{2}} = \sqrt{\dfrac{2}{4}} = \dfrac{\sqrt{2}}{\sqrt{4}} = \dfrac{\sqrt{2}}{2}$

37. $\sqrt{\dfrac{1}{8}} = \sqrt{\dfrac{1}{8} \cdot \dfrac{2}{2}} = \sqrt{\dfrac{2}{16}} = \dfrac{\sqrt{2}}{\sqrt{16}} = \dfrac{\sqrt{2}}{4}$

38. $\sqrt{\dfrac{5}{y}} = \sqrt{\dfrac{5}{y} \cdot \dfrac{y}{y}} = \sqrt{\dfrac{5y}{y^2}} = \dfrac{\sqrt{5y}}{\sqrt{y^2}} = \dfrac{\sqrt{5y}}{y}$

39. $\dfrac{2}{\sqrt{3}} = \dfrac{2}{\sqrt{3}} \cdot \dfrac{\sqrt{3}}{\sqrt{3}} = \dfrac{2\sqrt{3}}{3}$

40. $\dfrac{\sqrt{27}}{\sqrt{45}} = \sqrt{\dfrac{27}{45}} = \sqrt{\dfrac{3}{5}} = \sqrt{\dfrac{3}{5} \cdot \dfrac{5}{5}} = \sqrt{\dfrac{15}{25}} = \dfrac{\sqrt{15}}{\sqrt{25}} = \dfrac{\sqrt{15}}{5}$

41. $\dfrac{\sqrt{45x^2y}}{\sqrt{54y}} = \sqrt{\dfrac{45x^2y}{54y}} = \sqrt{\dfrac{5x^2}{6}} = \sqrt{\dfrac{5x^2}{6} \cdot \dfrac{6}{6}} = \sqrt{\dfrac{30x^2}{36}} = \dfrac{\sqrt{30x^2}}{\sqrt{36}} = \dfrac{x\sqrt{30}}{6}$

42.
$$\dfrac{4}{2 + \sqrt{3}} = \dfrac{4}{2 + \sqrt{3}} \cdot \dfrac{2 - \sqrt{3}}{2 - \sqrt{3}}$$
$$= \dfrac{4(2 - \sqrt{3})}{(2 + \sqrt{3})(2 - \sqrt{3})}$$
$$= \dfrac{8 - 4\sqrt{3}}{2^2 - (\sqrt{3})^2} = \dfrac{8 - 4\sqrt{3}}{4 - 3}$$
$$= \dfrac{8 - 4\sqrt{3}}{1} = 8 - 4\sqrt{3}$$

43. $10\sqrt{5} + 3\sqrt{5} = (10 + 3)\sqrt{5} = 13\sqrt{5}$

44.
$$\sqrt{80} - \sqrt{45} = \sqrt{16 \cdot 5} - \sqrt{9 \cdot 5}$$
$$= 4\sqrt{5} - 3\sqrt{5}$$
$$= (4 - 3)\sqrt{5}$$
$$= \sqrt{5}$$

45. $3\sqrt{2} - 5\sqrt{\dfrac{1}{2}} = 3\sqrt{2} - 5\sqrt{\dfrac{1}{2} \cdot \dfrac{2}{2}}$

$\qquad\qquad = 3\sqrt{2} - 5 \cdot \dfrac{\sqrt{2}}{2}$

$\qquad\qquad = 3\sqrt{2} - \dfrac{5}{2}\sqrt{2}$

$\qquad\qquad = \left(3 - \dfrac{5}{2}\right)\sqrt{2}$

$\qquad\qquad = \dfrac{1}{2}\sqrt{2}, \text{ or } \dfrac{\sqrt{2}}{2}$

46. $(2 + \sqrt{3})^2 = 2^2 + 2 \cdot 2 \cdot \sqrt{3} + (\sqrt{3})^2 = 4 + 4\sqrt{3} + 3 =$
$7 + 4\sqrt{3}$

47. $(2 + \sqrt{3})(2 - \sqrt{3}) = 2^2 - (\sqrt{3})^2 = 4 - 3 = 1$

48. $\qquad \sqrt{x - 3} = 7$

$\qquad (\sqrt{x - 3})^2 = 7^2$

$\qquad\qquad x - 3 = 49$

$\qquad\qquad\quad x = 52$

The number 52 checks. It is the solution.

49. $\qquad \sqrt{5x + 3} = \sqrt{2x - 1}$

$\qquad (\sqrt{5x + 3})^2 = (\sqrt{2x - 1})^2$

$\qquad\qquad 5x + 3 = 2x - 1$

$\qquad\qquad 3x + 3 = -1$

$\qquad\qquad\quad 3x = -4$

$\qquad\qquad\quad\; x = -\dfrac{4}{3}$

The number $-\dfrac{4}{3}$ does not check. There is no solution.

50. $\qquad\quad 1 + x = \sqrt{1 + 5x}$

$\qquad (1 + x)^2 = (\sqrt{1 + 5x})^2$

$\qquad 1 + 2x + x^2 = 1 + 5x$

$\qquad\qquad x^2 - 3x = 0$

$\qquad\qquad x(x - 3) = 0$

$\qquad x = 0 \;\; or \;\; x - 3 = 0$

$\qquad x = 0 \;\; or \qquad x = 3$

Both numbers check. The solutions are 0 and 3.

51. $\qquad\quad \sqrt{x} = \sqrt{x - 5} + 1$

$\qquad (\sqrt{x})^2 = (\sqrt{x - 5} + 1)^2$

$\qquad\quad x = x - 5 + 2\sqrt{x - 5} + 1$

$\qquad\quad x = x - 4 + 2\sqrt{x - 5}$

$\qquad\quad 4 = 2\sqrt{x - 5}$

$\qquad\quad 2 = \sqrt{x - 5}$

$\qquad\quad 2^2 = (\sqrt{x - 5})^2$

$\qquad\quad 4 = x - 5$

$\qquad\quad 9 = x$

The number 9 checks. It is the solution.

52. $\qquad a^2 + b^2 = c^2$

$\qquad 15^2 + b^2 = 25^2$

$\qquad 225 + b^2 = 625$

$\qquad\qquad b^2 = 400$

$\qquad\qquad\; b = 20$

53. $\qquad a^2 + b^2 = c^2$

$\qquad 1^2 + (\sqrt{2})^2 = c^2$

$\qquad\quad 1 + 2 = c^2$

$\qquad\qquad 3 = c^2$

$\qquad\quad \sqrt{3} = c \qquad$ Exact answer

$\qquad 1.732 \approx c \qquad$ Approximation

54. First we subtract to find the vertical distance of the descent:

$\qquad 30,000 \text{ ft} - 20,000 \text{ ft} = 10,000 \text{ ft}$

Let $c =$ the distance the plane travels during the descent. This is labeled "?" in the drawing in the text. We know that $10,000^2 + 50,000^2 = c^2$. We solve this equation.

$\qquad 10,000^2 + 50,000^2 = c^2$

$\qquad 100,000,000 + 2,500,000,000 = c^2$

$\qquad\qquad 2,600,000,000 = c^2$

$\qquad\qquad \sqrt{2,600,00,000} \text{ ft} = c$

$\qquad\qquad 50,990 \text{ ft} \approx c$

55. Let $d =$ the distance each brace reaches vertically. From the drawing in the text we see that $d^2 + 12^2 = 15^2$. We solve this equation.

$\qquad d^2 + 12^2 = 15^2$

$\qquad d^2 + 144 = 225$

$\qquad\qquad d^2 = 81$

$\qquad\qquad\; d = 9 \text{ ft}$

56. a) We substitute 200 for L in the formula.

$\qquad r = 2\sqrt{5 \cdot 200} = 2\sqrt{1000} \approx 63 \text{ mph}$

b) We substitute 90 for r and solve for L.

$\qquad\qquad 90 = 2\sqrt{5L}$

$\qquad\qquad 90^2 = (2\sqrt{5L})^2$

$\qquad\qquad 8100 = 4 \cdot 5L$

$\qquad\qquad 8100 = 20L$

$\qquad\quad 405 \text{ ft} = L$

57. *Discussion and Writing Exercise.* It is incorrect to take the square roots of the terms in the numerator individually; that is, $\sqrt{a + b}$ and $\sqrt{a} + \sqrt{b}$ are not equivalent. The following is correct:

$$\sqrt{\dfrac{9 + 100}{25}} = \dfrac{\sqrt{9 + 100}}{\sqrt{25}} = \dfrac{\sqrt{109}}{5}.$$

58. *Discussion and Writing Exercise.*

a) $\sqrt{5x^2} = \sqrt{5}\sqrt{x^2} = \sqrt{5} \cdot |x| = |x|\sqrt{5}$. The given statement is correct.

b) Let $b = 3$. Then $\sqrt{b^2 - 4} = \sqrt{3^2 - 4} = \sqrt{9 - 4} = \sqrt{5}$, but $b - 2 = 3 - 2 = 1$. The given statement is false.

c) Let $x = 3$. Then $\sqrt{x^2 + 16} = \sqrt{3^2 + 16} = \sqrt{9 + 16} = \sqrt{25} = 5$, but $x + 4 = 3 + 4 = 7$. The given statement is false.

59. After $\frac{1}{2}$ hr, the car traveling east at 50 mph has traveled $\frac{1}{2} \cdot 50$, or 25 mi and the car traveling south at 60 mph has traveled $\frac{1}{2} \cdot 60$, or 30 mph. These distances are the legs of a right triangle. The length of the hypotenuse of the triangle is the distance that separates the cars. Let $d =$ this distance. We know that $25^2 + 30^2 = d^2$. We solve this equation.

$$25^2 + 30^2 = d^2$$
$$625 + 900 = d^2$$
$$1525 = d^2$$
$$\sqrt{1525} \text{ mi} = d$$
$$39.051 \text{ mi} \approx d$$

60. $\sqrt{\sqrt{\sqrt{256}}} = \sqrt{\sqrt{16}} = \sqrt{4} = 2$

61.
$$A = \sqrt{a^2 + b^2}$$
$$A^2 = (\sqrt{a^2 + b^2})^2$$
$$A^2 = a^2 + b^2$$
$$A^2 - a^2 = b^2$$

If $b^2 = A^2 - a^2$, then $b = \sqrt{A^2 - a^2}$ or $b = -\sqrt{A^2 - a^2}$, so we have $b = \pm\sqrt{A^2 - a^2}$.

62. Using the drawing in the text, let $a =$ the hypotenuse of the triangle with legs 4 and x and let $b =$ the hypotenuse of the triangle with legs 9 and x. Then we have
$$4^2 + x^2 = a^2, \text{ or } 16 + x^2 = a^2$$
and
$$9^2 + x^2 = b^2, \text{ or } 81 + x^2 = b^2.$$
Note that a and b are also legs of the large triangle with hypotenuse $4 + 9$, or 13. Then we have
$$a^2 + b^2 = 13^2, \text{ or } a^2 + b^2 = 169.$$
Adding the two equations containing x, we have
$$16 + x^2 = a^2$$
$$\underline{81 + x^2 = b^2}$$
$$97 + 2x^2 = a^2 + b^2.$$
We substitute $97 + 2x^2$ for $a^2 + b^2$ in the equation pertaining to the large triangle.
$$a^2 + b^2 = 169$$
$$97 + 2x^2 = 169$$
$$2x^2 = 72$$
$$x^2 = 36$$
$$x = 6$$

Chapter 8 Test

1. The square roots of 81 are 9 and -9, because $9^2 = 81$ and $(-9)^2 = 81$.

2. $\sqrt{64} = 8$, taking the principal square root

3. $\sqrt{25} = 5$, so $-\sqrt{25} = -5$.

4. $\sqrt{116} \approx 10.770$

5. $\sqrt{87.4} \approx 9.349$, so $-\sqrt{87.4} \approx -9.349$.

6. $\sqrt{\dfrac{96 \cdot 38}{214.2}} \approx 4.127$

7. The radicand is the expression under the radical, $4 - y^3$.

8. Yes, because the radicand is nonnegative

9. No, because the radicand is negative

10. $\sqrt{a^2} = a$

11. $\sqrt{36y^2} = \sqrt{(6y)^2} = 6y$

12. $\sqrt{5}\sqrt{6} = \sqrt{5 \cdot 6} = \sqrt{30}$

13. $\sqrt{x - 8}\sqrt{x + 8} = \sqrt{(x - 8)(x + 8)} = \sqrt{x^2 - 64}$

14. $\sqrt{27} = \sqrt{9 \cdot 3} = \sqrt{9}\sqrt{3} = 3\sqrt{3}$

15. $\sqrt{25x - 25} = \sqrt{25(x - 1)} = \sqrt{25}\sqrt{x - 1} = 5\sqrt{x - 1}$

16. $\sqrt{t^5} = \sqrt{t^4 \cdot t} = \sqrt{t^4}\sqrt{t} = \sqrt{(t^2)^2}\sqrt{t} = t^2\sqrt{t}$

17. $\sqrt{5}\sqrt{10} = \sqrt{5 \cdot 10} = \sqrt{5 \cdot 2 \cdot 5} = \sqrt{5 \cdot 5}\sqrt{2} = 5\sqrt{2}$

18. $\sqrt{3ab}\sqrt{6ab^3} = \sqrt{3ab \cdot 6ab^3} = \sqrt{3 \cdot a \cdot b \cdot 2 \cdot 3 \cdot a \cdot b^2 \cdot b} = \sqrt{3 \cdot 3}\sqrt{a \cdot a}\sqrt{b \cdot b}\sqrt{b^2}\sqrt{2} = 3 \cdot a \cdot b \cdot b\sqrt{2} = 3ab^2\sqrt{2}$

19. $\sqrt{\dfrac{27}{12}} = \sqrt{\dfrac{9}{4}} = \dfrac{\sqrt{9}}{\sqrt{4}} = \dfrac{3}{2}$

20. $\sqrt{\dfrac{144}{a^2}} = \dfrac{\sqrt{144}}{\sqrt{a^2}} = \dfrac{12}{a}$

21. $\sqrt{\dfrac{2}{5}} = \sqrt{\dfrac{2}{5} \cdot \dfrac{5}{5}} = \sqrt{\dfrac{10}{25}} = \dfrac{\sqrt{10}}{\sqrt{25}} = \dfrac{\sqrt{10}}{5}$

22. $\sqrt{\dfrac{2x}{y}} = \sqrt{\dfrac{2x}{y} \cdot \dfrac{y}{y}} = \sqrt{\dfrac{2xy}{y^2}} = \dfrac{\sqrt{2xy}}{y}$

23. $\dfrac{\sqrt{27}}{\sqrt{32}} = \sqrt{\dfrac{27}{32}} = \sqrt{\dfrac{27}{32} \cdot \dfrac{2}{2}} = \sqrt{\dfrac{54}{64}} = \dfrac{\sqrt{54}}{\sqrt{64}} = \dfrac{\sqrt{9 \cdot 6}}{8} = \dfrac{3\sqrt{6}}{8}$

24. $\dfrac{\sqrt{35x}}{\sqrt{80xy^2}} = \sqrt{\dfrac{35x}{80xy^2}} = \sqrt{\dfrac{7}{16y^2}} = \dfrac{\sqrt{7}}{\sqrt{16y^2}} = \dfrac{\sqrt{7}}{4y}$

25. $3\sqrt{18} - 5\sqrt{18} = (3 - 5)\sqrt{18} = -2\sqrt{18} = -2\sqrt{9 \cdot 2} = -2 \cdot 3\sqrt{2} = -6\sqrt{2}$

26. $\sqrt{5} + \sqrt{\dfrac{1}{5}} = \sqrt{5} + \sqrt{\dfrac{1}{5} \cdot \dfrac{5}{5}} = \sqrt{5} + \sqrt{\dfrac{5}{25}} =$

$\sqrt{5} + \dfrac{\sqrt{5}}{5} = \left(1 + \dfrac{1}{5}\right)\sqrt{5} = \dfrac{6}{5}\sqrt{5}$, or $\dfrac{6\sqrt{5}}{5}$

27. $(4 - \sqrt{5})^2 = 4^2 - 2 \cdot 4 \cdot \sqrt{5} + (\sqrt{5})^2 = 16 - 8\sqrt{5} + 5 =$
$21 - 8\sqrt{5}$

28. $(4 - \sqrt{5})(4 + \sqrt{5}) = 4^2 - (\sqrt{5})^2 = 16 - 5 = 11$

29. $\dfrac{10}{4 - \sqrt{5}} = \dfrac{10}{4 - \sqrt{5}} \cdot \dfrac{4 + \sqrt{5}}{4 + \sqrt{5}} = \dfrac{10(4 + \sqrt{5})}{4^2 - (\sqrt{5})^2} = \dfrac{40 + 10\sqrt{5}}{16 - 5} =$

$\dfrac{40 + 10\sqrt{5}}{11}$

30. $a^2 + b^2 = c^2$

$8^2 + 4^2 = c^2$

$64 + 16 = c^2$

$80 = c^2$

$\sqrt{80} = c$ Exact answer

$8.944 \approx c$ Approximation

31. $\sqrt{3x} + 2 = 14$

$\sqrt{3x} = 12$

$(\sqrt{3x})^2 = 12^2$

$3x = 144$

$x = 48$

The number 48 checks. It is the solution.

32. $\sqrt{6x + 13} = x + 3$

$(\sqrt{6x + 13})^2 = (x + 3)^2$

$6x + 13 = x^2 + 6x + 9$

$0 = x^2 - 4$

$0 = (x + 2)(x - 2)$

$x + 2 = 0$ or $x - 2 = 0$

$x = -2$ or $x = 2$

Both numbers check. The solutions are -2 and 2.

33. $\sqrt{1 - x} + 1 = \sqrt{6 - x}$

$(\sqrt{1 - x} + 1)^2 = (\sqrt{6 - x})^2$

$1 - x + 2\sqrt{1 - x} + 1 = 6 - x$

$2 - x + 2\sqrt{1 - x} = 6 - x$

$2\sqrt{1 - x} = 4$

$\sqrt{1 - x} = 2$

$(\sqrt{1 - x})^2 = 2^2$

$1 - x = 4$

$-x = 3$

$x = -3$

The number -3 checks. It is the solution.

34. a) Substitute 28,000 for h in the formula.

$D = \sqrt{2 \cdot 28,000} = \sqrt{56,000} \approx 237$ mi

b) Substitute 261 for D and solve for h.

$261 = \sqrt{2h}$

$261^2 = (\sqrt{2h})^2$

$68,121 = 2h$

$34,060.5 = h$

The airplane is 34,060.5 ft high.

35. Let d = the length of a diagonal, in yd.

$a^2 + b^2 = c^2$

$60^2 + 110^2 = d^2$

$3600 + 12,100 = d^2$

$15,700 = d^2$

$\sqrt{15,700}$ yd $= d$

125.300 yd $\approx d$

36. $\sqrt{\sqrt{\sqrt{625}}} = \sqrt{\sqrt{25}} = \sqrt{5}$

37. $\sqrt{y^{16n}} = \sqrt{(y^{8n})^2} = y^{8n}$

Cumulative Review Chapters 1 - 8

1. $x^3 - x^2 + x - 1 = (-2)^3 - (-2)^2 + (-2) - 1$

$\qquad\qquad\qquad\quad = -8 - 4 - 2 - 1$

$\qquad\qquad\qquad\quad = -15$

2. $\qquad 2x^3 - 7 + \dfrac{3}{7}x^2 - 6x^3 - \dfrac{4}{7}x^2 + 5$

$\quad = (2 - 6)x^3 + \left(\dfrac{3}{7} - \dfrac{4}{7}\right)x^2 + (-7 + 5)$

$\quad = -4x^3 - \dfrac{1}{7}x^2 - 2$

3. $\dfrac{x - 6}{2x + 1}$

We set the denominator equal to 0 and solve for x.

$2x + 1 = 0$

$2x = -1$

$x = -\dfrac{1}{2}$

The expression is not defined for the replacement number $-\dfrac{1}{2}$.

4. No, because the radicand is negative

5. $(2 + \sqrt{3})(2 - \sqrt{3}) = 2^2 - (\sqrt{3})^2 = 4 - 3 = 1$

6. $\sqrt{196} = 14$, so $-\sqrt{196} = -14$.

7. $\sqrt{3}\sqrt{75} = \sqrt{3 \cdot 75} = \sqrt{3 \cdot 3 \cdot 25} = \sqrt{3 \cdot 3}\sqrt{25} = 3 \cdot 5 = 15$

8. $(1 - \sqrt{2})^2 = 1^2 - 2 \cdot 1 \cdot \sqrt{2} + (\sqrt{2})^2 = 1 - 2\sqrt{2} + 2 = 3 - 2\sqrt{2}$

9. $\dfrac{\sqrt{162}}{\sqrt{125}} = \dfrac{\sqrt{81 \cdot 2}}{\sqrt{25 \cdot 5}} = \dfrac{9\sqrt{2}}{5\sqrt{5}} = \dfrac{9\sqrt{2}}{5\sqrt{5}} \cdot \dfrac{\sqrt{5}}{\sqrt{5}} = \dfrac{9\sqrt{10}}{5 \cdot 5} = \dfrac{9\sqrt{10}}{25}$

10. $2\sqrt{45} + 3\sqrt{20} = 2\sqrt{9 \cdot 5} + 3\sqrt{4 \cdot 5}$

$\qquad\qquad\qquad = 2 \cdot 3\sqrt{5} + 3 \cdot 2\sqrt{5}$

$\qquad\qquad\qquad = 6\sqrt{5} + 6\sqrt{5}$

$\qquad\qquad\qquad = 12\sqrt{5}$

11. $(3x^4 - 2y^5)(3x^4 + 2y^5) = (3x^4)^2 - (2y^5)^2 = 9x^8 - 4y^{10}$

12. $(x^2 + 4)^2 = (x^2)^2 + 2 \cdot x^2 \cdot 4 + 4^2 = x^4 + 8x^2 + 16$

13. $\left(2x + \dfrac{1}{4}\right)\left(4x - \dfrac{1}{2}\right) = 2x \cdot 4x - 2x \cdot \dfrac{1}{2} + \dfrac{1}{4} \cdot 4x - \dfrac{1}{4} \dfrac{1}{2} =$

$8x^2 - x + x - \dfrac{1}{8} = 8x^2 - \dfrac{1}{8}$

14. $\dfrac{x}{2x-1} - \dfrac{3x+2}{1-2x} = \dfrac{x}{2x-1} - \dfrac{3x+2}{1-2x} \cdot \dfrac{-1}{-1}$

$\qquad\qquad\qquad = \dfrac{x}{2x-1} - \dfrac{-3x-2}{2x-1}$

$\qquad\qquad\qquad = \dfrac{x - (-3x-2)}{2x-1}$

$\qquad\qquad\qquad = \dfrac{x + 3x + 2}{2x-1}$

$\qquad\qquad\qquad = \dfrac{4x+2}{2x-1}$

15. $\quad (3x^2 - 2x^3) - (x^3 - 2x^2 + 5) + (3x^2 - 5x + 5)$

$= 3x^2 - 2x^3 - x^3 + 2x^2 - 5 + 3x^2 - 5x + 5$

$= -3x^3 + 8x^2 - 5x$

16. $\dfrac{2x+2}{3x-9} \cdot \dfrac{x^2-8x+15}{x^2-1} = \dfrac{(2x+2)(x^2-8x+15)}{(3x-9)(x^2-1)}$

$\qquad\qquad\qquad = \dfrac{2(x+1)(x-3)(x-5)}{3(x-3)(x+1)(x-1)}$

$\qquad\qquad\qquad = \dfrac{2\cancel{(x+1)}\cancel{(x-3)}(x-5)}{3\cancel{(x-3)}\cancel{(x+1)}(x-1)}$

$\qquad\qquad\qquad = \dfrac{2(x-5)}{3(x-1)}$

17. $\qquad \dfrac{2x^2-2}{2x^2+7x+3} \div \dfrac{4x-4}{2x^{-}5x-3}$

$= \dfrac{2x^2-2}{2x^2+7x+3} \cdot \dfrac{2x^2-5x-3}{4x-4}$

$= \dfrac{(2x^2-2)(2x^2-5x-3)}{(2x^2+7x+3)(4x-4)}$

$= \dfrac{2(x+1)(x-1)(2x+1)(x-3)}{(2x+1)(x+3)(4)(x-1)}$

$= \dfrac{\cancel{2}(x+1)\cancel{(x-1)}\cancel{(2x+1)}(x-3)}{\cancel{(2x+1)}(x+3)(\cancel{2}\cdot 2)\cancel{(x-1)}}$

$= \dfrac{(x+1)(x-3)}{2(x+3)}$

18.

$$
\begin{array}{r}
3x^2 + 4x + 9 \\
x-2 \overline{\smash{\big)}\ 3x^3 - 2x^2 + x - 5} \\
\underline{3x^3 - 6x^2} \\
4x^2 + x \\
\underline{4x^2 - 8x} \\
9x - 5 \\
\underline{9x - 18} \\
13
\end{array}
$$

The answer is $3x^2 + 4x + 9 + \dfrac{13}{x-2}$.

19. $\sqrt{2x^2 - 4x + 2} = \sqrt{2(x^2 - 2x + 1)} = \sqrt{2(x-1)^2} =$

$\sqrt{2}\sqrt{(x-1)^2} = \sqrt{2}(x-1)$

20. $x^{-9} \cdot x^{-3} = x^{-9+(-3)} = x^{-12} = \dfrac{1}{x^{12}}$

21. $\sqrt{\dfrac{50}{2x^8}} = \sqrt{\dfrac{25}{x^8}} = \dfrac{\sqrt{25}}{\sqrt{x^8}} = \dfrac{5}{x^4}$

22. $\dfrac{x - \dfrac{1}{x}}{1 - \dfrac{x-1}{2x}} = \dfrac{x - \dfrac{1}{x}}{1 - \dfrac{x-1}{2x}} \cdot \dfrac{2x}{2x}$

$\qquad = \dfrac{2x \cdot x - 2x \cdot \dfrac{1}{x}}{2x \cdot 1 - 2x\left(\dfrac{x-1}{2x}\right)}$

$\qquad = \dfrac{2x^2 - 2}{2x - (x-1)}$

$\qquad = \dfrac{2x^2 - 2}{2x - x + 1}$

$\qquad = \dfrac{2(x^2-1)}{x+1}$

$\qquad = \dfrac{2(x+1)(x-1)}{x+1}$

$\qquad = \dfrac{2\cancel{(x+1)}(x-1)}{\cancel{(x+1)} \cdot 1}$

$\qquad = 2(x-1)$

23. $\quad 3 - 12x^8 = 3(1 - 4x^8)$

$\qquad\qquad = 3(1^2 - (2x^4)^2)$

$\qquad\qquad = 3(1 + 2x^4)(1 - 2x^4)$

24. $12t - 4t^2 - 48t^4 = 4t(3 - t - 12t^3)$

25. $6x^2 - 28x + 16 = 2(3x^2 - 14x + 8) = 2(3x - 2)(x - 4)$

26. $\quad 4x^3 + 4x^2 - x - 1 = (4x^3 + 4x^2) + (-x - 1)$

$\qquad\qquad\qquad = 4x^2(x+1) - (x+1)$

$\qquad\qquad\qquad = (4x^2 - 1)(x+1)$

$\qquad\qquad\qquad = (2x+1)(2x-1)(x+1)$

27. $16x^4 - 56x^2 + 49 = (4x^2)^2 - 2 \cdot 4x^2 \cdot 7 + 7^2 = (4x^2 - 7)^2$

28. $x^2 + 3x - 180 = (x + 15)(x - 12)$

29. $\qquad\quad x^2 = -17x$

$\qquad x^2 + 17x = 0$

$\qquad x(x+17) = 0$

$\qquad x = 0 \;\; or \;\; x + 17 = 0$

$\qquad x = 0 \;\; or \qquad\quad x = -17$

The solutions are 0 and -17.

30. $\quad -3x < 30 + 2x$

$\quad\quad -5x < 30$

$\quad\quad \dfrac{-5x}{-5} > \dfrac{30}{-5}\quad$ Reversing the inequality symbol

$\quad\quad\quad x > -6$

The solution set is $\{x|x > -6\}$.

31. $\quad\quad \dfrac{1}{x} + \dfrac{2}{3} = \dfrac{1}{4}$, LCM $= 12x$

$\quad 12x\left(\dfrac{1}{x} + \dfrac{2}{3}\right) = 12x \cdot \dfrac{1}{4}$

$\quad 12x \cdot \dfrac{1}{x} + 12x \cdot \dfrac{2}{3} = 3x$

$\quad\quad\quad 12 + 8x = 3x$

$\quad\quad\quad\quad 12 = -5x$

$\quad\quad\quad -\dfrac{12}{5} = x$

The number $-\dfrac{12}{5}$ checks. It is the solution.

32. $\quad\quad x^2 - 30 = x$

$\quad\quad x^2 - x - 30 = 0$

$\quad (x - 6)(x + 5) = 0$

$\quad x - 6 = 0\ \ or\ \ x + 5 = 0$

$\quad\quad x = 6\ \ or\ \quad\ x = -5$

The solutions are 6 and -5.

33. $\quad -4(x + 5) \geq 2(x + 5) - 3$

$\quad\quad -4x - 20 \geq 2x + 10 - 3$

$\quad\quad -4x - 20 \geq 2x + 7$

$\quad\quad -6x - 20 \geq 7$

$\quad\quad\quad -6x \geq 27$

$\quad\quad\quad \dfrac{-6x}{-6} \leq \dfrac{27}{-6}\quad$ Reversing the inequality symbol

$\quad\quad\quad\quad x \leq -\dfrac{9}{2}$

The solution set is $\left\{x\,\middle|\,x \leq -\dfrac{9}{2}\right\}$.

34. $\quad\quad 2x^2 = 162$

$\quad\quad 2x^2 - 162 = 0$

$\quad 2(x^2 - 81) = 0$

$\quad\quad x^2 - 81 = 0$

$\quad (x + 9)(x - 9) = 0$

$\quad x + 9 = 0\ \ or\ \ x - 9 = 0$

$\quad\quad x = -9\ \ or\ \quad\ x = 9$

The solutions are -9 and 9.

35. $\quad \sqrt{2x - 1} + 5 = 14$

$\quad\quad \sqrt{2x - 1} = 9$

$\quad\quad (\sqrt{2x - 1})^2 = 9^2$

$\quad\quad\quad 2x - 1 = 81$

$\quad\quad\quad 2x = 82$

$\quad\quad\quad x = 41$

The number 41 checks. It is the solution.

36. $\quad\quad \sqrt{4x} + 1 = \sqrt{x} + 4$

$\quad\quad\quad \sqrt{4x} = \sqrt{x} + 3$

$\quad\quad (\sqrt{4x})^2 = (\sqrt{x} + 3)^2$

$\quad\quad\quad 4x = x + 6\sqrt{x} + 9$

$\quad\quad 3x - 9 = 6\sqrt{x}$

$\quad\quad x - 3 = 2\sqrt{x}\quad$ Dividing by 3

$\quad (x - 3)^2 = (2\sqrt{x})^2$

$\quad x^2 - 6x + 9 = 4x$

$\quad x^2 - 10x + 9 = 0$

$\quad (x - 1)(x - 9) = 0$

$\quad x - 1 = 0\ \ or\ \ x - 9 = 0$

$\quad\quad x = 1\ \ or\ \quad\ x = 9$

The number 1 does not check, but 9 does. The solution is 9.

37. $\quad\quad \dfrac{1}{4}x + \dfrac{2}{3}x = \dfrac{2}{3} - \dfrac{3}{4}x$, LCM is 12

$\quad 12\left(\dfrac{1}{4}x + \dfrac{2}{3}x\right) = 12\left(\dfrac{2}{3} - \dfrac{3}{4}x\right)$

$\quad 12 \cdot \dfrac{1}{4}x + 12 \cdot \dfrac{2}{3}x = 12 \cdot \dfrac{2}{3} - 12 \cdot \dfrac{3}{4}x$

$\quad\quad\quad 3x + 8x = 8 - 9x$

$\quad\quad\quad\quad 11x = 8 - 9x$

$\quad\quad\quad\quad 20x = 8$

$\quad\quad\quad\quad x = \dfrac{2}{5}$

The solution is $\dfrac{2}{5}$.

38. $\quad\quad \dfrac{x}{x - 1} - \dfrac{x}{x + 1} = \dfrac{1}{2x - 2}$

$\quad\quad \dfrac{x}{x - 1} - \dfrac{x}{x + 1} = \dfrac{1}{2(x - 1)}$,

$\quad\quad\quad\quad$ LCM is $2(x - 1)(x + 1)$

$\quad 2(x-1)(x+1)\left(\dfrac{x}{x-1} - \dfrac{x}{x+1}\right) = 2(x-1)(x+1) \cdot \dfrac{1}{2(x-1)}$

$\quad 2(x - 1)(x + 1) \cdot \dfrac{x}{x - 1} -$

$\quad 2(x - 1)(x + 1) \cdot \dfrac{x}{x + 1} = x + 1$

$\quad 2x(x + 1) - 2x(x - 1) = x + 1$

$\quad 2x^2 + 2x - 2x^2 + 2x = x + 1$

$\quad\quad\quad\quad 4x = x + 1$

$\quad\quad\quad\quad 3x = 1$

$\quad\quad\quad\quad x = \dfrac{1}{3}$

The number $\dfrac{1}{3}$ checks. It is the solution.

39. $\quad x = y + 3,\quad\quad (1)$

$\quad 3y - 4x = -13\quad (2)$

First substitute $y + 3$ for x in Equation (2) and solve for y.

$$3y - 4x = -13$$
$$3y - 4(y + 3) = -13$$
$$3y - 4y - 12 = -13$$
$$-y - 12 = -13$$
$$-y = -1$$
$$y = 1$$

Now substitute 1 for y in Equation (1) and find x.

$$x = y + 3 = 1 + 3 = 4$$

The ordered pair $(4, 1)$ checks in both equations. It is the solution.

40. $2x - 3y = 30,$
$5y - 2x = -46$

First we rewrite the second equation in the form $Ax + By = C$. Then we add.

$$\begin{array}{ll} 2x - 3y = 30, & (1) \\ \underline{-2x + 5y = -46} & (2) \\ 2y = -16 & \\ y = -8 & \end{array}$$

Now substitute -8 for y in one of the equations and solve for x.

$$2x - 3y = 30 \quad (1)$$
$$2x - 3(-8) = 30$$
$$2x + 24 = 30$$
$$2x = 6$$
$$x = 3$$

The ordered pair $(3, -8)$ checks in both equations. It is the solution.

41. $4A = pr + pq$
$4A = p(r + q)$

$$\frac{4A}{r + q} = p$$

42. $3y - 3x > -6$

First we graph the line $3y - 3x = -6$. The intercepts are $(0, -2)$ and $(2, 0)$. We draw a dashed line since the inequality symbol is $>$. Then we test a point that is not on the line. We use $(0, 0)$.

$$\begin{array}{c|c} 3y - 3x > -6 \\ \hline 3 \cdot 0 - 3 \cdot 0 \; ? \; -6 \\ 0 - 0 \\ 0 & \text{TRUE} \end{array}$$

We see that $(0, 0)$ is a solution of the inequality so we shade the half-plane that contains $(0, 0)$.

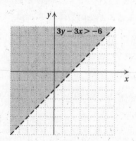

43. $x = 5$, or $x + 0 \cdot y = 5$

Any ordered pair $(5, y)$ is a solution of the equation. The graph is a vertical line with x-intercept $(5, 0)$.

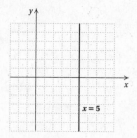

44. $2x - 6y = 12$

First we will find the intercepts. To find the y-intercept we let $x = 0$ and solve for y.

$$2 \cdot 0 - 6y = 12$$
$$-6y = 12$$
$$y = -2$$

The y-intercept is $(0, 2)$.

To find the x-intercept, let $y = 0$ and solve for x.

$$2x - 6 \cdot 0 = 12$$
$$2x = 12$$
$$x = 6$$

The x-intercept is $(6, 0)$.

We plot the intercepts and draw the graph.

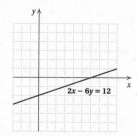

We can find a third point as a check. We let $x = 3$ and solve for y.

$$2 \cdot 3 - 6y = 12$$
$$6 - 6y = 12$$
$$-6y = 6$$
$$y = -1$$

The point $(3, -1)$ appears to lie on the graph, so the graph is probably correct.

45. First we find the slope.

$$m = \frac{-2 - 9}{1 - 5} = \frac{-11}{-4} = \frac{11}{4}$$

Then the equation is $y = \dfrac{11}{4}x + b$. Using the point $(5, 9)$, we substitute 5 for x and 9 for y and solve for b.

$$y = \frac{11}{4}x + b$$

$$9 = \frac{11}{4} \cdot 5 + b$$

$$9 = \frac{55}{4} + b$$

$$-\frac{19}{4} = b$$

Then the equation is $y = \frac{11}{4}x - \frac{19}{4}$.

46. First we solve for y.

$$5x - 3y = 9$$

$$-3y = -5x - 9$$

$$-\frac{1}{3}(-3y) = -\frac{1}{3}(-5x - 9)$$

$$y = \frac{5}{3}x + 3$$

Thus, the slope is $\frac{5}{3}$ and the y-intercept is $(0, 3)$.

47. When $x = -2$, $y = (-2)^3 - 4(-2) - 2 = -8 + 8 - 2 = -2$.
When $x = -1$, $y = (-1)^3 - 4(-1) - 2 = -1 + 4 - 2 = 1$.
When $x = 0$, $y = 0^3 - 4 \cdot 0 - 2 = 0 - 0 - 2 = -2$.
When $x = 1$, $y = 1^3 - 4 \cdot 1 - 2 = 1 - 4 - 2 = -5$.
When $x = 2$, $y = 2^3 - 4 \cdot 2 - 2 = 8 - 8 - 2 = -2$.

We could also estimate these values from the graph.

48. **Familiarize.** We let $h =$ the number of home runs Bonds would have in 162 games.

Translate. We translate to a proportion.

$$\text{Home runs} \rightarrow \frac{39}{88} = \frac{h}{162} \leftarrow \text{Home runs}$$
$$\text{Games} \rightarrow \qquad\qquad \leftarrow \text{Games}$$

Solve.

$$\frac{39}{88} = \frac{h}{162}$$

$$162 \cdot \frac{39}{88} = 162 \cdot \frac{h}{162}$$

$$72 \approx h$$

Check.

$\frac{39}{88} \approx 0.44$ and $\frac{72}{162} \approx 0.44$.

The ratios are the same, so the answer checks.

State. Bonds would hit about 72 home runs in a season of 162 games.

49. **Familiarize and Translate.** We have inverse variation between A and d, so an equation $A = k/d$, $k > 0$, applies.

Solve. First we find an equation of variation.

$$A = \frac{k}{d}$$

$$4 = \frac{k}{100}$$

$$400 = k$$

The equation of variation is $A = \frac{400}{d}$. Now find A when $d = 1000$.

$$A = \frac{400}{d}$$

$$A = \frac{400}{1000}$$

$$A = 0.4$$

Check. We could repeat the computation. Also, observe that, when the distance from the stage increased, the apparent size of the musicians decreased as we would expect with inverse variation.

State. If you were sitting 1000 ft from the stage, the musicians would appear to be 0.4 ft tall.

50. **Familiarize.** Let $h =$ the cost of one hamburger and $m =$ the cost of one milkshake.

Translate.

6 hamburgers plus 4 milkshakes cost $27.70.
$\quad 6h \qquad\quad + \qquad 4m \qquad\quad = \quad 27.70$

3 hamburgers plus 1 milkshake cost $11.35.
$\quad 3h \qquad\quad + \qquad 1 \cdot m \qquad = \quad 11.35$

After clearing decimals we have the following system of equations.

$$60h + 40m = 277, \quad (1)$$
$$300h + 100m = 1135 \quad (2)$$

Solve. We multiply Equation (1) by -5 and then add.

$$-300h - 200m = -1385$$
$$\underline{300h + 100m = 1135}$$
$$\qquad -100m = -250$$
$$\qquad\qquad m = 2.5$$

Now substitute 2.5 for m in one of the original equations and solve for h.

$$60h + 40m = 277 \quad (1)$$
$$60h + 40(2.5) = 277$$
$$60h + 100 = 277$$
$$60h = 177$$
$$h = 2.95$$

Check. If one hamburger costs $2.95 and one milkshake costs $2.50, then 6 hamburgers and 4 milkshakes cost $6(\$2.95) + 4(\$2.50) = \$17.70 + \$10.00 = \$27.70$ and 3 hamburgers and 1 milkshake cost $3(\$2.95) + \$2.50 = \$8.85 + \$2.50 = \$11.35$. The answer checks.

State. A hamburger costs $2.95 and a milkshake costs $2.50.

51. We first make a drawing. Let $h =$ the height of the top of the ladder.

We know that $4^2 + h^2 = 8^2$. We solve this equation.

$$4^2 + h^2 = 8^2$$
$$16 + h^2 = 64$$
$$h^2 = 48$$
$$h = \sqrt{48} \text{ m}$$
$$h \approx 6.928 \text{ m}$$

52. Familiarize. Let $x =$ the measure of the first angle. Then $2x =$ the measure of the second angle and $x + 2x - 48$, or $3x - 48 =$ the measure of the third angle. Recall that the sum of the measures of the angles of a triangle is $180°$.

Translate.

$$\underbrace{\text{The sum of the measures}}_{\downarrow} \quad \underset{\downarrow}{\text{is}} \quad \underset{\downarrow}{180°}.$$
$$x + 2x + (3x - 48) \quad = \quad 180$$

Solve.

$$x + 2x + (3x - 48) = 180$$
$$6x - 48 = 180$$
$$6x = 228$$
$$x = 38$$

If $x = 38$, then $2x = 2 \cdot 38 = 76$ and $3x - 48 = 3 \cdot 38 - 48 = 66$.

Check. If the angle measures are $38°$, $76°$, and $66°$, then the measure of the second angle is twice the measure of the first angle and the measure of the third angle is $48°$ less than the sum of the measures of the other two angles. Also, $38° + 76° + 66° = 180°$. The answer checks.

State. The angle measures are $38°$, $76°$, and $66°$.

53. Familiarize. Let $d =$ the number of defective resistors you would expect to find in a sample of 250.

Translate. We translate to a proportion.

$$\begin{array}{l} \text{Number defective} \rightarrow \\ \text{Sample size} \rightarrow \end{array} \frac{12}{150} = \frac{d}{250} \begin{array}{l} \leftarrow \text{Number defective} \\ \leftarrow \text{Sample size} \end{array}$$

Solve.

$$\frac{12}{150} = \frac{d}{250}$$
$$250 \cdot \frac{12}{150} = 250 \cdot \frac{d}{250}$$
$$20 = d$$

Check.

$$\frac{12}{150} = 0.08 \text{ and } \frac{20}{250} = 0.08.$$

The ratios are the same, so the answer is correct.

State. You would expect to find 20 defective resistors in a sample of 250.

54. Familiarize. Let $w =$ the width of the rectangle, in meters. Then $w + 3 =$ the length. Recall that the area of a rectangle is (length) × (width).

Translate.

$$\underbrace{\text{The area}}_{\downarrow} \quad \underset{\downarrow}{\text{is}} \quad \underset{\downarrow}{180 \text{ m}^2}.$$
$$(w + 3)w \quad = \quad 180$$

Solve.

$$(w + 3)w = 180$$
$$w^2 + 3w = 180$$
$$w^2 + 3w - 180 = 0$$
$$(w + 15)(w - 12) = 0$$
$$w + 15 = 0 \quad or \quad w - 12 = 0$$
$$w = -15 \quad or \quad \quad w = 12$$

Check. The width cannot be negative, so -15 cannot be a solution. If $w = 12$, then $w + 3 = 12 + 3 = 15$. If the width is 12 m and the length is 15 m, then the length is 3 m more than the width and the area is $15 \cdot 12$, or 180 m². The answer checks.

State. The length is 15 m and the width is 12 m.

55. Familiarize. Let $d =$ the number of dimes and $q =$ the number of quarters. Then the value of the dimes is $0.1d$ and the value of the quarters is $0.25q$.

Translate.

$$\underset{\downarrow}{\begin{array}{c} \text{Value} \\ \text{of dimes} \end{array}} \quad \underset{\downarrow}{\text{plus}} \quad \underset{\downarrow}{\begin{array}{c} \text{value of} \\ \text{quarters} \end{array}} \quad \underset{\downarrow}{\text{is}} \quad \underset{\downarrow}{\$19.00}.$$
$$0.1d \quad + \quad 0.25q \quad = \quad 19$$

$$\underset{\downarrow}{\underbrace{\begin{array}{c} \text{Number} \\ \text{of dimes} \end{array}}} \quad \underset{\downarrow}{\text{plus}} \quad \underset{\downarrow}{\underbrace{\begin{array}{c} \text{number} \\ \text{of quarters} \end{array}}} \quad \underset{\downarrow}{\text{is}} \quad \underset{\downarrow}{115}.$$
$$d \quad + \quad q \quad = \quad 115$$

After clearing decimals we have the following system of equations.

$$10d + 25q = 1900, \quad (1)$$
$$d + q = 115 \quad (2)$$

Solve. We multiply Equation (2) by -10 and then add.

$$10d + 25q = 1900$$
$$\underline{-10d - 10q = -1150}$$
$$15q = 750$$
$$q = 50$$

Now substitute 50 for q in one of the equations and solve for d.

$$d + q = 115 \quad (2)$$
$$d + 50 = 115$$
$$d = 65$$

Check. If there are 65 dimes and 50 quarters, there are $65 + 50$, or 115, coins in all. The value of the coins is $\$0.1(65) + \$0.25(50) = \$6.50 + \$12.50 = \$19.00$. The answer checks.

State. There are 65 dimes and 50 quarters.

56. Familiarize. Let $x =$ the amount originally invested. Using the formula for simple interest, $I = Prt$, we find that the simple interest for 1 yr is $x \cdot 4.5\% \cdot 1$, or $0.045x$.

Translate.

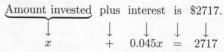

$$\underbrace{\text{Amount invested}}_{\downarrow} \quad \underset{}{\text{plus}} \quad \underset{\downarrow}{\text{interest}} \quad \underset{\downarrow}{\text{is}} \quad \underset{\downarrow}{\$2717}.$$
$$x \quad + \quad 0.045x \quad = \quad 2717$$

Solve.

$$x + 0.045x = 2717$$
$$1.045x = 2717$$
$$x = 2600$$

Check. 4.5% of \$2600 is 0.045(\$2600), or \$117, and \$2600 + \$117 = \$2717. The answer checks.

State. \$2600 was originally invested.

57. Familiarize. Let r = the faster speed. Then $r - 10$ = the slower speed. We organize the information in a table.

	Distance	Speed	Time
Faster trip	600	r	t_1
Slower trip	600	$r - 10$	t_2

Translate. We use the formula $d = rt$ in each row of the table to get two equations.

$$600 = rt_1 \qquad (1)$$
$$600 = (r - 10)t_2 \qquad (2)$$

Since $600 = rt_1$, we have $t_1 = \dfrac{600}{r}$. Also, since $600 = (r - 10)t_2$, we have $t_2 = \dfrac{600}{r - 10}$. We know that $t_2 = t_1 + 2$, so we can write the following equation:

$$\frac{600}{r - 10} = \frac{600}{r} + 2.$$

Solve.

$$\frac{600}{r - 10} = \frac{600}{r} + 2$$

$$r(r - 10) \cdot \frac{600}{r - 10} = r(r - 10)\left(\frac{600}{r} + 2\right)$$

$$600r = r(r - 10) \cdot \frac{600}{r} + r(r - 10)(2)$$

$$600r = 600(r - 10) + 2r(r - 10)$$

$$600r = 600r - 6000 + 2r^2 - 20r$$

$$0 = 2r^2 - 20r - 6000$$

$$0 = 2(r^2 - 10r - 3000)$$

$$0 = r^2 - 10r - 3000 \quad \text{Dividing by 2}$$

$$0 = (r + 50)(r - 60)$$

$$r + 50 = 0 \quad or \quad r - 60 = 0$$
$$r = -50 \quad or \qquad r = 60$$

Check. The speed cannot be negative, so -50 cannot be a solution. At a speed of 60 mph it takes 600/60, or 10 hr, to drive 60 miles. At a speed of $60 - 10$, or 50 mph, it takes 600/50, or 12 hr, to drive 600 mi. Since the second time is 2 hr longer than the first, the answer checks.

State. The speed going is 60 mph.

58. $\dfrac{\sqrt{27}}{\sqrt{45}} = \sqrt{\dfrac{27}{45}} = \sqrt{\dfrac{3}{5}} = \sqrt{\dfrac{3}{5} \cdot \dfrac{5}{5}} = \sqrt{\dfrac{15}{25}} = \dfrac{\sqrt{15}}{5}$

Answer (a) is correct.

59.
$$\frac{3}{x - 2} + \frac{6}{x^2 - 4} = \frac{5}{x + 2}$$

$$\frac{3}{x - 2} + \frac{6}{(x + 2)(x - 2)} = \frac{5}{x + 2},$$
$$\text{LCM is } (x + 2)(x - 2)$$

$$(x + 2)(x - 2)\left(\frac{3}{x - 2} + \frac{6}{(x + 2)(x - 2)}\right) =$$
$$(x + 2)(x - 2) \cdot \frac{5}{x + 2}$$

$$(x+2)(x-2) \cdot \frac{3}{x-2} + (x+2)(x-2) \cdot \frac{6}{(x+2)(x-2)} = 5(x-2)$$

$$3(x + 2) + 6 = 5(x - 2)$$

$$3x + 6 + 6 = 5x - 10$$

$$3x + 12 = 5x - 10$$

$$12 = 2x - 10$$

$$22 = 2x$$

$$11 = x$$

The number 11 checks, so answer (b) is correct.

60.
$$\frac{\dfrac{1}{x} + \dfrac{1}{y}}{\dfrac{x}{2} + \dfrac{y}{2}} = \frac{\dfrac{1}{x} \cdot \dfrac{y}{y} + \dfrac{1}{y} \cdot \dfrac{x}{x}}{\dfrac{x + y}{2}}$$

$$= \frac{\dfrac{y}{xy} + \dfrac{x}{xy}}{\dfrac{x + y}{2}}$$

$$= \frac{\dfrac{y + x}{xy}}{\dfrac{x + y}{2}}$$

$$= \frac{y + x}{xy} \cdot \frac{2}{x + y}$$

$$= \frac{(y + x)(2)}{xy(x + y)}$$

$$= \frac{(y + x)(2)}{xy(x + y)}$$

$$= \frac{2}{xy}$$

Answer (a) is correct.

61. $(4a^2b + 5c)(3a^2b - 2c) = 12a^4b^2 - 8a^2bc + 15a^2bc - 10c^2 = 12a^4b^2 + 7a^2bc - 10c^2$

Answer (e) is correct.

62. $|-3| = 3$ and -4 is to the left of 3 on the number line, so we have $-4 < |-3|$.

63. $|-4| = 4$, $|-3| = 3$, and 4 is to the right of 3 on the number line, so we have $|-4| > |-3|$.

64. Familiarize. Let x = the amount of pure water that should be added to make a 12% salt solution and let y = the total amount of the mixture. We organize the infor-

mation in a table.

Solution	Salt	Water	Mixture
Amount	200	x	y
Percent of salt	30%	0%	12%
Amount of salt	30% × 200, or 60	0% ×w, or 0	12% ×y, or 0.12y

Translate. From the first and last rows of the table we get two equations.

$$200 + x = y, \qquad (1)$$
$$60 + 0 = 0.12y, \text{ or } 60 = 0.12y \quad (2)$$

Solve. First we solve Equation (2) for y.

$$60 = 0.12y$$
$$500 = y$$

Substitute 500 for y in Equation (1) and solve for x.

$$200 + x = y$$
$$200 + x = 500$$
$$x = 300$$

Check. If 300 L of pure water is added, the mixture is $200 + 300$ or 500 L and it contains the 60 L of salt from the 200 L of the 30% salt solution. Since $60/500 = 0.12$, or 12%, the mixture is a 12% salt solution. The answer checks.

State. 300 L of pure water should be added.

65. $\quad \sqrt{x} + 1 = y, \quad (1)$
$\qquad \sqrt{x} + \sqrt{y} = 5 \quad (2)$

First solve Equation (1) for $\sqrt{x}$.

$$\sqrt{x} + 1 = y$$
$$\sqrt{x} = y - 1 \quad (3)$$

Now substitute $y - 1$ for $\sqrt{x}$ in Equation (2) and solve for y.

$$\sqrt{x} + \sqrt{y} = 5$$
$$y - 1 + \sqrt{y} = 5$$
$$\sqrt{y} = 6 - y$$
$$(\sqrt{y})^2 = (6 - y)^2$$
$$y = 36 - 12y + y^2$$
$$0 = y^2 - 13y + 36$$
$$0 = (y - 4)(y - 9)$$
$$y - 4 = 0 \quad or \quad y - 9 = 0$$
$$y = 4 \quad or \qquad y = 9$$

Substitute each value for y in Equation (3) and find x.

For 4: $\quad \sqrt{x} = 4 - 1$
$\qquad\qquad \sqrt{x} = 3$
$\qquad\qquad (\sqrt{x})^2 = 3^2$
$\qquad\qquad x = 9$

For 9: $\quad \sqrt{x} = 9 - 1$
$\qquad\qquad \sqrt{x} = 8$
$\qquad\qquad (\sqrt{x})^2 = 8^2$
$\qquad\qquad x = 64$

The ordered pair $(64, 9)$ does not check, but $(9, 4)$ does. The solution is $(9, 4)$.

66. First find the shortest distance from the restaurant to the corner of the building on which Vince's office is located. This is the hypotenuse of a right triangle with legs 500 ft and 200 ft.

$$a^2 + b^2 = c^2$$
$$500^2 + 200^2 = c^2$$
$$250,000 + 40,000 = c^2$$
$$290,000 = c^2$$
$$\sqrt{290,000} = c$$

Now find the shortest distance from Vince's office to the restaurant. This is the hypotenuse of a right triangle with legs $\sqrt{290,000}$ ft and 900 ft.

$$a^2 + b^2 = c^2$$
$$(\sqrt{290,000})^2 + 900^2 = c^2$$
$$290,000 + 810,000 = c^2$$
$$1,100,000 = c^2$$
$$\sqrt{1,100,000} = c$$
$$1049 \approx c$$

The shortest distance from the handset in Vince's office to the restaurant is about 1049 ft. Now we convert one quarter mile to feet to see if the handset is less than one quarter mile from the base unit.

$$\frac{1}{4} \text{ mi} = \frac{1}{4} \times 1 \text{ mi} = \frac{1}{4} \times 5280 \text{ ft} = 1328 \text{ ft}$$

We see that the handset can be used in the restaurant.

67. First we find the second leg of the triangle with hypotenuse 1 and one leg x. Let y = this leg.

$$x^2 + y^2 = 1$$
$$y^2 = 1 - x^2$$
$$y = \sqrt{1 - x^2}$$

Next we find the second leg of the triangle with hypotenuse $\sqrt{3}$ and one leg x. Let z = this leg.

$$x^2 + z^2 = (\sqrt{3})^2$$
$$x^2 + z^2 = 3$$
$$z^2 = 3 - x^2$$
$$z = \sqrt{3 - x^2}$$

Observe that the large triangle with legs $\sqrt{3}$ and 1 has hypotenuse of $y + z$. Then we have:

$$(\sqrt{3})^2 + 1^2 = (y+z)^2$$
$$(\sqrt{3})^2 + 1^2 = (\sqrt{1-x^2} + \sqrt{3-x^2})^2$$
$$3+1 = 1 - x^2 + 2\sqrt{1-x^2}\sqrt{3-x^2} + 3 - x^2$$
$$4 = 4 - 2x^2 + 2\sqrt{(1-x^2)(3-x^2)}$$
$$2x^2 = 2\sqrt{3-4x^2+x^4}$$
$$x^2 = \sqrt{3-4x^2+x^4}$$
$$(x^2)^2 = (\sqrt{3-4x^2+x^4})^2$$
$$x^4 = 3 - 4x^2 + x^4$$
$$0 = 3 - 4x^2$$
$$0 = (\sqrt{3}+2x)(\sqrt{3}-2x)$$

$$\sqrt{3}+2x = 0 \quad or \quad \sqrt{3}-2x = 0$$
$$2x = -\sqrt{3} \quad or \quad -2x = -\sqrt{3}$$
$$x = -\frac{\sqrt{3}}{2} \quad or \quad x = \frac{\sqrt{3}}{2}$$

Since x cannot be negative, we have $x = \dfrac{\sqrt{3}}{2} \approx 0.866$.

Chapter 9

Quadratic Equations

1. $x^2 - 3x + 2 = 0$

This equation is already in standard form.

$a = 1,\ b = -3,\ c = 2$

3. $\quad\quad 7x^2 = 4x - 3$

$7x^2 - 4x + 3 = 0 \quad\quad$ Standard form

$a = 7,\ b = -4,\ c = 3$

5. $\quad\quad\quad 5 = -2x^2 + 3x$

$2x^2 - 3x + 5 = 0 \quad\quad$ Standard form

$a = 2,\ b = -3,\ c = 5$

7. $x^2 + 5x = 0$

$x(x + 5) = 0$

$x = 0 \ \ or \ \ x + 5 = 0$

$x = 0 \ \ or \ \quad\quad x = -5$

The solutions are 0 and -5.

9. $3x^2 + 6x = 0$

$3x(x + 2) = 0$

$3x = 0 \ \ or \ \ x + 2 = 0$

$x = 0 \ \ or \ \quad\quad x = -2$

The solutions are 0 and -2.

11. $\quad\quad 5x^2 = 2x$

$5x^2 - 2x = 0$

$x(5x - 2) = 0$

$x = 0 \ \ or \ \ 5x - 2 = 0$

$x = 0 \ \ or \ \quad\quad 5x = 2$

$x = 0 \ \ or \ \quad\quad x = \dfrac{2}{5}$

The solutions are 0 and $\dfrac{2}{5}$.

13. $4x^2 + 4x = 0$

$4x(x + 1) = 0$

$4x = 0 \ \ or \ \ x + 1 = 0$

$x = 0 \ \ or \ \quad\quad x = -1$

The solutions are 0 and -1.

15. $0 = 10x^2 - 30x$

$0 = 10x(x - 3)$

$10x = 0 \ \ or \ \ x - 3 = 0$

$x = 0 \ \ or \ \quad\quad x = 3$

The solutions are 0 and 3.

17. $\quad\quad 11x = 55x^2$

$0 = 55x^2 - 11x$

$0 = 11x(5x - 1)$

$11x = 0 \ \ or \ \ 5x - 1 = 0$

$x = 0 \ \ or \ \quad\quad 5x = 1$

$x = 0 \ \ or \ \quad\quad x = \dfrac{1}{5}$

The solutions are 0 and $\dfrac{1}{5}$.

19. $\quad\quad 14t^2 = 3t$

$14t^2 - 3t = 0$

$t(14t - 3) = 0$

$t = 0 \ \ or \ \ 14t - 3 = 0$

$t = 0 \ \ or \ \quad\quad 14t = 3$

$t = 0 \ \ or \ \quad\quad t = \dfrac{3}{14}$

The solutions are 0 and $\dfrac{3}{14}$.

21. $5y^2 - 3y^2 = 72y + 9y$

$\quad\quad\quad 2y^2 = 81y$

$2y^2 - 81y = 0$

$y(2y - 81) = 0$

$y = 0 \ \ or \ \ 2y - 81 = 0$

$y = 0 \ \ or \ \quad\quad 2y = 81$

$y = 0 \ \ or \ \quad\quad y = \dfrac{81}{2}$

The solutions are 0 and $\dfrac{81}{2}$.

23. $x^2 + 8x - 48 = 0$

$(x + 12)(x - 4) = 0$

$x + 12 = 0 \quad\ \ or \ \ x - 4 = 0$

$\quad\quad x = -12 \ \ or \ \quad\quad x = 4$

The solutions are -12 and 4.

25. $5 + 6x + x^2 = 0$

$(5 + x)(1 + x) = 0$

$5 + x = 0 \ \ or \ \ 1 + x = 0$

$\quad\quad x = -5 \ \ or \ \quad\quad x = -1$

The solutions are -5 and -1.

27. $18 = 7p + p^2$

$0 = p^2 + 7p - 18$

$0 = (p + 9)(p - 2)$

$p + 9 = 0 \ \ or \ \ p - 2 = 0$

$\quad\ p = -9 \ \ or \ \quad\ p = 2$

The solutions are -9 and 2.

29.
$$-15 = -8y + y^2$$
$$0 = y^2 - 8y + 15$$
$$0 = (y - 5)(y - 3)$$
$$y - 5 = 0 \ \ or \ \ y - 3 = 0$$
$$y = 5 \ \ or \ \ \ \ y = 3$$
The solutions are 5 and 3.

31. $x^2 + 10x + 25 = 0$
$$(x + 5)(x + 5) = 0$$
$$x + 5 = 0 \ \ or \ \ x + 5 = 0$$
$$x = -5 \ \ or \ \ \ \ x = -5$$
The solution is -5.

33.
$$r^2 = 8r - 16$$
$$r^2 - 8r + 16 = 0$$
$$(r - 4)(r - 4) = 0$$
$$r - 4 = 0 \ \ or \ \ r - 4 = 0$$
$$r = 4 \ \ or \ \ \ \ r = 4$$
The solution is 4.

35.
$$6x^2 + x - 2 = 0$$
$$(3x + 2)(2x - 1) = 0$$
$$3x + 2 = 0 \ \ or \ \ 2x - 1 = 0$$
$$3x = -2 \ \ or \ \ \ \ 2x = 1$$
$$x = -\frac{2}{3} \ \ or \ \ \ \ x = \frac{1}{2}$$
The solutions are $-\frac{2}{3}$ and $\frac{1}{2}$.

37.
$$3a^2 = 10a + 8$$
$$3a^2 - 10a - 8 = 0$$
$$(3a + 2)(a - 4) = 0$$
$$3a + 2 = 0 \ \ or \ \ a - 4 = 0$$
$$3a = -2 \ \ or \ \ \ \ a = 4$$
$$a = -\frac{2}{3} \ \ or \ \ \ \ a = 4$$
The solutions are $-\frac{2}{3}$ and 4.

39.
$$6x^2 - 4x = 10$$
$$6x^2 - 4x - 10 = 0$$
$$2(3x^2 - 2x - 5) = 0$$
$$2(3x - 5)(x + 1) = 0$$
$$3x - 5 = 0 \ \ or \ \ x + 1 = 0$$
$$3x = 5 \ \ or \ \ \ \ x = -1$$
$$x = \frac{5}{3} \ \ or \ \ \ \ x = -1$$
The solutions are $\frac{5}{3}$ and -1.

41.
$$2t^2 + 12t = -10$$
$$2t^2 + 12t + 10 = 0$$
$$2(t^2 + 6t + 5) = 0$$
$$2(t + 5)(t + 1) = 0$$

$$t + 5 = 0 \ \ or \ \ t + 1 = 0$$
$$t = -5 \ \ or \ \ \ \ t = -1$$
The solutions are -5 and -1.

43.
$$t(t - 5) = 14$$
$$t^2 - 5t = 14$$
$$t^2 - 5t - 14 = 0$$
$$(t + 2)(t - 7) = 0$$
$$t + 2 = 0 \ \ or \ \ t - 7 = 0$$
$$t = -2 \ \ or \ \ \ \ t = 7$$
The solutions are -2 and 7.

45.
$$t(9 + t) = 4(2t + 5)$$
$$9t + t^2 = 8t + 20$$
$$t^2 + t - 20 = 0$$
$$(t + 5)(t - 4) = 0$$
$$t + 5 = 0 \ \ or \ \ t - 4 = 0$$
$$t = -5 \ \ or \ \ \ \ t = 4$$
The solutions are -5 and 4.

47. $16(p - 1) = p(p + 8)$
$$16p - 16 = p^2 + 8p$$
$$0 = p^2 - 8p + 16$$
$$0 = (p - 4)(p - 4)$$
$$p - 4 = 0 \ \ or \ \ p - 4 = 0$$
$$p = 4 \ \ or \ \ \ \ p = 4$$
The solution is 4.

49. $(t - 1)(t + 3) = t - 1$
$$t^2 + 2t - 3 = t - 1$$
$$t^2 + t - 2 = 0$$
$$(t + 2)(t - 1) = 0$$
$$t + 2 = 0 \ \ or \ \ t - 1 = 0$$
$$t = -2 \ \ or \ \ \ \ t = 1$$
The solutions are -2 and 1.

51.
$$\frac{24}{x - 2} + \frac{24}{x + 2} = 5$$
The LCM is $(x - 2)(x + 2)$.
$$(x - 2)(x + 2)\left(\frac{24}{x - 2} + \frac{24}{x + 2}\right) = (x - 2)(x + 2) \cdot 5$$
$$(x - 2)(x + 2) \cdot \frac{24}{x - 2} + (x - 2)(x + 2) \cdot \frac{24}{x + 2} = 5(x - 2)(x + 2)$$
$$24(x + 2) + 24(x - 2) = 5(x^2 - 4)$$
$$24x + 48 + 24x - 48 = 5x^2 - 20$$
$$48x = 5x^2 - 20$$
$$0 = 5x^2 - 48x - 20$$
$$0 = (5x + 2)(x - 10)$$

$$5x + 2 = 0 \quad or \quad x - 10 = 0$$
$$5x = -2 \quad or \qquad x = 10$$
$$x = -\frac{2}{5} \quad or \qquad x = 10$$

Both numbers check. The solutions are $-\frac{2}{5}$ and 10.

53.
$$\frac{1}{x} + \frac{1}{x+6} = \frac{1}{4}$$

The LCM is $4x(x+6)$.

$$4x(x+6)\left(\frac{1}{x} + \frac{1}{x+6}\right) = 4x(x+6) \cdot \frac{1}{4}$$

$$4x(x+6) \cdot \frac{1}{x} + 4x(x+6) \cdot \frac{1}{x+6} = x(x+6)$$
$$4(x+6) + 4x = x(x+6)$$
$$4x + 24 + 4x = x^2 + 6x$$
$$8x + 24 = x^2 + 6x$$
$$0 = x^2 - 2x - 24$$
$$0 = (x-6)(x+4)$$

$$x - 6 = 0 \quad or \quad x + 4 = 0$$
$$x = 6 \quad or \qquad x = -4$$

Both numbers check. The solutions are 6 and -4.

55.
$$1 + \frac{12}{x^2 - 4} = \frac{3}{x - 2}$$

The LCM is $(x+2)(x-2)$.

$$(x+2)(x-2)\left(1 + \frac{12}{(x+2)(x-2)}\right) =$$
$$(x+2)(x-2) \cdot \frac{3}{x-2}$$

$$(x+2)(x-2) \cdot 1 + (x+2)(x-2) \cdot \frac{12}{(x+2)(x-2)} =$$
$$3(x+2)$$

$$x^2 - 4 + 12 = 3x + 6$$
$$x^2 + 8 = 3x + 6$$
$$x^2 - 3x + 2 = 0$$
$$(x-2)(x-1) = 0$$

$$x - 2 = 0 \quad or \quad x - 1 = 0$$
$$x = 2 \quad or \qquad x = 1$$

The number 1 checks, but 2 does not. (It makes the denominators $x^2 - 4$ and $x - 2$ zero.) The solution is 1.

57.
$$\frac{r}{r-1} + \frac{2}{r^2 - 1} = \frac{8}{r+1}$$

The LCM is $(r-1)(r+1)$.

$$(r-1)(r+1)\left(\frac{r}{r-1} + \frac{2}{(r-1)(r+1)}\right) =$$
$$(r-1)(r+1) \cdot \frac{8}{r+1}$$

$$(r-1)(r+1) \cdot \frac{r}{r-1} + (r-1)(r+1) \cdot \frac{2}{(r-1)(r+1)} =$$
$$8(r-1)$$

$$r(r+1) + 2 = 8(r-1)$$
$$r^2 + r + 2 = 8r - 8$$
$$r^2 - 7r + 10 = 0$$
$$(r-5)(r-2) = 0$$

$$r - 5 = 0 \quad or \quad r - 2 = 0$$
$$r = 5 \quad or \qquad r = 2$$

Both numbers check. The solutions are 5 and 2.

59.
$$\frac{x-1}{1-x} = -\frac{x+8}{x-8}$$

The LCM is $(1-x)(x-8)$.

$$(1-x)(x-8) \cdot \frac{x-1}{1-x} = (1-x)(x-8)\left(-\frac{x+8}{x-8}\right)$$
$$(x-8)(x-1) = -(1-x)(x+8)$$
$$x^2 - 9x + 8 = -(x+8-x^2-8x)$$
$$x^2 - 9x + 8 = -(-x^2 - 7x + 8)$$
$$x^2 - 9x + 8 = x^2 + 7x - 8$$
$$16 = 16x$$
$$1 = x$$

The number 1 does not check. (It makes the denominator $1 - x$ zero.) There is no solution.

61.
$$\frac{5}{y+4} - \frac{3}{y-2} = 4$$

The LCM is $(y+4)(y-2)$.

$$(y+4)(y-2)\left(\frac{5}{y+4} - \frac{3}{y-2}\right) = (y+4)(y-2) \cdot 4$$
$$5(y-2) - 3(y+4) = 4(y^2 + 2y - 8)$$
$$5y - 10 - 3y - 12 = 4y^2 + 8y - 32$$
$$2y - 22 = 4y^2 + 8y - 32$$
$$0 = 4y^2 + 6y - 10$$
$$0 = 2(2y^2 + 3y - 5)$$
$$0 = 2(2y+5)(y-1)$$

$$2y + 5 = 0 \quad or \quad y - 1 = 0$$
$$2y = -5 \quad or \qquad y = 1$$
$$y = -\frac{5}{2} \quad or \qquad y = 1$$

The solutions are $-\frac{5}{2}$ and 1.

63. Familiarize. We will use the formula
$$d = \frac{n^2 - 3n}{2},$$

where d is the number of diagonals and n is the number of sides.

Translate. We substitute 10 for n.
$$d = \frac{10^2 - 3 \cdot 10}{2}$$

Solve. We do the computation.
$$d = \frac{10^2 - 3 \cdot 10}{2} = \frac{100 - 30}{2} = \frac{70}{2} = 35$$

Check. We can recheck our computation. We can also substitute 35 for d in the original formula and determine whether this yields $n = 10$. Our result checks.

State. A decagon has 35 diagonals.

65. Familiarize. We will use the formula
$$d = \frac{n^2 - 3n}{2},$$

where d is the number of diagonals and n is the number of sides.

Translate. We substitute 14 for d.

$$14 = \frac{n^2 - 3n}{2}$$

Solve. We solve the equation.

$$\frac{n^2 - 3n}{2} = 14$$

$$n^2 - 3n = 28 \quad \text{Multiplying by 2}$$

$$n^2 - 3n - 28 = 0$$

$$(n - 7)(n + 4) = 0$$

$$n - 7 = 0 \quad or \quad n + 4 = 0$$

$$n = 7 \quad or \qquad n = -4$$

Check. Since the number of sides cannot be negative, -4 cannot be a solution. To check 7, we substitute 7 for n in the original formula and determine if this yields $d = 14$. Our result checks.

State. The polygon has 7 sides.

67. Discussion and Writing Exercise

69. $\sqrt{64} = 8$, taking the principal square root

71. $\sqrt{8} = \sqrt{4 \cdot 2} = \sqrt{4}\sqrt{2} = 2\sqrt{2}$

73. $\sqrt{20} = \sqrt{4 \cdot 5} = \sqrt{4}\sqrt{5} = 2\sqrt{5}$

75. $\sqrt{405} = \sqrt{81 \cdot 5} = \sqrt{81}\sqrt{5} = 9\sqrt{5}$

77. 2.646

79. 1.528

81.
$$4m^2 - (m + 1)^2 = 0$$
$$4m^2 - (m^2 + 2m + 1) = 0$$
$$4m^2 - m^2 - 2m - 1 = 0$$
$$3m^2 - 2m - 1 = 0$$
$$(3m + 1)(m - 1) = 0$$
$$3m + 1 = 0 \quad or \quad m - 1 = 0$$
$$3m = -1 \quad or \qquad m = 1$$
$$m = -\frac{1}{3} \quad or \qquad m = 1$$

The solutions are $-\frac{1}{3}$ and 1.

83.
$$\sqrt{5}x^2 - x = 0$$
$$x(\sqrt{5}x - 1) = 0$$
$$x = 0 \quad or \quad \sqrt{5}x - 1 = 0$$
$$x = 0 \quad or \qquad \sqrt{5}x = 1$$
$$x = 0 \quad or \qquad x = \frac{1}{\sqrt{5}}, \text{ or } \frac{\sqrt{5}}{5}$$

The solutions are 0 and $\frac{\sqrt{5}}{5}$.

85. Graph $y_1 = 3x^2 - 7x$ and $y_2 = 20$. Then use the INTERSECT feature to find the first coordinate(s) of the point(s) of intersection. The solutions are 4 and approximately -1.7.

87. Graph $y_1 = 3x^2 + 8x$ and $y_2 = 12x + 15$. Then use the INTERSECT feature to find the first coordinate(s) of the point(s) of intersection. The solutions are 3 and approximately -1.7.

89. Graph $y_1 = (x - 2)^2 + 3(x - 2)$ and $y_2 = 4$. Then use the INTERSECT feature to find the first coordinate(s) of the point(s) of intersection. The solutions are -2 and 3.

91. Graph $y_1 = 16(x - 1)$ and $y_2 = x(x + 8)$. Then use the INTERSECT feature to find the first coordinate(s) of the point(s) of intersection. The solution is 4.

Exercise Set 9.2

1. $x^2 = 121$

$x = 11 \text{ or } x = -11$ Principle of square roots

The solutions are 11 and -11.

3. $5x^2 = 35$

$x^2 = 7$ Dividing by 5

$x = \sqrt{7} \text{ or } x = -\sqrt{7}$ Principle of square roots

The solutions are $\sqrt{7}$ and $-\sqrt{7}$.

5. $5x^2 = 3$

$$x^2 = \frac{3}{5}$$

$$x = \sqrt{\frac{3}{5}} \quad or \quad x = -\sqrt{\frac{3}{5}} \qquad \begin{array}{l}\text{Principle of} \\ \text{square roots}\end{array}$$

$$x = \sqrt{\frac{3}{5} \cdot \frac{5}{5}} \quad or \quad x = -\sqrt{\frac{3}{5} \cdot \frac{5}{5}} \qquad \begin{array}{l}\text{Rationalizing} \\ \text{denominators}\end{array}$$

$$x = \frac{\sqrt{15}}{5} \qquad or \quad x = -\frac{\sqrt{15}}{5}$$

The solutions are $\frac{\sqrt{15}}{5}$ and $-\frac{\sqrt{15}}{5}$.

7. $4x^2 - 25 = 0$

$$4x^2 = 25$$

$$x^2 = \frac{25}{4}$$

$$x = \frac{5}{2} \text{ or } x = -\frac{5}{2}$$

The solutions are $\frac{5}{2}$ and $-\frac{5}{2}$.

9. $3x^2 - 49 = 0$

$$3x^2 = 49$$

$$x^2 = \frac{49}{3}$$

$$x = \frac{7}{\sqrt{3}} \qquad or \quad x = -\frac{7}{\sqrt{3}}$$

$$x = \frac{7}{\sqrt{3}} \cdot \frac{\sqrt{3}}{\sqrt{3}} \quad or \quad x = -\frac{7}{\sqrt{3}} \cdot \frac{\sqrt{3}}{\sqrt{3}}$$

$$x = \frac{7\sqrt{3}}{3} \qquad or \quad x = -\frac{7\sqrt{3}}{3}$$

The solutions are $\frac{7\sqrt{3}}{3}$ and $-\frac{7\sqrt{3}}{3}$.

11. $4y^2 - 3 = 9$
$4y^2 = 12$
$y^2 = 3$
$y = \sqrt{3}$ *or* $y = -\sqrt{3}$
The solutions are $\sqrt{3}$ and $-\sqrt{3}$.

13. $49y^2 - 64 = 0$
$49y^2 = 64$
$y^2 = \dfrac{64}{49}$
$y = \dfrac{8}{7}$ *or* $y = -\dfrac{8}{7}$
The solutions are $\dfrac{8}{7}$ and $-\dfrac{8}{7}$.

15. $(x + 3)^2 = 16$
$x + 3 = 4$ *or* $x + 3 = -4$ Principle of square roots
$x = 1$ *or* $x = -7$
The solutions are 1 and -7.

17. $(x + 3)^2 = 21$
$x + 3 = \sqrt{21}$ *or* $x + 3 = -\sqrt{21}$ Principle of square roots
$x = -3 + \sqrt{21}$ *or* $x = -3 - \sqrt{21}$
The solutions are $-3 + \sqrt{21}$ and $-3 - \sqrt{21}$, or $-3 \pm \sqrt{21}$.

19. $(x + 13)^2 = 8$
$x + 13 = \sqrt{8}$ *or* $x + 13 = -\sqrt{8}$
$x + 13 = 2\sqrt{2}$ *or* $x + 13 = -2\sqrt{2}$
$x = -13 + 2\sqrt{2}$ *or* $x = -13 - 2\sqrt{2}$
The solutions are $-13 + 2\sqrt{2}$ and $-13 - 2\sqrt{2}$, or $-13 \pm 2\sqrt{2}$.

21. $(x - 7)^2 = 12$
$x - 7 = \sqrt{12}$ *or* $x - 7 = -\sqrt{12}$
$x - 7 = 2\sqrt{3}$ *or* $x - 7 = -2\sqrt{3}$
$x = 7 + 2\sqrt{3}$ *or* $x = 7 - 2\sqrt{3}$
The solutions are $7 + 2\sqrt{3}$ and $7 - 2\sqrt{3}$, or $7 \pm 2\sqrt{3}$.

23. $(x + 9)^2 = 34$
$x + 9 = \sqrt{34}$ *or* $x + 9 = -\sqrt{34}$
$x = -9 + \sqrt{34}$ *or* $x = -9 - \sqrt{34}$
The solutions are $-9 + \sqrt{34}$ and $-9 - \sqrt{34}$, or $-9 \pm \sqrt{34}$.

25. $\left(x + \dfrac{3}{2}\right)^2 = \dfrac{7}{2}$
$x + \dfrac{3}{2} = \sqrt{\dfrac{7}{2}}$ *or* $x + \dfrac{3}{2} = -\sqrt{\dfrac{7}{2}}$
$x = -\dfrac{3}{2} + \sqrt{\dfrac{7}{2}}$ *or* $x = -\dfrac{3}{2} - \sqrt{\dfrac{7}{2}}$
$x = -\dfrac{3}{2} + \sqrt{\dfrac{7}{2} \cdot \dfrac{2}{2}}$ *or* $x = -\dfrac{3}{2} - \sqrt{\dfrac{7}{2} \cdot \dfrac{2}{2}}$
$x = -\dfrac{3}{2} + \dfrac{\sqrt{14}}{2}$ *or* $x = -\dfrac{3}{2} - \dfrac{\sqrt{14}}{2}$
$x = \dfrac{-3 + \sqrt{14}}{2}$ *or* $x = \dfrac{-3 - \sqrt{14}}{2}$

The solutions are $\dfrac{-3 \pm \sqrt{14}}{2}$.

27. $x^2 - 6x + 9 = 64$
$(x - 3)^2 = 64$ Factoring the left side
$x - 3 = 8$ *or* $x - 3 = -8$ Principle of square roots
$x = 11$ *or* $x = -5$
The solutions are 11 and -5.

29. $x^2 + 14x + 49 = 64$
$(x + 7)^2 = 64$ Factoring the left side
$x + 7 = 8$ *or* $x + 7 = -8$ Principle of square roots
$x = 1$ *or* $x = -15$
The solutions are 1 and -15.

31. $x^2 - 6x - 16 = 0$
$x^2 - 6x \quad = 16$ Adding 16
$x^2 - 6x + \; 9 = 16 + 9$ Adding 9: $\left(\dfrac{-6}{2}\right)^2 =$
$(-3)^2 = 9$
$(x - 3)^2 = 25$
$x - 3 = 5$ *or* $x - 3 = -5$ Principle of square roots
$x = 8$ *or* $x = -2$
The solutions are 8 and -2.

33. $x^2 + 22x + \; 21 = 0$
$x^2 + 22x \qquad = -21$ Subtracting 21
$x^2 + 22x + 121 = -21 + 121$ Adding 121: $\left(\dfrac{22}{2}\right)^2 =$
$11^2 = 121$
$(x + 11)^2 = 100$
$x + 11 = 10$ *or* $x + 11 = -10$ Principle of square roots
$x = -1$ *or* $x = -21$
The solutions are -1 and -21.

35. $x^2 - 2x - 5 = 0$
$x^2 - 2x \quad = 5$
$x^2 - 2x + 1 = 5 + 1$ Adding 1: $\left(\dfrac{-2}{2}\right)^2 =$
$(-1)^2 = 1$
$(x - 1)^2 = 6$
$x - 1 = \sqrt{6}$ *or* $x - 1 = -\sqrt{6}$
$x = 1 + \sqrt{6}$ *or* $x = 1 - \sqrt{6}$
The solutions are $1 \pm \sqrt{6}$.

37. $x^2 - 22x + 102 = 0$
$x^2 - 22x \qquad = -102$
$x^2 - 22x + 121 = -102 + 121$ Adding 121:
$\left(\dfrac{-22}{2}\right)^2 = (-11)^2 = 121$
$(x - 11)^2 = 19$
$x - 11 = \sqrt{19}$ *or* $x - 11 = -\sqrt{19}$
$x = 11 + \sqrt{19}$ *or* $x = 11 - \sqrt{19}$
The solutions are $11 \pm \sqrt{19}$.

39. $x^2 + 10x - 4 = 0$

$x^2 + 10x = 4$

$x^2 + 10x + 25 = 4 + 25$ Adding 25: $\left(\dfrac{10}{2}\right)^2 =$

$5^2 = 25$

$(x + 5)^2 = 29$

$x + 5 = \sqrt{29} \quad or \quad x + 5 = -\sqrt{29}$

$x = -5 + \sqrt{29} \quad or \quad x = -5 - \sqrt{29}$

The solutions are $-5 \pm \sqrt{29}$.

41. $x^2 - 7x - 2 = 0$

$x^2 - 7x = 2$

$x^2 - 7x + \dfrac{49}{4} = 2 + \dfrac{49}{4}$ Adding $\dfrac{49}{4}$:

$\left(\dfrac{-7}{2}\right)^2 = \dfrac{49}{4}$

$\left(x - \dfrac{7}{2}\right)^2 = \dfrac{8}{4} + \dfrac{49}{4} = \dfrac{57}{4}$

$x - \dfrac{7}{2} = \dfrac{\sqrt{57}}{2} \quad or \quad x - \dfrac{7}{2} = -\dfrac{\sqrt{57}}{2}$

$x = \dfrac{7}{2} + \dfrac{\sqrt{57}}{2} \quad or \quad x = \dfrac{7}{2} - \dfrac{\sqrt{57}}{2}$

$x = \dfrac{7 + \sqrt{57}}{2} \quad or \quad x = \dfrac{7 - \sqrt{57}}{2}$

The solutions are $\dfrac{7 \pm \sqrt{57}}{2}$.

43. $x^2 + 3x - 28 = 0$

$x^2 + 3x = 28$

$x^2 + 3x + \dfrac{9}{4} = 28 + \dfrac{9}{4}$ Adding $\dfrac{9}{4}$: $\left(\dfrac{3}{2}\right)^2 = \dfrac{9}{4}$

$\left(x + \dfrac{3}{2}\right)^2 = \dfrac{121}{4}$

$x + \dfrac{3}{2} = \dfrac{11}{2} \quad or \quad x + \dfrac{3}{2} = -\dfrac{11}{2}$

$x = \dfrac{8}{2} \quad or \quad x = -\dfrac{14}{2}$

$x = 4 \quad or \quad x = -7$

The solutions are 4 and -7.

45. $x^2 + \dfrac{3}{2}x - \dfrac{1}{2} = 0$

$x^2 + \dfrac{3}{2}x = \dfrac{1}{2}$

$x^2 + \dfrac{3}{2}x + \dfrac{9}{16} = \dfrac{1}{2} + \dfrac{9}{16}$ Adding $\dfrac{9}{16}$: $\left(\dfrac{3/2}{2}\right)^2 =$

$\left(\dfrac{3}{4}\right)^2 = \dfrac{9}{16}$

$\left(x + \dfrac{3}{4}\right)^2 = \dfrac{17}{16}$

$x + \dfrac{3}{4} = \dfrac{\sqrt{17}}{4} \quad or \quad x + \dfrac{3}{4} = -\dfrac{\sqrt{17}}{4}$

$x = -\dfrac{3}{4} + \dfrac{\sqrt{17}}{4} \quad or \quad x = -\dfrac{3}{4} - \dfrac{\sqrt{17}}{4}$

$x = \dfrac{-3 + \sqrt{17}}{4} \quad or \quad x = \dfrac{-3 - \sqrt{17}}{4}$

The solutions are $\dfrac{-3 \pm \sqrt{17}}{4}$.

47. $2x^2 + 3x - 17 = 0$

$\dfrac{1}{2}(2x^2 + 3x - 17) = \dfrac{1}{2} \cdot 0$ Multiplying by $\dfrac{1}{2}$ to make the x^2-coefficient 1

$x^2 + \dfrac{3}{2}x - \dfrac{17}{2} = 0$

$x^2 + \dfrac{3}{2}x = \dfrac{17}{2}$

$x^2 + \dfrac{3}{2}x + \dfrac{9}{16} = \dfrac{17}{2} + \dfrac{9}{16}$ Adding $\dfrac{9}{16}$: $\left(\dfrac{3/2}{2}\right)^2 =$

$\left(\dfrac{3}{4}\right)^2 = \dfrac{9}{16}$

$\left(x + \dfrac{3}{4}\right)^2 = \dfrac{145}{16}$

$x + \dfrac{3}{4} = \dfrac{\sqrt{145}}{4} \quad or \quad x + \dfrac{3}{4} = -\dfrac{\sqrt{145}}{4}$

$x = \dfrac{-3 + \sqrt{145}}{4} \quad or \quad x = \dfrac{-3 - \sqrt{145}}{4}$

The solutions are $\dfrac{-3 \pm \sqrt{145}}{4}$.

49. $3x^2 + 4x - 1 = 0$

$\dfrac{1}{3}(3x^2 + 4x - 1) = \dfrac{1}{3} \cdot 0$

$x^2 + \dfrac{4}{3}x - \dfrac{1}{3} = 0$

$x^2 + \dfrac{4}{3}x = \dfrac{1}{3}$

$x^2 + \dfrac{4}{3}x + \dfrac{4}{9} = \dfrac{1}{3} + \dfrac{4}{9}$

$\left(x + \dfrac{2}{3}\right)^2 = \dfrac{7}{9}$

$x + \dfrac{2}{3} = \dfrac{\sqrt{7}}{3} \quad or \quad x + \dfrac{2}{3} = -\dfrac{\sqrt{7}}{3}$

$x = \dfrac{-2 + \sqrt{7}}{3} \quad or \quad x = -\dfrac{-2 - \sqrt{7}}{3}$

The solutions are $\dfrac{-2 \pm \sqrt{7}}{3}$.

51. $2x^2 = 9x + 5$

$2x^2 - 9x - 5 = 0$ Standard form

$\dfrac{1}{2}(2x^2 - 9x - 5) = \dfrac{1}{2} \cdot 0$

$x^2 - \dfrac{9}{2}x - \dfrac{5}{2} = 0$

$x^2 - \dfrac{9}{2}x = \dfrac{5}{2}$

$x^2 - \dfrac{9}{2}x + \dfrac{81}{16} = \dfrac{5}{2} + \dfrac{81}{16}$

$\left(x - \dfrac{9}{4}\right)^2 = \dfrac{121}{16}$

$$x - \frac{9}{4} = \frac{11}{4} \quad or \quad x - \frac{9}{4} = -\frac{11}{4}$$

$$x = \frac{20}{4} \quad or \qquad x = -\frac{2}{4}$$

$$x = 5 \quad or \qquad x = -\frac{1}{2}$$

The solutions are 5 and $-\frac{1}{2}$.

53.
$$6x^2 + 11x = 10$$
$$6x^2 + 11x - 10 = 0 \qquad \text{Standard form}$$
$$\frac{1}{6}(6x^2 + 11x - 10) = \frac{1}{6} \cdot 0$$
$$x^2 + \frac{11}{6}x - \frac{5}{3} = 0$$
$$x^2 + \frac{11}{6}x \qquad = \frac{5}{3}$$
$$x^2 + \frac{11}{6}x + \frac{121}{144} = \frac{5}{3} + \frac{121}{144}$$
$$\left(x + \frac{11}{12}\right)^2 = \frac{361}{144}$$

$$x + \frac{11}{12} = \frac{19}{12} \quad or \quad x + \frac{11}{12} = -\frac{19}{12}$$

$$x = \frac{8}{12} \quad or \qquad x = -\frac{30}{12}$$

$$x = \frac{2}{3} \quad or \qquad x = -\frac{5}{2}$$

The solutions are $\frac{2}{3}$ and $-\frac{5}{2}$.

55. Familiarize. We will use the formula $s = 16t^2$.

Translate. We substitute 1483 for s.
$$1483 = 16t^2$$

Solve. We solve the equation.
$$1483 = 16t^2$$
$$\frac{1483}{16} = t^2 \qquad \text{Solving for } t^2$$
$$92.6875 = t^2 \qquad \text{Dividing}$$
$$\sqrt{92.6875} = t \quad or \quad -\sqrt{92.6875} = t \qquad \text{Principle of square roots}$$
$$9.6 \approx t \quad or \qquad -9.6 \approx t \qquad \text{Using a calculator and rounding to the nearest tenth}$$

Check. The number -9.6 cannot be a solution, because time cannot be negative in this situation. We substitute 9.6 in the original equation.
$$s = 16(9.6)^2 = 16(92.16) = 1474.56$$

This is close. Remember that we approximated a solution. Thus we have a check.

State. It takes about 9.6 sec for an object to fall to the ground from the top of the Petronas Towers.

57. Familiarize. We will use the formula $s = 16t^2$.

Translate. We substitute 311 for s.
$$311 = 16t^2$$

Solve. We solve the equation.
$$311 = 16t^2$$
$$\frac{311}{16} = t^2 \qquad \text{Solving for } t^2$$
$$19.4375 = t^2 \qquad \text{Dividing}$$
$$\sqrt{19.4375} = t \quad or \quad -\sqrt{19.4375} = t \qquad \text{Principle of square roots}$$
$$4.4 \approx t \quad or \qquad -4.4 \approx t \qquad \text{Using a calculator and rounding to the nearest tenth}$$

Check. The number -4.4 cannot be a solution, because time cannot be negative in this situation. We substitute 4.4 in the original equation.
$$s = 16(4.4)^2 = 16(19.36) = 309.76$$

This is close. Remember that we approximated a solution. Thus we have a check.

State. The fall took approximately 4.4 sec.

59. Discussion and Writing Exercise

61. The <u>product</u> rule asserts when multiplying with exponential notation, if the bases are the same, we keep the base and add the exponents.

63. The number -5 is not the <u>principal square root</u> of 25.

65. The <u>quotient</u> rule asserts that when dividing with exponential notation, if the bases are the same, we keep the base and subtract the exponent of the denominator from the exponent of the numerator.

67. The <u>quotient</u> rule for radicals asserts that for any nonnegative radicand A and positive number B, $\frac{\sqrt{A}}{\sqrt{B}} = \sqrt{\frac{A}{B}}$.

69. $x^2 + bx + 36$

The trinomial is a square if the square of one-half the x-coefficient is equal to 36. Thus we have:
$$\left(\frac{b}{2}\right)^2 = 36$$
$$\frac{b^2}{4} = 36$$
$$b^2 = 144$$
$$b = 12 \quad or \quad b = -12 \qquad \text{Principle of square roots}$$

71. $x^2 + bx + 128$

The trinomial is a square if the square of one-half the x-coefficient is equal to 128. Thus we have:
$$\left(\frac{b}{2}\right)^2 = 128$$
$$\frac{b^2}{4} = 128$$
$$b^2 = 512$$
$$b = \sqrt{512} \quad or \quad b = -\sqrt{512}$$
$$b = 16\sqrt{2} \quad or \quad b = -16\sqrt{2}$$

73. $x^2 + bx + c$

The trinomial is a square if the square of one-half the x-coefficient is equal to c. Thus we have:

$$\left(\frac{b}{2}\right)^2 = c$$

$$\frac{b^2}{4} = c$$

$$b^2 = 4c$$

$$b = \sqrt{4c} \quad or \quad b = -\sqrt{4c}$$
$$b = 2\sqrt{c} \quad or \quad b = -2\sqrt{c}$$

75. $4.82x^2 = 12,000$

$$x^2 = \frac{12,000}{4.82}$$

$$x = \sqrt{\frac{12,000}{4.82}} \quad or \quad x = -\sqrt{\frac{12,000}{4.82}} \quad \text{Principle of square roots}$$

$$x \approx 49.896 \quad or \quad x \approx -49.896 \quad \text{Using a calculator and rounding}$$

The solutions are approximately 49.896 and -49.896.

77. $\dfrac{x}{9} = \dfrac{36}{4x}$, LCM is $36x$

$$36x \cdot \frac{x}{9} = 36x \cdot \frac{36}{4x} \quad \text{Multiplying by } 36x$$

$$4x^2 = 324$$

$$x^2 = 81$$

$$x = 9 \quad or \quad x = -9$$

Both numbers check. The solutions are 9 and -9.

Exercise Set 9.3

1.
$$x^2 - 4x = 21$$
$$x^2 - 4x - 21 = 0 \qquad \text{Standard form}$$

We can factor.
$$x^2 - 4x - 21 = 0$$
$$(x - 7)(x + 3) = 0$$

$$x - 7 = 0 \quad or \quad x + 3 = 0$$
$$x = 7 \quad or \qquad x = -3$$

The solutions are 7 and -3.

3.
$$x^2 = 6x - 9$$
$$x^2 - 6x + 9 = 0 \qquad \text{Standard form}$$

We can factor.
$$x^2 - 6x + 9 = 0$$
$$(x - 3)(x - 3) = 0$$

$$x - 3 = 0 \quad or \quad x - 3 = 0$$
$$x = 3 \quad or \qquad x = 3$$

The solution is 3.

5. $3y^2 - 2y - 8 = 0$

We can factor.
$$3y^2 - 2y - 8 = 0$$
$$(3y + 4)(y - 2) = 0$$

$$3y + 4 = 0 \quad or \quad y - 2 = 0$$
$$3y = -4 \quad or \qquad y = 2$$

$$y = -\frac{4}{3} \quad or \qquad y = 2$$

The solutions are $-\dfrac{4}{3}$ and 2.

7.
$$4x^2 + 4x = 15$$
$$4x^2 + 4x - 15 = 0 \qquad \text{Standard form}$$

We can factor.
$$4x^2 + 4x - 15 = 0$$
$$(2x - 3)(2x + 5) = 0$$

$$2x - 3 = 0 \quad or \quad 2x + 5 = 0$$
$$2x = 3 \quad or \qquad 2x = -5$$

$$x = \frac{3}{2} \quad or \qquad x = -\frac{5}{2}$$

The solutions are $\dfrac{3}{2}$ and $-\dfrac{5}{2}$.

9.
$$x^2 - 9 = 0 \qquad \text{Difference of squares}$$
$$(x + 3)(x - 3) = 0$$

$$x + 3 = 0 \quad or \quad x - 3 = 0$$
$$x = -3 \quad or \qquad x = 3$$

The solutions are -3 and 3.

11. $x^2 - 2x - 2 = 0$

$a = 1, \ b = -2, \ c = -2$

We use the quadratic formula.
$$x = \frac{-(-2) \pm \sqrt{(-2)^2 - 4 \cdot 1 \cdot (-2)}}{2 \cdot 1}$$

$$x = \frac{2 \pm \sqrt{4 + 8}}{2}$$

$$x = \frac{2 \pm \sqrt{12}}{2} = \frac{2 \pm \sqrt{4 \cdot 3}}{2}$$

$$x = \frac{2 \pm 2\sqrt{3}}{2} = \frac{2(1 \pm \sqrt{3})}{2}$$

$$x = 1 \pm \sqrt{3}$$

The solutions are $1 + \sqrt{3}$ and $1 - \sqrt{3}$, or $1 \pm \sqrt{3}$.

13. $y^2 - 10y + 22 = 0$

$a = 1, \ b = -10, \ c = 22$

We use the quadratic formula.
$$y = \frac{-(-10) \pm \sqrt{(-10)^2 - 4 \cdot 1 \cdot 22}}{2 \cdot 1}$$

$$y = \frac{10 \pm \sqrt{100 - 88}}{2}$$

$$y = \frac{10 \pm \sqrt{12}}{2} = \frac{10 \pm \sqrt{4 \cdot 3}}{2}$$

$$y = \frac{10 \pm 2\sqrt{3}}{2} = \frac{2(5 \pm \sqrt{3})}{2}$$

$$y = 5 \pm \sqrt{3}$$

The solutions are $5 + \sqrt{3}$ and $5 - \sqrt{3}$, or $5 \pm \sqrt{3}$.

15. $x^2 + 4x + 4 = 7$

$x^2 + 4x - 3 = 0$ Adding -7 to get standard

                                  form

$a = 1$, $b = 4$, $c = -3$

We use the quadratic formula.

$$x = \frac{-4 \pm \sqrt{4^2 - 4 \cdot 1 \cdot (-3)}}{2 \cdot 1} = \frac{-4 \pm \sqrt{16 + 12}}{2}$$

$$x = \frac{-4 \pm \sqrt{28}}{2} = \frac{-4 \pm \sqrt{4 \cdot 7}}{2}$$

$$x = \frac{-4 \pm 2\sqrt{7}}{2} = \frac{2(-2 \pm \sqrt{7})}{2}$$

$$x = -2 \pm \sqrt{7}$$

The solutions are $-2 + \sqrt{7}$ and $-2 - \sqrt{7}$, or $-2 \pm \sqrt{7}$.

17. $3x^2 + 8x + 2 = 0$

$a = 3$, $b = 8$, $c = 2$

We use the quadratic formula.

$$x = \frac{-8 \pm \sqrt{8^2 - 4 \cdot 3 \cdot 2}}{2 \cdot 3} = \frac{-8 \pm \sqrt{64 - 24}}{6}$$

$$x = \frac{-8 \pm \sqrt{40}}{6} = \frac{-8 \pm \sqrt{4 \cdot 10}}{6}$$

$$x = \frac{-8 \pm 2\sqrt{10}}{6} = \frac{2(-4 \pm \sqrt{10})}{2 \cdot 3}$$

$$x = \frac{-4 \pm \sqrt{10}}{3}$$

The solutions are $\dfrac{-4 + \sqrt{10}}{3}$ and $\dfrac{-4 - \sqrt{10}}{3}$, or

$\dfrac{-4 \pm \sqrt{10}}{3}$.

19. $2x^2 - 5x = 1$

$2x^2 - 5x - 1 = 0$ Adding -1 to get standard

                                   form

$a = 2$, $b = -5$, $c = -1$

We use the quadratic formula.

$$x = \frac{-(-5) \pm \sqrt{(-5)^2 - 4 \cdot 2 \cdot (-1)}}{2 \cdot 2} = \frac{5 \pm \sqrt{25 + 8}}{4}$$

$$x = \frac{5 \pm \sqrt{33}}{4}$$

The solutions are $\dfrac{5 + \sqrt{33}}{4}$ and $\dfrac{5 - \sqrt{33}}{4}$, or $\dfrac{5 \pm \sqrt{33}}{4}$.

21. $2y^2 - 2y - 1 = 0$

$a = 2$, $b = -2$, $c = -1$

We use the quadratic formula.

$$y = \frac{-(-2) \pm \sqrt{(-2)^2 - 4 \cdot 2 \cdot (-1)}}{2 \cdot 2} = \frac{2 \pm \sqrt{4 + 8}}{4}$$

$$y = \frac{2 \pm \sqrt{12}}{4} = \frac{2 \pm \sqrt{4 \cdot 3}}{4}$$

$$y = \frac{2 \pm 2\sqrt{3}}{4} = \frac{2(1 \pm \sqrt{3})}{2 \cdot 2}$$

$$y = \frac{1 \pm \sqrt{3}}{2}$$

The solutions are $\dfrac{1 + \sqrt{3}}{2}$ and $\dfrac{1 - \sqrt{3}}{2}$, or $\dfrac{1 \pm \sqrt{3}}{2}$.

23. $2t^2 + 6t + 5 = 0$

$a = 2$, $b = 6$, $c = 5$

We use the quadratic formula.

$$t = \frac{-6 \pm \sqrt{6^2 - 4 \cdot 2 \cdot 5}}{2 \cdot 2} = \frac{-6 \pm \sqrt{36 - 40}}{4}$$

$$t = \frac{-6 \pm \sqrt{-4}}{4}$$

Since square roots of negative numbers do not exist as real numbers, there are no real-number solutions.

25. $3x^2 = 5x + 4$

$3x^2 - 5x - 4 = 0$

$a = 3$, $b = -5$, $c = -4$

We use the quadratic formula.

$$x = \frac{-(-5) \pm \sqrt{(-5)^2 - 4 \cdot 3 \cdot (-4)}}{2 \cdot 3} = \frac{5 \pm \sqrt{25 + 48}}{6}$$

$$x = \frac{5 \pm \sqrt{73}}{6}$$

The solutions are $\dfrac{5 + \sqrt{73}}{6}$ and $\dfrac{5 - \sqrt{73}}{6}$, or $\dfrac{5 \pm \sqrt{73}}{6}$.

27. $2y^2 - 6y = 10$

$2y^2 - 6y - 10 = 0$

    $y^2 - 3y - 5 = 0$ Multiplying by $\dfrac{1}{2}$ to simplify

$a = 1$, $b = -3$, $c = -5$

We use the quadratic formula.

$$y = \frac{-(-3) \pm \sqrt{(-3)^2 - 4 \cdot 1 \cdot (-5)}}{2 \cdot 1} = \frac{3 \pm \sqrt{9 + 20}}{2}$$

$$y = \frac{3 \pm \sqrt{29}}{2}$$

The solutions are $\dfrac{3 + \sqrt{29}}{2}$ and $\dfrac{3 - \sqrt{29}}{2}$, or $\dfrac{3 \pm \sqrt{29}}{2}$.

29. $\dfrac{x^2}{x + 3} - \dfrac{5}{x + 3} = 0$,   LCM is $x + 3$

$$(x + 3)\left(\frac{x^2}{x + 3} - \frac{5}{x + 3}\right) = (x + 3) \cdot 0$$

$$x^2 - 5 = 0$$

$$x^2 = 5$$

$x = \sqrt{5}$ *or* $x = -\sqrt{5}$ Principle of square roots

Both numbers check. The solutions are $\sqrt{5}$ and $-\sqrt{5}$, or $\pm\sqrt{5}$.

31. $x + 2 = \dfrac{3}{x + 2}$

$(x + 2)(x + 2) = (x + 2) \cdot \dfrac{3}{x + 2}$ Clearing the fraction

$$x^2 + 4x + 4 = 3$$

$$x^2 + 4x + 1 = 0$$

$a = 1$, $b = 4$, $c = 1$

We use the quadratic formula.

$$x = \frac{-4 \pm \sqrt{4^2 - 4 \cdot 1 \cdot 1}}{2 \cdot 1} = \frac{-4 \pm \sqrt{16 - 4}}{2}$$

$$x = \frac{-4 \pm \sqrt{12}}{2} = \frac{-4 \pm \sqrt{4 \cdot 3}}{2}$$

$$x = \frac{-4 \pm 2\sqrt{3}}{2} = \frac{2(-2 \pm \sqrt{3})}{2}$$

$$x = -2 \pm \sqrt{3}$$

Both numbers check. The solutions are $-2 + \sqrt{3}$ and $-2 - \sqrt{3}$, or $-2 \pm \sqrt{3}$.

33. $\dfrac{1}{x} + \dfrac{1}{x+1} = \dfrac{1}{3}$, LCM is $3x(x+1)$

$$3x(x+1)\left(\frac{1}{x} + \frac{1}{x+1}\right) = 3x(x+1) \cdot \frac{1}{3}$$

$$3(x+1) + 3x = x(x+1)$$
$$3x + 3 + 3x = x^2 + x$$
$$6x + 3 = x^2 + x$$
$$0 = x^2 - 5x - 3$$

$a = 1$, $b = -5$, $c = -3$

We use the quadratic formula.

$$x = \frac{-(-5) \pm \sqrt{(-5)^2 - 4 \cdot 1 \cdot (-3)}}{2 \cdot 1} = \frac{5 \pm \sqrt{25 + 12}}{2}$$

$$x = \frac{5 \pm \sqrt{37}}{2}$$

The solutions are $\dfrac{5 + \sqrt{37}}{2}$ and $\dfrac{5 - \sqrt{37}}{2}$, or $\dfrac{5 \pm \sqrt{37}}{2}$.

35. $x^2 - 4x - 7 = 0$

$a = 1$, $b = -4$, $c = -7$

$$x = \frac{-(-4) \pm \sqrt{(-4)^2 - 4 \cdot 1 \cdot (-7)}}{2 \cdot 1}$$

$$x = \frac{4 \pm \sqrt{16 + 28}}{2} = \frac{4 \pm \sqrt{44}}{2}$$

$$x = \frac{4 \pm \sqrt{4 \cdot 11}}{2} = \frac{4 \pm 2\sqrt{11}}{2}$$

$$x = \frac{2(2 \pm \sqrt{11})}{2} = 2 \pm \sqrt{11}$$

Using a calculator, we have:

$2 + \sqrt{11} \approx 5.31662479 \approx 5.3$, and
$2 - \sqrt{11} \approx -1.31662479 \approx -1.3$.

The approximate solutions, to the nearest tenth, are 5.3 and -1.3.

37. $y^2 - 6y - 1 = 0$

$a = 1$, $b = -6$, $c = -1$

$$y = \frac{-(-6) \pm \sqrt{(-6)^2 - 4 \cdot 1 \cdot (-1)}}{2 \cdot 1}$$

$$y = \frac{6 \pm \sqrt{36 + 4}}{2} = \frac{6 \pm \sqrt{40}}{2}$$

$$y = \frac{6 \pm \sqrt{4 \cdot 10}}{2} = \frac{6 \pm 2\sqrt{10}}{2}$$

$$y = \frac{2(3 \pm \sqrt{10})}{2} = 3 \pm \sqrt{10}$$

Using a calculator, we have:

$3 + \sqrt{10} \approx 6.16227766 \approx 6.2$ and
$3 - \sqrt{10} \approx -0.1622776602 \approx -0.2$.

The approximate solutions, to the nearest tenth, are 6.2 and -0.2.

39. $4x^2 + 4x = 1$

$4x^2 + 4x - 1 = 0$ Standard form
$a = 4$, $b = 4$, $c = -1$

$$x = \frac{-4 \pm \sqrt{4^2 - 4 \cdot 4 \cdot (-1)}}{2 \cdot 4}$$

$$x = \frac{-4 \pm \sqrt{16 + 16}}{8} = \frac{-4 \pm \sqrt{32}}{8}$$

$$x = \frac{-4 \pm \sqrt{16 \cdot 2}}{8} = \frac{-4 \pm 4\sqrt{2}}{8}$$

$$x = \frac{4(-1 \pm \sqrt{2})}{4 \cdot 2} = \frac{-1 \pm \sqrt{2}}{2}$$

Using a calculator, we have:

$$\frac{-1 + \sqrt{2}}{2} \approx 0.2071067812 \approx 0.2 \text{ and}$$

$$\frac{-1 - \sqrt{2}}{2} \approx -1.207106781 \approx -1.2.$$

The approximate solutions, to the nearest tenth, are 0.2 and -1.2.

41. $3x^2 - 8x + 2 = 0$

$a = 3$, $b = -8$, $c = 2$

$$x = \frac{-(-8) \pm \sqrt{(-8)^2 - 4 \cdot 3 \cdot 2}}{2 \cdot 3}$$

$$x = \frac{8 \pm \sqrt{64 - 24}}{6} = \frac{8 \pm \sqrt{40}}{6}$$

$$x = \frac{8 \pm \sqrt{4 \cdot 10}}{6} = \frac{8 \pm 2\sqrt{10}}{6}$$

$$x = \frac{2(4 \pm \sqrt{10})}{2 \cdot 3} = \frac{4 \pm \sqrt{10}}{3}$$

Using a calculator, we have:

$$\frac{4 + \sqrt{10}}{3} \approx 2.387425887 \approx 2.4 \text{ and}$$

$$\frac{4 - \sqrt{10}}{3} \approx 0.2792407799 \approx 0.3.$$

The approximate solutions, to the nearest tenth, are 2.4 and 0.3.

43. Discussion and Writing Exercise

45. $\sqrt{40} - 2\sqrt{10} + \sqrt{90} = \sqrt{4 \cdot 10} - 2\sqrt{10} + \sqrt{9 \cdot 10}$
$$= \sqrt{4}\sqrt{10} - 2\sqrt{10} + \sqrt{9}\sqrt{10}$$
$$= 2\sqrt{10} - 2\sqrt{10} + 3\sqrt{10}$$
$$= (2 - 2 + 3)\sqrt{10}$$
$$= 3\sqrt{10}$$

47. $\sqrt{18} + \sqrt{50} - 3\sqrt{8} = \sqrt{9 \cdot 2} + \sqrt{25 \cdot 2} - 3\sqrt{4 \cdot 2}$

$\qquad = \sqrt{9}\sqrt{2} + \sqrt{25}\sqrt{2} - 3\sqrt{4}\sqrt{2}$

$\qquad = 3\sqrt{2} + 5\sqrt{2} - 3 \cdot 2\sqrt{2}$

$\qquad = 3\sqrt{2} + 5\sqrt{2} - 6\sqrt{2}$

$\qquad = (3 + 5 - 6)\sqrt{2}$

$\qquad = 2\sqrt{2}$

49. $\sqrt{80} = \sqrt{16 \cdot 5} = \sqrt{16}\sqrt{5} = 4\sqrt{5}$

51. $\sqrt{9000x^{10}} = \sqrt{900 \cdot 10 \cdot x^{10}} = \sqrt{900}\sqrt{x^{10}}\sqrt{10} = 30x^5\sqrt{10}$

53. $y = \dfrac{k}{x}$ Inverse variation

$235 = \dfrac{k}{0.6}$ Substituting 0.6 for x and 235 for y

$141 = k$ Constant of variation

$y = \dfrac{141}{x}$ Equation of variation

55. $5x + x(x-7) = 0$

$5x + x^2 - 7x = 0$

$x^2 - 2x = 0$ We can factor.

$x(x-2) = 0$

$x = 0 \ or \ x - 2 = 0$

$x = 0 \ or \ \quad\quad x = 2$

The solutions are 0 and 2.

57. $3 - x(x-3) = 4$

$3 - x^2 + 3x = 4$

$0 = x^2 - 3x + 1$ Standard form

$a = 1, \ b = -3, \ c = 1$

We use the quadratic formula.

$x = \dfrac{-(-3) \pm \sqrt{(-3)^2 - 4 \cdot 1 \cdot 1}}{2 \cdot 1} = \dfrac{3 \pm \sqrt{9-4}}{2}$

$x = \dfrac{3 \pm \sqrt{5}}{2}$

The solutions are $\dfrac{3+\sqrt{5}}{2}$ and $\dfrac{3-\sqrt{5}}{2}$, or $\dfrac{3 \pm \sqrt{5}}{2}$.

59. $(y+4)(y+3) = 15$

$y^2 + 7y + 12 = 15$

$y^2 + 7y - 3 = 0$ Standard form

$a = 1, \ b = 7, \ c = -3$

We use the quadratic formula.

$y = \dfrac{-7 \pm \sqrt{7^2 - 4 \cdot 1 \cdot (-3)}}{2 \cdot 1} = \dfrac{-7 \pm \sqrt{49+12}}{2}$

$y = \dfrac{-7 \pm \sqrt{61}}{2}$

The solutions are $\dfrac{-7+\sqrt{61}}{2}$ and $\dfrac{-7-\sqrt{61}}{2}$, or

$\dfrac{-7 \pm \sqrt{61}}{2}$.

61. $x^2 + (x+2)^2 = 7$

$x^2 + x^2 + 4x + 4 = 7$

$2x^2 + 4x + 4 = 7$

$2x^2 + 4x - 3 = 0$ Standard form

$a = 2, \ b = 4, \ c = -3$

We use the quadratic formula.

$x = \dfrac{-4 \pm \sqrt{4^2 - 4 \cdot 2 \cdot (-3)}}{2 \cdot 2} = \dfrac{-4 \pm \sqrt{16+24}}{4}$

$x = \dfrac{-4 \pm \sqrt{40}}{4} = \dfrac{-4 \pm \sqrt{4 \cdot 10}}{4}$

$x = \dfrac{-4 \pm 2\sqrt{10}}{4} = \dfrac{2(-2 \pm \sqrt{10})}{2 \cdot 2}$

$x = \dfrac{-2 \pm \sqrt{10}}{2}$

The solutions are $\dfrac{-2+\sqrt{10}}{2}$ and $\dfrac{-2\sqrt{10}}{2}$, or $\dfrac{-2 \pm \sqrt{10}}{2}$.

63.-69. Left to the student

Exercise Set 9.4

1. $q = \dfrac{VQ}{I}$

$I \cdot q = I \cdot \dfrac{VQ}{I}$ Multiplying by I

$Iq = VQ$ Simplifying

$I = \dfrac{VQ}{q}$ Dividing by q

3. $S = \dfrac{kmM}{d^2}$

$d^2 \cdot S = d^2 \cdot \dfrac{kmM}{d^2}$ Multiplying by d^2

$d^2 S = kmM$ Simplifying

$\dfrac{d^2 S}{kM} = m$ Dividing by kM

5. $S = \dfrac{kmM}{d^2}$

$d^2 \cdot S = d^2 \cdot \dfrac{kmM}{d^2}$ Multiplying by d^2

$d^2 S = kmM$ Simplifying

$d^2 = \dfrac{kmM}{S}$ Dividing by S

7. $T = \dfrac{10t}{W^2}$

$W^2 \cdot T = W^2 \cdot \dfrac{10t}{W^2}$ Multiplying by W^2

$W^2 T = 10t$

$W^2 = \dfrac{10t}{T}$ Dividing by T

$W = \sqrt{\dfrac{10t}{T}}$ Principle of square roots. Assume W is nonnegative.

9. $A = at + bt$

$A = t(a+b)$ Factoring

$\dfrac{A}{a+b} = t$ Dividing by $a+b$

11.
$$y = ax + bx + c$$
$$y - c = ax + bx \qquad \text{Subtracting } c$$
$$y - c = x(a + b) \qquad \text{Factoring}$$
$$\frac{y - c}{a + b} = x \qquad \text{Dividing by } a + b$$

13.
$$\frac{t}{a} + \frac{t}{b} = 1$$
$$ab\left(\frac{t}{a} + \frac{t}{b}\right) = ab \cdot 1 \qquad \text{Multiplying by } ab$$
$$ab \cdot \frac{t}{a} + ab \cdot \frac{t}{b} = ab$$
$$bt + at = ab$$
$$bt = ab - at \qquad \text{Subtracting } at$$
$$bt = a(b - t) \qquad \text{Factoring}$$
$$\frac{bt}{b - t} = a \qquad \text{Dividing by } b - t$$

15.
$$\frac{1}{p} + \frac{1}{q} = \frac{1}{f}$$
$$pqf\left(\frac{1}{p} + \frac{1}{q}\right) = pqf \cdot \frac{1}{f} \qquad \text{Multiplying by } pqf$$
$$pqf \cdot \frac{1}{p} + pqf \cdot \frac{1}{q} = pq$$
$$qf + pf = pq$$
$$qf = pq - pf \qquad \text{Subtracting } pf$$
$$qf = p(q - f) \qquad \text{Factoring}$$
$$\frac{qf}{q - f} = p \qquad \text{Dividing by } q - f$$

17.
$$A = \frac{1}{2}bh$$
$$2 \cdot A = 2 \cdot \frac{1}{2}bh \qquad \text{Multiplying by } 2$$
$$2A = bh$$
$$\frac{2A}{h} = b \qquad \text{Dividing by } h$$

19.
$$S = 2\pi r(r + h)$$
$$S = 2\pi r^2 + 2\pi rh \qquad \text{Removing parentheses}$$
$$S - 2\pi r^2 = 2\pi rh \qquad \text{Subtracting } 2\pi r^2$$
$$\frac{S - 2\pi r^2}{2\pi r} = h, \text{ or} \qquad \text{Dividing by } 2\pi r$$
$$\frac{S}{2\pi r} - r = h$$

21.
$$\frac{1}{R} = \frac{1}{r_1} + \frac{1}{r_2}$$
$$Rr_1 r_2 \cdot \frac{1}{R} = Rr_1 r_2 \left(\frac{1}{r_1} + \frac{1}{r_2}\right) \text{ Multiplying by } Rr_1 r_2$$
$$r_1 r_2 = Rr_1 r_2 \cdot \frac{1}{r_1} + Rr_1 r_2 \cdot \frac{1}{r_2}$$
$$r_1 r_2 = Rr_2 + Rr_1$$
$$r_1 r_2 = R(r_2 + r_1) \qquad \text{Factoring}$$
$$\frac{r_1 r_2}{r_2 + r_1} = R \qquad \text{Dividing by } r_2 + r_1$$

23.
$$P = 17\sqrt{Q}$$
$$\frac{P}{17} = \sqrt{Q} \qquad \text{Isolating the radical}$$
$$\left(\frac{P}{17}\right)^2 = (\sqrt{Q})^2 \qquad \text{Principle of squaring}$$
$$\frac{P^2}{289} = Q \qquad \text{Simplifying}$$

25.
$$v = \sqrt{\frac{2gE}{m}}$$
$$v^2 = \left(\sqrt{\frac{2gE}{m}}\right)^2 \qquad \text{Principle of squaring}$$
$$v^2 = \frac{2gE}{m}$$
$$mv^2 = 2gE \qquad \text{Multiping by } m$$
$$\frac{mv^2}{2g} = E \qquad \text{Dividing by } 2g$$

27.
$$S = 4\pi r^2$$
$$\frac{S}{4\pi} = r^2 \qquad \text{Dividing by } 4\pi$$
$$\sqrt{\frac{S}{4\pi}} = r \qquad \begin{array}{l}\text{Principle of square roots.} \\ \text{Assume } r \text{ is nonnegative.}\end{array}$$
$$\sqrt{\frac{1}{4} \cdot \frac{S}{\pi}} = r$$
$$\frac{1}{2}\sqrt{\frac{S}{\pi}} = r$$

29.
$$P = kA^2 + mA$$
$$0 = kA^2 + mA - P \qquad \text{Standard form}$$
$$a = k, \ b = m, \ c = -P$$
$$A = \frac{-b \pm \sqrt{b^2 - 4ac}}{2a} \qquad \text{Quadratic formula}$$
$$A = \frac{-m \pm \sqrt{m^2 - 4 \cdot k \cdot (-P)}}{2 \cdot k} \qquad \text{Substituting}$$
$$A = \frac{-m + \sqrt{m^2 + 4kP}}{2k} \qquad \text{Using the positive root}$$

31.
$$c^2 = a^2 + b^2$$
$$c^2 - b^2 = a^2$$
$$\sqrt{c^2 - b^2} = a \qquad \begin{array}{l}\text{Principle of square roots.} \\ \text{Assume } a \text{ is nonnegative.}\end{array}$$

33.
$$s = 16t^2$$
$$\frac{s}{16} = t^2$$
$$\sqrt{\frac{s}{16}} = t \qquad \begin{array}{l}\text{Principle of square roots.} \\ \text{Assume } t \text{ is nonnegative.}\end{array}$$
$$\frac{\sqrt{s}}{4} = t$$

35. $A = \pi r^2 + 2\pi rh$

$0 = \pi r^2 + 2\pi hr - A$

$a = \pi,\ b = 2\pi h,\ c = -A$

$r = \dfrac{-b \pm \sqrt{b^2 - 4ac}}{2a}$

$r = \dfrac{-2\pi h \pm \sqrt{(2\pi h)^2 - 4 \cdot \pi \cdot (-A)}}{2 \cdot \pi}$

$r = \dfrac{-2\pi h + \sqrt{4\pi^2 h^2 + 4\pi A}}{2\pi}$ Using the positive root

$r = \dfrac{-2\pi h + \sqrt{4(\pi^2 h^2 + \pi A)}}{2\pi}$

$r = \dfrac{-2\pi h + 2\sqrt{\pi^2 h^2 + \pi A}}{2\pi}$

$r = \dfrac{2\left(-\pi h + \sqrt{\pi^2 h^2 + \pi A}\right)}{2\pi}$

$r = \dfrac{-\pi h + \sqrt{\pi^2 h^2 + \pi A}}{\pi}$

37. $F = \dfrac{Av^2}{400}$

$400F = Av^2$ Multiplying by 400

$\dfrac{400F}{A} = v^2$ Dividing by A

$\sqrt{\dfrac{400F}{A}} = v$ Principle of square roots. Assume v is nonnegative.

$\sqrt{400 \cdot \dfrac{F}{A}} = v$

$20\sqrt{\dfrac{F}{a}} = v$

39. $c = \sqrt{a^2 + b^2}$

$c^2 = (\sqrt{a^2 + b^2})^2$ Principle of squaring

$c^2 = a^2 + b^2$

$c^2 - b^2 = a^2$

$\sqrt{c^2 - b^2} = a$ Principle of square roots. Assume a is nonnegative.

41. $h = \dfrac{a}{2}\sqrt{3}$

$2h = a\sqrt{3}$

$\dfrac{2h}{\sqrt{3}} = a$

$\dfrac{2h\sqrt{3}}{3} = a$ Rationalizing the denominator

43. $n = aT^2 - 4T + m$

$0 = aT^2 - 4T + m - n$

$a = a,\ b = -4,\ c = m - n$

$T = \dfrac{-b \pm \sqrt{b^2 - 4ac}}{2a}$

$T = \dfrac{-(-4) \pm \sqrt{(-4)^2 - 4 \cdot a \cdot (m-n)}}{2 \cdot a}$

$T = \dfrac{4 + \sqrt{16 - 4a(m-n)}}{2a}$ Using the positive root

$T = \dfrac{4 + \sqrt{4[4 - a(m-n)]}}{2a}$

$T = \dfrac{4 + 2\sqrt{4 - a(m-n)}}{2a}$

$T = \dfrac{2\left(2 + \sqrt{4 - a(m-n)}\right)}{2 \cdot a}$

$T = \dfrac{2 + \sqrt{4 - a(m-n)}}{a}$

45. $v = 2\sqrt{\dfrac{2kT}{\pi m}}$

$\dfrac{v}{2} = \sqrt{\dfrac{2kT}{\pi m}}$ Isolating the radical

$\left(\dfrac{v}{2}\right)^2 = \left(\sqrt{\dfrac{2kT}{\pi m}}\right)^2$ Principle of squaring

$\dfrac{v^2}{4} = \dfrac{2kT}{\pi m}$

$\dfrac{v^2}{4} \cdot \dfrac{\pi m}{2k} = \dfrac{2kT}{\pi m} \cdot \dfrac{\pi m}{2k}$ Multiplying by $\dfrac{\pi m}{2k}$

$\dfrac{v^2 \pi m}{8k} = T$

47. $3x^2 = d^2$

$x^2 = \dfrac{d^2}{3}$ Dividing by 3

$x = \dfrac{d}{\sqrt{3}}$ Principle of square roots. Assume x is nonnegative.

$x = \dfrac{d}{\sqrt{3}} \cdot \dfrac{\sqrt{3}}{\sqrt{3}}$ Rationalizing the denominator

$x = \dfrac{d\sqrt{3}}{3}$

49. $N = \dfrac{n^2 - n}{2}$

$2N = n^2 - n$ Multiplying by 2

$0 = n^2 - n - 2N$ Finding standard form

$a = 1,\ b = -1,\ c = -2N$

$n = \dfrac{-b \pm \sqrt{b^2 - 4ac}}{2a}$

$n = \dfrac{-(-1) \pm \sqrt{(-1)^2 - 4 \cdot 1 \cdot (-2N)}}{2 \cdot 1}$ Substituting

$n = \dfrac{1 + \sqrt{1 + 8N}}{2}$ Using the positive root

51.
$$S = \frac{a+b}{3b}$$
$$3b \cdot S = 3b \cdot \frac{a+b}{3b}$$
$$3bS = a + b$$
$$3bS - b = a$$
$$b(3S - 1) = a$$
$$b = \frac{a}{3S - 1}$$

53.
$$\frac{A - B}{AB} = Q$$
$$AB \cdot \frac{A - B}{AB} = AB \cdot Q$$
$$A - B = ABQ$$
$$A = ABQ + B$$
$$A = B(AQ + 1)$$
$$\frac{A}{AQ + 1} = B$$

55.
$$S = 180(n - 2)$$
$$S = 180n - 360$$
$$S + 360 = 180n$$
$$\frac{S + 360}{180} = n, \text{ or}$$
$$\frac{S}{180} + 2 = n$$

57.
$$A = P(1 + rt)$$
$$A = P + Prt$$
$$A - P = Prt$$
$$\frac{A - P}{Pr} = t$$

59.
$$\frac{A}{B} = \frac{C}{D}$$
$$BD \cdot \frac{A}{B} = BD \cdot \frac{C}{D}$$
$$AD = BC$$
$$D = \frac{BC}{A}$$

61.
$$C = \frac{Ka - b}{a}$$
$$a \cdot C = a \cdot \frac{Ka - b}{a}$$
$$aC = Ka - b$$
$$aC - Ka = -b$$
$$a(C - K) = -b$$
$$a = \frac{-b}{C - K}, \text{ or}$$
$$a = \frac{b}{K - C}$$

63. Discussion and Writing Exercise

65.
$$a^2 + b^2 = c^2 \qquad \text{Pythagorean equation}$$
$$4^2 + 7^2 = c^2 \qquad \text{Substituting}$$
$$16 + 49 = c^2$$
$$65 = c^2$$
$$\sqrt{65} = c \qquad \text{Exact answer}$$
$$8.062 \approx c \qquad \text{Approximate answer}$$

67.
$$a^2 + b^2 = c^2 \qquad \text{Pythagorean equation}$$
$$4^2 + 5^2 = c^2 \qquad \text{Substituting}$$
$$16 + 25 = c^2$$
$$41 = c^2$$
$$\sqrt{41} = c \qquad \text{Exact answer}$$
$$6.403 \approx c \qquad \text{Approximate answer}$$

69.
$$a^2 + b^2 = c^2 \qquad \text{Pythagorean equation}$$
$$2^2 + b^2 = (8\sqrt{17})^2 \qquad \text{Substituting}$$
$$4 + b^2 = 64 \cdot 17$$
$$4 + b^2 = 1088$$
$$b^2 = 1084$$
$$b = \sqrt{1084} \qquad \text{Exact answer}$$
$$b \approx 32.924 \qquad \text{Approximate answer}$$

71. We make a drawing. Let l = the length of the guy wire.

Then we use the Pythagorean equation.
$$10^2 + 18^2 = l^2$$
$$100 + 324 = l^2$$
$$424 = l^2$$
$$\sqrt{424} = l \qquad \text{Exact answer}$$
$$20.591 \approx l \qquad \text{Approximation}$$

The length of the guy wire is $\sqrt{424}$ ft ≈ 20.591 ft.

73. $\sqrt{3x} \cdot \sqrt{6x} = \sqrt{18x^2} = \sqrt{9 \cdot x^2 \cdot 2} = \sqrt{9}\sqrt{x^2}\sqrt{2} = 3x\sqrt{2}$

75. $3\sqrt{t} \cdot \sqrt{t} = 3\sqrt{t^2} = 3t$

77. a) $C = 2\pi r$
$$\frac{C}{2\pi} = r$$

b) $A = \pi r^2$
$$A = \pi \cdot \left(\frac{C}{2\pi}\right)^2 \qquad \text{Substituting } \frac{C}{2\pi} \text{ for } r$$
$$A = \pi \cdot \frac{C^2}{4\pi^2}$$
$$A = \frac{C^2}{4\pi}$$

79.
$$3ax^2 - x - 3ax + 1 = 0$$
$$3ax^2 + (-1 - 3a)x + 1 = 0$$
$$a = 3a, \ b = -1 - 3a, \ c = 1$$

$$x = \frac{-b \pm \sqrt{b^2 - 4ac}}{2a}$$

$$x = \frac{-(-1 - 3a) \pm \sqrt{(-1 - 3a)^2 - 4 \cdot 3a \cdot 1}}{2 \cdot 3a}$$

$$x = \frac{1 + 3a \pm \sqrt{1 + 6a + 9a^2 - 12a}}{6a}$$

$$x = \frac{1 + 3a \pm \sqrt{9a^2 - 6a + 1}}{6a}$$

$$x = \frac{1 + 3a \pm \sqrt{(3a - 1)^2}}{6a}$$

$$x = \frac{1 + 3a \pm (3a - 1)}{6a}$$

$$x = \frac{1 + 3a + 3a - 1}{6a} \quad or \quad x = \frac{1 + 3a - 3a + 1}{6a}$$

$$x = \frac{6a}{6a} \quad\quad\quad\quad or \quad x = \frac{2}{6a}$$

$$x = 1 \quad\quad\quad\quad\quad or \quad x = \frac{1}{3a}$$

The solutions are 1 and $\frac{1}{3a}$.

Exercise Set 9.5

1. **Familiarize.** Let $h =$ the height of the screen, in inches. Then $h + 27 =$ the width.

   Translate. We use the Pythagorean equation.

   $$h^2 + (h + 27)^2 = 70^2.$$

   Solve. We solve the equation.

   $$h^2 + (h + 27)^2 = 70^2$$
   $$h^2 + h^2 + 54h + 729 = 4900$$
   $$2h^2 + 54h + 729 = 4900$$
   $$2h^2 + 54h - 4171 = 0$$

   We use the quadratic formula with $a = 2$, $b = 54$, and $c = -4171$.

   $$h = \frac{-54 \pm \sqrt{54^2 - 4 \cdot 2 \cdot (-4171)}}{2 \cdot 2}$$

   $$= \frac{-54 \pm \sqrt{2916 + 33,368}}{4} = \frac{-54 \pm \sqrt{36,284}}{4}$$

   $$h = \frac{-54 - \sqrt{36,284}}{4} \quad or \quad h = \frac{-54 \pm \sqrt{36,284}}{4}$$

   $$h \approx -61 \quad\quad\quad or \quad h \approx 34$$

   Check. The height of the screen cannot be negative, so -61 cannot be a solution. If $h \approx 34$, then $h + 27 \approx 61$ and $34^2 + 61^2 = 4877 \approx 4900 = 70^2$. The answer checks.

   State. The width of the screen is about 61 in., and the height is about 34 in.

3. **Familiarize.** Using the labels on the drawing in the text we have $w =$ the width of the rectangle and $w + 3 =$ the length.

   Translate. Recall that area is length × width. Then we have

$(w + 3)(w) = 70.$

Solve. We solve the equation.

$$w^2 + 3w = 70$$
$$w^2 + 3w - 70 = 0$$
$$(w + 10)(w - 7) = 0$$

$$w + 10 = 0 \quad or \quad w - 7 = 0$$
$$w = -10 \quad or \quad\quad w = 7$$

Check. We know that -10 is not a solution of the original problem, because the width cannot be negative. When $w = 7$, then $w + 3 = 10$, and the area is $10 \cdot 7$, or 70. This checks.

State. The width of the rectangle is 7 ft, and the length is 10 ft.

5. **Familiarize.** Using the labels on the drawing in the text we have $s =$ the length of the shorter leg, in inches, and $s + 8 =$ the length of the longer leg.

   Translate. We use the Pythagorean equation.

   $$s^2 + (s + 8)^2 = (8\sqrt{13})^2$$

   Solve. We solve the equation.

   $$s^2 + (s + 8)^2 = (8\sqrt{13})^2$$
   $$s^2 + s^2 + 16s + 64 = 64 \cdot 13$$
   $$2s^2 + 16s + 64 = 832$$
   $$2s^2 + 16s - 768 = 0$$
   $$s^2 + 8s - 384 = 0 \quad\quad \text{Dividing by 2}$$
   $$(s + 24)(s - 16) = 0$$

   $$s + 24 = 0 \quad or \quad s - 16 = 0$$
   $$s = -24 \quad or \quad\quad s = 16$$

   Check. The length of a leg cannot be negative, so -24 cannot be a solution. If $s = 16$, then $s + 8 = 16 + 8 = 24$ and $16^2 + 24^2 = 832 = (8\sqrt{13})^2$. The answer checks.

   State. The lengths of the legs are 16 in. and 24 in.

7. **Familiarize.** We first make a drawing. We let x represent the length. Then $x - 4$ represents the width.

   Translate. The area is length × width. Thus, we have two expressions for the area of the rectangle: $x(x - 4)$ and 320. This gives us a translation.

   $$x(x - 4) = 320.$$

   Solve. We solve the equation.

   $$x^2 - 4x = 320$$
   $$x^2 - 4x - 320 = 0$$
   $$(x - 20)(x + 16) = 0$$

   $$x - 20 = 0 \quad or \quad x + 16 = 0$$
   $$x = 20 \quad or \quad\quad x = -16$$

   Check. Since the length of a side cannot be negative, -16 does not check. But 20 does check. If the length is 20, then

the width is $20 - 4$, or 16. The area is 20×16, or 320.
This checks.

State. The length is 20 cm, and the width is 16 cm.

9. **Familiarize.** We first make a drawing. We let x represent
the length of one leg. Then $x + 2$ represents the length of
the other leg.

Translate. We use the Pythagorean equation.
$$x^2 + (x+2)^2 = 8^2.$$

Solve. We solve the equation.
$$x^2 + x^2 + 4x + 4 = 64$$
$$2x^2 + 4x + 4 = 64$$
$$2x^2 + 4x - 60 = 0$$
$$x^2 + 2x - 30 = 0 \qquad \text{Dividing by 2}$$
$a = 1,\ b = 2,\ c = -30$
$$x = \frac{-2 \pm \sqrt{2^2 - 4 \cdot 1 \cdot (-30)}}{2 \cdot 1}$$
$$= \frac{-2 \pm \sqrt{4 + 120}}{2} = \frac{-2 \pm \sqrt{124}}{2}$$
$$= \frac{-2 \pm \sqrt{4 \cdot 31}}{2} = \frac{-2 \pm 2\sqrt{31}}{2}$$
$$= \frac{2(-1 \pm \sqrt{31})}{2} = -1 \pm \sqrt{31}$$

Using a calculator or Table 2 we find that $\sqrt{31} \approx 5.568$:
$$-1 + \sqrt{31} \approx -1 + 5.568 \quad or \quad -1 - \sqrt{31} \approx -1 - 5.568$$
$$\approx 4.6 \qquad\qquad or \qquad\qquad \approx -6.6$$

Check. Since the length of a leg cannot be negative, -6.6
does not check. But 4.6 does check. If the shorter leg is 4.6,
then the other leg is $4.6 + 2$, or 6.6. Then $4.6^2 + 6.6^2 =$
$21.16 + 43.56 = 64.72$ and using a calculator, $\sqrt{64.72} \approx$
$8.04 \approx 8$. Note that our check is not exact since we are
using an approximation.

State. One leg is about 4.6 m, and the other is about
6.6 m long.

11. **Familiarize.** We first make a drawing. We let x represent
the width and $x + 2$ the length.

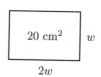

Translate. The area is length $\times$ width. We have two
expressions for the area of the rectangle: $(x+2)x$ and 20.
This gives us a translation.
$$(x+2)x = 20.$$

Solve. We solve the equation.

$$x^2 + 2x = 20$$
$$x^2 + 2x - 20 = 0$$
$a = 1,\ b = 2,\ c = -20$
$$x = \frac{-2 \pm \sqrt{2^2 - 4 \cdot 1 \cdot (-20)}}{2 \cdot 1}$$
$$= \frac{-2 \pm \sqrt{4 + 80}}{2} = \frac{-2 \pm \sqrt{84}}{2}$$
$$= \frac{-2 \pm \sqrt{4 \cdot 21}}{2} = \frac{-2 \pm 2\sqrt{21}}{2}$$
$$= \frac{2(-1 \pm \sqrt{21})}{2} = -1 \pm \sqrt{21}$$

Using a calculator or Table 2 we find that $\sqrt{21} \approx 4.583$:
$$-1 + \sqrt{21} \approx -1 + 4.583 \quad or \quad -1 - \sqrt{21} \approx -1 - 4.583$$
$$\approx 3.6 \qquad\qquad or \qquad\qquad \approx -5.6$$

Check. Since the length of a side cannot be negative,
-5.6 does not check. But 3.6 does check. If the width is
3.6, then the length is $3.6 + 2$, or 5.6. The area is $5.6(3.6)$,
or $20.16 \approx 20$. This checks.

State. The length is about 5.6 in., and the width is about
3.6 in.

13. **Familiarize.** We make a drawing and label it. We let
$w = $ the width of the rectangle and $2w = $ the length.

$$\boxed{20 \text{ cm}^2}\ w$$
$$2w$$

Translate. Recall that area = length $\times$ width. Then we
have
$$2w \cdot w = 20.$$

Solve. We solve the equation.
$$2w^2 = 20$$
$$w^2 = 10 \qquad \text{Dividing by 2}$$
$$w = \sqrt{10} \quad or \quad w = -\sqrt{10} \qquad \text{Principle of square roots}$$
$$w \approx 3.2 \quad or \quad w \approx -3.2$$

Check. We know that -3.2 is not a solution of the original
problem, because width cannot be negative. When $w \approx$
3.2, then $2w \approx 6.4$ and the area is about $(6.4)(3.2)$, or
20.48. This checks, although the check is not exact since
we used an approximation for $\sqrt{10}$.

State. The length is about 6.4 cm, and the width is about
3.2 cm.

15. **Familiarize.** Using the drawing in the text, we have $x =$
the thickness of the frame, $20 - 2x = $ the width of the
picture showing, and $25 - 2x = $ the length of the picture
showing.

Translate. Recall that area = length $\times$ width. Then we
have
$$(25 - 2x)(20 - 2x) = 266.$$

Solve. We solve the equation.

$$500 - 90x + 4x^2 = 266$$
$$4x^2 - 90x + 234 = 0$$
$$2x^2 - 45x + 117 = 0 \qquad \text{Dividing by 2}$$
$$(2x - 39)(x - 3) = 0$$

$$
\begin{array}{ccc}
2x - 39 = 0 & or & x - 3 = 0 \\
2x = 39 & or & x = 3 \\
x = 19.5 & or & x = 3
\end{array}
$$

Check. The number 19.5 cannot be a solution, because when $x = 19.5$ then $20 - 2x = -19$, and the width cannot be negative. When $x = 3$, then $20 - 2x = 20 - 2 \cdot 3$, or 14 and $25 - 2x = 25 - 2 \cdot 3$, or 19 and $19 \cdot 14 = 266$. This checks.

State. The thickness of the frame is 3 cm.

17. Familiarize. Referring to the drawing in the text, we complete the table.

	d	r	t
Upstream	40	$r - 3$	t_1
Downstream	40	$r + 3$	t_2
Total Time			14

Translate. Using $t = d/r$ and the rows of the table, we have
$$t_1 = \frac{40}{r - 3} \text{ and } t_2 = \frac{40}{r + 3}.$$
Since the total time is 14 hr, $t_1 + t_2 = 14$, and we have
$$\frac{40}{r - 3} + \frac{40}{r + 3} = 14.$$

Solve. We solve the equation. We multiply by $(r - 3)(r + 3)$, the LCM of the denominators.
$$(r-3)(r+3)\left(\frac{40}{r-3} + \frac{40}{r+3}\right) = (r-3)(r+3) \cdot 14$$
$$40(r+3) + 40(r-3) = 14(r^2 - 9)$$
$$40r + 120 + 40r - 120 = 14r^2 - 126$$
$$80r = 14r^2 - 126$$
$$0 = 14r^2 - 80r - 126$$
$$0 = 7r^2 - 40r - 63$$
$$0 = (7r + 9)(r - 7)$$

$$
\begin{array}{ccc}
7r + 9 = 0 & or & r - 7 = 0 \\
7r = -9 & or & r = 7 \\
r = -\dfrac{9}{7} & or & r = 7
\end{array}
$$

Check. Since speed cannot be negative, $-\dfrac{9}{7}$ cannot be a solution. If the speed of the boat is 7 km/h, the speed upstream is $7 - 3$, or 4 km/h, and the speed downstream is $7 + 3$, or 10 km/h. The time upstream is $\dfrac{40}{4}$, or 10 hr. The time downstream is $\dfrac{40}{10}$, or 4 hr. The total time is 14 hr. This checks.

State. The speed of the boat in still water is 7 km/h.

19. Familiarize. Let r represent the speed of the wind. Then the speed of the plane flying with the wind is $300 + r$ and the speed against the wind is $300 - r$. We fill in the table in the text.

	d	r	t
With wind	680	$300 + r$	t_1
Against wind	520	$300 - r$	t_2

Translate. Using $t = d/r$ and the rows of the table, we have
$$t_1 = \frac{680}{300 + r} \text{ and } t_2 = \frac{520}{300 - r}.$$
Since the total time is 4 hr, $t_1 + t_2 = 4$, and we have
$$\frac{680}{300 + r} + \frac{520}{300 - r} = 4.$$

Solve. We solve the equation. We multiply by $(300 + r)(300 - r)$, the LCM of the denominators.
$$(300+r)(300-r)\left(\frac{680}{300+r} + \frac{520}{300-r}\right) = (300+r)(300-r) \cdot 4$$
$$680(300 - r) + 520(300 + r) = 4(90,000 - r^2)$$
$$204,000 - 680r + 156,000 + 520r = 360,000 - 4r^2$$
$$360,000 - 160r = 360,000 - 4r^2$$
$$4r^2 - 160r = 0$$
$$4r(r - 40) = 0$$

$$
\begin{array}{ccc}
4r = 0 & or & r - 40 = 0 \\
r = 0 & or & r = 40
\end{array}
$$

Check. If $r = 0$, then the speed of the wind is 0 km/h. That is, there is no wind. In this case the plane travels 680 km in 680/300, or $2\frac{4}{15}$ hr, and it travels 520 km in 520/300, or $1\frac{11}{15}$ hr. The total time is $2\frac{4}{15} + 1\frac{11}{15}$, or 4 hr, so we have one solution. If the speed of the wind is 40 km/h, then the speed of the airplane with the wind is $300 + 40$, or 340 km/h, and the speed against the wind is $300 - 40$, or 260 km/h. The time with the wind is 680/340, or 2 hr, and the time against the wind is 520/260, or 2 hr. The total time is 2 hr + 2 hr, or 4 hr, so we have a second solution.

State. The speed of the wind is 0 km/h (There is no wind.) or 40 km/h.

21. Familiarize. We first make a drawing. We let r represent the speed of the current. Then $10 - r$ is the speed of the boat traveling upstream and $10 + r$ is the speed of the boat traveling downstream.

Upstream
$10 - r$ km/h

12 km

Downstream
$10 + r$ km/h

28 km

We summarize the information in a table.

	d	r	t
Upstream	12	$10-r$	t_1
Downstream	28	$10+r$	t_2
Total Time			4

Translate. Using $t = d/r$ and the rows of the table, we have

$$t_1 = \frac{12}{10-r} \text{ and } t_2 = \frac{28}{10+r}.$$

Since the total time is 4 hr, $t_1 + t_2 = 4$, and we have

$$\frac{12}{10-r} + \frac{28}{10+r} = 4.$$

Solve. We solve the equation. We multiply by $(10 - r)(10 + r)$, the LCM of the denominators.

$$(10-r)(10+r)\left(\frac{12}{10-r} + \frac{28}{10+r}\right) =$$
$$(10-r)(10+r) \cdot 4$$
$$12(10+r) + 28(10-r) = 4(100-r^2)$$
$$120 + 12r + 280 - 28r = 400 - 4r^2$$
$$400 - 16r = 400 - 4r^2$$
$$4r^2 - 16r = 0$$
$$4r(r-4) = 0$$

$$4r = 0 \quad or \quad r - 4 = 0$$
$$r = 0 \quad or \quad\quad\quad r = 4$$

Check. If $r = 0$, then the speed of the stream is 0 km/h. That is, the stream is still. In this case the boat travels 12 km in 12/10, or 1.2 hr, and it travels 28 km in 28/10, or 2.8 hr. The total time is 1.2 hr + 2.8 hr, or 4 hr, so we have one solution. If the speed of the current is 4 km/h, the speed upstream is $10 - 4$, or 6 km/h, and the speed downstream is $10 + 4$, or 14 km/h. The time upstream is 12/6, or 2 hr. The time downstream is 28/14, or 2 hr. The total time is 2 hr + 2 hr, or 4 hr. This checks also.

State. The speed of the stream is 0 km/h (The stream is still.) or 4 km/h.

23. Familiarize. We first make a drawing. We let r represent the speed of the boat in still water. Then $r - 4$ is the speed of the boat traveling upstream and $r + 4$ is the speed of the boat traveling downstream.

Upstream
$r - 4$ mph
←————————————————→
4 mi

Downstream
$r + 4$ mph
←————————————————————————→
12 mi

We summarize the information in a table.

	d	r	t
Upstream	4	$r-4$	t_1
Downstream	12	$r+4$	t_2
Total Time			2

Translate. Using $t = d/r$ and the rows of the table, we have

$$t_1 = \frac{4}{r-4} \text{ and } t_2 = \frac{12}{r+4}.$$

Since the total time is 2 hr, $t_1 + t_2 = 2$, and we have

$$\frac{4}{r-4} + \frac{12}{r+4} = 2.$$

Solve. We solve the equation. We multiply by $(r - 4)(r + 4)$, the LCM of the denominators.

$$(r-4)(r+4)\left(\frac{4}{r-4} + \frac{12}{r+4}\right) = (r-4)(r+4) \cdot 2$$
$$4(r+4) + 12(r-4) = 2(r^2-16)$$
$$4r + 16 + 12r - 48 = 2r^2 - 32$$
$$16r - 32 = 2r^2 - 32$$
$$0 = 2r^2 - 16r$$
$$0 = 2r(r-8)$$

$$2r = 0 \quad or \quad r - 8 = 0$$
$$r = 0 \quad or \quad\quad\quad r = 8$$

Check. If $r = 0$, then the speed upstream, $0 - 4$, would be negative. Since speed cannot be negative, 0 cannot be a solution. If the speed of the boat is 8 mph, the speed upstream is $8 - 4$, or 4 mph, and the speed downstream is $8 + 4$, or 12 mph. The time upstream is $\frac{4}{4}$, or 1 hr. The time downstream is $\frac{12}{12}$, or 1 hr. The total time is 2 hr. This checks.

State. The speed of the boat in still water is 8 mph.

25. Familiarize. We first make a drawing. We let r represent the speed of the stream. Then $9 - r$ represents the speed of the boat traveling upstream and $9 + r$ represents the speed of the boat traveling downstream.

Upstream
$9 - r$ km/h
←————————————————→
80 km

Downstream
$9 + r$ km/h
←————————————————→
80 km

We summarize the information in a table.

	d	r	t
Upstream	80	$9-r$	t_1
Downstream	80	$9+r$	t_2

Translate. Using $t = d/r$ and the rows of the table, we have

$$t_1 = \frac{80}{9-r} \text{ and } t_2 = \frac{80}{9+r}.$$

Since the total time is 18 hr, $t_1 + t_2 = 18$, and we have

$$\frac{80}{9-r} + \frac{80}{9+r} = 18.$$

Solve. We solve the equation. We multiply by $(9 - r)(9 + r)$, the LCM of the denominators.

$$(9-r)(9+r)\left(\frac{80}{9-r}+\frac{80}{9+r}\right) = (9-r)(9+r)\cdot 18$$
$$80(9+r) + 80(9-r) = 18(81-r^2)$$
$$720 + 80r + 720 - 80r = 1458 - 18r^2$$
$$1440 = 1458 - 18r^2$$
$$18r^2 = 18$$
$$r^2 = 1$$

$r = 1 \; or \; r = -1$ Principle of square roots

Check. Since speed cannot be negative, -1 cannot be a solution. If the speed of the stream is 1 km/h, the speed upstream is $9 - 1$, or 8 km/h, and the speed downstream is $9 + 1$, or 10 km/h. The time upstream is $\frac{80}{8}$, or 10 hr. The time downstream is $\frac{80}{10}$, or 8 hr. The total time is 18 hr. This checks.

State. The speed of the stream is 1 km/h.

27. Discussion and Writing Exercise

29. $\quad 5\sqrt{2} + \sqrt{18} = 5\sqrt{2} + \sqrt{9\cdot 2}$
$$= 5\sqrt{2} + \sqrt{9}\sqrt{2}$$
$$= 5\sqrt{2} + 3\sqrt{2}$$
$$= (5+3)\sqrt{2}$$
$$= 8\sqrt{2}$$

31. $\quad \sqrt{4x^3} - 7\sqrt{x} = \sqrt{4\cdot x^2 \cdot x} - 7\sqrt{x}$
$$= \sqrt{4}\sqrt{x^2}\sqrt{x} - 7\sqrt{x}$$
$$= 2x\sqrt{x} - 7\sqrt{x}$$
$$= (2x - 7)\sqrt{x}$$

33. $\quad \sqrt{2} + \sqrt{\frac{1}{2}} = \sqrt{2} + \sqrt{\frac{1}{2}\cdot\frac{2}{2}}$
$$= \sqrt{2} + \sqrt{\frac{2}{4}}$$
$$= \sqrt{2} + \frac{\sqrt{2}}{\sqrt{4}}$$
$$= \sqrt{2} + \frac{\sqrt{2}}{2}$$
$$= \left(1 + \frac{1}{2}\right)\sqrt{2}$$
$$= \frac{3}{2}\sqrt{2}, \text{or } \frac{3\sqrt{2}}{2}$$

35. $\quad \sqrt{24} + \sqrt{54} - \sqrt{48}$
$$= \sqrt{4\cdot 6} + \sqrt{9\cdot 6} - \sqrt{16\cdot 3}$$
$$= \sqrt{4}\cdot\sqrt{6} + \sqrt{9}\cdot\sqrt{6} - \sqrt{16}\cdot\sqrt{3}$$
$$= 2\sqrt{6} + 3\sqrt{6} - 4\sqrt{3}$$
$$= 5\sqrt{6} - 4\sqrt{3}$$

37. *Familiarize.* The radius of a 12-in. pizza is $\frac{12}{2}$, or 6 in. The radius of a d-in. pizza is $\frac{d}{2}$ in. The area of a circle is πr^2.

Translate.

Area of d-in. pizza	is	Area of 12-in. pizza	plus	Area of 12-in. pizza
↓	↓	↓	↓	↓
$\pi\left(\dfrac{d}{2}\right)^2$	$=$	$\pi\cdot 6^2$	$+$	$\pi\cdot 6^2$

Solve. We solve the equation.
$$\frac{d^2}{4}\pi = 36\pi + 36\pi$$
$$\frac{d^2}{4}\pi = 72\pi$$
$$\frac{d^2}{4} = 72 \quad \text{Dividing by } \pi$$
$$d^2 = 288$$

$d = \sqrt{288} \; or \; d = -\sqrt{288}$
$d = 12\sqrt{2} \; or \; d = -12\sqrt{2}$
$d \approx 16.97 \; or \; d \approx -16.97$ Using a calculator

Check. Since the diameter cannot be negative, -16.97 is not a solution. If $d = 12\sqrt{2}$, or 16.97, then $r = 6\sqrt{2}$ and the area is $\pi(6\sqrt{2})^2$, or 72π. The area of the two 12-in. pizzas is $2\cdot\pi\cdot 6^2$, or 72π. The value checks.

State. The diameter of the pizza should be $12\sqrt{2}$ in. $\approx$ 16.97 in.

The radius of a 16-in. pizza is $\frac{16}{2}$, or 8 in., so the area is $\pi(8)^2$, or 64π. We found that the area of two 12-in. pizzas is 72π and $72\pi > 64\pi$, so you get more to eat with two 12-in. pizzas than with a 16-in. pizza.

Exercise Set 9.6

1. $y = x^2 + 1$

We first find the vertex. The x-coordinate is
$$-\frac{b}{2a} = -\frac{0}{2\cdot 1} = 0.$$
We substitute into the equation to find the second coordinate of the vertex.
$$y = x^2 + 1 = 0^2 + 1 = 1$$
The vertex is (0,1). This is also the y-intercept. The line of symmetry is $x = 0$, the y-axis.

We choose some x-values on both sides of the vertex and graph the parabola.

When $x = 1$, $y = 1^2 + 1 = 1 + 1 = 2$.

When $x = -1$, $y = (-1)^2 + 1 = 1 + 1 = 2$.

When $x = 2$, $y = 2^2 + 1 = 4 + 1 = 5$.

When $x = -2$, $y = (-2)^2 + 1 = 4 + 1 = 5$.

x	y	
-2	5	
-1	2	
0	1	$\leftarrow$ Vertex
1	2	
2	5	
3	10	

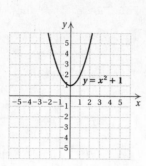

3. $y = -1 \cdot x^2$

Find the vertex. The x-coordinate is
$$-\frac{b}{2a} = -\frac{0}{2(-1)} = 0.$$

The y-coordinate is
$$y = -1 \cdot x^2 = -1 \cdot 0^2 = 0.$$

The vertex is $(0,0)$. This is also the y-intercept. The line of symmetry is $x = 0$, the y-axis.

Choose some x-values on both sides of the vertex and graph the parabola.

When $x = -2$, $y = -1 \cdot (-2)^2 = -1 \cdot 4 = -4$.

When $x = -1$, $y = -1 \cdot (-1)^2 = -1 \cdot 1 = -1$.

When $x = 1$, $y = -1 \cdot 1^2 = -1 \cdot 1 = -1$.

When $x = 2$, $y = -1 \cdot 2^2 = -1 \cdot 4 = -4$.

x	y	
0	0	$\leftarrow$ Vertex
-2	-4	
-1	-1	
1	-1	
2	-4	

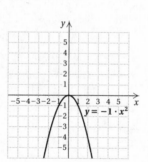

5. $y = -x^2 + 2x$

Find the vertex. The x-coordinate is
$$-\frac{b}{2a} = -\frac{2}{2(-1)} = -(-1) = 1.$$

The y-coordinate is
$$y = -x^2 + 2x = -(1)^2 + 2 \cdot 1 = -1 + 2 = 1.$$

The vertex is $(1,1)$.

We choose some x-values on both sides of the vertex and graph the parabola. We make sure we find y when $x = 0$. This gives us the y-intercept.

x	y	
1	1	$\leftarrow$ Vertex
0	0	$\leftarrow y$-intercept
-1	-3	
2	0	
3	-3	

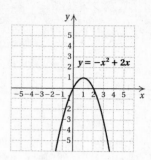

7. $y = 5 - x - x^2$, or $y = -x^2 - x + 5$

Find the vertex. The x-coordinate is
$$-\frac{b}{2a} = -\frac{-1}{2(-1)} = -\frac{1}{2}.$$

The y-coordinate is
$$y = 5 - x - x^2 = 5 - \left(-\frac{1}{2}\right) - \left(-\frac{1}{2}\right)^2 = 5 + \frac{1}{2} - \frac{1}{4} = \frac{21}{4}.$$

The vertex is $\left(-\frac{1}{2}, \frac{21}{4}\right)$.

We choose some x-values on both sides of the vertex and graph the parabola.

x	y	
$-\frac{1}{2}$	$\frac{21}{4}$	$\leftarrow$ Vertex
0	5	$\leftarrow y$-intercept
-1	5	
-2	3	
1	3	

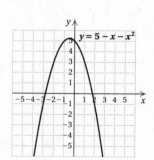

9. $y = x^2 - 2x + 1$

Find the vertex. The x-coordinate is
$$-\frac{b}{2a} = -\frac{-2}{2 \cdot 1} = -(-1) = 1.$$

The y-coordinate is
$$y = x^2 - 2x + 1 = 1^2 - 2 \cdot 1 + 1 = 1 - 2 + 1 = 0.$$

The vertex is $(1,0)$.

We choose some x-values on both sides of the vertex and graph the parabola.

x	y	
1	0	$\leftarrow$ Vertex
0	1	$\leftarrow y$-intercept
-1	4	
2	1	
3	4	

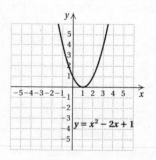

11. $y = -x^2 + 2x + 3$

Find the vertex. The x-coordinate is
$$-\frac{b}{2a} = -\frac{2}{2(-1)} = -(-1) = 1.$$
The y-coordinate is
$$y = -x^2 + 2x + 3 = -(1)^2 + 2 \cdot 1 + 3 = -1 + 2 + 3 = 4.$$
The vertex is $(1, 4)$.

We choose some x-values on both sides of the vertex and graph the parabola.

x	y	
1	4	←Vertex
0	3	←y-intercept
−1	0	
2	3	
3	0	

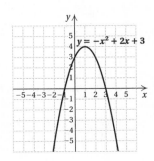

13. $y = -2x^2 - 4x + 1$

Find the vertex. The x-coordinate is
$$-\frac{b}{2a} = -\frac{-4}{2(-2)} = -1.$$
The y-coordinate is
$$y = -2x^2 - 4x + 1 = -2(-1)^2 - 4(-1) + 1 = -2 + 4 + 1 = 3.$$
The vertex is $(-1, 3)$.

We choose some x-values on both sides of the vertex and graph the parabola.

x	y	
−1	3	←Vertex
0	1	←y-intercept
1	−5	
−2	1	
−3	−5	

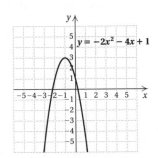

15. $y = 5 - x^2$, or $y = -x^2 + 5$

Find the vertex. The x-coordinate is
$$-\frac{b}{2a} = -\frac{0}{2(-1)} = 0.$$
The y-coordinate is
$$y = 5 - x^2 = 5 - 0^2 = 5.$$
The vertex is $(0, 5)$. This is also the y-intercept.

We choose some x-values on both sides of the vertex and graph the parabola.

x	y	
0	5	←Vertex
−1	4	
−2	1	
1	4	
2	1	

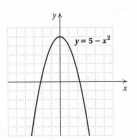

17. $y = \frac{1}{4}x^2$

Find the vertex. The x-coordinate is
$$-\frac{b}{2a} = -\frac{0}{2\left(\frac{1}{4}\right)} = 0.$$
The y-coordinate is
$$y = \frac{1}{4}x^2 = \frac{1}{4} \cdot 0^2 = 0.$$
The vertex is $(0, 0)$. This is also the y-intercept.

We choose some x-values on both sides of the vertex and graph the parabola.

x	y	
0	0	←Vertex
−2	1	
−4	4	
2	1	
4	4	

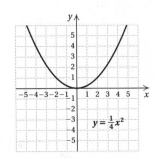

19. $y = -x^2 + x - 1$

Find the vertex. The x-coordinate is
$$-\frac{b}{2a} = -\frac{1}{2(-1)} = -\left(-\frac{1}{2}\right) = \frac{1}{2}.$$
The y-coordinate is
$$y = -x^2 + x - 1 = -\left(\frac{1}{2}\right)^2 + \frac{1}{2} - 1 = -\frac{1}{4} + \frac{1}{2} - 1 = -\frac{3}{4}.$$
The vertex is $\left(\frac{1}{2}, -\frac{3}{4}\right)$.

We choose some x-values on both sides of the vertex and graph the parabola.

x	y	
$\frac{1}{2}$	$-\frac{3}{4}$	←Vertex
0	−1	←y-intercept
−1	−3	
1	−1	
2	−3	

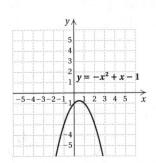

21. $y = -2x^2$

Find the vertex. The x-coordinate is
$$-\frac{b}{2a} = -\frac{0}{2(-2)} = 0.$$
The y-coordinate is
$$y = -2x^2 = -2 \cdot 0^2 = 0.$$
The vertex is $(0,0)$. This is also the y-intercept.

We choose some x-values on both sides of the vertex and graph the parabola.

x	y	
0	0	←Vertex
-1	-2	
-2	-8	
1	-2	
2	-8	

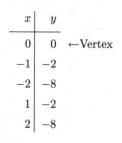

23. $y = x^2 - x - 6$

Find the vertex. The x-coordinate is
$$-\frac{b}{2a} = -\frac{-1}{2 \cdot 1} = -\left(-\frac{1}{2}\right) = \frac{1}{2}.$$
The y-coordinate is
$$y = x^2 - x - 6 = \left(\frac{1}{2}\right)^2 - \frac{1}{2} - 6 = \frac{1}{4} - \frac{1}{2} - 6 = -\frac{25}{4}.$$
The vertex is $\left(\frac{1}{2}, -\frac{25}{4}\right)$.

We choose some x-values on both sides of the vertex and graph the parabola.

x	y	
$\frac{1}{2}$	$-\frac{25}{4}$	←Vertex
0	-6	←y-intercept
-1	-4	
1	-6	
2	-4	

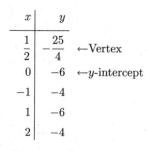

25. $y = x^2 - 2$

To find the x-intercepts we solve the equation $x^2 - 2 = 0$.
$$x^2 - 2 = 0$$
$$x^2 = 2$$
$x = \sqrt{2}$ or $x = -\sqrt{2}$ Principle of square roots

The x-intercepts are $(\sqrt{2}, 0)$ and $(-\sqrt{2}, 0)$.

27. $y = x^2 + 5x$

To find the x-intercepts we solve the equation $x^2 + 5x = 0$.
$$x^2 + 5x = 0$$
$$x(x + 5) = 0$$

$x = 0$ or $x + 5 = 0$
$x = 0$ or $x = -5$

The x-intercepts are $(0, 0)$ and $(-5, 0)$.

29. $y = 8 - x - x^2$

To find the x-intercepts we solve the equation $8 - x - x^2 = 0$.
$$8 - x - x^2 = 0$$
$$x^2 + x - 8 = 0 \qquad \text{Standard form}$$
$$a = 1, \ b = 1, \ c = -8$$
$$x = \frac{-1 \pm \sqrt{1^2 - 4 \cdot 1 \cdot (-8)}}{2 \cdot 1}$$
$$x = \frac{-1 \pm \sqrt{33}}{2}$$
The x-intercepts are $\left(\frac{-1 + \sqrt{33}}{2}, 0\right)$ and $\left(\frac{-1 - \sqrt{33}}{2}, 0\right)$.

31. $y = x^2 - 6x + 9$

To find the x-intercepts we solve the equation $x^2 - 6x + 9 = 0$.
$$x^2 - 6x + 9 = 0$$
$$(x - 3)(x - 3) = 0$$

$x - 3 = 0$ or $x - 3 = 0$
$x = 3$ or $x = 3$

The x-intercept is $(3, 0)$.

33. $y = -x^2 - 4x + 1$

To find the x-intercepts we solve the equation $-x^2 - 4x + 1 = 0$.
$$-x^2 - 4x + 1 = 0$$
$$x^2 + 4x - 1 = 0 \qquad \text{Standard form}$$
$$a = 1, \ b = 4, \ c = -1$$
$$x = \frac{-4 \pm \sqrt{4^2 - 4 \cdot 1 \cdot (-1)}}{2 \cdot 1}$$
$$x = \frac{-4 \pm \sqrt{20}}{2} = \frac{-4 \pm \sqrt{4 \cdot 5}}{2} = \frac{-4 \pm 2\sqrt{5}}{2}$$
$$x = \frac{2(-2 \pm \sqrt{5})}{2} = -2 \pm \sqrt{5}$$
The x-intercepts are $(-2 + \sqrt{5}, 0)$ and $(-2 - \sqrt{5}, 0)$.

35. $y = x^2 + 9$

To find the x-intercepts we solve the equation $x^2 + 9 = 0$.
$$x^2 + 9 = 0$$
$$x^2 = -9$$
The negative number -9 has no real-number square roots. Thus there are no x-intercepts.

37. Discussion and Writing Exercise

39.
$$\sqrt{8} + \sqrt{50} + \sqrt{98} + \sqrt{128}$$
$$= \sqrt{4 \cdot 2} + \sqrt{25 \cdot 2} + \sqrt{49 \cdot 2} + \sqrt{64 \cdot 2}$$
$$= 2\sqrt{2} + 5\sqrt{2} + 7\sqrt{2} + 8\sqrt{2}$$
$$= 22\sqrt{2}$$

41.
$$y = \frac{k}{x}$$

$$12.4 = \frac{k}{2.4} \quad \text{Substituting}$$

$$29.76 = k \quad \text{Variation constant}$$

$$y = \frac{29.76}{x} \quad \text{Equation of variation}$$

43. a) We substitute 128 for H and solve for t:
$$128 = -16t^2 + 96t$$
$$16t^2 - 96t + 128 = 0$$
$$16(t^2 - 6t + 8) = 0$$
$$16(t - 2)(t - 4) = 0$$

$$t - 2 = 0 \quad or \quad t - 4 = 0$$
$$t = 2 \quad or \quad t = 4$$

The projectile is 128 ft from the ground 2 sec after launch and again 4 sec after launch. The graph confirms this.

b) We find the first coordinate of the vertex of the function $H = -16t^2 + 96t$:
$$-\frac{b}{2a} = -\frac{96}{2(-16)} = -\frac{96}{-32} = -(-3) = 3$$

The projectile reaches its maximum height 3 sec after launch. The graph confirms this.

c) We substitute 0 for H and solve for t:
$$0 = -16t^2 + 96t$$
$$0 = -16t(t - 6)$$

$$-16t = 0 \quad or \quad t - 6 = 0$$
$$t = 0 \quad or \quad t = 6$$

At $t = 0$ sec the projectile has not yet been launched. Thus, we use $t = 6$. The projectile returns to the ground 6 sec after launch. The graph confirms this.

45. $y = x^2 + 2x - 3$

$a = 1, b = 2, c = -3$

$b^2 - 4ac = 2^2 - 4 \cdot 1 \cdot (-3) = 4 + 12 = 16$

Since $b^2 - 4ac = 16 > 0$, the equation $x^2 + 2x - 3 = 0$ has two real-number solutions.

47. $y = -0.02x^2 + 4.7x - 2300$

$a = -0.02, b = 4.7, c = -2300$

$b^2 - 4ac = (4.7)^2 - 4(-0.02)(-2300) = 22.09 - 184 = -161.91$

Since $b^2 - 4ac = -161.91 < 0$, the equation $-0.02x^2 + 4.7x - 2300 = 0$ has no real solutions.

Exercise Set 9.7

1. Yes; each member of the domain is matched to only one member of the range.

3. Yes; each member of the domain is matched to only one member of the range.

5. No; a member of the domain is matched to more than one member of the range. In fact, each member of the domain is matched to 3 members of the range.

7. Yes; each member of the domain is matched to only one member of the range.

9. This correspondence is a function, because each class member has only one seat number.

11. This correspondence is a function, because each shape has only one number for its area.

13. This correspondence is not a function, because it is reasonable to assume that at least one person has more than one aunt.

The correspondence is a relation, because it is reasonable to assume that each person has at least one aunt.

15. $f(x) = x + 5$
a) $f(4) = 4 + 5 = 9$
b) $f(7) = 7 + 5 = 12$
c) $f(-3) = -3 + 5 = 2$
d) $f(0) = 0 + 5 = 5$
e) $f(2.4) = 2.4 + 5 = 7.4$
f) $f\left(\frac{2}{3}\right) = \frac{2}{3} + 5 = 5\frac{2}{3}$

17. $h(p) = 3p$
a) $h(-7) = 3(-7) = -21$
b) $h(5) = 3 \cdot 5 = 15$
c) $h(14) = 3 \cdot 14 = 42$
d) $h(0) = 3 \cdot 0 = 0$
e) $h\left(\frac{2}{3}\right) = 3 \cdot \frac{2}{3} = \frac{6}{3} = 2$
f) $h(-54.2) = 3(-54.2) = -162.6$

19. $g(s) = 3s + 4$
a) $g(1) = 3 \cdot 1 + 4 = 3 + 4 = 7$
b) $g(-7) = 3(-7) + 4 = -21 + 4 = -17$
c) $g(6.7) = 3(6.7) + 4 = 20.1 + 4 = 24.1$
d) $g(0) = 3 \cdot 0 + 4 = 0 + 4 = 4$
e) $g(-10) = 3(-10) + 4 = -30 + 4 = -26$
f) $g\left(\frac{2}{3}\right) = 3 \cdot \frac{2}{3} + 4 = 2 + 4 = 6$

21. $f(x) = 2x^2 - 3x$
a) $f(0) = 2 \cdot 0^2 - 3 \cdot 0 = 0 - 0 = 0$
b) $f(-1) = 2(-1)^2 - 3(-1) = 2 + 3 = 5$
c) $f(2) = 2 \cdot 2^2 - 3 \cdot 2 = 8 - 6 = 2$
d) $f(10) = 2 \cdot 10^2 - 3 \cdot 10 = 200 - 30 = 170$
e) $f(-5) = 2(-5)^2 - 3(-5) = 50 + 15 = 65$
f) $f(-10) = 2(-10)^2 - 3(-10) = 200 + 30 = 230$

23. $f(x) = |x| + 1$

   a) $f(0) = |0| + 1 = 0 + 1 = 1$

   b) $f(-2) = |-2| + 1 = 2 + 1 = 3$

   c) $f(2) = |2| + 1 = 2 + 1 = 3$

   d) $f(-3) = |-3| + 1 = 3 + 1 = 4$

   e) $f(-10) = |-10| + 1 = 10 + 1 = 11$

   f) $f(22) = |22| + 1 = 22 + 1 = 23$

25. $f(x) = x^3$

   a) $f(0) = 0^3 = 0$

   b) $f(-1) = (-1)^3 = -1$

   c) $f(2) = 2^3 = 8$

   d) $f(10) = 10^3 = 1000$

   e) $f(-5) = (-5)^3 = -125$

   f) $f(-10) = (-10)^3 = -1000$

27. $F(x) = 2.75x + 71.48$

   a) $F(32) = 2.75(32) + 71.48$
$$= 88 + 71.48$$
$$= 159.48 \text{ cm}$$

   b) $F(30) = 2.75(30) + 71.48$
$$= 82.5 + 71.48$$
$$= 153.98 \text{ cm}$$

29. $\quad P(d) = 1 + \dfrac{d}{33}$

$$P(20) = 1 + \frac{20}{33} = 1\frac{20}{33} \text{ atm}$$

$$P(30) = 1 + \frac{30}{33} = 1\frac{10}{11} \text{ atm}$$

$$P(100) = 1 + \frac{100}{33} = 1 + 3\frac{1}{33} = 4\frac{1}{33} \text{ atm}$$

31. $\quad W(d) = 0.112d$

$\quad W(16) = 0.112(16) = 1.792 \text{ cm}$

$\quad W(25) = 0.112(25) = 2.8 \text{ cm}$

$\quad W(100) = 0.112(100) = 11.2 \text{ cm}$

33. Graph $f(x) = 3x - 1$

Make a list of function values in a table.

When $x = -1$, $f(-1) = 3(-1) - 1 = -3 - 1 = -4$.

When $x = 0$, $f(0) = 3 \cdot 0 - 1 = 0 - 1 = -1$.

When $x = 2$, $f(2) = 3 \cdot 2 - 1 = 6 - 1 = 5$.

x	$f(x)$
-1	-4
0	-1
2	5

Plot these points and connect them.

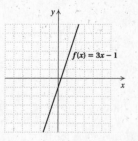

35. Graph $g(x) = -2x + 3$

Make a list of function values in a table.

When $x = -1$, $g(-1) = -2(-1) + 3 = 2 + 3 = 5$.

When $x = 0$, $g(0) = -2 \cdot 0 + 3 = 0 + 3 = 3$.

When $x = 3$, $g(3) = -2 \cdot 3 + 3 = -6 + 3 = -3$.

x	$g(x)$
-1	5
0	3
3	-3

Plot these points and connect them.

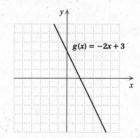

37. Graph $f(x) = \dfrac{1}{2}x + 1$.

Make a list of function values in a table.

When $x = -2$, $f(-2) = \dfrac{1}{2}(-2) + 1 = -1 + 1 = 0$.

When $x = 0$, $f(0) = \dfrac{1}{2} \cdot 0 + 1 = 0 + 1 = 1$.

When $x = 4$, $f(4) = \dfrac{1}{2} \cdot 4 + 1 = 2 + 1 = 3$.

x	$f(x)$
-2	0
0	1
4	3

Plot these points and connect them.

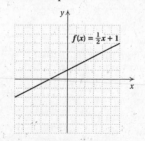

39. Graph $f(x) = 2 - |x|$.

Make a list of function values in a table.

When $x = -4$, $f(-4) = 2 - |-4| = 2 - 4 = -2$.

When $x = 0$, $f(0) = 2 - |0| = 2 - 0 = 2$.

When $x = 3$, $f(3) = 2 - |3| = 2 - 3 = -1$.

x	$f(x)$
-4	-2
0	2
3	-1

Plot these points and connect them.

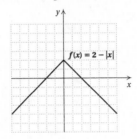

41. Graph $f(x) = x^2$.

Recall from Section 9.6 that the graph is a parabola. Make a list of function values in a table.

When $x = -2$, $f(-2) = (-2)^2 = 4$.

When $x = -1$, $f(-1) = (-1)^2 = 1$.

When $x = 0$, $f(0) = 0^2 = 0$.

When $x = 1$, $f(1) = 1^2 = 1$.

When $x = 2$, $f(2) = 2^2 = 4$.

x	$f(x)$
-2	4
-1	1
0	0
1	1
2	4

Plot these points and connect them.

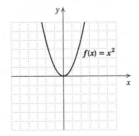

43. Graph $f(x) = x^2 - x - 2$.

Recall from Section 9.6 that the graph is a parabola. Make a list of function values in a table.

When $x = -1$, $f(-1) = (-1)^2 - (-1) - 2 = 1 + 1 - 2 = 0$.

When $x = 0$, $f(0) = 0^2 - 0 - 2 = -2$.

When $x = 1$, $f(1) = 1^2 - 1 - 2 = 1 - 1 - 2 = -2$.

When $x = 2$, $f(2) = 2^2 - 2 - 2 = 4 - 2 - 2 = 0$.

x	$f(x)$
-1	0
0	-2
1	-2
2	0

Plot these points and connect them.

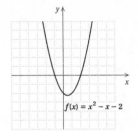

45. We can use the vertical line test:

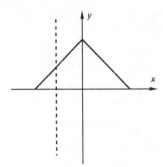

Visualize moving this vertical line across the graph. No vertical line will intersect the graph more than once. Thus, the graph is a graph of a function.

47. We can use the vertical line test:

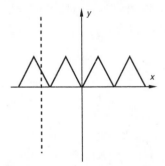

Visualize moving this vertical line across the graph. No vertical line will intersect the graph more than once. Thus, the graph is a graph of a function.

49. We can use the vertical line test.

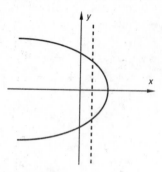

It is possible for a vertical line to intersect the graph more than once. Thus this is not the graph of a function.

51. We can use the vertical line test.

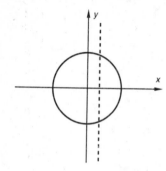

It is possible for a vertical line to intersect the graph more than once. Thus this is not a graph of a function.

53. Locate the point that is directly above 225. Then estimate its second coordinate by moving horizontally from the point to the vertical axis. The rate is about 75 per 10,000 men.

55. Discussion and Writing Exercise

57. The first equation is in slope-intercept form:

$$y = \frac{3}{4}x - 7, \; m = \frac{3}{4}$$

We write the second equation in slope-intercept form.

$$3x + 4y = 7$$
$$4y = -3x + 7$$
$$y = -\frac{3}{4}x + \frac{7}{4}, m = -\frac{3}{4}$$

Since the slopes are different, the equations do not represent parallel lines.

59. $2x - y = 6$, (1)
  $4x - 2y = 5$ (2)

We solve Equation (1) for y.

$2x - y = 6$ (1)

$2x - 6 = y$ Adding y and -6

Substitute $2x - 6$ for y in Equation (2) and solve for x.

$$4x - 2y = 5 \quad (2)$$
$$4x - 2(2x - 6) = 5$$
$$4x - 4x + 12 = 5$$
$$12 = 5$$

We get a false equation, so the system has no solution.

61. Graph $g(x) = x^3$.

Make a list of function values in a table. Then plot the points and connect them.

x	$g(x)$
-2	-8
-1	-1
0	0
1	1
2	8

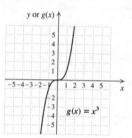

63. Graph $g(x) = |x| + x$.

Make a list of function values in a table. Then plot the points and connect them.

x	$f(x)$
-3	0
-2	0
-1	0
0	0
1	2
2	4
3	6

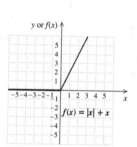

Chapter 9 Review Exercises

1. $8x^2 = 24$
  $x^2 = 3$
  $x = \sqrt{3}$ *or* $x = -\sqrt{3}$
  The solutions are $\sqrt{3}$ and $-\sqrt{3}$.

2. $40 = 5y^2$
  $8 = y^2$
  $y = \sqrt{8}$ *or* $y = -\sqrt{8}$
  $y = 2\sqrt{2}$ *or* $y = -2\sqrt{2}$
  The solutions are $2\sqrt{2}$ and $-2\sqrt{2}$.

3. $5x^2 - 8x + 3 = 0$
  $(5x - 3)(x - 1) = 0$
  $5x - 3 = 0$ *or* $x - 1 = 0$
  $5x = 3$ *or* $x = 1$
  $x = \frac{3}{5}$ *or* $x = 1$
  The solutions are $\frac{3}{5}$ and 1.

4.
$$3y^2 + 5y = 2$$
$$3y^2 + 5y - 2 = 0$$
$$(3y - 1)(y + 2) = 0$$
$$3y - 1 = 0 \quad or \quad y + 2 = 0$$
$$3y = 1 \quad or \quad y = -2$$
$$y = \frac{1}{3} \quad or \quad y = -2$$

The solutions are $\frac{1}{3}$ and -2.

5. $(x + 8)^2 = 13$
$$x + 8 = \sqrt{13} \quad or \quad x + 8 = -\sqrt{13}$$
$$x = -8 + \sqrt{13} \quad or \quad x = -8 - \sqrt{13}$$

The solutions are $-8 \pm \sqrt{13}$.

6. $9x^2 = 0$
$$x^2 = 0$$
$$x = 0$$

The solution is 0.

7. $5t^2 - 7t = 0$
$$t(5t - 7) = 0$$
$$t = 0 \quad or \quad 5t - 7 = 0$$
$$t = 0 \quad or \quad 5t = 7$$
$$t = 0 \quad or \quad t = \frac{7}{5}$$

The solutions are 0 and $\frac{7}{5}$.

8. $x^2 - 2x - 10 = 0$
$$a = 1, \ b = -2, \ c = -10$$
$$x = \frac{-(-2) \pm \sqrt{(-2)^2 - 4 \cdot 1 \cdot (-10)}}{2 \cdot 1}$$
$$x = \frac{2 \pm \sqrt{4 + 40}}{2}$$
$$x = \frac{2 \pm \sqrt{44}}{2} = \frac{2 \pm \sqrt{4 \cdot 11}}{2}$$
$$x = \frac{2 \pm 2\sqrt{11}}{2} = \frac{2(1 \pm \sqrt{11})}{2}$$
$$x = 1 \pm \sqrt{11}$$

The solutions are $1 \pm \sqrt{11}$.

9. $9x^2 - 6x - 9 = 0$
$$a = 9, \ b = -6, \ c = -9$$
$$x = \frac{-(-6) \pm \sqrt{(-6)^2 - 4 \cdot 9 \cdot (-9)}}{2 \cdot 9}$$
$$x = \frac{6 \pm \sqrt{36 + 324}}{18}$$
$$x = \frac{6 \pm \sqrt{360}}{18} = \frac{6 \pm \sqrt{36 \cdot 10}}{18}$$
$$x = \frac{6 \pm 6\sqrt{10}}{18} = \frac{6(1 \pm \sqrt{10})}{3 \cdot 6}$$
$$x = \frac{1 \pm \sqrt{10}}{3}$$

The solutions are $\frac{1 \pm \sqrt{10}}{3}$.

10.
$$x^2 + 6x = 9$$
$$x^2 + 6x - 9 = 0$$
$$a = 1, \ b = 6, \ c = -9$$
$$x = \frac{-6 \pm \sqrt{6^2 - 4 \cdot 1 \cdot (-9)}}{2 \cdot 1}$$
$$x = \frac{-6 \pm \sqrt{36 + 36}}{2}$$
$$x = \frac{-6 \pm \sqrt{72}}{2} = \frac{-6 \pm \sqrt{36 \cdot 2}}{2}$$
$$x = \frac{-6 \pm 6\sqrt{2}}{2} = \frac{2(-3 \pm 3\sqrt{2})}{2}$$
$$x = -3 \pm 3\sqrt{2}$$

The solutions are $-3 \pm 3\sqrt{2}$.

11.
$$1 + 4x^2 = 8x$$
$$4x^2 - 8x + 1 = 0$$
$$a = 4, \ b = -8, \ c = 1$$
$$x = \frac{-(-8) \pm \sqrt{(-8)^2 - 4 \cdot 4 \cdot 1}}{2 \cdot 4}$$
$$x = \frac{8 \pm \sqrt{64 - 16}}{8}$$
$$x = \frac{8 \pm \sqrt{48}}{8} = \frac{8 \pm \sqrt{16 \cdot 3}}{8}$$
$$x = \frac{8 \pm 4\sqrt{3}}{8} = \frac{4(2 \pm \sqrt{3})}{2 \cdot 4}$$
$$x = \frac{2 \pm \sqrt{3}}{2}$$

The solutions are $\frac{2 \pm \sqrt{3}}{2}$.

12. $6 + 3y = y^2$

$0 = y^2 - 3y - 6$

$a = 1,\ b = -3,\ c = -6$

$y = \dfrac{-(-3) \pm \sqrt{(-3)^2 - 4 \cdot 1 \cdot (-6)}}{2 \cdot 1}$

$y = \dfrac{3 \pm \sqrt{9 + 24}}{2} = \dfrac{3 \pm \sqrt{33}}{2}$

The solutions are $\dfrac{3 \pm \sqrt{33}}{2}$.

13. $3m = 4 + 5m^2$

$0 = 5m^2 - 3m + 4$

$a = 5,\ b = -3,\ c = 4$

$m = \dfrac{-(-3) \pm \sqrt{(-3)^2 - 4 \cdot 5 \cdot 4}}{2 \cdot 5}$

$m = \dfrac{3 \pm \sqrt{9 - 80}}{10} = \dfrac{3 \pm \sqrt{-71}}{10}$

Since the radicand is negative, there are no real-number solutions.

14. $3x^2 = 4x$

$3x^2 - 4x = 0$

$x(3x - 4) = 0$

$x = 0 \ \ or \ \ 3x - 4 = 0$

$x = 0 \ \ or \ \ \ \ \ 3x = 4$

$x = 0 \ \ or \ \ \ \ \ \ \ x = \dfrac{4}{3}$

The solutions are 0 and $\dfrac{4}{3}$.

15. $\dfrac{15}{x} - \dfrac{15}{x + 2} = 2$, LCM is $x(x + 2)$

$x(x+2)\left(\dfrac{15}{x} - \dfrac{15}{x+2}\right) = x(x+2)(2)$

$x(x+2) \cdot \dfrac{15}{x} - x(x+2) \cdot \dfrac{15}{x+2} = 2x(x+2)$

$15(x + 2) - 15x = 2x^2 + 4x$

$15x + 30 - 15x = 2x^2 + 4x$

$30 = 2x^2 + 4x$

$0 = 2x^2 + 4x - 30$

$0 = x^2 + 2x - 15$ Dividing by 2

$0 = (x + 5)(x - 3)$

$x + 5 = 0 \ \ or \ \ x - 3 = 0$

$x = -5 \ \ or \ \ \ \ \ x = 3$

Both numbers check. The solutions are -5 and 3.

16. $x + \dfrac{1}{x} = 2$, LCM is x

$x\left(x + \dfrac{1}{x}\right) = x \cdot 2$

$x \cdot x + x \cdot \dfrac{1}{x} = 2x$

$x^2 + 1 = 2x$

$x^2 - 2x + 1 = 0$

$(x - 1)^2 = 0$

$x - 1 = 0$

$x = 1$

The number 1 checks. It is the solution.

17. $x^2 - 5x + \ \ 2 = 0$

$x^2 - 5x \ \ \ \ \ \ = -2$

$x^2 - 5x + \dfrac{25}{4} = -2 + \dfrac{25}{4}$ Adding $\dfrac{25}{4}$: $\left(\dfrac{-5}{2}\right)^2 = \dfrac{25}{4}$

$\left(x - \dfrac{5}{2}\right)^2 = \dfrac{17}{4}$

$x - \dfrac{5}{2} = \dfrac{\sqrt{17}}{2} \ \ or \ \ x - \dfrac{5}{2} = -\dfrac{\sqrt{17}}{2}$

$x = \dfrac{5}{2} + \dfrac{\sqrt{17}}{2} \ \ or \ \ \ \ \ x = \dfrac{5}{2} - \dfrac{\sqrt{17}}{2}$

$x = \dfrac{5 + \sqrt{17}}{2} \ \ or \ \ \ \ \ x = \dfrac{5 - \sqrt{17}}{2}$

The solutions are $\dfrac{5 \pm \sqrt{17}}{2}$.

18. $3x^2 - 2x - \ \ 5 = 0$

$\dfrac{1}{3}(3x^2 - 2x - \ \ 5) = \dfrac{1}{3} \cdot 0$

$x^2 - \dfrac{2}{3}x - \dfrac{5}{3} = 0$

$x^2 - \dfrac{2}{3}x \ \ \ \ \ \ = \dfrac{5}{3}$

$x^2 - \dfrac{2}{3}x + \dfrac{1}{9} = \dfrac{5}{3} + \dfrac{1}{9}$ Adding $\dfrac{1}{9}$:

$\left[\dfrac{1}{2}\left(-\dfrac{2}{3}\right)\right]^2 = \left(-\dfrac{1}{3}\right)^2 = \dfrac{1}{9}$

$\left(x - \dfrac{1}{3}\right)^2 = \dfrac{16}{9}$

$x - \dfrac{1}{3} = \dfrac{4}{3} \ \ or \ \ x - \dfrac{1}{3} = -\dfrac{4}{3}$

$x = \dfrac{5}{3} \ \ or \ \ \ \ \ x = -1$

The solutions are $\dfrac{5}{3}$ and -1.

19. From Exercise 17, we know the solutions are $\dfrac{5 \pm \sqrt{17}}{2}$.

Using a calculator, we have

$\dfrac{5 + \sqrt{17}}{2} \approx 4.6$ and $\dfrac{5 - \sqrt{17}}{2} \approx 0.4$.

20. $4y^2 + 8y + 1 = 0$

$a = 4, b = 8, c = 1$

$y = \dfrac{-8 \pm \sqrt{8^2 - 4 \cdot 4 \cdot 1}}{2 \cdot 4}$

$y = \dfrac{-8 \pm \sqrt{64 - 16}}{8} = \dfrac{-8 \pm \sqrt{48}}{8}$

Using a calculator, we have $\dfrac{-8 + \sqrt{48}}{8} \approx -0.1$ and

$\dfrac{-8 - \sqrt{48}}{8} \approx -1.9$.

21.
$$V = \frac{1}{2}\sqrt{1 + \frac{T}{L}}$$
$$V^2 = \left(\frac{1}{2}\sqrt{1 + \frac{T}{L}}\right)^2$$
$$V^2 = \frac{1}{4}\left(1 + \frac{T}{L}\right)$$
$$4 \cdot V^2 = 4 \cdot \frac{1}{4}\left(1 + \frac{T}{L}\right)$$
$$4V^2 = 1 + \frac{T}{L}$$
$$4V^2 - 1 = \frac{T}{L}$$
$$L(4V^2 - 1) = L \cdot \frac{T}{L}$$
$$L(4V^2 - 1) = T$$

22. $y = 2 - x^2$, or $y = -x^2 + 2$

Find the vertex. The x-coordinate is

$-\dfrac{b}{2a} = -\dfrac{0}{2(-1)} = 0$.

The y-coordinate is

$y = 2 - x^2 = 2 - 0^2 = 2$.

The vertex is $(0, 2)$. This is also the y-intercept.

We choose some x-values on both sides of the vertex and graph the parabola.

x	y	
0	2	←Vertex
−1	1	
−2	−2	
1	1	
2	−2	

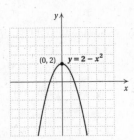

23. $y = x^2 - 4x - 2$

Find the vertex. The x-coordinate is

$-\dfrac{b}{2a} = -\dfrac{-4}{2 \cdot 1} = -(-2) = 2$.

The y-coordinate is

$y = 2^2 - 4 \cdot 2 - 2 = 4 - 8 - 2 = -6$.

The vertex is $(2, -6)$.

We choose some x-values on both sides of the vertex and graph the parabola.

x	y	
2	−6	← Vertex
0	−2	← y-intercept
−1	3	
3	5	
5	3	

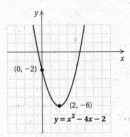

24. $y = 2 - x^2$

To find the x-intercepts we solve the equation $2 - x^2 = 0$.

$2 - x^2 = 0$

$2 = x^2$

$x = \sqrt{2}$ or $x = -\sqrt{2}$

The x-intercepts are $(\sqrt{2}, 0)$ and $(-\sqrt{2}, 0)$.

25. $y = x^2 - 4x - 2$

To find the x-intercepts we solve the equation $x^2 - 4x - 2 = 0$.

$x = \dfrac{-(-4) \pm \sqrt{(-4)^2 - 4 \cdot 1(-2)}}{2 \cdot 1}$

$x = \dfrac{4 \pm \sqrt{16 + 8}}{2} = \dfrac{4 + \sqrt{24}}{2}$

$x = \dfrac{4 \pm \sqrt{4 \cdot 6}}{2} = \dfrac{4 \pm 2\sqrt{6}}{2}$

$x = \dfrac{2(2 \pm \sqrt{6})}{2} = 2 \pm \sqrt{6}$

The x-intercepts are $(2 - \sqrt{6}, 0)$ and $(2 + \sqrt{6}, 0)$.

26. *Familiarize*. Using the labels on the drawing in the text, we let a and $a + 3$ represent the lengths of the legs, in cm.

Translate. We use the Pythagorean equation.

$a^2 + (a + 3)^2 = 5^2$

Solve.

$a^2 + (a + 3)^2 = 5^2$

$a^2 + a^2 + 6a + 9 = 25$

$2a^2 + 6a + 9 = 25$

$2a^2 + 6a - 16 = 0$

$a^2 + 3a - 8 = 0$ Dividing by 2

We use the quadratic formula.

$a = \dfrac{-3 \pm \sqrt{3^2 - 4 \cdot 1 \cdot (-8)}}{2 \cdot 1}$

$a = \dfrac{-3 \pm \sqrt{9 + 32}}{2} = \dfrac{-3 \pm \sqrt{41}}{2}$

Using a calculator, we have

$a = \dfrac{-3 - \sqrt{41}}{2} \approx -4.7$ and $a = \dfrac{-3 + \sqrt{41}}{2} \approx 1.7$.

Check. Since the length of a leg cannot be negative, -4.7 cannot be a solution. If $a \approx 1.7$, then $a + 3 \approx 4.7$ and $(1.7)^2 + (4.7)^4 = 24.98 \approx 25 = 5^2$. The answer checks.

State. The lengths of the legs are about 1.7 cm and 4.7 cm.

27. *Familiarize*. Using the labels on the drawing in the text, we let s and $s - 5$ represent the lengths of the legs, in ft.

Translate. We use the Pythagorean equation.

$$s^2 + (s - 5)^2 = 25^2$$

Solve.

$$s^2 + (s - 5)^2 = 25^2$$
$$s^2 + s^2 - 10s + 25 = 625$$
$$2s^2 - 10s + 25 = 625$$
$$2s^2 - 10s - 600 = 0$$
$$s^2 - 5s - 300 = 0 \quad \text{Dividing by 2}$$
$$(s - 20)(s + 15) = 0$$
$$s - 20 = 0 \quad or \quad s + 15 = 0$$
$$s = 20 \quad or \quad s = -15$$

Check. Since the length of a leg of the triangle cannot be negative, -15 cannot be a solution. If $s = 20$, then $s - 5 = 20 - 5 = 15$ and $20^2 + 15^2 = 625 = 25^2$. The answer checks.

State. The height of the ramp is 15 ft.

28. *Familiarize*. We will use the formula $s = 16t^2$.

Translate. We substitute 645 for s.

$$645 = 16t^2$$

Solve.

$$645 = 16t^2$$
$$\frac{645}{16} = t^2$$
$$40.3125 = t^2$$
$$t = \sqrt{40.3125} \quad or \quad t = -\sqrt{40.3125}$$
$$t \approx 6.3 \qquad or \quad t \approx -6.3$$

Check. Time cannot be negative in this application, so -6.3 cannot be a solution. We check 6.3.

$$16(6.3)^2 = 635.04 \approx 645$$

The answer is close, so we have a check. (Remember that we rounded the value of t.)

State. It would take about 6.3 sec for an object to fall to the ground from the top of Lake Point Towers.

29. $f(x) = 2x - 5$
$$f(2) = 2 \cdot 2 - 5 = 4 - 5 = -1$$
$$f(-1) = 2(-1) - 5 = -2 - 5 = -7$$
$$f(3.5) = 2(3.5) - 5 = 7 - 5 = 2$$

30. $g(x) = |x| - 1$
$$g(1) = |1| - 1 = 1 - 1 = 0$$
$$g(-1) = |-1| - 1 = 1 - 1 = 0$$
$$g(-20) = |-20| - 1 = 20 - 1 = 19$$

31. $C(p) = 15p$
$$C(180) = 15 \cdot 180 = 2700 \text{ calories}$$

32. $g(x) = 4 - x$

We find some function values.

When $x = -1$, $g(-1) = 4 - (-1) = 4 + 1 = 5$

When $x = 0$, $g(0) = 4 - 0 = 4$.

When $x = 3$, $g(3) = 4 - 3 = 1$.

x	$g(x)$
-1	5
0	4
3	1

Plot these points and connect them.

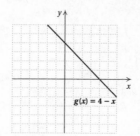

33. $f(x) = x^2 - 3$

Recall from Section 9.6 that the graph is a parabola. We find some function values.

When $x = -3$, $f(-3) = (-3)^2 - 3 = 9 - 3 = 6$.

When $x = -1$, $f(-1) = (-1)^2 - 3 = 1 - 3 = -2$.

When $x = 0$, $f(0) = 0^2 - 3 = 0 - 3 = -3$.

When $x = 1$, $f(1) = 1^2 - 3 = 1 - 3 = -2$.

When $x = 2$, $f(2) = 2^2 - 3 = 4 - 3 = 1$.

x	$f(x)$
-3	6
-1	-2
0	-3
1	-2
2	1

Plot these points and connect them.

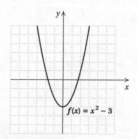

34. $h(x) = |x| - 5$

We find some function values.

When $x = -4$, $h(-4) = |-4| - 5 = 4 - 5 = -1$.

When $x = -2$, $h(-2) = |-2| - 5 = 2 - 5 = -3$.

When $x = 0$, $h(0) = |0| - 5 = 0 - 5 = -5$.

When $x = 1$, $h(1) = |1| - 5 = 1 - 5 = -4$.

When $x = 3$, $h(3) = |3| - 5 = 3 - 5 = -2$.

x	$h(x)$
-4	-1
-2	-3
0	-5
1	-4
3	-2

Plot these points and connect them.

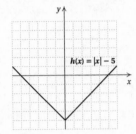

35. $f(x) = x^2 - 2x + 1$

Recall from Section 9.6 that the graph is a parabola.

When $x = -1$, $f(-1) = (-1)^2 - 2(-1) + 1 = 1 + 2 + 1 = 4$.

When $x = 0$, $f(0) = 0^2 - 2 \cdot 0 + 1 = 0 - 0 + 1 = 1$.

When $x = 1$, $f(1) = 1^2 - 2 \cdot 1 + 1 = 1 - 2 + 1 = 0$.

When $x = 2$, $f(2) = 2^2 - 2 \cdot 2 + 1 = 4 - 4 + 1 = 1$.

When $x = 3$, $f(3) = 3^2 - 2 \cdot 3 + 1 = 9 - 6 + 1 = 4$.

x	$f(x)$
-1	4
0	1
1	0
2	1
3	4

Plot these points and connect them.

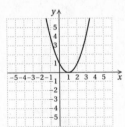

$f(x) = x^2 - 2x + 1$

36. It is possible for a vertical line to intersect the graph more than once, so this is not the graph of a function.

37. No vertical line will intersect the graph more than once, so this is the graph of a function.

38. *Discussion and Writing Exercise.*

Equation	Form	Example
Linear	Equivalent to $x = a$	$3x - 5 = 8$
Quadratic	$ax^2 + bx + c = 0$	$2x^2 - 3x + 1 = 0$
Rational	Contains one or more rational expressions	$\dfrac{x}{3} + \dfrac{4}{x-1} = 1$
Radical	Contains one or more radical expressions	$\sqrt{3x-1} = x - 7$
Systems of equations	$Ax + By = C$, $Dx + Ey = F$	$4x - 5y = 3$, $3x + 2y = 1$

39. *Discussion and Writing Exercise.*

a) The third line should be $x = 0$ *or* $x + 20 = 0$; the solution 0 is lost in the given procedure. Also, the last line should be $x = -20$.

b) The addition principle should be used at the outset to get 0 on one side of the equation. Since this was not done in the given procedure, the principle of zero products was not applied correctly.

40. *Familiarize.* Let $x = $ the first integer. Then $x + 1 = $ the second integer.

Translate. If the numbers are positive, then $(x + 1)^2$ is larger than x^2 and we have:

Square of larger number *minus* square of smaller number is 63.

$$(x+1)^2 - x^2 = 63$$

If the numbers are negative, then x^2 is larger than $(x + 1)^2$ and we have:

Square of smaller number *minus* square of larger number is 63.

$$x^2 - (x+1)^2 = 63$$

Solve. We solve each equation.

$$(x + 1)^2 - x^2 = 63$$
$$x^2 + 2x + 1 - x^2 = 63$$
$$2x + 1 = 63$$
$$2x = 62$$
$$x = 31$$

If $x = 31$, then $x + 1 = 31 + 1 = 32$.

$$x^2 - (x+1)^2 = 63$$
$$x^2 - (x^2 + 2x + 1) = 63$$
$$x^2 - x^2 - 2x - 1 = 63$$
$$-2x - 1 = 63$$
$$-2x = 64$$
$$x = -32$$

If $x = -32$, then $x + 1 = -32 + 1 = -31$.

Check. 31 and 32 are consecutive integers and $32^2 - 31^2 = 1024 - 961 = 63$. Also, -32 and -31 are consecutive integers and $(-32)^2 - (-31)^2 = 1024 - 961 = 63$. Both pairs of numbers check.

State. The integers are 31 and 32 or -32 and -31.

41. Familiarize. The area of a square with side s is s^2; the area of a circle with radius 5 in. is $\pi \cdot 5^2$, or 25π in^2.

Translate.

Area of square equals area of circle.

$$s^2 = 25\pi$$

Solve.
$$s^2 = 25\pi$$
$$s = 5\sqrt{\pi} \quad or \quad s = -5\sqrt{\pi}$$

Check. Since the length of a side of the square cannot be negative, $-5\sqrt{\pi}$ cannot be a solution. If the length of a side of the square is $5\sqrt{\pi}$, then the area of the square is $(5\sqrt{\pi})^2 = 25\pi$. Since this is also the area of the circle, $5\sqrt{\pi}$ checks.

State. $s = 5\sqrt{\pi}$ in. ≈ 8.9 in.

42. $x - 4\sqrt{x} - 5 = 0$

Let $u = \sqrt{x}$. Then $u^2 = x$. Substitute u for $\sqrt{x}$ and u^2 for x and solve for u.
$$u^2 - 4u - 5 = 0$$
$$(u+1)(u-5) = 0$$
$$u + 1 = 0 \quad or \quad u - 5 = 0$$
$$u = -1 \quad or \quad u = 5$$

Now we substitute $\sqrt{x}$ for u and solve for x.

$\sqrt{x} = -1$ has no real-number solutions.

If $u = 5$, then we have:
$$\sqrt{x} = 5$$
$$(\sqrt{x})^2 = 5^2$$
$$x = 25$$

The number 25 checks. It is the solution.

43. The graph of $y = (x+3)^2$ contains the points $(-4,1)$ and $(-2,1)$, so the solutions of $(x+3)^2 = 1$ are -4 and -2.

44. The graph of $y = (x+3)^2$ contains the points $(-5,4)$ and $(-1,4)$, so the solutions of $(x+3)^2 = 4$ are -5 and -1.

45. The graph of $y = (x+3)^2$ contains the points $(-6,9)$ and $(0,9)$, so the solutions of $(x+3)^2 = 9$ are -6 and 0.

46. The graph of $y = (x+3)^2$ contains the point $(-3,0)$, so the solution of $(x+3)^2 = 0$ is -3.

Chapter 9 Test

1. $7x^2 = 35$
$$x^2 = 5$$
$$x = \sqrt{5} \quad or \quad x = -\sqrt{5}$$
The solutions are $\sqrt{5}$ and $-\sqrt{5}$.

2. $7x^2 + 8x = 0$
$$x(7x + 8) = 0$$

$$x = 0 \quad or \quad 7x + 8 = 0$$
$$x = 0 \quad or \quad 7x = -8$$
$$x = 0 \quad or \quad x = -\frac{8}{7}$$
The solutions are 0 and $-\frac{8}{7}$.

3. $48 = t^2 + 2t$
$$0 = t^2 + 2t - 48$$
$$0 = (t+8)(t-6)$$
$$t + 8 = 0 \quad or \quad t - 6 = 0$$
$$t = -8 \quad or \quad t = 6$$
The solutions are -8 and 6.

4. $3y^2 - 5y = 2$
$$3y^2 - 5y - 2 = 0$$
$$(3y+1)(y-2) = 0$$
$$3y + 1 = 0 \quad or \quad y - 2 = 0$$
$$3y = -1 \quad or \quad y = 2$$
$$y = -\frac{1}{3} \quad or \quad y = 2$$
The solutions are $-\frac{1}{3}$ and 2.

5. $(x-8)^2 = 13$
$$x - 8 = \sqrt{13} \quad or \quad x - 8 = -\sqrt{13}$$
$$x = 8 + \sqrt{13} \quad or \quad x = 8 - \sqrt{13}$$
The solutions are $8 \pm \sqrt{13}$.

6. $x^2 = x + 3$
$$x^2 - x - 3 = 0$$
$$a = 1, b = -1, c = -3$$
$$x = \frac{-(-1) \pm \sqrt{(-1)^2 - 4 \cdot 1 \cdot (-3)}}{2 \cdot 1}$$
$$x = \frac{1 \pm \sqrt{1 + 12}}{2}$$
$$x = \frac{1 \pm \sqrt{13}}{2}$$
The solutions are $\frac{1 \pm \sqrt{13}}{2}$.

7. $m^2 - 3m = 7$
$$m^2 - 3m - 7 = 0$$
$$a = 1, b = -3, c = -7$$
$$m = \frac{-(-3) \pm \sqrt{(-3)^2 - 4 \cdot 1 \cdot (-7)}}{2 \cdot 1}$$
$$m = \frac{3 \pm \sqrt{9 + 28}}{2}$$
$$m = \frac{3 \pm \sqrt{37}}{2}$$
The solutions are $\frac{3 \pm \sqrt{37}}{2}$.

8. $10 = 4x + x^2$

$0 = x^2 + 4x - 10$

$a = 1,\ b = 4,\ c = -10$

$x = \dfrac{-4 \pm \sqrt{4^2 - 4 \cdot 1 \cdot (-10)}}{2 \cdot 1}$

$x = \dfrac{-4 \pm \sqrt{16 + 40}}{2} = \dfrac{-4 \pm \sqrt{56}}{2}$

$x = \dfrac{-4 \pm \sqrt{4 \cdot 14}}{2} = \dfrac{-4 \pm 2\sqrt{14}}{2}$

$x = \dfrac{2(-2 \pm \sqrt{14})}{2} = -2 \pm \sqrt{14}$

The solutions are $-2 \pm \sqrt{14}$.

9. $3x^2 - 7x + 1 = 0$

$a = 3,\ b = -7,\ c = 1$

$x = \dfrac{-(-7) \pm \sqrt{(-7)^2 - 4 \cdot 3 \cdot 1}}{2 \cdot 3}$

$x = \dfrac{7 \pm \sqrt{49 - 12}}{6} = \dfrac{7 \pm \sqrt{37}}{6}$

The solutions are $\dfrac{7 \pm \sqrt{37}}{6}$.

10. $x - \dfrac{2}{x} = 1$, LCM is x

$x\left(x - \dfrac{2}{x}\right) = x \cdot 1$

$x \cdot x - x \cdot \dfrac{2}{x} = x$

$x^2 - 2 = x$

$x^2 - x - 2 = 0$

$(x - 2)(x + 1) = 0$

$x - 2 = 0\ \ or\ \ x + 1 = 0$

$x = 2\ \ or\ \ \ \ \ \ \ x = -1$

Both numbers check. The solutions are 2 and -1.

11. $\dfrac{4}{x} - \dfrac{4}{x + 2} = 1$, LCM is $x(x+2)$

$x(x + 2)\left(\dfrac{4}{x} - \dfrac{4}{x + 2}\right) = x(x + 2) \cdot 1$

$x(x + 2) \cdot \dfrac{4}{x} - x(x + 2) \cdot \dfrac{4}{x + 2} = x(x + 2)$

$4(x + 2) - 4x = x^2 + 2x$

$4x + 8 - 4x = x^2 + 2x$

$8 = x^2 + 2x$

$0 = x^2 + 2x - 8$

$0 = (x + 4)(x - 2)$

$x + 4 = 0\ \ or\ \ x - 2 = 0$

$x = -4\ \ or\ \ \ \ \ \ \ x = 2$

Both numbers check. The solutions are -4 and 2.

12. $x^2 - 4x - 10 = 0$

$x^2 - 4x\ \ \ \ \ \ = 10$

$x^2 - 4x +\ \ \ 4 = 10 + 4$ Adding 4: $\left(\dfrac{-4}{2}\right)^2 = (-2)^2 = 4$

$(x - 2)^2 = 14$

$x - 2 = \sqrt{14}\ \ \ \ or\ \ x - 2 = -\sqrt{14}$

$x = 2 + \sqrt{14}\ \ or\ \ \ \ \ \ \ x = 2 - \sqrt{14}$

The solutions are $2 \pm \sqrt{14}$.

13. From Exercise 12 we know that the solutions of the equation are $2 \pm \sqrt{14}$.

Using a calculator, we have

$2 - \sqrt{14} \approx -1.7$ and $2 + \sqrt{14} \approx 5.7$.

14. $d = an^2 + bn$

$0 = an^2 + bn - d$

We will use the quadratic formula with $a = a$, $b = b$, and $c = -d$.

$n = \dfrac{-b \pm \sqrt{b^2 - 4 \cdot a \cdot (-d)}}{2 \cdot a}$

$n = \dfrac{-b + \sqrt{b^2 + 4ad}}{2a}$ Using the positive square root

15. To find the x-intercepts we solve the following equation.

$-x^2 + x + 5 = 0$

$x^2 - x - 5 = 0$ Standard form

$a = 1,\ b = -1,\ c = -5$

$x = \dfrac{-(-1) \pm \sqrt{(-1)^2 - 4 \cdot 1 \cdot (-5)}}{2 \cdot 1}$

$x = \dfrac{1 \pm \sqrt{1 + 20}}{2} = \dfrac{1 \pm \sqrt{21}}{2}$

The x-intercepts are $\left(\dfrac{1 - \sqrt{21}}{2}, 0\right)$ and $\left(\dfrac{1 + \sqrt{21}}{2}, 0\right)$.

16. $y = 4 - x^2$, or $y = -x^2 + 4$

Find the vertex. The x-coordinate is

$-\dfrac{b}{2a} = -\dfrac{0}{2(-1)} = 0.$

The y-coordinate is

$y = 4 - x^2 = 4 - 0^2 = 4.$

The vertex is $(0, 4)$. This is also the y-intercept.

We choose some x-values on both sides of the vertex and graph the parabola.

x	y	
0	4	←Vertex
-1	3	
-2	0	
1	3	
2	0	

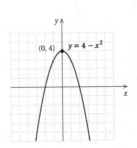

17. $y = -x^2 + x + 5$

Find the vertex. The x-coordinate is

$$-\frac{b}{2a} = -\frac{1}{2(-1)} = -\left(-\frac{1}{2}\right) = \frac{1}{2}.$$

The y-coordinate is

$$y = -\left(\frac{1}{2}\right)^2 + \frac{1}{2} + 5 = -\frac{1}{4} + \frac{1}{2} + 5 = \frac{21}{4}.$$

The vertex is $\left(\frac{1}{2}, \frac{21}{4}\right)$.

We choose some x-values on both sides of the vertex and graph the parabola.

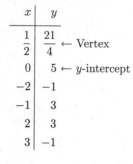

x	y
$\frac{1}{2}$	$\frac{21}{4}$ $\leftarrow$ Vertex
0	5 $\leftarrow$ y-intercept
-2	-1
-1	3
2	3
3	-1

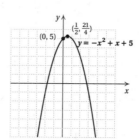

18. $f(x) = \frac{1}{2}x + 1$

$$f(0) = \frac{1}{2} \cdot 0 + 1 = 0 + 1 = 1$$

$$f(1) = \frac{1}{2} \cdot 1 + 1 = \frac{1}{2} + 1 = 1\frac{1}{2}$$

$$f(2) = \frac{1}{2} \cdot 2 + 1 = 1 + 1 = 2$$

19. $g(t) = -2|t| + 3$

$$g(-1) = -2|-1| + 3 = -2 \cdot 1 + 3 = -2 + 3 = 1$$

$$g(0) = -2|0| + 3 = -2 \cdot 0 + 3 = 0 + 3 = 3$$

$$g(3) = -2|3| + 3 = -2 \cdot 3 + 3 = -6 + 3 = -3$$

20. *Familiarize.* Using the labels on the drawing in the text, we let l and $l-4$ represent the length and width of the rug, respectively, in meters. Recall that the area of a rectangle is (length) $\times$ (width).

Translate.

$$\underbrace{\text{The area}}_{l(l-4)} \ \underbrace{\text{is}}_{=} \ \underbrace{16.25 \text{ m}^2}_{16.25}.$$

Solve.

$$l(l-4) = 16.25$$
$$l^2 - 4l = 16.25$$
$$l^2 - 4l - 16.25 = 0$$

The factorization of $l^2 - 4l - 16.25$ is not readily apparent so we will use the quadratic formula with $a = 1$, $b = -4$, and $c = -16.25$.

$$l = \frac{-(-4) \pm \sqrt{(-4)^2 - 4 \cdot 1 \cdot (-16.25)}}{2 \cdot 1}$$

$$l = \frac{4 \pm \sqrt{16 + 65}}{2} = \frac{4 \pm \sqrt{81}}{2}$$

$$l = \frac{4 \pm 9}{2}$$

$$l = \frac{4 - 9}{2} = \frac{-5}{2} = -\frac{5}{2}, \text{ or } -2.5$$

or

$$l = \frac{4 + 9}{2} = \frac{13}{2} = 6.5$$

Check. Since the length cannot be negative, -2.5 cannot be a solution. If $l = 6.5$, then $l - 4 = 6.5 - 4 = 2.5$ and $6.5(2.5) = 16.25$. Thus, the width is 4 m less than the length and the area is 16.25 m^2. The answer checks.

State. The length of the rug is 6.5 m, and the width is 2.5 m.

21. *Familiarize.* Let $r =$ the speed of the boat in still water. Then $r - 2 =$ the speed upstream and $r + 2 =$ the speed downstream. We organize the information in a table.

	d	r	t
Upstream	44	$r - 2$	t_1
Downstream	52	$r + 2$	t_2

Translate. Using $t = d/r$ and the rows of the table, we have

$$t_1 = \frac{44}{r-2} \text{ and } t_2 = \frac{52}{r+2}.$$

Since the total time is 4 hr, $t_1 + t_2 = 4$, and we have

$$\frac{44}{r-2} + \frac{52}{r+2} = 4.$$

Solve. We solve the equation. We multiply by $(r-2)(r+2)$, the LCM of the denominators.

$$(r-2)(r+2)\left(\frac{44}{r-2} + \frac{52}{r+2}\right) = (r-2)(r+2) \cdot 4$$
$$44(r+2) + 52(r-2) = 4(r^2 - 4)$$
$$44r + 88 + 52r - 104 = 4r^2 - 16$$
$$96r - 16 = 4r^2 - 16$$
$$0 = 4r^2 - 96r$$
$$0 = 4r(r - 24)$$

$$4r = 0 \ \ or \ \ r - 24 = 0$$
$$r = 0 \ \ or \ \ \ \ \ \ \ \ r = 24$$

Check. The boat cannot travel upstream if its speed in still water is 0 km/h. If the speed of the boat in still water is 24 km/h, then it travels at a speed of $24 - 2$, or 22 km/h, upstream and $24 + 2$, or 26 km/h, downstream. At 22 km/h the boat travels 44 km in 44/22, or 2 hr. At 26 km/h it travels 52 km in 52/26, or 2 hr. The total time is $2 + 2$, or 4 hr, so the answer checks.

State. The speed of the boat in still water is 24 km/h.

22. In 2010, $t = 2010 - 1940 = 70$.

$$R(70) = 30.18 - 0.06(70) = 30.18 - 4.2 = 25.98$$

We predict that the record will be 25.98 min in 2010.

23. $h(x) = x - 4$

We find some function values.

When $x = -1$, $h(x) = -1 - 4 = -5$.

When $x = 2$, $h(x) = 2 - 4 = -2$.

When $x = 5$, $h(x) = 5 - 4 = 1$.

x	$h(x)$
-1	-5
2	-2
5	1

Plot these points and connect them.

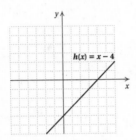

24. $g(x) = x^2 - 4$

Recall from Section 9.6 that the graph is a parabola. We find some function values.

When $x = -3$, $g(x) = (-3)^2 - 4 = 9 - 4 = 5$.

When $x = -1$, $g(x) = (-1)^2 - 4 = 1 - 4 = -3$.

When $x = 0$, $g(x) = 0^2 - 4 = 0 - 4 = -4$.

When $x = 2$, $g(2) = 2^2 - 4 = 4 - 4 = 0$.

When $x = 3$, $g(3) = 3^2 - 4 = 9 - 4 = 5$.

x	$g(x)$
-3	5
-1	-3
0	-4
2	0
3	5

Plot these points and connect them.

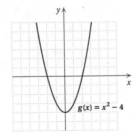

25. No vertical line will intersect the graph more than once, so this is the graph of a function.

26. It is possible for a vertical line to intersect the graph more than once, so this is not the graph of a function.

27. *Familiarize*. We make a drawing. Let $s =$ the length of a side of the square, in feet. Then $s + 5 =$ the length of a diagonal.

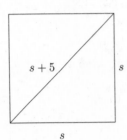

Translate. We use the Pythagorean equation.

$$s^2 + s^2 = (s + 5)^2$$

Solve.

$$s^2 + s^2 = (s + 5)^2$$
$$2s^2 = s^2 + 10s + 25$$
$$s^2 - 10s - 25 = 0$$

We use the quadratic formula with $a = 1$, $b = -10$, and $c = -25$.

$$s = \frac{-(-10) \pm \sqrt{(-10)^2 - 4 \cdot 1 \cdot (-25)}}{2 \cdot 1}$$

$$s = \frac{10 \pm \sqrt{100 + 100}}{2} = \frac{10 \pm \sqrt{200}}{2}$$

$$s = \frac{10 \pm \sqrt{100 \cdot 2}}{2} = \frac{10 \pm 10\sqrt{2}}{2}$$

$$s = \frac{2(5 \pm 5\sqrt{2})}{2} = 5 \pm 5\sqrt{2}$$

Check. Since $5 - 5\sqrt{2}$ is negative, it cannot be the length of a side of the square. If the length of a side is $5 + 5\sqrt{2}$ ft, then $(5+5\sqrt{2})^2 + (5+5\sqrt{2})^2 = 25 + 50\sqrt{2} + 50 + 25 + 50\sqrt{2} + 50 = 150 + 100\sqrt{2}$. The length of a diagonal is $5 + 5\sqrt{2} + 5$ ft, or $10 + 5\sqrt{2}$, and $(10 + 5\sqrt{2})^2 = 100 + 100\sqrt{2} + 50 = 150 + 100\sqrt{2}$. Since these lengths satisfy the Pythagorean equation, the answer checks.

State. The length of a side of the square is $5 + 5\sqrt{2}$ ft.

28. $x - y = 2$,   (1)

    $xy = 4$ (2)

Solve Equation (1) for y.

$$x - y = 2$$
$$-y = -x + 2$$
$$y = x - 2$$

Substitute $x - 2$ for y in Equation (2) and solve for x.

$$xy = 4$$
$$x(x - 2) = 4$$
$$x^2 - 2x = 4$$
$$x^2 - 2x - 4 = 0$$

We use the quadratic formula with $a = 1$, $b = -2$, and $c = -4$.

$$x = \frac{-(-2) \pm \sqrt{(-2)^2 - 4 \cdot 1 \cdot (-4)}}{2 \cdot 1}$$

$$x = \frac{2 \pm \sqrt{4 + 16}}{2} = \frac{2 \pm \sqrt{20}}{2}$$

$$x = \frac{2 \pm \sqrt{4 \cdot 5}}{2} = \frac{2 \pm 2\sqrt{5}}{2}$$

$$x = \frac{2(1 \pm \sqrt{5})}{2} = 1 \pm \sqrt{5}$$

We are asked to find only x, so we stop here.

Cumulative Review Chapters 1 - 9

1. x^3 means $x \cdot x \cdot x$.

2. $(x-3)^2 + 5 = (10-3)^2 + 5 = 7^2 + 5 = 49 + 5 = 54$

3. First we divide to find decimal notation for $\frac{3}{11}$.

```
     0.2 7 2 7
11 |3.0 0 0 0
     2 2
     ‾‾‾
       8 0
       7 7
       ‾‾‾
         3 0
         2 2
         ‾‾‾
           8 0
           7 7
           ‾‾‾
             3
```

We have a repeating decimal.

$\frac{3}{11} = 0.\overline{27}$, so $-\frac{3}{11} = -0.\overline{27}$.

4. $15 = 3 \cdot 5$

$48 = 2 \cdot 2 \cdot 2 \cdot 2 \cdot 3$

LCM $= 2 \cdot 2 \cdot 2 \cdot 2 \cdot 3 \cdot 5$, or 240

5. This distance of -7 from 0 is 7, so $|-7| = 7$.

6. $-6 + 12 + (-4) + 7 = 6 + (-4) + 7$
$$= 2 + 7$$
$$= 9$$

7. $2.8 - (-12.2) = 2.8 + 12.2 = 15$

8. $-\frac{3}{8} \div \frac{5}{2} = -\frac{3}{8} \cdot \frac{2}{5} = -\frac{3 \cdot 2}{8 \cdot 5} = -\frac{3 \cdot \cancel{2}}{\cancel{2} \cdot 4 \cdot 5} = -\frac{3}{20}$

9. $13 \cdot 6 \div 3 \cdot 2 \div 13 = 78 \div 3 \cdot 2 \div 13$
$$= 26 \cdot 2 \div 13$$
$$= 52 \div 13$$
$$= 4$$

10. $4m + 9 - (6m + 13) = 4m + 9 - 6m - 13 = -2m - 4$

11. $3x = -24$
$$\frac{3x}{3} = \frac{-24}{3}$$
$$x = -8$$

The solution is -8.

12. $3x + 7 = 2x - 5$
$$3x + 7 - 2x = 2x - 5 - 2x$$
$$x + 7 = -5$$
$$x + 7 - 7 = -5 - 7$$
$$x = -12$$

The solution is -12.

13. $3(y-1) - 2(y+2) = 0$
$$3y - 3 - 2y - 4 = 0$$
$$y - 7 = 0$$
$$y = 7$$

The solution is 7.

14. $x^2 - 8x + 15 = 0$
$$(x-3)(x-5) = 0$$
$$x - 3 = 0 \quad or \quad x - 5 = 0$$
$$x = 3 \quad or \quad x = 5$$

The solutions are 3 and 5.

15. $y - x = 1, \quad (1)$
$y = 3 - x \quad (2)$

Substitute $3 - x$ for y in Equation (1) and solve for x.

$$y - x = 1$$
$$3 - x - x = 1$$
$$3 - 2x = 1$$
$$-2x = -2$$
$$x = 1$$

Substitute 1 for x in one of the original equations and find y. We use Equation (2).

$$y = 3 - x = 3 - 1 = 2$$

The ordered pair $(1, 2)$ checks in both equations. It is the solution.

16. $x + y = 17, \quad (1)$
$\underline{x - y = 7 \quad (2)}$
$2x \quad\quad = 24 \quad$ Adding
$x = 12$

Substitute 12 for x in one of the original equations and solve for y.

$$x + y = 17 \quad (1)$$
$$12 + y = 17$$
$$y = 5$$

The ordered pair $(12, 5)$ checks in both equations. It is the solution.

17. $4x - 3y = 3, \quad (1)$
$3x - 2y = 4 \quad (2)$

First we multiply Equation (1) by 2 and Equation (2) by -3 and then add.

$$8x - 6y = 6$$
$$\underline{-9x + 6y = -12}$$
$$-x \quad\quad = -6$$
$$x = 6$$

Substitute 6 for x in one of the original equations and solve for y.

$$3x - 2y = 4 \qquad (2)$$
$$3 \cdot 6 - 2y = 4$$
$$18 - 2y = 4$$
$$-2y = -14$$
$$y = 7$$

The ordered pair $(6, 7)$ checks in both equations. It is the solution.

18.
$$x^2 - x - 6 = 0$$
$$(x - 3)(x + 2) = 0$$
$$x - 3 = 0 \quad or \quad x + 2 = 0$$
$$x = 3 \quad or \qquad x = -2$$

The solutions are 3 and -2.

19.
$$x^2 + 3x = 5$$
$$x^2 + 3x - 5 = 0$$

We will use the quadratic formula with $a = 1$, $b = 3$, and $c = -5$.

$$x = \frac{-3 \pm \sqrt{3^2 - 4 \cdot 1 \cdot (-5)}}{2 \cdot 1}$$
$$x = \frac{-3 \pm \sqrt{9 + 20}}{2} = \frac{-3 \pm \sqrt{29}}{2}$$

20.
$$3 - x = \sqrt{x^2 - 3}$$
$$(3 - x)^2 = (\sqrt{x^2 - 3})^2$$
$$9 - 6x + x^2 = x^2 - 3$$
$$9 - 6x = -3$$
$$-6x = -12$$
$$x = 2$$

The number 2 checks. It is the solution.

21.
$$5 - 9x \leq 19 + 5x$$
$$5 - 14x \leq 19$$
$$-14x \leq 14$$
$$\frac{-14x}{-14} \geq \frac{14}{-14} \qquad \text{Reversing the inequality symbol}$$
$$x \geq -1$$

The solution set is $\{x | x \geq -1\}$.

22.
$$-\frac{7}{8}x + 7 = \frac{3}{8}x - 3, \text{ LCM is 8}$$
$$8\left(-\frac{7}{8}x + 7\right) = 8\left(\frac{3}{8}x - 3\right)$$
$$8\left(-\frac{7}{8}x\right) + 8 \cdot 7 = 8 \cdot \frac{3}{8}x - 8 \cdot 3$$
$$-7x + 56 = 3x - 24$$
$$-10x + 56 = -24$$
$$-10x = -80$$
$$x = 8$$

The solution is 8.

23.
$$0.6x - 1.8 = 1.2x$$
$$10(0.6x - 1.8) = 10(1.2x) \quad \text{Clearing decimals}$$
$$6x - 18 = 12x$$
$$-18 = 6x$$
$$-3 = x$$

The solution is -3.

24.
$$-3x > 24$$
$$\frac{-3x}{-3} < \frac{24}{-3} \quad \text{Reversing the inequality symbol}$$
$$x < -8$$

The solution set is $\{x | x < -8\}$.

25.
$$23 - 19y - 3y \geq -12$$
$$23 - 22y \geq -12$$
$$-22y \geq -35$$
$$\frac{-22y}{-22} \leq \frac{-35}{-22} \quad \text{Reversing the inequality symbol}$$
$$y \leq \frac{35}{22}$$

The solution set is $\left\{y \left| y \leq \frac{35}{22}\right.\right\}$.

26.
$$3y^2 = 30$$
$$y^2 = 10$$
$$y = \sqrt{10} \text{ or } y = -\sqrt{10}$$

The solutions are $\sqrt{10}$ and $-\sqrt{10}$.

27. $(x - 3)^2 = 6$
$$x - 3 = \sqrt{6} \qquad or \quad x - 3 = -\sqrt{6}$$
$$x = 3 + \sqrt{6} \quad or \qquad x = 3 - \sqrt{6}$$

The solutions are $3 \pm \sqrt{6}$.

28.
$$\frac{6x - 2}{2x - 1} = \frac{9x}{3x + 1},$$
$$\text{LCM is } (2x - 1)(3x + 1)$$
$$(2x-1)(3x+1) \cdot \frac{6x-2}{2x-1} = (2x - 1)(3x + 1) \cdot \frac{9x}{3x + 1}$$
$$(3x + 1)(6x - 2) = 9x(2x - 1)$$
$$18x^2 - 2 = 18x^2 - 9x$$
$$-2 = -9x$$
$$\frac{2}{9} = x$$

The number $\frac{2}{9}$ checks. It is the solution.

29.
$$\frac{2x}{x+1} = 2 - \frac{5}{2x}, \text{ LCM is } 2x(x+1)$$

$$2x(x+1) \cdot \frac{2x}{x+1} = 2x(x+1)\left(2 - \frac{5}{2x}\right)$$

$$2x \cdot 2x = 2x(x+1)(2) - 2x(x+1) \cdot \frac{5}{2x}$$

$$4x^2 = 4x(x+1) - 5(x+1)$$

$$4x^2 = 4x^2 + 4x - 5x - 5$$

$$4x^2 = 4x^2 - x - 5$$

$$0 = -x - 5$$

$$x = -5$$

The number -5 checks. It is the solution.

30.
$$\frac{2x}{x+3} + \frac{6}{x} + 7 = \frac{18}{x^2 + 3x}$$

$$\frac{2x}{x+3} + \frac{6}{x} + 7 = \frac{18}{x(x+3)}, \text{ LCM is } x(x+3)$$

$$x(x+3)\left(\frac{2x}{x+3} + \frac{6}{x} + 7\right) = x(x+3) \cdot \frac{18}{x(x+3)}$$

$$2x \cdot x + 6(x+3) + 7x(x+3) = 18$$

$$2x^2 + 6x + 18 + 7x^2 + 21x = 18$$

$$9x^2 + 27x + 18 = 18$$

$$9x^2 + 27x = 0$$

$$9x(x+3) = 0$$

$$9x = 0 \quad or \quad x+3 = 0$$

$$x = 0 \quad or \quad \quad x = -3$$

Since both 0 and -3 make a denominator 0 in the original equation, there is no solution.

31.
$$\sqrt{x+9} = \sqrt{2x-3}$$

$$(\sqrt{x+9})^2 = (\sqrt{2x-3})^2$$

$$x + 9 = 2x - 3$$

$$-x + 9 = -3$$

$$-x = -12$$

$$x = 12$$

The number 12 checks. It is the solution.

32.
$$A = \frac{4b}{t}$$

$$t \cdot A = t \cdot \frac{4b}{t}$$

$$At = 4b$$

$$\frac{At}{4} = b$$

33.
$$\frac{1}{t} = \frac{1}{m} - \frac{1}{n}$$

$$mnt \cdot \frac{1}{t} = mnt \cdot \left(\frac{1}{m} - \frac{1}{n}\right)$$

$$mn = mnt \cdot \frac{1}{m} - mnt \cdot \frac{1}{n}$$

$$mn = nt - mt$$

$$mn + mt = nt$$

$$m(n+t) = nt$$

$$m = \frac{nt}{n+t}$$

34.
$$r = \sqrt{\frac{A}{\pi}}$$

$$r^2 = \left(\sqrt{\frac{A}{\pi}}\right)^2$$

$$r^2 = \frac{A}{\pi}$$

$$\pi \cdot r^2 = \pi \cdot \frac{A}{\pi}$$

$$\pi r^2 = A$$

35.
$$y = ax^2 - bx$$

$$0 = ax^2 - bx - y$$

We will use the quadratic formula with $a = a$, $b = -b$, and $c = -y$.

$$x = \frac{-(-b) \pm \sqrt{(-b)^2 - 4 \cdot a \cdot (-y)}}{2 \cdot a}$$

$$x = \frac{b + \sqrt{b^2 + 4ay}}{2a} \quad \text{Taking the positive square root}$$

36. $x^{-6} \cdot x^2 = x^{-6+2} = x^{-4} = \dfrac{1}{x^4}$

37. $\dfrac{y^3}{y^{-4}} = y^{3-(-4)} = y^{3+4} = y^7$

38. $(2y^6)^2 = 2^2(y^6)^2 = 4y^{6 \cdot 2} = 4y^{12}$

39. $2x - 3 + 5x^3 - 2x^3 + 7x^3 + x = 10x^3 + 3x - 3$

40. $(4x^3 + 3x^2 - 5) + (3x^3 - 5x^2 + 4x - 12) = 7x^3 - 2x^2 + 4x - 17$

41. $(6x^2 - 4x + 1) - (-2x^2 + 7) = 6x^2 - 4x + 1 + 2x^2 - 7 = 8x^2 - 4x - 6$

42. $-2y^2(4y^2 - 3y + 1) = -2y^2 \cdot 4y^2 - (-2y^2)(3y) - 2y^2 \cdot 1 = -8y^4 - (-6y^3) - 2y^2 = -8y^4 + 6y^3 - 2y^2$

43. $(2t-3)(3t^2 - 4t + 2) = (2t-3) \cdot 3t^2 - (2t-3) \cdot 4t + (2t-3) \cdot 2$
$$= 6t^3 - 9t^2 - (8t^2 - 12t) + 4t - 6$$
$$= 6t^3 - 9t^2 - 8t^2 + 12t + 4t - 6$$
$$= 6t^3 - 17t^2 + 16t - 6$$

44. $\left(t - \dfrac{1}{4}\right)\left(t + \dfrac{1}{4}\right) = t^2 - \left(\dfrac{1}{4}\right)^2 = t^2 - \dfrac{1}{16}$

45. $(3m - 2)^2 = (3m)^2 - 2 \cdot 3m \cdot 2 + 2^2 = 9m^2 - 12m + 4$

46. $(15x^2y^3 + 10xy^2 + 5) - (5xy^2 - x^2y^2 - 2)$
$= 15x^2y^3 + 10xy^2 + 5 - 5xy^2 + x^2y^2 + 2$
$= 15x^2y^3 + x^2y^2 + 5xy^2 + 7$

47. $(x^2 - 0.2y)(x^2 + 0.2y) = (x^2)^2 - (0.2y)^2 = x^4 - 0.04y^2$

48. $(3p + 4q^2)^2 = (3p)^2 + 2 \cdot 3p \cdot 4q^2 + (4q^2)^2 =$
$9p^2 + 24pq^2 + 16q^4$

49. $\dfrac{4}{2x - 6} \cdot \dfrac{x - 3}{x + 3} = \dfrac{4(x + 3)}{(2x - 6)(x + 3)}$
$= \dfrac{4(x + 3)}{2(x - 3)(x + 3)}$
$= \dfrac{2 \cdot 2 \cdot \cancel{(x + 3)}}{2(x - 3)\cancel{(x + 3)}}$
$= \dfrac{2}{x - 3}$

50. $\dfrac{3a^4}{a^2 - 1} \div \dfrac{2a^3}{a^2 - 2a + 1} = \dfrac{3a^4}{a^2 - 1} \cdot \dfrac{a^2 - 2a + 1}{2a^3}$
$= \dfrac{3a^4(a^2 - 2a + 1)}{(a^2 - 1)(2a^3)}$
$= \dfrac{3 \cdot a^3 \cdot a \cdot (a - 1)(a - 1)}{(a + 1)(a - 1) \cdot 2 \cdot a^3}$
$= \dfrac{3 \cdot \cancel{a^3} \cdot a \cancel{(a - 1)}(a - 1)}{(a + 1)\cancel{(a - 1)} \cdot 2 \cdot \cancel{a^3}}$
$= \dfrac{3a(a - 1)}{2(a + 1)}$

51. $\dfrac{3}{3x - 1} + \dfrac{4}{5x} = \dfrac{3}{3x - 1} \cdot \dfrac{5x}{5x} + \dfrac{4}{5x} \cdot \dfrac{3x - 1}{3x - 1}$
$= \dfrac{3 \cdot 5x}{5x(3x - 1)} + \dfrac{4(3x - 1)}{5x(3x - 1)}$
$= \dfrac{15x}{5x(3x - 1)} + \dfrac{12x - 4}{5x(3x - 1)}$
$= \dfrac{15x + 12x - 4}{5x(3x - 1)}$
$= \dfrac{27x - 4}{5x(3x - 1)}$

52. $\dfrac{2}{x^2 - 16} - \dfrac{x - 3}{x^2 - 9x + 20}$
$= \dfrac{2}{(x + 4)(x - 4)} - \dfrac{x - 3}{(x - 4)(x - 5)}$
$= \dfrac{2}{(x + 4)(x - 4)} \cdot \dfrac{x - 5}{x - 5} - \dfrac{x - 3}{(x - 4)(x - 5)} \cdot \dfrac{x + 4}{x + 4}$
$= \dfrac{2(x - 5)}{(x + 4)(x - 4)(x - 5)} - \dfrac{(x - 3)(x + 4)}{(x + 4)(x - 4)(x - 5)}$
$= \dfrac{2x - 10}{(x + 4)(x - 4)(x - 5)} - \dfrac{x^2 + x - 12}{(x + 4)(x - 4)(x - 5)}$
$= \dfrac{2x - 10 - (x^2 + x - 12)}{(x + 4)(x - 4)(x - 5)}$
$= \dfrac{2x - 10 - x^2 - x + 12}{(x + 4)(x - 4)(x - 5)}$
$= \dfrac{-x^2 + x + 2}{(x + 4)(x - 4)(x - 5)}$

53. $8x^2 - 4x = 4x \cdot 2x - 4x \cdot 1 = 4x(2x - 1)$

54. $25x^2 - 4 = (5x)^2 - 2^2 = (5x + 2)(5x - 2)$

55. $6y^2 - 5y - 6 = (3y + 2)(2y - 3)$ Using trial and error

56. $m^2 - 8m + 16 = m^2 - 2 \cdot m \cdot 4 + 4^2 = (m - 4)^2$

57. $x^3 - 8x^2 - 5x + 40 = (x^3 - 8x^2) + (-5x + 40)$
$= x^2(x - 8) - 5(x - 8)$
$= (x^2 - 5)(x - 8)$

58. $3a^4 + 6a^2 - 72 = 3(a^4 + 2a^2 - 24)$
$= 3(a^2 + 6)(a^2 - 4)$
$= 3(a^2 + 6)(a + 2)(a - 2)$

59. $16x^4 - 1 = (4x^2)^2 - 1^2$
$= (4x^2 + 1)(4x^2 - 1)$
$= (4x^2 + 1)[(2x)^2 - 1^2]$
$= (4x^2 + 1)(2x + 1)(2x - 1)$

60. $49a^2b^2 - 4 = (7ab)^2 - 2^2 = (7ab + 2)(7ab - 2)$

61. $9x^2 + 30xy + 25y^2 = (3x)^2 + 2 \cdot 3x \cdot 5y + (5y)^2 =$
$(3x + 5y)^2$

62. $2ac - 6ab - 3db + dc = (2ac - 6ab) + (-3db + dc)$
$= 2a(c - 3b) + d(-3b + c)$
$= (2a + d)(c - 3b)$

63. $15x^2 + 14xy - 8y^2 = (5x - 2y)(3x + 4y)$

64. $\dfrac{\dfrac{3}{x} + \dfrac{1}{2x}}{\dfrac{1}{3x} - \dfrac{3}{4x}} = \dfrac{\dfrac{3}{x} + \dfrac{1}{2x}}{\dfrac{1}{3x} - \dfrac{3}{4x}} \cdot \dfrac{12x}{12x}$
$= \dfrac{\dfrac{3}{x} \cdot 12x + \dfrac{1}{2x} \cdot 12x}{\dfrac{1}{3x} \cdot 12x - \dfrac{3}{4x} \cdot 12x}$
$= \dfrac{36 + 6}{4 - 9}$
$= \dfrac{42}{-5} = -\dfrac{42}{5}$

65. $\sqrt{49} = 7$ Taking the principal square root

66. $\sqrt{625} = 25$, so $-\sqrt{625} = -25$.

67. $\sqrt{64x^2} = \sqrt{(8x)^2} = 8x$

68. $\sqrt{a + b}\sqrt{a - b} = \sqrt{(a + b)(a - b)} = \sqrt{a^2 - b^2}$

69. $\sqrt{32ab}\sqrt{6a^4b^2} = \sqrt{32ab \cdot 6a^4b^2} =$
$\sqrt{16 \cdot 2 \cdot a \cdot b \cdot 2 \cdot 3 \cdot a^4 \cdot b^2} = \sqrt{16}\sqrt{2 \cdot 2}\sqrt{a^4}\sqrt{b^2}\sqrt{3ab} =$
$4 \cdot 2 \cdot a^2 \cdot b\sqrt{3ab} = 8a^2b\sqrt{3ab}$

70. $\sqrt{150} = \sqrt{25 \cdot 6} = \sqrt{25}\sqrt{6} = 5\sqrt{6}$

71. $\sqrt{243x^3y^2} = \sqrt{81 \cdot 3 \cdot x^2 \cdot x \cdot y^2} = \sqrt{81}\sqrt{x^2}\sqrt{y^2}\sqrt{3x} =$
$9xy\sqrt{3x}$

72. $\sqrt{\dfrac{100}{81}} = \dfrac{\sqrt{100}}{\sqrt{81}} = \dfrac{10}{9}$

73. $\sqrt{\dfrac{64}{x^2}} = \dfrac{\sqrt{64}}{\sqrt{x^2}} = \dfrac{8}{x}$

74. $4\sqrt{12} + 2\sqrt{12} = 6\sqrt{12} = 6\sqrt{4 \cdot 3} = 6\sqrt{4}\sqrt{3} =$
$6 \cdot 2\sqrt{3} = 12\sqrt{3}$

75. $\dfrac{\sqrt{72}}{\sqrt{45}} = \sqrt{\dfrac{72}{45}} = \sqrt{\dfrac{8}{5}} = \sqrt{\dfrac{8}{5} \cdot \dfrac{5}{5}} = \dfrac{\sqrt{40}}{5} =$
$\dfrac{\sqrt{4 \cdot 10}}{5} = \dfrac{2\sqrt{10}}{5}$

76. $a^2 + b^2 = c^2$
$9^2 + b^2 = 41^2$
$81 + b^2 = 1681$
$b^2 = 1600$
$b = 40$

77. $y = \dfrac{1}{3}x - 2$

We choose some values of x and find the corresponding y-values. We use multiples of 3 to avoid fractions.

When $x = 3$, $y = \dfrac{1}{3}(-3) - 2 = -1 - 2 = -3$.

When $x = 0$, $y = \dfrac{1}{3} \cdot 0 - 2 = 0 - 2 = -2$.

When $x = 3$, $y = \dfrac{1}{3} \cdot 3 - 2 = 1 - 2 = -1$.

x	y
-3	-3
0	-2
3	-1

We plot the points and draw and label the graph.

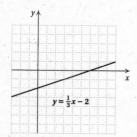

78. $2x + 3y = -6$

First we will find the intercepts. To find the y-intercept let $x = 0$ and solve for y.
$2 \cdot 0 + 3y = -6$
$3y = -6$
$y = -2$

The y-intercept is $(0, -2)$.

To find the x-intercept let $y = 0$ and solve for x.

$2x + 3 \cdot 0 = -6$
$2x = -6$
$x = -3$

The x-intercept is $(-3, 0)$.

Plot these points and draw the graph.

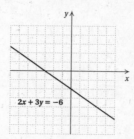

We can use a third point as a check. When $x = 3$ we have:
$2 \cdot 3 + 3y = -6$
$6 + 3y = -6$
$3y = -12$
$y = -4$

It appears that the point $(3, -4)$ is on the graph, so the graph is probably correct.

79. $y = -3$, or $y = 0 \cdot x - 3$

No matter what value we choose for x, y is always -3. The graph is a horizontal line with y-intercept $(0, -3)$.

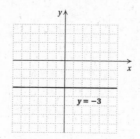

80. $x \geq -3$

First we graph the line $x = -3$. We draw a solid line since the inequality symbol is $\geq$. Then we test a point not on the line. We use $(0, 0)$.

$$\begin{array}{c} x \geq -3 \\ \hline 0 \ ? \ -3 \quad \text{TRUE} \end{array}$$

Since $(0, 0)$ is a solution of the inequality we shade the half-plane that contains (0.0).

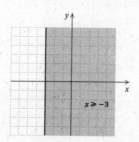

81. $4x - 3y > 12$

First we graph the line $4x - 3y = 12$. The intercepts are $(0, -4)$ and $(3, 0)$. We draw a dashed line since the inequality symbol is $>$. Now we test a point not on the line. We use $(0, 0)$.

$$\begin{array}{c|c} \underline{4x - 3y > 12} \\ 4 \cdot 0 - 3 \cdot 0 \; ? \; 12 \\ 0 \; \big| \quad \text{FALSE} \end{array}$$

Since $(0, 0)$ is not a solution we shade the half-plane that does not contain $(0, 0)$.

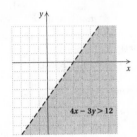

82. $y = x^2 + 2x + 1$

Find the vertex. The x-coordinate is

$$-\frac{b}{2a} = -\frac{2}{2 \cdot 1} = -1.$$

The y-coordinate is

$$y = x^2 + 2x + 1 = (-1)^2 + 2(-1) + 1 = 1 - 2 + 1 = 0.$$

The vertex is $(-1, 0)$.

We choose some x-values on both sides of the vertex and graph the parabola.

x	y	
-1	0	$\leftarrow$ Vertex
0	1	$\leftarrow$ y-intercept
-3	4	
-2	1	
1	4	

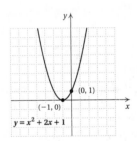

83. $9x^2 - 12x - 2 = 0$

$$\frac{1}{9}(9x^2 - 12x - 2) = \frac{1}{9} \cdot 0$$

$$x^2 - \frac{4}{3}x - \frac{2}{9} = 0$$

$$x^2 - \frac{4}{3}x \qquad = \frac{2}{9}$$

$$x^2 - \frac{4}{3}x + \frac{4}{9} = \frac{2}{9} + \frac{4}{9} \quad \text{Adding: } \frac{4}{9}:$$

$$\left[\frac{1}{2}\left(-\frac{4}{3}\right)\right]^2 = \left(-\frac{2}{3}\right)^2 = \frac{4}{9}$$

$$\left(x - \frac{2}{3}\right)^2 = \frac{6}{9}$$

$$x - \frac{2}{3} = \frac{\sqrt{6}}{3} \qquad or \qquad x - \frac{2}{3} = -\frac{\sqrt{6}}{3}$$

$$x = \frac{2}{3} + \frac{\sqrt{6}}{3} \quad or \qquad x = \frac{2}{3} - \frac{\sqrt{6}}{3}$$

$$x = \frac{2 + \sqrt{6}}{3} \quad or \qquad x = \frac{2 - \sqrt{6}}{3}$$

The solutions are $\dfrac{2 \pm \sqrt{6}}{3}$.

84.
$$4x^2 = 4x + 1$$

$$4x^2 - 4x - 1 = 0$$

$$a = 4, \; b = -4, \; c = -1$$

$$x = \frac{-(-4) \pm \sqrt{(-4)^2 - 4 \cdot 4 \cdot (-1)}}{2 \cdot 4}$$

$$x = \frac{4 \pm \sqrt{16 + 16}}{8} = \frac{4 \pm \sqrt{32}}{8}$$

Using a calculator, we have

$$\frac{4 - \sqrt{32}}{8} \approx -0.2 \text{ and } \frac{4 + \sqrt{32}}{8} \approx 1.2.$$

85. $\underbrace{\text{What percent}}_{\downarrow \atop p}$ of 52 is 13?
$$\begin{array}{ccccc} & \downarrow & \downarrow & \downarrow & \downarrow \\ & \cdot & 52 & = & 13 \end{array}$$

To solve the equation we divide both sides by 52 and then convert to percent notation.

$$p \cdot 52 = 13$$

$$\frac{p \cdot 52}{52} = \frac{13}{52}$$

$$p = 0.25 = 25\%$$

Thus, 25% of 52 is 13.

86. 12 is 20% of $\underbrace{\text{what number?}}_{\downarrow}$

$$\begin{array}{cccccc} \downarrow & \downarrow & \downarrow & \downarrow & & \downarrow \\ 12 & = & 20\% & \times & & b, \text{ or} \\ 12 & = & 0.2 & \times & & b \end{array}$$

To solve the equation we divide both sides by 0.2.

$$12 = 0.2 \times b$$

$$\frac{12}{0.2} = \frac{0.2 \times b}{0.2}$$

$$60 = b$$

Thus, 12 is 20% of 60.

87. *Familiarize*. We will use the work principal, substituting 8 for a and 10 for b.

Translate.
$$\frac{t}{8} + \frac{t}{10} = 1$$

Solve. First we multiply by 40 to clear the fractions.

$$\frac{t}{8} + \frac{t}{10} = 1$$

$$40\left(\frac{t}{8} + \frac{t}{10}\right) = 40 \cdot 1$$

$$40 \cdot \frac{t}{8} + 40 \cdot \frac{t}{10} = 40$$

$$5t + 4t = 40$$

$$9t = 40$$

$$t = \frac{40}{9}, \text{ or } 4\frac{4}{9}$$

Check. In $\frac{40}{9}$ hr crew A does $\frac{40}{9} \cdot \frac{1}{8}$, or $\frac{5}{9}$, of the job and crew B does $\frac{40}{9} \cdot \frac{1}{10}$, or $\frac{4}{9}$, of the job. Together they do $\frac{5}{9} + \frac{4}{9} = 1$ entire job. The answer checks.

State. It would take the crews $4\frac{4}{9}$ hr to do the job working together.

88. **Familiarize.** Using the labels on the drawing in the text, we let $w =$ the width of the screen, in feet, and $w + 4 =$ the length. Recall that the area of a rectangle is (length) $\times$ (width).

Translate. The area is 96 ft^2, so we have

$$(w + 4)w = 96.$$

Solve.

$$(w + 4)w = 96$$

$$w^2 + 4w = 96$$

$$w^2 + 4w - 96 = 0$$

$$(w + 12)(w - 8) = 0$$

$$w + 12 = 0 \quad or \quad w - 8 = 0$$

$$w = -12 \quad or \quad w = 8$$

Check. The width cannot be negative, so -12 cannot be a solution. If $w = 8$, then $w + 4 = 12$. The length is 4 ft more than the width and the area is $12 \cdot 8$, or 96 ft^2. The answer checks.

State. The length of the screen is 12 ft, and the width is 8 ft.

89. **Familiarize.** Let $r =$ the speed of the stream, in km/h. Then the boat travels upstream at a speed of $8 - r$ and it travels downstream at a speed of $8 + r$. We organize the information in a table.

	d	r	t
Upstream	60	$8 - r$	t_1
Downstream	60	$8 + r$	t_2

Translate. Using $t = d/r$ and the rows of the table we have

$$t_1 = \frac{60}{8 - r} \text{ and } t_2 = \frac{60}{8 + r}.$$

Since the total time is 16 hr, $t_1 + t_2 = 16$, and we have

$$\frac{60}{8 - r} + \frac{60}{8 + r} = 16.$$

Solve. First we multiply by $(8 - r)(8 + r)$, the LCM of the denominators.

$$(8 - r)(8 + r)\left(\frac{60}{8 - r} + \frac{60}{8 + r}\right) = (8 - r)(8 + r) \cdot 16$$

$$60(8 + r) + 60(8 - r) = 16(64 - r^2)$$

$$480 + 60r + 480 - 60r = 1024 - 16r^2$$

$$960 = 1024 - 16r^2$$

$$-64 = -16r^2$$

$$4 = r^2$$

$$r = 2 \quad or \quad r = -2$$

Check. The speed of the current cannot be negative, so -2 cannot be a solution. If the speed of the current is 2 km/h, then the boat travels upstream at $8 - 2$, or 6 km/h. At this speed it will travel 60 km in $60/6$, or 10 hr. The boat travels downstream at $8 + 2$, or 10 km/h. At this speed it will travel 60 km in $60/10$, or 6 hr. The total time is $10 + 6$, or 16 hr, so the answer checks.

State. The speed of the current is 2 km/h.

90. **Familiarize.** Let $w =$ the width of the floor, in meters. Then $w + 7 =$ the length.

Translate. We use the Pythagorean equation.

$$a^2 + b^2 = c^2$$

$$w^2 + (w + 7)^2 = 13^2$$

Solve.

$$w^2 + (w + 7)^2 = 13^2$$

$$w^2 + w^2 + 14w + 49 = 169$$

$$2w^2 + 14w + 49 = 169$$

$$2w^2 + 14w - 120 = 0$$

$$w^2 + 7w - 60 = 0 \quad \text{Dividing by 2}$$

$$(w + 12)(w - 5) = 0$$

$$w + 12 = 0 \quad or \quad w - 5 = 0$$

$$w = -12 \quad or \quad w = 5$$

Check. The width cannot be negative, so -12 cannot be a solution. If $w = 5$, then $w + 7 = 12$ and $5^2 + 12^2 = 25 + 144 = 169 = 13^2$. The answer checks.

State. The length of the floor is 12 m.

91. **Familiarize.** Let $x =$ the smaller odd integer. Then $x + 2 =$ the larger odd integer.

Translate.

The sum of the squares of the integers is 74.

$$\underbrace{x^2 + (x + 2)^2}_{} = 74$$

Solve.

$$x^2 + (x + 2)^2 = 74$$

$$x^2 + x^2 + 4x + 4 = 74$$

$$2x^2 + 4x + 4 = 74$$

$$2x^2 + 4x - 70 = 0$$

$$x^2 + 2x - 35 = 0 \quad \text{Dividing by 2}$$

$$(x + 7)(x - 5) = 0$$

$$x + 7 = 0 \quad or \quad x - 5 = 0$$
$$x = -7 \quad or \quad x = 5$$

Check. If $x = -7$, then $x + 2 = -7 + 2 = -5$ and $(-7)^2 + (-5)^2 = 49 + 25 = 74$. If $x = 5$, then $x + 2 = 5 + 2 = 7$ and $5^2 + 7^2 = 25 + 49 = 74$. The answer checks.

State. The numbers are -7 and -5 or 5 and 7.

92. Familiarize. Let $x =$ the amount of solution A and $y =$ the amount of solution B to be used, in liters. We organize the information in a table.

Solution	A	B	Mixture
Amount	x	y	60
Percent of alcohol	75%	50%	$66\frac{2}{3}$%
Amount of alcohol	75%x	50%y	$66\frac{2}{3}$% $\times$ 60, or 40

Translate. The first and last rows of the table give us two equations:

$$x + y = 60 \text{ and}$$
$$75\%x + 50\%y = 40, \text{ or } 0.75x + 0.5y = 40.$$

After clearing decimals we have the following system of equations.

$$x + y = 60, \quad (1)$$
$$75x + 50y = 4000 \quad (2)$$

Solve. First we multiply Equation (1) by -50 and then add.

$$-50x - 50y = -3000$$
$$\underline{75x + 50y = 4000}$$
$$25x \quad\quad = 1000$$
$$x = 40$$

Substitute 40 for x and one of the equations and solve for y.

$$x + y = 60 \quad (1)$$
$$40 + y = 60$$
$$y = 20$$

Check. If 40 L of solution A and 20 L of solution B are used, the solution consists of $40 + 20$, or 60 L. The amount of alcohol in the solution is $0.75(40) + 0.5(20) = 30 + 10 = 40$ L. The answer checks.

State. 40 L of solution A and 20 L of solution B should be used.

93. Familiarize. We will use the formula $s = 16t^2$.

Translate. Substitute 984 for s in the formula.

$$s = 16t^2$$
$$984 = 16t^2$$

Solve.

$$984 = 16t^2$$
$$\frac{984}{16} = t^2$$

$$\sqrt{\frac{984}{16}} = t \quad or \quad -\sqrt{\frac{984}{16}} = t$$
$$7.8 \approx t \quad or \quad -7.8 \approx t$$

Check. The time cannot be negative in this application, so -7.8 cannot be a solution. If $t \approx 7.8$, then $16(7.8)^2 = 973.44 \approx 984$. The answer checks. (Remember that we rounded the value of t.)

State. It would take an object about 7.8 sec to fall to the ground from the top of the Eiffel Tower.

94. Familiarize and Translate. Let $p =$ the amount of the pay and $h =$ the number of hours worked. We have direct variation between p and h so an equation $p = kh$, $k > 0$, applies.

Solve. First we find an equation of variation.

$$p = kh$$
$$242.52 = k \cdot 43$$
$$5.64 = k$$

The equation of variation is $p = 5.64h$. Note that the variation constant k is the hourly rate at which the student is paid.

Now we find the value of p when $h = 80$.

$$p = 5.64h$$
$$p = 5.64(80) = 451.20$$

Check. We can repeat the computations. Also note that, when the number of hours increased, the pay also increased. This is what we expect with direct variation.

State. The student would earn $451.20 for working 80 hr. The variation constant is the amount earned per hour.

95. Familiarize. Let $c =$ the number of cars entering the city each morning.

Translate. We reword and translate to an equation.

$$3654 \text{ is } \underbrace{\text{three-fifths}} \text{ of } \underbrace{\text{what number?}}$$
$$\downarrow \quad \downarrow \quad\quad \downarrow \quad\quad \downarrow \quad\quad \downarrow$$
$$3654 = \quad\quad \frac{3}{5} \quad \cdot \quad\quad c$$

Solve.

$$3654 = \frac{3}{5} \cdot c$$
$$\frac{5}{3} \cdot 3654 = c$$
$$6090 = c$$

Check. $\frac{3}{5} \cdot 6090 = \frac{3 \cdot 6090}{5} = \frac{3 \cdot \cancel{5} \cdot 1218}{\cancel{5} \cdot 1} = 3654.$

The answer checks.

State. 6090 cars enter the city each morning.

96. Familiarize. Let $x =$ the number of pounds of the more expensive nuts and $y =$ the number of pounds of the less expensive nuts to be used. We organize the information in a table.

	More expensive	Less expensive	Mixture
Cost	$3.30	$2.40	$2.70
Amount	x	y	42
Total cost	$3.3x$	$2.4y$	$2.70(42)$, or $113.40

Translate. The last two rows of the table give us two equations.

$$x + y = 42,$$
$$3.3x + 24y = 113.40$$

After clearing decimals we have the following system of equations.

$$x + y = 42, \qquad (1)$$
$$33x + 24y = 1134 \quad (2)$$

Solve. First we multiply Equation (1) by -24 and then add.

$$-24x - 24y = -1008$$
$$\underline{33x + 24y = 1134}$$
$$9x \qquad = 126$$
$$x = 14$$

Now substitute 14 for x in one of the equations and solve for y.

$$x + y = 42 \quad (1)$$
$$14 + y = 42$$
$$y = 28$$

Check. If 14 lb of the more expensive nuts and 28 lb of the less expensive nuts are used, then the mixture weighs $14 + 28$, or 42 lb. The total cost of the mixture is $3.30(14) + $2.40(28) = $46.20 + $67.20 = $113.40. The answer checks.

State. 14 lb of the $3.30 per pound nuts and 28 lb of the $2.40 per pound nuts should be used.

97. Familiarize. Let r the speed of the plane in still air. Then its speed with the wind is $r + 20$ and the speed against the wind is $r - 20$. We organize the information in a table.

	Distance	Speed	Time
With the wind	d	$r + 20$	3
Against the wind	d	$r - 20$	4

Translate. Using $d = rt$ in each row of the table we have

$$d = (r + 20)3, \quad (1)$$
$$d = (r - 20)4. \quad (2)$$

Solve. We substitute $(r + 20)3$ for d in Equation (2) and solve for r.

$$(r + 20)3 = (r - 20)4$$
$$3r + 60 = 4r - 80$$
$$60 = r - 80$$
$$140 = r$$

Check. If the speed of the plane in still air is 140 mph, then it travels at $140 + 20$, or 160 mph, with the wind. At this speed, in 3 hr it travels $160 \cdot 3$, or 480 mi. The plane travels at $140 - 20$, or 120 mph, against the wind. At this speed, in 4 hr it travels $120 \cdot 4$, or 480 mi. The distances are the same, so the answer checks.

State. The speed of the plane in still air is 140 mph.

98. The solutions of the equation are the first coordinates of the x-intercepts of the graph. We see that they are -3 and 2.

99. We let $y = 0$ and solve for x.

$$x^2 + 4x + 1 = 0$$
$$a = 1, \ b = 4, \ c = 1$$
$$x = \frac{-4 \pm \sqrt{4^2 - 4 \cdot 1 \cdot 1}}{2 \cdot 1}$$
$$x = \frac{-4 \pm \sqrt{16 - 4}}{2} = \frac{-4 \pm \sqrt{12}}{2}$$
$$x = \frac{-4 \pm \sqrt{4 \cdot 3}}{2} = \frac{-4 \pm 2\sqrt{3}}{2}$$
$$x = \frac{2(-2 \pm \sqrt{3})}{2} = -2 \pm \sqrt{3}$$

The x-intercepts are $(-2 - \sqrt{3}, 0)$ and $(-2 + \sqrt{3}, 0)$.

100. We solve for y to write the equation in the form $y = mx + b$.

$$-6x + 3y = -24$$
$$3y = 6x - 24$$
$$\frac{1}{3} \cdot 3y = \frac{1}{3}(6x - 24)$$
$$y = 2x - 8$$

The slope is 2 and the y-intercept is $(0, -8)$.

101. First we solve each equation for y.

$$y - x = 4 \qquad 3y + x = 8$$
$$y = x + 4 \qquad 3y = -x + 8$$
$$y = -\frac{1}{3}x + \frac{8}{3}$$

The slopes are not the same, so the lines are not parallel. The product of the slopes is $1\left(-\frac{1}{3}\right) = -\frac{1}{3} \neq -1$, so the lines are not perpendicular. Thus, the lines are neither parallel nor perpendicular.

102. $m - \dfrac{y_2 - y_1}{x_2 - x_1} = \dfrac{9 - (-6)}{-4 - (-5)} = \dfrac{15}{1} = 15$

103.
$$y = kx$$
$$100 = k \cdot 10$$
$$10 = k$$

The equation of variation is $y = 10x$.

When $x = 64$, we have:

$$y = 10x$$
$$y = 10 \cdot 64$$
$$y = 640$$

104.
$$y = \frac{k}{x}$$
$$100 = \frac{k}{10}$$
$$1000 = k$$

The equation of variation is $y = \dfrac{1000}{x}$.

When $x = 125$, we have:

$$y = \frac{1000}{x}$$
$$y = \frac{1000}{125}$$
$$y = 8$$

105. No vertical line will intersect the graph more than once, so this is the graph of a function.

106. It is possible for a vertical line to intersect the graph more than once, so this is not the graph of a function.

107. $f(x) = x^2 + x - 2$

We find some function values. Recall from Section 9.6 that the graph is a parabola.

x	$f(x)$
-3	4
-2	0
0	-2
1	0
2	4

We plot these points and draw the graph.

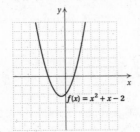

108. $g(x) = |x + 2|$

We find some function values, plot the points, and draw the graph.

x	$f(x)$
-5	3
-3	1
-2	0
0	2
3	5

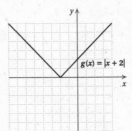

109. $f(x) = 2x^2 + 7x - 4$

$$f(0) = 2 \cdot 0^2 + 7 \cdot 0 - 4 = 0 + 0 - 4 = -4$$

$$f(-4) = 2(-4)^2 + 7(-4) - 4 = 2 \cdot 16 - 28 - 4 = 32 - 28 - 4 = 0$$

$$f\left(\frac{1}{2}\right) = 2\left(\frac{1}{2}\right)^2 + 7 \cdot \frac{1}{2} - 4 = 2 \cdot \frac{1}{4} + \frac{7}{2} - 4 = \frac{1}{2} + \frac{7}{2} - 4 = 0$$

110. *Familiarize*. Let $r =$ the speed of the wind, in mph. Then the airplane travels at a speed of $300 - r$ against the wind and at $300 + r$ with the wind. We organize the information in a table.

	Distance	Speed	Time
Against wind	408	$300 - r$	t_1
With wind	492	$300 + r$	t_2

Translate. Using $t = d/r$ in each row of the table we have

$$t_1 = \frac{408}{300 - r} \text{ and } t_2 = \frac{492}{300 + r}.$$

Since the total time is 3 hr, $t_1 + t_2 = 3$ and we have

$$\frac{408}{300 - r} + \frac{492}{300 + r} = 3$$

Solve. First we multiply by $(300 - r)(300 + r)$, the LCM of the denominators.

$$(300 - r)(300 + r)\left(\frac{408}{300 - r} + \frac{492}{300 + r}\right) = (300 - r)(300 + r) \cdot 3$$

$$408(300 + r) + 492(300 - r) = 3(90,000 - r^2)$$

$$122,400 + 408r + 147,600 - 492r = 270,000 - 3r^2$$

$$270,000 - 84r = 270,000 - 3r^2$$

$$0 = 84r - 3r^2$$

$$0 = 3r(28 - r)$$

$$3r = 0 \quad or \quad 28 - r = 0$$
$$r = 0 \quad or \quad 28 = r$$

Check. Since we are assuming there is some wind, 0 cannot be a solution. If the speed of the wind is 28 mph, then the plane travels at a speed of $300 - 28$, or 272 mph, against the wind. At this speed, it travels 408 mi in 408/272, or 1.5 hr. With the wind the plane travels at $300 + 28$, or 328 mph. At this speed, it travels 492 mi in 492/328, or 1.5 hr. The total time is $1.5 + 1.5$, or 3 hr. The answer checks.

State. The speed of the wind is 28 mph. This is between 22 and 29 mph, so answer (c) is correct.

111. $2x^2 + 6x + 5 = 4$

$2x^2 + 6x + 1 = 0$

$a = 2,\ b = 6,\ c = 1$

$x = \dfrac{-6 \pm \sqrt{6^2 - 4 \cdot 2 \cdot 1}}{2 \cdot 2}$

$x = \dfrac{-6 \pm \sqrt{36 - 8}}{4} = \dfrac{-6 \pm \sqrt{28}}{4}$

$x = \dfrac{-6 \pm \sqrt{4 \cdot 7}}{4} = \dfrac{-6 \pm 2\sqrt{7}}{4}$

$x = \dfrac{2(-3 \pm \sqrt{7})}{2 \cdot 2} = \dfrac{-3 \pm \sqrt{7}}{2}$

Answer (d) is correct.

112. $\quad S = \dfrac{a+b}{3b}$

$3bS = a + b \quad$ Multiplying by $3b$

$3bS - b = a$

$b(3S - 1) = a$

$b = \dfrac{a}{3S - 1}$

Answer (d) is correct.

113. $3x - 4y = 12$

We find the intercepts. To find the y-intercept, let $x = 0$ and solve for y.

$3 \cdot 0 - 4y = 12$

$-4y = 12$

$y = -3$

The y-intercept is $(0, -3)$.

To find the x-intercept, let $y = 0$ and solve for x.

$3x - 4 \cdot 0 = 12$

$3x = 12$

$x = 4$

The x-intercept is $(4, 0)$.

Graph (c) is the only graph with these intercepts, so (c) is the correct answer.

114. $|x| = 12$

The solutions are the numbers whose distance from 0 is 12. They are -12 and 12.

115. $\sqrt{\sqrt{\sqrt{81}}} = \sqrt{\sqrt{9}} = \sqrt{3}$

116. $x^2 - bx + 225 = x^2 - bx + 15^2$

If the trinomial is a square, then:

$b = 2 \cdot 1 \cdot 15 \ \ or \ \ b = -2 \cdot 1 \cdot 15$

$b = 30 \qquad\quad or \quad b = -30$

117. Let $y =$ the second leg of the small right triangle with hypotenuse $\sqrt{2}$ and one leg x.

$x^2 + y^2 = (\sqrt{2})^2$

$x^2 + y^2 = 2$

$y^2 = 2 - x^2$

$y = \sqrt{2 - x^2}$

Let $z =$ the second leg of the small right triangle with hypotenuse 1 and one leg x.

$x^2 + z^2 = 1^2$

$x^2 + z^2 = 1$

$z^2 = 1 - x^2$

$z = \sqrt{1 - x^2}$

Then $y + z =$ the hypotenuse of the large right triangle with legs $\sqrt{2}$ and 1.

$(\sqrt{2})^2 + 1^2 = (y + z)^2$

$2 + 1 = (\sqrt{2 - x^2} + \sqrt{1 - x^2})^2$

$3 = 2 - x^2 + 2\sqrt{2 - x^2}\sqrt{1 - x^2} + 1 - x^2$

$3 = -2x^2 + 3 + 2\sqrt{(2 - x^2)(1 - x^2)}$

$2x^2 = 2\sqrt{2 - 3x^2 + x^4}$

$x^2 = \sqrt{2 - 3x^2 + x^4} \qquad$ Dividing by 2

$(x^2)^2 = (\sqrt{2 - 3x^2 + x^4})^2$

$x^4 = 2 - 3x^2 + x^4$

$0 = 2 - 3x^2$

$3x^2 = 2$

$x^2 = \dfrac{2}{3}$

$x = \sqrt{\dfrac{2}{3}} \qquad$ Taking the positive square root

$x = \sqrt{\dfrac{2}{3} \cdot \dfrac{3}{3}} = \dfrac{\sqrt{6}}{3}$

118. Since $(x - 3)(x + 3) = x^2 - 9$, the expressions are equivalent.

119. When $x = 0$, $\dfrac{x + 3}{3} = \dfrac{0 + 3}{3} = 1 \neq 0$, so the expressions are not equivalent.

120. When $x = 1$, $(x + 5)^2 = (1 + 5)^2 = 6^2 = 36$ and $x^2 + 25 = 1^2 + 25 = 1 + 25 = 26$. Since $36 \neq 26$, the expressions are not equivalent.

121. When $x = 3$, $\sqrt{x^2 + 16} = \sqrt{3^2 + 16} = \sqrt{9 + 16} = \sqrt{25} = 5$ and $x + 4 = 3 + 4 = 7$. Since $5 \neq 7$, the expressions are not equivalent.

122. By definition, $\sqrt{x^2} = |x|$ so the expressions are equivalent.